Twelfth Edition

Human Genetics

Concepts and Applications

Ricki Lewis

Genetic Counselor
CareNet Medical Group
Schenectady, New York

Adjunct Assistant Professor of
Medical Education
Alden March Bioethics Institute
Albany Medical College

Writer, Medscape Medical News
    Genetic Literacy Project

Blogger, Public Library of Science

Mc
Graw
Hill
Education

HUMAN GENETICS: CONCEPTS AND APPLICATIONS, TWELFTH EDITION

Published by McGraw-Hill Education, 2 Penn Plaza, New York, NY 10121. Copyright © 2018 by McGraw-Hill Education. All rights reserved. Printed in the United States of America. Previous editions © 2015, 2012, and 2010. No part of this publication may be reproduced or distributed in any form or by any means, or stored in a database or retrieval system, without the prior written consent of McGraw-Hill Education, including, but not limited to, in any network or other electronic storage or transmission, or broadcast for distance learning.

Some ancillaries, including electronic and print components, may not be available to customers outside the United States.

This book is printed on acid-free paper.

3 4 5 6 7 8 9 LKV 21 20 19 18

ISBN 978-1-259-70093-4
MHID 1-259-70093-3

Portfolio Manager: *Justin Wyatt, Ph.D.*
Senior Product Developer: *Donna Nemmers*
Senior Marketing Manager: *Kelly Brown*
Content Project Managers: *Jane Mohr, Jessica Portz, Christina Nelson, and Sandra Schnee*
Senior Buyer: *Sandy Ludovissy*
Senior Designer: *Tara McDermott*
Content Licensing Specialists: *Lori Hancock and Lori Slattery*
Cover Image: *©Geri Lavrov/Getty Images (woman); ©Adrian Neal/Getty Images (helix model)*
Compositor: *SPi Global*
Typeface: *10/12 STIXMathJax Main*
Printer: *LSC Communications*

All credits appearing on page or at the end of the book are considered to be an extension of the copyright page.

Library of Congress Cataloging-in-Publication Data

Names: Lewis, Ricki.
Title: Human genetics : concepts and applications / Ricki Lewis, genetic
  counselor, CareNet Medical Group, Schenectady, New York; adjunct assistant
  professor of medical education, Alden March Bioethics Institute, Albany
  Medical College; writer, Medscape Medical News, Genetic Literacy Project;
  blogger, Public Library of Science.
Description: Twelfth edition. | New York, NY : McGraw-Hill Education, [2018]
  | Includes bibliographical references and index.
Identifiers: LCCN 2017016868 | ISBN 9781259700934 (alk. paper)
Subjects: LCSH: Human genetics–Textbooks.
Classification: LCC QH431 .L41855 2018 | DDC 611/.01816—dc23 LC record available at
https://lccn.loc.gov/2017016868

The Internet addresses listed in the text were accurate at the time of publication. The inclusion of a website does not indicate an endorsement by the authors or McGraw-Hill Education, and McGraw-Hill Education does not guarantee the accuracy of the information presented at these sites.

mheducation.com/highered

About the Author

Courtesy of Dr. Wendy Josephs

Ricki Lewis has built an eclectic career in communicating the excitement of genetics and genomics, combining skills as a geneticist and a journalist. She currently writes the popular weekly blog, DNA Science, at Public Library of Science (http://blogs.plos.org/dnascience/) and contributes frequent articles to Medscape Medical News and the Genetic Literacy Project. Dr. Lewis has authored or coauthored several university-level textbooks and is the author of the narrative nonfiction book, *The Forever Fix: Gene Therapy and the Boy Who Saved It*, as well as an essay collection, a novel, and a short "basics" book on human genetics. She teaches an online course on "Genethics" for the Alden March Bioethics Institute of Albany Medical College and is a genetic counselor for a private medical practice. Her passion is rare genetic diseases; she writes often about affected families who are pioneering DNA-based treatments.

Dedicated to the families who live with genetic diseases, the health care providers who help them, and the researchers who develop new tests and treatments.

Brief Contents

Contents

PART 1 Introduction 1

CHAPTER 1
What Is in a Human Genome? 1

CHAPTER 2
Cells 15

CHAPTER 3
Meiosis, Development, and Aging 40

PART 2 Transmission Genetics 66

CHAPTER 4
Single-Gene Inheritance 66

Preface

Human Genetics Touches Us All

More than a million people have had their genomes sequenced, most of them since the last edition of this book was published in 2014. When I wrote the first edition, the idea to sequence "the" human genome was just becoming reality. The growing field of genomics, of considering all of our genes, is now revealing that we are much more alike than different, yet those differences among 3 million of our 3.2 billion DNA building blocks hold clues to our variation and diversity. It has been a privilege to chronicle the evolution of human genetics, from an academic subfield of life science and a minor medical specialty to a growing body of knowledge that will affect us all.

The twelfth edition opens with the hypothetical "Eve's Genome" and ends with "Do You Want Your Genome Sequenced?" In between, the text touches on what exome and genome sequencing have revealed about single-gene diseases so rare that they affect only a single family to clues to such common and complex conditions as intellectual disability and autism. Exome and genome sequencing are also important in such varied areas as understanding our origins, solving crimes, and tracking epidemics. In short, DNA sequencing will affect most of us.

As the cost of genome sequencing plummets, we all may be able to look to our genomes for echoes of our pasts and hints of our futures—if we so choose. We may also learn what we can do to counter our inherited tendencies and susceptibilities. Genetic knowledge is informative and empowering. This book shows you how and why this is true.

Ricki Lewis

What Sets This Book Apart

Current Content

The exciting narrative writing style, with clear explanations of concepts and mechanisms propelled by stories, historical asides, and descriptions of new technologies reflects Dr. Lewis's eclectic experience as a health and science writer, blogger, professor, and genetic counselor, along with her expertise in genetics. Updates to this edition include

- Children benefiting from genetic technologies
- Cannabidiol to treat genetic seizure disorders
- "Variants of uncertain significance" as test results
- DNA profiling and the Srebrenica genocide
- Steel syndrome in Harlem
- Archaic humans
- Chimeric antigen receptor technology
- Genome editing, gene drives, and synthetic genomes
- Learning from the genomes of the deceased

Connections and Context

For human genome sequence information to be useful, we need to discover all of the ways that genes interact. The patterns with which different parts of the genome touch in a cell's nucleus serves as a metaphor for the new edition of this book. Originally conceived as two-thirds "concepts" followed by one-third "applications," the book has evolved as has the science, with the tentacles of technology no longer constrained to that final third, but touching other chapters, in which the science of genetics becomes applied:

- The very first illustration, figure 1.1, depicts DNA wound around proteins to form a nucleosome, the unit of chromatin. Part of the figure repeats as an inset in figure 11.6, which zooms in on the molecular events as nucleosomes open and close during gene expression.
- The "diagnostic odyssey" of young Millie McWilliams is told in Clinical Connection 1.1. Millie appears again in figure 4.18, in the context of genome sequencing of parent-child trios to track the genetic causes of rare diseases.
- The cell cycle first appears as figure 2.12, then again as figure 2.15 but with the checkpoints added. In figure 18.3, the cell cycle appears yet again in the context of the photo of dividing cancer cells next to it.
- The journey from fertilized ovum to cleavage embryo, then to implantation in the uterine lining, is depicted in figure 3.15. It appears again in figure 21.3 to orient the stages and places of assisted reproductive technologies such as *in vitro* fertilization.

- A recurring representation of different-colored shapes moving in and out of an ancestral "population" traces the forces of evolutionary change throughout chapter 15: nonrandom mating (figure 15.2), migration (figure 15.3), genetic drift (figure 15.5), mutation (figure 15.7), and natural selection (figure 15.8), and then all together in figure 15.14.
- Table 20.1 defines and describes all types of genetic testing, with references to their mentions in previous chapters.
- Table 22.2 reviews genomics coverage in other chapters.

The historical roots from which today's genetic technologies emerged appear in *A Glimpse of History* boxes throughout the book.

The chapter and unit organization remain from the eleventh edition, with a few meaningful moves of material to more logical places. The essay on mitochondrial transfer that appeared in the last edition in the context of assisted reproductive technologies in chapter 21 is now with mitochondria, in chapter 5 as a *Bioethics* box. The "diseaseome" that was at the end of chapter 1 is now in chapter 11, in the context of gene expression. Examples of exome and genome sequencing are threads throughout that knit the ongoing transition from genetics to genomics. Chapter 1 is now more molecular in focus because today even grade-schoolers are familiar with DNA. New subheads throughout the book ease understanding and studying.

Changes in terminology reflect the bigger picture of today's genetics. "Abnormal" is now the less judgmental "atypical." Use of the general term "gene variant" clarifies the fuzziness of the distinction between "mutation" and "polymorphism." Both refer to changes in the DNA sequence, but in the past, "mutation" has been considered a rare genetic change and "polymorphism" a more common one. "Gene variant" is a better general term since genome sequencing has revealed that some mutations have no effects in certain individuals—again, due to gene-gene interactions, many as yet unknown.

The Lewis Guided Learning System

Each chapter begins with two views of the content. *Learning Outcomes* embedded in the table of contents guide the student in mastering material. *The Big Picture* encapsulates the overall theme of the chapter. The opening essay and figure grab attention. Content flows logically through three to five major sections per chapter that are peppered with high-interest boxed readings (*Clinical Connections, Bioethics, A Glimpse of History*, and *Technology Timelines*). End-of-chapter pedagogy progresses from straight recall to applied and creative questions and challenges, including a question based on the chapter opener. The *Clinical Connections* and *Bioethics* boxes have their own question sets. *Key Concepts Questions* after each major section reinforce learning.

Dynamic Art

Outstanding photographs and dimensional illustrations, vibrantly colored, are featured throughout *Human Genetics: Concepts and Applications*. Figure types include process figures with numbered steps, micro to macro representations, and the combination of art and photos to relate stylized drawings to real-life structures.

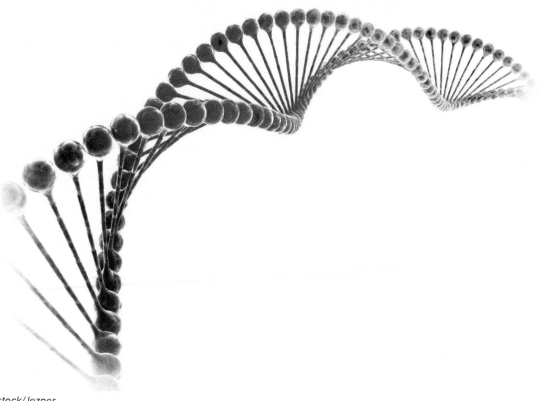

Shutterstock/Jezper

New to This Edition

Highlights in the new edition update information and discoveries, ease learning, and conceptually connect chapters. Updates include:

Chapter 1 What Is in a Human Genome?
- How a precision medicine program is integrating human genome information with environmental factors to dissect health and disease, on a population level

Chapter 4 Single-Gene Inheritance
- How genome analysis provides a new view of Mendel's laws

Chapter 8 Genetics of Behavior
- Schizophrenia arises from excess synaptic pruning
- Syndromes that include autism

Chapter 9 DNA Structure and Replication
- More subheads ease learning
- The "loop-ome" brings genes together

Chapter 10 Gene Action: From DNA to Protein
- Why proteins are important

Chapter 12 Gene Mutation
- More subheads to distinguish mutations, polymorphisms, and gene variants
- Figures and discussion on somatic mosaicism
- The famous painting of the "blue people" of Kentucky

Chapter 13 Chromosomes
- Less history, more new technology

Chapter 14 Constant Allele Frequencies and DNA Forensics
- DNA profiling confirms genocide

Chapter 15 Changing Allele Frequencies
- Steel syndrome in East Harlem—how considering population substructure improves health care

Chapter 16 Human Ancestry and Evolution
- Gene flow among archaic and modern humans

Chapter 18 Cancer Genetics and Genomics
- The "3 strikes" to cancer
- Chimeric antigen receptor technology
- Liquid biopsy

Chapter 19 DNA Technologies
- Genome editing and gene drives

Chapter 20 Genetic Testing and Treatment
- Genome editing in research and the clinic
- Gene therapy successes

Chapter 22 Genomics
- Sequencing genomes of the deceased
- Synthetic genomes

New Figures

- 1.1 Levels of genetics
- 1.3 Gene to protein to person
- 1.4 A mutation can alter a protein, causing symptoms
- 1.8 Precision medicine
- 2.21 The human microbiome
- 3.23 Zika virus causes birth defects
- 4.18 Parent and child trios
- 5.9 Ragged red fibers in mitochondrial disease
- 8.9 Schizophrenia and overactive synaptic pruning
- 9.14 DNA looping
- 11.7 Open or closed chromatin
- 12.11 The blue people of Kentucky
- 16.6 Gene flow among archaic and anatomically modern humans
- 16.9 The dystrophin gene
- 19.11 CRISPR-Cas9 genome editing
- 19.12 Gene drives

New Tables

- 3.2 Paternal Age Effect Conditions
- 8.3 Famous People Who Had Autistic Behaviors
- 8.4 Genetic Syndromes That Include Autism
- 16.1 Neanderthal Genes in Modern Human Genomes
- 19.4 Genome Editing Techniques
- 19.5 Applications of CRISPR-Cas9 Genome Editing
- 20.2 Genes Associated with Athletic Characteristics
- 20.3 Pharmacogenetics
- 22.2 Genomics Coverage in Other Chapters

ACKNOWLEDGMENTS

Human Genetics: Concepts and Applications, Twelfth Edition, would not have been possible without the assistance of the editorial and production team: senior brand manager Justin Wyatt, product development director Rose Koos, marketing manager Kelly Brown, content licensing specialists Lori Slattery and Lori Hancock, designer Tara McDermott, senior product developer Donna Nemmers, project managers Jane Mohr and Jessica Portz, copyeditor Carey Lange, and proofreaders Jane Hoover and Sue Nodine. I am also grateful to the dozens of colleagues who have reviewed this textbook over the years and have contributed to its currency and accuracy. Many thanks to the instructors and students who have reached out to me through the years with helpful suggestions and support. Special thanks to my friends in the rare disease community who have shared their stories, and to Jonathan Monkemeyer and David Bachinsky for helpful Facebook posts. As always, many thanks to my wonderful husband Larry for his support and encouragement and to my three daughters and their partners, Emmanuel the future doctor in Africa, my cats, and Cliff the hippo.

This book continually evolves thanks to input from instructors and students. Please let me know your thoughts and suggestions for improvement. (rickilewis54@gmail.com)

Applying Human Genetics

Chapter Openers

Courtesy of the Gavin R. Stevens Foundation

Source: Centers for Disease Control and Prevention (CDC)

A GLIMPSE OF HISTORY

The Human Touch

Clinical Connections

© Clinic for Special Children, 2013

Technology Timelines

Courtesy of Lori Sames.
Photo by Dr. Wendy Josephs

Bioethics

The Lewis Guided Learning System

Learning Outcomes preview major chapter topics in an inquiry-based format according to numbered sections.

The Big Picture encapsulates chapter content at the start.

Chapter Openers vividly relate content to real life.

Key Concepts Questions follow each numbered section.

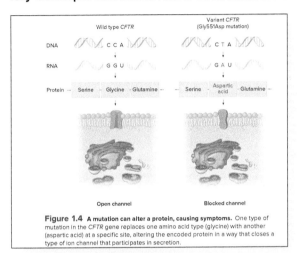

Figure 1.4 A mutation can alter a protein, causing symptoms. One type of mutation in the *CFTR* gene replaces one amino acid type (glycine) with another (aspartic acid) at a specific site, altering the encoded protein in a way that closes a type of ion channel that participates in secretion.

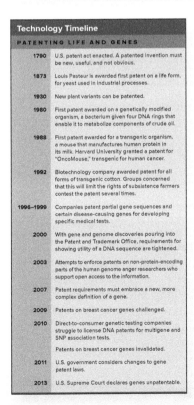

Technology Timeline

PATENTING LIFE AND GENES

1790	U.S. patent act enacted. A patented invention must be new, useful, and not obvious.
1873	Louis Pasteur is awarded first patent on a life form, for yeast used in industrial processes.
1930	New plant variants can be patented.
1980	First patent awarded on a genetically modified organism, a bacterium given four DNA rings that enable it to metabolize components of crude oil.
1988	First patent awarded for a transgenic organism, a mouse that manufactures human protein in its milk. Harvard University granted a patent for "OncoMouse," transgenic for human cancer.
1992	Biotechnology company awarded patent for all forms of transgenic cotton. Groups concerned that this will limit the rights of subsistence farmers contest the patent several times.
1996–1999	Companies patent partial gene sequences and certain disease-causing genes for developing specific medical tests.
2000	With gene and genome discoveries pouring into the Patent and Trademark Office, requirements for showing utility of a DNA sequence are tightened.
2003	Attempts to enforce patents on non-protein-encoding parts of the human genome anger researchers who support open access to the information.
2007	Patent requirements must embrace a new, more complex definition of a gene.
2009	Patents on breast cancer genes challenged.
2010	Direct-to-consumer genetic testing companies struggle to license DNA patents for multigene and SNP association tests. Patents on breast cancer genes invalidated.
2011	U.S. government considers changes to gene patent laws.
2013	U.S. Supreme Court declares genes unpatentable.

In-Chapter Review Tools, such as Key Concepts Questions, summary tables, and timelines of major discoveries, are handy tools for reference and study. Most boldfaced terms are consistent in the chapters, summaries, and glossary.

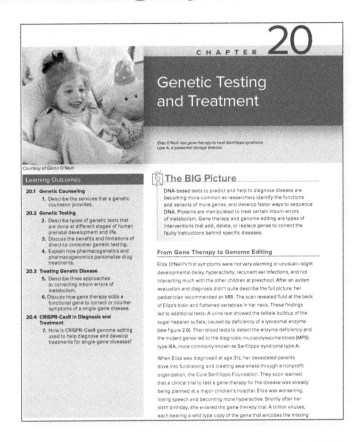

CHAPTER 20

Genetic Testing and Treatment

Eliza O'Neill had gene therapy to treat Sanfilippo syndrome type A, a lysosomal storage disease.

Courtesy of Glenn O'Neill

Learning Outcomes

20.1 Genetic Counseling
1. Describe the services that a genetic counselor provides.

20.2 Genetic Testing
2. Describe types of genetic tests that are done at different stages of human prenatal development and life.
3. Discuss the benefits and limitations of direct-to-consumer genetic testing.
4. Explain how pharmacogenetics and pharmacogenomics personalize drug treatments.

20.3 Treating Genetic Disease
5. Describe three approaches to correcting inborn errors of metabolism.
6. Discuss how gene therapy adds a functional gene to correct or counter symptoms of a single-gene disease.

20.4 CRISPR-Cas9 in Diagnosis and Treatment
7. How is CRISPR-Cas9 genome editing used to help diagnose and develop treatments for single-gene diseases?

The BIG Picture

DNA-based tests to predict and help to diagnose disease are becoming more common as researchers identify the functions and variants of more genes, and develop faster ways to sequence DNA. Proteins are manipulated to treat certain inborn errors of metabolism. Gene therapy and genome editing are types of interventions that add, delete, or replace genes to correct the faulty instructions behind specific diseases.

From Gene Therapy to Genome Editing

Eliza O'Neill's first symptoms were not very alarming or unusual—slight developmental delay, hyperactivity, recurrent ear infections, and not interacting much with the other children at preschool. After an autism evaluation and diagnosis didn't quite describe the full picture, her pediatrician recommended an MRI. The scan revealed fluid at the back of Eliza's brain and flattened vertebrae in her neck. These findings led to additional tests. A urine test showed the telltale buildup of the sugar heparan sulfate, caused by deficiency of a lysosomal enzyme (see figure 2.6). Then blood tests to detect the enzyme deficiency and the mutant genes led to the diagnosis: mucopolysaccharidosis (MPS) type IIIA, more commonly known as Sanfilippo syndrome type A.

When Eliza was diagnosed at age 3½, her devastated parents dove into fundraising and creating awareness through a nonprofit organization, the Cure Sanfilippo Foundation. They soon learned that a clinical trial to test a gene therapy for the disease was already being planned at a major children's hospital. Eliza was worsening, losing speech and becoming more hyperactive. Shortly after her sixth birthday, she entered the gene therapy trial. A trillion viruses, each bearing a wild type copy of the gene that encodes the missing

---

Bioethics

Designer Babies: Is Prenatal Genetic Testing Eugenic?

Modern genetics is sometimes compared to eugenics because genetic technologies may affect reproductive choices and can influence which alleles are passed to the next generation. However, medical genetics and eugenics differ in their intent. Eugenics aims to allow only people with certain "valuable" genotypes to reproduce, for the supposed benefit of the population as a whole. The goal of medical genetics, in contrast, is to prevent and alleviate suffering in individuals and families. But the once-clear line between eugenics and genetics is starting to blur as access to prenatal DNA testing widens, the scope of testing broadens to exomes and genomes, and sequencing cost plummets.

For decades prenatal genetic testing has focused on detecting the most common aneuploids—extra or missing chromosomes—or single-gene diseases known to occur in a family. As chapter 13 describes, chorionic villus sampling and amniocentesis have been used to visualize fetal chromosomes. In 2011 it became possible to collect, sequence, and overlap small pieces of placenta-derived cell-free DNA in the circulation of a pregnant woman and reconstruct a full genome sequence (see figure 13.10). Then, the sequence can be analyzed for genotypes that cause or increase the risk of developing known diseases. This knowledge would theoretically allow a quality control of sorts in terms of which fetuses, with which characteristics, complete development. Such extensive analysis is not (yet) commercially available, but is done as part of research protocols.

Preimplantation genetic diagnosis (see figure 21.6) may provide a form of selection because it checks the genes of early embryos and chooses those with certain genotypes to continue development. Finally, use of technologies such as genome editing (see section 20.4) enable manipulation of a fertilized ovum's DNA, although such germline intervention is generally prohibited.

The ability to sequence genomes has the potential to extend prenatal investigation from the more common chromosomal conditions to many aspects of a future individual's health and perhaps other characteristics, such as personality traits, appearance, and intelligence. Figure 15.C takes a simple view of a complex idea — altering the frequency of inherited traits in a future human population.

Figure 15C Designer babies. Will widespread use of genetic technologies to create or select perfect children have eugenic effects? © Finn Brandt/Getty Images

Questions for Discussion

1. Is the lower birth rate of people with trisomy 21 Down syndrome a sign of eugenics (see Bioethics in chapter 13)? Cite a reason for your answer.
2. Is genetic manipulation to enhance an individual a eugenic measure?
3. Do you think that eugenics should be distinguished from medical genetics based on intent, or can widespread genetic testing to prevent disease have an effect on the population that is essentially eugenic?
4. Tens of thousands of years ago, humans with very poor eyesight were likely not to have survived to reproductive age. Is wearing corrective lenses a eugenic measure? Why or why not?

Bioethics and **Clinical Connection** boxes include Questions for Discussion.

Genome Sequencing Ends a Child's "Diagnostic Odyssey"

Millie McWilliams was born September 2, 2005. At first, Millie seemed healthy, lifting her head and rolling over when most babies do. "But around 6 months, her head became shaky, like an infant's. Then she stopped saying 'dada,'" recalled her mother Angela.

By her first birthday, Millie couldn't crawl or sit, and her head shaking had become a strange, constant swaying. She had bouts of irritability and vomiting and the peculiar habit of biting her hands and fingers. In genetic diseases, odd habits and certain facial features can be clues. None of the many tests, scans, and biopsies that Millie underwent led to a diagnosis.

By age 6, Millie had lost the ability to speak, was intellectually disabled, and confined to a wheelchair, able to crawl only a few feet. Today she requires intensive home-based therapies. "She likes to look at what she wants, with an intense stare," said Angela. She loves country music and Beyoncé, and every once in awhile something funny will happen and she'll break into a big smile.

Millie's pediatrician, Dr. Sarah Soden, suggested that genome sequencing, already being done at the medical center where Millie receives care, might reveal the worsening symptoms (**figure 1A**). So the little girl and her parents had their genomes sequenced in December 2011. Dr. Soden's team identified a suspicious mutation, but the gene had never been linked to a childhood disease.

In February 2013, a medical journal published a report about four children with mutations in this gene who had symptoms strikingly like those of Millie. An answer had finally emerged: Millie has Bainbridge-Ropers syndrome. Even her facial structures—arched eyebrows, flared nostrils, and a high forehead—matched, as well as the hand-biting symptom.

Millie is missing two DNA bases in the gene *ASXL3*. DNA bases are "read" three at a time to indicate the amino acids in a protein, so missing two bases garbled the code, leading to tiny, nonfunctional proteins for that particular gene. Somehow the glitch caused the symptoms. Because Millie's father Earl and Angela do not have the mutation, it originated in either a sperm or an egg that went on to become Millie.

Figure 1A Dr. Sarah Soden examines Millie McWilliams. Genome sequencing identified the cause of Millie's intellectual disability, lack of mobility, and even her hand-biting. *Courtesy of Children's Mercy Kansas City*

So far a few dozen individuals have been diagnosed with Bainbridge-Ropers syndrome, and families have formed a support group and Facebook page. Although there is no treatment yet, the families are happy to have an answer, because sometimes parents blame themselves. Said Angela, "It was a relief to finally put a name on it and figure out what was actually going on with her, and then to understand that other families have this too. I've been able to read about her diagnosis and what other kids are going through."

Questions for Discussion

1. Millie has a younger brother and an older sister. Why don't they have Bainbridge-Ropers syndrome?
2. Would exome sequencing have discovered Millie's mutation?
3. Find a Facebook page for families that have members with a specific genetic disease and list topics that parents of affected children discuss.
4. Do you think it is valuable to have a diagnosis of a condition that has no treatment? Why or why not?

Clinical Connection boxes discuss how genetics and genomics impact health and health care.

Summary

7.1 Genes and the Environment Mold Traits

1. **Multifactorial** traits reflect influences of the environment and genes. A **polygenic** trait is determined by more than one gene and varies continuously in expression.
2. Single-gene traits are rare. For most traits, many genes contribute to a small, but not necessarily equal, degree.

7.2 Polygenic Traits Are Continuously Varying

3. Genes that contribute to polygenic traits are called **quantitative trait loci.** The frequency distribution of phenotypes for a polygenic trait forms a bell curve.

7.3 Methods to Investigate Multifactorial Traits

4. **Empiric risk** measures the likelihood that a multifactorial trait will recur based on **incidence**. The risk rises with genetic closeness, severity, and number of affected relatives.
5. **Heritability** estimates the proportion of variation in a multifactorial trait due to genetics in a particular population at a particular time. The **coefficient of relatedness** is the proportion of genes that two people related in a certain way share.
6. Characteristics shared by adopted people and their biological parents are mostly inherited, whereas similarities between adopted people and their adoptive parents reflect environmental influences.
7. **Concordance** measures the frequency of expression of a trait in both members of MZ or DZ twin pairs. The more influence genes exert over a trait, the higher the differences in concordance between MZ and DZ twins.
8. **Genome-wide association studies** correlate patterns of genetic markers (**single nucleotide polymorphisms** and/or **copy number variants**) to increased disease risk. They may use a **cohort study** to follow a large group over time, or a **case-control study** on matched pairs.
9. An **affected sibling pair study** identifies homozygous regions that may include genes of interest. **Homozygosity mapping** identifies mutations in genome regions that are homozygous because the parents shared recent ancestors.

7.4 A Closer Look: Body Weight

10. Leptin and associated proteins affect appetite. Fat cells secrete leptin in response to eating, which decreases appetite.
11. Populations that switch to a high-fat, high-calorie diet and a less-active lifestyle reveal effects of the environment on body weight.

Review Questions

1. Explain how Mendel's laws apply to multifactorial traits.
2. Choose a single-gene disease and describe how environmental factors may affect the phenotype.
3. Explain the difference between a Mendelian multifactorial trait and a polygenic multifactorial trait.
4. Do all genes that contribute to a polygenic trait do so to the same degree?
5. Explain why the curves shown in figures 7.2, 7.3, and 7.4 have the same bell shape, even though they represent different traits.
6. How can skin color have a different heritability at different times of the year?
7. Explain how the twins in figure 7.4 can have such different skin colors.
8. In a large, diverse population, why are medium brown skin colors more common than very white or very black skin?
9. Which has a greater heritability—eye color or height? State a reason for your answer.
10. Describe the type of information resulting from a(n)
    a. empiric risk determination.
    b. twin study.
    c. adoption study.
    d. genome-wide association study.
11. Name three types of proteins that affect cardiovascular functioning and three that affect body weight.
12. What is a limitation of a genome-wide association study?
13. Explain how genome sequencing may ultimately make genome-wide association studies unnecessary.

Applied Questions

1. "Heritability" is often used in the media to refer to the degree to which a trait is inherited. How is this definition different from the scientific one?
2. Would you take a drug that was prescribed to you based on your race? Cite a reason for your answer.
3. The incidence of obesity in the United States has doubled over the past two decades. Is this due more to genetic or environmental factors? Cite a reason for your answer.
4. One way to calculate heritability is to double the difference between the concordance values for MZ versus DZ twins. For multiple sclerosis, concordance for MZ twins is 30 percent, and for DZ twins, 3 percent. What is the heritability? What does the heritability suggest about the

Each chapter ends with a point-by-point **Summary.**

Review Questions assess content knowledge.

Applied Questions help students develop problem-solving skills. The first question in this section relates back to the chapter opener.

Forensics Focus

1. Establishing time of death is critical information in a murder investigation. Forensic entomologists can estimate the "postmortem interval" (PMI), or the time at which insects began to deposit eggs on the corpse, by sampling larvae of specific insect species and consulting developmental charts to determine the stage. The investigators then count the hours backwards to estimate the PMI. Blowflies are often used for this purpose, but their three larval stages look remarkably alike in shape and color, and development rate varies with environmental conditions. With luck, researchers can count back 6 hours from the developmental time for the largest larvae to estimate the time of death.

   In many cases, a window of 6 hours is not precise enough to narrow down suspects when the victim visited several places and interacted with many people in the hours before death. Suggest a way that gene expression profiling might be used to more precisely define the PMI and extrapolate a probable time of death.

Case Studies and Research Results

1. Kabuki syndrome is named for the resemblance of an affected individual to a performer wearing the dramatic makeup used in traditional Japanese theater called Kabuki. The face has long lashes, arched eyebrows, flared eyelids, a flat nose tip, and large earlobes. The syndrome is associated with many symptoms, including developmental delay and intellectual disability, seizures, a small head (microcephaly), weak muscle tone, fleshy fingertips, cleft palate, short stature, hearing loss, and heart problems. Both genes associated with the condition result in too many regions of closed chromatin. Drugs that inhibit histone deacetylases (enzymes that remove acetyl groups from histone proteins) are effective. Explain how the drugs work.

2. To make a "reprogrammed" induced pluripotent stem (iPS) cell (see figure 2.20), researchers expose fibroblasts taken from skin to "cocktails" that include transcription factors. The fibroblasts divide and give rise to iPS cells, which, when exposed to other transcription factors, divide and yield daughter cells that specialize in distinctive ways that make them different from the original fibroblasts. How do transcription factors orchestrate these changes in cell type?

3. Using an enzyme called DNAse 1, researchers can determine which parts of the genome are in the "open chromatin" configuration in a particular cell. How could this technique be used to develop a new cancer treatment?

Forensics Focus questions probe the use of genetic information in criminal investigations.

Case Studies and Research Results use stories based on accounts in medical and scientific journals; real clinical cases; posters and reports from professional meetings; interviews with researchers; and fiction to ask students to analyze data and predict results.

Dynamic Art Program

Multilevel Perspective

Illustrations depicting complex structures show microscopic and macroscopic views to help students see relationships among increasingly detailed drawings.

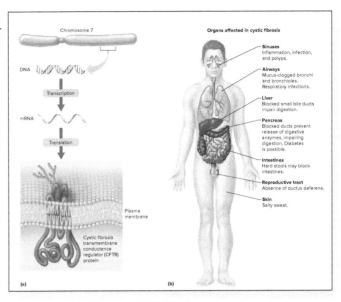

Figure 1.3 From gene to protein to person.

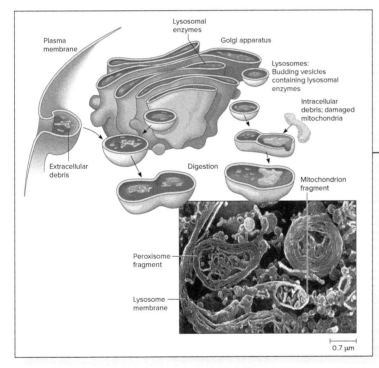

Figure 2.6 Lysosomes are trash centers.
© *Prof. P. Motta & T. Naguro/SPL/Science Source*

Combination Art

Drawings of structures are paired with micrographs to provide the best of both perspectives: the realism of photos and the explanatory clarity of line drawings.

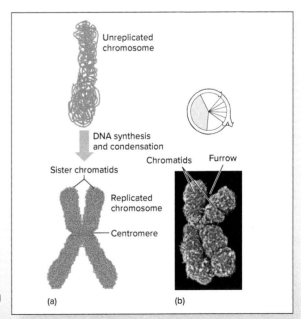

Figure 2.13 Replicated and unreplicated chromosomes. *(b):* © *SPL/Science Source*

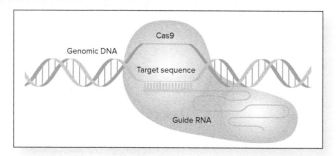

Figure 19.11 CRISPR-Cas9.

New Technologies

Genome editing can replace mutant genes with wild type alleles to counter disease or "drive out" pest populations.

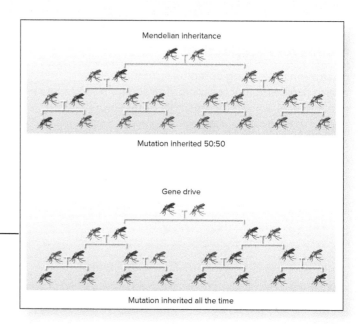

Figure 19.12 A gene drive.

Figure 4A Regular exercise helps many people who have cystic fibrosis. *Courtesy of Evan Scully*

Clinical Coverage

Evan Scully controls his cystic fibrosis with an exercise schedule of running, gym workouts, and yoga.

Process Figures

Complex processes are broken down into a series of numbered smaller steps that are easy to follow. Here, cancer evolves from an initial breakthrough "driver" mutation through additional mutations as the tumor expands and invades healthy tissue (figure 18.10).

Cancer trigger: inherited mutation or environmental insult that causes mutation

Epithelial cell

Nucleus

Oncogene turned on
or
Tumor suppressor gene turned off

1 BREAKTHROUGH: *First driver mutation* Cancer begins in a single cell when an oncogene is turned on or a tumor suppressor gene is turned off, lifting controls on cell division.

Loss of cell division control
Loss of specialization

2 EXPANSION: *Second driver mutation* Cancer cells divide more frequently and can survive in an environment with scarce nutrients, oxygen, and growth factors.

3 INVASION: *Further mutations in multiple pathways* The concerous tumor grows and spreads locally, and then attracts blood vessels and enters the bloodstream.

To other tissues

Capillary   Tumor cell

Blood vessel

Tumor

Figure 18.10 The "three strikes" of cancer.

 **connect**®

McGraw-Hill Connect® is a highly reliable, easy-to-use homework and learning management solution that utilizes learning science and award-winning adaptive tools to improve student results.

Homework and Adaptive Learning

- Connect's assignments help students contextualize what they've learned through application, so they can better understand the material and think critically.

- Connect will create a personalized study path customized to individual student needs through SmartBook®.

- SmartBook helps students study more efficiently by delivering an interactive reading experience through adaptive highlighting and review.

Connect's Impact on Retention Rates, Pass Rates, and Average Exam Scores

without Connect   with Connect

Using **Connect** improves retention rates by **19.8%**, passing rates by **12.7%**, and exam scores by **9.1%**.

Over **7 billion questions** have been answered, making McGraw-Hill Education products more intelligent, reliable, and precise.

73% of instructors who use **Connect** require it; instructor satisfaction **increases** by 28% when **Connect** is required.

Quality Content and Learning Resources

- Connect content is authored by the world's best subject matter experts, and is available to your class through a simple and intuitive interface.

- The Connect eBook makes it easy for students to access their reading material on smartphones and tablets. They can study on the go and don't need internet access to use the eBook as a reference, with full functionality.

- Multimedia content such as videos, simulations, and games drive student engagement and critical thinking skills.

Robust Analytics and Reporting

- Connect Insight® generates easy-to-read reports on individual students, the class as a whole, and on specific assignments.

- The Connect Insight dashboard delivers data on performance, study behavior, and effort. Instructors can quickly identify students who struggle and focus on material that the class has yet to master.

- Connect automatically grades assignments and quizzes, providing easy-to-read reports on individual and class performance.

Impact on Final Course Grade Distribution

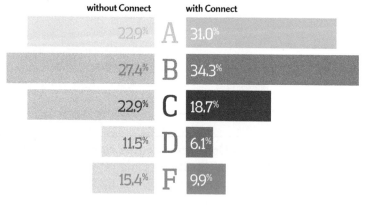

without Connect		with Connect
22.9%	A	31.0%
27.4%	B	34.3%
22.9%	C	18.7%
11.5%	D	6.1%
15.4%	F	9.9%

More students earn **As** and **Bs** when they use **Connect**.

Trusted Service and Support

- Connect integrates with your LMS to provide single sign-on and automatic syncing of grades. Integration with Blackboard®, D2L®, and Canvas also provides automatic syncing of the course calendar and assignment-level linking.

- Connect offers comprehensive service, support, and training throughout every phase of your implementation.

- If you're looking for some guidance on how to use Connect, or want to learn tips and tricks from super users, you can find tutorials as you work. Our Digital Faculty Consultants and Student Ambassadors offer insight into how to achieve the results you want with Connect.

© Tom Grill/Corbis RF

What Is in a Human Genome?

A child's genome holds information on her health, where she came from, and what she might experience and achieve—but the environment is very important too in guiding who she is and will become.

Learning Outcomes

1.1 Introducing Genes and Genomes

1. Explain what genetics is and what it is not.
2. Distinguish between gene and genome.
3. Define *bioethics*.

1.2 Levels of Genetics and Genomics

4. List the levels of genetics.
5. Explain how DNA is maintained and how it provides the information to construct a protein.
6. Explain how a mutation can cause a disease.
7. Define *exome*.
8. Distinguish between Mendelian and multifactorial traits.
9. Explain how genetics underlies evolution.

1.3 Applications of Genetics and Genomics

10. List some practical uses of DNA information.
11. Explain how DNA information can be considered with other types of information to learn about maintaining health and treating disease.
12. Distinguish between traditional breeding and genetically modifying organisms.
13. Describe a situation in which exome sequencing can be useful.

1.4 A Global Perspective on Genomes

14. Explain how investigating genomes extends beyond interest in ourselves.

 The BIG Picture

The human genome is a vast store of information encoded in the sequence of building blocks of the molecule deoxyribonucleic acid (DNA). Genetic information affects our health and traits, and holds clues to how we are biologically related to one another.

Eve's Genome

A baby is born. Drops of blood from her heel are placed into a small device that sends personal information into her electronic medical record. The device deciphers the entire sequence of DNA building blocks wound into the nucleus of a white blood cell. This is Eve's genome. Past, present, and future are encoded in nature's master informational molecule, deoxyribonucleic acid, or DNA—with room for environmental influences.

Eve's genome indicates overall good genetic health. She has a mild clotting disease that the nurse suspected when two gauze patches were needed to stop the bleeding from the heel stick. Two rare variants of the gene that causes cystic fibrosis (CF) mean that Eve is susceptible to certain respiratory infections and sensitive to irritants, but her parents knew that from prenatal testing. Fortunately the family lives in a rural area far from pollution, and Eve will have to avoid irritants such as smoke and dust.

The inherited traits that will emerge as Eve grows and develops range from interesting to important. Her hair will darken and curl, and genes that contribute to bone development indicate that she'll have a small nose, broad forehead, and chiseled cheekbones. If she follows a healthy diet, she'll be as tall as her parents. On the serious side, Eve has inherited a mutation in a gene that greatly raises her risk of developing certain types of cancers. Her genes predict a healthy heart, but she might develop diabetes unless she exercises regularly and limits carbohydrates in her diet.

Many traits are difficult to predict because of environmental influences, including experiences. What will Eve's personality be like? How intelligent will she be? How will she react to stress? What will be her passions?

Genome sequencing also reveals clues to Eve's past, which is of special interest to her father, who was adopted. She has gene variants common among the Eastern European population of her mother's origin, and others that match people from Morocco. Is that her father's heritage? Eve is the beautiful consequence of a mix of her parents' genomes, receiving half of her genetic material from each.

Do you want to know the information in your genome?

1.1 Introducing Genes and Genomes

Genetics is the study of inherited traits and their variation. Genetics is not the same as genealogy, which considers relationships but not traits. Because some genetic tests can predict illness, genetics has also been compared to fortunetelling. However, genetics is a life science. **Heredity** is the transmission of traits and biological information between generations, and genetics is the study of how traits are transmitted.

Inherited traits range from obvious physical characteristics, such as freckles and red hair, to many aspects of health, including disease. Talents, quirks, personality traits, and other difficult-to-define characteristics might appear to be inherited if they affect several family members, but may reflect shared genetic and environmental influences. Attributing some traits to genetics, such as sense of humor or whether or not one votes, are oversimplifications. These connections are associations, not causes.

Over the past decade, genetics has exploded from a mostly academic discipline and a minor medical specialty dealing with rare diseases, to the new basis of some fields, such as oncology (cancer care). Genetics is a part of everyday discussion. Personal genetic information is accessible, and we are learning the contribution of genes to the most common traits and diseases. Many health care providers are learning how to integrate DNA information into clinical practice.

Like all sciences, genetics has its own vocabulary. Some technical terms and expressions may be familiar, but actually have precise scientific definitions. Conversely, the language of genetics sometimes enters casual conversation. *"It's in her DNA,"* for example, usually means an inborn trait, not a specific DNA sequence. The terms and concepts introduced in this chapter are explained and explored in detail in subsequent chapters.

Genes are the units of heredity. Genes are biochemical instructions that tell **cells**, the basic units of life, how to manufacture certain proteins. These proteins, in turn, impart or control the characteristics that create much of our individuality. A gene consists of the long molecule **deoxyribonucleic acid (DNA)**. The DNA transmits information in its sequence of four types of building blocks, which function like an alphabet.

The complete set of genetic instructions characteristic of an organism, including protein-encoding genes and other DNA sequences, constitutes a **genome**. Nearly all of our cells contain two copies of the genome. Following a multi-year, international effort, researchers published the deciphered sequences of the first human genomes, in 2003. However, scientists are still analyzing what all of our genes do, and how genes interact and respond to environmental stimuli. Only a tiny fraction of the 3.2 billion building blocks of our genetic instructions determines the most interesting parts of ourselves—our differences. Analyzing and comparing genomes, which constitute the field of **genomics**, reveal how closely related we are to each other and to other species.

Genetics directly affects our lives and those of our relatives, including our descendants. Principles of genetics also touch history, politics, economics, sociology, anthropology, art, the law, athletics, and psychology. Genetic questions force us to wrestle with concepts of benefit and risk, even tapping our deepest feelings about right and wrong. A field of study called **bioethics** began in the 1970s to address moral issues and controversies that arise in applying medical technology. Bioethicists today confront concerns that arise from new *genetic* technology, such as privacy, use of genetic information, and discrimination. Essays throughout this book address bioethical issues.

Key Concepts Questions 1.1

1. Distinguish between genetics and heredity.
2. Distinguish between a gene and a genome.
3. Describe the type of information that the DNA sequence of a gene encodes.
4. Define *bioethics*.

1.2 Levels of Genetics and Genomics

Genetics considers the transmission of information at several levels. It begins with the molecular level and broadens through cells, tissues and organs, individuals, families, and finally to populations and the evolution of species (**figure 1.1**).

Instructions and Information: DNA

DNA resembles a spiral staircase or double helix. The "rails," or backbone, consist of alternating chemical groups (sugars and phosphates) and are the same in all DNA molecules.

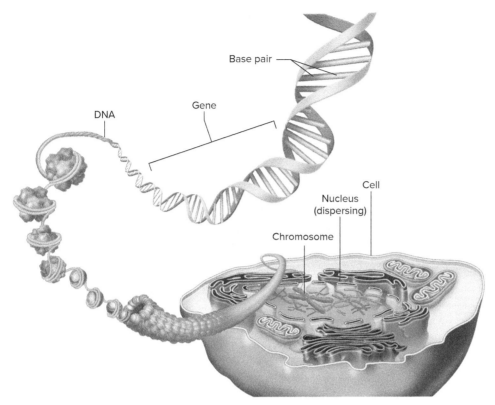

Figure 1.1 **Levels of genetics.** Genetics can be considered at several levels, from DNA, to genes, to chromosomes, to genomes, to the more familiar individuals, families, and populations.

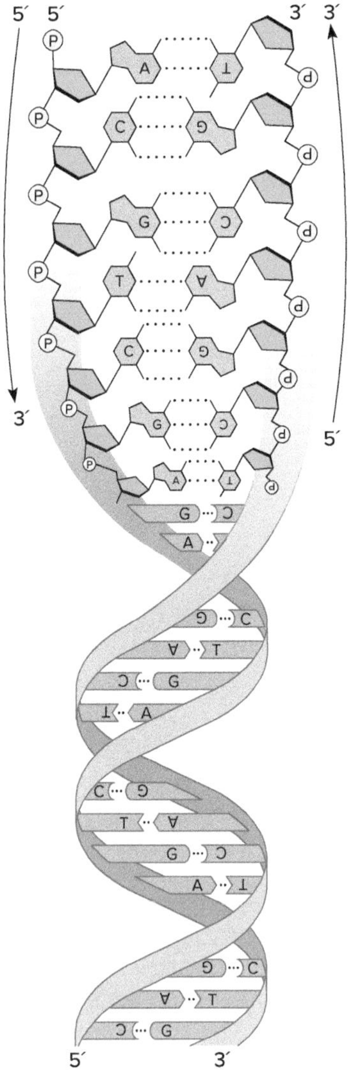

Figure 1.2 **The DNA double helix.** The 5' and 3' labels indicate the head-to-tail organization of the DNA double helix. A, C, T, and G are bases. S stands for sugar and P for phosphate. The green five-sided shapes represent the sugars.

The two strands of the double helix are oriented in opposite directions, like two snakes biting each other's tails. The "steps" of the DNA double helix are pairs of the four types of building blocks, or **nitrogenous bases**: **adenine** (A) and **thymine** (T), which attract each other, and **cytosine** (C) and **guanine** (G), which attract each other (**figure 1.2**). The information that a DNA molecule imparts is in the sequences of A, T, C, and G.

The chemical structure of DNA gives the molecule two key abilities that are essential for the basis of life: DNA can both perpetuate itself when a cell divides and provide information to manufacture specific proteins. Each set of three consecutive DNA bases is a code for a particular amino acid, and amino acids are the building blocks of proteins.

Accessing genetic information occurs in three processes: replication of DNA, transcription of RNA from the information in DNA, and translation of protein from RNA. Chapter 10 discusses these complex processes in detail.

In **DNA replication**, the chains of the double helix untwist and separate, and then each half builds a new partner chain from free DNA bases. A and T attract and C and G attract. Then **transcription** copies the sequence of part of one strand of a DNA molecule into a related molecule, messenger **ribonucleic acid (RNA)**. In **translation**, each three RNA bases in a row attract another type of RNA that functions as a connector, bringing in a particular amino acid. The amino acids align and link like snap beads, forming a protein. The inherited disease cystic fibrosis (CF) illustrates how proteins provide the

traits associated with genes. The protein that is abnormal in CF works like a selective doorway in cells lining the airways and certain other body parts, thickening secretions when it doesn't work properly (**figure 1.3**).

A change in a gene, or **mutation**, can have an effect at the whole-person level, such as causing a disease. **Figure 1.4** depicts the effect of a mutation in the gene that causes CF when mutant, which is called *CFTR* (cystic fibrosis transmembrane conductance regulator). A change of a "C" in the DNA sequence at a specific location in the gene to a "T" inserts the amino acid aspartic acid rather than the amino acid glycine as the protein forms. The resulting protein cannot open to the cell's surface, removing channels for certain salt components, causing the symptoms described in figure 1.3.

The human genome has about 20,325 protein-encoding genes, and these DNA sequences comprise the **exome**. Protein-encoding genes account for only about 1.5 percent of the human genome, yet this portion accounts for about 85 percent of

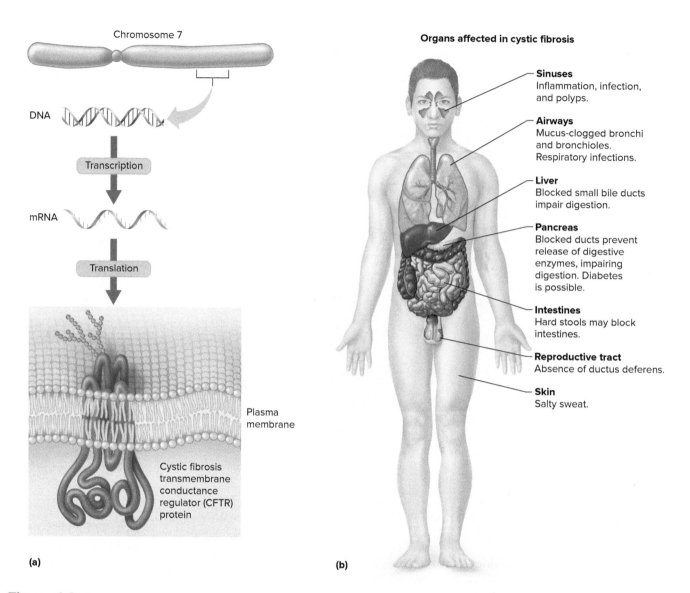

(a)

Chromosome 7

DNA

Transcription

mRNA

Translation

Plasma
membrane

Cystic fibrosis
transmembrane
conductance
regulator (CFTR)
protein

(b)

Organs affected in cystic fibrosis

Sinuses
Inflammation, infection,
and polyps.

Airways
Mucus-clogged bronchi
and bronchioles.
Respiratory infections.

Liver
Blocked small bile ducts
impair digestion.

Pancreas
Blocked ducts prevent
release of digestive
enzymes, impairing
digestion. Diabetes
is possible.

Intestines
Hard stools may block
intestines.

Reproductive tract
Absence of ductus deferens.

Skin
Salty sweat.

Figure 1.3 **From gene to protein to person.** **(a)** The gene encoding the CFTR protein, causing cystic fibrosis when in a variant form (a mutation), is part of the seventh largest chromosome. CFTR normally folds into a channel that regulates the flow of salt components (ions) into and out of cells lining the respiratory tract, pancreas, intestines, and elsewhere. **(b)** Cystic fibrosis causes several symptoms.

(Source: Data from "Reverse genetics and cystic fibrosis" by M. C. Iannuzi and F. S. Collins. *American Journal of Respiratory Cellular and Molecular Biology* 2:309–316 [1990].)

known genetic diseases. The rest of the genome includes many DNA sequences that assist in protein synthesis or turn protein-encoding genes on or off. The ongoing effort to understand what individual genes do is termed *annotation*.

The same protein-encoding gene may vary slightly in DNA base sequence from person to person. These gene variants are called **alleles**. The changes in DNA sequence that distinguish alleles arise by mutation. (The word "mutation" is also used as a noun to refer to the changed gene.) Once a gene mutates, the change is passed on when the cell that contains it divides. If the change is in a sperm or egg cell that becomes a fertilized egg, it is passed to the next generation.

Some mutations cause disease, and others provide variation such as freckled skin. Mutations can also help. One rare mutation makes a person's cells unable to manufacture a surface protein that binds HIV. These people are resistant to HIV

infection. Mutations that have no detectable effect because they do not change the encoded protein in a way that affects its function are sometimes called gene variants. They are a little like a minor spelling error that does not obscure the meaning of a sentence.

The DNA sequences of the human genome are dispersed among 23 structures called **chromosomes**. When a cell is dividing, the chromosomes wind up so tightly that they can be seen under a microscope when stained, appearing rod shaped. The DNA of a chromosome is continuous, but it includes hundreds of genes, plus other sequences.

A human **somatic cell** (non-sex cell) has 23 pairs of chromosomes. Twenty-two of these 23 pairs are **autosomes**, which do not differ between the sexes. The autosomes are numbered from 1 to 22, with 1 being the largest. The other two chromosomes, the X and the Y, are **sex chromosomes**.

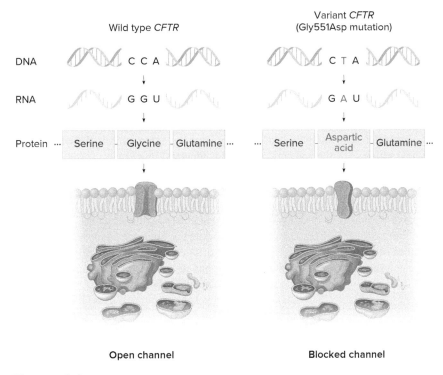

DNA ··· C C A ···   Wild type *CFTR*

DNA ··· C T A ···   Variant *CFTR* (Gly551Asp mutation)

RNA ··· G G U ···

RNA ··· G A U ···

Protein ··· Serine – Glycine – Glutamine ···

Protein ··· Serine – Aspartic acid – Glutamine ···

Open channel

Blocked channel

Figure 1.4 **A mutation can alter a protein, causing symptoms.** One type of mutation in the *CFTR* gene replaces one amino acid type (glycine) with another (aspartic acid) at a specific site, altering the encoded protein in a way that closes a type of ion channel that participates in secretion.

The Y chromosome bears genes that determine maleness. In humans, a female has two X chromosomes and a male has one X and one Y chromosome. Charts called **karyotypes** display the chromosome pairs from largest to smallest.

To summarize, a human somatic cell has two complete sets of genetic information (genomes). The protein-encoding genes are scattered among 3.2 billion DNA bases in each set of 23 chromosomes.

A trait caused predominantly by a single gene is termed Mendelian. Most characteristics are **multifactorial traits**, which means that they are determined by one or more genes and environmental factors (**figure 1.5**). The more factors that contribute to a trait or illness—inherited or environmental—the more difficult it is to predict the risk of occurrence in a particular family member. The bone-thinning condition osteoporosis illustrates the various factors that can contribute to a disease. Several genes elevate osteoporosis risk by conferring susceptibility to fractures, but so do smoking, lack of weight-bearing exercise, and a calcium-poor diet.

Environmental effects on gene action counter the idea of "genetic determinism," that "we are our genes." This idea may be harmful or helpful. As part of social policy, genetic determinism can be disastrous. Assuming that one group of people is genetically less intelligent than another can lower expectations and even lead to fewer educational opportunities for people perceived as inferior. Environment, in fact, has a large impact on intellectual development. On the other hand, knowing the genetic contribution to a trait can provide information that can help a health care provider select a treatment most likely to work, with minimal adverse effects.

The Body: Cells, Tissues, and Organs

A human body consists of approximately 30 trillion cells. All somatic cells except red blood cells contain two copies of the genome, but cells differ in appearance and activities because they use only some of their genes. Which genes a cell uses at any given time depends upon environmental conditions inside and outside the body.

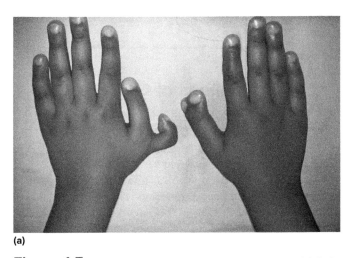

(a)

(b)

Figure 1.5 **Mendelian versus multifactorial traits.** **(a)** Polydactyly—extra fingers and/or toes—is a Mendelian trait (single gene). **(b)** Hair color is multifactorial, controlled by at least three genes plus environmental influences, such as the bleaching effects of sun exposure. *(a): © Lester V. Bergman/Getty Images; (b): © Steve Mason/Getty RF*

Like the Internet, a genome contains a wealth of information, but only some of it is needed in a particular cell under particular circumstances. The use, or "expression," of different subsets of genes to manufacture proteins drives the **differentiation**, or specialization, of distinctive cell types. An adipose cell is filled with fat, but not the contractile proteins of muscle cells. Both cell types, however, have two complete genomes. Groups of differentiated cells assemble and interact with each other and the nonliving materials that they secrete to form aggregates called **tissues**. Table 2.1 lists the four basic tissue types, which are composed of more than 290 types of cells.

Tissues intertwine and layer to form organs, which connect into organ systems. The stomach, for example, is a sac made of muscle that also has a lining of epithelial tissue, nervous tissue, and a supply of blood, which is a type of connective tissue. Many organs include rare, unspecialized **stem cells**. A stem cell can divide to yield another stem cell and a cell that differentiates. Stem cells provide a reserve supply of cells that enable an organ to grow and repair damage.

Relationships: From Individuals to Families

Two terms distinguish the alleles that are *present* in an individual from the alleles that are *expressed*. The **genotype** refers to the underlying instructions (alleles present), whereas the **phenotype** is the visible trait, biochemical change, or effect on health (alleles expressed). Alleles are further distinguished by how many copies are necessary to affect the phenotype. A **dominant** allele has an effect when present in just one copy (on one chromosome), whereas a **recessive** allele must be present on both chromosomes of a pair to be expressed.

Individuals are genetically connected into families. A person has approximately half of his or her gene variants in common with each parent, sibling, and offspring, and one-quarter with each grandparent. First cousins share one-eighth of their gene variants. Charts called **pedigrees** depict the members of a family and indicate which individuals have particular inherited traits.

The Bigger Picture: From Populations to Evolution

Above the family level of genetic organization is the population. In a strict biological sense, a population is a group of individuals that can have healthy offspring together. In a genetic sense, a population is a large collection of alleles, distinguished by their frequencies. People from a Swedish population, for example, would have a greater frequency of alleles that specify light hair and skin than people from a population in Nigeria, who tend to have dark hair and skin. All the alleles in a population constitute the **gene pool**. (An individual does not have a gene pool.)

Population genetics is applied in health care, forensics, and other fields. It is also the basis of evolution, which is defined as changing allele frequencies in populations. These small-scale genetic changes underlie the species distinctions we most often associate with evolution.

Comparing DNA sequences for individual genes, or the amino acid sequences of the proteins that the genes encode, can reveal how closely related different types of organisms are. The assumption is that the more similar the DNA sequences are, the more recently two species diverged from a shared ancestor, and the more closely related they are. This is a more plausible explanation than two species having evolved similar or identical gene sequences coincidentally. The same logic applies to family patterns of inherited traits. It is more likely that a brother and sister share approximately half of their gene variants because they have the same parents than that half of their genetic material is identical by chance.

More information is available in full genome sequences than in single genes. Humans, for example, share more than 98 percent of the DNA sequence with chimpanzees. Our genomes differ from theirs more in gene organization and in the number of copies of genes. Learning the functions of the human-specific genes may explain the differences between us and them—such as our sparse body hair and use of spoken language. Figure 16.8 highlights some of our distinctively human traits.

At the genome level, we are much more like each other genetically than are other mammals. Chimpanzees are more distinct from each other than we are! The most genetically diverse modern people are from Africa, where humanity arose. The gene variants among different modern ethnic groups include subsets of our ancestral African gene pool.

Key Concepts Questions 1.2

1. List and define the levels of genetic information.
2. Explain how DNA carries and maintains information.
3. Explain how a mutation can cause a disease.
4. Explain how a gene can exist in more than one form.
5. Distinguish between Mendelian and multifactorial traits.
6. Explain how gene expression underlies composition of the human body.
7. Distinguish between genotype and phenotype; dominant and recessive.
8. Explain how comparing DNA sequences can clarify evolutionary relationships.

1.3 Applications of Genetics and Genomics

Genetics is impacting many areas of our lives, from health care choices, to what we eat and wear, to unraveling our pasts and guiding our futures. "Citizen scientists" are discovering genetic information about themselves while helping researchers compile databases that will help many.

Thinking about genetics evokes fear, hope, anger, and wonder, depending upon context and circumstance. Following are a few eclectic uses of DNA information, then glimpses of applications of genetics and genomics that are explored more fully in subsequent chapters:

- Identifying which of several pets produced feces, so a stool sample can be brought to a veterinarian to diagnose the sick animal
- Predicting shelf life of fruits and vegetables, detecting spoiled meat, identifying allergens, and indicating degree of fermentation in cheese
- Identifying victims of human trafficking at transportation centers by comparing the DNA of suspected victims to DNA from concerned relatives
- Detecting disease-causing mutations or abnormal chromosome numbers in a fetus from DNA in a pregnant woman's blood
- Performing rapid diagnosis of an infectious disease on the battlefield
- Creating a tree of life depiction of how all species are related
- Selecting crops and show animals for breeding
- Choosing people to date
- Detecting tiny amounts of DNA in fur, feathers, or feces of rare or elusive species to sequence their genomes and learn more about them (**figure 1.6**)

Establishing Identity

A technique called **DNA profiling** compares DNA sequences among individuals to establish or rule out identity, relationships, or ancestry. The premise is that the more DNA sequences two individuals share, the more closely related they are.

DNA profiling has varied applications, in humans and other species. The term is most often used in the context of forensic science, which is the collecting of physical evidence of a crime. Comparing DNA collected at crime scenes to DNA in samples from suspects often leads to convictions, and also to reversing convictions erroneously made using other forms of evidence.

DNA profiling is useful in identifying victims of natural disasters, such as violent storms and earthquakes. In happier circumstances, DNA profiles maintained in databases assist adopted individuals in locating blood relatives and children of sperm donors in finding their biological fathers and half-siblings.

Another use of DNA profiling is to analyze food, because foods were once organisms, which have species-specific DNA sequences. Analyzing DNA sequences revealed horsemeat in meatballs sold at a restaurant chain, cheap fish sold as gourmet varieties, and worms in cans of sardines.

Illuminating History

DNA analysis is a time machine of sorts. It can connect past to present, from determining family relationships to establishing geographic origins of specific populations. DNA evidence is perhaps most interesting when it contradicts findings from anthropology and history. Consider three examples, from most recent to most ancient.

DNA analysis confirmed that Thomas Jefferson had children with his slave Sally Hemings. The president was near Hemings 9 months before each of her seven children was born, and the children resembled him. Male descendants of Sally Hemings share an unusual Y chromosome sequence with the president's male relatives. His only son with his wife died in infancy, so researchers deduced the sequence of the president's Y chromosome from descendants of his uncle. Today the extended family holds reunions (**figure 1.7**).

DNA testing can provide views into past epidemics of infectious diseases by detecting genes of the pathogens. For example, analysis of DNA in the mummy of the Egyptian king Tutankhamun, who died in 1323 B.C.E. at age 19, revealed DNA

(a)

(b)

Figure 1.6 DNA is used to study endangered and difficult-to-capture animals. The field of conservation genomics collects and analyzes DNA from fur, feathers, and feces. The approach is used to study the American pika **(a)** and a rare subspecies of panther **(b)** that lives in isolated areas of southeast Russia and northeast China. *(a): Source: Jim Peaco/National Park Service; (b): © Destinyweddingstudio/Shutterstock*

Figure 1.7 **DNA reveals and clarifies history.** After DNA evidence showed that Thomas Jefferson likely fathered children with his slave Sally Hemings, confirming gossip of the time, descendants of both sides of the family met, and continue to do so. © Leslie Close/AP Images

from the microorganism that causes malaria. He likely died from complications of malaria following a leg fracture from weakened bones, rather than from intricate murder plots, a kick from a horse, or a fall from a chariot, as had been thought. His tomb included a cane and drugs, supporting the diagnosis based on DNA evidence.

The Basque people, who have a distinct language, live on the coastal border of France and Spain. The 700,000 modern Basques had long been thought to descend from hunter-gatherers who lived in the area about 7,500 years ago, before the first farmers arrived. DNA told a different story. Researchers compared the genome sequences of bones from eight Basque farmers who had lived in a cave in northern Spain from 5,500 to 3,500 years ago to genomes from other skeletons representing several European hunter-gatherers and early farming groups, as well as to modern Europeans. While the ancient farmers had genomes representing many groups, including those of hunter-gatherers, the Basques indeed have a unique genome—but one that descends from the earliest farmers, not from hunter-gatherers. Apparently their uniqueness today is due to their self-imposed isolation as the rest of Europe interbred.

Precision Medicine

In several nations, people are volunteering to have their genomes sequenced to learn more about health and disease. The DNA data are considered along with other types of information that can impact health, such as environmental exposures, exercise, diet, lifestyle factors, and the many microbes that live in and on the human body, collectively termed the **microbiome** (**figure 1.8**). In the United States, a precision medicine initiative is tracking 1 million volunteers who are having their genomes sequenced and providing as much information about their lives as possible.

On a smaller scale, a precision medicine approach consults DNA information to select drugs that are most likely to work and least likely to have side effects in a particular individual. This strategy, called **pharmacogenetics**, is already used to guide prescription of more than 150 drugs. Some highly effective new drugs that collectively treat a variety of conditions, from cystic fibrosis to cancers, are targeted to patients who have specific mutations in specific genes.

Healthy people have much to contribute to precision medicine by helping researchers identify gene variant combinations that contribute to wellness and longevity. Consider one participant's story:

My grandmother, mother, and myself all look at least 15 years younger than we are. My mother and I have absolutely no health concerns and have vital signs that are also indicative of people 15 to 20 years younger. I'm 46 and my mother is 64. I know there are world record holders who have amazingly long life spans, too, and other families with characteristics of slower aging, like my own. Maybe there is something in our DNA that delays aging?

The initiative in the United States, and efforts elsewhere such as the United Kingdom's 100,000 Genomes Project, are involving the general population in collecting information that will improve health care for many. For example, imagine identifying a group of people who all have a mutation known to cause a specific disease, but some of them do not have the disease. Something in the environment or that the protected people are doing might explain their health, and the finding used to develop treatments.

Genetic Modification

Genetic modification means altering a gene or genome in a way that does not occur in nature, such as giving a carrot a gene from a green bean that isn't part of the carrot genome. Traditional agriculture and animal breeding do not result in "genetically modified organisms" (GMOs) because they select traits within one species.

In health care, GMOs in the form of bacteria bearing human genes have provided many drugs, such as insulin and clotting factors, since the 1970s. Foods are genetically modified to be more nutritious, easier to cultivate, or able to grow in the presence of herbicides and pesticides. In the United States, more than 90 percent of the crops of corn, soybeans, and cotton are GMOs and the public has been eating them for more than a quarter century. However, some GMO foods have been failures. The "Flavr Savr" tomato, for example, was genetically modified to have a longer shelf life, but it had a terrible taste!

Genetically modified crops and drugs have been available for many years. A newer technology, **genome editing**, can replace, remove, or add specific genes into the cells of any organism (**figure 1.9**). Researchers are using genome-editing techniques experimentally to alter individual somatic cells growing in laboratory glassware, and genome editing may one day be used to treat certain inherited diseases. However, in the meantime, many researchers have agreed not to edit the genomes of human gametes (sperm or eggs) or fertilized eggs, which would create a genetically modified human. The most talked-about genome-editing tool is CRISPR-Cas9, but similar technologies

Figure 1.9 **Genome editing.** CRISPR-Cas9 is one genome-editing technique. It uses a protein (Cas9) that functions much like a wrench, along with RNA molecules (CRISPRs) that guide the tool to a specific site in a genome. Genome editing enables researchers to add, delete, or replace specific DNA sequences in the cells of any type of organism. *Ernesto del Aguilo, III, NHGRI*

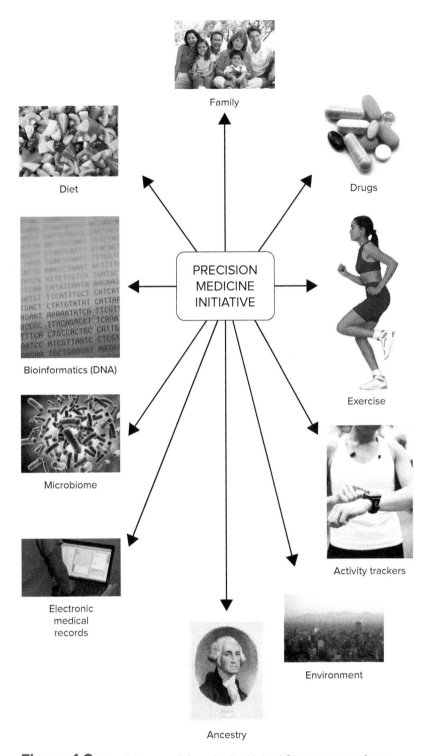

Figure 1.8 **Precision medicine.** In the United States, researchers are analyzing the genome sequences of 1 million citizens with other sources of "big data," such as electronic medical records, diet, exercise, family history and ancestry, the microbiome, and environmental exposures. *(Family):* © *Stockbroker/Alamy; (Drugs):* © *Dan Wilkie/Getty Images; (Exercise):* © *Glyn Jones/Corbis; (Activity trackers):* © *Disuke Martis /Getty Images; (Environment):* © *Phototreat/Getty Images; (Ancestry): Courtesy Yale University Library; (Electronic medical records):* © *McGraw-Hill Higher Education, Inc.; (Microbiome):* © *Disuke Martis/Getty Images; (Bioinformatics (DNA)):* © *Getty/Stockphoto; (Diet):* © *Babaz/Shutterstock*

have been used since 2009, and the idea to edit the genome has been discussed since the 1980s.

Exome Sequencing

Exome sequencing determines the order of the DNA bases of all parts of the genome that encode proteins—that is, about 20,325 genes. The information is compared to databases that list many gene variants (alleles) and their associations with specific phenotypes, such as diseases. Exome sequencing is valuable after more conventional tests, such as tests for single gene diseases and chromosome abnormalities, do not explain a person's symptoms.

The information in an exome sequence can be lifesaving. For example, it showed that a 2-year-old boy who had severe eating difficulties and was very thin had a mutation known to cause Marfan syndrome, although he didn't have the characteristic long limbs and fingers that might have alerted a doctor. However, one symptom of Marfan syndrome is an enlarged aorta, the largest artery, and often this is not obvious until the aorta bursts and the person dies. After exome sequencing revealed a Marfan mutation, an ultrasound scan indeed showed a bulge in the wall of the aorta near the boy's heart, which could have resulted in his sudden death. Doctors successfully patched the bulging blood vessel.

Exome sequencing is particularly valuable in identifying extremely rare diseases—swiftly. In the past, parents of children with very unusual symptoms referred to the multiyear effort to find a physician who recognized the disease as a "diagnostic odyssey," in reference to Homer's epic poem about Greek hero Odysseus' 10-year journey home. Most diagnostic odysseys for genetic diseases took 5 years or longer. With exome as well as full genome sequencing, diagnosis can take just hours. **Clinical Connection 1.1** describes how genome sequencing led to diagnosis of a rare genetic disease in a child.

Genome Sequencing Ends a Child's "Diagnostic Odyssey"

Millie McWilliams was born September 2, 2005. At first, Millie seemed healthy, lifting her head and rolling over when most babies do. "But around 6 months, her head became shaky, like an infant's. Then she stopped saying 'dada'," recalled her mother Angela.

By her first birthday, Millie couldn't crawl or sit, and her head shaking had become a strange, constant swaying. She had bouts of irritability and vomiting and the peculiar habit of biting her hands and fingers. In genetic diseases, odd habits and certain facial features can be clues. None of the many tests, scans, and biopsies that Millie underwent led to a diagnosis.

By age 6, Millie had lost the ability to speak, was intellectually disabled, and confined to a wheelchair, able to crawl only a few feet. Today she requires intensive home-based therapies. But Millie can communicate with her parents. "She likes to look at what she wants, with an intense stare," said Angela. She loves country music and Beyoncé, and every once in awhile something funny will happen and she'll break into a big smile.

Millie's pediatrician, Dr. Sarah Soden, suggested that genome sequencing, already being done at the medical center where Millie receives care, might explain the worsening symptoms (**figure 1A**). So the little girl and her parents had their genomes sequenced in December 2011. Dr. Soden's team identified a suspicious mutation, but the gene had never been linked to a childhood disease.

In February 2013, a medical journal published a report about four children with mutations in this gene who had symptoms strikingly like those of Millie. An answer had finally emerged: Millie has Bainbridge-Ropers syndrome. Even her facial structures—arched eyebrows, flared nostrils, and a high forehead—matched, as well as the hand-biting symptom.

Millie is missing two DNA bases in the gene *ASXL3*. DNA bases are "read" three at a time to indicate the amino acids in a protein, so missing two bases garbled the code, leading to tiny, nonfunctional proteins for that particular gene. Somehow the glitch caused the symptoms. Because Millie's father Earl and Angela do not have the mutation, it originated in either a sperm or an egg that went on to become Millie.

Figure 1A Dr. Sarah Soden examines Millie McWilliams. Genome sequencing identified the cause of Millie's intellectual disability, lack of mobility, and even her hand-biting. *Courtesy of Children's Mercy Kansas City*

So far a few dozen individuals have been diagnosed with Bainbridge-Ropers syndrome, and families have formed a support group and Facebook page. Although there is no treatment yet, the families are happy to have an answer, because sometimes parents blame themselves. Said Angela, "It was a relief to finally put a name on it and figure out what was actually going on with her, and then to understand that other families have this too. I've been able to read about her diagnosis and what other kids are going through."

Questions for Discussion

1. Millie has a younger brother and an older sister. Why don't they have Bainbridge-Ropers syndrome?

2. Would exome sequencing have discovered Millie's mutation?

3. Find a Facebook page for families that have members with a specific genetic disease and list topics that parents of affected children discuss.

4. Do you think it is valuable to have a diagnosis of a condition that has no treatment? Why or why not?

Key Concepts Questions 1.3

1. List some uses of DNA information in identifying individuals and in clarifying relationships in history.

2. Explain how precision medicine uses more than just DNA information and helps people other than those with genetic diseases.

3. Distinguish genetic modification from traditional breeding.

4. Describe the role of exome sequencing in solving medical mysteries.

1.4 A Global Perspective on Genomes

We share the planet with many thousands of other species. We aren't familiar with many of Earth's residents because we can't observe their habitats, or we can't grow them in laboratories. **Metagenomics** is a field that describes much of the invisible living world by sequencing all of the DNA in a habitat, such as soil, an insect's gut, garbage, or a volume of captured air over a city. A project that involves swabbing various surfaces in the New York City subway system is painting a portrait of an "urban microbiome." Metagenomics studies are showing how species interact, and may yield information useful in developing new drugs or energy sources.

Metagenomics researchers collect and sequence DNA, then consult databases of known genomes to imagine what the organisms might be like. The first metagenomics project described life in the Sargasso Sea. This 2-million-square-mile area off the coast of Bermuda was thought to lack life beneath its thick cover of seaweed, which is so abundant that Christopher Columbus thought he'd reached land when his ships came upon it. Boats have been lost in the Sargasso Sea, which includes the area known as the Bermuda Triangle. Researchers collected more than a billion DNA molecules from the depths, representing about 1,800 microbial species and including more than a million previously unknown genes.

Perhaps the most interesting subject of metagenomics is the human body, which is home to trillions of bacteria. Section 2.5 discusses the human microbiome, the collection of life within us.

Genetics is a special branch of life science because it affects us intimately. Equal access to genetic tests and treatments, misuse of genetic information, and abuse of genetics to intentionally cause harm are compelling social issues that parallel scientific progress.

Genetics and genomics are spawning technologies that may vastly improve quality of life. But at first, tests and treatments are costly and not widely available. Whereas some people in economically and politically stable nations may be able to afford genetic tests or exome or genome sequencing, or even take genetic tests for curiosity, poor people in other nations struggle to survive, often lacking basic vaccines and medicines. In an African nation where two of five children suffer from AIDS and many die from other infectious diseases, newborn screening for rare single-gene diseases may seem impractical. However, genetic diseases weaken people so that they become more susceptible to infectious diseases, which they can pass to others.

Recognizing that human genome information can ultimately benefit everyone, organizations such as the United Nations, World Health Organization, and World Bank are discussing how nations can share new diagnostic tests and therapeutics that arise from genome information about ourselves and the microbes that make us sick. **Bioethics** discusses instances when genetic testing can threaten privacy.

Key Concepts Questions 1.4

1. Explain how metagenomics explores genetics beyond the human body.
2. Name social issues that arise from technologies based on genetics.

Bioethics

Genetic Testing and Privacy

The field of bioethics began in the 1950s to address issues raised by medical experimentation during World War II. Bioethics initially centered on informed consent, paternalism, autonomy, allocation of scarce medical resources, justice, and definitions of life and death. Today, the field covers medical and biotechnologies and the dilemmas they present. Genetic testing is a key issue in bioethics because its informational nature affects privacy, yet collecting genetic information from many individuals is required to make discoveries of clinical importance that could affect many (**figure 1B**). Consider these situations.

Testing Incoming Freshmen

Today students in genetics classes at some colleges take genetic tests. Medical students are having their exomes or genomes sequenced to learn how to eventually use genetic information in providing patient care. But a situation in 2010 at a college indicated difficulties that could arise from pressuring people to take genetic tests.

In summer 2010, incoming freshmen at the University of California, Berkeley, received, along with class schedules and

Figure 1B **Genetic privacy.** How can we protect DNA sequence data? © *Jane Ades, NHGRI*

dorm assignments, kits to send in DNA samples to test for three genes that control three supposedly harmless traits. Participation

(Continued)

was voluntary, and because the intent was to gather data, informed consent was not required. However, after genetics groups, bioethicists, policy analysts, and consumer groups protested, the Department of Public Health ruled that the tests provided personal medical information and should be conducted by licensed medical labs. Because this quintupled the cost, the university changed the program to collect aggregate data, rather than individually identified results.

The three genetic tests were to detect lactose intolerance, alcohol metabolism, and folic acid metabolism. The alcohol test detects variants of a gene that cause a facial flush, nausea, and heart palpitations after drinking, particularly in East Asians—who made up a significant part of the freshman class. Certain mutations in this gene raise the risk of developing esophageal cancer, and so test results may be useful, but they could also encourage drinking.

The Military

A new recruit hopes that the DNA sample given when military service begins is never used—it is stored to identify remains. Until recently, genetic tests have only been performed for two specific illnesses that could endanger soldiers under certain environmental conditions. Carriers of sickle cell disease can develop painful blocked circulation at high altitudes, and carriers of G6PD deficiency react badly to anti-malaria medication. Carriers wear red bands on their arms to alert officers to keep the soldiers from harmful situations. In the future, the military may use genetic information to identify soldiers at risk for such conditions as depression and posttraumatic stress disorder. Deployments can be tailored to personal risks, minimizing suffering.

Research Study Participants

When genetic studies considered only a few genes, peoples' identities were protected, because there were many more people than genotypes. That is, a person was unlikely to be the only one to have a particular genotype. That is no longer true. Because studies now probe a million or more pieces of genetic information, an individual's genotype can be traced to a particular group being investigated. The more ways that we can detect that people vary, the easier it is to identify any one of them. It is like adding digits to a zip code or a new area code to a phone number to increase the pool of identifiers.

Clever consulting of information other than genotypes can identify individuals. If a child's DNA information is in a study and his or her name is in a database that includes a rare disease name and a hometown, comparing these sources can match a name to a DNA sequence. **Bioethics** in chapter 16 describes a clever combination of information from a Google search and from a genealogical database that graduate students used to identify participants in an experiment.

Questions for Discussion

1. If a genetic test reveals a mutation that could harm a blood relative, should the first person's privacy be sacrificed to inform the second person?

2. Under what circumstances, if any, should a government mandate any type of genetic testing?

3. Some student athletes have died of complications from being carriers of sickle cell disease. What are the risks and benefits of testing student athletes for sickle cell disease carrier status?

4. Do you think that passenger screening at airports should include DNA scans?

5. Exome sequencing to identify a mutation that could cause a particular set of symptoms in a patient can reveal another genetic condition that has not yet been detected. Under what circumstances, if any, do you think patients should receive such "secondary findings"? **Bioethics** in chapter 20 explores whether or not secondary findings provide too much information.

Summary

1.1 Introducing Genes and Genomes

1. **Genes** are the instructions to manufacture proteins, which determine inherited traits. Genes are composed of **deoxyribonucleic acid** (**DNA**).

2. A **genome** is a complete set of genetic information. A **cell**, the basic unit of life, contains two genomes. **Genomics** compares and analyzes the functions of many genes. **Bioethics** addresses issues and controversies that arise in applying medical technology and using genetic information.

1.2 Levels of Genetics and Genomics

3. The **nitrogenous base** (A, C, T, G) sequences of genes encode proteins. **Ribonucleic acid (RNA)** molecules carry DNA sequence information so that it can be utilized. **DNA replication** maintains the genetic information. **Transcription** copies DNA information into RNA, and **translation** assembles amino acids into proteins.

4. A **mutation** is a change in a gene that can cause a disease if it alters the amino acid sequence of the specified protein. Mutation also refers to the process of change.

5. Much of the genome does not encode proteins. The part that does is called the **exome**.

6. Variants of a gene are called **alleles**. They are inherited or arise by mutation. Alleles may differ slightly from one another, but encode the same protein.

7. **Chromosomes** consist of DNA and protein. A **somatic cell** in humans has 23 chromosome pairs. The 22 types of **autosomes** do not include genes that specify sex. The X and Y **sex chromosomes** bear genes that determine sex. **Karyotypes** are chromosome charts.

8. Single genes determine Mendelian traits. **Multifactorial traits** reflect the influence of one or more genes and the environment.

9. Cells undergo **differentiation** by expressing subsets of genes. **Stem cells** divide to yield other stem cells and cells that differentiate. **Tissues** are groups of cells with a shared function.

10. The **phenotype** is the gene's expression. An allele combination constitutes the **genotype**. Alleles may be **dominant** (exerting an effect in a single copy) or **recessive** (requiring two copies for expression).

11. **Pedigrees** are diagrams used to study traits in families.

12. Genetic populations are defined by their collections of alleles, termed the **gene pool**. Genome comparisons among species reveal evolutionary relationships.

1.3 Applications of Genetics and Genomics

13. Comparing DNA sequences can provide information on identity, relationships, and history.

14. Precision medicine considers DNA information along with other types of information to learn about health and to develop new treatments for diseases.

15. Genetically modified organisms have genes that are added, removed, or replaced, possibly from other species.

1.4 A Global Perspective on Genomes

16. In **metagenomics**, DNA collected from specific habitats, including the human body, is sequenced to learn more about microbiomes.

Review Questions

1. Place the following terms in size order, from largest to smallest, based on the structures or concepts they represent.
   a. chromosome
   b. gene pool
   c. gene
   d. DNA
   e. genome

2. Provide an example of a bioethical issue.

3. Explain how DNA encodes information.

4. Distinguish between
   a. an autosome and a sex chromosome.
   b. genotype and phenotype.
   c. DNA and RNA.
   d. recessive and dominant traits.
   e. pedigrees and karyotypes.
   f. gene and genome.
   g. exome and genome.

5. Explain how all humans have the same genes, but vary genetically.

6. Explain how all cells in a person's body have the same genome, but are of hundreds of different types that look and function differently.

7. What is the assumption behind the comparison of DNA sequences to deduce that two individuals, or two species, are related?

8. Cite two examples of DNA sequencing used to identify an individual organism.

9. List three types of information being considered in a precision medicine initiative.

10. Define *genome editing*.

Applied Questions

1. If and when you or your child have genome sequencing done, what types of information would you like to receive?

2. Until recently, the Basque people of Europe were thought to have descended directly from hunter-gatherers, based on language similarities. Why is shared DNA information a more reliable indicator of ancestry than shared language?

3. Why might people accept drugs that are manufactured in genetically modified organisms, but not "GM" foods?

4. Would you want to carry your exome or genome information in your smartphone? Cite a reason for your answer.

5. Describe both a frivolous use and a serious use of DNA testing.

6. Explain why exome sequencing can be almost as valuable as genome sequencing.

7. What do you want to learn, if anything, about your own genome?

Forensics Focus

1. It is springtime and Romeo, a fluffy black-and-white Maine Coon cat, bolts outdoors. Hours later he returns, consumes a can of seafood cat food and a few liver treats, then grabs some fried eggplant from the counter but spits it out. He licks his fur for awhile, nonchalantly swallowing wads of hair, then falls into a deep sleep. The next morning, Romeo's humans awaken to the unmistakable sounds of a cat vomiting. They poke around in the material to deduce what may be disturbing Romeo's digestion, but it is too slimy to tell. Describe how DNA testing might be used to discover the source of Romeo's discomfort.

2. DNA samples are taken from inside the cheeks of people arrested on suspicion of having committed serious crimes. Only a small portion of the genome information is recorded (13 repeated sequences that do not encode proteins; see chapter 14) and the DNA itself is discarded. Explain why sampling this part of the genome does not indicate anything about the person's health, ancestry, or traits.

3. Researchers in Canada published a report on 468 legal cases that cited "genetic predisposition" to explain physical injury or criminal behavior due to mental illness or substance abuse. (In 86 cases the genetic information didn't harm the plaintiff; in 134 cases it did; and in 248 cases the legal significance of the genetic findings was unclear.) Under what circumstances, if any, do you think genetic information might be useful in a legal case, and from whose point of view?

4. Make up a crime scene scenario (or find a real one) in which DNA from a nonhuman provided critical evidence.

Case Studies and Research Results

1. Many large-scale genome sequencing projects focus on individuals who presently have similar diagnoses, such as adults with cardiovascular disease or children with developmental delay, autism, and intellectual disability. In contrast, the Genomic Postmortem Research Project is sequencing the genomes of 300 deceased individuals who had been patients at a large clinic that kept meticulous electronic medical records. What type of information might be gleaned from this genome information?

2. An artist travels around New York City collecting hairs, fingernails, discarded paper cups, gum, cigarette butts, and other trash; then extracts the DNA and uses a computer program to construct a face from the genetic information. The artist then prints the results in three dimensions, producing a sculpture that she displays in a gallery along with the original DNA-bearing evidence. The algorithm considers 50 inherited traits that affect facial features, including information on ancestry. What would you do if you wandered into a gallery and found a sculpture of yourself or someone you know? How private is the DNA on a wad of discarded chewing gum?

3. A study collected DNA from children at a state fair, to identify genes that contribute to normal health and development. If you were a parent of a child at the fair, what questions would you have asked before donating your child's DNA?

4. What might be some problems or challenges that might arise in implementing a large-scale project that entails sequencing genomes (see figure 1.8)?

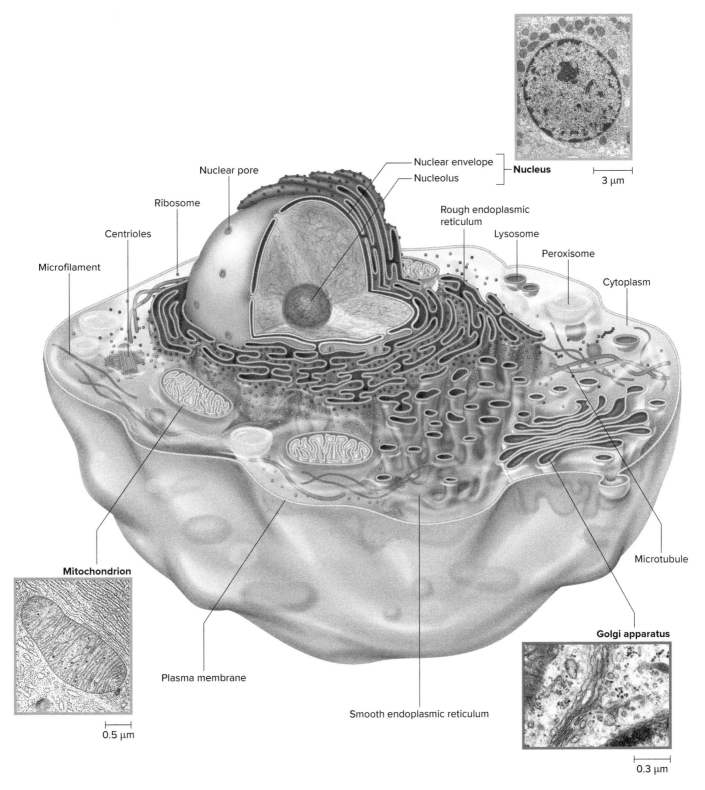

Nuclear pore

Ribosome

Centrioles

Microfilament

Nuclear envelope

Nucleolus

Nucleus

3 μm

Rough endoplasmic
reticulum

Lysosome

Peroxisome

Cytoplasm

Mitochondrion

0.5 μm

Plasma membrane

Smooth endoplasmic reticulum

Microtubule

Golgi apparatus

0.3 μm

Figure 2.3 Generalized animal cell. Organelles provide specialized functions for the cell. Most of these structures are transparent; colors are used here to distinguish them. Different cell types have different numbers of organelles. All cell types have a single nucleus, except for red blood cells, which expel their nuclei as they mature. *(top): © David M. Phillips/The Population Council/Photo Researchers/Science Source; (left): © K.R. Porter/Science Source; (right): © EM Research Services, Newcastle University RF*

The remainder of the cell—that is, everything but the nucleus, organelles, and the outer boundary, or **plasma membrane**—is **cytoplasm**. (The plasma membrane is also called the cell membrane.) Other cellular components include stored proteins, carbohydrates, and lipids; pigment molecules; and various other small chemicals. (The cytoplasm is called

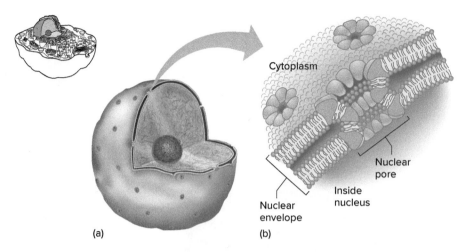

Figure 2.4 **The nucleus is the genetic headquarters.** **(a)** The largest structure in a typical human cell, the nucleus lies within two membrane layers that make up the nuclear envelope. **(b)** Nuclear pores allow specific molecules to move in and out of the nucleus through the envelope.

cytosol when these other parts are removed.) We now take a closer look at three cellular functions: secretion, digestion inside cells, and energy production.

Secretion—The Eukaryotic Production Line

Organelles interact in ways that coordinate basic life functions and provide the characteristics of specialized cell types. Secretion, which is the release of a substance from a cell, illustrates one way that organelles function together.

Secretion begins when the body sends a biochemical message to a cell to begin producing a particular substance. For example, when a newborn suckles the mother's breast, her brain releases hormones that signal cells in her breast to rapidly increase the production of the complex mixture that makes up milk (**figure 2.5**). In response, information in certain genes is copied into molecules of **messenger RNA (mRNA)**, which then exit the nucleus (see steps 1 and 2 in figure 2.5). In the cytoplasm, the mRNAs, with the help of ribosomes and another type of RNA called **transfer RNA (tRNA)**, direct the manufacture of milk proteins. These include nutritive proteins called caseins, antibodies that protect against infection, and enzymes.

Most protein synthesis occurs on a maze of interconnected membranous tubules and sacs called the **endoplasmic reticulum (ER)** (see step 3 in figure 2.5). The ER winds from the nucleus envelope outward to the plasma membrane, forming a vast tubular network that transports molecules from one part of a cell to another. The ER nearest the nucleus, which is flattened and studded with ribosomes, is called rough ER, because the ribosomes make it appear fuzzy when viewed under an electron microscope. Protein synthesis begins on the rough ER when messenger RNA attaches to the ribosomes. Amino acids from the cytoplasm are then linked, following the instructions in the mRNA's sequence, to form particular proteins that will either exit the cell or become part of membranes (step 3, figure 2.5).

Proteins are also synthesized on ribosomes not associated with the ER. These proteins remain in the cytoplasm.

The ER acts as a quality control center for the cell. Its chemical environment enables the forming protein to start folding into the three-dimensional shape necessary for its specific function. Misfolded proteins are pulled out of the ER and degraded, much as an obviously defective toy might be pulled from an assembly line at a toy factory and discarded. Misfolded proteins can cause disease, as discussed further in chapter 10.

As the rough ER winds out toward the plasma membrane, the ribosomes become fewer and the tubules widen, forming a section called smooth ER. Here, lipids are made and added to the proteins arriving from the rough ER (step 4, figure 2.5). The lipids and proteins are transported until the tubules of the smooth ER narrow and end. Then the proteins exit the ER in membrane-bounded, saclike organelles called **vesicles**, which pinch off from the tubular endings of the membrane. Lipids are exported without a vesicle, because a vesicle is itself made of lipid.

A loaded vesicle takes its contents to the next stop in the secretory production line, a **Golgi apparatus** (step 5, figure 2.5), which looks like a stack of pancakes. This processing center is a column of four to six interconnected flat, membrane-enclosed sacs. Here, sugars, such as the milk sugar lactose, are made. Some sugars attach to proteins to form glycoproteins or to lipids to form glycolipids, which become parts of plasma membranes. Proteins finish folding in the Golgi apparatus and become active. Some cell types have just a few Golgi apparatuses, but those that secrete may have hundreds.

The components of complex secretions, such as milk, are temporarily stored in Golgi apparatuses. Droplets of proteins and sugars then bud off in vesicles that move outward to the plasma membrane, fleetingly becoming part of it until they are secreted to the cell's exterior. Lipids exit the plasma membrane directly, taking bits of it with them (step 6, figure 2.5).

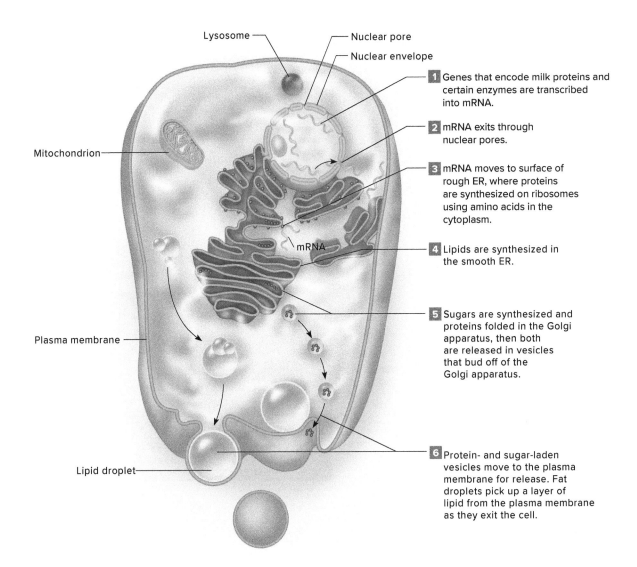

Figure 2.5 **Secretion: making milk.** Milk production and secretion illustrate organelle functions and interactions in a cell from a mammary gland: (1) through (6) indicate the order in which organelles participate in this process. Lipids are secreted in separate droplets from proteins and their attached sugars. This cell is highly simplified.

In the breast, epithelial (lining) cells called lactocytes ("milk cells") form tubules, into which they secrete the components of milk. When the baby suckles, contractile cells squeeze the milk through the tubules and out of holes in the nipples.

A Golgi apparatus processes many types of proteins. These products are sorted when they exit the Golgi into different types of vesicles, a little like merchandise leaving a warehouse being packaged into different types of containers, depending upon the destination. Molecular tags target vesicles to specific locations in the cell, or indicate that the contents are to be released to outside the cell.

A type of transport of molecules between cells uses vesicles called **exosomes** that bud from one cell and then travel to, merge with, and empty their contents into other cells. Exosomes are only 30 to 100 nanometers (billionths of a meter) in diameter. They may carry proteins, lipids, and RNA, and have been identified in many cell types. Exosomes remove debris,

transport immune system molecules, and provide a vast communication network among cells.

Intracellular Digestion—Lysosomes and Peroxisomes

Just as clutter and garbage accumulate in an apartment, debris builds up in cells. Organelles called **lysosomes** ("bodies that cut") handle the garbage. Lysosomes are membrane-bounded sacs that contain enzymes that dismantle bacterial remnants, worn-out organelles, and other material such as excess cholesterol (**figure 2.6**). The enzymes also break down some digested nutrients into forms that the cell can use.

Lysosomes fuse with vesicles carrying debris from outside or within the cell, and the lysosomal enzymes then degrade the contents. The cell's disposing of its own trash is called **autophagy** ("eating self"). For example, a type of vesicle that

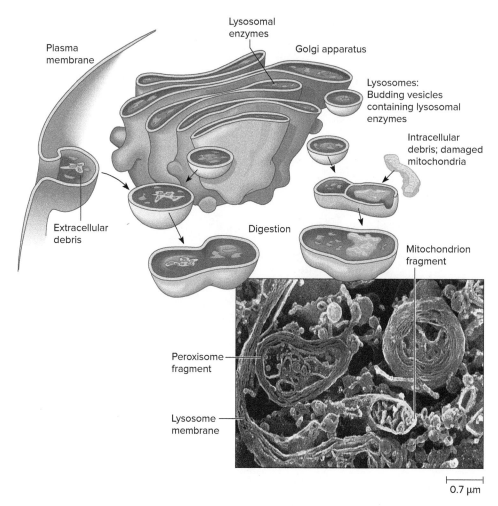

Plasma membrane

Lysosomal enzymes

Golgi apparatus

Lysosomes: Budding vesicles containing lysosomal enzymes

Intracellular debris; damaged mitochondria

Extracellular debris

Digestion

Mitochondrion fragment

Peroxisome fragment

Lysosome membrane

0.7 µm

Figure 2.6 Lysosomes are trash centers. Lysosomes fuse with vesicles or damaged organelles, activating the enzymes within to recycle the molecules. Lysosomal enzymes also dismantle bacterial remnants. These enzymes require a highly acidic environment to function. © *Prof. P. Motta & T. Naguro/SPL/Science Source*

forms from the plasma membrane, called an **endosome**, ferries extra low-density lipoprotein (LDL) cholesterol to lysosomes. A loaded lysosome moves toward the plasma membrane and fuses with it, releasing its digested contents to the outside. Lysosomes maintain the highly acidic environment that their enzymes require to function, without harming other cell parts that acids could destroy.

Cells differ in their number of lysosomes. Cells called macrophages that move about and engulf bacteria have many lysosomes. Liver cells require many lysosomes to break down cholesterol, toxins, and drugs.

All lysosomes contain forty-three types of digestive enzymes, which must be maintained in balance. Absence or malfunction of an enzyme causes a "lysosomal storage disease," in which the molecule that the missing or abnormal enzyme normally degrades accumulates. The lysosomes swell, crowding other organelles and interfering with the cell's functions. In Tay-Sachs disease, for example, an enzyme is deficient that normally breaks down lipids in the cells that surround nerve cells. As the nervous system becomes buried in lipid, the infant begins to lose skills, such as sight, hearing, and mobility. Death

is typically within 3 years. Even before birth, the lysosomes of affected cells swell. These diseases affect about 10,000 people worldwide. (See table 20.4 and figure 20.4.)

Peroxisomes are sacs with single outer membranes that are studded with several types of proteins and that house enzymes that perform a variety of functions. The enzymes catalyze reactions that break down certain lipids and rare biochemicals, synthesize bile acids used in fat digestion, and detoxify compounds that result from exposure to oxygen free radicals. Peroxisomes are large and abundant in liver and kidney cells, which handle toxins.

The 1992 film *Lorenzo's Oil* recounted the true story of a child with adrenoleukodystrophy, caused by an absent peroxisomal enzyme. A type of lipid called a very-long-chain fatty acid built up in Lorenzo's brain and spinal cord. Early symptoms of ALD include low blood sugar, skin darkening, muscle weakness, visual loss, altered behavior and cognition, and irregular heartbeat. The patient eventually loses control over the limbs and usually dies within a few years. Lorenzo's parents devised a combination of edible oils that diverts another enzyme to compensate for the missing one. Lorenzo lived

30 years, which may have been due to the oil, or to the excellent supportive care that he received. Providing a functional copy of the gene, called gene therapy, can halt progression of the disease. Gene therapy for ALD is being tested on newborns who have the disease. (Chapter 20 discusses gene therapy.)

Energy Production—Mitochondria

Cells require continual energy to power the chemical reactions of life. Organelles called **mitochondria** provide energy by breaking the chemical bonds that hold together the nutrient molecules in food.

A mitochondrion has an outer membrane similar to those in the ER and Golgi apparatus and an inner membrane that forms folds called cristae (**figure 2.7**). These folds hold enzymes that catalyze the biochemical reactions that release energy from nutrient molecules. The freed energy is captured and stored in the bonds that hold together a molecule called adenosine triphosphate (ATP). In this way, ATP functions a little like an energy debit card.

A cell may have a few hundred to tens of thousands of mitochondria, depending upon activity level. A typical liver cell has about 1,700 mitochondria, but a muscle cell, with its very high energy requirements, may have 10,000. A major symptom of diseases that affect mitochondria is fatigue. Mitochondria contain a small amount of DNA different in sequence from the DNA in the nucleus (see figure 5.10). Chapter 5 discusses mitochondrial inheritance and disease, and chapter 16 describes how mitochondrial genes provide insights into early human migrations.

Table 2.2 summarizes the structures and functions of organelles.

Biological Membranes

Just as the character of a community is molded by the people who enter and leave it, the special characteristics of different cell types are shaped in part by the substances that enter and

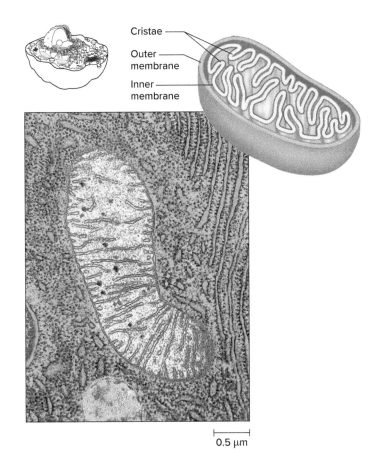

Figure 2.7 A mitochondrion extracts energy. Cristae, which are infoldings of the inner membrane of a mitochondrion, increase the available surface area containing enzymes for energy reactions. © *Bill Longcore/Photo Researchers/Science Source*

leave them. The plasma membrane completely surrounds the cell and monitors the movements of molecules in and out. How the chemicals that comprise the plasma membrane associate with each other determines which substances can enter or leave the cell.

Table 2.2	Structures and Functions of Organelles	
Organelle	**Structure**	**Function**
Endoplasmic reticulum	Membrane network; rough ER has ribosomes, smooth ER does not	Site of protein synthesis and folding; lipid synthesis
Golgi apparatus	Stacks of membrane-enclosed sacs	Site where sugars are made and linked into starches or joined to lipids or proteins; proteins finish folding; secretions stored
Lysosome	Sac containing digestive enzymes	Degrades debris; recycles cell contents
Mitochondrion	Two membranes; inner membrane enzyme-studded	Releases energy from nutrients
Nucleus	Porous, double-membraned sac containing DNA	Separates DNA within cell
Peroxisome	Sac containing enzymes	Breaks down and detoxifies various molecules
Ribosome	Two associated globular subunits of RNA and protein	Scaffold and catalyst for protein synthesis
Vesicle	Membrane-bounded sac	Temporarily stores or transports substances

Membranes also form the outer boundaries of mitochondria, lysosomes, peroxisomes, and vesicles, and completely comprise other organelles, such as the endoplasmic reticulum and Golgi apparatus. A cell's membranes are more than simple coverings, and the molecules embedded in them have specific functions.

Membrane Structure

A biological membrane has a distinctive structure. It is a double layer (bilayer) of molecules called phospholipids. A phospholipid is a fat molecule with attached phosphate groups. A phosphate group (PO_4) is a phosphorus atom bonded to four oxygen atoms. A phospholipid is depicted as a head with two parallel tails, shown in the inset to **figure 2.8**.

The tendency of lipid molecules to self-assemble into sheets makes the formation of biological membranes possible. The molecules form sheets because their ends react oppositely to water: The phosphate end of a phospholipid is attracted to water, and thus is hydrophilic ("water loving"); the other end, which consists of two chains of fatty acids, moves away from water, and is therefore hydrophobic ("water fearing"). Because of these forces, phospholipid molecules in water spontaneously form bilayers. Their hydrophilic surfaces are exposed to the watery exterior and interior of the cell. The hydrophobic surfaces face each other on the inside of the bilayer, away from water, and block entry and exit to most substances that dissolve in water. However, certain molecules can cross the membrane through proteins that form passageways, or when temporarily joined to a "carrier" protein.

Some membrane proteins form channels for ions (atoms or molecules with electrical charge). "Channelopathies" are diseases that stem from faulty ion channels. For example, figure 1.4 depicts the ion channel that is abnormal in cystic fibrosis. Another type of channelopathy can cause either extreme pain in response to no apparent stimulus, or lack of pain. A famous case involved a boy who became a performer, walking on hot coals and stabbing himself to entertain crowds in Pakistan. Researchers are investigating these pain syndromes to develop new painkillers.

Proteins are embedded in the phospholipid bilayer of biological membranes. Some proteins traverse the entire structure, while others extend from one or both faces (figure 2.8). The phospholipid bilayer is oily, and some proteins move within it like ships on a sea.

The Plasma Membrane Enables Cell-to-Cell Communication

The proteins, glycoproteins, and glycolipids that extend from a plasma membrane create surface topographies that are important in a cell's functions and to its interactions with other cells. The surfaces of a person's cells indicate not only that they are part of that particular body, but also that they are part of a specific organ and type of tissue.

Many molecules that extend from the plasma membrane are receptors, which are structures that have indentations or other shapes that fit and hold specific molecules outside the cell. The molecule that binds to the receptor, called the ligand, may set into motion a cascade of chemical reactions inside the cell that carries out a particular activity, such as dividing.

In a cellular communication process called **signal transduction**, a series of molecules that are part of the plasma membrane form pathways that detect signals from outside the cell and transmit them inward, where yet other molecules orchestrate the cell's response. In a different process called **cellular adhesion**, the plasma membrane helps cells attach to certain other cells. These cell-to-cell connections are important in forming tissues.

Faulty signal transduction or cellular adhesion can harm health. For example, cancer is due to cells' failure to recognize or react to signals to cease dividing. Cancer cells also have abnormal cellular adhesion, which enables them to invade healthy tissue.

The Cytoskeleton

The **cytoskeleton** is a meshwork of protein rods and tubules that serves as the cell's architecture, positioning organelles and providing overall three-dimensional shapes. The proteins of the cytoskeleton are continually broken down and built up as a cell performs specific activities. Some cytoskeletal elements function as rails that transport cellular contents; other parts, called motor molecules, power the movement of organelles along these rails as they convert chemical energy into mechanical energy.

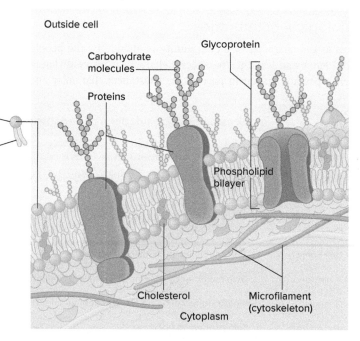

Figure 2.8 Anatomy of a plasma membrane. Mobile proteins are embedded throughout a phospholipid bilayer. Carbohydrates jut from the membrane's outer face. The inset (left) shows a phospholipid molecule.

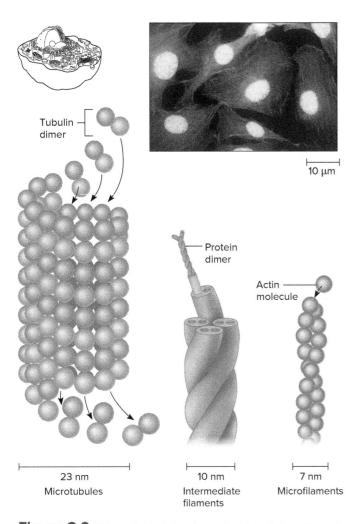

Figure 2.9 **The cytoskeleton is made of protein rods and tubules.** The three major components of the cytoskeleton are microtubules, intermediate filaments, and microfilaments. Using special stains, the cytoskeletons in the cells in the inset appear orange under the microscope. (The abbreviation nm stands for nanometer, which is a billionth of a meter.) *(Photo): © Dr. Gopal Murti/Science Source*

The cytoskeleton includes three major types of elements—**microtubules**, **microfilaments**, and **intermediate filaments** (**figure 2.9**). They are distinguished by protein type, diameter, and how they aggregate into larger structures. Other proteins connect these components, creating a framework that provides the cell's strength and ability to resist force and maintain shape.

Long, hollow microtubules provide many cellular movements. A microtubule is composed of pairs (dimers) of a protein, called tubulin, assembled into a hollow tube. Adding or removing tubulin molecules changes the length of the microtubule.

Cells contain both formed microtubules and individual tubulin molecules. When the cell requires microtubules to carry out a specific function—cell division, for example—free tubulin dimers self-assemble into more microtubules. After the cell divides, some of the microtubules fall apart into individual

tubulin dimers, replenishing the cell's supply of building blocks. Cells perpetually build up and break down microtubules.

Microtubules maintain cellular organization and enable transport of substances within the cell. Abnormal microtubules cause several neurodegenerative diseases. For example, they may prevent neurotransmitters from reaching the synapses between neurons or between neurons and muscle cells.

Microtubules form hairlike structures called cilia, from the Latin meaning "cells' eyelashes" (**figure 2.10**). Cilia are of two types: motile cilia that move, and primary cilia that do not move but serve a sensory function.

Motile cilia have one more pair of microtubules than primary microtubules. Coordinated movement of motile cilia generates a wave that moves the cell or propels substances along its surface. The coordinated motion of many cilia moving a substance resembles "crowd surfing"—when a performer jumps into the audience and is then held aloft and passed back up to the stage. Similarly, motile cilia pass inhaled particles up and out of respiratory tubules and move egg cells in the female reproductive tract. Because motile cilia are so widespread, defects in them can cause multiple symptoms. One such ciliopathy—"sick cilia disease"—is Bardet-Biedl syndrome, which causes obesity, visual loss, diabetes, cognitive impairment, and extra fingers and/or toes. There are several types of Bardet-Biedl syndrome, and the condition affects both motile and primary cilia.

Many types of cells have primary cilia, which do not move and serve as antennae, sensing signals from outside cells and passing them to specific locations inside cells. Primary cilia sense light entering the eyes, urine leaving the kidney tubules, blood flowing from vessels in the heart, and pressure on cartilage. Although these cilia do not move, they stimulate some cells to move, such as those that form organs in an embryo, and cells that help wounds to heal. Absence of primary cilia can harm health, such as in polycystic kidney disease. Cells may have many motile cilia but usually have only one cilium that does not move.

Microfilaments are long, thin rods composed of many molecules of the protein actin. They are solid and narrower than microtubules, enable cells to withstand stretching and compression, and help anchor one cell to another. Microfilaments provide many other functions in the cell through proteins that interact with actin. When any of these proteins is

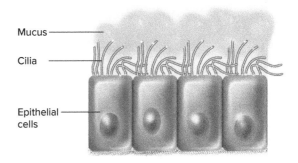

Figure 2.10 **Cilia are built of microtubules.** Motile cilia move secretions such as mucus on the cell surfaces of the respiratory tubes. In contrast, primary cilia do not move, functioning like antennae to sense changes. This figure shows only motile cilia.

A Girl with Giant Axons

Hannah Sames's cells do not make the protein gigaxonin, which normally breaks down intermediate filaments and recycles their components. In Hannah's hair cells (**figure 2B**), intermediate filaments made of keratin proteins build up, kinking the strands. In her neurons, different types of intermediate filament proteins accumulate, swelling the long nerve extensions (axons) that send messages to her muscles. Hannah is one of a few dozen people in the world known to have giant axonal neuropathy (GAN).

When Hannah was born on March 5, 2004, her parents and sisters were charmed that her hair was kinky while theirs was stick-straight. All seemed well until Hannah was 2 years, 5 months old, and her grandmother noticed her left arch rolling inward. Doctors thought she would outgrow it, but by her third birthday, both arches were involved and her gait had become awkward. Still physicians were not alarmed. Then Hannah's aunt showed phone video of Hannah walking to a physical therapist she worked with, who thought Hannah's gait resembled that of a child with muscular dystrophy.

With the idea of muscular dystrophy mentioned to the pediatrician, Hannah finally had genetic testing for the more common inherited childhood neurological conditions, but results were all normal. Then, at a visit with a pediatric neurologist, the doctor took one look at Hannah and reached for a large textbook high on a shelf, flipped through it, and showed Hannah's parents, Lori and Matt, a photo of a skinny boy with kinky hair and a high forehead and braces that went just below the knee—he had GAN, and he looked exactly like Hannah.

A test for the gigaxonin gene confirmed the diagnosis—Hannah had two identical "deletion" mutations—she did not have the gene at all. Lori and Matt were devastated when a genetic counselor told them that there were no treatments and no research on the extremely rare disease. The parents would not accept that, and within days had formed Hannah's Hope Foundation, which funded development of gene therapy (see section 20.3), the first into the spinal cord.

Figure 2B **Hannah Sames has giant axonal neuropathy, which affects intermediate filaments.** Her beautiful curls are one of the symptoms, but she prefers today to straighten her hair. *Courtesy, Lori Sames. Photo by Dr. Wendy Josephs*

Hannah received functional gigaxonin genes just after her twelfth birthday. By that time, she was using a wheelchair and beginning to have trouble speaking and seeing. If all goes well, the gene therapy will halt the progression of the disease. Meanwhile, Hannah is working hard to strengthen her muscles. Gene therapy into the spinal cord may be developed to treat other, more common conditions, and perhaps spinal cord injuries.

Questions for Discussion

1. Why is the experimental gene therapy for GAN important, even though only a few people have this disease?
2. What is unusual about intermediate filaments, compared to microtubules and microfilaments?
3. Did the pediatric neurologist base his diagnosis of Hannah on her genotype or on her phenotype?

---

absent or abnormal, a genetic disease results. The actin part of the cytoskeleton is abnormal in Alzheimer disease and in some diseases of heart muscle.

Intermediate filaments have diameters intermediate between those of microtubules and microfilaments, and unlike those other components of the cytoskeleton, intermediate filaments are composed of different types of proteins in different cell types. However, all intermediate filaments consist of paired proteins entwined to form nested coiled rods. Intermediate filaments are scarce in many cell types but are abundant in nerve cells and skin cells. **Clinical Connection 2.2** describes how abnormal intermediate filaments affect several cell types in a girl who has a rare disease called giant axonal neuropathy.

Key Concepts Questions 2.2

1. Distinguish prokaryotic from eukaryotic cells.
2. List the chemical constituents of cells.
3. Explain the general functions of organelles.
4. Describe the organelles that interact as a cell secretes, degrades debris, and acquires energy.
5. Explain how the structure of the plasma membrane enables its functions.
6. List the components of the cytoskeleton and describe some cytoskeletal functions.

2.3 Cell Division and Death

In a human body, new cells form as old ones die, at different rates in different tissues. Growth, development, maintaining health, and healing from disease or injury require an intricate interplay between the rates of **mitosis** and **cytokinesis**, which divide the DNA and the rest of the cell, respectively, and **apoptosis**, a form of cell death.

An adult human body consists of about 30 trillion cells, and billions are replaced daily. Cells must die as part of normal development, molding organs much as a sculptor carefully removes clay to shape the desired object. Apoptosis carves fingers and toes, for example, from weblike structures that telescope out from an embryo's developing form (**figure 2.11**). Apoptosis is a Greek word for "leaves falling from a tree." It is a precise, genetically programmed sequence of events that is a normal part of development.

The Cell Cycle

Many cell divisions enable a fertilized egg to develop into a many-trillion-celled organism. A series of events called the **cell cycle** describes the sequence of activities as a cell prepares for and undergoes division.

Cell cycle rate varies in different tissues at different times. A cell lining the small intestine's inner wall may divide frequently, throughout life, whereas a neuron in the brain may never divide. A cell in the deepest skin layer may divide as long as a person lives, and even divide a few times after a person dies! Frequent mitosis enables the embryo and fetus to grow rapidly. By birth, the mitotic rate slows dramatically. Later,

mitosis maintains the numbers and positions of specialized cells in tissues and organs.

The cell cycle is continual, but we describe it with stages. The two major stages are **interphase** (not dividing) and mitosis (dividing) (**figure 2.12**). In mitosis, a cell duplicates its chromosomes, then in cytokinesis it apportions one set of chromosomes, along with organelles, into each of two resulting cells, called daughter cells. Mitosis maintains the set of 23 chromosome pairs characteristic of a human somatic cell. Another form of cell division, **meiosis**, produces sperm or eggs, which have half the amount of genetic material that somatic cells do (or 23 single chromosomes, comprising one copy of the genome). Chapter 3 discusses meiosis in the context of development.

Interphase—A Time of Great Activity

During interphase a cell continues the basic biochemical functions of life, while also replicating its DNA and some organelles. Interphase is divided into two gap (G_1 and G_2) **phases** and one synthesis (**S**) **phase**. In addition, a cell can exit the cell cycle at G_1 to enter a quiet phase called G_0. A cell in G_0 is alive and maintains its specialized characteristics but does not replicate its DNA or divide. From G_0, a cell may also proceed to mitosis and divide, or die. Apoptosis may ensue if the cell's

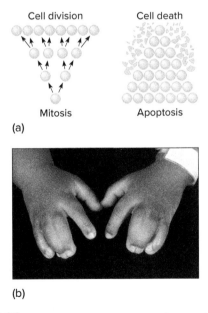

Figure 2.11 **Mitosis and apoptosis mold a body.** **(a)** Cell numbers increase from mitosis and decrease from apoptosis. **(b)** In the embryo, apoptosis normally carves fingers and toes from webbed structures. In syndactyly, apoptosis fails to carve digits, and webbing persists, as it does in these hands. *(b): © Mediscan/Medical-on-Line/Alamy*

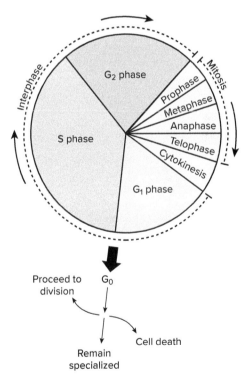

Figure 2.12 **The cell cycle.** The cell cycle is divided into interphase, when cellular components are replicated, and mitosis, when the cell distributes its contents into two daughter cells. Interphase is divided into G_1 and G_2, when the cell duplicates specific molecules and structures, and S phase, when it replicates DNA. Mitosis is divided into four stages plus cytokinesis, when the cells separate. G_0 is a "time-out" when a cell "decides" which course of action to follow.

DNA is so damaged that cancer might result. G_0, then, is when a cell's fate is either decided or put on hold.

During G_1, which follows mitosis, the cell resumes synthesis of proteins, lipids, and carbohydrates. These molecules will contribute to building the extra plasma membrane required to surround the two new cells that form from the original one. G_1 is the period of the cell cycle that varies the most in duration among different cell types. Slowly dividing cells, such as those in the liver, may exit at G_1 and enter G_0, where they remain for years. In contrast, the rapidly dividing cells in bone marrow speed through G_1 in 16 to 24 hours. Cells of the early embryo may skip G_1 entirely.

During S phase, the cell replicates its entire genome. This happens simultaneously from several starting points, to get the enormous job done. After DNA replication, each chromosome consists of two copies of the genome joined at an area called the **centromere**. In most human cells, S phase takes 8 to 10 hours. Many proteins are also synthesized during this phase, including those that form the mitotic **spindle**, which will pull the chromosomes apart. Microtubules form structures called **centrioles** near the nucleus. Centriole microtubules join with other proteins and are oriented at right angles to each other, forming paired, oblong structures called **centrosomes** that organize other microtubules into the spindle.

G_2 occurs after the DNA has been replicated but before mitosis begins. More proteins are synthesized during this phase. Membranes are assembled from molecules made during G_1 and are stored as small, empty vesicles beneath the plasma membrane. These vesicles will merge with the plasma membrane to make enough of a boundary to enclose the two daughter cells.

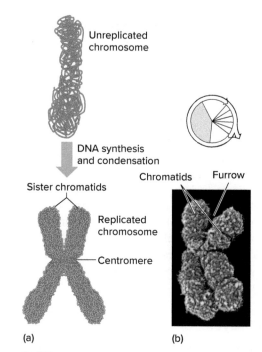

Figure 2.13 **Replicated and unreplicated chromosomes.** Chromosomes are replicated during S phase, before mitosis begins. Two genetically identical chromatids of a replicated chromosome join at the centromere **(a)**. The photomicrograph **(b)** shows a replicated human chromosome. *(b): © SPL/ Science Source*

Mitosis—The Cell Divides

As mitosis begins, the replicated chromosomes are condensed enough to be visible, when stained, under a microscope. The long strands of chromosomal material in replicated chromosomes are called **chromatids**, and if attached at a centromere, they are called sister chromatids. The space between sister chromatids is called a furrow (**figure 2.13**). At a certain point during mitosis, a replicated chromosome's centromere splits. This allows its sister chromatids to separate into two individual chromosomes. So a chromosome can be in an unreplicated form, in a replicated form consisting of two sister chromatids, or unwound and not visible when stained. Although the centromere of a replicated chromosome appears as a constriction, its DNA is replicated.

During **prophase**, the first stage of mitosis, DNA coils tightly. This coiling shortens and thickens the chromosomes,

Figure 2.14 **Mitosis in a human cell.** Replicated chromosomes separate and are distributed into two cells from one. In a separate process called cytokinesis, the cytoplasm and other cellular structures distribute and pinch off into two daughter cells. (Not all chromosome pairs are depicted.) *(Photos): © Ed Reschke*

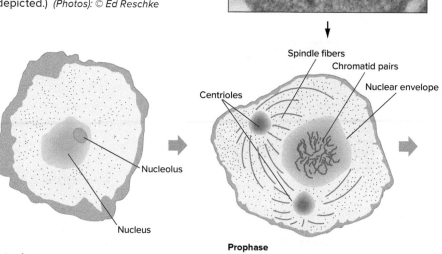

Interphase
Chromosomes are uncondensed.

Prophase
Condensed chromosomes take up stain. The spindle assembles, centrioles appear, and the nuclear envelope breaks down.

easing their separation (**figure 2.14**). Microtubules assemble from tubulin building blocks in the cytoplasm, forming the spindles. Toward the end of prophase, the nuclear membrane breaks down. The nucleolus is no longer visible.

Metaphase follows prophase. Chromosomes attach to the spindle at their centromeres and align along the center of the cell, which is called the equator. Metaphase chromosomes are under great tension, but they appear motionless because they are pulled with equal force on both sides, like a tug-of-war rope pulled taut.

Next, during **anaphase**, the plasma membrane indents at the center, where the metaphase chromosomes line up. A band of microfilaments forms on the inside face of the plasma membrane, constricting the cell down the middle. Then the centromeres part, which relieves the tension and releases one chromatid from each pair to move to opposite ends of the cell—like a tug-of-war rope breaking in the middle and the participants falling into two groups. Microtubule movements stretch the dividing cell. During the very brief anaphase, a cell fleetingly contains twice the normal number of chromosomes because each chromatid becomes an independently moving chromosome, but the cell has not yet physically divided.

In **telophase**, the final stage of mitosis, the cell looks like a dumbbell with a set of chromosomes at each end. The spindle falls apart, and nucleoli and the membranes around the nuclei re-form at each end of the elongated cell. Division of the genetic material is now complete. Next, during cytokinesis,

organelles and macromolecules are distributed between the two daughter cells. Finally, the microfilament band contracts like a drawstring, separating the newly formed cells.

Control of the Cell Cycle

Control of mitosis is a daunting task. Quadrillions of mitoses occur in a lifetime, and not at random. When and where a somatic cell divides is crucial to health. Too little mitosis, and an injury goes unrepaired; too much, and an abnormal growth forms.

Groups of interacting proteins function at specific times in the cell cycle, called checkpoints, and they ensure that chromosomes are correctly replicated and apportioned into daughter cells (**figure 2.15**). A "DNA damage checkpoint" temporarily pauses the cell cycle while special proteins repair damaged DNA. An "apoptosis checkpoint" turns on as mitosis begins. During this checkpoint, proteins called survivins override signals telling the cell to die, so that mitosis (division) rather than apoptosis (death) occurs. Later during mitosis, the "spindle assembly checkpoint" oversees construction of the spindle and the binding of chromosomes to it.

Cells obey an internal "clock" that tells them approximately how many times to divide. Mammalian cells grown (cultured) in a dish divide about 40 to 60 times. The mitotic clock ticks down with time. A connective tissue cell from a fetus, for example, will divide about 50 more times. A similar cell from an adult divides only 14 to 29 more times.

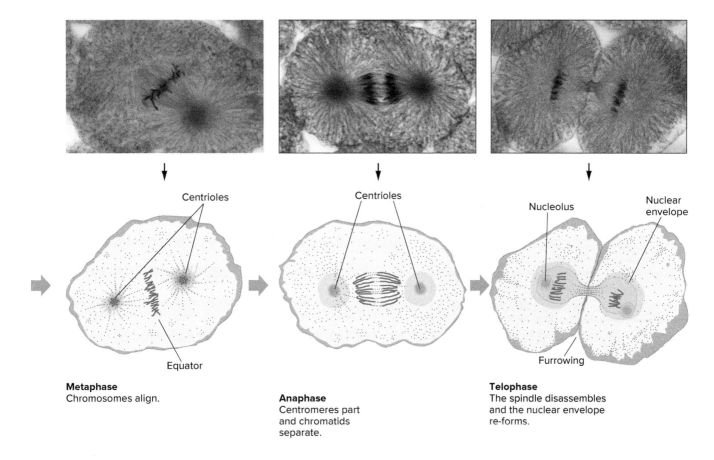

Metaphase
Chromosomes align.

Anaphase
Centromeres part and chromatids separate.

Telophase
The spindle disassembles and the nuclear envelope re-forms.

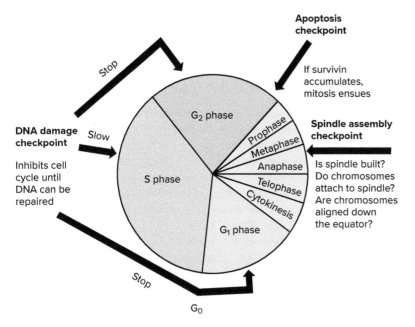

Figure 2.15 Cell cycle checkpoints. Checkpoints ensure that mitotic events occur in the correct sequence. Many types of cancer result from faulty checkpoints.

How can a cell "know" how many divisions remain? The answer lies in the chromosome tips, called **telomeres** (**figure 2.16**). Telomeres function like cellular fuses that burn down as pieces are lost from the ends. Telomeres consist of hundreds to thousands of repeats of a specific six DNA-base sequence (TTAGGG, see figure 13.2). At each mitosis, the telomeres lose 50 to 200 endmost bases, gradually shortening the chromosome. After approximately 50 divisions, a critical length of telomere DNA is lost, which signals mitosis to stop. The cell may remain alive but not divide again, or it may die.

Not all cells have shortening telomeres. In eggs and sperm, in cancer cells, and in a few types of normal cells that must continually supply new cells (such as bone marrow cells), an enzyme called telomerase keeps chromosome tips long. However, most cells do not produce telomerase, and their chromosomes gradually shrink.

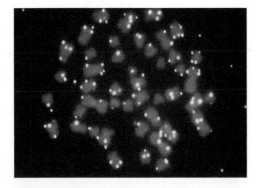

Figure 2.16 Telomeres. Fluorescent tags mark the telomeres in this human cell. © Health Protection Agency/Science Source

The rate of telomere shortening provides a cellular "clock." It not only counts down a cell's remaining life span, but may also sense environmental stimuli. Chronic stress, obesity, lack of exercise, and elevated blood sugar are associated with accelerated telomere shortening.

Factors from outside the cell can affect a cell's mitotic clock. Crowding can slow or halt mitosis. Normal cells growing in culture stop dividing when they form a one-cell-thick layer lining the container. This limitation to division is called contact inhibition. If the layer tears, the cells that border the tear grow and divide, filling in the gap. The cells stop dividing once the space is filled. Perhaps a similar mechanism in the body limits mitosis.

Hormones and growth factors are chemical signals that control the cell cycle from outside. A **hormone** is made in a gland and transported in the bloodstream to another part of the body, where it exerts a specific effect. Hormones secreted in the brain, for example, signal the cells lining a young woman's uterus to build up each month by mitosis in preparation for possible pregnancy. A growth factor acts more locally. Epidermal growth factor (EGF), for example, stimulates cell division in the skin beneath a scab. Certain cancer drugs work by plugging growth factor receptors on cancer cells, blocking the signals to divide.

Two types of proteins, the cyclins and kinases, interact inside cells, activating the genes whose products carry out mitosis. The two types of proteins form pairs. Cyclin levels fluctuate regularly throughout the cell cycle, while kinase levels stay the same. A certain number of cyclin-kinase pairs turn on the genes that trigger mitosis. Then, as mitosis begins, enzymes degrade the cyclin. The cycle starts again as cyclin begins to build up during the next interphase. A successful cancer drug called imatinib (Gleevec), discussed in Clinical Connection 18.1, targets a kinase.

Apoptosis

Apoptosis rapidly and neatly dismantles a cell into membrane-enclosed pieces that a phagocyte (a cell that engulfs and destroys another) can mop up. It is a little like packing the contents of a messy room into garbage bags, then disposing of it all. In contrast is necrosis, a form of cell death associated with inflammation and damage, rather than an orderly, contained destruction.

Like mitosis, apoptosis is a continuous process. It begins when a "death receptor" on the cell's plasma membrane receives a signal to die. Within seconds, enzymes called caspases are activated inside the doomed cell, stimulating each other and snipping apart various cell components. These killer enzymes:

- destroy enzymes that replicate and repair DNA;
- activate enzymes that cut DNA into similarly sized pieces;

- tear apart the cytoskeleton, collapsing the nucleus and condensing the DNA within;
- cause mitochondria to release molecules that increase caspase activity and end the energy supply;
- abolish the cell's ability to adhere to other cells; and
- attract phagocytes that dismantle the cell remnants.

A dying cell has a characteristic appearance (**figure 2.17**). It rounds up as contacts with other cells are cut off, and the plasma membrane undulates, forming bulges called blebs. The nucleus bursts, releasing DNA pieces, as mitochondria decompose. Then the cell shatters. Almost instantly, pieces of membrane encapsulate the cell fragments, which prevents inflammation. Within an hour, the cell is gone.

From the embryo onward through development, mitosis and apoptosis are synchronized and, as a result, tissue neither overgrows nor shrinks. In this way, a child's liver retains its shape as she grows into adulthood, yet enlarges. During early development, mitosis and apoptosis orchestrate the ebb and flow of cell numbers as new structures form. Later, these processes protect—mitosis produces new skin to heal a scraped knee; apoptosis peels away sunburnt skin cells that might otherwise become cancerous. Cancer is a profound derangement of the balance between cell division and cell death. In cancer, mitosis occurs too frequently or too many times, or apoptosis happens too infrequently. Chapter 18 discusses cancer in detail.

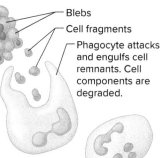

Death receptor on doomed cell binds signal molecule. Caspases are activated within.

Caspases destroy various proteins and other cell components. Cell undulates.

Blebs
Cell fragments
Phagocyte attacks and engulfs cell remnants. Cell components are degraded.

Figure 2.17 Death of a cell. A cell undergoing apoptosis loses its shape, forms blebs, and falls apart. Caspases destroy the cell's insides. Phagocytes digest the remains. Note the blebs on the dying liver cells in the top photograph. Sunburn peeling is one example of apoptosis. *(top): © David McCarthy/Photo Researchers/ Science Source; (bottom): © Peter Skinner/Photo Researchers/Science Source*

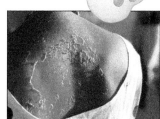

Key Concepts Questions 2.3

1. Explain why mitosis and apoptosis must be balanced.
2. Describe the phases of the cell cycle.
3. List the stages and the major events of mitosis.
4. Discuss how chromosome structure and terminology change during the cell cycle.
5. Explain control of the cell cycle.
6. List the events of apoptosis.

2.4 Stem Cells

A growing body must maintain the proportions and numbers of different cell types. Considering the composition of a human body by cell type, red blood cells are by far the most abundant. It is strange to realize that the cell types we are most aware of—fat and muscle cells—comprise only 0.2 percent or less of the total cell count. This is because the cells of fat and muscle tissue are much larger than red blood cells.

Bodies grow and heal thanks to cells that retain the ability to divide, generating new cells like themselves and cells that will specialize. **Stem cells** and **progenitor cells** renew tissues so that as the body grows or loses cells to apoptosis, injury, and disease, new cells are produced that take their places.

Cell Lineages

A stem cell divides by mitosis to yield either two daughter cells that are stem cells like itself, or one that is a stem cell and one that is a partially specialized progenitor cell (**figure 2.18**). The characteristic of **self-renewal** is what makes a stem cell a stem cell—its ability to continue the lineage of cells that can divide to give rise to another cell like itself. In contrast to a stem cell, a progenitor cell cannot self-renew, and its daughters specialize as any of a restricted number of cell types. A fully differentiated cell, such as a mature blood cell, descends from a sequence of increasingly specialized progenitor cell intermediates, each one less like a stem cell and more like a blood cell. Our more than 290 differentiated cell types develop

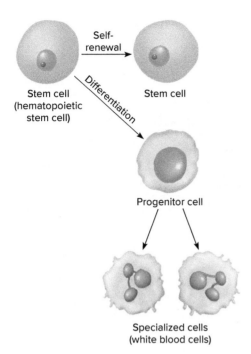

Figure 2.18 **Stem cells and progenitor cells.** A stem cell is less specialized than the progenitor cell that descends from it by mitosis. Stem cells provide raw material for producing specialized cells, while retaining the ability to generate new cells. A hematopoietic stem cell in the bone marrow produces progenitors whose daughter cells specialize as types of blood cells.

from sequences, called lineages, of stem, progenitor, and their increasingly differentiated daughter cells. **Figure 2.19** shows parts of a few cell lineages.

Stem cells and progenitor cells are described in terms of developmental potential—that is, according to the number of possible fates of their daughter cells. A fertilized ovum is the ultimate stem cell. It is totipotent, which means that it can give rise to every cell type, including cells of the membranes that support the embryo. Other stem cells and progenitor cells are pluripotent: Their daughter cells have fewer possible fates. Some are multipotent: Their daughter cells have only a few developmental "choices." This is a little like a college freshman's consideration of many majors, compared to a junior's more narrowed focus in selecting courses.

As stem cell descendants specialize, they express some genes and ignore others. An immature bone cell forms from a progenitor cell by manufacturing mineral-binding proteins and enzymes. In contrast, an immature muscle cell forms from a muscle progenitor cell that accumulates contractile proteins. The bone cell does not produce muscle proteins, nor does the muscle cell produce bone proteins. All cells, however, synthesize proteins for basic "housekeeping" functions, such as energy acquisition and protein synthesis.

Many, if not all, of the organs in an adult human body have stem or progenitor cells. These cells can divide when injury or illness occurs and generate new cells to replace damaged ones. Stem cells in the adult may have been set aside in the embryo

or fetus as repositories of future healing. Some stem cells, such as those from bone marrow, can travel to and replace damaged or dead cells elsewhere in the body, in response to signals that are released in injury or disease. Because every cell except red blood cells contains all of an individual's DNA, any cell type, given appropriate signals, can in theory become any other. This concept is the basis of some technologies that use stem cells.

Researchers study stem cells to learn more about biology and to develop treatments for a variety of diseases and injuries—not just inherited conditions. These cells come from both donors' and patients' own bodies, as **figure 2.20** illustrates. The cells can be mass produced in laboratory glassware, and if they originate from a patient's cell or nucleus, they are a genetic match.

Stem Cell Sources

There are three general sources of human stem cells: **embryonic stem (ES) cells**, **induced pluripotent stem (iPS) cells**, and "adult" stem cells (**Table 2.3**).

Embryonic stem (ES) cells form in a laboratory dish from certain cells taken from a very early embryo, at a stage called the inner cell mass (ICM) (see figure 3.15). Some ICM cells, under certain conditions, become pluripotent and can self-renew—that is, they function as stem cells.

The ICM cells used to derive human ES cells can come from two sources: "leftover" embryos from fertility clinics that would otherwise be destroyed, and somatic cell nuclear transfer, a process in which a nucleus from a somatic cell (such as a skin fibroblast) is transferred to an egg cell that has had its own nucleus removed. An embryo is grown, and its ICM cells are used to make ES cells. Researchers can also derive ES cells from somatic cell nuclear transfer into egg cells directly, without using embryos. Somatic cell nuclear transfer is popularly called "cloning" because it copies the nucleus donor's genome. It was made famous by Dolly the sheep, whose original nucleus came from a mammary gland cell of a 6-year-old ewe. The age of the donor may be why Dolly did not live as long as a sheep conceived the natural way. **Bioethics** in chapter 3 discusses "Why a Clone Is Not an Exact Duplicate."

Induced pluripotent stem (iPS) cells are somatic cells that are "reprogrammed" to differentiate into any of several cell types. Reprogramming a cell may take it back through developmental time to an ES cell-like state. Then the cell divides and gives rise to cells that specialize as different, desired cell types. Or, cells can be reprogrammed directly into another cell type. Reprogramming instructions may be in the form of DNA, RNA, or other chemical factors. Induced pluripotent stem cells do not require cells from an embryo, and they are a genetic match to the person of origin. For now, iPS cells are a research tool and will not be used to treat disease until researchers learn more about how they function in a human body.

To return to the college major analogy, reprogramming a cell is like a senior in college deciding to change major. A French major wanting to become an engineer would have to start over, taking very different courses. But a biology major wanting to become a chemistry major would not need to start

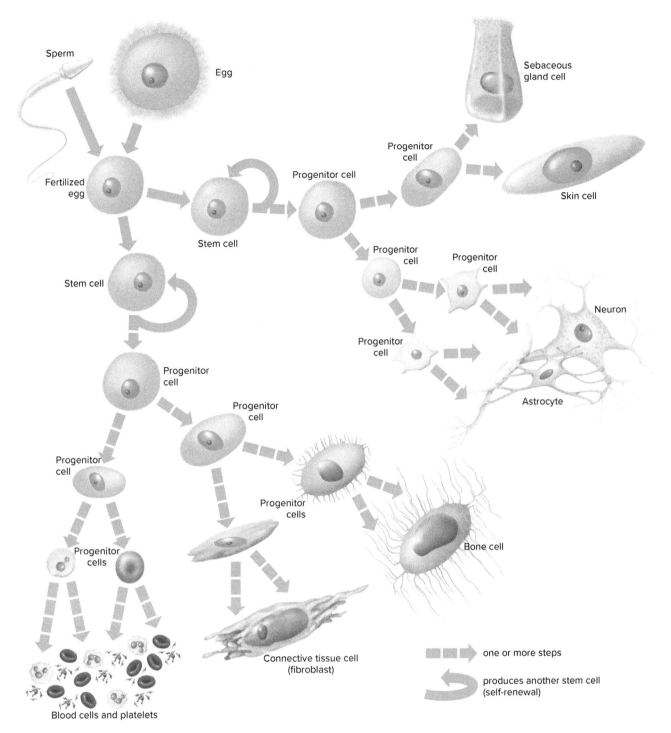

Figure 2.19 **Pathways to cell specialization.** All cells in the human body descend from stem cells, through the processes of mitosis and differentiation. The differentiated cells on the bottom are all connective tissues (blood, connective tissue, and bone), but the blood cells are more closely related to each other than they are to the other two cell types. On the upper right, the skin and sebaceous (oil) gland cells share a recent progenitor, and both share a more distant progenitor with neurons and supportive astrocytes. Imagine how complex the illustration would be if it depicted all 290-plus types of cells in a human body!

from scratch because many of the same courses apply to both majors. So it is for stem cells. Taking a skin cell from a man with heart disease and deriving a healthy heart muscle cell might require taking that initial cell back to an ES or iPS state, because these cells come from very different lineages.

But turning a skeletal muscle cell into a smooth muscle cell requires fewer steps backwards because the two cell types differentiate from the same type of progenitor cell.

A third source of stem cells are those that naturally are part of the body, called "adult" stem cells. They are more

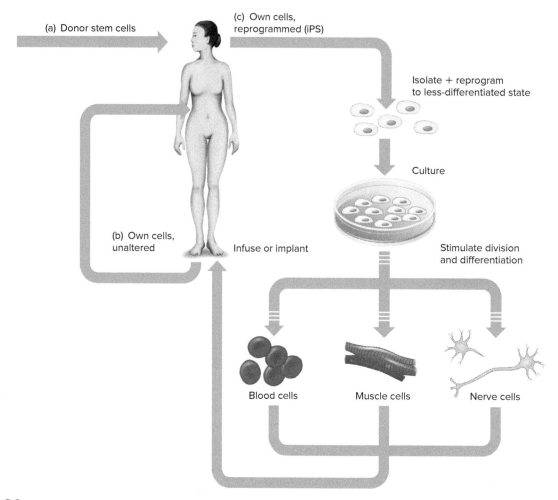

Figure 2.20 Using stem cells to heal. **(a)** Stem cells from donors (bone marrow or umbilical cord blood) are already in use. **(b)** A person's cells may be used, unaltered, to replace damaged tissue, such as bone marrow. **(c)** It is possible to "reprogram" a person's cells in culture, taking them back to a less specialized state and then nurturing them to differentiate as a needed cell type.

Table 2.3	Stem Cell Sources
Stem Cell Type	**Source**
Embryonic stem cell	Inner cell mass of very early embryo; somatic cell nuclear transfer into egg cell
Induced pluripotent stem cell	Genes or other chemicals reprogram somatic cell nucleus; no embryos required
"Adult" stem cell	Somatic cells that normally function as stem cells, from any stage of development from fertilized ovum through elderly

accurately called tissue-specific or somatic stem cells because they are found in the tissues of embryos, fetuses, and children, and not just in adult bodies. Adult stem cells self-renew, but most give rise only to a few types of specialized daughter cells. Researchers are still discovering niches of adult stem cells in the body. Many potentially valuable adult stem cells are discarded as medical waste, such as fat, placenta, and brain tissue.

Stem Cell Applications

Stem cells are being used in four basic ways. In the first, drug discovery and development, stem cell cultures supply the human cells that are affected in a particular disease, which may be difficult or impossible to culture in a laboratory. Drugs are tested on these cells. Liver and heart cells derived from stem cells are particularly useful in testing drugs for side effects, because the liver detoxifies many drugs, and many drugs harm the heart. Using stem cells in drug development can minimize the need to experiment on animals and can eliminate candidate drugs with adverse effects before they are tested on people.

A second application of stem cells growing in culture is to observe the earliest signs of a disease, which may begin long before symptoms are noted. The first disease for which human iPS cells were derived was amyotrophic lateral sclerosis (ALS), also known as Lou Gehrig's disease. In ALS, motor neurons that enable a person to move gradually fail. A few years after the first signs of weakness or stumbling, death comes from failure of the respiratory muscles. ALS had been difficult to study because motor neurons do not survive for long in laboratory glassware because they do not divide. However, iPS cells

Banking Stem Cells: When Is It Necessary?

The parents-to-be were very excited by the company's promise: *Bank your baby's cord blood stem cells and benefit from breakthroughs. Be prepared for the unknowns in life.*

The website profiled children saved from certain diseases using stored umbilical cord blood. The statistics were persuasive: More than 70 diseases are currently treatable with cord blood transplants, and 10,000 procedures have already been done.

With testimonials like that, it is little wonder that parents collectively spend more than $100 million per year to store cord blood. The ads and statistics are accurate but misleading, because of what they *don't* say. Most people never actually use the umbilical cord blood stem cells that they store. The scientific reasons go beyond the fact that treatable diseases are very rare. In addition, cord blood stem cells are not nearly as pluripotent as some other stem cells, limiting their applicability. Perhaps the most compelling reason that stem cell banks are rarely used beyond initial storage is logic: For a person with an inherited disease, *healthy* stem cells are required—not his or her own, which could cause the disease all over again because the mutation is in every cell. The patient needs a well-matched donor, such as a healthy sibling.

Commercial cord blood banks may charge more than $1,000 for the initial collection plus an annual fee. However, the U.S. National Institutes of Health and organizations in many other nations have supported not-for-profit banks for years, and these may not charge fees. Donations of cord blood to these facilities are not to help the donors directly, but to help whomever can use the cells.

Commercial stem cell banks are not just for newborns. One company banks tiny embryonic-like stem cells for an initial charge of $7,500 and a $750 annual fee, to enable people to donate and store their own stem cells when young and healthy for personal use in times of unexpected medical need. The cells come from a person's blood and, in fact, one day may be very useful, but the research has yet to be done supporting use of the cells in specific treatments.

Questions for Discussion

1. Storing stem cells is not regulated by the U.S. government the way that a drug or a surgical procedure is because it is a service that will be helpful for treatments not yet invented. Do you think such banks should be regulated, and if so, by whom and how?

2. What information should companies offering to store stem cells post on their websites?

3. Do you think that advertisements for cord blood storage services that have quotes and anecdotal reports, but do not mention that most people who receive stem cell transplants do not receive their own cells, are deceptive? Or do you think it is the responsibility of the consumer to discover this information?

4. Several companies store stem cells extracted from baby teeth, although a use for such stem cells has not yet been found. (The use of DNA from a tooth in the chapter opener was to diagnose a child posthumously.) Suggest a different way to obtain stem cells that have the genome of a living child.

---

derived from fibroblasts in patients' skin are reprogrammed in culture to become ALS motor neurons, providing an endless source of the hard-to-culture cells from the abundant fibroblasts. Researchers worldwide are establishing cell banks to store iPS cells corresponding to a variety of diseases.

The third application of stem cells is to create tissues and organs for use in implants and transplants, or to study. This approach is not new—the oldest such treatment, a bone marrow transplant, has been available for more than 60 years. Many other uses of adult stem cells, delivered as implants, transplants, or infusions into the bloodstream, are being tested. A patient's own bone marrow stem cells, for example, can be removed, isolated, perhaps altered, grown, bathed in selected factors, and infused back into the patient, where they follow natural signals to damaged tissues.

The fourth application of stem cells became clear with the creation of iPS cells. It might be possible to introduce reprogramming proteins directly into the body to stimulate stem cells in their natural niches. Once we understand the signals, we might not need the cells. The applications of stem cells seem limited only by our imaginations, but some companies take advantage of what the public does not know about the science. Bioethicists call this practice "stem cell tourism." Bioethics discusses stem cell banking.

Key Concepts Questions 2.4

1. Explain why different cell types are not present in the body in equal numbers.
2. State the two general characteristics of stem cells.
3. Distinguish stem cells from progenitor cells.
4. Define *cell lineage*.
5. Describe the functions of stem cells throughout life.
6. Distinguish among embryonic stem cells, induced pluripotent stem cells, and adult stem cells.
7. Describe how stem cells might be used in health care.

2.5 The Human Microbiome

Many of the cells within a human body are not actually human—our bodies are vast ecosystems for microscopic life. The average person consists of about 30 trillion cells, but a healthy human body also includes about 39 trillion cells that are bacterial, fungal, or protozoan, as well as many viruses. The cells within and on us that are not actually *of* us constitute the human **microbiome**.

The term *microbiome* borrows from the ecological term *biome* to indicate all of the species in an area. Trillions of bacteria can fit into a human body because their cells, which are prokaryotic, are much smaller than ours. The Human Microbiome Project has revealed our bacterial residents, although the first to notice our bacteria was inventor of the light microscope Antonie van Leeuwenhoek, who in 1676 described the "animalcules" in his mouth. Today, researchers look at DNA sequences to identify the bacterial species that inhabit the human mouth. Knowledge of the human microbiome will be increasingly applied to improving health, once we learn whether a particular collection of microbes causes or contributes to an illness or is a result of it.

Each of us has a shared "core microbiome" of bacterial species, but also many other species that reflect our differing genomes, environments, habits, ages, diets, and health. Different body parts house distinctive communities of microbes, altered by experience. A circumcised penis has a different microbiome than an uncircumcised one, for example, and the vaginal microbiome of a mother differs from that of a woman who hasn't had a child. Different communities of bacteria inhabit a person's armpits, groin, navel, bottoms of the feet, spaces between fingers, and the space between the buttocks. The microbiome changes with experience and environmental exposures, as **figure 2.21** shows.

The 10 trillion bacteria that make up the human "gut microbiome" have long been known to help us digest certain foods. The mouth alone houses more than 600 species of bacteria. Analysis of their genomes yields practical information. For example, the genome sequence of the bacterium *Treponema denticola* reveals how it survives amid the films other bacteria form in the mouth and how it causes gum disease.

The more distant end of our digestive tract is easy to study in feces. The large intestine alone houses 6,800 bacterial species. One of the first microbiome studies examined soiled diapers from babies during their first year—one the child of the chief investigator—chronicling the establishment of the gut bacterial community. Newborns start out with clean intestines. After various bacteria come and go, very similar species remain from baby to baby by the first birthday.

Illness can alter the bacterial populations within us, and we use knowledge of our gut microbiome to improve health. Probiotics are live microorganisms, such as bacteria and yeasts, that, when ingested, confer a health benefit. For example, certain *Lactobacillus* strains added to yogurt help protect against *Salmonella* foodborne infection.

Probiotics typically consist of one or several microbial species. In contrast, "fecal transplantation" is a treatment based on altering the microbiome that replaces hundreds of bacterial species at once. In the procedure, people with recurrent infection from *Clostridium difficile*, which causes severe diarrhea, receive feces from a healthy donor. The bacteria transferred in the donated feces reconstitute a healthy gut microbiome, treating the infection. Fecal transplantation has been performed by enema in cattle for a century, and since 1958 in humans. The material, fresh or frozen, is introduced by enema, although clinical trials are testing delivery via a tube through the nose to the small intestine, or in a capsule. Do not try this yourself!

Researchers have been studying the human microbiome for a short time, but already have learned or confirmed the following facts.

1. Certain skin bacteria cause acne, but others keep skin clear.
2. Circumcision protects against certain viral infections, including HIV.
3. Lowered blood sugar following weight-loss surgery is partly due to a changed gut microbiome.
4. An altered microbiome hastens starvation in malnourished children.
5. Antibiotics temporarily alter the gut microbiome.

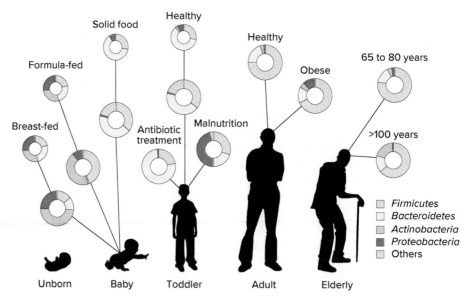

Figure 2.21 **The human microbiome.** Our habits, experiences, and environmental exposures influence the makeup of our microbiomes. The colors indicate different species of bacteria. *Image courtesy: US National Library of Medicine. Image source: Ottman N. Smidt H, de Vos WM and Belzer C (2012) The function of our microbiota: who is out there and what do they do? Front. Cell. Inf. Microbio, 2:104 doi: 10.3389/fcimb.2012.00104.*

6. The birth of agriculture 10,000 years ago introduced the bacteria that cause dental caries.

7. Microbiome imbalances may contribute to or cause asthma, cancers, obesity, psoriasis, Crohn's disease, gum disease, and other conditions.

8. Babies born by Cesarean section (surgically) have different microbiomes than babies born vaginally.

Our genomes have evolved along with our microbiomes. Certain modern practices, such as use of antibiotic drugs, sanitation, and use of antimicrobial soap, may affect the composition of our personal microbiomes.

Summary

2.1 Introducing Cells

1. Cells are the basic units of life and comprise the human body. Inherited traits and illnesses can be understood at the cellular and molecular levels.

2. All cells share certain features, but they are also specialized because they express different subsets of genes.

3. **Somatic** (body) **cells** are **diploid**, and **germ cells** (sperm and egg cells) are **haploid**. **Stem cells** produce new cells.

2.2 Cell Components

4. A **prokaryotic cell** does not have a **nucleus** or other **organelles**. A **eukaryotic cell** has organelles, including a nucleus.

5. Cells consist primarily of water and several types of macromolecules: **carbohydrates**, **lipids**, **proteins**, and **nucleic acids**.

6. Organelles sequester related biochemical reactions, improving the efficiency of life functions and protecting the cell. The cell also consists of **cytoplasm** and certain chemicals.

7. The nucleus contains DNA and a **nucleolus**, a site of ribosome synthesis. **Ribosomes** provide scaffolds for protein synthesis; they exist free in the cytoplasm or complexed with the rough **endoplasmic reticulum (ER)**.

8. In secretion, the rough ER is where protein synthesis begins. Smooth ER is the site of lipid synthesis, transport, and packaging; the **Golgi apparatus** packages secretions into **vesicles**, which exit through the **plasma membrane**. **Exosomes** ferry molecules between cells.

9. **Lysosomes** contain enzymes that dismantle debris (autophagy), and **peroxisomes** house enzymes that perform a variety of functions.

10. Enzymes in **mitochondria** extract energy from nutrients.

11. The plasma membrane is a protein-studded phospholipid bilayer. It controls which substances exit and enter the cell (**signal transduction**) and how the cell interacts with other cells (**cellular adhesion**).

12. The **cytoskeleton** is a protein framework of hollow **microtubules**, made of tubulin, and solid **microfilaments**, which consist of actin. **Intermediate filaments** are made of more than one protein type.

2.3 Cell Division and Death

13. Coordination of cell division (**mitosis**) and cell death (**apoptosis**) maintains cell numbers, enabling structures to enlarge during growth and development but preventing abnormal growth.

14. The **cell cycle** indicates whether a cell is dividing (mitosis) or not (**interphase**). Interphase consists of G_1 and G_2 **phases**, when proteins and lipids are produced, and a synthesis **(S) phase**, when DNA is replicated.

15. Mitosis proceeds in four stages. In **prophase**, replicated chromosomes consisting of two sister **chromatids** condense, the **spindle** assembles, the nuclear membrane breaks down, and the nucleolus is no longer visible. In **metaphase**, replicated chromosomes align along the center of the cell. In **anaphase**, the **centromeres** part, equally dividing the now unreplicated chromosomes into two daughter cells. In **telophase**, the new cells separate. **Cytokinesis** apportions organelles and other components into the daughter cells.

16. Internal and external factors control the cell cycle. Checkpoints are times when proteins regulate the cell cycle. **Telomere** (chromosome tip) length determines how many more times mitosis will occur. Crowding, **hormones**, and growth factors signal cells from the outside; the interactions of cyclins and kinases trigger mitosis from inside.

17. In apoptosis, a receptor on the plasma membrane receives a death signal, which activates caspases that tear apart the cell in an orderly fashion. Membrane surrounds the pieces, preventing inflammation.

2.4 Stem Cells

18. **Stem cells** self-renew, producing daughter cells that are also stem cells, and they also yield daughter cells that specialize. **Progenitor cells** give rise to more specialized daughter cells but do not **self-renew**.

19. A fertilized ovum is totipotent. Some stem cells are pluripotent and some are multipotent. Cells are connected through lineages.

20. The three sources of human stem cells are **embryonic stem (ES) cells**, **induced pluripotent stem (iPS) cells**, and adult stem cells.

21. Stem cell technology enables researchers to observe the origins of diseases and to devise new types of treatments.

2.5 The Human Microbiome

22. About 60 percent of the cells in a human body are microorganisms. They are our **microbiome**.

23. Different body parts house different communities of microbes.

24. Genes interact with the microbiome and with environmental factors to mold most traits.

Review Questions

1. The terms used in cell biology can be quite colorful. Explain how the following technical terms describe the things they name (you might have to look some word parts up, or just guess). An example is organelle = "little organ."
   a. haploid
   b. eukaryote
   c. cytoplasm
   d. lactocyte
   e. lysosome
   f. autophagy
   g. peroxisome
   h. telomere
   i. microbiome

2. Match each organelle to its function.

Organelle	Function
a. lysosome	1. lipid synthesis
b. rough ER	2. houses DNA
c. nucleus	3. energy extraction
d. smooth ER	4. dismantles debris
e. Golgi apparatus	5. detoxification
f. mitochondrion	6. protein synthesis
g. peroxisome	7. processes secretions

3. Name and describe a disease that affects each of the following organelles or structures, using information from the text or elsewhere.
   a. lysosome
   b. peroxisome
   c. mitochondrion
   d. cytoskeleton
   e. ion channel

4. Identify an organelle whose name is an eponym (a person's name) rather than a descriptive term.

5. What advantage does compartmentalization provide to a large and complex cell?

6. Give two examples of how the plasma membrane functions as more than just a covering of the cell's insides.

7. List four types of controls on cell cycle rate.

8. How can all of a person's cells contain exactly the same genetic material, yet specialize as bone cells, nerve cells, muscle cells, and connective tissue cells?

9. Distinguish between
   a. a bacterial cell and a eukaryotic cell.
   b. interphase and mitosis.
   c. mitosis and apoptosis.
   d. rough ER and smooth ER.
   e. microtubules and microfilaments.
   f. a stem cell and a progenitor cell.
   g. totipotent and pluripotent.

10. How are intermediate filaments similar to microtubules and microfilaments, and how are they different?

11. List the events of each stage of mitosis.

12. How is a progenitor cell like a stem cell yet also not like a stem cell?

13. How is the common definition of a stem cell as a cell that "turns into any cell type" incorrect?

14. Distinguish among ES cells, iPS cells, and adult stem cells, and state the pros and cons of working with each to develop a therapy.

15. Define *microbiome*.

16. "I heard about a great new organic, gluten-free, nonfat, low-calorie, high-energy way to remove all of the bacteria in our intestines!" enthused the young woman in the health food store to her friend. Is this a good idea or not? Cite a reason for your answer.

17. State one or more ways the human genome and human microbiome are similar and one or more ways they differ.

Applied Questions

1. The chapter opener describes diagnosing Rett syndrome, which affects neurons in the brain, from DNA extracted from a saved baby tooth. Explain why a mutation detected in one cell type can be used to help diagnose an inherited condition that affects a different cell type.

2. How might abnormalities in each of the following contribute to cancer?
   a. cellular adhesion
   b. signal transduction
   c. balance between mitosis and apoptosis
   d. cell cycle control
   e. telomerase activity
   f. an activated stem cell

3. In which organelle would a defect cause fatigue on a whole-body level?

4. If you wanted to create a synthetic organelle to test new drugs for toxicity, which natural organelle's function would you try to replicate?

5. An inherited form of migraine is caused by a mutation in a gene *(SCN1A)* that encodes a sodium channel in neurons. What is a sodium channel, and in which cell structure is it located?

6. In a form of Parkinson disease, a protein called alpha synuclein accumulates because of impaired autophagy. Which organelle is implicated in this form of the disease?

7. Why wouldn't a cell in an embryo likely be in phase G_0?

8. Explain how replacing an enzyme or implanting umbilical cord stem cells from a donor can treat a lysosomal storage disease.

9. Describe two ways to derive stem cells without using human embryos.

10. Two roommates regularly purchase 100-calorie packages of cookies. After a semester of each roommate eating one package a night, and following similar diet and exercise plans otherwise, one roommate has gained 12 pounds, but the other's weight has stayed the same. If genetics does not account for the difference in weight gain, what other factor discussed in the chapter might?

11. Stem cells from bone marrow send membrane bubbles containing proteins to fibroblasts (connective tissue cells) near the site of an injury, stimulating them to divide and migrate to the wound, where they promote healing and form scar tissue. What are the membrane bubbles called?

Forensics Focus

1. A man who owned a company that provides human tissues was sentenced to serve many years in prison for trafficking in body parts taken, without consent, from dismembered corpses from funeral homes in Pennsylvania, New Jersey, and New York. Thousands of parts from hundreds of bodies were used in surgical procedures, including hip replacements and dental implants. The most commonly used product was a paste made of bone. Many family members of the tissue sources testified at the trials. Explain why cells from bone tissue can be matched to blood or cheek lining cells from blood relatives.

Case Studies and Research Results

1. The boy is described as a "Michelin tire baby," the nickname for "congenital skin crease Kunze type." He has rings of creases on his arms and legs, unusual facial features, intellectual disability, slow growth, and a cleft palate. He has a mutation in the gene *TUBB,* and his condition is termed a "tubulinopathy." What part of his cells is affected?

2. In the past, researchers studied different cell types in bulk, because it is easier to dismantle them from part of an organ. A newer field, called single-cell genomics, instead sequences genomes of individual cells. This is possible using microfluidic techniques. What is an advantage of describing a cell based on analysis of the genome of a single cell compared to pooling data from cells extracted from a tissue?

3. Studies show that women experiencing chronic stress, such as from caring for a severely disabled child, have telomeres that shorten at an accelerated rate. Suggest a study that would address the question of whether men have a similar reaction to chronic stress.

4. The editors of a journal about stem cells wrote, "iPSCs will no doubt provide the lens through which we'll come to understand the richness and flexibility of cellular identity." What do they mean?

5. Researchers isolated stem cells from fat removed from people undergoing liposuction, a procedure to remove fat. The stem cells can give rise to muscle, fat, bone, and cartilage cells.
   a. Are the stem cells totipotent or pluripotent?
   b. Are these stem cells ES cells, iPS cells, or adult stem cells?

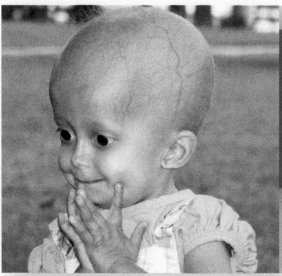

Courtesy of the Progeria Research Foundation

CHAPTER 3

Meiosis, Development, and Aging

This child has progeria, a rare genetic syndrome in which some signs of aging are greatly accelerated. A worldwide effort to find all affected children has enabled testing of a drug that, so far, improves some symptoms, extends life, and may lead to treatments of common diseases in the general aging population.

Learning Outcomes

3.1 The Reproductive System
1. Describe the structures of the male and female reproductive systems.

3.2 Meiosis
2. Explain why meiosis is necessary to reproduce.
3. Summarize the events of meiosis.

3.3 Gametes Mature
4. List the steps in sperm and oocyte formation.

3.4 Prenatal Development
5. Describe early prenatal development.
6. Explain how the embryo differs from the fetus.

3.5 Birth Defects
7. Define *critical period*.
8. List some teratogens.

3.6 Maturation and Aging
9. Describe common diseases that begin in adulthood.
10. Explain how rapid aging syndromes occur.

 ## The BIG Picture

Our reproductive systems enable us to start a new generation. First our genetic material must be halved, so it can combine with that of a partner to reconstitute a full diploid genome. Then genetic programs operate as the initial cell divides and its daughter cells specialize. The forming tissues fold into organs and the organs interact, slowly building a new human body. Mutations can cause disease at any stage of development or age.

Progress for Progeria

A child with Hutchinson-Gilford progeria syndrome (HGPS) appears old. Shortly after birth, weight gain begins to slow and hair to thin. The gums remain smooth, as teeth do not erupt. Joints stiffen and bones weaken. Skin wrinkles as the child's chubbiness melts away too quickly, and a cherubic toddler becomes increasingly birdlike in appearance.

Beneath the child's toughening skin, blood vessels stiffen with premature atherosclerosis, fat pockets shrink, and connective tissue hardens. Inside cells, chromosome tips whittle down at an accelerated pace, marking time too quickly. However, some organs remain healthy, and intellect is spared. The child usually dies during adolescence, from atherosclerosis that may cause a heart attack or stroke.

About 350 to 400 children worldwide are known to have this strange illness, first described in 1886. When researchers in 2003 identified the mutant gene that causes HGPS, they realized, from the nature of the molecular defect, that several existing drugs might alleviate certain symptoms.

Researchers tested a failed cancer drug (lonafarnib), tracking children's weight to assess whether the drug had an effect. These children usually stop growing at age 3, but 9 of 25 children in the initial clinical trial had a greater than 50 percent increase in the rate of weight gain. The artery stiffening and blockages of atherosclerosis not only stopped progressing, but reversed! For some children, hearing improved and bones became stronger. More than 70 children have now participated in the clinical trial, and so far the drug has increased life span by at least 1.6 years.

The Progeria Research Foundation is trying to locate all of the children with progeria in the world, and involve them in testing several promising drugs, both old and new. The drugs discovered to help in progeria might also be useful in people who develop the symptoms as part of advancing age.

3.1 The Reproductive System

Genes orchestrate our physiology from a few days after conception through adulthood. Expression of specific sets of genes sculpts the differentiated cells that interact, aggregate, and fold, forming the organs of the body. Abnormal gene functioning can affect health at all stages of development. Certain single-gene mutations act before birth, causing broken bones, dwarfism, cancer, and other conditions. Some mutant genes exert their effects during childhood, appearing as early developmental delays or loss of skills. Inherited forms of heart disease and breast cancer can appear in early or middle adulthood, which is earlier than multifactorial forms of these conditions. Pattern baldness is a common single-gene trait that may not become obvious until well into adulthood. This chapter explores the stages of the human life cycle that form the backdrop against which genes function.

The first cell that leads to development of a new individual forms when a **sperm** from a male and an **oocyte** (also called an egg) from a female join. Sperm and oocytes are **gametes**, or sex cells. They mix genetic material from past generations. Because we have thousands of genes, some with many variants, each person (except for identical twins) has a unique combination of inherited traits.

Sperm and oocytes are produced in the reproductive system. The reproductive organs are organized similarly in the male and female. Each system has

- paired structures, called **gonads**, where the sperm and oocytes are manufactured;
- tubular structures that transport these cells; and
- hormones and secretions that control reproduction.

The Male

Sperm cells develop within a 125-meter-long network of seminiferous tubules, which one researcher describes as "a massive tangle of spaghetti." The seminiferous tubules are packed into paired, oval organs called **testes** (testicles) (**figure 3.1**). The testes are the male gonads. They lie outside the abdomen within a sac, the scrotum. This location keeps the testes cooler than the rest of the body, which is necessary for sperm to develop. Leading from each testis is a tightly coiled tube, the epididymis, in which sperm cells mature and are stored. Each epididymis continues into another tube, the ductus deferens. Each ductus deferens bends behind the bladder and joins the urethra, which is the tube that carries sperm and urine out of the body through the penis.

Along the sperm's path, three glands add secretions. The ductus deferentia pass through the prostate gland, which produces a thin, milky, acidic fluid that activates the sperm to swim. Opening into the ductus deferens is a duct from the

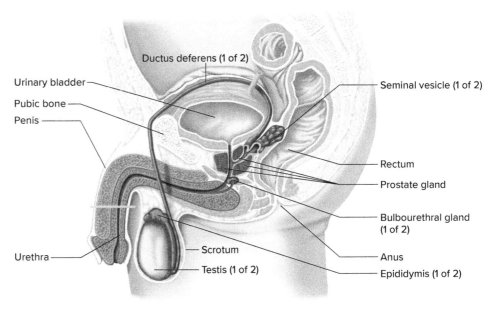

Figure 3.1 **The human male reproductive system.** Sperm cells are manufactured in the seminiferous tubules, which wind tightly within the testes and descend into the scrotum. The prostate gland, seminal vesicles, and bulbourethral glands add secretions to the sperm cells to form seminal fluid. Sperm mature and are stored in the epididymis and exit through the ductus deferens. The paired ductus deferentia join in the urethra, which transports seminal fluid from the body.

seminal vesicles, which secrete fructose (an energy-rich sugar) and hormonelike prostaglandins, which may stimulate contractions in the female that help sperm and oocyte meet. The bulbourethral glands, each about the size of a pea, join the urethra where it passes through the body wall. They secrete an alkaline mucus that coats the urethra before sperm are released. All of these secretions combine to form the seminal fluid that carries sperm. The fluid is alkaline.

During sexual arousal, the penis becomes erect, which enables it to penetrate and deposit sperm in the female reproductive tract. At the peak of sexual stimulation, a pleasurable sensation called orgasm occurs, accompanied by rhythmic muscular contractions that eject the sperm from each ductus deferens through the urethra and out the penis. The discharge of sperm from the penis, called ejaculation, delivers 200 million to 600 million sperm cells.

The Female

The female sex cells develop in paired organs in the abdomen called **ovaries (figure 3.2)**, which are the female gonads. Within each ovary of a newborn girl are about a million immature oocytes. Each individual oocyte nestles within nourishing follicle cells, and each ovary houses oocytes in different stages of development. After puberty, about once a month, one ovary releases the most mature oocyte. Beating cilia sweep the mature oocyte into the fingerlike projections of one of two uterine (also called fallopian) tubes. The uterine tube carries the oocyte into a muscular, saclike organ called the uterus, or womb.

The oocyte released from the ovary may encounter a sperm, usually in a uterine tube. If the sperm enters the oocyte and the DNA of the two gametes merges into a new nucleus, the result is a fertilized ovum. After a day, this first cell divides while moving through the uterine tube. It settles into the lining of the uterus, where it may continue to divide and develop into an embryo. If fertilization does not occur, the oocyte, along with much of the uterine lining, is shed as the menstrual flow. Hormones coordinate the monthly menstrual cycle.

The lower end of the uterus narrows and leads to the cervix, which opens into the tubelike vagina. The vaginal opening is protected on the outside by two pairs of fleshy folds. At the upper juncture of both pairs is the 2-centimeter-long clitoris, which is anatomically similar to the penis. Rubbing the clitoris triggers female orgasm. Hormones control the cycle of oocyte maturation and the preparation of the uterus to nurture a fertilized ovum.

3.2 Meiosis

Gametes form from special cells, called germline cells, in a type of cell division called **meiosis** that halves the chromosome number. A further process, maturation, sculpts the distinctive characteristics of sperm and oocyte. The organelle-packed oocyte has 90,000 times the volume of the streamlined sperm.

Gametes contribute 23 different chromosomes, constituting one copy of the genome, to a fertilized ovum. In contrast, somatic cells contain 23 *pairs* of chromosomes, or 46 in total. One member of each pair comes from the person's mother and one comes from the father. The chromosome pairs are called **homologous pairs**, or *homologs* for short. Homologs have the same genes in the same order but may carry different alleles, or variants, of the same gene. Recall from chapter 2 that gametes are **haploid** (1n), which means that they have only one of each type of chromosome, and somatic cells are **diploid** (2n), with two copies of each chromosome type.

It is the meeting of sperm and oocyte that restores two full genomes. Without meiosis, the sperm and oocyte would each contain 46 chromosomes, and the fertilized ovum would have twice the normal number of chromosomes,

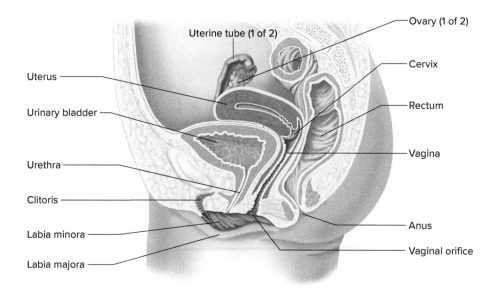

Figure 3.2 The human female reproductive system. Oocytes mature in the paired ovaries. Once a month after puberty, an ovary releases one oocyte, which is drawn into a nearby uterine tube. If a sperm fertilizes the oocyte in the uterine tube, the fertilized ovum continues into the uterus, where for 9 months it divides and develops. If the oocyte is not fertilized, the body expels it, along with the built-up uterine lining, as the menstrual flow.

Labels in figure: Uterine tube (1 of 2); Ovary (1 of 2); Cervix; Rectum; Vagina; Anus; Vaginal orifice; Uterus; Urinary bladder; Urethra; Clitoris; Labia minora; Labia majora

or 92. Such a genetically overloaded cell, called a polyploid, usually undergoes only a few cell divisions before stopping.

In addition to producing gametes, meiosis mixes up trait combinations. For example, a person may produce one gamete containing alleles encoding green eyes and freckles, yet another gamete with alleles encoding brown eyes and no freckles. Meiosis explains why siblings differ genetically from each other and from their parents.

In a much broader sense, meiosis, as the mechanism of sexual reproduction, provides genetic diversity. A population of sexually reproducing organisms is made up of individuals with different genotypes and phenotypes. Genetic diversity may enable a population to survive an environmental challenge. A population of asexually reproducing organisms, such as bacteria or genetically identical crops, consists of individuals with the same genome sequence. Should a new threat arise, such as an infectious disease that kills only organisms with a certain genotype, then the entire asexual population could perish. In a sexually reproducing population, individuals that inherited a certain combination of genes may survive. This differential survival of certain genotypes is the basis of evolution by natural selection, discussed in chapter 15.

Meiosis is actually two divisions of the genetic material. The first division is called a **reduction division** (or meiosis I) because it reduces the number of replicated chromosomes from 46 to 23. The second division, called an equational division (or meiosis II), produces four cells from the two cells formed in the first division by splitting the replicated chromosomes. **Figure 3.3** shows an overview of meiosis. The colors represent the contributions of the two parents, whereas size distinguishes different chromosomes. Only two types of chromosomes are depicted due to space constraints. In the body there would be 23 pairs of replicated chromosomes as meiosis begins.

As in mitosis, meiosis occurs after an interphase period when DNA is replicated (doubled) (**table 3.1**). After interphase, prophase I (so called because it is the prophase of meiosis I) begins as the replicated chromosomes condense and become

Table 3.1	Comparison of Mitosis and Meiosis	
Mitosis	**Meiosis**
One division	Two divisions
Two daughter cells per cycle	Four daughter cells per cycle
Daughter cells genetically identical	Daughter cells genetically different
Chromosome number of daughter cells same as that of parent cell (2n)	Chromosome number of daughter cells half that of parent cell (1n)
Occurs in somatic cells	Occurs in germline cells
Occurs throughout life cycle	In humans, completes after sexual maturity
Used for growth, repair, and asexual reproduction	Used for sexual reproduction, producing new gene combinations

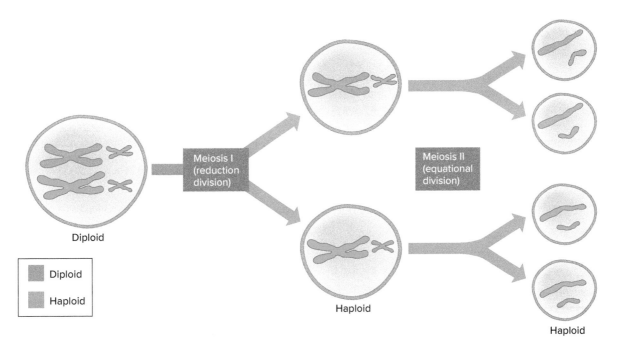

Figure 3.3 Overview of meiosis. Meiosis gives rise to haploid gametes. This simplified illustration follows the fate of two chromosome pairs rather than the true 23 pairs. In actuality, the first meiotic division reduces the number of chromosomes to 23, all in the replicated form. In the second meiotic division, the cells essentially undergo mitosis. The result of the two meiotic divisions (in this illustration and in reality) is four haploid cells. In this figure, homologous pairs of chromosomes are indicated by size, and parental origin of chromosomes by color.

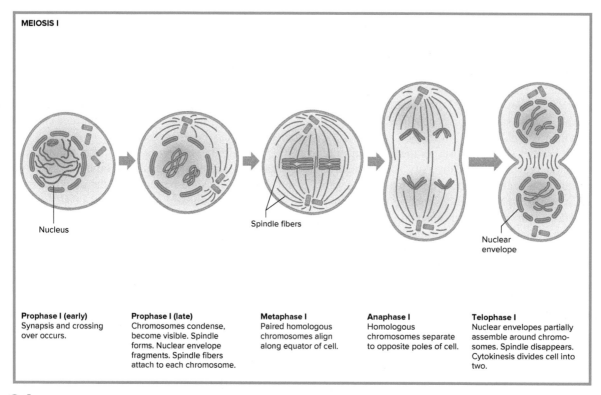

MEIOSIS I

Prophase I (early)
Synapsis and crossing over occurs.

Prophase I (late)
Chromosomes condense, become visible. Spindle forms. Nuclear envelope fragments. Spindle fibers attach to each chromosome.

Metaphase I
Paired homologous chromosomes align along equator of cell.

Anaphase I
Homologous chromosomes separate to opposite poles of cell.

Telophase I
Nuclear envelopes partially assemble around chromosomes. Spindle disappears. Cytokinesis divides cell into two.

Nucleus

Spindle fibers

Nuclear envelope

Figure 3.4 **Meiosis I.** The reduction division of meiosis I halves the number of replicated chromosomes from 46 to 23 in a human cell. (The figure depicts a representative two chromosome types.)

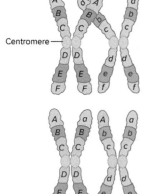

Homologous pair of chromosomes (schematized)

Centromere

Figure 3.5 **Crossing over recombines genes.** Crossing over generates genetic diversity by recombining genes, mixing parental traits. The capital and lowercase forms of the same letter represent different variants (alleles) of the same gene. A chromosome has hundreds to thousands of genes.

visible when stained (**figure 3.4**). A spindle forms. Toward the middle of prophase I, the homologs line up next to one another, gene by gene, in an event called synapsis. A mixture of RNA and protein holds the chromosome pairs together.

During synapsis, the homologs exchange parts, or **cross over (figure 3.5**). All four chromatids that comprise each homologous chromosome pair align as exchanges occur. After crossing over, each homolog bears some genes from each parent. Prior to this, all of the genes on a homolog were derived from one parent. New gene combinations arise from crossing over when the parents carry different alleles. Toward the end of prophase I, the synapsed chromosomes separate but remain attached at a few points along their lengths.

To understand how crossing over mixes trait combinations, consider a simplified example. Suppose that homologs carry genes for hair color, eye color, and finger length. One of the chromosomes carries alleles for blond hair, blue eyes, and short fingers. Its homolog carries alleles for black hair, brown eyes, and long fingers. After crossing over, one of the chromosomes might bear alleles for blond hair, brown eyes, and long fingers, and the other might bear alleles for black hair, blue eyes, and short fingers.

Meiosis continues in metaphase I, when the homologs align down the center of the cell. Each member of a homologous pair attaches to a spindle fiber at an opposite pole. The pattern in which the chromosomes align during metaphase I generates genetic diversity. For each homologous pair, the pole the maternally or paternally derived member goes to is random. It is a little like the number of different ways that 23 boys and 23 girls can line up in opposite-sex pairs.

The greater the number of chromosomes, the greater the genetic diversity generated in metaphase I. For two pairs of homologs, four (2^2) different metaphase alignments are possible. For three pairs of homologs, eight (2^3) different alignments can occur. Our 23 chromosome pairs can line up in 8,388,608 (2^{23}) different ways. This random alignment of chromosomes causes **independent assortment** of the genes that they carry. Independent assortment means that the fate of a gene on one chromosome is not influenced by a gene on a different chromosome (**figure 3.6**). Independent assortment accounts for a basic law of inheritance discussed in chapter 4.

Homologs separate in anaphase I and move to opposite poles by telophase I. These movements establish a haploid set

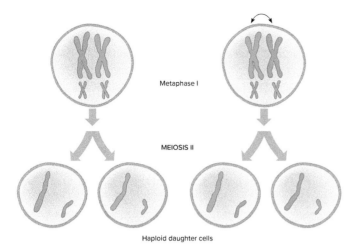

Metaphase I

MEIOSIS II

Haploid daughter cells

Figure 3.6 **Independent assortment.** The pattern in which homologs randomly align during metaphase I determines the combination of maternally and paternally derived chromosomes in the daughter cells. Two pairs of chromosomes can align in two ways to produce four possibilities in the daughter cells. The potential variability that meiosis generates for all 23 chromosome pairs, especially considering crossing over, is great.

of still-replicated chromosomes at each end of the stretched-out cell. Unlike in mitosis, the centromeres of each homolog in meiosis I remain together. During a second interphase, chromosomes unfold into thin threads. Proteins are manufactured, but DNA is not replicated a second time. The single DNA

replication, followed by the double division of meiosis, halves the chromosome number.

Prophase II marks the start of the second meiotic division (**figure 3.7**). The chromosomes are again condensed and visible. In metaphase II, the replicated chromosomes align down the center of the cell. In anaphase II, the centromeres part, and the newly formed chromosomes, each now in the unreplicated form, move to opposite poles. In telophase II, nuclear envelopes form around the four nuclei, which then separate into individual cells. The net result of meiosis is four haploid cells, each carrying a new assortment of genes and chromosomes that hold a single copy of the genome.

Meiosis generates astounding genetic variety. Any of a person's more than 8 million possible combinations of chromosomes can meet with any of the more than 8 million combinations of a partner, raising potential variability to more than 70 trillion ($8,388,608^2$) genetically unique individuals! Crossing over contributes almost limitless genetic diversity.

Key Concepts Questions 3.2

1. Distinguish between haploid and diploid.

2. Explain how meiosis maintains the chromosome number over generations and mixes gene combinations.

3. Discuss the two mechanisms that generate genotypic diversity.

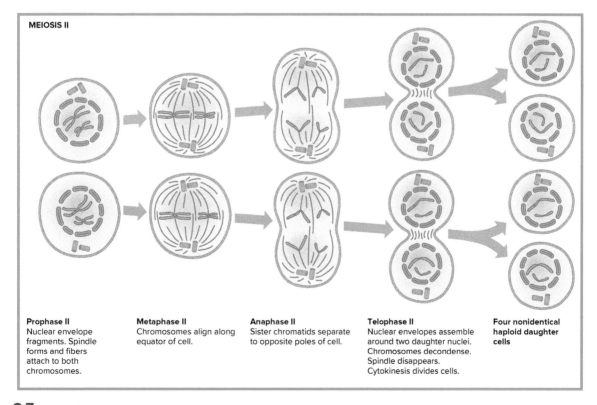

MEIOSIS II

Prophase II
Nuclear envelope fragments. Spindle forms and fibers attach to both chromosomes.

Metaphase II
Chromosomes align along equator of cell.

Anaphase II
Sister chromatids separate to opposite poles of cell.

Telophase II
Nuclear envelopes assemble around two daughter nuclei. Chromosomes decondense. Spindle disappears. Cytokinesis divides cells.

Four nonidentical haploid daughter cells

Figure 3.7 **Meiosis II.** The equational division of meiosis II pulls apart the replicated chromosomes, producing 46 chromosomes in the unreplicated form. (This figure depicts two representative chromosome types.)

3.3 Gametes Mature

Meiosis happens in both sexes, but further steps elaborate the very different-looking sperm and oocyte. Different distributions of cell components create the distinctions between sperm and oocytes. The forming gametes of the maturing male and female proceed through similar stages, but with sex-specific terminology and vastly different timetables. A male begins manufacturing sperm at puberty and continues throughout life, whereas a female begins meiosis when she is a fetus. Meiosis in the female completes only if a sperm fertilizes an oocyte.

Sperm Form

Spermatogenesis, the formation of sperm cells, begins in a diploid stem cell called a **spermatogonium (figure 3.8)**. This cell divides mitotically, yielding two daughter cells. One cell continues to specialize into a mature sperm. The other daughter cell remains a stem cell, able to self-renew and continually produce more sperm.

Bridges of cytoplasm attach several spermatogonia, and their daughter cells enter meiosis together. As these spermatogonia mature, they accumulate cytoplasm and replicate their DNA, becoming primary spermatocytes.

During reduction division (meiosis I), each primary spermatocyte divides, forming two equal-sized haploid cells called secondary spermatocytes. In meiosis II, each secondary spermatocyte divides to yield two equal-sized spermatids. Each spermatid then develops the characteristic sperm tail, or flagellum.

The base of the tail has many mitochondria, which will split ATP molecules to release energy that will propel the sperm inside the female reproductive tract. After spermatid differentiation, some of the cytoplasm connecting the cells falls away, leaving mature, tadpole-shaped spermatozoa (singular, *spermatozoon*), or sperm. **Figure 3.9** presents an anatomical view showing the stages of spermatogenesis within the seminiferous tubules.

A sperm, which is a mere 0.006 centimeter (0.0023 inch) long, must travel about 18 centimeters (7 inches) to reach an oocyte. Each sperm cell consists of a tail, body or midpiece, and a head region (**figure 3.10**). A membrane-covered area on

---

A GLIMPSE OF HISTORY

Sperm have fascinated biologists for centuries. Antonie van Leeuwenhoek was the first to view human sperm under a microscope, in 1678, concluding that they were parasites in semen. By 1685, he had modified his view, writing that sperm contain a preformed human, called a homunculus, and are seeds requiring nurturing in a female to start a new life. The illustration in **figure 3A** was drawn by Dutch histologist Niklaas Hartsoeker in 1694 and represents the once-popular homunculus hypothesis.

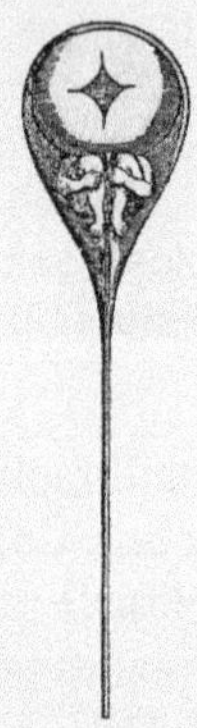

Figure 3A A sperm cell was once thought to house a tiny human. Illustration by Nicolaas Hartsoeker, from Essai de Dioptrique, 1695.

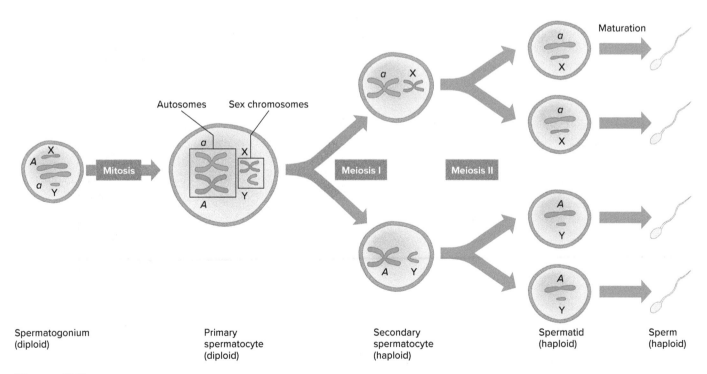

Figure 3.8 Sperm formation (spermatogenesis). Primary spermatocytes have the normal diploid number of 23 chromosome pairs. The large pair of chromosomes represents autosomes (non-sex chromosomes). The X and Y chromosomes are sex chromosomes.

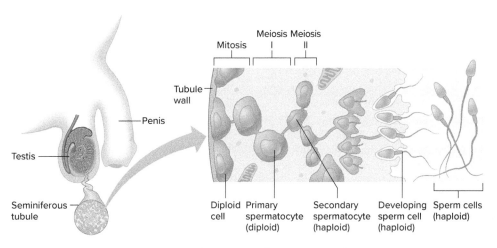

Figure 3.9 Meiosis and maturation produce sperm cells. Diploid cells divide through mitosis in the linings of the seminiferous tubules. Some daughter cells then undergo meiosis, producing haploid secondary spermatocytes, which differentiate into mature sperm cells.

the front end, the acrosome, contains enzymes that help the sperm cell penetrate the protective layers around the oocyte. Within the large sperm head, DNA wraps around proteins. The sperm's DNA at this time is genetically inactive. A male manufactures trillions of sperm in his lifetime. Although many of these will come close to an oocyte, few will actually touch one.

Meiosis in the male has built-in protections that help prevent sperm from causing some birth defects. Spermatogonia that are exposed to toxins tend to be so damaged that they never mature into sperm. More mature sperm cells exposed to toxins are often so damaged that they cannot swim.

Oocytes Form

Meiosis in the female, called **oogenesis** (egg making), begins with a diploid cell, an **oogonium**. Unlike male cells, oogonia are not attached to each other. Instead, follicle cells surround each oogonium. As each oogonium grows, cytoplasm accumulates, DNA replicates, and the cell becomes a primary oocyte. The ensuing meiotic division in oogenesis, unlike the male pathway, produces cells of different sizes.

In meiosis I, the primary oocyte divides into two cells: a small cell with very little cytoplasm, called a first **polar body**, and a much larger cell called a secondary oocyte (**figure 3.11**). Each cell is haploid, with the chromosomes in replicated form. In meiosis II, the tiny first polar body may divide to yield two polar bodies of equal size, with unreplicated chromosomes, or

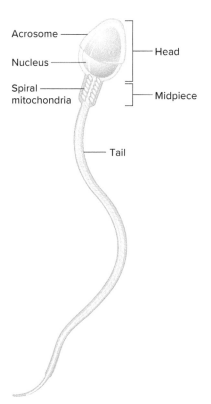

Figure 3.10 Sperm. A sperm has distinct regions that assist in delivering DNA to an oocyte.

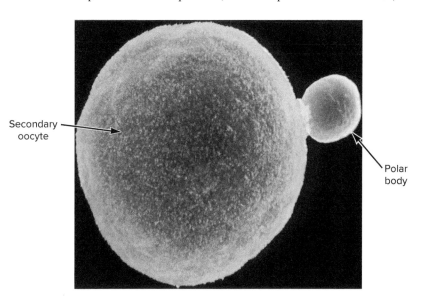

Figure 3.11 Meiosis in a female produces a secondary oocyte and a polar body. Unequal division and apportioning of cell parts enable the cell destined to become a fertilized ovum to accumulate most of the cytoplasm and organelles from the primary oocyte, but with only one genome copy. The oocyte receives most of the cytoplasm that would have gone into the meiotic product that became the polar body if the division had been equal. © *Prof. P.M. Motta/Univ. "La Sapienza", Rome/Photo Researchers/Science Source*

the first polar body may decompose. The secondary oocyte, however, divides unequally in meiosis II to produce another small polar body, with unreplicated chromosomes, and the mature egg cell, or ovum, which contains a large volume of cytoplasm. **Figure 3.12** summarizes meiosis in the female, and **figure 3.13** provides an anatomical view of the process.

Most of the cytoplasm among the four meiotic products in the female ends up in only one cell, the ovum. The woman's body absorbs the polar bodies, which normally play no further role in development. Rarely, a sperm fertilizes a polar body. When this happens, the woman's hormones respond as if she is pregnant, but a disorganized clump of cells that is not an embryo grows for a few weeks, and then leaves the woman's body. This event is a type of miscarriage called a "blighted ovum."

Before birth, a female has a million or so oocytes arrested in prophase I. (This means that when your grandmother was pregnant with your mother, the oocyte that would be fertilized and eventually become you was already there.) By puberty, about 400,000 oocytes remain. After puberty, meiosis I continues in one or several oocytes each month, but halts again at metaphase II. In response to specific hormonal cues each month, one ovary releases a secondary oocyte; this event is ovulation. The oocyte drops into a uterine tube, where waving cilia move it toward the uterus. Along the way, if a sperm penetrates the oocyte membrane, then female meiosis completes, and a fertilized ovum forms. If the secondary oocyte is not fertilized, it degenerates and leaves the body in the menstrual flow, without meiosis completed.

A female ovulates about 400 oocytes between puberty and menopause. Most oocytes degrade, because fertilization is so rare. Furthermore, only one in three of the oocytes that do meet and merge with a sperm cell will continue to grow, divide, and specialize to eventually form a new individual.

The diminishing number of oocytes present as a female ages was thought to be due to a dwindling original supply of the cells. However, researchers have discovered that human ovaries contain oocyte-producing stem cells, so it is not clear why the number falls.

Meiosis and Mutations

The gametes of older people are more likely to have new mutations (that is, not *inherited* mutations) than the gametes of younger people. Older women are at higher risk of producing oocytes that have an extra or missing chromosome. If fertilized, such oocytes lead to offspring with the chromosomal conditions described in chapter 13. Older men are also more likely to produce gametes that have genetic errors, but sperm tend to have single-gene mutations rather than chromosome-level changes. The "paternal age effect" usually causes dominant single-gene diseases. That is, only one copy of the mutant gene causes the condition.

The different timetables of meiosis explain why sperm introduce single-gene conditions whereas oocytes originate chromosome imbalances. In females, oocytes exist on the brink of meiosis I for years, and the cells complete meiosis II only if they are fertilized. Mistakes occur when gametes are active, and that is when they are distributing their chromosomes. If a homologous pair doesn't separate, the result could be an oocyte with an extra or missing chromosome.

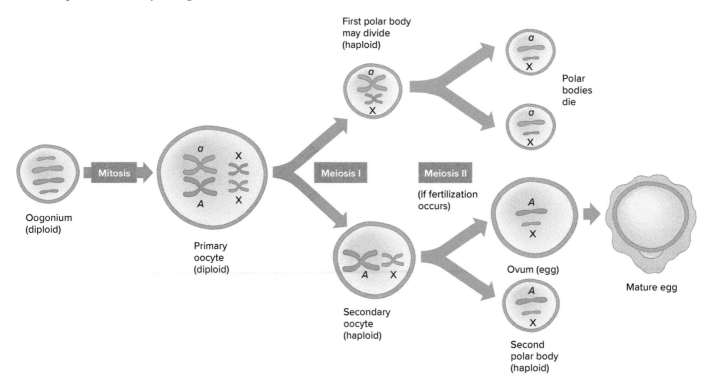

Figure 3.12 Ovum formation (oogenesis). Primary oocytes have the diploid number of 23 chromosome pairs. Meiosis in females concentrates most of the cytoplasm into one large cell called an oocyte (or egg). The other products of meiosis, called polar bodies, contain the other three sets of chromosomes and normally degenerate.

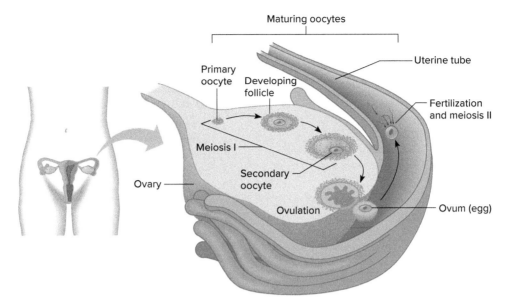

Figure 3.13 **The making of oocytes.** Oocytes develop within the ovary in follicles. An ovary contains many oocytes in various stages of maturation. After puberty, the most mature oocyte in one ovary bursts out each month, in an event called ovulation.

Sperm develop in only 74 days. "Paternal age effect" conditions arise from stem cells in the testis that divide every 16 days, from puberty on, offering many opportunities for DNA replication (discussed in chapter 9) to make a mistake, generating a dominant mutation. **Table 3.2** describes some paternal age effect conditions. They occur in genes of the fibroblast growth factor receptor (FGFR) family and affect skeletal growth. Mutations in *FGFR* genes that arise in the testicles as a man ages skew meiosis so that more spermatogonia give rise to cells with the mutation than to cells without it. The result is a mosaic cell population, or what has been called a "testicular time bomb."

The maternal age effect in causing chromosomal imbalances has been recognized since the nineteenth century, when physicians noticed that babies with trisomy 21 Down syndrome

tended to be the youngest children in large families. The paternal age effect has only recently been recognized. Researchers examined testes from elderly men who had died and donated their bodies to science. Each testis was divided into 6 slices of 32 pieces, and probed for mutations that cause multiple endocrine neoplasia (an inherited cancer syndrome). Mutations were found in discrete sections of the testes, indicating that stem cells perpetuated the mutation as they divided. The older the man at death, the more testis cells had the mutations.

Table 3.2	Paternal Age Effect Conditions
Disease	**Phenotype**
Achondroplasia	Short-limbed dwarfism (see figure 5.1a)
Crouzan syndrome	Premature fusion of skull bones in infancy, causing wide-spaced and bulging eyes, beaked nose, short upper lip, small upper jaw, and jutting lower jaw
Multiple endocrine neoplasia 2	Cancers of thyroid, parathyroid, and adrenal glands
Pfeiffer syndrome	Premature fusion of skull bones in infancy, short and fused fingers and toes
Thanatophoric dysplasia	Severe short-limbed dwarfism

Key Concepts Questions 3.3

1. Explain the steps of sperm and oocyte formation and specialization.
2. Describe how the timetables of spermatogenesis and oogenesis differ.
3. Explain how paternal and maternal age effect conditions differ.

3.4 Prenatal Development

A prenatal human is considered an **embryo** for the first 8 weeks, when rudiments of all body parts form. During the first week, the embryo is in a "preimplantation" stage because it has not yet settled into the uterine lining. Some biologists consider a prenatal human to be an embryo when it begins to develop tissue layers, at about 2 weeks.

Prenatal development after the eighth week is the fetal period, when structures grow and specialize. From the start of the ninth week until birth, the prenatal human organism is a **fetus**.

Why a Clone Is Not an Exact Duplicate

Cloning creates a genetic replica of an individual. In contrast, normal reproduction and development combine genetic material from two individuals. In fiction, scientists have cloned Nazis, politicians, dinosaurs, children, and organ donors. Real scientists have cloned sheep, mice, cats, pigs, dogs, monkeys, and amphibians.

Identical multiples, such as twins and triplets, are natural clones. Armadillos naturally always give birth to identical quadruplets (**figure 3B**). A technique called "somatic cell nuclear transfer," or just "nuclear transfer," is used to create clones. It transfers a nucleus from a somatic cell into an oocyte whose nucleus has been removed.

Cloning cannot produce an exact replica of a person, for several reasons:

- **Premature cellular aging.** In some species, telomeres of chromosomes in the donor nucleus are shorter than those in the recipient cell (see figure 2.16).
- **Altered gene expression.** In normal development, for some genes, one copy is turned off, depending upon which parent transmits it. This phenomenon is called genomic imprinting (see section 6.5). In cloning, genes in a donor nucleus skip passing through a sperm or oocyte, and thus are not imprinted. Lack of imprinting causes cloned animals to be unusually large.
- **More mutations.** DNA from a donor cell has had years to accumulate mutations. A mutation might not be noticeable in one of millions of somatic cells in a body but it could be devastating if that nucleus is used to program development of a new individual.
- **X inactivation.** At a certain time in early prenatal development in female mammals, one X chromosome is inactivated. Whether that X chromosome is from the mother or the father occurs at random in each cell, creating an overall mosaic pattern of expression for genes on the X chromosome. The pattern of X inactivation of a female clone would probably not match that of her nucleus donor,

because X inactivation normally occurs in the embryo, not in the first cell (see section 6.4).
- **Mitochondrial DNA.** A clone's mitochondria descend from the recipient oocyte, not from the donor cell, because mitochondria are in the cytoplasm, not the nucleus (see figure 5.10).

The environment is another powerful factor in why a clone isn't an identical copy. For example, coat color patterns differ in cloned calves and cats. When the animals were embryos, cells destined to produce pigment moved in a unique way in each individual, producing different color patterns. In humans, experience, nutrition, stress, exposure to infectious disease, and many other factors join our genes in molding who we are.

Questions for Discussion

1. Which of your characteristics do you think could not be duplicated in a clone, and why?
2. What might be a reason to clone humans?
3. What are potential dangers of cloning humans? Of cloning pets?
4. Do human clones exist naturally?

Figure 3B Armadillos are clones. These armored mammals always give birth to four genetically identical offspring—that is, quadruplet clones. © *Bianca Lavies/National Geographic Creative*

Sperm and Oocyte Meet at Fertilization

Hundreds of millions of sperm cells are deposited in the vagina during sexual intercourse. A sperm cell can survive in a woman's body for up to 3 days, but the oocyte can only be fertilized in the 12 to 24 hours after ovulation.

The woman's body helps sperm reach an oocyte. A process in the female called capacitation chemically activates sperm, and the oocyte secretes a chemical that attracts sperm. Contractions of the female's muscles and moving of sperm tails propel the sperm. Still, only 200 or so sperm get near the oocyte.

A sperm first contacts a covering of follicle cells, called the corona radiata, that guards a secondary oocyte. The sperm's acrosome then bursts, releasing enzymes that bore through a protective layer of glycoprotein (the zona pellucida) beneath the corona radiata. Fertilization, or conception, begins when the outer membranes of the sperm and secondary oocyte meet and a protein on the sperm head contacts a different type of protein on the oocyte (**figure 3.14**). A wave of electricity spreads physical and chemical changes across the oocyte surface, which keep other sperm out. More than one sperm can enter an oocyte, but the resulting cell has too much genetic material for development to follow.

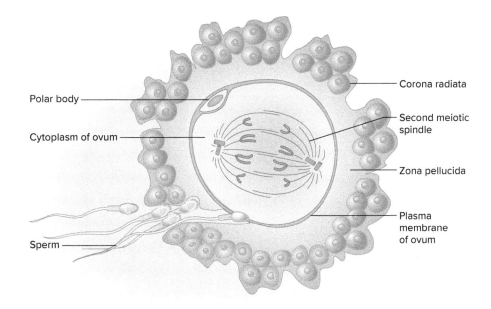

Polar body

Cytoplasm of ovum

Sperm

Corona radiata

Second meiotic spindle

Zona pellucida

Plasma membrane of ovum

Figure 3.14 **Fertilization.** Fertilization by a sperm cell induces the oocyte (arrested in metaphase II) to complete meiosis. Before fertilization occurs, the sperm's acrosome bursts, spilling enzymes that help the sperm's nucleus enter the oocyte.

Usually only the sperm's head enters the oocyte. Within 12 hours of the sperm's penetration, the ovum's nuclear membrane disassembles, and the two sets of chromosomes, called pronuclei, approach one another. Within each pronucleus, DNA replicates. Fertilization completes when the two genetic packages meet and merge, forming the genetic instructions for a new individual. The fertilized ovum is called a **zygote**. The **Bioethics** reading discusses cloning, which is a way to start development without a fertilized egg.

The Embryo Cleaves and Implants

A day after fertilization, the zygote divides by mitosis, beginning a period of frequent cell division called **cleavage** (**figure 3.15**). The resulting early cells are called **blastomeres**. When the blastomeres form a solid ball of sixteen or more cells, the embryo is called a **morula** (Latin for "mulberry," which it resembles).

During cleavage, organelles and molecules from the secondary oocyte's cytoplasm still control cellular activities, but some of the embryo's genes begin to function. The ball of cells hollows out, and its center fills with fluid, creating a **blastocyst**. (The term "cyst" refers to the fluid-filled center.) Some of the cells form a clump on the inside lining called the **inner cell mass**. Its formation is the first event that distinguishes cells from each other by their relative positions. The inner cell mass continues developing, forming the embryo.

A week after conception, the blastocyst nestles into the uterine lining. This event, called implantation, takes about a week. As implantation begins, the outermost cells of the blastocyst, called the trophoblast, secrete human chorionic gonadotropin (hCG), a hormone that prevents menstruation. This hormone detected in a woman's urine or blood is one sign of pregnancy.

Today's home tests for the pregnancy hormone hCG cap a long legacy of ways to tell if a woman was expecting. Etchings on an Egyptian papyrus from 1350 B.C.E. show a woman whose menstrual period was late urinating on wheat and barley seeds. If wheat sprouted it would be a girl, if barley grew it was a boy, and if nothing happened, then the woman wasn't with child. The effect on seeds might have been due to estrogens in the urine. In the ensuing centuries, "piss prophets" attempted to divine the pregnant state from the shades and textures of urine.

In the early twentieth century, researchers seeking ways to confirm pregnancy measured levels of the hormones that wax and wane during the menstrual cycle and the body parts that they affect. They injected the urine of possibly pregnant women into rabbits, rats, mice, or frogs, leading to the once-famous announcement of impending parenthood, *"The rabbit died!"* Actually, all the rabbits injected with human urine died, as doctors probed the ovaries for the swelling that would indicate presence of the telltale human pregnancy hormone.

In the 1970s, increasing interest in women's health sped development of a pregnancy test for the active part of the hCG molecule using chemicals in a test tube, rather than sacrificing rabbits. At first doctors sent urine to labs for analysis. Then in 1978 came the first at-home "early pregnancy test," sold at drug stores and revolutionary at the time. The thin blue line that indicated pregnancy took much of the mystery and anxiety out of confirming what a woman may have already strongly suspected.

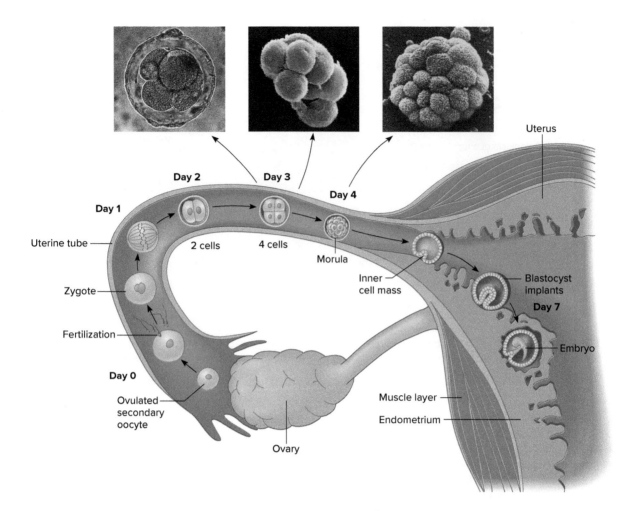

Figure 3.15 **Cleavage: From ovulation to implantation.** The zygote forms in the uterine tube when a sperm nucleus fuses with the nucleus of an oocyte. The first divisions proceed while the zygote moves toward the uterus. By day 7, the zygote, now called a blastocyst, begins to implant in the uterine lining. *(left): © Petit Format/Nestle/Photo Researchers/Science Source; (middle): © P.M. Motta & J. Van Blerkom/SPL/Science Source; (right): © Petit Format/Nestle/Photo Researchers/Science Source*

The Embryo Forms

During the second week of prenatal development, a space called the amniotic cavity forms between the inner cell mass and the outer cells anchored to the uterine lining. Then the inner cell mass flattens into a two-layered embryonic disc. The layer nearest the amniotic cavity is the **ectoderm**; the inner layer, closer to the blastocyst cavity, is the **endoderm**. Shortly after, a third layer, the **mesoderm**, forms in the middle. This three-layered, curved, sandwich-like structure is called the primordial embryo, or the **gastrula** (**figure 3.16**).

The cells that form the layers of the primordial embryo, called **primary germ layers**, become "determined," or fated, to develop as specific cell types. The cells specialize by the progressive switching off of the expression of genes important and active in the early embryo as other genes begin to be expressed. Gene expression shuts off when a small molecule called a methyl (CH_3; a carbon bonded to three hydrogen atoms) binds certain genes and proteins. In the early embryo, methyls bind the proteins (histones) that hold the long DNA molecules in a coil (see figure 9.13 and figure 11.5). In the later embryo,

methyls bind to and silence specific sets of genes. The pattern of methyl group binding that guides the specialization of the embryo establishes the epigenome. The term **epigenetic** refers to changes to DNA expression, not to DNA nucleotide base sequence. Each layer of the embryo retains stem cells as the organism develops. Under certain conditions, stem cells may produce daughter cells that can specialize as many cell types.

Each primary germ layer gives rise to certain structures. Cells in the ectoderm become skin, nervous tissue, or parts of certain glands. Endoderm cells form parts of the liver and pancreas and the linings of many organs. The middle layer of the embryo, the mesoderm, forms many structures, including muscle, connective tissues, the reproductive organs, and the kidneys.

Genes called homeotics control how the embryo develops parts in the right places. Mutations in these genes cause some very interesting conditions, including forms of intellectual disability, autism, blood cancer, and blindness. **Clinical Connection 3.1** describes a human disease that results from mutations in homeotic genes. **Table 3.3** summarizes the stages of early human prenatal development.

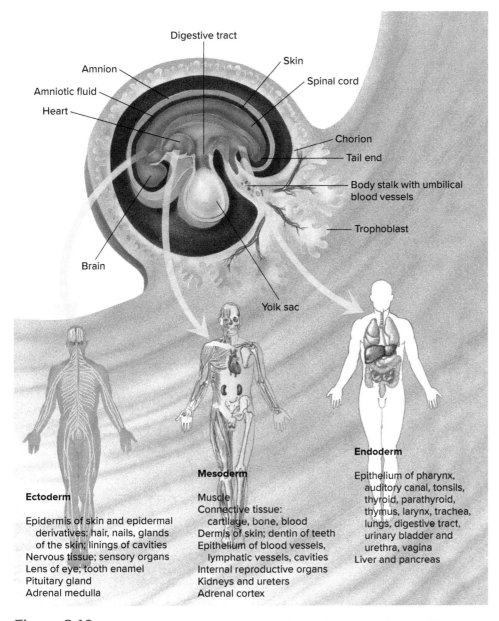

Digestive tract

Amnion

Amniotic fluid

Heart

Skin

Spinal cord

Chorion

Tail end

Body stalk with umbilical blood vessels

Trophoblast

Brain

Yolk sac

Ectoderm

Epidermis of skin and epidermal
   derivatives; hair, nails, glands
   of the skin; linings of cavities
Nervous tissue; sensory organs
Lens of eye; tooth enamel
Pituitary gland
Adrenal medulla

Mesoderm

Muscle
Connective tissue:
   cartilage, bone, blood
Dermis of skin; dentin of teeth
Epithelium of blood vessels,
   lymphatic vessels, cavities
Internal reproductive organs
Kidneys and ureters
Adrenal cortex

Endoderm

Epithelium of pharynx,
   auditory canal, tonsils,
   thyroid, parathyroid,
   thymus, larynx, trachea,
   lungs, digestive tract,
   urinary bladder and
   urethra, vagina
Liver and pancreas

Figure 3.16 **The primordial embryo.** When the three primary germ layers of the embryo form at gastrulation, many cells become "fated" to follow a specific developmental pathway.

Table 3.3	Stages and Events of Early Human Prenatal Development	
Stage	**Time Period**	**Principal Events**
Fertilized ovum	12 to 24 hours following ovulation	Oocyte fertilized; zygote has 23 pairs of chromosomes and is genetically distinct
Cleavage	30 hours to third day	Mitosis increases cell number
Morula	Third to fourth day	Solid ball of cells
Blastocyst	Fifth day through second week	Hollowed fluid-filled ball forms trophoblast (outside) and inner cell mass, which implants and flattens to form embryonic disc
Gastrula	End of second week	Primary germ layers form

Supportive Structures Form

As an embryo develops, structures form that support and protect it. These include chorionic villi, the placenta, the yolk sac, the allantois, the umbilical cord, and the amniotic sac.

By the third week after conception, fingerlike outgrowths called chorionic villi extend from the area of the embryonic disc close to the uterine wall. The villi project into pools of the woman's blood. Her blood system and the embryo's are separate, but nutrients and oxygen diffuse across the chorionic villi from her circulation to the embryo. Wastes leave the embryo's circulation and enter the woman's circulation, to be excreted.

By 10 weeks, the placenta is fully formed from the chorionic villi. The placenta links the woman and her fetus for the rest of the pregnancy. It secretes hormones that maintain pregnancy and sends nutrients to the fetus.

Other structures nurture the developing embryo. The yolk sac manufactures blood cells, as does the allantois, a membrane surrounding the embryo that gives rise to the umbilical blood vessels. The umbilical cord forms around these vessels and attaches to the center of the placenta. Toward the end of the embryonic period, the yolk sac shrinks, as the amniotic sac swells with fluid that cushions the embryo and maintains a constant temperature and pressure. The amniotic fluid contains fetal urine and a few fetal cells.

Two of the supportive structures that develop during pregnancy provide material for prenatal tests, discussed in chapters 13 and 20. **Chorionic villus sampling** examines chromosomes from cells snipped off the chorionic villi at 10 weeks. Because the villi cells and the embryo's cells come from the same fertilized ovum, an abnormal chromosome in villi cells should also be in the embryo. In **amniocentesis**, a sample of amniotic fluid is taken and fetal cells in it are examined for biochemical, genetic, and chromosomal anomalies. However, chorionic villus sampling and amniocentesis may be replaced with collection and analysis of fetal DNA that is in the pregnant woman's bloodstream. This DNA is "cell free"—that is, outside blood cells, in the liquid part of the blood. Eight weeks after conception, about 10 percent of the cell-free DNA in the maternal circulation comes from the fetus or the placenta.

The umbilical cord is another prenatal structure that has medical applications. In addition to blood stem cells mentioned in **Bioethics** in chapter 2, the cord yields bone, fat, nerve, and cartilage cells. Stem cells from the cord itself are used to treat a respiratory disease of newborns. The stem cells give rise to lung cells that secrete surfactant, which is the chemical that inflates the microscopic air sacs, and the cell type that exchanges oxygen for carbon dioxide. Stem cells from umbilical cords are abundant, easy to obtain and manipulate, and can give rise to almost any cell type.

Multiples

Twins and other multiples arise early in development. Fraternal, or **dizygotic (DZ)**, **twins** result when two sperm fertilize two oocytes. This can happen if ovulation occurs in two ovaries in the same month, or if two oocytes leave the same ovary and are both fertilized. DZ twins are no more alike than any two siblings, although they share a very early environment in the uterus. The tendency to ovulate two oocytes in a month, leading to conception of DZ twins, can run in families.

Identical, or **monozygotic (MZ)**, **twins** descend from a single fertilized ovum and therefore are genetically identical. They are natural clones. Three types of MZ twins can form, depending upon when the fertilized ovum or very early embryo splits (**figure 3.17**). This difference in timing determines which supportive structures the twins share. About a third of all MZ twins have completely separate chorions and amnions, and

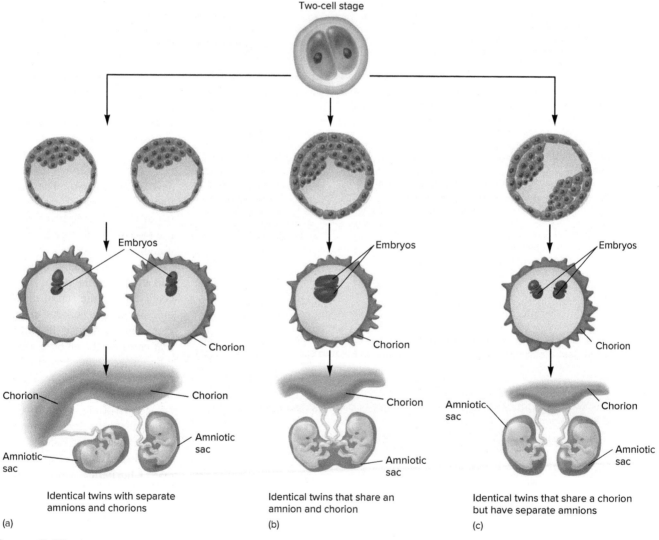

Figure 3.17 Types of identical twins. Identical (MZ) twins originate at three points in development. **(a)** In about one-third of identical twins, separation of cells into two groups occurs before the trophoblast forms on day 5. These twins have separate chorions and amnions. **(b)** About 1 percent of identical twins share a single amnion and chorion, because the tissue splits into two groups after these structures have already formed. **(c)** In about two-thirds of identical twins, the split occurs after day 5 but before day 9. These twins share a chorion but have separate amnions. Fraternal (DZ) twins result from two sperm fertilizing two secondary oocytes. These twins develop their own amniotic sacs, yolk sacs, allantois, placentas, and umbilical cords.

When an Arm Is Really a Leg: Homeotic Mutations

Flipping the X ray showed Stefan Mundlos, MD, that his hunch was correct—the patient's arms were odd-looking and stiff because the elbows were actually knees! The condition, Liebenberg syndrome, had been described in 1973 among members of a five-generation white South African family (**figure 3C**). Four males and six females had stiff elbows and wrists and short fingers that looked misplaced. A trait that affects both sexes in every generation is autosomal dominant inheritance—each child of a person with the unusual limbs has a 50:50 chance of having the condition, too.

In 2000, a medical journal described a second family with Liebenberg syndrome. Several members had restricted movements because they couldn't bend their misshapen elbows. Then in 2010, a report appeared on identical twin girls with stiff elbows and long arms, with fingers that looked like toes.

In 2012, Dr. Mundlos noted that the muscles and tendons of the elbows, as well as the bones of the arms, weren't quite right in his patient. The doctor, an expert in the comparative anatomy of limb bones of different animals, realized that the stiff elbows were acting like knees. The human elbow joint hinges and rotates, but the knee extends the lower leg straight out. Then an X-ray scan of the patient's arm fell to the floor. He suddenly realized that the entire limb looked like a leg.

Genes that switch body parts are termed *homeotic*. They are well studied in organisms as evolutionarily diverse as fruit flies, flowering plants, and mice, affecting the positions of larval segments, petals, legs, and much more. Assignment of body parts begins in the early embryo, when cells look alike but are already fated to become specific structures. Gradients (increasing or decreasing concentrations) of "morphogen" proteins in an embryo program a particular region to develop a certain way. Mix up the messages, and an antenna becomes a leg, or an elbow a knee.

Homeotic genes include a 180-base-long DNA sequence, called the homeobox, which enables the encoded protein to bind other proteins that turn on sets of other genes, crafting an embryo, section by section. Homeotic genes line up on their chromosomes in the precise order in which they're deployed in development, like chapters in an instruction manual to build a body.

The human genome has four clusters of homeotic genes, and mutations in them cause disease. In certain lymphomas, a homeotic mutation sends white blood cells along the wrong developmental pathway, resulting in too many of some blood cell types and too few of others. The abnormal ears, nose, mouth, and throat of DiGeorge (22q11.2 deletion) syndrome echo the abnormalities in *Antennapedia,* a fruit fly mutant that has legs on

its head. Extra and fused fingers and various bony alterations also stem from homeotic mutations.

The search for the mutation behind the arm-to-leg Liebenberg phenotype began with abnormal chromosomes. Affected members of the three known families were each missing 134 DNA bases in the same part of the fifth largest chromosome. The researchers identified a gene called *PITX1* that controls other genes that oversee limb development. In the Liebenberg families, the missing DNA places an "enhancer" gene near *PITX1*, altering its expression in a way that mixes up developmental signals so that the forming arm instead becomes a leg.

Questions for Discussion

1. What is the genotype and phenotype of Liebenberg syndrome?
2. How can homeotic mutations be seen in such different species as humans, mice, fruit flies, and flowering plants?
3. Explain the molecular basis of a homeotic mutation and the resulting phenotype.
4. Describe a human disease other than Liebenberg syndrome that results from a homeotic mutation.

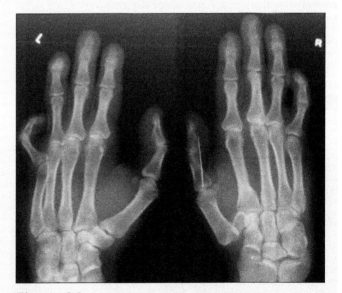

Figure 3C **The hands of a person with Liebenberg syndrome resemble feet; arms resemble legs.** *Photo courtesy of Dr. Malte Spielmann*

about two-thirds share a chorion but have separate amnions. Slightly fewer than 1 percent of MZ twins share both amnion and chorion. These differences may expose the different types of MZ twins to slightly different uterine environments. For example, if one chorion develops more attachment sites to the maternal circulation, one twin may receive more nutrients and gain more weight than the other.

In 1 in 50,000 to 100,000 pregnancies, an embryo divides into twins after the point at which the two groups of cells can develop as two individuals, between days 13 and 15. The result is conjoined or "Siamese" twins. The latter name comes from Chang and Eng Bunker, who were born in Thailand, then called Siam, in 1811. They were joined by a band of tissue from the navel to the breastbone, and could easily have been separated today. Chang and Eng lived for 63 years, attached. They fathered 22 children and divided each week between their wives.

For Abigail and Brittany Hensel, shown in **figure 3.18**, the separation occurred after day 9 of development, but before day 14. Biologists know this because the shared organs have derivatives of ectoderm, mesoderm, and endoderm; that is, when the early embryo divided incompletely, the three primary germ layers had not yet fully sorted themselves out into two bodies. The Hensels are extremely rare "incomplete twins." They are "dicephalic," which means that they have two heads. Despite the fact that anatomically they are not as distinct as Chang and Eng were from each other, the two young women are very much individuals.

Each twin has her own neck, head, heart, stomach, gall-bladder, and lungs. Each has one leg and one arm, and a third arm between their heads was surgically removed. Each twin also has her own nervous system. The twins share a large liver, a single bloodstream, a large ribcage and diaphragm muscle, and all organs below the navel (colon, bladder, pelvis, and reproductive organs). They have three kidneys. Because at birth Abigail and Brittany were strong and healthy, doctors suggested surgery to separate them. But their parents, aware of other cases where only one child survived separation, declined surgery.

The Hensel twins have learned to work together, from crawling as babies to biking and driving a car today. As teens, Abigail and Brittany enjoyed kickball, volleyball, and basketball, and developed distinctive tastes in clothing and in food. They graduated from college and had their own reality TV series. Today they are elementary school teachers.

MZ twins occur in 3 to 4 pregnancies per 1,000 births worldwide. In North America, twins occur in about 1 in 81 pregnancies, which means that 1 in 40 of us is a twin. However, not all twins survive to be born. One study of twins detected early in pregnancy showed that up to 70 percent of the eventual births are of a single child. This is called the "vanishing twin" phenomenon.

The Embryo Develops

As prenatal development proceeds, different rates of cell division in different parts of the embryo fold the forming tissues into intricate patterns. In a process called embryonic induction, the specialization of one group of cells causes adjacent groups of cells to specialize. Gradually, these changes mold the three primary germ layers into organs and organ systems. Organogenesis is the transformation of the simple three layers of the embryo into distinct organs. During these weeks, the embryo is sensitive to environmental influences, such as exposure to toxins and viruses.

During the third week of prenatal development, a band called the primitive streak appears along the back of the embryo. Some nations designate day 14 of prenatal development and primitive streak formation as the point beyond which they ban research on the human embryo. The reasoning is that the primitive streak is the first sign of a nervous system and day 14 is when implantation completes.

The primitive streak gradually elongates to form an axis that other structures organize around as they develop. The primitive streak eventually gives rise to connective tissue progenitor cells and the notochord, which is a structure that forms the basic framework of the skeleton. The notochord induces a sheet of overlying ectoderm to fold into the hollow **neural tube**, which develops into the brain and spinal cord (central nervous system).

If the neural tube does not completely close by day 28, a neural tube defect (NTD) occurs, in which parts of the brain or spinal cord protrude from the open head or spine, and body parts below the defect cannot move. Surgery can correct some NTDs. Lack of the B vitamin folic acid can cause NTDs in embryos with a genetic susceptibility. For this reason, the U.S. government adds the vitamin to grains, and pregnant women take supplements. A blood test during the fifteenth week of pregnancy detects a substance from the fetus's liver called alpha fetoprotein (AFP) that moves at an abnormally rapid rate into the woman's circulation if there is an open NTD. Using hints from more than 200 genes associated with NTDs in other

Figure 3.18 Conjoined twins. Abigail and Brittany Hensel are the result of incomplete twinning during the first 2 weeks of prenatal development. *Courtesy of Brittany and Abigail Hensel*

vertebrates, investigators are searching the genome sequences of people with NTDs to identify patterns of gene variants that could contribute to risk for or cause these conditions.

Appearance of the neural tube marks the beginning of organ development. Shortly after, a reddish bulge containing the heart appears. The heart begins to beat around day 18, and this is easily detectable by day 22. Soon the central nervous system starts to form.

The fourth week of embryonic existence is one of spectacularly rapid growth and differentiation (**figure 3.19**). Arms and legs begin to extend from small buds on the torso. Blood cells form and fill primitive blood vessels. Immature lungs and kidneys begin to develop.

By the fifth and sixth weeks, the embryo's head appears too large for the rest of its body. Limbs end in platelike structures with tiny ridges, and gradually apoptosis sculpts the fingers and toes. The eyes are open, but they do not yet have lids or irises. By the seventh and eighth weeks, a skeleton composed of cartilage forms. The embryo is now about the length and weight of a paper clip. At the end of 8 weeks of gestation, the prenatal human has tiny versions of all structures that will be present at birth. It is now a fetus.

The Fetus Grows

During the fetal period, body proportions approach those of a newborn. Initially, the ears lie low and the eyes are widely spaced. Bone begins to replace the softer cartilage. As nerve and muscle functions become coordinated, the fetus moves.

Sex is determined at conception, when a sperm bearing an X or Y chromosome meets an oocyte, which always carries an X chromosome. An individual with two X chromosomes is a female, and one with an X and a Y is a male. A gene on the Y chromosome, called *SRY* (for "sex-determining region of the Y"), determines maleness.

Anatomical differences between the sexes appear at week 6, after the *SRY* gene is expressed in males. Male hormones stimulate male reproductive organs and glands to differentiate from existing, indifferent structures. In a female, the indifferent structures of the early embryo develop as female organs and glands, under the control of other genes. Differences may be noticeable on ultrasound scans by 12 to 15 weeks. Chapter 6 discusses sexual development further.

By week 12, the fetus sucks its thumb, kicks, makes fists and faces, and has the beginnings of teeth. It breathes amniotic fluid in and out, and urinates and defecates into it. The first trimester (3 months) of pregnancy ends.

By the fourth month, the fetus has hair, eyebrows, lashes, nipples, and nails (**figure 3.20**). By 18 weeks, the vocal cords have formed, but the fetus makes no sound because it doesn't breathe air. By the end of the fifth month, the fetus curls into a head-to-knees position. It weighs about 454 grams (1 pound). During the sixth month, the skin appears wrinkled because there isn't much fat beneath it, and turns pink as capillaries fill with blood. By the end of the second trimester, the fetus kicks and jabs and may even hiccup. It is now about 23 centimeters (9 inches) long.

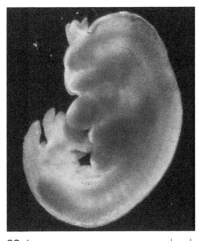
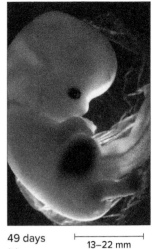

28 days (a)   4–6 mm   49 days (b)   13–22 mm

Figure 3.19 **Human embryos at (a) 28 days and (b) 49 days.** *(a): © Petit Format/Nestle/Photo Researchers/Science Source; (b): © Petit Format/SPL/Photo Researchers/Science Source*

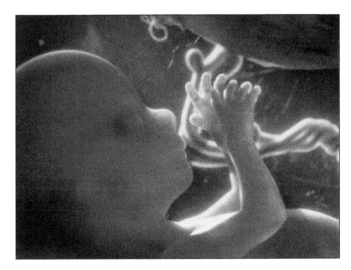

Figure 3.20 **A 16-week fetus.** A fetus at this stage has hair, eyebrows, lashes, and other features that look human. *© Nestle/Petit Format/Science Source*

In the final trimester, fetal brain cells rapidly link into networks as organs elaborate and grow. A layer of fat forms beneath the skin. The digestive and respiratory systems mature last, which is why infants born prematurely often have difficulty digesting milk and breathing. Approximately 266 days after a single sperm burrowed its way into an oocyte, a baby is ready to be born.

The birth of a healthy baby is against the odds. Of every 100 secondary oocytes exposed to sperm, 84 are fertilized. Of these 84, only 69 implant in the uterus, 42 survive 1 week or longer, 37 survive 6 weeks or longer, and 31 are born alive. Of the fertilized ova that do not survive, about half have chromosomal abnormalities that cause problems too severe for development to proceed.

1. Describe the events of fertilization.

2. Distinguish among the zygote, morula, and blastocyst.

3. What happens to the inner cell mass?

4. Explain how some genes are silenced as development proceeds.

5. Explain the significance of the formation of primary germ layers.

6. Describe the supportive structures that enable an embryo to develop.

7. Explain how twins arise.

8. Distinguish the embryo from the fetus.

9. List the events of the embryonic and fetal periods.

3.5 Birth Defects

Certain genetic abnormalities or toxic exposures can affect development in an embryo or fetus, causing birth defects. Only a genetic birth defect can be passed to future generations. Although development can be abnormal in many ways, 97 percent of newborns appear healthy at birth.

The Critical Period

The specific nature of a birth defect reflects the structures developing when the damage occurs. The time when genetic abnormalities, toxic substances, or viruses can alter a specific structure is its **critical period (figure 3.21)**. Some body parts, such as fingers and toes, are sensitive for short periods of time. In contrast, the brain is sensitive throughout prenatal development, and connections between nerve cells continue to change throughout life. Because of the brain's continuous critical period, many birth defect syndromes include learning disabilities or intellectual disability.

About two-thirds of all birth defects arise from a disruption during the embryonic period. More subtle defects, such as learning disabilities, that become noticeable only after infancy, are often caused by interventions during the fetal period. A disruption in the first trimester might cause severe intellectual disability; in the seventh month of pregnancy, it might cause difficulty in learning to read.

Some birth defects are caused by a mutation that acts at a specific point in prenatal development. In a rare inherited condition called phocomelia, for example, a mutation halts limb development from the third to the fifth week of the embryonic period, severely stunting arms and legs. The risk that a genetically caused birth defect will affect a particular family member can be calculated.

Many birth defects are caused by toxic substances the pregnant woman encounters. These environmentally caused problems will not affect other family members unless they, too, are exposed to the environmental trigger. Chemicals or other agents that cause birth defects are called **teratogens** (Greek for "monster-causing"). While it is best to avoid teratogens while pregnant, some women may need to continue to take potentially teratogenic drugs to maintain their own health.

Teratogens

Most drugs are not teratogens. Whether or not exposure to a particular drug causes birth defects may depend upon a woman's genes. For example, certain variants of a gene that control the body's use of an amino acid called homocysteine affect whether or not the medication valproic acid causes birth defects. Valproic acid is used to prevent seizures and symptoms of bipolar disorder. Rarely, it can cause NTDs, heart defects, hernias, and clubfoot. Women who have this gene variant (*MTHFR C677T*) can use a different drug to treat seizures or bipolar disorder when they try to conceive.

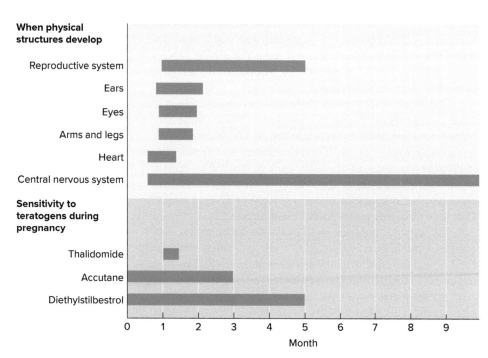

Figure 3.21 Critical periods of development. The nature of a birth defect resulting from drug exposure depends upon which structures were developing at the time of exposure. Isotretinoin (Accutane) is an acne medication that causes cleft palate and eye, brain, and heart defects. Diethylstilbestrol (DES) was used in the 1950s to prevent miscarriage. It caused vaginal cancer in some "DES daughters." Thalidomide was used to prevent morning sickness.

Thalidomide

The idea that the placenta protects the embryo and fetus was tragically disproven between 1957 and 1961, when 10,000 children were born in Europe with what seemed, at first, to be phocomelia. This genetic disease is rare, and therefore couldn't be the cause of the sudden problem. Instead, the mothers had all taken a mild tranquilizer, thalidomide, to alleviate nausea early in pregnancy, during the critical period for limb formation. Many "thalidomide babies" were born with incomplete or missing legs and arms.

The United States was spared from the thalidomide disaster because an astute government physician noted the drug's adverse effects on laboratory monkeys. Still, several "thalidomide babies" were born in South America in 1994, where pregnant women were given the drug. In spite of its teratogenic effects, thalidomide is used to treat leprosy, AIDS, and certain blood and bone marrow cancers.

Alcohol

A pregnant woman who has just one or two alcoholic drinks a day, or perhaps a large amount at a single crucial time, risks a fetal alcohol spectrum disorder in her unborn child. An analysis of 1,728 people with these disorders showed that more than 90 percent have behavior problems, 80 percent have communication challenges, 70 percent have abnormal cognition, and about 50 percent have hyperactivity and short attention span. People with fetal alcohol spectrum disorders are also more likely to have hearing and/or visual loss.

Tests for variants of genes that encode proteins that regulate alcohol metabolism may be able to identify women whose fetuses are at elevated risk for developing fetal alcohol spectrum disorders. Until these tests are validated, pregnant women are advised to avoid all alcoholic beverages.

The most severe manifestation of alcohol exposure before birth is fetal alcohol syndrome (FAS). Most children with FAS have small heads and flat faces (**figure 3.22**). Growth is slow

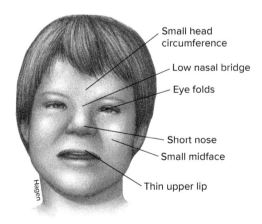

Figure 3.22 Fetal alcohol syndrome. Some children whose mothers drank alcohol during pregnancy have characteristic flat faces.

Labels in figure:
- Small head circumference
- Low nasal bridge
- Eye folds
- Short nose
- Small midface
- Thin upper lip

before and after birth. Teens and young adults are short and have small heads. More than 80 percent of them retain the facial characteristics of a young child with FAS.

The long-term cognitive effects of prenatal alcohol exposure are more severe than the physical vestiges. Intellectual impairment ranges from minor learning disabilities to intellectual disability. Many adults with FAS function at early grade-school level. They never develop social and communication skills and cannot understand the consequences of actions, form friendships, take initiative, and interpret social cues. A person with FAS may have the cognitive symptoms without the facial characteristics, so considering alcohol consumption is important for diagnosis.

Greek philosopher Aristotle noticed problems in children of women with alcoholism more than 23 centuries ago. In the United States today, 1 to 3 of every 1,000 infants has FAS, meaning 2,000 to 6,000 affected children are born each year. Many more children have milder "alcohol-related effects." A fetus of a woman with active alcoholism has a 30 to 45 percent chance of harm from alcohol exposure.

Nutrients

Certain nutrients ingested in large amounts, particularly vitamins, act as drugs. The acne medicine isotretinoin (Accutane) is a vitamin A derivative that causes spontaneous abortion and defects of the heart, nervous system, and face in exposed embryos. Another vitamin A–based drug, used to treat psoriasis, as well as excesses of vitamin A itself, also cause birth defects. Vitamin A and its derivatives may accumulate enough to harm embryos because they are stored in body fat for up to 3 years.

Excess vitamin C can harm a fetus that becomes accustomed to the large amounts the woman takes. After birth, when the vitamin supply suddenly plummets, the baby may develop symptoms of vitamin C deficiency, such as bruising and becoming infected easily.

Malnutrition threatens a fetus. The opening essay to chapter 11 describes the effects of starvation on embryos during the bleak "Dutch Hunger Winter" of 1944–1945. Poor nutrition later in pregnancy affects the development of the placenta and can cause low birth weight, short stature, tooth decay, delayed sexual development, and learning disabilities.

Viral Infection

Viruses are small enough to cross the placenta and reach a fetus. Some viruses that cause mild symptoms in an adult, such as the chickenpox virus, may devastate a fetus. Men can transmit viral infections to an embryo or fetus during sexual intercourse. Zika virus, for example, is carried in semen. This virus was discovered in Uganda in 1947, but came to the world's attention in 2015, after it was associated with a dramatic increase in the incidence of the birth defect microcephaly in Brazil (**figure 3.23**). About 11 percent of women exposed to the virus during the first trimester of pregnancy have a child with microcephaly or another brain abnormality, whether or

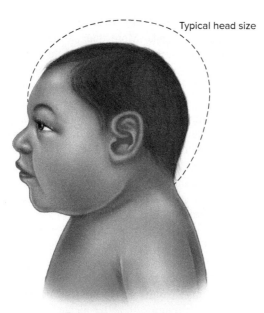
Typical head size

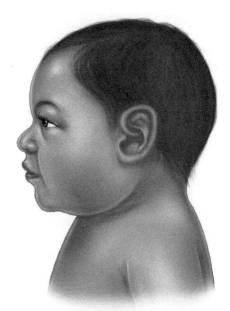

Baby with microcephaly | Baby with typical head size

Figure 3.23 **Zika virus causes birth defects.** Exposure of a fetus to Zika virus can cause an abnormally small brain (microcephaly) that has scattered calcifications, inflammation, and no characteristic infoldings.

not the women had symptoms of viral infection. Mosquitoes spread Zika virus, which causes a mild illness of fever, rash, joint pain, and red eyes in some adults.

HIV can reach a fetus through the placenta or infect a newborn via blood contact during birth. The risk of transmission is significantly reduced if a pregnant woman takes anti-HIV drugs. Fetuses of HIV-infected women are at higher risk for low birth weight, prematurity, and stillbirth if the woman's health is failing.

German measles (rubella) is a well-known viral teratogen. In the United States, in the early 1960s, an epidemic of the usually mild illness caused 20,000 birth defects and 30,000 stillbirths. Exposure during the first trimester of pregnancy caused cataracts, deafness, and heart defects. Exposure during the second or third trimester of pregnancy caused learning disabilities, speech and hearing problems, and type 1 diabetes mellitus. Incidence of these problems, called congenital rubella syndrome, has dropped markedly since vaccination eliminated the disease in the United States. Similar problems may emerge among children exposed to Zika virus in the uterus as they near school age.

Key Concepts Questions 3.5

1. What is the critical period?
2. Explain why most birth defects that develop during the embryonic period are more severe than problems that arise during fetal development.
3. Define *teratogen* and give an example of one.

3.6 Maturation and Aging

"Aging" means moving through the life cycle, despite advertisements for products that promise to reverse the process. As we age, the limited life spans of cells are reflected in the waxing and waning of biological structures and functions. Although some aspects of our anatomy and physiology peak very early—such as the number of brain cells or hearing acuity, which do so in childhood—age 30 seems to be a turning point for decline. Some researchers estimate that, after this age, the human body becomes functionally less efficient by about 0.8 percent each year.

Can we slow aging? In the quest to extend life, people have sampled everything from turtle soup to owl meat to human blood. More recently, attention has turned to a component of red wine, dark chocolate, and raspberries called resveratrol. It is a type of enzyme called a **sirtuin** that regulates energy use in cells by altering the expression of certain sets of genes. Through their effect on energy metabolism, the sirtuins seem to prevent or delay several diseases that are more common in the aged, such as heart disease and neurodegenerative conditions. Levels of sirtuins are lower in brain cells from people who have died of Alzheimer disease. However, clinical trials testing synthetic sirtuins in dosages up to the equivalent of a thousand bottles of red wine drunk at once indicate safety, but had no effect on Alzheimer disease progression.

Many diseases that begin in adulthood, or are associated with aging, have genetic components. These diseases tend to be multifactorial, because it takes many years for environmental exposures to alter gene expression in ways that noticeably affect health. Following is a closer look at how genes may impact health throughout life.

Adult-Onset Inherited Diseases

Human prenatal development is a highly regulated program of genetic switches that are turned on in specific body parts at specific times. Environmental factors can affect how certain genes are expressed before birth in ways that create risks that appear much later. For example, adaptations that enable a fetus to grow despite near-starvation become risk factors for certain common conditions of adulthood, when conserving calories is no longer needed. Such diseases include coronary artery disease, obesity, stroke, hypertension, schizophrenia, and type 2 diabetes mellitus. A malnourished fetus has intrauterine growth retardation (IUGR), and though born on time, is quite small. Premature infants, in contrast, are small but are born early, and are not predisposed to conditions resulting from IUGR.

More than 100 studies correlate low birth weight due to IUGR with increased incidence of cardiovascular disease and diabetes later in life. Much of the data come from war records because enough time has elapsed to study the effects of prenatal malnutrition as people age. The introduction to chapter 11 describes how prenatal nutrient deficiency may set the stage for later schizophrenia.

How can poor nutrition before birth cause disease decades later? Perhaps to survive, the starving fetus redirects its circulation to protect vital organs, as muscle mass and hormone production change to conserve energy. Growth-retarded babies have too little muscle tissue, and because muscle is the primary site of insulin action, glucose metabolism changes. Thinness at birth, and the accelerated weight gain in childhood that often compensates, increases risk for coronary heart disease and type 2 diabetes much later.

In contrast to the delayed effects of fetal malnutrition, symptoms of single-gene diseases can begin anytime. Most inherited conditions that affect children are recessive. Even a fetus can have symptoms of inherited disease, such as the broken bones of osteogenesis imperfecta ("brittle bone disease").

Most dominantly inherited conditions start to affect health in early to middle adulthood. In polycystic kidney disease, for example, cysts that may have been present but undetected in the kidneys during one's twenties begin causing bloody urine, high blood pressure, and abdominal pain in the thirties. The joint destruction of osteoarthritis may begin in one's thirties, but not be painful for 20 years.

Mutations in single genes cause from 5 to 10 percent of Alzheimer disease cases, producing initial symptoms in the forties and fifties. Noninherited Alzheimer disease typically begins later in life.

Whatever the age of onset, Alzheimer disease starts gradually. Mental function declines steadily for 3 to 10 years following the first symptoms of depression and short-term memory loss. Cognitive skills ebb. A person may forget how to put one foot in front of the other, or, confused and forgetful, wander away. Finally, the patient cannot recognize loved ones and can no longer perform basic functions such as speaking or eating.

The brains of Alzheimer disease patients contain deposits of a protein, called beta amyloid, in learning and memory centers. The brains of people with Alzheimer disease also contain structures called neurofibrillary tangles, which consist of a protein called tau. Tau binds to and disrupts microtubules in nerve cell branches, destroying the shape of the cell, impairing its ability to communicate. Clinical Connection 5.1 discusses Alzheimer disease further.

Disorders That Resemble Accelerated Aging

Genes control aging both passively (as structures break down) and actively (by initiating new activities). A group of "premature aging" inherited disorders is rare, but may hold clues to how genes control aging in all of us, as discussed in the chapter opener.

Disorders that accelerate aging and shorten life, or that make a person appear older than they are but do not affect lifespan, are termed progeroid. The most severe such disorders are the progerias, which shorten life span. Most accelerated aging conditions are caused by the inability of cells to adequately repair DNA, which is discussed in section 12.7. With poor DNA repair, mutations that would ordinarily be corrected persist. Over time, the accumulation of mutations destabilizes the entire genomes of somatic cells, and more mutations occur. Changes associated with aging ensue.

Table 3.4 lists the more common progeroid syndromes. They vary in severity. People with Rothmund-Thomson syndrome, for example, may have a normal life span, but develop gray hair or baldness, cataracts, cancers, and osteoporosis at a young age. Werner syndrome becomes apparent before age 20, causing death before age 50 from diseases associated with aging. Young adults with Werner syndrome develop atherosclerosis, type 2 diabetes, hair graying and loss, osteoporosis, cataracts, and wrinkled skin. They are short because they skip the growth spurt of adolescence.

Hutchinson-Gilford progeria syndrome is the most severe rapid aging condition. In addition to the symptoms the chapter opener describes are abnormalities at the cellular level. Normal cells growing in culture divide about 50 times before dying. Cells from Hutchinson-Gilford progeria syndrome patients die in culture after only 10 to 30 divisions.

Table 3.4	Premature Aging Syndromes	
Disease	**Incidence**	**Average Life Span**
Ataxia telangiectasia	1/60,000	20
Cockayne syndrome	1/100,000	20
Hutchinson-Gilford progeria syndrome	<1/1,000,000	13
Rothmund-Thomson syndrome	<1/100,000	Normal
Trichothiodystrophy	<1/100,000	10
Werner syndrome	<1/100,000	50

Hutchinson-Gilford progeria is caused by a single DNA base change in the gene that encodes the protein lamin A. That base is a site that determines how parts of the protein are cut and joined, and when it is altered, the protein is missing 50 amino acids. The shortened protein is called progerin. It remains stuck to the endoplasmic reticulum, instead of being transported into the nucleus through the nuclear pores, as happens to normal lamin A protein. Progerin travels through the tubules of the ER to the nuclear membrane. This stresses the nuclear membrane, causing it to bubble or "bleb" inward, altering the way the nuclear lamina (the layer on the inside face) binds the chromatin (DNA complexed with protein) within. Somehow, disturbing the chromatin hampers DNA repair, allowing mutations associated with the signs of aging to occur. Several drugs block the molecule that holds progerin to the ER. Lonafarnib, the drug being tested on children with progeria, blocks formation of progerin. Rapamycin, a second drug, works in a different way, enabling cells to remove the toxic progerin.

In stem cells from bone marrow of patients with Hutchinson-Gilford progeria, progerin shifts the activities of certain genes in ways that promote bone formation and suppress fat deposition, sculpting the skeletal appearance. The molecular basis of progeria suggests that robust DNA repair contributes to longevity. The cells of these children lack that protection, and the mutations that age us all accumulate much faster.

Genes and Longevity

Aging reflects gene activity plus a lifetime of environmental influences (**figure 3.24**). Families and genetically isolated populations with many aged members have a lucky collection of gene variants plus shared environmental influences such as good nutrition, excellent health care, and devoted relatives. Genome comparisons among people who've passed their hundredth birthdays to those who have died of the common illnesses of older age are revealing genes that influence longevity.

Figure 3.24 **Analyzing the genomes of older people can provide clues to health for everyone.** © Wealan Pollard/age fotostock RF

People who have lived past 100 years are called centenarians. Most enjoy excellent health and are socially active, then succumb rapidly to diseases that typically claim people decades earlier. Centenarians fall into three broad groups—about 20 percent of them never get the diseases that kill most people; 40 percent get these diseases, but at much older ages; and the other 40 percent live with and survive the more common diseases of aging: heart disease, stroke, cancers, type 2 diabetes, and dementias. Those past 110 are called supercentenarians; about 80 are known worldwide.

Although the environment seems to play an important role in the deaths of people ages 60 to 85, past that age, genetic effects predominate. That is, someone who dies at age 68 of lung cancer can probably blame a lifetime of cigarette smoking. But a smoker who dies at age 101 of the same disease probably had gene variants that protected against lung cancer. Centenarians have higher levels of large lipoproteins that carry cholesterol (high-density lipoprotein [HDL]) than other people, which researchers estimate adds 20 years of life.

Children and siblings of centenarians tend to be long-lived as well, supporting the idea that longevity is inherited. Brothers of centenarians are 17 times as likely to live past age 100 as the average man, and sisters are 8.5 times as likely. Unhealthy habits among some centenarians hint that genes are protecting them. One researcher suggests that the saying, "The older you get, the sicker you get" be replaced with "The older you get, the healthier you've been."

Centenarians have inherited two types of gene variants—those that directly protect them and wild type alleles of genes that, when mutant, cause disease.

Types of genes that affect longevity control:

- immune system functioning;
- insulin secretion and glucose metabolism;
- response to stress;
- the cell cycle;
- DNA repair;
- lipid (including cholesterol) metabolism;
- nutrient metabolism; and
- production of antioxidant enzymes.

Researchers estimate that at least 130 genes have variants that influence how long a person is likely to live.

One well-studied aging gene encodes the growth hormone receptor, which is a protein that enables a cell to respond to signals to enlarge, a key part of early development. People who inherit two recessive mutations in the gene have Laron syndrome. They are very short—adults stand under 4 feet tall—but they do not develop cancer or diabetes. Because of the near absence of these common diseases of aging, people with Laron syndrome live long lives. In El Oro Province in southern Ecuador, more than 100 people have Laron syndrome. They descend from Eastern European Jewish people who fled to the area to escape the Spanish Inquisition in the fifteenth century, and avoiding churches led them to the unoccupied, rural region.

An isolated population such as the one in El Oro, where people mate with close relatives and produce offspring, creates a situation in which mutations are overrepresented. However,

Table 3.5	Longevity Genes	
Gene	**Protects Against**	**Population Studied**
Apolipoprotein C3 (*APOC-3*)	Hypertension, diabetes	Ashkenazi Jews
	Cardiovascular disease	Amish
Bitter taste receptor (*TAW2R16*)	Poisoning, digestive problems	Calabria, Italy
Cholesteryl ester transfer protein (*CETP*)	Cardiovascular disease	Ashkenazi Jews
Forkhead box O3 (*FOXO3*)	Cancer, cardiovascular disease	Japanese-Americans
Growth hormone receptor (*GHR*)	Diabetes, cancer	Ecuador, Israel
Uncoupling proteins (*UCP 2, 3, 4*)	Oxidative damage, poor energy use	Calabria, Italy

the genomes are relatively uniform, making gene variants easier to identify than in more outbred populations. Laron syndrome somehow protects these people from two major diseases of older age. **Table 3.5** lists other single genes that are associated with longevity in isolated populations. Environmental factors are also important in longevity. The very-long-lived people of Calabria, an area of southern Italy, for example, eat mostly fruits and vegetables, which may contribute to their excellent health.

Several research projects are collecting and curating genome sequences from specific populations, many of them genetically more uniform than most, like the "little people" of southern Ecuador. The longest study is the New England Centenarian Study, which began in 1988 to amass information on the families of the oldest citizens then known in the United States. Another study is investigating genomes of nursing home residents, and a program in California is probing the genomes of the "wellderly." So far, these people share never having had heart disease and never having smoked. Several very-long-lived people have had cancer, indicating that cancers are survivable.

Researchers at the University of Pittsburgh have identified places in the genome that harbor "successful aging genes" with variants that preserve cognition. Other studies are looking at genes implicated in the diseases that kill most of us for variants that long-lived siblings share. Considered together, perhaps these studies will provide information that will help the majority of us who have not been fortunate enough to have inherited longevity gene variants.

Key Concepts Questions 3.6

1. Explain how starvation before birth sets the stage for later disease.

2. How do recessive and dominant conditions differ in terms of the human life cycle?

3. Describe a single-gene disorder that speeds aspects of aging.

4. Explain how genes control aging.

Summary

3.1 The Reproductive System

1. The male and female reproductive systems include paired **gonads** and networks of tubes in which **sperm** and **oocytes** are made.

2. Male **gametes** originate in seminiferous tubules within the **testes**, then pass through the epididymis and ductus deferentia, where they mature before exiting the body through the urethra during sexual intercourse. The prostate gland, the seminal vesicles, and the bulbourethral glands add secretions.

3. Female gametes originate in the **ovaries**. Each month after puberty, one ovary releases an oocyte into a uterine tube. The oocyte then moves to the uterus for implantation (if fertilized) or expulsion.

3.2 Meiosis

4. **Meiosis** reduces the chromosome number in gametes from **diploid** to **haploid**, maintaining the chromosome number between generations. Meiosis ensures genetic variability by **independently assorting** combinations of genes into gametes as **homologous pairs** of chromosomes randomly align and **cross over**.

5. Meiosis I, a **reduction division**, halves the number of chromosomes. Meiosis II, an **equational division**, produces four cells from the two that result from meiosis I, without another DNA replication.

6. Crossing over occurs during prophase I. It mixes up paternally and maternally derived genes.

7. Chromosomes segregate and independently assort in metaphase I, which determines the distribution of genes from each parent.

3.3 Gametes Mature

8. **Spermatogenesis** begins with **spermatogonia**, which accumulate cytoplasm and replicate their DNA, becoming primary spermatocytes. After meiosis I, the cells become haploid secondary spermatocytes. In meiosis II, the

secondary spermatocytes divide, each yielding two spermatids, which then differentiate into spermatozoa.

9. In **oogenesis**, some **oogonia** grow and replicate their DNA, becoming primary oocytes. In meiosis I, the primary oocyte divides to yield one large secondary oocyte and a small **polar body**. In meiosis II, the secondary oocyte divides to yield the large ovum and another polar body. Female meiosis is completed at fertilization.

10. Genetic errors in sperm from older men are usually dominant single-gene mutations. Genetic errors in oocytes from older women are typically extra or absent chromosomes.

3.4 Prenatal Development

11. In the female, sperm are capacitated and drawn toward a secondary oocyte. One sperm burrows through the oocyte's protective layers with acrosomal enzymes. Fertilization occurs when the sperm and oocyte fuse and their DNA combines in one nucleus, forming the **zygote**. Electrochemical changes in the egg surface block additional sperm from entering. **Cleavage** begins and a 16-celled **morula** forms. Between days 3 and 6, the morula arrives at the uterus and hollows, forming a **blastocyst** made up of **blastomeres**. The trophoblast and **inner cell mass** form. Around day 6 or 7, the blastocyst implants, and trophoblast cells secrete hCG, which prevents menstruation.

12. During the second week, the amniotic cavity forms as the inner cell mass flattens. **Ectoderm** and **endoderm** form, and then **mesoderm** appears, establishing the **primary germ layers** of the **gastrula**. **Epigenetic** effects oversee cell differentiation as cells in each germ layer begin to develop into specific organs.

13. During the third week, the placenta, yolk sac, allantois, and umbilical cord begin to form as the amniotic cavity swells with fluid.

14. **Monozygotic (MZ) twins** result when one fertilized ovum splits. **Dizygotic (DZ) twins** result from two fertilized ova.

15. Organs form throughout the embryonic period. The primitive streak, notochord and **neural tube**, arm and leg buds, heart, facial features, and skeleton develop.

16. At the eighth week, the **embryo** becomes a **fetus**, with all structures present but not fully grown. Organ rudiments grow and specialize. The developing organism moves and reacts. In the last trimester, the brain develops rapidly, and fat is deposited beneath the skin. The digestive and respiratory systems mature last.

3.5 Birth Defects

17. Birth defects can result from a mutation or an environmental influence.

18. A substance that causes birth defects is a **teratogen**. Environmentally caused birth defects are not transmitted to future generations.

19. The time when a structure is sensitive to damage from an abnormal gene or environmental intervention is its **critical period**.

3.6 Maturation and Aging

20. Genes cause or predispose us to illness throughout life. Single-gene diseases that strike early tend to be recessive; most adult-onset single-gene conditions are dominant.

21. Malnutrition before birth can alter gene expression in ways that cause illness later.

22. Progeroid syndromes are single-gene disorders that speed aging-associated changes. The progeria subtype shortens life.

23. Long life is due to genetics and environmental influences. Researchers identify longevity genes by studying genetically isolated populations with many long-lived members.

Review Questions

1. How many sets of human chromosomes are in each of the following cell types?
   a. an oogonium
   b. a primary spermatocyte
   c. a spermatid
   d. a cell from either sex during anaphase of meiosis I
   e. a cell from either sex during anaphase of meiosis II
   f. a secondary oocyte
   g. a polar body derived from a primary oocyte

2. List the structures and functions of the male and female reproductive systems.

3. A dog has 39 pairs of chromosomes. Considering only the random alignment of chromosomes, how many genetically different puppies are possible when two dogs mate? Is this number an underestimate or overestimate of the actual total? Why?

4. How does meiosis differ from mitosis?

5. What do oogenesis and spermatogenesis have in common, and how do they differ?

6. How does gamete maturation differ in the male and female?

7. Why is it necessary for spermatogenesis and oogenesis to generate stem cells?

8. Describe the events of fertilization.

9. Define *epigenome*. How does the epigenome differ from the gene variants that make up the genome of an embryo?

10. Write the time sequence in which the following structures begin to develop: notochord, gastrula, inner cell mass, fetus, zygote, morula.

11. Why does exposure to teratogens produce more severe health effects in an embryo than in a fetus?

12. The same birth defect syndrome can be caused by a mutant gene or exposure to a teratogen. How do the consequences of each cause differ for future generations?

13. List three teratogens, and explain how they disrupt prenatal development.

14. How are sirtuins thought to extend life?

15. Describe two ways that children with Hutchinson-Gilford progeria syndrome age rapidly.

16. Cite two pieces of evidence that genes control aging.

Applied Questions

1. Progeria is among the rarest of rare diseases. Some people question the utility of developing treatments for extremely uncommon conditions. Suggest one way that finding a treatment for one specific aspect of progeria might help many others.

2. Up to what stage, if any, do you think it is ethical to experiment on a prenatal human? Cite reasons for your answer.

3. Under a microscope, a first and second polar body look alike. What structure would distinguish them?

4. Armadillos always give birth to identical quadruplets. Are the offspring clones?

5. The morning-after pill prevents pregnancy if taken up to 5 days after having unprotected sex. Based on the stages of prenatal development, how does the pill prevent pregnancy?

6. In about 1 in 200 pregnancies, a sperm fertilizes a polar body instead of an oocyte. A mass of tissue that is not an embryo develops. Why can't a polar body support the development of an entire embryo?

7. "Congratulations! You have a 97 percent chance of having a healthy baby!" says Jill to Janna, who has just read the results of her early pregnancy test. How is Jill's encouragement statistically incorrect?

8. Should a woman be held legally responsible if she drinks alcohol, smokes, or abuses drugs during pregnancy and it harms her child? Should liability apply to all substances that can harm a fetus, or only to those that are illegal?

9. Explain how a man can transmit Zika virus infection to offspring.

10. The antidepressant drugs fluoxetine and paroxetine increase the risk of certain birth defects affecting the heart two- to threefold, but the absolute risk is quite low—that is, few babies are affected. What should a health care provider consider in deciding whether to advise a particular pregnant patient to switch to a similar drug that is not associated with these increased risks?

11. What are possible benefits and dangers of predicting how long a person will live from analyzing his or her genome sequence?

Case Studies and Research Results

1. A mutation in the "testis-expressed 11" gene causes male infertility. The protein that the gene encodes is normally made in spermatids. Explain how the mutation may cause infertility.

2. Human embryonic stem cells can be derived and cultured from an 8-celled cleavage embryo and from a cell of an inner cell mass. Explain the difference between these stages of human prenatal development.

3. Victor, a 34-year-old artist, was killed in a car accident. He and his wife Emma hadn't started a family yet, but planned to soon. The morning after the accident, Emma asked if some of her husband's sperm could be collected and frozen, for her to use to have a child. Do you think that this "postmortem sperm retrieval" should be done? For further information skip ahead to **Bioethics** in chapter 21, "Removing and Using Gametes After Death."

4. A pediatrician with training in anatomy and embryology was examining the jaw of a little girl with auriculocondylar syndrome. The child had tiny ears, a small head and mouth, and an unusual jaw. The doctor realized from studying X rays that the lower jaw looked exactly like an upper jaw—and the girl's mother looked just like the child. What type of mutation does this family have? Describe in general terms how it causes the phenotype.

5. A few children have been reported who have thin sparse hair, bony facial features, little body fat, and appear aged. What further information would help to confirm a diagnosis of a progeroid syndrome?

6. The television program *Orphan Black* is about a young woman and how she interacts with her clones—more than a dozen.

   a. Describe how the clones might have been created using somatic cell nuclear transfer.

   b. Wikipedia describes the clones of *Orphan Black* as being made using *in vitro* fertilization (IVF), in which sperm fertilize oocytes in a laboratory dish. Considering your knowledge of meiosis, explain how IVF would in fact *not* be a way to create clones.

   c. How might you manipulate a very early embryo to produce clones?

7. Hendrikje van Andel-Schipper died in 2005 at age 115. In 2014, researchers identified 450 mutations in white blood cells from a blood sample taken at her autopsy. "Henny" had lived a healthy life, dying of a stomach cancer she didn't even know she had. She never had a blood disease. What is the significance of the discovery that her white blood cells had so many mutations?

8. A study of 202 individuals over the age of 90 living in Calabria in southern Italy is comparing the siblings of these people to their spouses. How could this strategy reveal the genetic underpinnings of longevity?

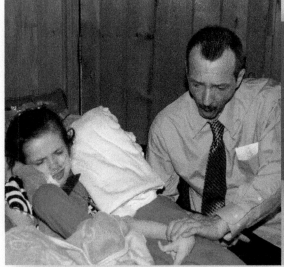

Courtesy of Jane Mervar

CHAPTER 4

Single-Gene Inheritance

Karli Mervar and her father Karl suffered from Huntington disease at the same time. Karl's gaunt look results from the constant movements that are part of the disease—his cells used energy rapidly.

Learning Outcomes

4.1 Following the Inheritance of One Gene

1. List the characteristics that distinguish single-gene diseases from other types of diseases.
2. Describe how Mendel deduced that recessive traits seem to disappear in hybrids.
3. Define and distinguish *heterozygote* and *homozygote*; *dominant* and *recessive*; *phenotype* and *genotype*.
4. Explain how the law of segregation reflects the events of meiosis.
5. Indicate how a Punnett square is used to track inheritance patterns.

4.2 Single-Gene Inheritance Is Rare

6. Explain how a gene alone may not solely determine a trait.
7. Distinguish between autosomal recessive and autosomal dominant inheritance.

4.3 Following the Inheritance of More Than One Gene

8. Explain how Mendel's experiments followed the inheritance of more than one gene.
9. Explain how the law of independent assortment reflects the events of meiosis.

4.4 Pedigree Analysis

10. Explain how pedigrees show single-gene transmission.
11. Explain how exome and genome sequencing can reveal whether a sick child with healthy parents inherited two autosomal recessive mutations or has a new, dominant mutation.

 The BIG Picture

The laws underlying the inheritance of single-gene traits and illnesses have their origins in experiments using pea plants that Gregor Mendel conducted more than a century ago. Mendel's laws reverberate in the interpretation of DNA tests done today, from investigating single genes to sequencing exomes and genomes, which allows geneticists to distinguish inherited mutations from new mutations.

Juvenile Huntington Disease: The Cruel Mutation

Huntington disease (HD) is known as an illness that typically begins in early adulthood, causing uncontrollable movements and changes in behavior and thinking (cognition), with death 15 to 20 years later. HD is dominant, which means that each child of an affected individual need inherit only one copy of the mutant gene to develop the disease. However, 10 percent of people who have HD are under age 20. They have juvenile Huntington disease.

The situation in the Mervar family is unusual in that an adult (Karl) and his young children were or are affected. Karl's first symptoms started when he and his wife Jane's youngest daughter, Karli, began to have trouble paying attention in school. Karl had become abusive, paranoid, and unstable on his feet, due to the disease. By age 5 Karli's left side stiffened and her movements slowed. Soon she could no longer skip, hop, or jump, and she developed a racing heartbeat, itching, pain, and unintelligible speech. Karl was diagnosed 6 weeks after Karli in 2002. He was 35, she just 6. Karli's disease progressed rapidly. By age 9 she required a feeding tube, suffered seizures, and

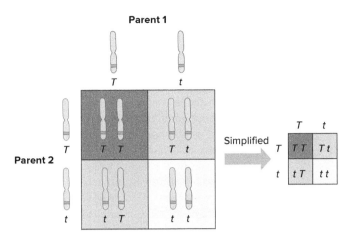

Figure 4.4 **A Punnett square.** A Punnett square illustrates how alleles combine in offspring. The different types of gametes of one parent are listed along the top of the square, with those of the other parent listed on the left-hand side. Each compartment displays the genotype that results when gametes that correspond to that compartment join.

4.2 Single-Gene Inheritance Is Rare

Mendel's first law addresses traits and illnesses caused by single genes, which are also called Mendelian or monofactorial. Single-gene diseases, such as sickle cell disease and muscular dystrophy, are rare compared to infectious diseases, cancer, and multifactorial diseases. The actions of at least one gene and the environment cause multifactorial diseases. Many single-gene diseases affect fewer than 1 in 10,000 individuals.

Single-gene inheritance is usually more complicated than a pea plant having green or yellow peas because many phenotypes associated with single genes are influenced by other genes as well as by environmental factors. That is, the single gene controls trait transmission, but other genes and the environment affect the degree of the trait or severity of the illness. This complexity is

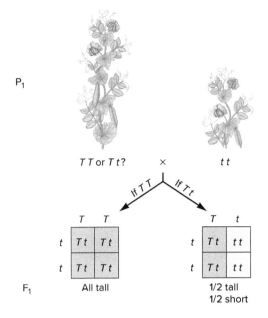

Figure 4.5 **Test cross.** Breeding a tall pea plant with homozygous recessive short plants reveals whether the tall plant is true-breeding (*TT*) or non-true-breeding (*Tt*). Punnett squares usually indicate only the alleles.

one reason why analyzing exomes and genomes, as well as gene expression "connectomes" as depicted in figure 11.4, is important for fleshing out phenotypes. Eye color provides a good example of how our view of single-gene traits has evolved with increasing knowledge of our genomes.

Eye Color

Most people have brown eyes; blue and green eyes are almost exclusively in people of European ancestry. The color of the iris is due to melanin pigments, which come in two forms—the dark brown/black eumelanin, and the red-yellow pheomelanin. In the eye, cells called melanocytes produce melanin, which is stored in structures called melanosomes in the outermost layer of the iris. People differ in the amount of melanin and number of melanosomes, but have about the same number of melanocytes in their eyes.

Nuances of eye color—light versus dark brown, clear blue versus greenish or hazel—arise from the distinctive peaks and valleys at the back of the iris. Thicker regions darken appearance of the pigments, rendering brown eyes nearly black in some parts and blue eyes closer to purple. The bluest eyes have thin irises with very little pigment. The effect of the iris surface on color is a little like the visual effect of a rough-textured canvas on paint.

A single gene on chromosome 15, *OCA2*, confers eye color by controlling melanin synthesis. If this gene is missing, albinism results, causing pale skin and red eyes (**figure 4.6**; see also figure 4.15). A recessive allele of this gene confers blue color and a dominant allele confers brown. But inheritance of eye color is more complicated than this. Near the *OCA2* gene on chromosome 15 is a second gene, *HERC2*, that controls expression of

OCA2

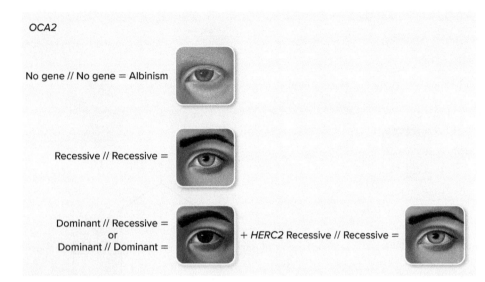

No gene // No gene = Albinism

Recessive // Recessive =

Dominant // Recessive =
or   + HERC2 Recessive // Recessive =
Dominant // Dominant =

Figure 4.6 **Eye color.** At least two genes determine eye color in humans. Inheriting two particular recessive alleles in *HERC2* abolishes the effect of *OCA2,* causing blue eyes. The two genes are near each other on chromosome 15.

the *OCA2* gene. A recessive allele of *HERC2* abolishes the control over *OCA2,* and blue eyes result. A person must inherit two copies of the recessive allele in *HERC2* to have blue eyes.

If blue eye color is the disruption of a "normal" function, why has it persisted? A clue comes from evolution. The *HERC2* gene is found in many species, indicating that it is ancient and important, because it has persisted. Perhaps mutations in *HERC2* arose long ago among hunter-gatherers in Europe, and the unusual individuals with the pale eyes were, for whatever reason, more desirable as sexual partners. Over time, this sexual selection would have increased the proportion of the population with blue eyes. Other explanations are possible.

Modes of Inheritance

Modes of inheritance are rules that explain the common patterns of single-gene transmission, and are derived from Mendel's laws. Knowing mode of inheritance makes it possible to calculate the probability that a particular couple will have a child who inherits a particular condition. The way that a trait is passed depends on whether the gene that determines it is on an autosome or on a sex chromosome, and whether the allele is recessive or dominant. **Table 4.2** compares two of the four modes of inheritance, which apply to transmission of genes

on the autosomes (chromosomes 1 through 22). Table 6.2 covers traits transmitted through the sex chromosomes (X and Y).

In autosomal dominant inheritance, a trait can appear in either sex because an autosome carries the gene. If a child has the trait, at least one parent also has it. Autosomal dominant traits do not skip generations because if no offspring inherit the mutation in one generation, transmission stops. Huntington disease is an autosomal dominant condition. The Punnett square in **figure 4.7** depicts inheritance of an autosomal dominant trait or condition. Many autosomal dominant diseases do not cause symptoms until adulthood.

In autosomal recessive inheritance, a trait can appear in either sex. Affected individuals have a homozygous recessive genotype, whereas in heterozygotes (carriers) the wild type allele masks expression of the mutant allele. A person with cystic fibrosis, for example, inherits a mutant allele from each carrier parent.

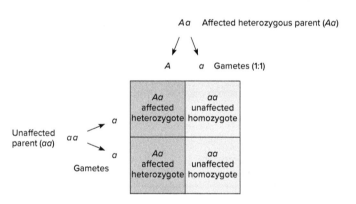

Figure 4.7 **Autosomal dominant inheritance.** When one parent has an autosomal dominant condition and the other does not, each offspring has a 50 percent probability of inheriting the mutant allele and the condition. In the family from the chapter opener, Karl was the "*Aa*" parent and Jane the "*aa*" parent. Each of their daughters faced the 1 in 2 probability of inheriting Huntington disease.

Table 4.2	Comparison of Autosomal Dominant and Autosomal Recessive inheritance
Autosomal Dominant	**Autosomal Recessive**
Males and females affected, with equal frequency	Males and females affected, with equal frequency
Successive generations affected until no one inherits the mutation	Can skip generations
Affected individual has an affected parent, unless he or she has a *de novo* mutation	Affected individual has parents who are affected or are carriers (heterozygotes)

Mendel's first law can be used to calculate the probability that an individual will have either of two phenotypes. The probabilities of each possible genotype are added. For example, the chance that a child whose parents are both carriers of cystic fibrosis will *not* have the condition is the sum of the probability that she has inherited two normal alleles (1/4) plus the chance that she herself is a heterozygote (1/2), or 3/4 in total. Note that this also equals 1 minus the probability that she is homozygous recessive and has the condition.

The ratios that Mendel's first law predicts for autosomal recessive inheritance apply to each offspring anew. If a couple has a child with an autosomal recessive illness, each of their future children faces the same 25 percent risk of inheriting the condition.

Most autosomal recessive conditions appear unexpectedly in families, because they are transmitted silently, through heterozygotes (carriers). **Clinical Connection 4.1** discusses one of them, cystic fibrosis. However, a situation in which an autosomal recessive condition is more likely to recur is when blood relatives have children together. The higher risk of having a child with a particular autosomal recessive condition is because the related parents may carry the same alleles inherited from an ancestor that they have in common, such as a great-grandparent. Marriage between relatives introduces **consanguinity**, which means "shared blood"—a figurative description, because genes are not passed in blood. Alleles inherited from shared ancestors are said to be "identical by descent."

Consanguinity is part of many cultures. For example, marriage between first cousins occurs in about a third of the Pakistani community in England, a population that researchers have been following. In this group, consanguinity doubles the birth defect rate, but it is still low—less than 4 percent of births. Consanguinity may happen unknowingly in close-knit communities where people do not realize they are related, and among families using the same sperm donor.

Logic explains why consanguinity raises risk of inheriting autosomal recessive diseases. An unrelated man and woman have eight different grandparents, but first cousins have only six, because they share one pair through their parents, who are siblings (see figure 4.14c). That is, the probability of two relatives inheriting the same disease-causing recessive allele is greater than that of two unrelated people having the same allele by chance. However, genome-wide studies show that cousins tend to inherit different parts of the shared ancestor's genome, which explains why these populations are healthier than might be expected given the high frequency of consanguinity.

The nature of the phenotype is important when evaluating the transmission of single-gene traits. Some diseases are too severe for people to live long enough or feel well enough to have children. For example, each adult sibling of a person who is a known carrier of Tay-Sachs disease has a two-thirds chance of also being a carrier because only two genotypes are possible for an adult—homozygous for the wild type allele or a carrier who inherits the mutant allele from either parent. A homozygous recessive individual for this brain disease would not have survived childhood.

Geneticists who study human traits and illnesses can hardly set up crosses as Mendel did to demonstrate modes of inheritance, but they can pool information from families whose members have the same trait or illness to deduce the mode of inheritance. Consider a simplified example of 50 couples in whom both partners are carriers of sickle cell disease. If 100 children are born, about 25 of them would be expected to have sickle cell disease. Of the remaining 75, theoretically 50 would be carriers like their parents, and the remaining 25 would have two wild type alleles.

Solving a Problem in Following a Single Gene

Using Mendel's laws to predict phenotypes and genotypes requires a careful reading of the problem to organize relevant information. Common sense is useful, too. The following general steps can help to solve a problem based on the inheritance of a single-gene trait:

1. List all possible genotypes and phenotypes for the trait.
2. Determine the genotypes of the individuals in the first (P_1) generation. Use information about those individuals' parents.
3. After deducing genotypes, derive the possible alleles in gametes each individual produces.
4. Unite these gametes in all combinations to reveal all possible genotypes. Calculate ratios for the F_1 generation.
5. To extend predictions to the F_2 generation, use the genotypes of the specified F_1 individuals and repeat steps 3 and 4.

As an example, consider curly hair. If C is the dominant allele, conferring curliness, and c is the recessive allele, then CC and Cc genotypes confer curly hair. A person with a cc genotype has straight hair.

Wendy has beautiful curls, and her husband Rick has straight hair. Wendy's father is bald, but once had curly hair, and her mother has stick-straight hair. What is the probability that Wendy and Rick's child will have straight hair? Steps 1 through 5 solve the problem:

1. State possible genotypes: CC, Cc = curly; cc = straight.
2. Determine genotypes: Rick must be cc, because his hair is straight. Wendy must be Cc, because her mother has straight hair and therefore gave her a c allele.
3. Determine gametes: Rick's sperm carry only c. Half of Wendy's oocytes carry C, and half carry c.
4. Unite the gametes:

5. Conclusion: Each child of Wendy and Rick's has a 50 percent chance of having curly hair (Cc) and a 50 percent chance of having straight hair (cc).

"65 Roses": Progress in Treating Cystic Fibrosis

Young children who cannot pronounce the name of their disease call it "65 roses." Cystic fibrosis (CF), although still a serious illness, is a genetic disease success story thanks to a new class of drugs that correct the underlying problem in protein folding that causes many cases.

Recall from figures 1.3 and 1.4 that CF affects ion channels that control chloride movement out of cells in certain organs. In the lungs, thick, sticky mucus accumulates and creates an environment hospitable to certain bacteria that are uncommon in healthy lungs. A clogged pancreas prevents digestive secretions from reaching the intestines, impairing nutrient absorption. A child with CF has trouble breathing and maintaining weight.

Life with severe CF is challenging. In summertime, a child must avoid water from hoses, which harbor lung-loving *Pseudomonas* bacteria. A bacterium called *Burkholderia cepacia* easily spreads in summer camps. Cookouts spew lung-irritating particulates. Too much chlorine in pools irritates lungs whereas too little invites bacterial infection. New infections arise, too. In the past few years, multidrug-resistant *Mycobacterium abscessus*, related to the pathogen that causes tuberculosis, has affected up to 10 percent of CF patients in the United States and Europe. Other bacteria that can't grow in a laboratory infect CF patients. Sequenced bacterial DNA in lung fluid from children with CF reveals more than 60 species of bacteria.

CF is inherited from two carrier parents, and affects about 30,000 people in the United States and about 70,000 worldwide. Many other people may have cases so mild that they are not diagnosed with CF but instead with frequent respiratory infections. More than 2,000 variants are known in the cystic fibrosis transmembrane regulator (*CFTR*) gene, which encodes the chloride channel protein. About half of the variants are known to affect health. Today pregnant women are tested for the more common mutations to see if they are carriers, and if they are, their partners are tested. If both parents-to-be are carriers, then the fetus has a 25 percent chance of having inherited CF and may be tested as well. In most states all newborns receive a genetic test for CF. Before genetic testing became available, diagnosis typically took months and was based on observing "failure to thrive," salty sweat, and foul-smelling stools.

When CF was recognized in 1938, life expectancy was only 5 years; today median survival is about age 40, with many patients living longer, thanks to several types of treatments. Inhaled antibiotics control respiratory infections and daily "bronchial drainage" exercises shake mucus from the lungs (**figure 4A**). A vibrating vest worn for half-hour periods two to four times a day loosens mucus. Digestive enzymes mixed into soft foods enhance nutrient absorption, although some patients require feeding tubes.

Figure 4A **Regular exercise helps many people who have cystic fibrosis.** Evan Scully, who has CF, is shown here running 8 miles along the Great Wall of China during the World Athletics Championships in August 2015. He attributes his relatively good health to being active. "I do a 3-day exercise cycle—about an hour running on day 1, gym workout on day 2, and yoga and movement on day 3." When he was running competitively for Ireland, he did 120 miles a week! *Courtesy of Evan Scully*

New drugs to treat CF work in various ways: correcting misfolded CFTR protein, restoring liquid on airway surfaces, breaking up mucus, improving nutrition, and fighting inflammation and infection. A drug called ivacaftor (Kalydeco) that became available in 2012 at first helped only the 5 percent of patients with a mutation called G551D, in whom the ion channel proteins reach the cell membrane but once there, remain closed, like locked gates. The drug binds the proteins in a way that opens the channels. Kalydeco given with a second drug helps other people with CF whose mutations cause dismantling of CFTR in the endoplasmic reticulum and people in whom the protein reaches the plasma membrane and actually folds into an ion channel but doesn't stay there long enough to work. A third new drug enables protein synthesis to ignore mutations that shorten the protein.

Questions for Discussion

1. Which parts of the cell are affected in CF?
2. If a child with CF has two parents who do not have the disease, what is the risk that a future sibling will inherit CF?
3. How does the drug Kalydeco work?
4. For some people who have taken Kalydeco, the first sign that the drug is working is that their flatulence no longer smells foul. What characteristic of the disease might this observation indicate is improving?

On the Meaning of Dominance and Recessiveness

Knowing whether an allele is dominant or recessive is critical in determining the risk of inheriting a particular condition (phenotype). Dominance and recessiveness arise from the genotype, and reflect the characteristics or abundance of a protein.

Mendel based his definitions of dominance and recessiveness on what he could see—one allele masking the effect of the second one. Today we can often add a cellular or molecular explanation.

The basis of an inborn error of metabolism is easy to picture. These diseases are typically recessive because the half normal amount of the enzyme that a carrier produces is usually sufficient to maintain health. The one normal allele, therefore, compensates for the mutant one, to which it is dominant.

A recessive trait is said to arise from a "loss-of-function" because the recessive allele usually prevents the production or activity of the normal protein. In contrast, some dominantly inherited diseases are said to be due to a "gain-of-function," because they result from the action of an abnormal protein that interferes with the function of the normal protein. Huntington disease results from a gain-of-function in which the dominant mutant allele encodes an abnormally long protein that prevents the normal protein from functioning in certain brain cells. Huntington disease is a gain-of-function because individuals who are missing one copy of the gene do not have the illness. That is, the protein encoded by the mutant HD allele must be abnormal, not absent, to cause the disease. The gain-of-function nature of HD is why people with one mutant allele have the same phenotype as the rare individuals with two mutant alleles (**figure 4.8**).

Recessive diseases tend to be more severe, and produce symptoms earlier, than dominant diseases. Disease-causing recessive alleles remain in populations because healthy heterozygotes pass them to future generations. In contrast, if a dominant mutation arises that harms health early in life, people who have the allele are either too ill or do not survive long enough to reproduce. The allele eventually becomes rare in the population unless it arises anew by mutation in the gametes of a person who does not have the disease. Dominant diseases whose symptoms do not appear until adulthood, or that do not severely disrupt health, remain in a population because they do not prevent a person from having children and passing on the mutation.

Under certain circumstances, for some genes, a heterozygous individual (a carrier) can develop symptoms. This is the case for sickle cell disease. Carriers can develop a life-threatening breakdown of muscle if exposed to the combination of environmental heat, intense physical activity, and dehydration. Several college athletes died from these symptoms, prompting sports authorities to begin testing athletes for sickle cell disease carrier status. **Case Studies and Research Results** at the chapter's end discusses this problem.

4.3 Following the Inheritance of More Than One Gene

In a second set of experiments, Mendel examined the inheritance of two traits at a time. Today, his classic experiments on the inheritance of more than one trait are more relevant than ever, thanks to technologies that consider many genes simultaneously.

Mendel's Second Law

The law of **independent assortment** states that for two genes on different chromosomes, the inheritance of one gene does not influence the chance of inheriting the other gene. The two genes are said to "independently assort" because they are packaged into gametes at random (**figure 4.9**). Two genes that are far apart on the same chromosome also appear to independently assort, because so many crossovers take place between them that it is as if they are part of separate chromosomes (see figure 3.5).

Mendel looked at seed shape, which was either round or wrinkled (determined by the R gene), and seed color, which was either yellow or green (determined by the Y gene). When he crossed true-breeding plants that had round, yellow seeds

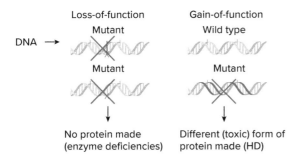

Figure 4.8 **A mutation can confer a loss-of-function or a gain-of-function.** A loss-of-function mutation is typically recessive and results in a deficit or absence of the gene's protein product. A gain-of-function mutation is typically dominant and alters the encoded protein, introducing a new activity.

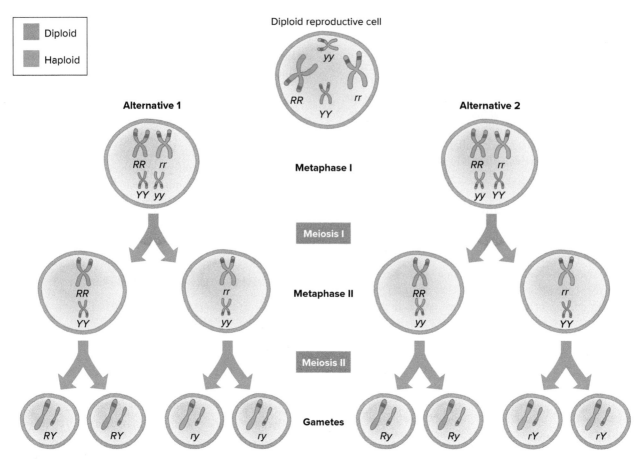

Figure 4.9 Mendel's second law—independent assortment. The independent assortment of genes carried on different chromosomes results from the random alignment of chromosome pairs during metaphase of meiosis I. An individual of genotype *RrYy*, for example, manufactures four types of gametes, containing the dominant alleles of both genes (*RY*), the recessive alleles of both genes (*ry*), and a dominant allele of one with a recessive allele of the other (*Ry* or *rY*). The allele combination depends upon which chromosomes are packaged together in a gamete—and this happens at random.

to true-breeding plants that had wrinkled, green seeds, all the progeny had round, yellow seeds (**figure 4.10**). These offspring were double heterozygotes, or dihybrids, of genotype *RrYy*. From their appearance, Mendel deduced that round is dominant to wrinkled, and yellow to green.

Next, he self-crossed the dihybrid plants in a **dihybrid cross**, so named because two genes and traits are followed. Mendel found four types of seeds in the next, third generation: 315 plants with round, yellow seeds; 108 plants with round, green seeds; 101 plants with wrinkled, yellow seeds; and 32 plants with wrinkled, green seeds. These classes appeared in a ratio of 9:3:3:1.

Mendel then crossed each plant from the third generation to plants with wrinkled, green seeds (genotype *rryy*). These test crosses established whether each plant in the third generation was true-breeding for both genes (genotypes *RRYY* or *rryy*), true-breeding for one gene but heterozygous for the other (genotypes *RRYy*, *RrYY*, *rrYy*, or *Rryy*), or heterozygous for both genes (genotype *RrYy*). Mendel could explain the 9:3:3:1 proportion of progeny classes only if one gene does not influence transmission of the other. Each parent would produce

equal numbers of four different types of gametes: *RY*, *Ry*, *rY*, and *ry*. Each of these combinations has one gene for each trait. A Punnett square for this cross shows that the four types of seeds:

1. round, yellow (*RRYY*, *RrYY*, *RRYy*, and *RrYy*),
2. round, green (*RRyy* and *Rryy*),
3. wrinkled, yellow (*rrYY* and *rrYy*), and
4. wrinkled, green (*rryy*)

are present in the ratio 9:3:3:1, just as Mendel found.

Solving a Problem in Following Multiple Genes

A Punnett square for three genes has 64 boxes; for four genes, 256 boxes. An easier way to predict genotypes and phenotypes in multigene crosses is to use the mathematical laws of probability that are the basis of Punnett squares. Probability predicts the likelihood of an event.

An application of probability theory called the product rule can predict the chance that parents with known genotypes

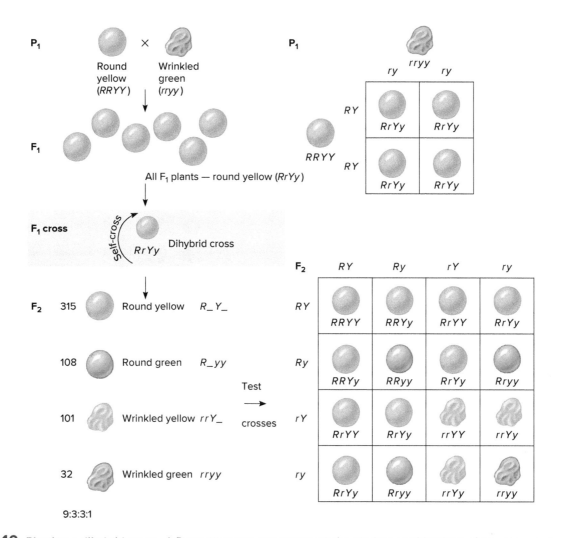

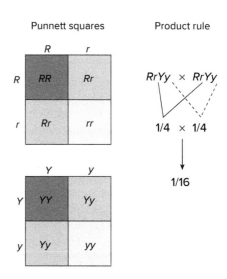

Figure 4.10 **Plotting a dihybrid cross.** A Punnett square can represent the random combinations of gametes produced by dihybrid individuals. An underline in a genotype (in the F₂ generation) indicates that either a dominant or a recessive allele is possible. The numbers in the F₂ generation are Mendel's experimental data. Test crosses with *rryy* plants revealed the genotypes of the F₂ generation, depicted in the 16-box Punnett square.

can produce offspring of a particular genotype. The product rule states that the chance that two independent events will both occur equals the product of the chances that either event will occur alone. Consider the probability of obtaining a plant with wrinkled, green peas (genotype *rryy*) from dihybrid (*RrYy*) parents. Do the reasoning for one gene at a time, then multiply the results (**figure 4.11**).

A Punnett square depicting a cross of two *Rr* plants indicates that the probability of producing *rr* progeny is 25 percent, or 1/4. Similarly, the chance of two *Yy* plants producing a *yy* plant is 1/4. Therefore, the chance of dihybrid parents (*RrYy*) producing homozygous recessive (*rryy*) offspring is 1/4 multiplied by 1/4, or 1/16. Now consult the 16-box Punnett square for Mendel's dihybrid cross again (see figure 4.10). Only one of the 16 boxes is *rryy*, just as the product rule predicts. **Figure 4.12** shows how these tools can be used to predict offspring genotypes and phenotypes for three human traits simultaneously.

Figure 4.11 **The product rule.**

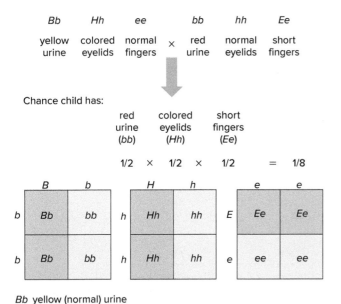

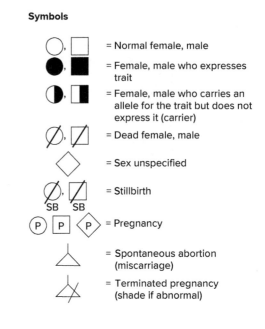

$$\frac{1}{2} \times \frac{1}{2} \times \frac{1}{2} = \frac{1}{8}$$

Bb yellow (normal) urine
bb beeturia (red urine after eating beets)
Hh colored eyelids
hh normal eyelids
Ee brachydactyly (short fingers)
ee normal fingers

Figure 4.12 **Using probability to track three traits in people.** A man has normal urine, colored eyelids, and normal fingers. His partner has red urine after she eats beets, and has normal eyelids and short fingers. The chance that their child will have red urine after eating beets, colored eyelids, and short fingers is 1/8.

Key Concepts Questions 4.3

1. How is Mendel's second law relevant today?

2. Explain how meiotic events underlie independent assortment for two or more genes.

3. Explain how Mendel demonstrated that genes on different chromosomes independently assort.

4. How are Punnett squares and probability used to follow inheritance of more than one trait?

4.4 Pedigree Analysis

For genetics researchers and genetic counselors, the bigger the family the better—the more children in a generation, the easier it is to deduce a mode of inheritance. Charts called **pedigrees** are used to display family relationships and depict which relatives have specific phenotypes and, sometimes, genotypes. A human pedigree serves the same purpose as one for purebred dogs or cats or thoroughbred horses—it represents relationships and sometimes traits. A pedigree in genetics differs from a family tree in genealogy, and from a genogram in social work, in that it indicates inherited diseases or traits as well as relationships and ancestry. Pedigrees may also include molecular data, test results, and information on variants of multiple genes.

In a pedigree, lines link shapes. Vertical lines represent generations, and horizontal lines that connect two shapes

at their centers depict partners. Shapes connected by vertical lines that are joined horizontally represent siblings. Squares indicate males; circles, females; and diamonds, individuals of unspecified sex. Roman numerals designate generations. Arabic numerals or names indicate individuals. **Figure 4.13** shows these and other commonly used pedigree symbols. Colored or shaded shapes indicate individuals who express a trait, and half-filled shapes are known carriers.

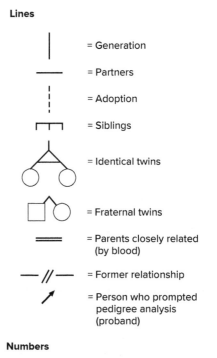

Numbers

Roman numerals = generations

Arabic numerals = individuals in a generation

Figure 4.13 **Pedigree components.** Symbols representing individuals are connected to form pedigree charts, which display the inheritance patterns of traits.

Pedigrees Then and Now

The earliest pedigrees were strictly genealogical, not indicating traits. **Figure 4.14***a* shows such a pedigree for a highly inbred branch of the ancient Egyptian royal family tree. The term *pedigree* arose in the fifteenth century, from the French *pie de grue*, which means "crane's foot." Pedigrees at that time, typically depicting large families, showed parents linked by curved lines to their many offspring. The overall diagram often resembled a bird's foot.

One of the first pedigrees to trace an inherited illness was an extensive family tree of several European royal families, indicating which members had the clotting disease hemophilia B, also called factor IX deficiency (see figure 6.7). The mutant gene probably originated in Queen Victoria of England in the nineteenth century. In 1845, a genealogist named Pliny Earle constructed a pedigree of a family with colorblindness using musical notation—half notes for unaffected females, quarter notes for colorblind females, and filled-in and squared-off notes to represent the many colorblind males. In the early twentieth century, eugenicists tried to use pedigrees to show

that traits such as criminality, feeblemindedness, and promiscuity were the consequence of faulty genes (see Section 15.7).

Today, pedigrees are important both for helping families identify the risk of transmitting an inherited illness and as starting points for identifying and describing, or annotating, a gene from the human genome sequence. People who have kept meticulous family records, such as the Mormons and the Amish, are invaluable in helping researchers follow the inheritance of particular genes. (See Clinical Connection 15.1.) Very large pedigrees can provide information on many individuals with a particular rare disease. The researchers then search affected individuals' DNA to identify a specific sequence they have all inherited that is not found in healthy family members. Within this section of a chromosome lies the causative mutation. Discovery of a mutation that causes an early-onset form of Alzheimer disease, for example, took researchers to a remote village in Colombia, where the original mutation present today in a 1,000-plus-member family came from a Spanish settler who had arrived in the seventeenth century.

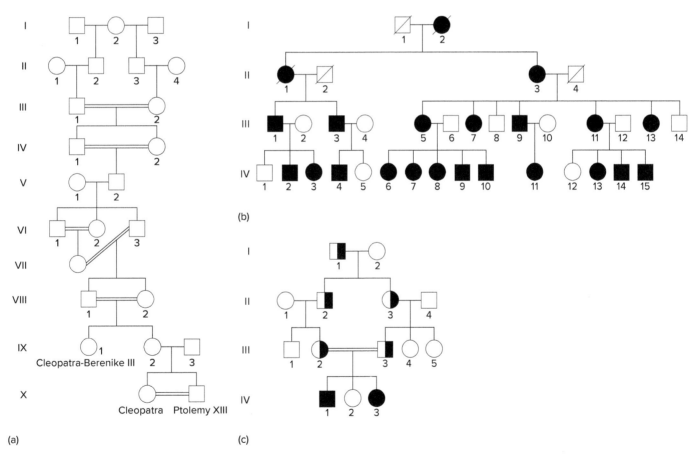

Figure 4.14 Unusual pedigrees. (a) A partial pedigree of Egypt's Ptolemy dynasty shows only genealogy, not traits. It appears almost ladderlike because of the extensive consanguinity. From 323 B.C. to Cleopatra's death in 30 B.C., the family experienced one pairing between cousins related through half-brothers (generation III), four brother-sister pairings (generations IV, VI, VIII, and X), and an uncle-niece relationship (generations VI and VII). Cleopatra married her brother, Ptolemy XIII, when he was 10 years old! These marriage patterns were an attempt to preserve the royal blood. **(b)** A family with polydactyly (extra fingers and toes) extends laterally, with many children, because the phenotype doesn't affect reproduction. **(c)** This pedigree shows marriage of first cousins. They share one set of grandparents and, therefore, risk passing on the same recessive alleles to offspring.

Pedigrees Display Mendel's Laws

Visual learners can easily "see" a mode of inheritance in a pedigree. Consider a pedigree for an autosomal recessive trait, albinism. Homozygous recessive individuals in the third (F₂) generation lack an enzyme necessary to manufacture the pigment melanin and, as a result, display pale hair and skin color (**figure 4.15**). Their parents are inferred to be heterozygotes (carriers). One partner from each pair of grandparents must also be a carrier. For some diseases, carrier status is detectable with a blood or urine test that shows half the normal amount of the gene product.

Recall that an autosomal dominant trait does not skip generations and can affect both sexes. A typical pedigree for an autosomal dominant trait has some squares and circles filled in to indicate affected individuals in each generation. Figure 4.14*b* is a pedigree for an autosomal dominant trait, extra fingers and toes (polydactyly), which is shown in figure 1.5*a*.

A pedigree may be inconclusive, which means that either autosomal recessive or autosomal dominant inheritance can explain the pattern of filled-in symbols. **Figure 4.16** shows one such pedigree, for a type of hair loss called alopecia areata. According to the pedigree, this trait can be passed in an autosomal dominant mode because it affects both males and females and is present in every generation. However, the pedigree can also depict autosomal recessive inheritance if the individuals represented by unfilled symbols are carriers. Inconclusive pedigrees tend to arise when families are small and the trait is not severe enough to impair fertility.

Pedigrees may be difficult to construct or interpret for reasons that have nothing to do with DNA. People may hesitate to supply information because the symptoms embarrass them. Families with adoption, children born out of wedlock, serial relationships, blended families, and use of assisted reproductive technologies (see chapter 21) may not fit easily into the rules of pedigree construction. Many people cannot trace their families back far enough to reveal a mode of inheritance.

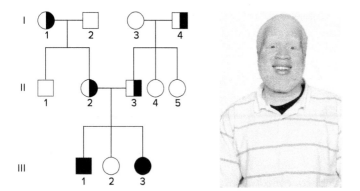

Figure 4.15 **Albinism is autosomal recessive.** Albinism affects males and females and can skip generations, as it does in generations I and II in this pedigree. Homozygous recessive individuals lack an enzyme needed to produce melanin, which colors the eyes, skin, and hair. © *Paul Buarns/Getty Images RF*

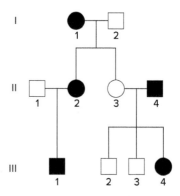

Figure 4.16 **An inconclusive pedigree.** This pedigree could account for an autosomal dominant trait or an autosomal recessive trait that does not prevent affected individuals from having children. (Unfilled symbols could represent carriers.)

Solving a Problem in Conditional Probability

Pedigrees and Punnett squares can be used to trace a conditional probability, which is when an offspring's genotype depends on the parents' genotypes, which may not be obvious from their phenotypes. For example, if a person has an autosomal recessive trait or condition, his or her parents are inferred to be carriers (unless a new mutation has arisen in the affected individual).

The family represented in **figure 4.17** illustrates conditional probability. Deshawn has sickle cell disease. His unaffected parents, Kizzy and Ike, must each be heterozygotes (carriers). Deshawn's sister, Taneesha, is also healthy, and she is expecting her first child. Taneesha's husband, Antoine, has no family history of sickle cell disease. What is the probability that Taneesha's child will inherit her mutant allele and be a carrier?

Taneesha's concern raises two questions. First, what is the probability that she is a carrier? Because she is the product of a monohybrid cross, and she is not homozygous recessive (sick), she has a 2 in 3 chance of being a carrier. If so, the chance that she will transmit the mutant allele is 1 in 2, because she has two copies of the gene, and only one allele goes into each gamete. To calculate the overall risk to her child, multiply the probability that she is a carrier (2/3) by the chance that if she is, she will transmit the mutant allele (1/2). The result is 1/3.

Exome and Genome Sequencing Clarify Pedigrees

An increasingly common clinical scenario occurs when a child has a syndrome (a collection of symptoms) not seen before, and both parents are unaffected. Two explanations are possible. Either both parents are carriers (heterozygotes) for recessive mutations in the same gene for a condition that has not yet been identified, or a *de novo* (new) dominant mutation arose spontaneously in the sick child, in a sperm or an oocyte. The distinction is important to a family because a mutation inherited from carrier parents could be transmitted to siblings of

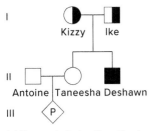

(a) Taneesha's brother, Deshawn, has sickle cell disease.

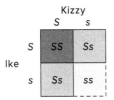

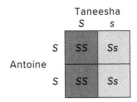

(b) Probability that Taneesha is a carrier: $^2/_3$

Taneesha
S   s

	S	s
S	SS	Ss
S	SS	Ss

Antoine

(c) If Taneesha is a carrier, chance that fetus is a carrier: $^1/_2$

Total probability $= ^2/_3 \times ^1/_2 = ^1/_3$

Figure 4.17 **Making predictions.** Taneesha's brother Deshawn has sickle cell disease **(a).** Taneesha wonders if her fetus has inherited the sickle cell allele. First, she must calculate the chance that she is a carrier. The Punnett square in **(b)** shows that this risk is 2 in 3. (She must be genotype *SS* or *Ss* but cannot be *ss* because she does not have the disease.) The risk that the fetus is a carrier, assuming that the father is not a carrier, is half Taneesha's risk of being a carrier, or 1 in 3 **(c).**

the affected child, according to Mendel's first law. In contrast, a mutation originating in a single gamete would *not* affect a sibling, unless the parent was a mosaic and has other gametes with the mutation. However, a *de novo* dominant mutation *can* be passed on to the *next* generation after the affected child, according to Mendel's first law, with a probability of 1 in 2. Chapter 12 discusses *de novo* mutations further.

Exome and genome sequencing of parents-child "trios" can identify a mutant gene in the affected child and distinguish whether the condition is autosomal recessive and inherited, or autosomal dominant and new. Algorithms compare the DNA sequences of the affected child and unaffected parents, searching for variants of "candidate genes" whose function, or malfunction, might explain the specific symptoms. The challenge is a little like finding a needle in a haystack. Exome sequencing typically detects 70 to 175 *de novo* gene variants, but fewer

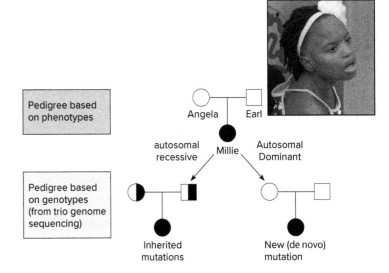

Figure 4.18 **Parent and child trios that have their exomes or genomes sequenced can enable geneticists to distinguish Mendelian inheritance of a recessive mutation from carrier parents from a new, dominant mutation in the child.**
A pedigree based on phenotypes only reveals that two healthy parents have a sick child. The pedigree based on genotypes can distinguish inheritance of a mutation from carrier parents from a *de novo* mutation in the child. *Courtesy Children's Mercy Kansas City*

than 3 and often none of them change the encoded protein in a way that can affect the phenotype. The numbers of identified gene variants are higher for sequencing the entire genome.

When Millie McWilliams, described in Clinical Connection 1.1, and her parents had their genomes sequenced, the analysis revealed her *de novo* mutation in the *ASXL3* gene on chromosome 18. She is missing two bases in the gene, which causes the rare Bainbridge-Ropers syndrome. Her parents are wild type for this gene (**figure 4.18**). The case of Millie and thousands of others with genetic syndromes, many of them children, illustrate the marriage of genome sequencing and the more traditional genetic tools of observing families and charting their traits—all based on Mendel's laws of inheritance.

Key Concepts Questions 4.4

1. How do pedigrees depict family relationships and transmission of inherited traits?

2. How do pedigrees and Punnett squares apply Mendel's laws to predict the recurrence risks of inherited diseases or traits?

3. Explain how exome and genome sequencing can be used to distinguish autosomal recessive inheritance in a "parents-child trio" from a new dominant mutation affecting the child.

Summary

4.1 Following the Inheritance of One Gene

1. Single-gene diseases differ from other types of diseases in the ability to predict risk of occurrence in a particular family member, possibility of presymptomatic testing, increased prevalence in specific populations, and the possibility of correcting or compensating for the underlying mutation.

2. Modes of inheritance are patterns of the probability that people related in a certain way will inherit a trait or illness.

3. Mendel's laws, based on pea plant crosses, derive from how chromosomes act during meiosis. The laws apply to all diploid organisms.

4. Mendel used statistics to investigate why some traits vanish in hybrids. The law of **segregation** states that alleles of a gene go into separate gametes during meiosis. Mendel demonstrated segregation of seven traits in pea plants using **monohybrid crosses**.

5. A diploid individual with two identical alleles of a gene is a **homozygote**. A **heterozygote** has two different alleles of a gene. A gene may have many alleles.

6. A **dominant** allele masks the expression of a **recessive** allele.

7. An individual may be homozygous dominant, homozygous recessive, or heterozygous for a given pair of alleles.

8. The alleles for a gene in an individual constitute the **genotype**, and their expression the **phenotype**.

9. The most common allele in a population is the **wild type**, and variants are **mutant**. **Mutation** is a change in a gene.

10. When Mendel crossed two true-breeding plants, then bred the resulting hybrids to each other, the two variants of the trait appeared in a 3:1 phenotypic ratio. Crossing these progeny revealed a genotypic ratio of 1:2:1.

11. A **Punnett square** follows the transmission of alleles and is based on probability.

4.2 Single-Gene Inheritance Is Rare

12. Eye color illustrates how a single-gene trait can be affected by other genes.

13. An **autosomal dominant** trait affects males and females and does not skip generations.

14. An **autosomal recessive** trait affects males or females and may skip generations.

15. Autosomal recessive conditions are more likely in families where there is **consanguinity**. Recessive diseases tend to be more severe and cause symptoms earlier than dominant diseases.

16. Genetic problems can be solved by tracing alleles as gametes form and then combine in a new individual.

17. Dominance and recessiveness reflect how alleles affect the abundance or activity of the gene's protein product. Loss-of-function mutations lead to a missing protein. Gain-of-function mutations alter a protein's function.

4.3 Following the Inheritance of More Than One Gene

18. Mendel's second law, the law of **independent assortment**, follows transmission of two or more genes on different chromosomes.

19. **Dihybrid crosses** showed that a random assortment of maternally and paternally derived chromosomes during meiosis yields gametes with different gene combinations.

20. The product rule states that the chance that two independent genetic events occur equals the product of the probabilities that each event will occur. The product rule is useful in following the inheritance of genes on different chromosomes.

4.4 Pedigree Analysis

21. A **pedigree** is a chart that depicts family relationships and patterns of inheritance for particular traits. A pedigree can be inconclusive if a dominant trait or illness does not impair fertility. An individual with a *de novo* dominant mutation or two autosomal recessive alleles inherited from carrier parents can appear the same in a pedigree.

22. Pedigrees and **Punnett squares** can predict the phenotypic and genotypic classes of offspring in cases of conditional probability in which knowing a parent's genotype restricts the possible genotypic classes of an offspring.

23. Comparing exome or genome sequences in parents-child trios when the child has an unrecognized syndrome and the parents are unaffected can distinguish autosomal recessive inheritance from the parents from a *de novo* dominant mutation in the child.

Review Questions

1. List four ways that single-gene diseases such as Huntington disease and cystic fibrosis differ from an infectious disease, such as malaria or influenza.

2. How does meiosis explain Mendel's laws?

3. Describe two ways to inherit blue eyes.

4. Compare how an autosomal recessive condition affects a family to how an autosomal dominant condition does so.

5. Discuss how Mendel derived the two laws of inheritance without knowing about chromosomes.

6. Distinguish between
   a. autosomal recessive and autosomal dominant inheritance.
   b. Mendel's first and second laws.
   c. a homozygote and a heterozygote.
   d. a monohybrid and a dihybrid cross.
   e. a loss-of-function mutation and a gain-of-function mutation.
   f. a Punnett square and a pedigree.

7. Explain how a dominant disease can occur if its symptoms prevent reproduction.

8. Why would Mendel's results for the dihybrid cross have been different if the genes for the traits he followed were near each other on the same chromosome?

9. Why are extremely rare autosomal recessive diseases more likely to appear in families in which blood relatives have children together?

10. How does the pedigree of the ancient Egyptian royal family in figure 4.14*a* differ from a pedigree a genetic counselor might use today?

11. What are possible genotypes of the parents of a person who has two HD mutations?

12. What is the probability that two individuals with an autosomal recessive trait, such as albinism, will have a child with the same genotype and phenotype as they have?

13. Explain why discovering a *de novo* mutation in a child who has a syndrome does not pose an elevated risk for her siblings, but does to her children.

Applied Questions

1. Calculate the probability that all three children in the Mervar family would inherit Huntington disease.

2. Two couples chat in the waiting room of a guidance counselor at an elementary school in Flint, Michigan. Each family has six children. Some of them are having trouble keeping up at school, with worsening memory and fatigue and irritability. The affected children have physical complaints too. One family lives in a home built in 2012. The other lives in an apartment building that was built before World War II. What questions would you ask, and what information would you need, to determine whether the cause of the children's problems is due to an inherited disease? What might be a specific environmental explanation?

3. Predict the phenotypic and genotypic ratios for crossing the following pea plants:

    a. short × short

    b. short × true-breeding tall

    c. true-breeding tall × true-breeding tall

4. What are the genotypes of the pea plants that would have to be bred to yield one plant with restricted pods for every three plants with inflated pods?

5. If pea plants with all white seed coats are crossed, what are the possible phenotypes of their progeny?

6. Pea plants with restricted yellow pods are crossed to plants that are true-breeding for inflated green pods and the F_1 plants are crossed. Derive the phenotypic and genotypic ratios for the F_2 generation.

7. More than 100 genes cause deafness when mutant. What is the most likely mode of inheritance in families in which all children and the parents were born deaf?

8. The MacDonalds raise Labrador retrievers. In one litter, two of eight puppies, a male and a female, have a condition called exercise-induced collapse. After about 15 minutes of intense exercise, the dogs wobble about, develop a fever, and their hind legs collapse. The parents are healthy. What is the mode of inheritance?

9. Chands syndrome is autosomal recessive and causes curly hair, underdeveloped nails, and abnormally shaped eyelids. In the following pedigree, which individuals must be carriers?

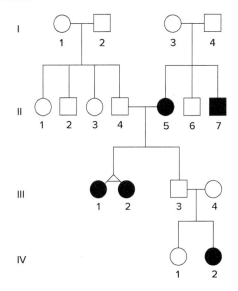

Chands syndrome

10. Lorenzo has a double row of eyelashes, which he inherited from his mother as a dominant trait. His maternal grandfather is the only other relative to have it. Fatima, who has normal eyelashes, marries Lorenzo. Their first child, Nicola, has normal eyelashes. Now Fatima is pregnant again and hopes for a child with double eyelashes. What chance does the child have of inheriting double eyelashes? Draw a pedigree of this family.

11. In peeling skin syndrome, the outer skin layers fall off on the hands and feet. The pedigrees depict three affected families. What do the families share that might explain the appearance of this otherwise rare condition?

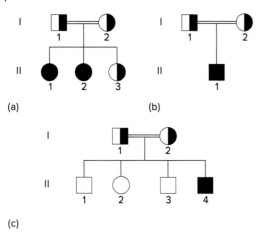

(a)

(b)

(c)

Peeling skin syndrome

12. The child referenced in figure 4.12 who has red urine after eating beets, colored eyelids, and short fingers, is genotype *bbHhEe*. The genes for these traits are on different chromosomes. If he has children with a woman who is a trihybrid, what are the expected genotypic and phenotypic ratios for their offspring?

Forensics Focus

1. A woman, desperate to complete her family tree for an upcoming family reunion, cornered a stranger in a fast-food restaurant. Her genealogical research had identified him as a distant cousin, and she needed his DNA. He refused to cooperate, looked scared, and ran off. The woman took his discarded coffee cup and collected DNA from traces of saliva, which she sent on a swab to a DNA ancestry testing company. (This is a true story.)

   a. Do you think the woman was justified in her action? Why or why not?

   b. What are the strengths and limitations of using genealogical information (family records, word-of-mouth) versus DNA testing to construct a pedigree?

2. A young woman walking to her car in a parking lot late at night was attacked and raped. She recalled that the rapist wore a green silk shirt, a belt with a metal buckle that left a bruise on her abdomen, jeans, and running shoes. She also remembered that he had white skin, a shaved head, and startling blue eyes. She recalled footsteps approaching and then someone yanking the man off her, but her head was let go and hit the pavement, knocking her unconscious.

   A few weeks after the rape, a naked male body washed up in a nearby river, and police found a belt buckle farther downstream that matched the pattern of the woman's bruise. The body, however, was headless. How can police determine the eye color of the corpse to help in identifying or ruling him out as the rapist?

Case Studies and Research Results

1. Hannah Sames (see Clinical Connection 2.2) inherited two identical mutations that delete part of the gigaxonin gene. A boy who received gene therapy to treat his hereditary blindness (see section 20.3) has two different recessive mutations in the *RPE65* gene. Which child is a compound heterozygote?

2. In 2006, a 19-year-old freshman at Rice University collapsed during strenuous football practice. He died the next morning following acute exertional rhabdomyolysis, a complication of being a carrier (heterozygote) for sickle cell disease in which muscles break down. The student's parents sued Rice University and the National Collegiate Athletic Association (NCAA). In response, the NCAA mandated sickle cell carrier testing for all Division 1 student-athletes. However, a student can sign a waiver to opt out of sickle cell testing, absolving the NCAA of liability. In some states, students may already know their carrier status because they were tested as newborns. Dehydration and heat exposure are also risk factors for the deadly muscle condition in anyone, even people who are not carriers for or have sickle cell disease. Drinking enough water and resting can help prevent the muscle problem. The genetic disease is more prevalent among people of African ancestry, but anyone can be a carrier. The NCAA estimates that requiring testing in all Division 1 schools would identify 300 to 400 carriers each year and save one or two lives.

   a. Was the NCAA response to test all Division 1 students a good idea? State a reason for your answer.

   b. How might the NCAA improve its plan to test all Division 1 athletes?

   c. Suggest a way to prevent deaths from exertional rhabdomyolysis other than genetic testing.

3. When Peter and Martha were 24 and expecting their first child, they learned that Peter's mother, who was adopted, had early signs of Huntington disease (HD) (see chapter opener). A genetic counselor explained the mode of inheritance and said that Peter could take a "predictive" genetic test to find out if he had inherited the dominant mutation. Peter did not want the information about himself, but Martha did not want to have a child who would inherit HD. Martha requested that the fetus be tested but that Peter not be tested. The genetic counselor explained that people under age 18 were discouraged from having predictive testing. Peter was against testing the fetus, pointing out that symptoms do not begin until adulthood, and a treatment might be available 20 or so years in the future.

   a. Why do you think geneticists advise against testing people under 18 years of age? Do you agree or disagree with this practice?

   b. What is the mode of inheritance of HD?

   c. What is the risk that Peter's sister Kate, who is 19, inherited the mutation?

   d. If the fetus could be tested, how might this pose a problem for Peter?

4. When Dr. Hugh Rienhoff, an internist and clinical geneticist, first saw his newborn daughter Bea in 2003, he knew that her long feet, clenched fingers, poor muscle tone, widely spaced eyes, and facial birthmark might be due to a genetic syndrome. As Bea got older but did not gain much weight, appearing thin with birdlike legs, her father recognized signs of two well-known connective tissue diseases, but single-gene tests ruled them out. No one else in the family had any of Bea's characteristics. When pediatricians couldn't put a name to Bea's condition, her father bought second-hand DNA sequencing equipment and searched for an answer himself. "It was eerie examining her DNA, as though I were peering through a powerful microscope looking deep into my daughter while she patiently lay on the microscope stage, looking up, hoping for answers," he wrote.

   After ruling out still other conditions, Dr. Rienhoff turned to exome sequencing. Bea indeed has a mutation in the gene that encodes transforming growth factor β-3, blocking her cells from receiving certain growth signals. The gene is similar to the genes behind the two conditions that Dr. Rienhoff had initially suspected. The parents and brothers are wild type for this gene.

   Today Bea is bright and athletic and looks more lean and lanky than having a medical condition. She will soon be old enough to start thinking about having children someday. What is the likelihood that she can pass her syndrome on to her children?

CHAPTER 5

Beyond Mendel's Laws

Gavin Stevens has a form of Leber congenital amaurosis that was the eighteenth to be discovered, using exome sequencing. He cannot see, but is an amazing musician. Gavin bounced to hiphop as a toddler, could play anything he heard on the piano before he turned 2, and today attends the Academy of Music for the Blind in Los Angeles, where he sings opera.

Courtesy of the Gavin R. Stevens Foundation

Learning Outcomes

5.1 When Gene Expression Appears to Alter Mendelian Ratios

1. Explain the effect of lethal alleles on Mendelian ratios.
2. State how DNA structure underlies multiple gene variants.
3. Distinguish among complete dominance, incomplete dominance, and codominance.
4. Distinguish epistasis from allele interactions.
5. Describe how penetrance, expressivity, and pleiotropy affect gene expression.
6. Explain how a phenocopy can appear to be inherited.

5.2 Mitochondrial Genes

7. Describe the mode of inheritance of a mitochondrial trait.
8. Explain how mitochondrial DNA differs from nuclear DNA.

5.3 Linkage

9. Explain how linked traits are inherited differently from Mendelian traits.
10. Discuss the basis of linkage in meiosis.
11. Explain how linkage is the basis of genetic maps and genome-wide association studies.

The BIG Picture

The modes of inheritance that Gregor Mendel's elegant experiments on peas revealed can be obscured when genes have many variants, interact with each other or the environment, are in mitochondria, or are linked on the same chromosome.

Mutations in Different Genes Cause Blindness

"Genetic heterogeneity" refers to mutations in different genes that cause the same symptoms. This technical concept came to life when, on a sunny summer Saturday in Philadelphia, families with a form of visual loss called Leber congenital amaurosis (LCA) gathered to hear researchers talk about gene therapy that can enable people with one type of the condition to see. At least 21 genes cause LCA.

Jennifer Pletcher was at the conference. Her daughter Finley has a mutation in the gene *RDH12*. Jennifer had met a few LCA families on Facebook, but they had mutations in different genes. At the conference, one of the speakers asked attendees to stand when their mutation was called. When the speaker announced "*RDH12*," Jennifer's family stood, and soon three other families did the same. According to Jennifer, this awesome experience led to friendships with these families, and they now keep in contact with each other about changes in their children.

Not all were able to stand and be counted among those with known mutations. Troy and Jennifer Stevens were there to learn which mutation their then-2-year-old son Gavin had. The parents were to meet with the head of the lab testing their DNA, but Gavin's mutation was not among the known mutations. So parents and child had their exomes sequenced. They hoped that knowing the gene behind the condition might one day lead to developing a gene therapy, which adds a functional gene. Exome sequencing indeed discovered Gavin's mutation, and the foundation his parents started is funding gene therapy research. While Gavin awaits treatment, he reads and writes Braille fluently and attends the Academy of Music for the Blind in California. He can play several instruments and is a masterful singer with amazing stage presence.

5.1 When Gene Expression Appears to Alter Mendelian Ratios

Single genes seldom completely control a phenotype in the way that Mendel's experiments with peas suggested. Genes interact with each other and with environmental influences. When transmission patterns of a visible trait do not exactly fit autosomal recessive or autosomal dominant inheritance, Mendel's laws are still operating. The underlying genotypic ratios persist, but other factors affect the phenotypes.

This chapter considers three general phenomena that seem to be exceptions to Mendel's laws, but are actually not: gene expression, mitochondrial inheritance, and linkage. In several circumstances, phenotypic ratios appear to contradict Mendel's laws, but they do not. Chapter 11 revisits gene expression from a molecular point of view.

Lethal Allele Combinations

A genotype (allele combination) that causes death is, by strict definition, lethal. Death from genetic disease can occur at any stage of development or life. Tay-Sachs disease is lethal by age 3 or 4, whereas Huntington disease may not be lethal until late middle age. In a population and evolutionary sense, a lethal genotype has a more specific meaning—it causes death before the individual can reproduce, which prevents passage of mutations to the next generation.

In organisms used in experiments, such as fruit flies, pea plants, or mice, lethal allele combinations remove an expected progeny class following a specific cross. For example, in a cross of heterozygous flies carrying lethal alleles in the same gene, homozygous recessive progeny die as embryos, leaving only heterozygous and homozygous dominant adult fly offspring. In humans, early-acting lethal alleles cause spontaneous abortion. When both parents carry a recessive lethal allele for the same gene, each pregnancy has a 25 percent chance of spontaneously aborting—that is, 25 percent of embryos are homozygous

recessive. They do not develop further, and therefore this genotype is not seen in any person.

An example of a lethal genotype in humans is achondroplastic dwarfism, which has the distinct phenotype of a long trunk, short limbs, and a large head bearing a flat face (**figure 5.1**). It is an autosomal dominant trait, but is most often the result of a spontaneous (new) mutation. Each child of two people with achondroplasia has a one in four chance of inheriting both mutant alleles; however, because such homozygotes are not seen, this genotype is presumed to be lethal. Each child therefore faces a 2/3 probability of having achondroplasia and a 1/3 probability of being of normal height, illustrating conditional probability. Homozygotes for achondroplasia mutations in other species cannot breathe because the lungs do not have room to inflate. The mutation is in the gene that encodes a receptor for a growth factor. Without the receptor, growth is severely stunted.

Multiple Alleles

A person has two alleles for any autosomal gene—one part of each homologous chromosome. However, a gene can exist in more than two allelic forms in a population because it can mutate in many ways. That is, the sequence of hundreds of DNA bases that makes up a gene can be altered in many ways, just as mistakes can occur anywhere in a written sentence. Different allele combinations can produce variations in the phenotype. The more alleles, the more variations of the phenotype are possible. An individual with two different mutant alleles for the same gene is called a *compound heterozygote.*

For some inherited diseases, knowing the genotype enables a physician to predict the general course of an illness, or which of several symptoms are likely to develop. This is the case for cystic fibrosis (CF) (see figures 1.3 and 1.4 and Clinical Connection 4.1). Some CF genotypes cause frequent, severe respiratory infections; highly congested lungs; and poor weight gain. A different CF genotype increases susceptibility to bronchitis, sinusitis, and pneumonia, and another causes only male infertility.

Different Dominance Relationships

In complete dominance, only one allele is expressed; the other is not. In **incomplete dominance**, the heterozygous phenotype is intermediate between that of either homozygote.

Enzyme deficiencies in which a threshold level is necessary for health illustrate both complete and incomplete dominance. For example, Tay-Sachs disease displays complete dominance because the heterozygote (carrier) is as healthy as a homozygous dominant individual. However, the heterozygote has an intermediate level of enzyme between the homozygous dominant (full enzyme level) and homozygous recessive (no enzyme). Half the normal amount of enzyme is sufficient for health, which is why at the whole-person level, the wild type allele is completely dominant.

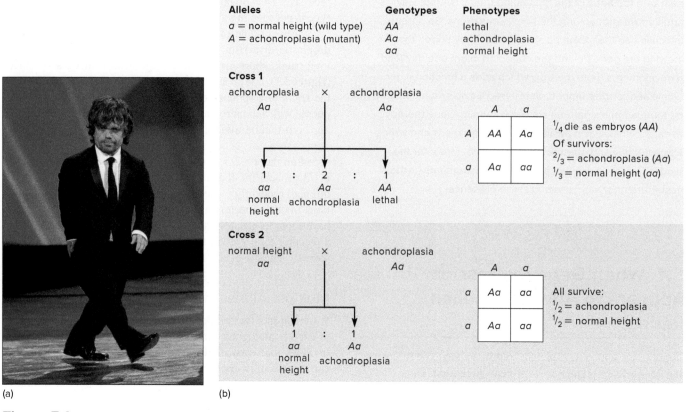

Alleles
a = normal height (wild type)
A = achondroplasia (mutant)

Genotypes
AA
Aa
aa

Phenotypes
lethal
achondroplasia
normal height

Cross 1

achondroplasia × achondroplasia
Aa   Aa

1   :   2   :   1
aa   Aa   AA
normal achondroplasia lethal
height

	A	a
A	AA	Aa
a	Aa	aa

¹/₄ die as embryos (AA)

Of survivors:
²/₃ = achondroplasia (Aa)
¹/₃ = normal height (aa)

Cross 2

normal height × achondroplasia
aa   Aa

1   :   1
aa   Aa
normal achondroplasia
height

	A	a
a	Aa	aa
a	Aa	aa

All survive:
¹/₂ = achondroplasia
¹/₂ = normal height

(a)   (b)

Figure 5.1 **Lethal alleles.** **(a)** A person with achondroplasia has inherited a mutant dominant allele. Inheriting two such alleles is lethal to embryos. **(b)** If two people with achondroplasia have children, each child has a 2/3 chance of inheriting achondroplasia and 1/3 of being normal height. This is a conditional probability because one in four conceptions does not survive. *(a): © Kevin Winter/ Getty Images*

Familial hypercholesterolemia (FH) is an example of incomplete dominance that can be observed in carriers on both the molecular and whole-body levels. A person with two disease-causing alleles does not have receptors on liver cells that take up the low-density lipoprotein (LDL) form of cholesterol from the bloodstream, so it builds up. A person with one disease-causing allele has half the normal number of receptors. Someone with two wild type alleles has the normal number of receptors. **Figure 5.2** shows how measurement of plasma cholesterol reflects these three genotypes. The phenotypes parallel the number of receptors—individuals with two mutant alleles die in childhood of heart attacks, those with one mutant allele may suffer heart attacks in young adulthood, and those with two wild type alleles do not develop this inherited form of heart disease.

Different alleles that are both expressed in a heterozygote are **codominant**. The ABO blood group system is based on the expression of codominant alleles.

Blood types are determined by the patterns of molecules on the surfaces of red blood cells. Most of these molecules are proteins embedded in the plasma membrane with attached sugars that extend from the cell surface. The sugar is an **antigen**, a molecule that the immune system recognizes and responds to. People in blood group A have an allele that encodes an enzyme that adds a piece to a certain sugar attached to the plasma membrane,

producing antigen A. In people with blood type B, the allele and its encoded enzyme are slightly different, which places a different piece on the sugar, producing antigen B. People in blood group AB have both antigen types. Blood group O results from yet a third allele of this gene. It is missing just one DNA nucleotide, but this changes the encoded enzyme in a way that removes the sugar chain from its final piece (**figure 5.3**). As a result, type O red blood cells do not have either A or B antigens.

The A and B alleles are codominant, and both are completely dominant to O. Considering the genotypes reveals how these interactions occur. In the past, ABO blood types have been described as variants of a gene called "I," which stands for isoagglutinin. The three alleles are I^A, I^B, and i. People with blood type A have antigen A on the surfaces of their red blood cells, and may be of genotype $I^A I^A$ or $I^A i$. People with blood type B have antigen B on their red blood cell surfaces, and may be of genotype $I^B I^B$ or $I^B i$. People with the rare blood type AB have both antigens A and B on their cell surfaces, and are genotype $I^A I^B$. People with blood type O have neither antigen, and are genotype ii.

Fiction plots often misuse ABO blood type terminology, assuming that a child's ABO type must match that of one parent. This is not true because a person with type A or B blood can be heterozygous. A person who is genotype $I^A i$ and

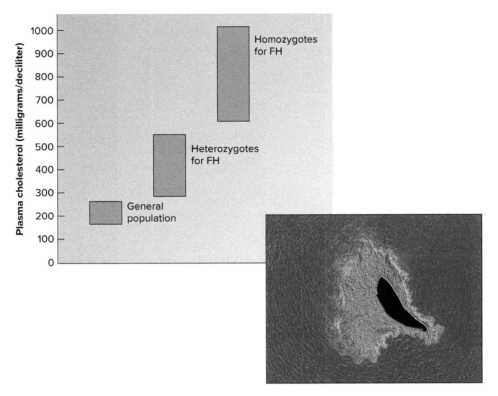

Figure 5.2 Incomplete dominance. A heterozygote for familial hypercholesterolemia (FH) has approximately half the normal number of cell surface receptors in the liver for LDL cholesterol. An individual with two mutant alleles has the severe form of FH, with liver cells that totally lack the receptors. As a result, serum cholesterol level is very high. The photograph shows an artery blocked with cholesterol-rich plaque. Cholesterol is also deposited in joints and many other body parts. © *Alfred Pasieka/Science Photo Library/Science Source*

a person who is $I^B i$ can jointly produce offspring of any ABO genotype or phenotype, as **figure 5.4** illustrates.

Epistasis

Mendel's laws can appear not to operate when one gene masks or otherwise affects the phenotype of another. This phenomenon is called **epistasis**. It refers to interaction between *different* genes, not between the alleles of the same gene. A gene that affects expression of another is called a modifier gene.

In epistasis, the blocked gene is expressed (transcribed into RNA) normally, but the protein product of the modifier gene inactivates it, removes a structure needed for it to contribute to the phenotype, or otherwise counteracts its effects. A familiar epistatic interaction is albinism, in which one gene blocks the action of genes that confer color (see figure 4.15). Albinism is seen in many species.

A blood type called the Bombay phenotype also illustrates epistasis. It results from an interaction between a gene called *H* and the *I* gene that confers ABO blood type. The *H* gene controls the placement of a molecule to which antigens A and B attach on red blood cell surfaces. A person of genotype *hh* can't make that molecule, so the A and B antigens cannot attach to red blood cell surfaces. The A and B antigens fall off and the person tests as type O, but may be any ABO genotype.

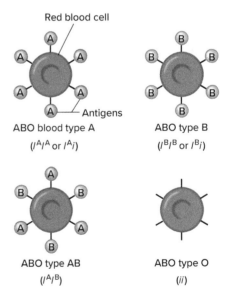

Figure 5.3 ABO blood types illustrate codominance. ABO blood types are based on antigens on red blood cell surfaces. This depiction greatly exaggerates the size of the A and B antigens. Genotypes are in parentheses.

Epistasis can explain why siblings who inherit the same disease can suffer to differing degrees. One study examined siblings who both inherited spinal muscular atrophy (SMA)

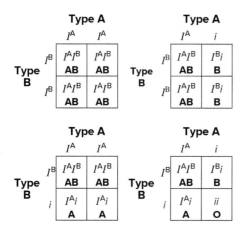

Figure 5.4 **Codominance.** The I^A and I^B alleles of the *I* gene are codominant, but they follow Mendel's law of segregation. These Punnett squares show the genotypes that could result when a person with type A blood has children with a person with type B blood.

type 1, in which nerves cannot signal muscles. The muscles weaken and atrophy, usually proving fatal in early childhood. The mutation encodes an abnormal protein that shortens axons, which are the extensions on nerve cells that send messages. Some siblings who inherited the SMA genotype, however, never developed symptoms. They can thank a variant of another gene, *plastin 3*, which increases production of the cytoskeletal protein actin that extends axons. Because the healthy siblings inherited the ability to make extra long axons, the axon-shortening effects of SMA were not harmful.

Penetrance and Expressivity

The same genotype can produce different degrees of a phenotype in different individuals because of influences of other genes, as well as environmental influences such as nutrition, exposure to toxins, and stress. This is why two individuals who have the same *CF* genotype may have different clinical experiences. One person may be much sicker because she also inherited gene variants predisposing her to develop asthma and respiratory allergies. Two terms describe the degrees of expression of a single gene. **Penetrance** refers to the percentage of individuals who have a particular genotype who have the associated phenotype. **Expressivity** refers to variability in severity of a phenotype, or the extent to which the gene is expressed.

An allele combination that produces a phenotype in everyone who inherits it is completely penetrant. Huntington disease (see the opener to chapter 4) is nearly completely penetrant. Almost all people who inherit the mutant allele will develop symptoms if they live long enough. Complete penetrance is rare.

A genotype is incompletely penetrant if some individuals do not express the phenotype (have no symptoms). Polydactyly (see figure 1.5) is incompletely penetrant. Some people who inherit the dominant allele have more than five digits on a hand or foot. Yet others who must have inherited the allele because they have an affected parent and child have ten fingers and ten toes. Penetrance is described numerically. If 80 of 100 people who inherit the dominant polydactyly allele have extra digits, the genotype is 80 percent penetrant.

A phenotype is variably expressive if symptoms vary in intensity among different people. One person with polydactyly might have an extra digit on both hands and a foot, but another might have just one extra fingertip. Polydactyly is both incompletely penetrant and variably expressive.

Pleiotropy

A single-gene disease with many symptoms, or a gene that controls several functions or has more than one effect, is termed **pleiotropic**. Such conditions can be difficult to trace through families because people with different subsets of symptoms may appear to have different diseases.

On a molecular level, pleiotropy occurs when a single protein affects different body parts, participates in more than one biochemical reaction, or has different effects in different amounts. Consider Marfan syndrome. The most common form of this autosomal dominant condition is a defect in an elastic connective tissue protein called fibrillin. The protein is abundant in the lens of the eye, in the aorta (the largest artery in the body, leading from the heart), and in bones of the limbs, fingers, and ribs. The symptoms are lens dislocation, long limbs, spindly fingers, and a caved-in chest (**figure 5.5**). The most serious symptom is a weakening in the aorta, which can suddenly burst. If the weakening is detected early, a synthetic graft can replace the section of artery wall and save the person's life.

Genetic Heterogeneity

Mutations in different genes that produce the same phenotype lie behind **genetic heterogeneity**. It can occur when genes encode enzymes or other proteins that are part of the same biochemical pathway, or when proteins affect the same body part, such as the visual loss conditions described in the chapter opener. Genetic heterogeneity may make it appear that Mendel's laws are not operating, even though they are. The different forms of Leber congenital amaurosis arise because there are many ways that a mutation can disrupt the functioning of the rods and cones, the cells that provide vision (**figure 5.6**). If a man who is homozygous recessive for a mutation in one of the genes that causes the condition has a child with a woman who is homozygous recessive for a different gene, then the child would not inherit either form of blindness because he or she would be heterozygous for both genes.

Discovering additional genes that can cause a known disease is happening more often as the human genome is analyzed, and such discoveries can have practical repercussions. Consider osteogenesis imperfecta, in which abnormal collagen causes fragile bones. Before a second causative gene was discovered, some parents of children who were brought to the hospital with frequent fractures and who did not have a mutation in the one known gene were accused of child abuse. Today eight genetically distinct forms of the disease are recognized.

Figure 5.5 **Marfan syndrome is pleiotropic.** Jonathan Larson died just before the opening of the Broadway play that he wrote, *Rent,* when his aorta tore apart. He had gone to two New York City hospitals complaining of chest pain, but no one recognized the symptoms of Marfan syndrome. Mr. Larson's long thin face is characteristic of the syndrome. © *Columbia University, ho/AP Images*

Clinical Connection 5.1 describes a common disease whose familial forms are genetically heterogeneic—Alzheimer disease.

Phenocopies

An environmentally caused trait that appears to be inherited is a **phenocopy**. Such a trait can either produce symptoms that resemble those of a known single-gene disease or mimic inheritance patterns by affecting certain relatives. For example, the limb birth defect caused by the drug thalidomide, discussed in chapter 3, is a phenocopy of the rare inherited illness phocomelia. Physicians recognized the environmental disaster when they began seeing many children born with what looked like phocomelia. A birth defect caused by exposure to a widely used teratogen was more likely than a sudden increase in incidence of a rare inherited disease.

An infection can be a phenocopy if it affects more than one family member. Children who have AIDS may have

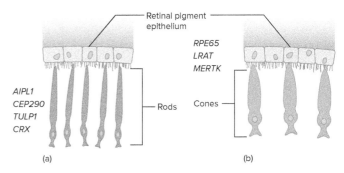

Figure 5.6 **Many routes to blindness.** Mutations in more than 100 genes cause degeneration of the retina. This multilayered structure at the back of the eye includes the rods and cones, which are cells that signal incoming light to the brain (photoreceptors). This illustration indicates seven of these genes (in italics). Their encoded proteins have various functions: activating vitamin A, cleaning up debris, providing energy, and maintaining the functioning of the rods and cones.

parents who also have the disease, but these children acquired AIDS by viral infection, not by inheriting a mutation.

Common symptoms may resemble those of an inherited condition, but be due to an environmental situation. For example, an underweight child who has frequent colds may show some signs of cystic fibrosis, but may instead suffer from malnutrition.

The Human Genome Sequence Adds Perspective

As researchers describe more genes from the human genome, some of the phenomena just discussed are turning out not to be rare, as had once been thought. Terms such as *epistasis* and *genetic heterogeneity* are beginning to overlap and blur. For example, most people who have Marfan syndrome have a mutation in the fibrillin gene. However, some people with the syndrome instead have a mutation in the gene that encodes the transforming growth factor beta receptor (TGFβR). Fibrillin and TGFβR are part of the same biochemical pathway. So, Marfan syndrome fits the definition of genetic heterogeneity because mutations in different genes cause identical symptoms. Yet the forms of the syndrome are also epistatic because a variant TGFβR blocks the activity of fibrillin.

Gene interactions also underlie penetrance and expressivity because even genes that do not directly interact, in space or time, can affect each other's expression. This is the case for Huntington disease, in which cells in a certain part of the brain die. Siblings who inherit the same HD mutation may differ in the number of cells that they have in the affected brain area because of variants of other genes that affected the division rate of neural stem cells in the brain during embryonic development. As a result, an individual who inherits HD, but also extra brain cells, might develop symptoms much later in life than a brother or sister who does not have such a built-in reserve supply. If the delay is long enough that death comes from another cause, the HD mutation would then be nonpenetrant.

Technologies that reveal gene expression patterns in different tissues are painting detailed portraits of pleiotropy, showing that inherited diseases may affect more tissues or organs than are obvious from symptoms. Finally, more cases of genetic heterogeneity are being discovered as researchers identify genes with redundant or overlapping functions.

Table 5.1 summarizes phenomena that appear to alter single-gene inheritance. Our definitions and designations are changing as improving technology enables us to describe and differentiate diseases in greater detail. Phenomena such as variable expressivity, incomplete penetrance, epistasis, pleiotropy, and genetic heterogeneity, once considered unusual characteristics of single genes, may turn out to be the norm.

Clinical Connection 5.1

The Roots of Familial Alzheimer Disease

"What's that on that tiny screen?" asked 82-year-old Ginny for the fifth time in half an hour as her son glanced down at his phone.

"Mom, don't you know you've asked me that several times since I got here? It's my wallpaper, you know, my cat Trouser," he said, showing her the photo of the orange cat on his mobile phone and trying not to become impatient.

"No, I've never asked you that before." She paused, looking puzzled. "What did you say his name is?"

In the months following that conversation, Ginny's short-term memory declined further. She could rarely concentrate long enough to finish reading an article or follow a conversation. In the grocery store, she couldn't find items she'd been buying for decades. Aware of her growing deficits, she became depressed. Finally, her son suggested she have a complete neurological exam. By the time she could see a physician, other signs had emerged. Ginny couldn't recall her zip code or the name of the small town where she grew up. Sometimes she couldn't remember where things belonged—she put a cantaloupe in the bathtub.

Ginny had "mild cognitive impairment" (MCI). It could be the start of Alzheimer disease, which affects 26 million people worldwide. After ruling out a vitamin deficiency and lack of oxygen to the brain as causes of Ginny's forgetfulness, a neurologist started her on a drug to slow breakdown of the neurotransmitter acetylcholine in the brain. The doctor also prescribed an antidepressant, which revived Ginny enough so that she was more willing to leave her apartment. If the MCI progressed to Alzheimer disease, Ginny would continue to lose thinking, reasoning, learning, and communicating abilities, and one day would not recognize her loved ones. Eventually, she would no longer speak and smile, and would cease walking, not because her legs wouldn't function, but because she would forget how to walk. Yet the haze would sometimes seem to lift from her eyes and the old Ginny would return for a precious moment or two.

In Alzheimer disease, certain brain parts—the amygdala (seat of emotion) and hippocampus (the memory center)—become

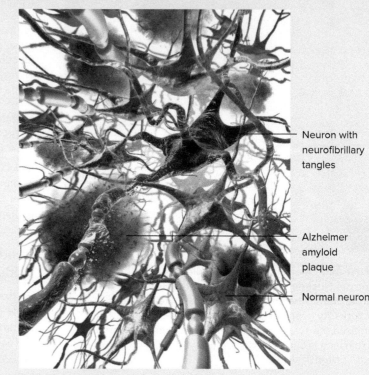

Neuron with neurofibrillary tangles

Alzheimer amyloid plaque

Normal neuron

Figure 5A **In Alzheimer disease, amyloid plaques form outside neurons and neurofibrillary tangles form inside the cells.** © Marc Phares/Science Source

buried in two types of protein. Amyloid precursor protein normally converts iron into a safe form. In Alzheimer disease, the protein is cut into unusually sized amyloid beta peptides, which aggregate to form "plaques" outside brain cells because cells cannot remove the material fast enough. At the same time, unsafe iron accumulates inside brain neurons, and zinc builds up in the plaques outside them. Ginny's brain scan showed plaques accumulating. The second type of protein, called tau, was building up, too, clumping into "tangles" inside brain neurons (**figure 5A**). The telltale plaques and tangles, present to a lesser extent in everyone, could cause the symptoms of Alzheimer disease or result from them. Either way, their abundance in a cerebrospinal fluid test, combined with Ginny's cognitive problems, strongly suggested Alzheimer disease.

(Continued)

Table 5A — Genes Associated with Alzheimer Disease

Causative Gene	Chromosome	Mechanism
Amyloid precursor protein (*APP*)	21	Unusually sized pieces aggregate outside brain cells.
Calcium homeostasis modulator 1 (*CALHM1*)	10	Controls amyloid cutting.
Presenilin 1 (*PSEN1*)	14	Forms part of secretase (enzyme) that cuts APP.
Presenilin 2 (*PSEN2*)	1	Forms part of secretase (enzyme) that cuts APP.
Risk Gene		
Apolipoprotein E4 (*APOE4*)	19	Unusually sized pieces add phosphates to tau protein, making it accumulate and impairing microtubule binding.
Triggering receptor expressed on myeloid cells (*TREM*)	6	Promotes brain inflammation.

Fewer than 1 percent of Alzheimer cases are familial (inherited), caused by mutations in any of several genes that affect amyloid accumulation or clearance (**table 5A**). However, some variants of genes associated with Alzheimer disease are actually protective. One variant of the gene *APOE* increases the risk of developing a late-onset form threefold in a heterozygote and fifteenfold in a homozygote, yet a different variant *lowers* risk. Other genes increase susceptibility but raise risk to such a small degree that they have little predictive value.

Questions for Discussion

1. Why is Alzheimer disease difficult to diagnose, especially in the early stage?
2. How is the genetics of Alzheimer disease complex?
3. Describe the protein abnormalities that occur in Alzheimer disease.
4. How common are inherited forms of Alzheimer disease?

Table 5.1 — Factors That Alter Single-Gene Phenotypic Ratios

Phenomenon	Effect on Phenotype	Example
Lethal alleles	A phenotypic class does not survive to reproduce.	Achondroplasia
Multiple alleles	Many variants or degrees of a phenotype are possible.	Cystic fibrosis
Incomplete dominance	A heterozygote's phenotype is intermediate between those of the two homozygotes.	Familial hypercholesterolemia
Codominance	A heterozygote's phenotype is distinct from and not intermediate between those of the two homozygotes.	ABO blood types
Epistasis	One gene masks or otherwise affects another's phenotype.	Bombay phenotype
Penetrance	Some individuals with a particular genotype do not have the associated phenotype.	Polydactyly
Expressivity	A genotype is associated with a phenotype of varying intensity.	Polydactyly
Pleiotropy	The phenotype includes many symptoms, with different subsets in different individuals.	Marfan syndrome
Phenocopy	An environmentally caused condition has symptoms and a recurrence pattern similar to those of a known inherited trait.	Infection
Genetic heterogeneity	Genotypes of different genes cause the same phenotype.	Osteogenesis imperfecta

Gregor Mendel derived the two laws of inheritance working with traits conferred by genes located on different chromosomes in the nucleus. When genes are on the same chromosomes, however, the associated traits may not appear in Mendelian ratios. The rest of this chapter considers two types of gene transmission that are not Mendelian—mitochondrial inheritance and linkage.

5.2 Mitochondrial Genes

Mitochondria are the cellular organelles that house the reactions that derive energy from nutrients (see figure 2.7). Each of the hundreds to thousands of mitochondria in each human cell contains several copies of a "mini-chromosome" that carries just 37 genes. Genes encoded in mitochondrial DNA (mtDNA) act in the mitochondrion, but the organelle also requires the activities of certain genes from the nucleus.

The inheritance patterns and mutation rates for mitochondrial genes differ from those for genes in the nucleus. Rather than being transmitted equally from both parents, mitochondrial genes are maternally inherited. They are passed only from an individual's mother because the sperm head, which enters an oocyte at fertilization, does not include mitochondria, which are found in the sperm midsection, where they provide energy for moving the tail. In the rare instances when mitochondria from sperm enter an oocyte, they are usually selectively destroyed early in development. Pedigrees that follow mitochondrial genes therefore show a woman passing the trait to all her children, whereas a male cannot pass the trait to any of his (**figure 5.7**).

DNA in the mitochondria differs functionally from DNA in the nucleus in several ways (**figure 5.8**):

- MtDNA does not cross over.
- It mutates faster than DNA in the nucleus because fewer types of DNA repair are available and DNA-damaging oxygen free radicals are produced in the energy reactions.
- Mitochondrial genes are not wrapped in proteins.
- Mitochondrial genes are not "interrupted" by DNA sequences that do not encode protein.
- A cell has many mitochondria and each mitochondrion contains several copies of the mitochondrial genome.
- Mitochondria with different alleles for the same gene can reside in the same cell.

Mitochondrial Diseases

Mitochondrial genes encode proteins that participate in protein synthesis and energy production. Twenty-four of the 37 genes encode RNA molecules (22 transfer RNAs and 2 ribosomal RNAs) that help assemble proteins. The other 13 mitochondrial genes encode proteins that function in cellular respiration, which is the process that uses energy from digested nutrients to synthesize ATP, the biological energy molecule.

A class of diseases results from mutations in mitochondrial genes. They are called mitochondrial myopathies and have specific names, but news reports often lump them together as "mitochondrial disease." Symptoms arise from tissues whose cells normally have many mitochondria, such as skeletal muscle, and include great fatigue, weak and flaccid muscles, and intolerance to exercise. Skeletal muscle fibers appear "red and ragged" when stained and viewed under a light microscope, their abundant abnormal mitochondria visible beneath the plasma membrane (**figure 5.9**). Diseases considered to be mitochondrial may also result from mutations in nuclear genes that encode proteins that are essential for mitochondrial function.

A mutation in a mitochondrial gene that encodes a tRNA or rRNA can be devastating because it impairs the organelle's general ability to manufacture proteins. Consider what happened to Lindzy, a once active and articulate dental hygienist. In her forties, Lindzy gradually began to slow down at work. She heard a buzzing in her ears and developed difficulty talking and walking. Then her memory began to fade in and out, she became lost easily in familiar places, and her conversation made no sense. Her condition worsened,

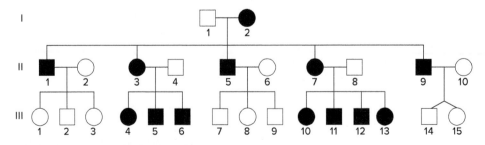

Figure 5.7 Inheritance of mitochondrial genes. Mothers pass mitochondrial genes to all offspring. Fathers do not transmit mitochondrial genes because sperm only rarely contribute mitochondria to fertilized ova. If mitochondria from a male do enter, they are destroyed.

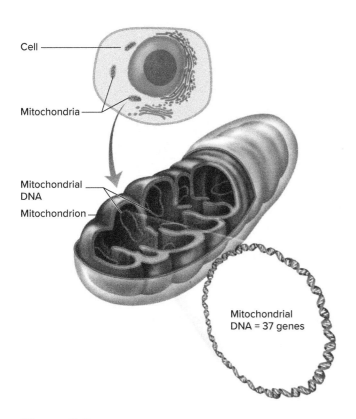

Figure 5.8 **Mitochondrial DNA.** A mitochondrion contains several rings of DNA. Different alleles can reside on different copies of the mitochondrial "mini-chromosome." A typical cell has hundreds to thousands of mitochondria, each of which has many copies of its mini-chromosome.

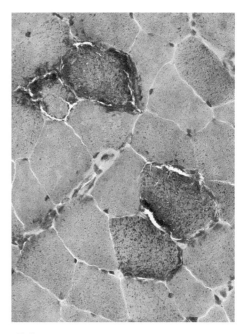

Figure 5.9 **Red-ragged fibers are a hallmark of a mitochondrial disease.** These skeletal muscle cells are from an individual with MELAS (mitochondrial myopathy encephalopathy lactic acidosis syndrome). *Courtesy of Alan Pestronk, MD*

and she developed diabetes, seizures, and pneumonia and became deaf and demented. She was finally diagnosed with MELAS, which stands for "mitochondrial myopathy encephalopathy lactic acidosis syndrome." Her muscle cells had the telltale red-ragged fibers. Lindzy died. Her son and daughter will likely develop the condition because they inherited her mitochondria.

About 1 in 200 people has a mutation in a mitochondrial gene that could cause disease. However, mitochondrial diseases are rare, affecting about 1 in 6,500 people, apparently because of a weeding-out process during egg formation. A mitochondrial mutation may disrupt energy acquisition so greatly in an oocyte that it cannot survive.

Theoretically, a woman with a mitochondrial disease can avoid transmitting it to her children if her mitochondria can be replaced with healthy mitochondria from a donor. **Bioethics** describes two ways to do this, and notes the controversy over creating "three-parent" children.

Heteroplasmy

The fact that a cell contains many mitochondria makes possible a condition called **heteroplasmy**, in which a mutation is in some mitochondrial chromosomes, but not others. At each cell division, the mitochondria are distributed at random into daughter cells. Over time, the chromosomes within a mitochondrion tend to be all wild type or all mutant for any particular gene, but different mitochondria can have different alleles predominating. As an oocyte matures, the number of mitochondria drops from about 100,000 to 100 or fewer. If the woman is heteroplasmic for a mutation, by chance, she can produce an oocyte that has mostly mitochondria that are wild type, mostly mitochondria that have the mutation, or anything in between (**figure 5.10**). In this way, a woman who does not have a mitochondrial disease, because the mitochondria bearing the mutation are either rare or not abundant in affected cell types, can nevertheless pass the associated condition to a child. Therefore, mitochondrial inheritance is both complex and unpredictable.

Heteroplasmy has several consequences on phenotypes. Expressivity may vary widely among siblings, depending upon how many mutation-bearing mitochondria were in the oocyte that became each individual. Severity of symptoms reflects which tissues have cells whose mitochondria bear the mutation. This is the case for a family with Leigh syndrome, which affects the enzyme that directly produces ATP. Two boys died of the severe form of the disease, because the brain regions that control movement rapidly degenerated. Another sibling was blind and had central nervous system degeneration. Several relatives, however, suffered only mild impairment of their peripheral vision. The more severely affected family members had more brain cells that received the mutation-bearing mitochondria.

The most severe mitochondrial illnesses are heteroplasmic. This is presumably because *homo*plasmy—when all mitochondria bear the mutant allele—too severely impairs protein

Replacing Mitochondria

The tiny mitochondrial genome of 16,569 DNA base pairs can cause severe illness when any of its 37 genes, which encode a few proteins, rRNAs, and tRNAs, is mutant. A way to prevent a mitochondrial disease, theoretically, is to intervene at conception and replace the mitochondria in an oocyte from a woman who has a mitochondrial mutation with mitochondria that do not have mutations.

Researchers replace mitochondria in several ways: They (1) transfer the male and female pronuclei from an *in vitro* fertilized ovum into an oocyte donated from a healthy woman that has had its nucleus removed, so that the pronuclei share a cytoplasm with donor mitochondria; (2) transfer the female pronucleus into a donated oocyte with its nucleus removed and then fertilize that cell *in vitro* or (3) transfer the spindle apparatus, to which mitochondria cling, from a donor cell to a fertilized ovum. However it happens, the resulting first cell has two haploid genomes plus the small amount of mitochondrial DNA from the donor, a third individual. In 2016 a boy was born using the second method, and so far is healthy. Two siblings had died from Leigh syndrome.

Mitochondrial replacement has worked in monkeys but in other animals has led to developmental problems, including accelerated aging, poor growth, lowered fertility, and early death. Some researchers caution that the rarity of mitochondrial diseases, compared to the high mutation rate of mitochondrial DNA, suggests that embryos that inherit the diseases are naturally weeded out. Another complicating factor may be epigenetic effects from a cytoplasm on a nucleus from a different individual.

When the U.S. Food and Drug Administration held a meeting to discuss the science behind mitochondrial manipulation before the first child was born from such a procedure, a public comment session dealt with the **bioethics** of creating a "three-parent" embryo. Another objection came from a young woman who has suffered with a mitochondrial disease for more than a decade. She suggested that research regarding mitochondrial disease try to help those who are severely affected by it.

Questions for Discussion

1. Do you think that a person resulting from mitochondrial replacement would actually have three parents, as many popular accounts of the technology claim? If so, is creation of such an individual unethical, and why?
2. Is mitochondrial replacement necessary?
3. What are the potential risks and benefits of mitochondrial replacement?

---

synthesis or energy production for embryonic development to complete. Often, severe heteroplasmic mitochondrial diseases do not produce symptoms until adulthood, because it takes many cell divisions, and therefore years, for a cell to receive enough mitochondria bearing mutant alleles to cause symptoms.

A Glimpse of History describes the effect of heteroplasmy on solving a royal mystery.

About one in ten of the DNA bases in the mitochondrial genome are heteroplasmic—that is, they can differ within an individual. The mutations that generate these single-base

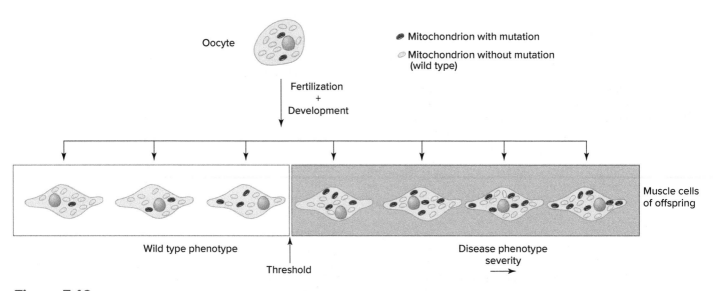

Figure 5.10 **Mitochondrial inheritance.** Mitochondria and their genes are passed only from the mother. Cells have many mitochondria. If an oocyte is heteroplasmic, differing numbers of copies of a mitochondrial mutation may be transmitted. The phenotype arises from the proportion of mitochondria bearing the mutation.

On a July night in 1918, Tsar Nicholas II of Russia and his family, the royal Romanovs, were shot, their bodies damaged with acid, and buried. In another July, in 1991, two amateur historians found the grave. DNA testing revealed the number of buried people, Y chromosome sequences distinguished males from females, and mitochondrial DNA sequences identified a mother (the Tsarina) and three daughters. When researchers consulted the DNA of modern-day relatives to link the remains to the royal family, they encountered a problem: The remains of the suspected Tsar and his living great-grandniece Xenia differed at nucleotide position 16169 in the mitochondrial DNA. More puzzling, retesting the remains showed that at this site in the mitochondrial genome, the purported Tsar had in some samples thymine (T), yet in others cytosine (C). In yet another July, in 1994, researchers dug up the body of Nicholas's brother, the Grand Duke Georgij Romanov, and solved the mystery. Mitochondrial DNA nucleotide position 16169 in bone cells from the Grand Duke also had either T or C. The family had heteroplasmy.

variations in mtDNA probably occur all the time, but only those that originate early in development have a chance to accumulate enough to be detectable.

Mitochondrial DNA Reveals the Past

Interest in mtDNA extends beyond the medical. In forensics it is used to link suspects to crimes. The forensic usefulness of mtDNA is based on the fact that it is more likely than nuclear DNA to remain after extensive damage, because cells have many copies of it.

Analysis of mtDNA can also identify war dead and support or challenge historical records. Sequencing mtDNA identified the son of Marie Antoinette and Louis XVI, who supposedly died in prison at age 10. In 1845, the boy was given a royal burial, but some people thought the buried child was an imposter. His heart had been stolen at the autopsy, and through a series of bizarre events, wound up, dried out, in the possession of the royal family. A few years ago, researchers compared mtDNA sequences from cells in the boy's heart to corresponding sequences in heart and hair cells from Marie Antoinette (her decapitated body identified by her fancy underwear), two of her sisters, and living relatives Queen Anne of Romania and her brother. The mtDNA evidence showed that the buried boy was indeed the prince, Louis XVII. Chapter 16 discusses how researchers consult mtDNA sequences to reconstruct ancient migration patterns.

5.3 Linkage

The genes responsible for the traits that Mendel studied in pea plants were on different chromosomes. When genes are close to each other on the same chromosome, they usually do not segregate at random during meiosis and therefore their expression does not support Mendel's predictions. Instead, genes close on a chromosome are packaged into the same gametes and are said to be "linked" (**figure 5.11**). Linkage has this precise meaning in genetics. The term is popularly used to mean any association between two events or observations.

Linkage in genetics refers to the transmission of genes on the same chromosome. Linked genes do *not* assort independently and do *not* produce Mendelian ratios for crosses tracking two or more genes. Understanding and using linkage as a mapping tool helped to identify many disease-causing genes before genome sequencing became possible.

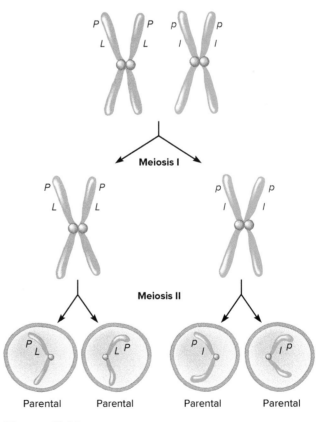

Figure 5.11 **Inheritance of linked genes.** Genes linked closely to one another are usually inherited together when the chromosome is packaged into a gamete.

Discovery in Pea Plants

William Bateson and R. C. Punnett first observed the unexpected ratios indicating linkage in the early 1900s, again in pea plants. They crossed true-breeding plants with purple flowers and long pollen grains (genotype *PPLL*) to true-breeding plants with red flowers and round pollen grains (genotype *ppll*). The plants in the next generation, of genotype *PpLl*, were then self-crossed. But this dihybrid cross did not yield the expected 9:3:3:1 phenotypic ratio that Mendel's second law predicts (**figure 5.12**).

Bateson and Punnett noticed that two types of third-generation peas, those with the parental phenotypes *P_L_* and *ppll*, were more abundant than predicted, while the other two progeny classes, *ppL_* and *P_ll*, were less common (the blank indicates that the allele can be dominant or recessive). The more prevalent parental allele combinations, Bateson and Punnett hypothesized, could be due to genes that are transmitted on the same chromosome and that therefore do not separate during meiosis. The two less common offspring classes could also be explained by a meiotic event, crossing over. Recall that crossing over is an exchange between homologs that mixes up maternal and paternal gene combinations without disturbing the sequence of genes on the chromosome (**figure 5.13**).

Progeny that exhibit this mixing of maternal and paternal alleles on a single chromosome are called **recombinant**. *Parental* and *recombinant* are context-dependent terms. Had the parents in Bateson and Punnett's crosses been of genotypes *ppL_* and *P_ll*, then *P_L_* and *ppll* would be recombinant rather than parental classes.

Two other terms describe the configurations of linked genes in dihybrids. Consider a pea plant with genotype *PpLl*. These alleles can be part of the chromosomes in either of two ways. If the two dominant alleles are on one chromosome and the two recessive alleles on the other, the genes are in *"cis."* In the opposite configuration, with one dominant and one recessive allele on each chromosome, the genes are in *"trans"* (**figure 5.14**). Whether alleles in a dihybrid are in *cis* or *trans* is important in distinguishing recombinant from parental progeny classes in specific crosses and in clinical cases.

Linkage Maps

As Bateson and Punnett were discovering linkage in peas, geneticist Thomas Hunt Morgan and his coworkers in Columbia University's "fly room" were doing the same using the fruit fly, *Drosophila melanogaster.* They assigned genes to relative positions on chromosomes and compared progeny class sizes to assess whether traits were linked. The pairs of traits fell into four groups. Within each group, crossed dihybrids did not produce offspring classes according to Mendel's second law. Also, the number of linkage groups—four—matched the number of chromosome pairs in the fly. Coincidence? No. The traits fell into four groups because their genes are inherited together on the same chromosome.

The genius of the work on linkage in fruit flies was twofold. First, the researchers used test crosses (see figure 4.5) to follow parental versus recombinant progeny. Second, the

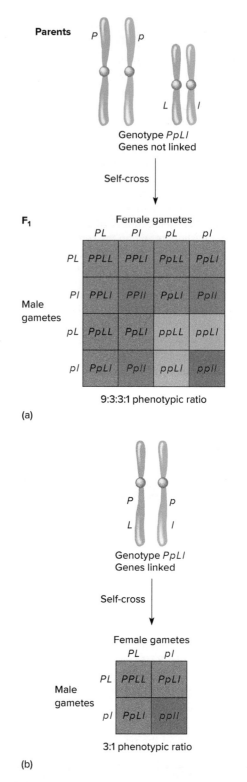

Figure 5.12 Expected results of a dihybrid cross. (a) Unlinked genes assort independently. The gametes represent all possible allele combinations. The expected phenotypic ratio of a dihybrid cross is 9:3:3:1. **(b)** If genes are linked, only two allele combinations are expected in the gametes. The phenotypic ratio is 3:1, the same as for a monohybrid cross.

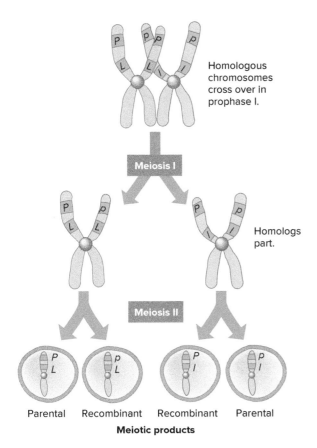

Homologous chromosomes cross over in prophase I.

Meiosis I

Homologs part.

Meiosis II

Parental   Recombinant   Recombinant   Parental

Meiotic products

Figure 5.13 Crossing over disrupts linkage. The linkage between two genes may be interrupted if the chromosome they are on crosses over with its homolog between the two genes. Crossing over packages recombinant groupings of the genes into gametes.

fly room investigators translated their data into actual maps depicting positions of genes on chromosomes.

The idea for a genetic map grew out of Morgan's observation that the sizes of the recombinant classes varied for different genes. In 1911 he hypothesized that the farther apart

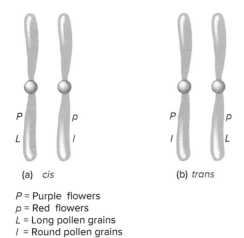

(a) *cis*   (b) *trans*

P = Purple flowers
p = Red flowers
L = Long pollen grains
l = Round pollen grains

Figure 5.14 Allele configuration is important. Parental chromosomes can be distinguished from recombinant chromosomes only if the allele configuration of the two genes is known—they are either in *cis* **(a)** or in *trans* **(b)**.

two genes are on a chromosome, the more likely they are to cross over simply because more physical distance separates them (**figure 5.15**). He proposed that the size of a recombinant class is directly proportional to the distance between the genes on the chromosome they share. Then undergraduate student Alfred Sturtevant devised a way to represent the correlation between crossover frequency and the distance between genes as a **linkage map**. These diagrams showed the order of genes on chromosomes and the relative distances between them. The distance was represented using "map units" called centimorgans, where 1 centimorgan equals 1 percent recombination. These units are still used to estimate genetic distance along a chromosome. Linkage maps of human genes were important in the initial sequencing of the human genome.

The frequency of a crossover between any two linked genes is inferred from the proportion of offspring from a cross that are recombinant. Frequency of recombination is based on the percentage of meiotic divisions that break the linkage between—that is, separate—two parental alleles. Genes at opposite ends of the same chromosome cross over often, generating a large recombinant class. Genes lying very close on the chromosome would only rarely be separated by a crossover. The probability that genes on opposite ends of a chromosome cross over approaches the probability that, if on different chromosomes, they would independently assort—about 50 percent. **Figure 5.16** illustrates this distinction.

The situation with linked genes is like a street lined with stores on both sides. There are more places to cross the street between stores at opposite ends on opposite sides than between two stores in the middle of the block on opposite sides of the street. Similarly, more crossovers, or progeny with recombinant genotypes, are seen when two genes are farther apart on the same chromosome.

Geneticists at Columbia University mapped several genes on all four fruit fly chromosomes, while researchers in other labs assigned genes to the human X chromosome. Localizing genes on the X chromosome was easier than doing so on the autosomes, because X-linked traits follow an inheritance pattern that is distinct from the one all autosomal genes follow. In human males, with their single X chromosome, recessive alleles on the X are expressed and observable. Chapter 6 returns to this point.

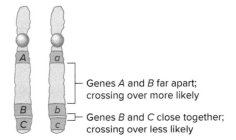

Genes *A* and *B* far apart; crossing over more likely

Genes *B* and *C* close together; crossing over less likely

Figure 5.15 Breaking linkage. Crossing over is more likely between widely spaced linked genes *A* and *B,* or between *A* and *C,* than between more closely spaced linked genes *B* and *C* because there is more room for an exchange to occur.

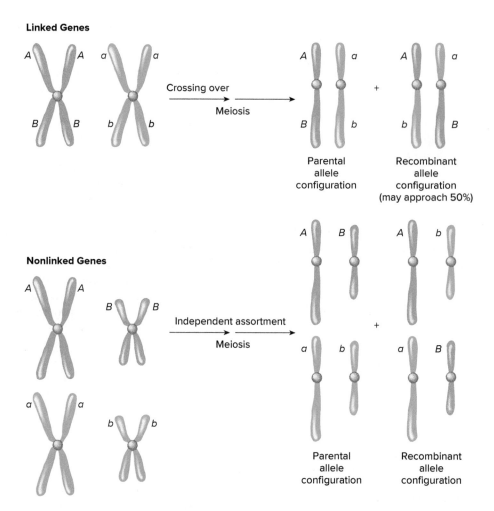

Linked Genes

Crossing over
Meiosis

Parental allele configuration

Recombinant allele configuration (may approach 50%)

Nonlinked Genes

Independent assortment
Meiosis

Parental allele configuration

Recombinant allele configuration

Figure 5.16 Linkage versus nonlinkage (independent assortment). When two genes are widely separated on a chromosome, the likelihood of a crossover is so great that the recombinant class may approach 50 percent—which may appear to be the result of independent assortment.

By 1950, geneticists began to think about mapping genes on the 22 human autosomes. To start, a gene must be matched to its chromosome. This became possible when geneticists identified people with a particular inherited condition or trait and an unusual chromosome.

In 1968, researcher R. P. Donohue was looking at chromosomes in his own white blood cells when he noticed a dark area consistently located near the centromere of one member of his largest chromosome pair (chromosome 1). He examined chromosomes from several family members for the dark area, noting also whether each family member had a blood type called Duffy. (Recall that blood types refer to the patterns of sugars on red blood cell surfaces.) Donohue found that the Duffy blood type was linked to the chromosome variant. That is, he could predict a relative's Duffy blood type by whether or not the chromosome had the dark area. This was the first assignment of a trait in humans to an autosome.

Finding a chromosomal variation linked to a family trait like Donohue did was unusual. More often, researchers mapped genes in experimental organisms, such as fruit flies, by calculating percent recombination (crossovers) between two genes with known locations on a chromosome. However, because human parents do not have hundreds of offspring, nor do they produce a new generation every 10 days, getting enough information to establish linkage relationships for us requires observing the same traits in many families and pooling the results. Today, even though we can sequence human genomes, linkage remains a powerful tool to track disease-associated genes.

Solving Linkage Problems Uses Logic

To illustrate determining the degree of linkage by percent recombination using several families, consider the traits of Rh blood type and a form of anemia called elliptocytosis. An Rh-positive phenotype corresponds to genotypes *RR* or *Rr*. (This is simplified.) The anemia corresponds to genotypes *EE* or *Ee*.

In 100 one-child families, one parent is Rh negative with no anemia (*rree*), and the other parent is Rh positive with anemia (*RrEe*), and the *R* and *E* (or *r* and *e*) alleles are in *cis*. Of the 100 offspring, 96 have parental genotypes (*re/re* or *RE/re*) and four are recombinants for these two genes (*Re/re* or *rE/re*). Percent recombination is therefore 4 percent, and the two linked genes are 4 centimorgans apart.

Another pair of linked genes in humans is the gene that underlies nail-patella syndrome, a rare autosomal dominant disease that causes absent or underdeveloped fingernails and toenails and painful arthritis in the knee and elbow joints, and the *I* gene that determines the ABO blood type, on chromosome 9. These two genes are 10 map units apart, which geneticists determined by pooling information from many families. The information is used to predict genotypes and phenotypes in offspring, as in the following example.

Greg and Susan each have nail-patella syndrome. Greg has type A blood. Susan has type B blood. What is the chance that their child inherits normal nails and knees and type O blood? A genetic counselor deduces their allele configurations using information about the couple's parents (**figure 5.17**).

Greg's mother has nail-patella syndrome and type A blood. His father has normal nails and type O blood. Therefore, Greg must have inherited the dominant nail-patella syndrome allele (*N*) and the I^A allele from his mother, on the same chromosome. We know this because Greg has type A blood and his father has type O blood—therefore, he couldn't have gotten the I^A allele from his father. Greg's other chromosome 9 must carry the alleles *n* and *i*. His alleles are therefore in *cis.*

Susan's mother has nail-patella syndrome and type O blood, and so Susan inherited *N* and *i* on the same chromosome. Because her father has normal nails and type B blood, her homolog from him bears alleles *n* and I^B. Her alleles are in *trans.*

Determining the probability that Susan and Greg's child could have normal nails and knees and type O blood is the easiest question the couple could ask. The only way this genotype can arise from theirs is if an *ni* sperm (which occurs with a frequency of 45 percent, based on pooled data) fertilizes an *ni* oocyte (which occurs 5 percent of the time). The result—according to the product rule—is a 2.25 percent chance of producing a child with the *nnii* genotype.

Calculating other genotypes for their offspring is more complicated, because more combinations of sperm and oocytes could account for them. For example, a child with nail-patella syndrome and type AB blood could arise from all combinations that include I^A and I^B as well as at least one *N* allele (assuming that *NN* has the same phenotype as *Nn*).

The Rh blood type and elliptocytosis, and nail-patella syndrome and ABO blood type, are examples of linked gene pairs. A linkage map begins to emerge when percent recombination is known between all possible pairs of three or more linked genes, just as a road map with more landmarks provides more information on distance and direction. Consider genes *x, y,* and *z* (**figure 5.18**). If the percent recombination between *x* and *y* is 10, between *x* and *z* is 4, and between *z* and *y* is 6, then the order of the genes on the chromosome is *x-z-y,* the only order that accounts for the percent recombination data. It is a little like deriving a geographical map from knowing the distances between cities.

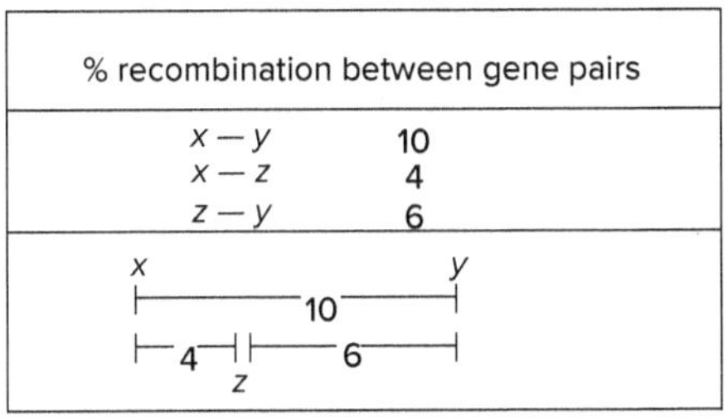

Figure 5.18 Recombination mapping. If we know the percent recombination between all possible pairs of three genes, we can determine their relative positions on the chromosome.

	Greg		Susan
Phenotype	nail-patella syndrome, type A blood		nail-patella syndrome, type B blood
Genotype	$NnI^{A}i$		$NnI^{B}_$
Allele configuration	$\frac{N\ I^A}{n\ i}$		$\frac{N\ i}{n\ I^B}$
Gametes:	sperm	frequency	oocytes
Parental	$N\ I^A$	45%	$N\ i$
	$n\ i$	45%	$n\ I^B$
Recombinants	$N\ i$	5%	$N\ I^B$
	$n\ I^A$	5%	$n\ i$

N = nail-patella syndrome
n = normal

Figure 5.17 Inheritance of nail-patella syndrome. Greg inherited the *N* and I^A alleles from his mother; that is why the alleles are on the same chromosome. His *n* and *i* alleles must therefore be on the homolog. Susan inherited alleles *N* and *i* from her mother, and *n* and I^B from her father. Population-based probabilities are used to calculate the likelihood of phenotypes in the offspring of this couple. Note that in this case, map distances are known and are used to predict outcomes.

Genetic maps derived from percent recombination between linked genes accurately reflect the order of the genes on the chromosome, but the distances are estimates because crossing over is not equally likely across the genome. Some DNA sequences are nearly always inherited together, like two inseparable friends. This nonrandom association between DNA sequences is called **linkage disequilibrium (LD)**. The human genome consists of many LD blocks, where stretches of alleles stick together, interspersed with areas where crossing over is prevalent. Chapters 7, 15, and 16 discuss the use of LD blocks, called **haplotypes**, to track genes in populations.

From Linkage to Genome-Wide Associations

The first human genes mapped to their chromosomes encoded blood proteins, because these were easy to study. In 1980, researchers began using DNA sequences near genes of interest as landmarks called **genetic markers**. These markers need not encode proteins that cause a phenotype. They might be DNA sequence differences that alter where a DNA cutting enzyme cuts, differing numbers of short repeated sequences of DNA with no obvious function, or single sites where the base varies among individuals. The term "genetic marker" is used popularly to mean any DNA sequence that is associated with a phenotype, usually one affecting health.

Computers tally how often genes and their markers are inherited together. The "tightness" of linkage between a marker and a gene of interest is represented as a LOD score, where LOD stands for "logarithm of the odds." A LOD score indicates the likelihood that particular crossover frequency data indicate linkage, rather than the inheritance of two alleles by chance. The higher the LOD score, the closer the two genes on the chromosome.

A LOD score of 3 or greater signifies linkage. It means that the observed data are 1,000 (10^3) times more likely to have occurred if the two DNA sequences (a disease-causing allele and its marker) are linked than if they reside on different chromosomes and just happen to often be inherited together by chance. It is somewhat like deciding whether two coins tossed together 1,000 times always come up both heads or both tails by chance, or because they are taped together side by side in that position, as are linked genes. If the coins land with the same sides up in all 1,000 trials, it indicates they are likely taped.

Before sequencing of the first human genomes, genetic markers were used to predict which individuals in some families were most likely to have inherited a particular disease, before symptoms began. Such indirect analyses are no longer necessary, because tests directly detect disease-causing genes.

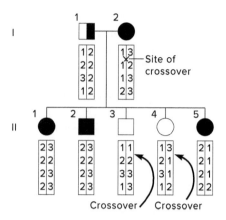

Figure 5.19 Haplotypes. The numbers in bars beneath pedigree symbols enable researchers to track specific chromosome segments with markers. Disruptions of a marker sequence indicate crossover sites.

However, genetic markers are still used to distinguish parts of chromosomes. In pedigrees, designations for markers linked into haplotypes are sometimes placed beneath the symbols to further describe chromosomes. In the family with cystic fibrosis depicted in **figure 5.19**, each set of numbers beneath a symbol represents a haplotype. Knowing the haplotype of individual II–2 reveals which chromosome in parent I–1 contributed the mutant allele. Because Mr. II–2 received haplotype 3233 from his affected mother, his other haplotype, 2222, came from his father. Because Mr. II–2 is affected and his father is not, the father must be a heterozygote, and 2222 must be the haplotype linked to the mutant *CFTR* allele.

Today, entire human genomes can be sequenced in under a day. However, researchers are still filling in the orders of genes on chromosomes and determining gene functions and interactions.

Key Concepts Questions 5.3

1. Why are linked genes inherited in different patterns than unlinked genes?

2. What is the relationship between crossover frequency and relative positions of genes on chromosomes?

3. How were the first human linkage maps constructed?

4. What is linkage disequilibrium?

5. What are genetic markers, LOD scores, and haplotypes?

Summary

5.1 When Gene Expression Appears to Alter Mendelian Ratios

1. Homozygosity for lethal recessive alleles stops development before birth, eliminating an offspring class.

2. A gene can have multiple alleles because its DNA sequence can vary in many ways. Different allele combinations may produce different expressions of the phenotype.

3. Heterozygotes of **incompletely dominant** alleles have phenotypes intermediate between those associated with the two homozygotes. **Codominant** alleles are both expressed in the phenotype.

4. In **epistasis**, one gene affects the phenotype of another.

5. An incompletely **penetrant** genotype is not expressed in all individuals who inherit it. Phenotypes that vary in intensity among individuals are **variably expressive**.

6. **Pleiotropic** genes have several expressions due to functioning in shared pathways, processes, or structures.

7. In **genetic heterogeneity**, two or more genes specify the same phenotype.

8. A **phenocopy** is a characteristic that appears to be inherited but is environmentally caused.

9. Sequencing of many human genomes is revealing many examples of epistasis, nonpenetrance, pleiotropy, and genetic heterogeneity as genes are identified and their interactions deciphered.

5.2 Mitochondrial Genes

10. Cells have many mitochondria, and each mitochondrion has many copies of the mitochondrial genome.

11. Only females transmit mitochondrial genes; males can inherit such a trait but cannot pass it on.

12. Mitochondrial chromosomes do not cross over, and mtDNA mutates more frequently than nuclear DNA.

13. The thirty-seven mitochondrial genes encode tRNA, rRNA, or proteins involved in protein synthesis or energy reactions.

14. Many mitochondrial diseases are due to **heteroplasmy**, in which mitochondria in a single cell have mutations in different alleles.

5.3 Linkage

15. Genes on the same chromosome are **linked** and, unlike genes that independently assort, produce many individuals with parental genotypes and a few with **recombinant** genotypes.

16. **Linkage maps** depict the order of genes on a chromosome and the distances between them. Researchers can examine a group of known linked DNA sequences (a **haplotype**) to follow the inheritance of certain chromosomes.

17. We can predict the probabilities that certain genotypes will appear in progeny if we know crossover frequencies from pooled data and whether linked alleles are in *cis* or in *trans*.

18. Genetic linkage maps assign distances between linked genes based on crossover frequencies.

19. **Genetic markers** are DNA sequences near a gene of interest that are inherited with it due to linkage.

Review Questions

1. Explain how each of the following phenomena can disrupt Mendelian phenotypic ratios.

   a. lethal alleles
   b. multiple alleles
   c. incomplete dominance
   d. codominance
   e. epistasis
   f. incomplete penetrance
   g. variable expressivity
   h. pleiotropy
   i. a phenocopy
   j. genetic heterogeneity

2. How does the relationship between dominant and recessive alleles of a gene differ from epistasis?

3. Why can transmission of an autosomal dominant trait with incomplete penetrance look like autosomal recessive inheritance?

4. How does inheritance of ABO blood types exhibit both complete dominance and codominance?

5. How could two people with albinism have a child who has normal skin pigment?

6. Name two diseases discussed in the chapter that demonstrate genetic heterogeneity.

7. Explain ways that genome sequencing is showing that phenomena that we once thought were exceptions to Mendel's laws and quite rare are actually more common.

8. How can epistasis explain incomplete penetrance?

9. The lung condition emphysema may be caused by lack of an enzyme or by smoking. In which situation is the condition a phenocopy?

10. List three ways that mtDNA differs from DNA in a cell's nucleus.

11. Describe why inheritance of mitochondrial DNA and linkage do not follow Mendel's laws.

12. Explain how a child conceived with donor mitochondria can be described as having three genetic parents.

13. Explain how a pedigree for a maternally inherited trait differs from one for an autosomal dominant trait.

14. If researchers could study pairs of human genes as easily as they can study pairs of genes in fruit flies, how many linkage groups would they detect?

15. The popular media often use words that have precise meanings in genetics, but more general common meanings. Provide the two types of meanings for "linked" and "marker."

Applied Questions

1. For each of the diseases described in situations (a) through (i), indicate which of the following phenomena (A–H) is at work. More than one may apply.

   A. lethal alleles
   B. multiple alleles
   C. epistasis
   D. incomplete penetrance
   E. variable expressivity
   F. pleiotropy
   G. a phenocopy
   H. genetic heterogeneity

   a. A woman with severe neurofibromatosis type 1 has brown spots on her skin and several large tumors beneath her skin. A genetic test shows that her son has the disease-causing autosomal dominant allele, but he has no symptoms.

   b. A man would have a widow's peak, if he wasn't bald.

   c. A man and woman have six children. They also had two stillbirths—fetuses that stopped developing shortly before birth.

   d. Mutations in a gene that encodes a muscle protein called titin cause 22 percent of cases of inherited dilated cardiomyopathy, a form of heart disease. Other single genes cause the other cases.

   e. A woman with dark brown skin uses a bleaching cream with a chemical that darkens her fingertips and ears, making her look like she has the inherited disease alkaptonuria.

   f. In Labrador retrievers, the B allele confers black coat color and the b allele brown coat color. The E gene controls the expression of the B gene. If a dog inherits the E allele, the coat is golden no matter what the B genotype is. A dog of genotype ee expresses the B (black) phenotype.

   g. Two parents are heterozygous for genes that cause albinism, but each gene specifies a different enzyme in the biochemical pathway for skin pigment synthesis. Their children thus do not face a 25 percent risk of having albinism.

   h. All children with Aicardi syndrome have defects in their retinas that look like white craters (chorioretinal lacunae), but only some of the 4,000 recognized cases also have absence of the corpus callosum (the band of nerve fibers that joins the halves of the brain).

   i. Two young children in a family have very decayed teeth. Their parents think it is genetic, but the true cause is a babysitter who puts them to sleep with juice bottles in their mouths.

2. If many family studies of a particular autosomal recessive condition show fewer affected individuals than Mendel's law predicts, the explanation may be either incomplete penetrance or lethal alleles. How might considering haplotypes help to determine which of these two possibilities is the cause?

3. A man who has type O blood has a child with a woman who has type A blood. The woman's mother has AB blood, and her father, type O. What is the probability that the child has each of the following blood types?

   a. O
   b. A
   c. B
   d. AB

4. Enzymes are used in blood banks to remove the A and B antigens from blood types A and B, making the blood type O. Does this process alter the phenotype or the genotype?

5. Ataxia-oculomotor apraxia syndrome, which impairs the ability to feel and move the limbs, usually begins in early adulthood. The molecular basis of the disease is impairment of ATP production in mitochondria, but the mutant gene is in the nuclei of the cells. Would this disease be inherited in a Mendelian fashion? Explain your answer.

6. What is the chance that Greg and Susan, the couple with nail-patella syndrome, could have a child with normal nails and type AB blood?

7. What is a danger of taking action in response to the results of a genetic test for the variant of the APOE4 gene associated with a high risk of Alzheimer disease?

8. In prosopagnosia, a person has "face blindness"—he or she cannot identify individuals by their faces. It is inherited as an autosomal dominant trait and affects people to different degrees. Some individuals learn early in life to identify people by voice or style of dress, and so appear not to have the condition. Only a small percentage of cases are inherited; most are the result of stroke or brain injury. Which of the following does face blindness demonstrate? Explain your choices.

   a. incomplete penetrance
   b. variable expressivity
   c. pleiotropy
   d. a phenocopy

9. Many people who have the "iron overload" disease hereditary hemochromatosis are homozygous for a variant of the *C282Y* gene. How would you determine the penetrance of this condition?

10. A Martian creature called a gazook has 17 chromosome pairs. On the largest chromosome are genes for three traits—round or square eyeballs (*R* or *r*); a hairy or smooth tail (*H* or *h*); and 9 or 11 toes (*T* or *t*). Round eyeballs, hairy tail, and 9 toes are dominant to square eyeballs, smooth tail, and 11 toes. A trihybrid male has offspring with a female who has square eyeballs, a smooth tail, and 11 toes on each of her three feet.

She gives birth to 100 little gazooks, who have the following phenotypes:

- 40 have round eyeballs, a hairy tail, and 9 toes
- 40 have square eyeballs, a smooth tail, and 11 toes
- 6 have round eyeballs, a hairy tail, and 11 toes
- 6 have square eyeballs, a smooth tail, and 9 toes
- 4 have round eyeballs, a smooth tail, and 11 toes
- 4 have square eyeballs, a hairy tail, and 9 toes

a. Draw the allele configurations of the parents.

b. Identify the parental and recombinant progeny classes.

c. What is the crossover frequency between the *R* and *T* genes?

Forensics Focus

1. "Earthquake McGoon" was 32 years old when the plane he was piloting over North Vietnam was hit by groundfire on May 6, 1954. Of the five others aboard, only two survived. McGoon, actually named James B. McGovern, was well known for his flying in World War II, and for his jolliness. Remains of a man about his height and age at death were discovered in late 2002 but could not be identified by dental records. However, DNA sampled from a leg bone enabled forensic scientists to identify him. Name the type of DNA likely analyzed, and describe the further information that was needed to make the identification.

Case Studies and Research Results

1. Suzanne is deaf. In early childhood, she began having fainting spells, especially when she became excited. When she fainted while opening Christmas gifts, her parents took her to the hospital, where doctors said that there wasn't a problem. As the spells continued, Suzanne became able to predict the attacks, telling her parents that her head hurt beforehand. Her parents took her to a neurologist, who checked Suzanne's heart and diagnosed long QT syndrome with deafness, which is a severe form of inherited heartbeat irregularity. Ten different genes can cause long QT syndrome. The doctor told them of a case from 1856: A young girl, called at school to face the headmaster for an infraction, became so agitated that she dropped dead. The parents were not surprised; they had lost two other children to great excitement.

   The family visited a medical geneticist, who discovered that each parent had a mild heartbeat irregularity that did not produce symptoms. Suzanne's parents had normal hearing. Her younger brother Frank was also hearing impaired and suffered night terrors but had so far not fainted during the day. Like Suzanne, he had the full syndrome. Vanessa, still a baby, was also tested. She did not have either form of the family's illness; her heartbeat was normal.

   Today, Suzanne and Frank are treated with beta blocker drugs, and each has an implantable defibrillator to correct a potentially fatal heartbeat. Suzanne's diagnosis may have saved her brother's life.

   a. Which of the following applies to the condition in this family?

      i. genetic heterogeneity

      ii. pleiotropy

      iii. variable expressivity

      iv. incomplete dominance

      v. a phenocopy

   b. How is the inheritance pattern of this form of long QT syndrome consistent with that of familial hypercholesterolemia?

   c. Explain how it is possible that Vanessa did not inherit either the serious or the asymptomatic form of the disease.

   d. Do the treatments for the disease affect the genotype or the phenotype?

2. Max, born in 2004, was well for the first few weeks of his life, but then he became difficult to awaken, stopped gaining weight and then began to lose it, and then developed seizures. His pediatrician suspected that Max had cobalamin C deficiency, an inability to use vitamin B12, also called cobalamin. But Max did not have a mutation in what—at the time—was the only known causative gene, on an autosome. When Max was 9, researchers sequenced his exome and discovered that he had a mutation in a gene on the X chromosome that controls a transcription factor that in turn controls use of the vitamin. What phenomenon described in section 5.1 does this case illustrate?

3. A 5-year-old girl who tired very easily and was losing strength was diagnosed with a mitochondrial disease. Genetic testing revealed two mutations in a gene that is on chromosome 1, but no mutations in mitochondrial genes. Explain how the risk to other children in this family differs from the risk of inheriting a mitochondrial disease caused by mutation in a mitochondrial gene.

© Blend/Image Source

CHAPTER 6

Matters of Sex

In 2015, the government of China officially ended the 35-year one-child policy that has profoundly skewed the sex ratio, with effects on population composition that will persist for years.

Learning Outcomes

6.1 Our Sexual Selves

1. Describe the factors that contribute to whether we are and feel male or female.
2. Distinguish between the X and Y chromosomes.
3. Discuss how manipulating the sex ratio can affect societies.

6.2 Traits Inherited on Sex Chromosomes

4. Distinguish between Y linkage and X linkage.
5. Compare and contrast X-linked recessive inheritance and X-linked dominant inheritance.

6.3 Sex-Limited and Sex-Influenced Traits

6. Discuss the inheritance pattern of a trait that appears in only one sex.
7. Define *sex-influenced trait*.

6.4 X Inactivation

8. Explain why X inactivation is necessary.
9. Explain how X inactivation is an epigenetic change.
10. Discuss how X inactivation affects the phenotype in female mammals.

6.5 Parent-of-Origin Effects

11. Explain the chemical basis of silencing the genetic contribution from one parent.
12. Explain how differences in the timetables of sperm and oocyte formation can lead to parent-of-origin effects.

 ## The BIG Picture

Sex affects our lives in many ways. Which sex chromosomes we are dealt at conception sets the developmental program for maleness or femaleness, but gene expression before and after birth greatly influences how that program unfolds.

An End to China's One-Child Policy

China's one-child policy began with a letter from the Communist Party in September 1980 announcing a "voluntary" program to limit family size, to counter overpopulation. Couples having only one child enjoyed financial incentives from the government that were revoked if they had a second child.

Prenatal diagnosis was widely used to identify XX fetuses and many families wanting their only child to be a boy did not continue or report pregnancies carrying girls. Forced pregnancy terminations and sterilizations and, rarely, infanticide of girl babies, all contributed to an unnaturally skewed sex ratio favoring males. The reasoning was societal: A son would care for his aging parents, but a daughter would care for her in-laws.

China's one-child policy prevented hundreds of millions of births. By 2005, 121 boys were being born for every 100 girls. In 2007, a committee of experts asked the government to end the policy because it created "social problems and personality diseases," but the end did not come until 2015.

Today, many children in China have few siblings, cousins, aunts, or uncles. In rural "bachelor villages," young men cannot find brides. Coddled only-children, called "little emperors," now dominate the younger generation, and studies have shown them to be more self-centered and less cooperative than people who grew up with siblings. Some job ads indicate "no single children." The Chinese government now awards housing subsidies and scholarships to families that have girls, in the hope that a more natural sex ratio will return. That will take time.

The effects of altering the sex ratio will echo in China for years. By 2020, at least 20 million men will lack female partners. By 2050, 25 percent of the population will be older than 65, most of them male, with few young people to care for them. China's one-child policy has been called the world's most radical social experiment.

6.1 Our Sexual Selves

Maleness or femaleness is determined at conception, when she inherits two X chromosomes or he inherits an X and a Y (**figure 6.1**). Another level of sexual identity comes from the control that hormones exert over the development of reproductive structures. Finally, both biological factors and social cues influence sexual feelings, including the strong sense of whether we are female or male—our gender identity.

Sexual Development

Gender differences become apparent around the ninth week of prenatal development. During the fifth week, all embryos develop two unspecialized gonads, which are organs that will develop as either testes or ovaries. Each such "indifferent"

gonad forms near two sets of ducts that offer two developmental options. If one set of tubes, called the Müllerian ducts, continues to develop, female sexual structures form. If the other set, the Wolffian ducts, persists, male sexual structures form.

The option to follow either developmental pathway occurs during the sixth week, depending upon the sex chromosomes present and the actions of certain genes. If a gene on the Y chromosome called *SRY* (for "sex-determining region of the Y") is activated, hormones steer development along the male route. In the absence of *SRY* activation, a female develops (**figure 6.2**).

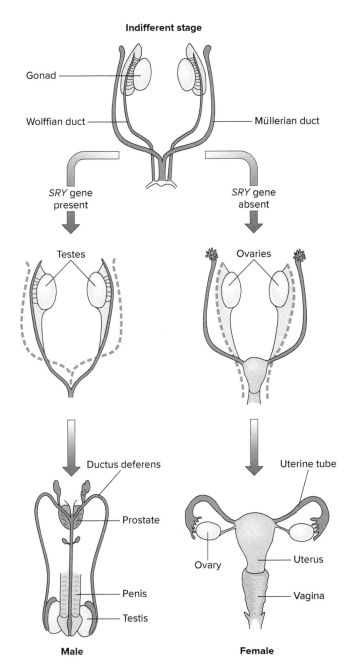

Figure 6.2 **Male or female?** The paired duct systems in the early human embryo may develop into male *or* female reproductive organs. The red tubes represent female structures and the blue tubes, male structures.

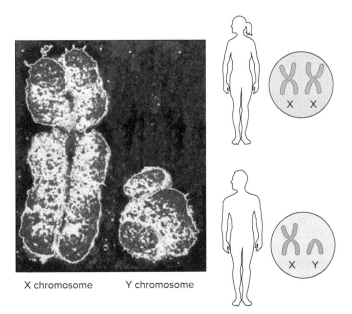

Figure 6.1 **The sex chromosomes.** A human female has two X chromosomes and a human male has one X chromosome and one Y chromosome. © *Biophoto Associates/ Photo Researchers/Science Source*

Femaleness was once considered a "default" option in human development, defined as the absence of maleness. Sex determination is more accurately described as a fate imposed on ambiguous precursor structures. Several genes in addition to *SRY* guide early development of reproductive structures. A mutation in the gene *Wnt4,* for example, causes an XX female to have high levels of male sex hormones and lack a vaginal canal and uterus. Ovaries do not develop properly and secondary sex characteristics do not appear. Hence, *Wnt4* is essential for development and maturation as a female.

Sex Chromosomes

The sex with two different sex chromosomes is called the **heterogametic** sex, and the other, with two of the same sex chromosome, is the **homogametic** sex. In humans, males are heterogametic (XY) and females are homogametic (XX). Some other species are different. In birds and snakes, for example, males are ZZ (homogametic) and females are ZW (heterogametic).

The sex chromosomes differ in size and capacity. The human X chromosome has more than 1,500 genes and is much larger than the Y chromosome, which has 231 protein-encoding genes. In meiosis in a male, the X and Y chromosomes act as if they are a pair of homologs.

Identifying genes on the human Y chromosome has been difficult. Before human genomes were easily sequenced, researchers inferred the functions and locations of Y-linked genes by observing how men missing parts of the chromosome differed from normal. Creating linkage maps for the Y chromosome was not possible, because it does not cross over along all of its length and because of its highly unusual organization. About 95 percent of the chromosome harbors 22 male-specific genes, and many DNA segments here are palindromes. In written languages, palindromes are sequences of letters that read the same in both directions—"Madam, I'm Adam," for example. This symmetry in a DNA sequence destabilizes DNA replication so that during meiosis, sections of a Y chromosome attract each other, which can loop out parts in between. Such deletions in the Y account for some cases of male infertility.

The Y chromosome has a very short arm and a long arm (**figure 6.3**). At both tips are regions, called PAR1 and PAR2, which are termed "pseudoautosomal" because they can cross over with counterparts on the X chromosome. The pseudoautosomal regions comprise only 5 percent of the Y chromosome and include 63 pseudoautosomal genes that encode proteins that function in both sexes. These genes control bone growth, cell division, immunity, signal transduction, the synthesis of hormones and receptors, fertility, and energy metabolism.

Most of the Y chromosome is the male-specific region, or MSY, which lies between the two pseudoautosomal regions. Many of the genes in the MSY are essential to fertility, including *SRY*. (**A Glimpse of History** describes the discovery of the sex-determining gene *SRY*.) The MSY consists of three classes of DNA sequences. About 10 to 15 percent of the MSY sequence is almost identical to parts of the X chromosome. Protein-encoding genes are scarce here. Another 20 percent of the MSY is somewhat similar to X chromosome sequences,

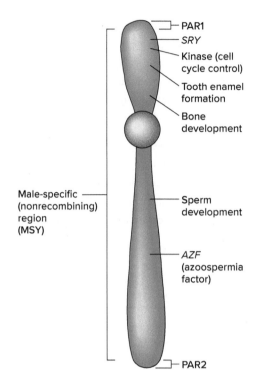

Figure 6.3 **Anatomy of the Y chromosome.** The Y chromosome has two pseudoautosomal regions (PAR1 and PAR2) and a large central area (MSY) that comprises about 95 percent of the chromosome and has only 22 genes. *SRY* determines male sex. *AZF* encodes a protein essential to producing sperm; mutations in it cause infertility.

and may be remnants of an ancient autosome that gave rise to the X chromosome. The rest of the MSY is riddled with palindromes. The MSY has many repeats and the specified proteins combine in different ways, which is one reason males may have 20, 21, or 22 genes in the MSY.

The Phenotype Forms

The *SRY* gene encodes an important type of protein called a **transcription factor**, which controls the expression of other genes. The *SRY* transcription factor stimulates male

---

A GLIMPSE OF HISTORY

The Y chromosome was first seen under a light microscope in 1923, and researchers soon recognized its association with maleness. For many years, they sought to identify the gene or genes that determine sex. Important clues came from two interesting types of people—men who have two X chromosomes (XX male syndrome), and women who have one X and one Y chromosome (XY female syndrome). The XX males actually have a small piece of a Y chromosome, and the XY females do not have a small part of the Y chromosome. The part of the Y chromosome present in the XX males is the same part that is missing in the XY females. In 1990, researchers found the *SRY* gene here.

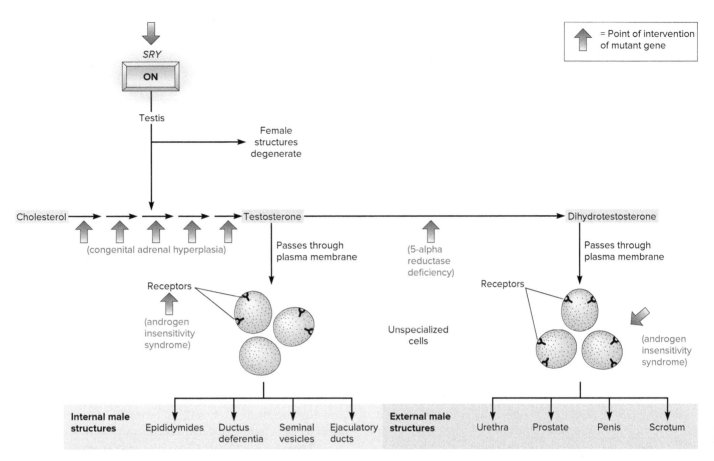

Figure 6.4 Mutations that affect male sexual development. In normal male prenatal development, activation of a set of genes beginning with *SRY* sends signals that destroy female rudiments, while activating the biochemical pathway that produces testosterone and dihydrotestosterone, which promote the development of male structures. The green arrows indicate where mutations disrupt normal sexual development. Diseases are identified in red.

development by sending signals to the indifferent gonads that destroy potential female structures while stimulating development of male structures.

Prenatal sexual development is a multistep process, and mutations can intervene at several points (**figure 6.4**). For example, in androgen insensitivity syndrome, a mutation in a gene on the X chromosome blocks formation of receptors for androgens (the male sex hormones testosterone and dihydrotestosterone [DHT]). As a result, cells in early reproductive structures do not receive signals to develop as male. The person looks female, but is XY.

Several terms are used to describe individuals whose genetic/chromosomal sex and physical structures, both internal and external, are not consistent with one gender. *Hermaphroditism* is an older and more general term for an individual with male and female sexual structures. *Intersex* refers to individuals whose internal structures are inconsistent with external structures, or whose genitalia are ambiguous. It is the preferred term outside of medical circles. *Pseudohermaphroditism* refers to the presence of both female and male structures but at different life stages.

Living with pseudohermaphroditism can be confusing. Consider people who have 5-alpha reductase deficiency, which is autosomal recessive. Affected individuals have a normal Y chromosome, a wild type *SRY* gene, and testes. The internal male reproductive tract develops and internal female structures do not, so the male anatomy is present on the inside. But the child looks like a girl on the outside. Without 5-alpha reductase, which normally catalyzes the reaction of testosterone to form DHT, a penis cannot form. At puberty, when the adrenal glands, which sit atop the kidneys, start to produce testosterone, this XY individual, who thought she was female, starts to exhibit the signs of maleness—deepening voice, growth of facial hair, and the sculpting of muscles into a masculine physique. Instead of developing breasts and menstruating, the clitoris enlarges into a penis. Usually sperm production is normal. XX individuals with 5-alpha reductase deficiency look female.

The degree to which pseudohermaphroditism disturbs the individual depends as much on society as it does on genetics. In the Dominican Republic in the 1970s, 22 young girls reached the age of puberty and began to transform into boys. They had a form of 5-alpha reductase deficiency that was fairly common in the population due to consanguinity (blood relatives having children). The parents were happy that they had sons after all, and so these special adolescents were given their own gender name—*guevedoces,* for "penis at age 12." They were fully accepted as whatever they wanted to be. This isn't always the case. A realistic novel, *Middlesex,* tells the story of a young man with this condition who grew up as a female.

In a more common form of pseudohermaphroditism, congenital adrenal hyperplasia (CAH) due to 21-hydroxylase deficiency, an enzyme block causes testosterone and DHT to accumulate. It is autosomal recessive, and both males and females are affected. The higher levels of androgens cause precocious puberty in males or male secondary sex characteristics to develop in females. A boy may enter puberty as early as 3 years of age, with well-developed musculature, small testes, and an enlarged penis. At birth, a girl may have a swollen clitoris that appears to be a small penis. The person is female internally, but as puberty begins the voice deepens, facial hair grows, and menstruation never happens.

Prenatal tests that detect chromosomal sex have changed the way that pseudohermaphroditism is diagnosed, from phenotype to genotype. Before, the condition was detected only after puberty, when a girl began to look like a boy. Today pseudohermaphroditism is suspected when a prenatal chromosome check reveals an X and a Y chromosome, but the newborn looks like a girl.

A transgender person has a gender identity that does not match the gender assigned at birth based on genitalia. The genetic or physical basis of transgender is not known. Some transgender individuals have surgery to better match their physical characteristics to their gender identity. Chromosomal sex remains binary (female or male), but gender identity is a more "fluid" phenomenon.

Is Homosexuality Inherited?

No one really knows why we have feelings of belonging to one gender or the other, or of being attracted to the same or opposite gender, but these feelings are intense. In homosexuality, a person's phenotype and genotype are consistent, and physical attraction is toward members of the same sex. Homosexuality is seen in all human cultures and has been observed for thousands of years. It has been documented in more than 500 animal species.

Homosexuality reflects complex input from genes and the environment. The genetic influence may be seen in the strong feelings that homosexual individuals have as young children, long before they know of the existence or meaning of the term. Other evidence comes from identical twins, who are more likely to both be homosexual than are both members of fraternal twin pairs.

Experiments in the 1990s identified genetic markers on the X chromosome that were more often identical among pairs of homosexual brothers than among other pairs of brothers. This finding led to the idea that a single gene, or a few genes, dictates sexual preference, but these results could not be confirmed. Further studies on twins have indicated what many people have long suspected—that the roots of homosexuality are not simple.

Twin studies compare a trait between identical and fraternal twin pairs, to estimate the rough proportion of a trait that can be attributed to genes. Chapter 7 discusses this approach further.

Table 6.1	Sexual Identity	
Level	**Events**	**Timing**
Chromosomal/genetic	XY = male XX = female	Fertilization
Gonadal sex	Undifferentiated structure begins to develop as testis or ovary	6 weeks after fertilization
Phenotypic sex	Development of external and internal reproductive structures continues as male or female in response to hormones	8 weeks after fertilization, at puberty
Gender identity	Strong feelings of being male or female develop	From childhood, possibly earlier
Sexual orientation	Attraction to same or opposite sex	From childhood

Courtesy Professor Jennifer A. Marshall-Graves, Australian National University.

Such a study done on all adult twins in Sweden found that in males, genetics contributes about 35 percent to homosexuality, whereas among females the genetic contribution is about 18 percent. Homosexuality likely reflects the input of many genes and environmental factors, and may arise in a variety of ways.

Table 6.1 summarizes the components of sexual identity.

Sex Ratio

Mendel's law of segregation predicts that populations should have approximately equal numbers of male and female newborns. That is, male meiosis should yield equal numbers of X-bearing and Y-bearing sperm. After birth, societal and environmental factors may favor survival of one gender over the other.

The proportion of males to females in a human population is called the **sex ratio**. It is calculated as the number of males divided by the number of females multiplied by 1,000, for people of a particular age. (Some organizations describe sex ratio based on 1.0.) A sex ratio of equal numbers of males and females would be designated 1,000. The sex ratio at conception is called the primary sex ratio. In the United States for the past six decades, newborn boys have slightly outnumbered newborn girls, with the primary sex ratio averaging 1,050. Boys are slightly in excess because Y-bearing sperm weigh slightly less than X-bearing sperm, and so they may reach the oocyte faster and be more likely to enter.

The sex ratio at birth is termed secondary, and at maturity it is called tertiary. It can change markedly with age due to illnesses and environmental factors that affect the sexes differently, such as participating in wars and giving birth. When a society intentionally alters the sex ratio, results can be drastic, as the chapter opener discusses.

At the other end of the human life cycle, the sex ratio favors females in most populations. For people over the age of 65 in the United States, for example, the sex ratio is 720, meaning that there are 720 men for every 1,000 women. The ratio among older people is the result of diseases that are more likely to be fatal in men, as well as behaviors that may shorten their life spans compared to those of women.

Key Concepts Questions 6.1

1. Distinguish the sex chromosome constitutions of human females and males.
2. Describe how male and female structures develop from "indifferent" forms.
3. Explain how the structure of the Y chromosome has made analysis of its genes challenging.
4. Describe conditions that disrupt sexual development.
5. Explain how genes and the environment contribute to homosexuality.
6. Define *sex ratio*.

6.2 Traits Inherited on Sex Chromosomes

Genes on the Y chromosome are **Y-linked** and genes on the X chromosome are **X-linked**. Y-linked traits are rare because the chromosome has few genes, and many of its genes have counterparts on the X chromosome. Y-linked traits are passed from male to male, because a female does not have a Y chromosome. No other Y-linked traits besides infertility (which obviously can't be passed on) are yet clearly defined. Some traits at first attributed to the Y chromosome are actually due to genes that have been inserted into that chromosome from other chromosomes, such as a deafness gene.

A disproportionate number of X-linked genes cause illness when mutant. The chromosome includes 4 percent of all the genes in the human genome, but accounts for about 10 percent of Mendelian (single-gene) diseases.

Genes on the X chromosome have different patterns of expression in females and males because a female has two X chromosomes and a male just one. In females, X-linked traits are passed just like autosomal traits—that is, two copies are required for expression of a recessive allele and one copy for a dominant allele. In males, however, a single copy of an X-linked allele causes expression of the trait or illness because there is no copy of the gene on a second X chromosome to mask the effect. A man inherits an X-linked trait only from his mother. The human male is considered **hemizygous** for X-linked traits, because he has only one set of X-linked genes.

Understanding how sex chromosomes are inherited is important in predicting phenotypes and genotypes in offspring. A male inherits his Y chromosome from his father and his X chromosome from his mother (**figure 6.5**). A female inherits one X chromosome from each parent. If a mother is heterozygous for a particular X-linked gene, her son or daughter has a 50 percent chance of inheriting either allele from her. X-linked

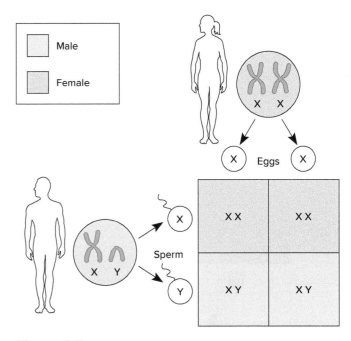

Figure 6.5 **Sex determination in humans.** An oocyte has a single X chromosome. A sperm cell has either an X or a Y chromosome. If a Y-bearing sperm cell with a functional *SRY* gene fertilizes an oocyte, then the zygote is a male (XY). If an X-bearing sperm cell fertilizes an oocyte, then the zygote is a female (XX).

traits are always passed on the X chromosome from a mother to a son or from either parent to a daughter. There is no male-to-male transmission of X-linked traits.

X-Linked Recessive Inheritance

An X-linked recessive trait is expressed in females if the causative allele is present in two copies. Many times, an X-linked trait passes from an unaffected heterozygous mother to an affected son. **Table 6.2** summarizes the transmission of an X-linked recessive trait compared to transmission of an X-linked dominant trait, discussed in the next section.

Table 6.2	Comparison of X-Linked Recessive and X-Linked Dominant Inheritance
X-Linked Recessive Trait	**X-Linked Dominant Trait**
Always expressed in the male	Expressed in females in one copy
Expressed in a female homozygote and very rarely in a female heterozygote	Much more severe effects in males
Affected male inherits trait from heterozygote or homozygote mother	High rates of miscarriage due to early lethality in males
Affected female inherits trait from affected father and affected or heterozygote mother	Passed from male to all daughters but to no sons

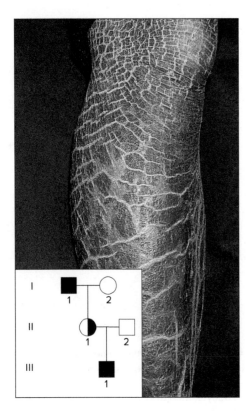

Figure 6.6 **An X-linked recessive trait.** Ichthyosis is transmitted as an X-linked recessive trait. A grandfather and grandson were affected in this family. *Courtesy of Dr. Mark A. Crowe*

mutation or she inherited it. Either way, she passed the mutation to carrier daughters and one mildly affected son.

For many years, historians assumed that the royal families of Europe suffered from the more common form of hemophilia, type A. However, researchers in 2009 tested bits of bone from Tsarina Alexandra and Crown Prince Alexei for the wild type and mutant alleles for the genes that cause hemophilia A as well as B. Both genes are on the X chromosome. The result: Mother Alexandra was a carrier of the rarer hemophilia B, and her son Crown Prince Alexei had the disease, which had been known from documents describing his severe bleeding episodes.

The transmission pattern of hemophilia B is consistent with the criteria for an X-linked recessive trait listed in table 6.2. A daughter can inherit an X-linked recessive disease or trait if her father is affected and her mother is a carrier, because the daughter inherits one mutation-bearing X chromosome from each parent. Without a biochemical test, though, an unaffected woman will not know she is a carrier for an X-linked recessive trait unless she has an affected son.

A woman whose brother has hemophilia B has a 1 in 2 risk of being a carrier. Both her parents are healthy, but her mother must be a carrier because her brother is affected. Her risk is the chance that she has inherited the X chromosome bearing the hemophilia allele from her mother. The chance of the woman conceiving a son is 1 in 2, and of that son inheriting hemophilia is 1 in 2. Using the product rule, the risk that she is a carrier and will have a son with hemophilia, out of all the possible children she can conceive, is $1/2 \times 1/2 \times 1/2$, or $1/8$.

X-Linked Dominant Inheritance

Dominant X-linked conditions and traits are rare. Again, gene expression differs between the sexes (see table 6.2). A female who inherits a dominant X-linked allele or in whom the mutation originates has the associated trait or illness, but a male who inherits the allele is usually more severely affected because he has no other allele to mask its effect.

The children of a man who has a normal X chromosome and a woman who has a dominant mutation on one X chromosome face the risks summarized in **figure 6.8**. However, for a severe condition such as Rett syndrome (see the chapter 2 opener), females may be too disabled to have children. Rett syndrome and similar X-linked dominant conditions affect only girls, because sons would inherit the X chromosome bearing the mutation from their mothers.

An X-linked dominant condition that could be mild enough to be passed to children is incontinentia pigmenti (IP; see figure 6.8b). In affected females, swirls of skin pigment arise where melanin penetrates the deeper skin layers. A newborn girl with IP has yellow, pus-filled vesicles on her limbs that come and go over the first few weeks. Then the lesions become warty and eventually become brown splotches that may remain for life, although they fade with time. Males who inherit the mutant allele from their heterozygous mothers do not survive to be born, which is why women with the disease who become pregnant have a miscarriage rate of about 25 percent.

If an X-linked condition is not lethal, a man may be healthy enough to transmit it to offspring. Consider the small family depicted in the pedigree in **figure 6.6**. A middle-aged man who had rough, brown, scaly skin did not realize his condition was inherited until his daughter had a son. By a year of age, the boy's skin resembled his grandfather's. In the condition, called ichthyosis, an enzyme deficiency blocks removal of cholesterol from skin cells. The upper skin layer cannot peel off as it normally does, and appears brown and scaly. A test of the daughter's skin cells revealed that she produced half the normal amount of the enzyme, indicating that she was a carrier.

Colorblindness is another X-linked recessive trait that does not hamper the ability of a man to have children. About 8 percent of males of European ancestry are colorblind, as are 4 percent of males of African descent. Only 0.4 percent of females in both groups are colorblind. **Clinical Connection 6.1** looks at this interesting trait.

Figure 6.7 shows part of an extensive pedigree for another X-linked recessive trait, the blood-clotting disease hemophilia B. It is also known as Christmas disease and factor IX ("FIX") deficiency. Part (a) depicts the combination of pedigree symbols and a Punnett square to trace transmission of the trait. Dominant and recessive alleles are indicated by superscripts to the X and Y chromosomes. Part (b) shows how the mutant allele arose in one of Queen Victoria's X chromosomes and then passed to other members of the royal families of England, Germany, Spain, and Russia. The queen either had a new

Colorblindness and Tetrachromacy

English chemist John Dalton and his brother saw things differently from most people. Sealing wax that appeared red to other people was as green as a leaf to the Dalton brothers. Pink wildflowers were blue. The Dalton brothers had X-linked recessive colorblindness.

Curious about the cause of his colorblindness, John Dalton asked his physician, Joseph Ransome, to dissect his eyes after he died. When that happened, Ransome snipped off the back of one eye, removing the retina, where the cone cells that provide color vision are nestled among the more abundant rod cells that impart black-and-white vision. Because Ransome could see red and green normally when he peered through the back of his friend's disembodied eyeball, he concluded that it was not an abnormal filter in front of the eye that altered color vision. He stored the eyes in dry air, enabling researchers at the London Institute of Ophthalmology to analyze DNA in Dalton's eyeballs in 1994. Dalton's remaining retina lacked one of the three types of photopigments that enable cone cells to capture certain wavelengths of light.

Color Vision Basics

Our three types of cone cells correspond to three types of photopigments, making us "trichromats." Primates other than apes, such as monkeys, are dichromats. An object appears colored because it reflects certain wavelengths of light, and each photopigment captures a specific range of wavelengths. The brain interprets the information as a visual perception, much as an artist mixes the three primary colors to create many hues.

Each photopigment has a vitamin A–derived portion called retinal and a protein portion called opsin. The three types of opsins correspond to short, middle, and long wavelengths of light. Mutations in opsin genes cause three types of colorblindness. A gene on chromosome 7 encodes shortwave opsins, and mutations in it produce rare "blue" colorblindness. Dalton had deuteranopia (green colorblindness); his eyes lacked the middle-wavelength opsin. The third type, protanopia (red colorblindness), lacks long-wavelength opsin. Deuteranopia and protanopia are X-linked.

A Molecular View

Jeremy Nathans of Johns Hopkins University also personally contributed to understanding color vision. He used a cow version of a protein called rhodopsin that provides black-and-white vision to identify the human version of the gene, then used the human rhodopsin gene to search his own DNA. He found three genes with similar sequences: one on chromosome 7 and two on the X chromosome.

Nathans can see colors, but his opsin genes provided a clue to how colorblindness arises. His X chromosome has one red opsin gene and two green genes, instead of the normal one of each. Because the red and green genes have similar sequences, they can misalign during meiosis in the female (**figure 6A**). The resulting oocytes have either two or none of one opsin gene type. An oocyte missing an opsin gene would, when fertilized by a Y-bearing sperm, give rise to a colorblind male.

A mutation in an opsin gene can also be beneficial. In some women mispairing during meiosis yields an X chromosome that

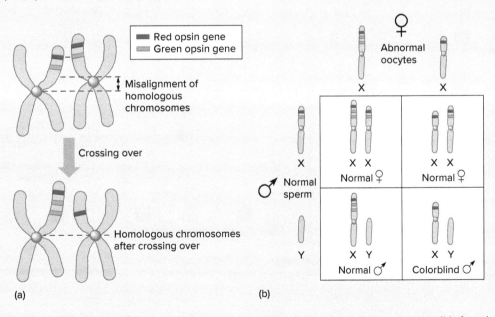

(a) (b)

Figure 6A **How colorblindness arises.** (a) The sequence similarities among the opsin genes responsible for color vision may cause chromosomes to misalign during meiosis in the female. Offspring may inherit too many, or too few, opsin genes. A son inheriting an X chromosome missing an opsin gene will be colorblind. A daughter would have to inherit two mutations to be colorblind. (b) A missing gene causes X-linked colorblindness.

(Continued)

has three opsins: the normal red and green, as well as another that has part red and part green DNA sequences. Combined with her chromosome 7 blue opsin, she has four opsin genes, and is termed a tetrachromat. If she is also artistic and has a good vocabulary (both environmental influences), she may perceive colors that others cannot and be able to describe them. One tetrachromat, for example, can see undertones that ruin the color of clothing, becomes frustrated with automated paint mixers because they are not precise enough, and even detected changes in the sky just before tornado-strength winds began. A cosmetic company hired her to help design products.

Questions for Discussion

1. Discuss how the three types of colorblindness differ in genotype and phenotype.
2. Explain how Dr. Nathans discovered the mutations that cause colorblindness.
3. How does a tetrachromat differ from a trichromat?

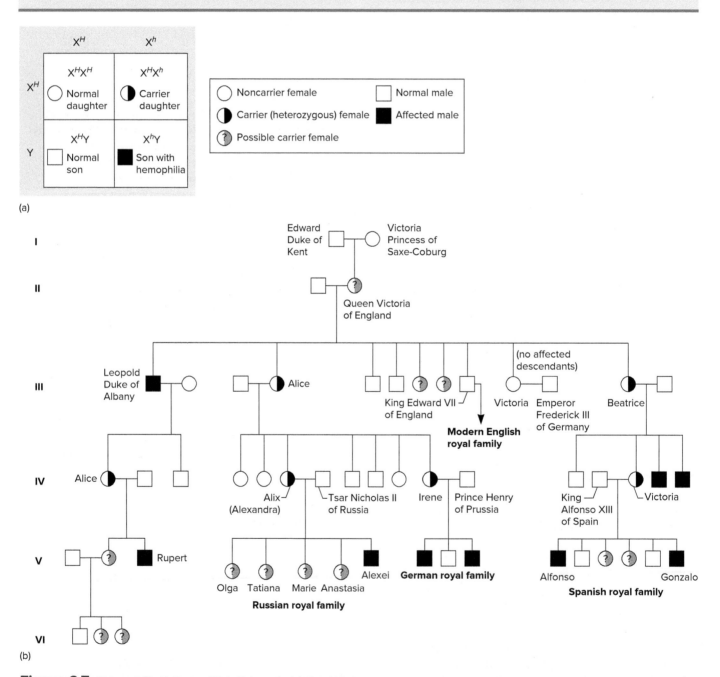

Figure 6.7 **Hemophilia B (factor IX deficiency).** **(a)** This X-linked recessive disease usually passes from a heterozygous woman (designated $X^H X^h$, where h is the hemophilia-causing allele) to heterozygous daughters or hemizygous sons. The father has normal blood clotting. **(b)** The disease has appeared in the royal families of England, Germany, Spain, and Russia. The modern royal family in England does not carry hemophilia.

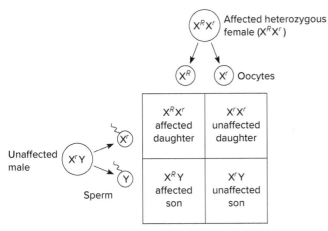

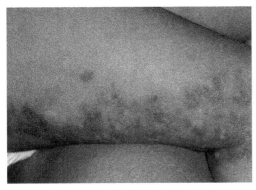

(a)

(b)

Figure 6.8 **X-linked dominant inheritance.** **(a)** A female who has an X-linked dominant trait has a 1 in 2 probability of passing it to her offspring, male or female. Males are generally more severely affected than females. **(b)** Note the characteristic patchy pigmentation on the leg of a girl who has incontinentia pigmenti. *(b): Courtesy of Richard Alan Lewis M.D., M.S., Baylor College of Medicine*

Solving a Problem of X-Linked Inheritance

Mendel's first law (segregation) applies to genes on the X chromosome. The same logic is used to solve problems of X-linked inheritance as to trace traits transmitted on autosomes, but considering the X and Y chromosomes in Punnett squares. Follow these steps:

1. Look at the pattern of inheritance. Different frequencies of affected males and females in each generation may suggest X linkage. For an X-linked recessive trait:
   - An affected male has a carrier mother.
   - An unaffected female with an affected brother has a 50 percent (1 in 2) chance of being a carrier.
   - An affected female has a carrier or affected mother *and* an affected father.
   - A carrier (female) has a carrier or affected mother or an affected father.

For an X-linked dominant trait:
   - There may be no affected males, because they die early.
   - An affected female has an affected mother.
2. Draw the pedigree.
3. List all genotypes and phenotypes and their probabilities.
4. Assign genotypes and phenotypes to the parents. Consider clues in the phenotypes of relatives.
5. Determine how alleles separate into gametes for the genes of interest on the X and Y chromosomes.
6. Unite the gametes in a Punnett square.
7. Determine the phenotypic and genotypic ratios for the F_1 generation.
8. To predict further generations, use the F_1 genotypes and repeat steps 4 through 6.

Consider as an example Kallmann syndrome, which causes a poor or absent sense of smell and small testes or ovaries. It is X-linked recessive. Tanisha does not have Kallmann syndrome, but her brother Jamal and her maternal cousin Malcolm (her mother's sister's child) have it. Tanisha's and Malcolm's parents are unaffected, as is Tanisha's husband Sam. Tanisha and Sam wish to know the risk that a son will inherit the condition. Sam has no affected relatives.

Solution

Mode of inheritance: The trait is X-linked recessive because males are affected through carrier mothers.

K = wild type; k = Kallmann syndrome

Genotypes	Phenotypes
X^KX^K, X^KX^k, X^KY	normal
X^kX^k, X^kY	affected

Individual	Genotype	Phenotype	Probability
Tanisha	X^KX^k or X^KX^K	normal (carrier)	50% each
Jamal	X^kY	affected	100%
Malcolm	X^kY	affected	100%
Sam	X^KY	normal	100%

Tanisha's gametes

if she is a carrier:	X^K	X^k
Sam's gametes:	X^K	Y

Punnett square

	X^K	X^k
X^K	X^KX^K	X^KX^k
Y	X^KY	X^kY

Interpretation: If Tanisha is a carrier, the probability that their son will have Kallmann syndrome is 50 percent, or 1 in 2. This is a conditional probability. The chance that any son will have the condition is actually 1 in 4, because Tanisha also has a 50 percent chance of being genotype X^KX^K and therefore not a carrier.

6.3 Sex-Limited and Sex-Influenced Traits

An X-linked recessive trait generally is more prevalent in males than females. Other situations, however, can affect gene expression in the sexes differently.

Sex-Limited Traits

A **sex-limited trait** affects a structure or function of the body that is present in only males or only females (**figure 6.9**). The gene for such a trait may be X-linked or autosomal.

Understanding sex-limited inheritance is important in animal breeding. For example, a New Zealand cow named Marge has a mutation that makes her milk very low in saturated fat; she is founding a commercial herd. Males play their part by transmitting the mutation, even though they do not make milk. In humans, beard growth is sex-limited. A woman does not grow a beard, because she does not manufacture the hormones required for facial hair growth. She can, however, pass to her sons the genes specifying heavy beard growth.

Curiously, a father-to-be can affect the health of the mother-to-be. Although an inherited disease that causes symptoms associated with pregnancy is obviously sex-limited, the male genome may play a role by contributing to the development of supportive structures, such as the placenta. This is the case for preeclampsia, a sudden rise in blood pressure late in pregnancy. It kills 50,000 women worldwide each year. A study of 1.7 million pregnancies in Norway found that if a man's first wife had preeclampsia, his second wife had double the risk of developing it. Another study found that women whose mothers-in-law developed preeclampsia when pregnant with the women's husbands had twice the rate of developing the condition themselves. Perhaps a gene from the father affects the placenta in a way that elevates the pregnant woman's blood pressure.

Sex-Influenced Traits

With a **sex-influenced trait**, an allele is dominant in one sex but recessive in the other. Such a gene may be X-linked or autosomal. The difference in expression can be caused by hormonal differences between the sexes. For example, an autosomal gene for hair growth pattern has two alleles, one that produces hair all over the head and another that causes pattern baldness. The baldness allele is dominant in males but recessive in females, which is why more men than women are bald. A heterozygous male is bald, but a heterozygous female is not. A bald woman is homozygous recessive. Even a bald woman tends to have some wisps of hair, whereas an affected male may be completely hairless on the top of his head.

(a)　　　(b)

Figure 6.9　Sex-limited traits. Milk composition **(a)** and beard growth **(b)** are examples of sex-limited traits. *(a): Source: Bob Nichols USDA Natural Resources Conservation Services; (b): © Ariel Skelly/Getty Images*

6.4 X Inactivation

Females have two alleles for every gene on the X chromosomes, whereas males have only one. In mammals, a mechanism called **X inactivation** balances this apparent inequality in the expression of genes on the X chromosome.

Equaling Out the Sexes

By the time the embryo of a female mammal consists of eight cells, about 75 percent of the genes on one X chromosome in each cell are inactivated, and the remaining 25 percent are expressed to different degrees in different females. Which X chromosome is mostly turned off in each cell—the one inherited from the mother or the one from the father—is usually random. As a result, a female mammal expresses the X chromosome genes inherited from her father in some cells and those from her mother in others. She is, therefore, a mosaic for expression of most genes on the X chromosome (**figure 6.10**).

The gene *XIST* in part of the X chromosome called the X inactivation center controls X inactivation. *XIST* encodes an RNA that binds to a specific site on the same (inactivated) X chromosome. From this site to the chromosome tip, most of the X chromosome is then silenced. Figure 13.14 shows how *XIST* is used to turn off the extra chromosome that causes most cases of Down syndrome.

Once an X chromosome is inactivated in one cell, all its daughter cells have the same X chromosome shut off. Because the inactivation occurs early in development, the adult female has patches of tissue that differ in their expression of X-linked genes. With each cell in her body having only one active X chromosome, she is roughly equivalent to the male in gene expression.

X inactivation can alter the phenotype (gene expression), but not the genotype. The change is reversed in germline cells destined to become oocytes, and this is why a fertilized ovum does not have an inactivated X chromosome. X inactivation is an example of an **epigenetic** change—one that is passed from one cell generation to the next but that does not alter the DNA base sequence.

We can observe X inactivation at the cellular level because the turned-off X chromosome absorbs a stain much faster than the active one. This differential staining occurs because inactivated DNA has chemical methyl (CH_3) groups bound to it that prevent it from being transcribed into RNA and also enable it to absorb stain.

X inactivation can be used to check the sex of an individual. The nucleus of a cell of a female, during interphase, has one dark-staining X chromosome called a Barr body. A cell from a male has no Barr body because his one X chromosome remains active.

Effect of X Inactivation on the Phenotype

The consequences of X inactivation on the phenotype can be interesting. For homozygous X-linked genotypes, X inactivation has no effect. Whichever X chromosome is turned off, a

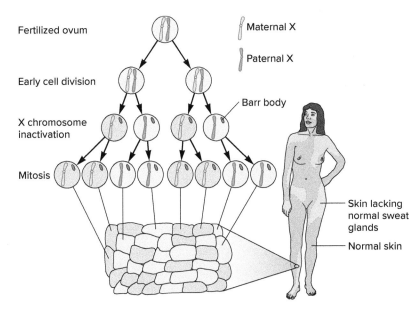

Figure 6.10 X inactivation. A female is a mosaic for expression of genes on the X chromosome because of the random inactivation of either the maternal or paternal X in each cell early in prenatal development. In anhidrotic ectodermal dysplasia, a woman has patches of skin that do not have sweat glands and hair. (Colors distinguish cells with the inactivated X; they do not depict skin color.)

copy of the same allele is left to be expressed. For heterozygotes, however, X inactivation leads to expression of one allele or the other. This doesn't affect health if enough cells express the functional gene product. However, traits that affect skin or fur reveal X inactivation. The swirls of skin color in incontinentia pigmenti (IP) patients reflect patterns of X inactivation in skin cells (see figure 6.8b). Where the normal allele for melanin pigment is shut off, pale swirls develop. Where pigment is produced, brown swirls result.

A familiar example of X inactivation is the coat colors of tortoiseshell and calico cats. An X-linked gene confers brownish-black (dominant) or yellow-orange (recessive) color. A female cat heterozygous for this gene has patches of each color, forming a tortoiseshell pattern that arises from different cells expressing either of the two alleles (**figure 6.11**). The earlier the X inactivation, the larger the patches, because more cell divisions can occur after the event, producing more daughter cells. White patches may form due to epistasis by an autosomal gene that shuts off pigment synthesis. A cat with colored patches against such a white background is a calico. Tortoiseshell and calico cats are nearly always female. A male can have one of these coat patterns only if he inherits an extra X chromosome.

X inactivation has clinical implications. A female who is heterozygous for an X-linked recessive gene can express the associated condition if the normal allele is inactivated in the tissues that the illness affects. Consider a carrier of hemophilia A (factor VIII deficiency). If the X chromosome carrying the normal allele for the clotting factor is turned off in her liver, where clotting factors are made, then the blood will clot slowly enough to cause mild hemophilia. (Luckily for her, slowed clotting time also greatly reduces her risk of cardiovascular disease caused by blood clots blocking circulation.)

Figure 6.11 Visualizing X inactivation. X inactivation is obvious in a calico cat. X inactivation is rarely observable in humans because most cells do not remain together during development, as a cat's skin cells do. © *Animal Attraction/OS50/ Getty RF*

A carrier of an X-linked trait who expresses the phenotype, such as a heterozygote for hemophilia A or B, is called a **manifesting heterozygote**. Whether or not a manifesting heterozygote results from X inactivation depends upon how adept cells are at sharing. Two lysosomal storage diseases, which are deficiencies of specific enzymes that normally dismantle cellular debris in lysosomes, illustrate how cell-cell interactions mediate effects of X inactivation in manifesting heterozygotes. In carriers of Hunter syndrome (also called mucopolysaccharidosis II), cells that make the enzyme readily send it to neighboring cells that do not, essentially correcting the defect in cells that can't make the enzyme. Carriers of Hunter syndrome do not have symptoms because cells get enough enzyme. Boys with Hunter syndrome are deaf, intellectually disabled, have dwarfism and abnormal facial features, heart damage, and an enlarged liver and spleen. In contrast, in Fabry disease, cells do not readily release the enzyme alpha-galactosidase A, so a female who is a heterozygote may have cells in the affected organs that lack the enzyme. She may develop mild symptoms of this disease that causes skin lesions, abdominal pain, and kidney failure in boys.

In humans, X inactivation can be used to identify carriers of some X-linked diseases. This is the case for Lesch-Nyhan syndrome, in which an affected boy has cerebral palsy; bites his fingers, toes, and lips to the point of mutilation; is intellectually disabled; and passes painful urinary stones. Mutation results in defective or absent HGPRT, an enzyme. A woman who carries Lesch-Nyhan syndrome can be detected when hairs from widely separated parts of her head are tested for HGPRT. (Hair is used for the test because it is accessible and produces the enzyme.) If some hairs contain HGPRT but others do not, she is a carrier. The hair cells that lack the enzyme have turned off the X chromosome that carries the normal allele; the hair cells that manufacture the normal enzyme have turned off the X chromosome that carries the disease-causing allele. The woman is healthy because her brain has enough HGPRT, but each son has a 50 percent chance of inheriting the disease.

X inactivation affects the severity of Rett syndrome, the X-linked dominant disease discussed in the chapter 2 opener. Ninety-nine percent of cases arise anew, from mutations in X-bearing sperm cells. Rarely, Rett syndrome may be inherited from a woman who has a very mild case because, by chance, the X chromosomes bearing the mutation are silenced in her brain cells.

Theoretically, X inactivation evens out the sexes for expression of X-linked genes. In actuality, however, a female may *not* be equivalent, in gene expression, to a male because she has two cell populations, whereas a male has only one. One of a female's two cell populations has the X she inherited from her father active, and the other has the X chromosome she inherited from her mother active. For heterozygous X-linked genes, she has some cells that manufacture the protein encoded by one allele, and some cells that produce the protein specified by the other allele. Although the alleles of most heterozygous genes are about equally represented, sometimes X inactivation can be skewed. In such a case, most cells express the X inherited from the same parent. This unequal X inactivation pattern can happen if the two X chromosomes have different alleles for a gene that controls cell division rate, giving certain cells a survival advantage.

Key Concepts Questions 6.4

1. Explain how X inactivation compensates for differences between males and females in the numbers of copies of genes on the X chromosome.
2. When in prenatal development does X inactivation begin?
3. Discuss why and how the effects of X inactivation are noticeable in heterozygotes.

6.5 Parent-of-Origin Effects

In Mendel's pea experiments, it didn't matter whether a trait came from the male or female parent. For certain genes in mammals, however, parental origin influences age of onset of a disease or symptom severity. Two mechanisms of parent-of-origin effects are genomic imprinting and differences between the developmental timetables of sperm and oocytes.

Case Studies and Research Results

1. For each case description, identify the principle at work from the list that follows. More than one answer may apply to a given case.

   A. Y-linked inheritance
   B. X-linked recessive
   C. X-linked dominant inheritance
   D. sex-limited inheritance
   E. sex-influenced inheritance
   F. X inactivation or manifesting heterozygote
   G. imprinting abnormality

   a. A child with Russell-Silver syndrome displays poor growth, a large head, a characteristic triangular face, and digestive problems. In some cases, a gene on chromosome 11 that is normally methylated in the father is not methylated.

   b. Six-year-old LeQuan inherited Fabry disease from his mother, who is a heterozygote for the causative allele. The gene, on the X chromosome, encodes a lysosomal enzyme. LeQuan would die before age 50 of heart failure, kidney failure, or a stroke, but fortunately he can be treated with twice-monthly infusions of the enzyme. His mother, Echinecea, recently began experiencing recurrent fevers, a burning pain in her hands and feet, a rash, and sensitivity to cold. She is experiencing mild Fabry disease.

   c. The Chandler family has many male members who have a form of retinitis pigmentosa (RP) in which the cells that capture light energy in the retina degenerate, causing gradual visual loss. Several female members of the family presumed to be carriers because they have affected sons are tested for RP genes on chromosomes 1, 3, 6, and the X, but are found to not carry these genes. Years ago, Rachel married her cousin Ross, who has the family's form of RP. They had six children. The three sons are all affected, but their daughters all have normal vision.

   d. Simon's mother and his mother's sister are breast cancer survivors, and their mother died of the disease. Simon's sister Maureen has a genetic test and learns that she has inherited the *BRCA1* mutation that increases risk of developing certain cancers. Simon has two daughters but doesn't want to be tested because he thinks a man cannot transmit a trait that affects a body part that is more developed in females.

   e. Tribbles are extraterrestrial mammals that long ago invaded a starship on the television program *Star Trek*. A gene called *frizzled* causes kinky hair in female tribbles who inherit just one allele. However, two mutant alleles must be inherited for a male tribble to have kinky hair. The gene is on an autosome.

   f. Devon died at age 16 of Lowe syndrome. He had slight intellectual disability, visual problems (cataracts and glaucoma), seizures, poor muscle tone, and progressive kidney failure, which was ultimately fatal.

   His sister Lily is pregnant, and wonders whether she is a carrier of the disease that killed her brother. She remembers a doctor saying that her mother Drucilla was a carrier. Lily's physician determines that she is a carrier because she has cataracts, which is a clouding of the lenses. It has not yet affected her vision. When a prenatal test reveals that Lily's fetus is a female, her doctor tells her not to worry about Lowe syndrome.

   g. Mating among Texas field crickets depends upon females responding to a male mating call. The sounds must arrive at a particular frequency to excite the females, who do not sing back in response. However, females can pass on an allele that confers the "correct" frequency of singing.

   h. When Winthrop was a baby, he was diagnosed with "failure to thrive." At 14 months of age, he suddenly took an interest in food, and his parents couldn't feed him fast enough. By age 4, Winthrop was obese, with disturbing behavior. He was so hungry that after he'd eaten his meal and everyone else's leftovers, he'd hunt through the garbage for more. Finally a psychiatrist who had a background in genetics diagnosed Prader-Willi syndrome. Testing showed that the allele for the Prader-Willi gene that Winthrop had inherited from his father was abnormally methylated.

   i. Certain breeds of dogs have cryptorchidism, in which the testicles do not descend into the scrotum. The trait is passed through females.

2. Reginald has mild hemophilia A that he can control by taking a clotting factor. He marries Lydia, whom he met at the hospital where he and Lydia's brother, Marvin, receive the same treatment. Lydia and Marvin's mother and father, Emma and Clyde, do not have hemophilia. What is the probability that Reginald and Lydia's son will inherit hemophilia A?

3. Harold works in a fish market, but the odor does not bother him because he has anosmia, an X-linked recessive lack of the sense of smell. Harold's wife, Shirley, has a normal sense of smell. Harold's sister, Maude, also has a normal sense of smell, as does her husband, Phil, and daughter, Marsha, but their identical twin boys, Alvin and Simon, cannot detect odors. Harold and Maude's parents, Edgar and Florence, can smell normally. Draw a pedigree for this family, indicating people who must be carriers of the anosmia mutation.

4. When a South African sprinter won the 800-meter race at the World Championships in 2009, with a time significantly improved over past events, the International Association of Athletics Federations asked her to take a chromosome test. The implication of the much-publicized request was that she was really a male. For reasons of privacy, the results of the test were never released. Which gene would have been tested to determine the sprinter's chromosomal sex? Is this the same as gender?

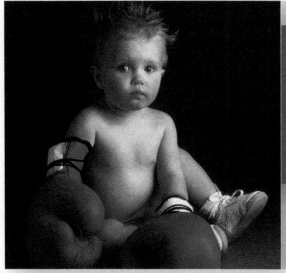

CHAPTER 7

© Corbis RF

Multifactorial Traits

Should parents submit their children's DNA for genetic testing to determine which sports the youngsters should play?

The BIG Picture

Who we are and how we feel arise from an intricate interplay between our genes and environmental influences. Understanding genetic contributions to traits and illnesses can suggest how we can alter our environments to improve our lives.

The Complex Genetics of Athletics

A website offers a single-gene test that purportedly provides parents and coaches early information on a child's future success in team or individual sports requiring speed/power or endurance. The test is for variants of a gene that encodes a protein called actinin 3 that is expressed in skeletal muscle. One genotype is overrepresented among elite sprinters, another among world-class endurance runners. The test takes a simplistic view of a complex trait.

Athletic ability is multifactorial—many genes as well as environmental influences determine at which activities or sports a person may excel. Any single gene is unlikely to have a great influence. Environmental factors include exposure to pollution and toxins, as well as opportunities to participate in sports. The ability to work well on a team cannot be reduced to a simple string of DNA letters that encodes a protein.

The idea to market tests for athletic genes may have come from knowledge of rare mutations in other genes that bestow great physical prowess. Members of a German family with a mutation in a "double muscle" gene are amazing weight lifters; a Scandinavian family of Olympic skiers has a mutation that increases the number of red blood cells. Genes also influence metabolic rate, bone mineral density, fat storage, glucose use, and lung function. Table 20.2 describes genes associated with athletic characteristics.

Genetic testing to predict athletic success illustrates genetic determinism—the idea that our genes alone determine who we are or what we can do. Using such test results to choose a child's sport can stress a child with no interest in competing, or discourage a child who loves a particular sport.

7.1 Genes and the Environment Mold Traits

A woman who is a prolific writer has a daughter who becomes a successful novelist. An overweight man and woman have obese children. A man whose father suffers from alcoholism has the same problem. Are these characteristics—writing talent, obesity, and alcoholism—inherited or learned? Such traits, and nearly all others, are not the result of an "either/ or" mechanism, but reflect the input of many genes as well as environmental influences. Even single-gene diseases are modified by environmental factors and/or other genes. A child with cystic fibrosis, for example, has inherited a single-gene disease, but the severity of her symptoms reflects which variant (or variants) of the gene she has, other genes that affect her immune system, the pathogens to which she is exposed, and the quality of the air she breathes (see Clinical Connection 4.1). This chapter considers characteristics that represent input from many genes, and the tools used to study them.

A trait can be described as either single-gene (Mendelian or monogenic) or **polygenic**. As its name implies, a polygenic trait reflects the activities of more than one gene. Both single-gene and polygenic traits can also be **multifactorial**, which means they are influenced by the environment. Lung cancer is a multifactorial trait (**figure 7.1**). Variants of genes that increase the risk of becoming addicted to nicotine and of developing cancer come into play—but may not ever be expressed if a person never smokes and breathes only fresh air. A diet high in carbohydrates that have a high glycemic index (raise blood glucose level rapidly) elevates lung cancer risk among non-Hispanic whites. Purely polygenic traits—those not influenced by the environment at all—are rare. Eye color, discussed in chapter 4, is close to being purely polygenic.

Polygenic multifactorial traits include common ones, such as height, skin color, body weight, many illnesses, and behavioral conditions and tendencies. Behavioral traits are

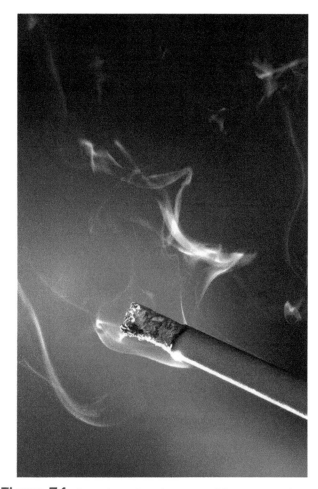

Figure 7.1 **Genetic and environmental factors contribute to lung cancer risk.** Mutations raise lung cancer risk in several ways: impairing DNA repair, promoting inflammation, blocking detoxification of carcinogens, keeping telomeres long, and promoting addiction. These genetic risk factors interact with each other and with environmental influences, such as smoking, breathing polluted air, and diet. © *Brand X Pictures RF*

not inherently different from other types of traits; they involve the functioning of the brain, rather than another organ. Chapter 8 discusses them. A more popular term for multifactorial is "complex," but multifactorial is more precise and is not confused with the general definition of "complex." The genes underlying a multifactorial trait are not more complicated than others. They follow Mendel's laws, but contribute only partly to a trait and are therefore more difficult to track.

A polygenic multifactorial condition reflects additive contributions of several genes. Each gene confers a degree of susceptibility, but the input of these genes is not necessarily equal. While a rare allele may exert a large influence, several common alleles may each contribute only slightly to a trait. For example, three genes contribute significantly to the risk of developing type 2 diabetes mellitus, but other genes exert smaller effects.

Different genes may contribute different parts of a phenotype that was once thought to be due to the actions of a single gene. This is the case for migraine. A gene on chromosome 1 contributes sensitivity to sound; a gene on chromosome 5 produces the pulsating headache and sensitivity to light; and a gene on chromosome 8 is associated with nausea and vomiting. In addition, environmental influences trigger migraine in some people, such as eating certain foods, stress, or certain weather conditions. **Clinical Connection 7.1** takes a closer look at risk factors for heart disease.

Key Concepts Questions 7.1

1. Explain how polygenic traits differ from single-gene (Mendelian) traits.
2. Define *multifactorial traits*.
3. Are the contributions of individual genes to a polygenic multifactorial trait necessarily equal?

Clinical Connection 7.1

Many Genes Control Heart Health

Many types of cells and processes must interact for the heart and blood vessels (the cardiovascular system) to circulate blood, and many genes maintain the system. The environment has strong effects, too. For example, intake of vitamin K, necessary for blood to clot, influences the severity of single-gene clotting diseases. Cardiovascular disease affects one in three individuals.

Genes control the heart and blood vessels in several ways: transporting lipids; blood clotting; blood pressure; and how well white blood cells stick to blood vessel walls. Lipids can only circulate when bound to proteins, forming lipoproteins. Several genes encode the protein parts of lipoproteins, which are called apolipoproteins. Some types of lipoproteins carry lipids in the blood to tissues, where they are used, and other types of lipoproteins take lipids to the liver, where they are broken down into biochemicals that the body can excrete. An allele of the apolipoprotein E gene (*E4*) increases the risk of a heart attack threefold in people who smoke.

Maintaining a healthy cardiovascular system requires a balance between having enough lipids inside cells and not having an excess outside cells. Genes specify enzymes that process lipids, the proteins that transport them, and receptor proteins that admit lipids into cells. Atherosclerosis is the buildup of fatty material, called plaque, on the interior linings of arteries.

An enzyme, lipoprotein lipase, lines the walls of the smallest blood vessels, where it breaks down fat packets released from the small intestine and liver. Lipoprotein lipase is activated by high-density lipoproteins (HDLs), and it breaks down low-density lipoproteins (LDLs). Because low LDL levels are associated with a healthy cardiovascular system in many families, low LDL is used as a biomarker of heart health. High HDL was once widely used as a biomarker for health, until cardiologists recognized what geneticists had long known—that in some families with no heart problems, HDL is low, usually a danger sign. People with elevated LDL levels that diet and exercise cannot control may take statin drugs, which block the enzyme the liver requires to synthesize cholesterol.

The fluidity of the blood is also critical to health. Overly active clotting factors or extra sticky white blood cells (leukocytes) can block blood flow, usually in blood vessels in the heart or in the legs. Poor clotting causes dangerous bleeding. Because clotting factors are proteins, clotting is genetically controlled.

Elevated cholesterol in the blood illustrates the multifactorial nature of cardiovascular disease. People who inherit one mutant allele of the LDL receptor gene typically suffer heart attacks in early adulthood, and the rare individuals who inherit two mutant alleles die much earlier (see figure 5.2). But high cholesterol can arise from the actions of at least 12 other genes.

Genetic test panels detect alleles of genes that contribute to risk of developing cardiovascular disease. More than 50 genes regulate blood pressure, and more than 95 contribute to inherited variation in blood cholesterol and triglyceride levels. Tests of gene expression can indicate which cholesterol-lowering drugs are most likely to be effective and tolerable for a particular individual. Computer analysis of multigene tests accounts for controllable environmental factors, such as exercising, not smoking, and maintaining a healthy weight (**table 7A**). **Figure 7A** shows an artery blocked by fatty plaque. Diet, regular exercise, and medication can help to counter an inherited tendency to deposit cholesterol-rich material on the interior linings of arteries.

Table 7A	Risk Factors for Cardiovascular Disease	
Uncontrollable		**Uncontrollable**
Age		Fatty diet
Male sex		Hypertension
Genes		Smoking
Lipid metabolism		High serum cholesterol
Apolipoproteins		Stress
Lipoprotein lipase		Diabetes
Blood clotting		
Fibrinogen		
Clotting factors		
Inflammation		
C-reactive protein		
Homocysteine metabolism		
Leukocyte adhesion		

(Continued)

Figure 7A **Cardiovascular disease.** Accumulation of fatty plaque blocks circulation of the blood through arteries. © The McGraw-Hill Companies, Inc.

7.2 Polygenic Traits Are Continuously Varying

For a polygenic trait, the combined action of many genes often produces a "shades of gray" or "continuously varying" phenotype, also called a quantitative trait. DNA sequences that contribute to polygenic traits are called **quantitative trait loci (QTLs)**. A multifactorial trait is continuously varying if it is also polygenic. That is, it is the multigene component of the trait that contributes the continuing variation of the phenotype. The individual genes that confer a polygenic trait follow Mendel's laws, but together they do not produce single-gene phenotypic ratios. They all contribute to the phenotype, but without being dominant or recessive to each other. Single-gene traits are instead discrete or qualitative, often providing an "all-or-none" phenotype such as "normal" versus "affected."

A polygenic trait varies in populations, as our many nuances of hair color, body weight, and cholesterol levels demonstrate. Some genes contribute more to a polygenic trait than others. Within genes, alleles can have differing impacts depending upon exactly how they alter an encoded protein and how common they are in a population. For example, a mutation in the LDL receptor gene greatly raises blood serum cholesterol level. But because fewer than 1 percent of individuals in most populations have this mutation, it contributes very little to variation in cholesterol level in a population; however, the mutation has a large impact on the person who has it.

Although the expression of a polygenic trait is continuous, we can categorize individuals into classes and calculate the frequencies of the classes. When we do this and plot the frequency for each phenotype class, a bell-shaped curve results. Even when different numbers of genes affect the trait, the curve takes the same shape, as the following examples show.

Fingerprint Patterns

Skin on the fingertips is folded into raised patterns called dermal ridges that align to form loops, whorls, and arches. This pattern is a fingerprint. A technique called dermatoglyphics ("skin writing") compares the number of ridges that comprise these patterns to identify and distinguish individuals (**figure 7.2**).

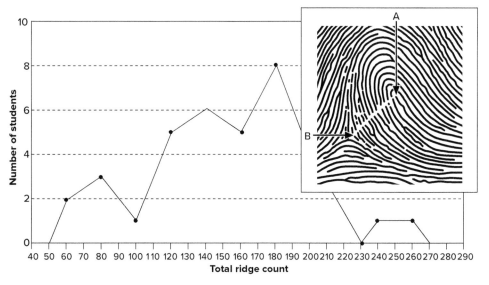

Figure 7.2 **Anatomy of a fingerprint.** Total ridge counts for a number of individuals, plotted on a bar graph, form an approximately bell-shaped curve. The number of ridges between landmark points A and B on this loop pattern is 12. Total ridge count includes the number of ridges on all fingers. *Data from Gordon Mendenhall, Thomas Mertens, and Jon Hendrix, "Fingerprint Ridge Count" in The American Biology Teacher, vol. 51, no. 4, April 1989, pp. 204–206.*

Dermatoglyphics is part of genetics because certain diseases (such as Down syndrome) produce unusual ridge patterns. Forensic fingerprint analysis is also an application of dermatoglyphics.

Genes largely determine the number of ridges in a fingerprint, but the environment can affect them, too. During weeks 6 through 13 of prenatal development, the ridge pattern can be altered as the fetus touches the finger and toe pads to the wall of the amniotic sac. This early environmental effect explains why the fingerprints of identical twins, who share all genes, are in some cases not exactly alike.

A fingerprint is quantified with a measurement called a total ridge count, which tallies the numbers of ridges in whorls, loops, or arches. The average total ridge count in a male is 145, and in a female, 126. Plotting total ridge count yields the bell curve of a continuously varying trait.

Height

The effect of the environment on height is obvious—people who do not eat enough do not reach their genetic potential for height. Students lined up according to height, but raised in two different decades and under different circumstances, illustrate the effects of genes and the environment on this continuously varying trait. Students from 1920 are on average considerably shorter than students from recent years. The tallest individuals from 1920 were 5′9″, whereas the tallest now are 6′5″. The difference in heights between then and now is attributed to improved diet and better overall health and perhaps the fact that many men were killed in the first world war, in the years before the first heights were measured. At least 50 genes affect height. **Figure 7.3** shows a bell curve for height.

Skin Color and Race

More than 100 genes affect pigmentation in skin, hair, and the irises. Melanin pigments color the skin to different degrees in different individuals. In the skin, as in the iris (see figure 4.6), melanocytes are cells that contain melanin in packets called melanosomes. Melanocytes extend between the tilelike skin cells, distributing pigment granules through the skin layers.

Figure 7.3 **The inheritance of height.** Genetics students at the University of Notre Dame lined up by height in inches, revealing the continuously varying nature of height. © *The McGraw-Hill Companies, Inc./Photo by David Hyde and Wayne Falda*

Some melanin exits the melanocytes and enters the hardened cells in the skin's upper layers. Here the melanin breaks into pieces, and as the skin cells are pushed up toward the skin's surface as stem cells beneath them divide, the bits of melanin provide color. The pigment protects against DNA damage from ultraviolet radiation. Exposure to sunlight increases melanin synthesis. **Figure 7.4***a* shows a three-gene model for human skin color. This is an oversimplification, but it illustrates how several genes can contribute to a variable trait.

We all have about the same number of melanocytes per unit area of skin. People have different skin colors because they vary in melanosome number, size, and density of pigment distribution. Different skin colors arise from the number and distribution of melanin pieces in cells in the uppermost skin layers.

Skin color is one physical trait that is used to distinguish race. The definition of race based largely on skin color is more a social construct than a biological concept, because skin color is only one of thousands of traits whose frequencies vary in different populations. From a genetic perspective, when referring to nonhumans, races are groups within species that are distinguished by different allele frequencies. Humans are actually a lot less variable in appearance than other mammals. We may classify people by skin color because it is an obvious visible way to distinguish individuals, but this trait is *not* a reliable indicator of ancestry.

The concept of race based on skin color falls apart when considering many genes. That is, two people with very dark skin may be less alike than either is to another person with very

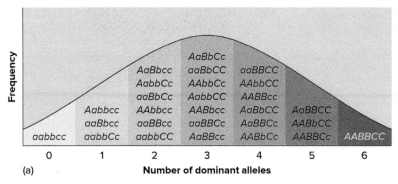

Figure 7.4 **Variations in skin color. (a)** A model of three genes, with two alleles each, can explain broad hues of human skin. In actuality, this trait likely involves many more than three genes. **(b)** Humans come in a great variety of skin colors. Skin color genes can assort in interesting ways. These beautiful young ladies, Marcia and Millie, are twins! Their father is Jamaican with dark skin and tight dark curls, and their mother is European with fair skin and golden-brown hair. *(b): © SWNS.com*

light skin. For example, sub-Saharan Africans and Australian aborigines have dark skin, but are dissimilar in other inherited characteristics. Their dark skins may reflect adaptation to life in a sunny, tropical climate rather than recent shared ancestry. Overall, 93 percent of varying inherited traits are no more common in people of one skin color than any other.

Testing DNA indicates that it makes more biological sense to classify people by ancestry than by skin color. In a sociology class, 100 students had their DNA tested for percent contribution from "European white," "black African," "Asian," and "Native American" gene variants. Many students were surprised. A light-skinned black man learned that genetically he is approximately half black, half white. Another student who considered herself black was 58 percent white European. The U.S. Census Bureau, in recognition of the complexity of classifying people into races based on skin color, began to allow "mixed race" as a category in 2000.

In a genetic sense the concept of race based on skin color has little meaning, but in a practical sense, racial groups *do* have different incidences of certain diseases, because of the tendency to choose partners within a group, which retains certain alleles. However, racial differences in disease prevalence may also result from social inequities, such as access to good nutrition or health care. Observations that populations of particular races have a higher incidence of certain illnesses have led to "race-based prescribing." For example, certain hypertension and heart disease drugs are marketed to African Americans, because this group has a higher incidence of these conditions than do people in other groups. But on the individual level a white person might be denied a drug that would work, or a black person given one that doesn't, if the treatment decision is based on a trait not directly related to how the body responds to a drug.

It is more accurate to prescribe drugs based on personal genotypes that determine whether or not a particular drug will work or have side effects, than by the color of a person's skin. For example, researchers cataloged twenty-three markers for genes that control drug metabolism in 354 people representing blacks (Bantu, Ethiopian, and Afro-Caribbean), whites (Norwegian, Armenian, and Ashkenazi Jews), and Asians (Chinese and New Guinean). The genetic markers fell into four distinct groups that predict which of several blood thinners, chemotherapies, and painkillers will be effective—and these response groups did not at all match the groups based on skin color. Chapter 20 continues discussion of selecting drugs based on genotype.

Key Concepts Questions 7.2

1. Define *quantitative trait loci.*
2. What does "continuously varying" mean?
3. Explain how a bell curve describes the distribution of phenotype classes of any polygenic trait.
4. Explain how fingerprint patterns, height, and skin color are multifactorial traits.

7.3 Methods to Investigate Multifactorial Traits

Predicting recurrence risks for polygenic traits is much more challenging than doing so for single-gene traits. Researchers use several strategies to investigate these traits.

Empiric Risk

Using Mendel's laws, it is possible to predict the risk that a single-gene trait will recur in a family from knowing the mode of inheritance—such as autosomal dominant or recessive. To predict the chance that a polygenic multifactorial trait will occur in a particular individual, geneticists use **empiric risk**, which is based on incidence in a specific population. **Incidence** is the rate at which a certain event occurs, such as the number of new cases of a disease diagnosed per year in a population of known size. **Prevalence** is the proportion or number of individuals in a population who have a particular disease at a specific time, such as during one year.

Empiric risk is not a calculation, but a population statistic based on observation. The population might be broad, such as an ethnic group or community, or genetically more well defined, such as families that have cystic fibrosis. Empiric risk increases with severity of the disease, number of affected family members, and how closely related a person is to affected individuals. For example, empiric risk is used to predict the likelihood of a child being born with a neural tube defect (NTD). In the United States, the overall population risk of carrying a fetus with an NTD is about 1 in 1,000 (0.1 percent). For people of English, Irish, or Scottish ancestry, the risk is about 3 in 1,000. However, if a sibling has an NTD, for any ethnic group, the risk of recurrence increases to 3 percent, and if two siblings are affected, the risk for a third child is even greater.

If a trait has an inherited component, then it makes sense that the closer the relationship between two individuals, one of whom has the trait, the greater the probability that the second individual has the trait, too, because they share more genes. Studies of empiric risk support this logic. **Table 7.1** summarizes empiric risks for relatives of individuals with cleft lip (**figure 7.5**).

Because empiric risk is based solely on observation, it is useful to derive risks for diseases with poorly understood transmission patterns. For example, certain multifactorial diseases affect one sex more often than the other. Pyloric stenosis, an overgrowth of muscle at the juncture between the stomach and the small intestine, is five times more common among males than females. The condition must be corrected surgically shortly after birth, or the newborn will be unable to digest foods. Empiric data show that the risk of recurrence for the brother of an affected brother is 3.8 percent, but the risk for the brother of an affected sister is 9.2 percent. Empiric risk is based on real-world observations, even if the cause of the illness or why it is more common in one gender than another is not known.

Table 7.1	Empiric Risk of Recurrence for Cleft Lip	
Relationship to Affected Person	**Empiric Risk of Recurrence**	
Identical twin	40.0%	
Sibling	4.1%	
Child	3.5%	
Niece/nephew	0.8%	
First cousin	0.3%	
General population risk (no affected relatives)	0.1%	

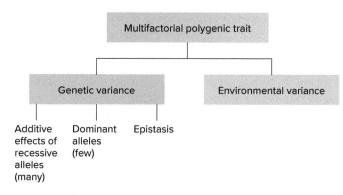

Figure 7.6 Heritability estimates the genetic contribution to the variability of a trait. Observed variance in a multifactorial polygenic trait or illness reflects genetic and environmental contributions.

Figure 7.5 Cleft lip. Cleft lip is more likely in a person who has a relative with the condition. Corrective surgery is highly effective. *Source: Centers for Disease Control and Prevention (CDC)*

Heritability

Charles Darwin noted that some of the variation of a trait is due to inborn differences in populations, and some to differences in environmental influences. A measurement called **heritability**, designated H, estimates the proportion of the phenotypic variation for a trait that is due to genetic differences in a certain population at a certain time. The distinction between empiric risk and heritability is that empiric risk could be affected by nongenetic influences, whereas heritability focuses on the genetic component of the variation in a trait. Heritability refers to the degree of *variation* in a trait due to genetics, and not to the proportion of the trait itself attributed to genes.

Figure 7.6 outlines the factors that contribute to observed variation in a trait. Heritability equals 1.0 for a trait whose variability is completely the result of gene action, such as in a population of laboratory mice who share the same environment. Without environmental variability, genetic differences alone determine expression of the trait in the population. Variability of most traits, however, is due to differences among genes and environmental components. **Table 7.2** lists some traits and their heritabilities.

Heritability changes as the environment changes. For example, the heritability of skin color is higher in the winter months, when sun exposure is less likely to increase melanin synthesis. The same trait may be highly heritable in two populations, but certain variants much more common in one group due to long-term environmental differences. Populations in equatorial Africa, for example, have darker skin than sun-deprived Scandinavians.

Researchers use several statistical methods to estimate heritability. One way is to compare the actual proportion of pairs of people related in a certain manner who share a particular trait to the expected proportion of pairs that would share it if it were inherited in a Mendelian fashion. The expected proportion is derived by knowing the blood relationships of the individuals and using a measurement called the **coefficient of relatedness**, which is the proportion of genes that two people related in a certain way share (**table 7.3**).

Table 7.2	Heritabilities for Some Human Traits
Trait	**Heritability**
Clubfoot	0.8
Height	0.8
Blood pressure	0.6
Body mass index	0.4–0.7
Verbal aptitude	0.7
Mathematical aptitude	0.3
Spelling aptitude	0.5
Total fingerprint ridge count	0.9
Intelligence	0.5–0.8
Total serum cholesterol	0.6

Table 7.3	Coefficient of Relatedness for Pairs of Relatives	
Relationship	**Degree of Relationship**	**Percent Shared Genes (Coefficient of Relatedness)**
Sibling to sibling	1°	50% (1/2)
Parent to child	1°	50% (1/2)
Uncle/aunt to niece/nephew	2°	25% (1/4)
Grandparent to grandchild	2°	25% (1/4)
First cousin to first cousin	3°	12.5% (1/8)

A parent and child share 50 percent of their genes because of the mechanism of meiosis. Siblings share on average 50 percent of their genes because they have a 50 percent chance of inheriting each allele for a gene from each parent. Genetic counselors use the designations primary (1°), secondary (2°), and tertiary (3°) relatives when calculating risks (see table 7.3; **figure 7.7**). For extended or complicated pedigrees, the value of 1 in 2 (or 50 percent) between siblings and between parent-child pairs can be used to trace and calculate the percentage of genes shared between people related in other ways.

If the heritability of a trait is very high, then of a group of 100 sibling pairs, nearly 50 would be expected to have the

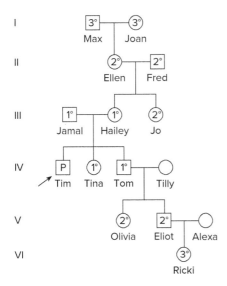

Figure 7.7 Tracing relatives. Tim has an inherited illness. A genetic counselor drew this pedigree to explain the approximate percentage of genes Tim shares with relatives. This information can be used to alert certain relatives to their risks. ("P" is the proband, or affected individual who initiated the study. See table 7.3 for definitions of 1°, 2°, and 3° relationships.)

same phenotype, because siblings share on average 50 percent of their genes. Height is a trait for which heritability reflects the environmental influence of nutrition. Of 100 sibling pairs in a population, for example, 40 might be the same number of inches tall. Heritability for height among this group of sibling pairs is 0.40/0.50, or 80 percent, which is the observed phenotypic variation divided by the expected phenotypic variation if environment had no influence.

Genetic variance for a polygenic trait is mostly due to the additive effects of recessive alleles of different genes. For some traits, a few dominant alleles can greatly influence the phenotype, but because they are rare, they do not contribute greatly to heritability. This is the case for heart disease caused by a faulty LDL receptor. Heritabilities for some traits or diseases may be underestimated if only mutations that change DNA base sequences are compared, without also considering **copy number variants (CNVs)**, which are differences in the numbers of copies of a DNA sequence.

Epistasis (interaction between alleles of different genes) can also influence heritability. To account for the fact that different genes affect a phenotype to differing degrees, geneticists calculate a "narrow" heritability that considers only additive recessive effects and a "broad" heritability that also considers the effects of rare dominant alleles and epistasis. For LDL cholesterol level, for example, the narrow heritability is 0.36 but the broad heritability is 0.96, indicating that a rare dominant allele has a large impact.

Understanding multifactorial inheritance is important in agriculture. A breeder needs to know whether genetic or environmental influences contribute to variability in such traits as birth weight, milk yield, and egg hatchability. It is also valuable to know whether the genetic influences are additive or epistatic. The breeder can control the environment by adjusting the conditions under which animals are raised and crops grown and control genetic effects by setting up crosses between particular individuals.

Studying multifactorial traits in humans is difficult because information must be obtained from many families. Two special types of people, however, can help geneticists to tease apart the genetic and environmental components of the variability of multifactorial traits—adopted individuals and twins.

Adopted Individuals

An adopted person typically shares environmental influences, but not many gene variants, with the adoptive family. Conversely, adopted individuals share genes, but not the exact environment, with their biological parents. Therefore, similarities between adopted people and adoptive parents reflect mostly environmental influences, whereas similarities between adoptees and their biological parents reflect mostly genetic influences. Information on both sets of parents can indicate how heredity and the environment contribute to a trait.

Adoption studies go back many years. An early investigation used a database of all adopted children in Denmark and their families from 1924 to 1947. One study examined correlations between causes of death among biological and adoptive parents and adopted children. If a biological parent died of infection before age 50, the child he or she gave up for adoption was five times more likely to die of infection at a young age than a similar person in the general population. This may be because inherited variants in immune system genes increase susceptibility to certain infections. In support of this hypothesis, the risk that an adopted individual would die young from infection did not correlate with adoptive parents' deaths from infection before age 50. Researchers concluded that genetics mostly determines length of life, but there are also environmental influences. For example, if adoptive parents died before age 50 of cardiovascular disease, their adopted children were three times as likely to die of heart and blood vessel disease as a person in the general population. What environmental factor might explain this finding?

Twins

Studies that use twins to separate the genetic from the environmental contribution to a phenotype provide more meaningful information than studies of adopted individuals.

A trait that occurs more frequently in both members of identical (monozygotic, or MZ) twin pairs than in both members of fraternal (dizygotic, or DZ) twin pairs is at least partly controlled by heredity. Geneticists calculate the **concordance** of a trait as the percentage of pairs in which both twins express the trait among pairs of twins in whom at least one has the trait. Twins who differ in a trait are said to be discordant for it.

In one study, 142 MZ twin pairs and 142 DZ twin pairs took a "distorted tunes test," in which 26 familiar songs were played, each with at least one note altered. A person was considered "tune deaf" if he or she failed to detect the mistakes in three or more tunes. Concordance for "tune deafness" was 67 percent for MZ twins but only 44 percent for DZ twins, indicating a considerable inherited component in the ability to accurately perceive musical pitch. **Table 7.4** compares twin types for a variety of hard-to-measure traits. (Figure 3.17 shows how DZ and MZ twins arise.)

Diseases caused by single genes that approach 100 percent penetrance, whether dominant or recessive, also approach 100 percent concordance in MZ twins. That is, if one identical twin has the disease, so does the other. However, among DZ twins, concordance generally is 50 percent for a dominant trait and 25 percent for a recessive trait. These are the Mendelian values that apply to any two nontwin siblings. For a polygenic trait with little environmental input, concordance values for

Table 7.4	Concordance Values for Some Traits in Twins	
Trait	**MZ (Identical) Twins**	**DZ (Fraternal) Twins**
Acne	14%	14%
Alzheimer disease	78%	39%
Anorexia nervosa	55%	7%
Autism	90%	4.5%
Bipolar disorder	33%–80%	0%–8%
Cleft lip with or without cleft palate	40%	3%–6%
Hypertension	62%	48%
Schizophrenia	40%–50%	10%

MZ twins are significantly greater than for DZ twins. A trait molded mostly by the environment exhibits similar concordance values for both types of twins.

Comparing twin types assumes that both types of twins share similar experiences. In fact, MZ twins are often closer emotionally than DZ twins. This discrepancy between the closeness of the two types of twins can lead to misleading results. A study from the 1940s, for example, concluded that tuberculosis is inherited because concordance among MZ twins was higher than among DZ twins. Actually, the infectious disease more readily passed between MZ twins because their parents kept them closer. However, we do inherit susceptibilities to some infectious diseases. MZ twins would share such genes, whereas DZ twins would only be as likely as any sibling pairs to do so.

A more informative way to assess the genetic component of a multifactorial trait is to study MZ twins who were separated at birth, then raised in different environments. The Minnesota Twin Family Study at the University of Minnesota has used this "twins reared apart" approach. Since 1989, thousands of sets of twins and triplets who were separated at birth have visited the laboratories of Thomas Bouchard there. For a week or more, the twins and triplets are tested for physical and behavioral traits, including 24 blood types, handedness, direction of hair growth, fingerprint pattern, height, weight, organ system function, intelligence, allergies, and dental patterns. The participants provide DNA samples. Researchers videotape facial expressions and body movements in different circumstances and probe participants' fears, interests, and superstitions.

Twins and triplets separated at birth provide natural experiments for distinguishing nature from nurture. Many of their common traits can be attributed to genetics, especially if their environments have been very different. Their differences tend to come from differences in upbringing, because their genes are identical (MZ twins and triplets) or similar (DZ twins and triplets).

Some MZ twins separated at birth and reunited later are remarkably similar, even when they grow up in very different adoptive families (**figure 7.8**). Idiosyncrasies are particularly striking. One pair of twins who met for the first time when they were in their thirties responded identically to questions; each paused for 30 seconds, rotated a gold necklace she was wearing three times, and then answered the question. Coincidence, or genetics?

The "twins reared apart" approach is not an ideal way to separate nature from nurture. MZ twins and other multiples share an environment in the uterus and possibly in early infancy that may affect later development. Siblings, whether adoptive or biological, do not always share identical home environments. Differences in sex, general health, school and peer experiences, temperament, and personality affect each individual's perception of such environmental influences as parental affection and discipline.

Genome-Wide Association Studies

A **genome-wide association study (GWAS)** compares many landmarks (genetic markers) across the genome between two large groups of people—one with a particular trait or disease and one without it. Identifying parts of the genome that are much more common among the people with the trait or illness can lead researchers to genes that contribute to the phenotype.

Genome-wide association studies use several types of genetic markers (**table 7.5**). **Single nucleotide polymorphisms (SNPs)** and copy number variants (CNVs) describe the DNA

Table 7.5	Types of Information Used in Genome-Wide Association Studies
Marker Type	**Definition**
SNP	A single nucleotide polymorphism is a site in the genome that has a different DNA base in >1% of a population.
CNV	A copy number variant is a tandemly repeated DNA sequence, such as CGTA CGTA CGTA.
Gene expression	The pattern of genes that are overexpressed and/or underexpressed in people with a particular trait or disease.
	Epigenetic signature of methyl groups binding DNA.

base sequence. A SNP is a site in the genome that has a different DNA base in at least 1 percent of a population (**figure 7.9**). A CNV is a DNA sequence that repeats a different number of times in different individuals (**figure 7.10**). A CNV does not provide information in the same way as a gene that

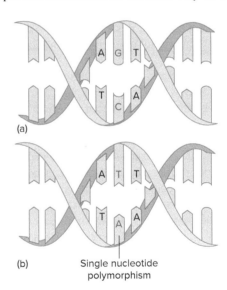

(a)

(b)   Single nucleotide
           polymorphism

Figure 7.9 SNPs are sites of variability. The DNA base pair in red is a SNP—a site that differs in more than 1 percent of a population. (The percentage may change as SNPs are identified in more individuals.) In this case, 98% of individuals might have GC at the indicated position **(a)**, and 2% have TA **(b)**.

Figure 7.8 Identical twins have much in common. In addition to physical traits, MZ twins may share tastes, preferences, and behaviors. Studying MZ twins separated at birth and reunited is a way to assess which traits are inherited.
© Bob Kreisel /Alamy

GATTACA	Allele 1
GATTACAGATTACA	Allele 2
GATTACAGATTACAGATTACA	Allele 3
GATTACAGATTACAGATTACAGATTACA	Allele 4

Figure 7.10 Copy number variants provide a different type of genetic information. For copy number variants, different numbers of repeats of a short DNA base sequence are considered to be different alleles.

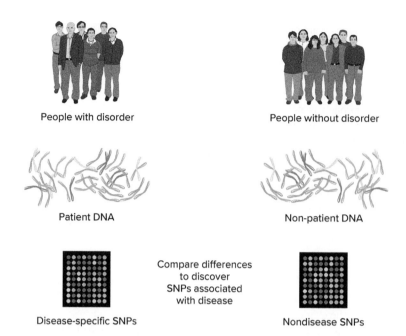

People with disorder

People without disorder

Patient DNA

Non-patient DNA

Disease-specific SNPs

Compare differences
to discover
SNPs associated
with disease

Nondisease SNPs

Figure 7.11 **Tracking genes in groups.** Genome-wide association studies seek DNA sequence variants that are shared with much greater frequency among individuals with the same illness or trait than among others. The squares are DNA microarrays, which display short, labeled DNA pieces (see figure 19.7). Different patterns indicate different alleles.

encodes protein does, but it is another way to distinguish individuals. CNVs are useful in forensic applications, discussed in chapter 14.

Gene expression patterns are also used in genome-wide association studies. These patterns indicate which proteins are overproduced or underproduced in people with the trait or illness, compared to unaffected controls. Yet another way to compare genomes is by the sites to which methyl (CH_3) groups bind, shutting off gene expression. This is an epigenetic change because it doesn't affect the DNA base sequence.

To achieve statistical significance, a genome-wide association study must include at least 100,000 markers. It is the association of markers to a trait or disease that is informative (**figure 7.11**). Typically, genome-wide association studies use a million or more SNPs, grouped into half a million or so haplotypes. A specific "tag SNP" is used to identify a haplotype.

A genome-wide association study is a stepwise focusing in on parts of the genome responsible to some degree for a trait (**figure 7.12**). In general, a group of people with the same condition or trait and a control group have their DNA isolated and genotyped for the 500,000 tag SNPs. Statistical algorithms identify the uniquely shared SNPs among the group of individuals with the trait or disease. Repeating the process for additional populations narrows the SNPs and strengthens the association. It is important to validate a SNP association in different population groups, to be certain that it is the trait of interest that is being tracked, and not another part of the genome that members of one population share due to their common ancestry.

Several study designs are used in genome-wide association studies (**table 7.6**). In a cohort study, researchers follow a large group of individuals over time and measure many aspects of their health. The most famous is the Framingham Heart Study, which began tracking thousands of people and their descendants in Massachusetts in 1968. Nine thousand of them are participating in a genome-wide association study.

In a case-control study, each individual in one group is matched to an individual in another group who shares as many characteristics as possible, such as age, sex, activity level, and

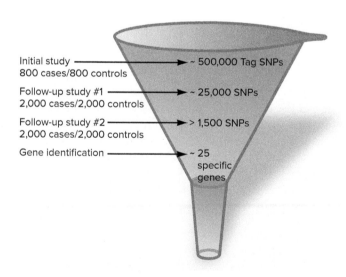

Initial study
800 cases/800 controls
~ 500,000 Tag SNPs

Follow-up study #1
2,000 cases/2,000 controls
~ 25,000 SNPs

Follow-up study #2
2,000 cases/2,000 controls
> 1,500 SNPs

Gene identification
~ 25 specific genes

Figure 7.12 **A stepwise approach to gene discovery.** Genome-wide association study results must be validated in several different populations. Further research is necessary to go beyond association to demonstrate correlation and cause.

© Sami Sarkis/Photographer's Choice/Getty Images RF

CHAPTER 8

Genetics of Behavior

Child abuse and PTSD. *Inheriting a variant of a particular gene may influence how a person handles memories of child abuse throughout her or his lifetime.*

Learning Outcomes

8.1 Genes and Behavior

1. Identify the physical basis of behavioral traits in the brain.
2. Explain how genes can affect behavior.

8.2 Sleep

3. Describe how studies on dogs led to discovery of a narcolepsy gene in humans.
4. Explain how mutations in the *period 2* gene disrupt sleep-wake cycles.

8.3 Intelligence and Intellectual Disability

5. Describe development of intelligence tests.
6. List causes of intellectual disability.

8.4 Drug Addiction

7. State the two identifying characteristics of drug addiction.
8. Discuss genetic and neurological evidence that drug addiction is a biological phenomenon.
9. Explain how people can be addicted to drugs that are derived from plants.
10. Describe two types of receptors that are implicated in drug addiction.

8.5 Mood Disorders

11. Explain how genes control neurotransmitter abnormalities that lie behind major depressive disorder.
12. Explain how the genetics of bipolar disorder is complex.

8.6 Schizophrenia

13. Explain how schizophrenia differs from the mood disorders.

14. Discuss how synaptic pruning can unite environmental and genetic contributions to elevating the risk of developing schizophrenia.

8.7 Autism

15. List factors that increase the risk of developing autism.
16. Explain three ways to use exome or genome sequence data to discover genes that contribute to autism.
17. Discuss the role of neurexins and neuroligins in causing some cases of autism.

The BIG Picture

Genes that affect brain structure and function give rise to such behavioral traits and tendencies as sleep patterns, addictions, and intelligence. Variants of such genes, with environmental factors, contribute to or cause mood disorders, schizophrenia, and autism. Once considered part of psychology and psychiatry, conditions that affect how we think and communicate are becoming the province of neurology and genetics.

Posttraumatic Stress Disorder: Who Is Predisposed?

A young woman who was physically and emotionally abused as a child becomes a warm, caring mother with a successful career. Another woman with a similar upbringing grows into an anxious adult who suffers from posttraumatic stress disorder (PTSD).

PTSD is a reaction to traumatic physical harm, or the perceived threat of harm, which persists long after the triggering event has passed. The person suffers from anxiety, flashbacks, and avoidance of memories of the event, resulting from inappropriate activation of the stress hormones that flooded the bloodstream during the original trauma. Often associated with combat experiences, PTSD also follows health scares, natural disasters, crimes, and other disturbing events. PTSD is rare.

Many factors contribute to the development of PTSD. One study of 2,000 people who had been severely traumatized from abuse, either as children or adults, found that those abused as children who developed PTSD were significantly more likely to have a specific variant of a gene called *FKBP5* than other individuals. The effect is both genetic and epigenetic. In people with the gene variant who were abused as children and developed PTSD, methyl groups on the gene are absent—an effect not seen in the people who did not develop PTSD. As a result of the decreased methylation, receptors for stress hormones (glucocorticoids) malfunction, and the body cannot react and adapt to stress.

The study found this epigenetic change in white blood cells, but other experiments on cells growing in the laboratory showed that it also occurs and persists in neural progenitor cells that differentiate into brain nerve cells that will last a lifetime. In a person, the epigenetic change happened at a critical period in development, setting the stage in a child for an inability to fully use stress hormones later in life. The result: PTSD.

8.1 Genes and Behavior

Behavior is a complex continuum of emotions, moods, intelligence, and personality that drives how we function on a daily basis. We are, to an extent, defined and judged by our many behaviors. They control how we communicate, cope with negative feelings, and react to stress. Behavioral disorders are common, with wide-ranging and sometimes overlapping symptoms. Past chapters have referred to inherited *diseases,* which are conditions for which the precise cause is known. Behavioral conditions are termed *disorders* because many causes are possible, and they may not be known.

Many of our behaviors occur in response to environmental factors, but

how we respond may have genetic underpinnings, as the chapter opener describes for flashing back to abuse during childhood. Understanding the biology behind behavior can help scientists develop treatments for disorders that have behavioral symptoms.

Behavioral genetics considers nervous system function and variation, including the hard-to-define qualities of mood and mind. The human brain weighs about 3 pounds and resembles a giant gray walnut, but with the appearance and consistency of pudding. It consists of 100 billion nerve cells, or **neurons**, and at least a trillion other cells called **neuroglia**, which support and nurture the neurons.

Brain neurons connect and interact in complex networks that control all of the body. Branches from each of the 100 billion neurons in the brain form close associations, called **synapses**, with 1,000 to 10,000 other neurons. Neurons communicate across these tiny spaces using chemical signals called **neurotransmitters**. Networked neurons oversee broad functions such as sensation and perception, memory, reasoning, and muscular movements.

Genes affect behavior by encoding diverse proteins that control the production and distribution of neurotransmitters. **Figure 8.1** indicates the points where genes act in sending and receiving nervous system information. Enzymes oversee

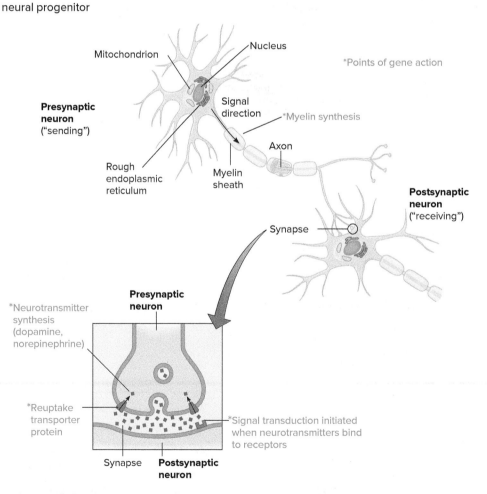

Figure 8.1 Neurotransmission. Many genes that affect behavior encode proteins that affect neurotransmission, which is sending a chemical signal from one neuron to another across a synapse.

the synthesis of neurotransmitters and their transport from the sending (presynaptic) neuron across the synapse to receptors on the receiving (postsynaptic) neuron. Proteins called transporters ferry neurotransmitters from sending to receiving neurons, and proteins also form the subunits of the receptors that bind neurotransmitters. Genes control the synthesis of myelin, a substance consisting of fats and proteins that coats neuron extensions called axons. Myelin, which accumulates in and causes the plasma membranes of neuroglia to balloon out, forms a sheath that insulates the neuron, which speeds neurotransmission. Myelin-rich neuroglia wrap around an axon like a bandage wound around a finger.

Many behavioral traits and disorders, like other characteristics, reflect a major influence from a single gene, perhaps one whose protein product takes part directly in neurotransmission, but also small inputs from variants of hundreds of other genes. Inheriting certain subsets of variants of these genes makes an individual susceptible to developing a certain disorder in the presence of a particular environmental stimulus. People who have different behavioral disorders, such as depression, addiction, and schizophrenia, may share underlying gene variants. Human behavior is genetically quite complex!

Deciphering the genetic components of behavioral traits and diseases uses traditional empiric risk determination, adoptee and twin study data, and exome and genome sequencing

to identify responsible gene variants. Considering genes may make it possible to subtype behavioral conditions in a biologically meaningful way so that diagnoses will be personalized and effective treatments begun sooner.

Table 8.1 lists the 12-month prevalence of some behavioral disorders, which refers to people who have had symptoms in the year prior to being asked. In contrast, lifetime prevalence indicates the proportion of individuals who report symptoms at any time over a lifetime.

Key Concepts Questions 8.1

1. What is behavioral genetics?
2. What is the cellular basis of behavior?
3. How do genes influence behavior?

8.2 Sleep

Sleep has been called "a vital behavior of unknown function," and, indeed, without sleep, animals die. We spend a third of our lives in this mysterious state.

Genes influence sleep characteristics. When asked about sleep duration, schedule, quality, nap habits, and whether they are "night owls" or "morning people," MZ (identical) twins report significantly more in common than do DZ twins. This is true even for MZ twins separated at birth. Twin studies of brain wave patterns through four of the five stages of sleep confirm a hereditary influence. The fifth stage, rapid eye movement or REM sleep, is associated with dreaming and therefore may indicate the input of experience more than genes.

Narcolepsy

Researchers discovered the first gene related to sleep in 1999, for a condition called "narcolepsy with cataplexy" in dogs. Mutations in several genes cause narcolepsy in humans.

A person (or dog) with narcolepsy falls asleep suddenly several times a day. Extreme daytime sleepiness greatly disrupts daily activities. People with narcolepsy have a tenfold higher rate of car accidents. Another symptom is sleep paralysis, which is the inability to move for a few minutes after awakening. The most dramatic manifestation of narcolepsy is cataplexy. During these short and sudden episodes of muscle weakness, the jaw sags, the head drops, knees buckle, and the person falls to the ground. This often occurs during a bout of laughter or excitement. Narcolepsy with cataplexy affects only 0.02 to 0.06 percent of the general populations of North America and Europe, but the fact that it is much more common in certain families suggests a genetic component.

Studies on dogs led the way to discovery of one human narcolepsy gene. In 1999, researchers discovered mutations in a gene that encodes a receptor for a neuropeptide (a short protein

Table 8.1	12-Month Prevalence of Some Behavioral Disorders in the U.S. Adult Population
Condition	**Prevalence (%)**
Anxiety	18.0
Social phobia	6.8
Posttraumatic stress disorder	3.5
Generalized anxiety disorder	3.1
Obsessive compulsive disorder	1.0
Panic disorder	2.7
Autism spectrum disorders (children)	1.5
Eating disorders	3.0
Anorexia nervosa	female 0.9, male 0.3
Binge eating	1.2
Bulimia nervosa	0.3
Mood disorders	9.5
Major depressive disorder	6.7
Bipolar disorder	2.6
Schizophrenia	1.1

Data from the National Institute of Mental Health

Figure 8.2 Letting sleeping dogs lie. These Doberman pinschers have inherited narcolepsy. They suddenly fall into a short but deep sleep while playing. Research on dogs with narcolepsy led to the discovery of a gene that affects sleep in humans and development of a drug to treat insomnia.
© Stanford University Center for Narcolepsy

that functions as a neurotransmitter) called hypocretin. In Doberman pinschers and Labrador retrievers, the receptor does not reach the cell surfaces of certain brain cells, preventing the cells from receiving signals to promote a state of awakeness. **Figure 8.2** shows a still frame of a film of narcoleptic dogs playing. Suddenly, they all collapse! A minute later, they get up and resume their antics. To induce a narcoleptic episode in puppies, researchers let them play with each other. Feeding older dogs meat excites them so much that they can take a while to finish a meal because they fall down in delight so often. Getting narcoleptic dogs to breed is difficult, too, for sex is even more exciting than play or food!

The hypocretin receptor gene turned out to be the same as a gene discovered a year earlier that encodes a neuropeptide that controls eating, called orexin. Neurons that produce the neuropeptide in different parts of the brain account for the two effects. Researchers study hypocretin/orexin and the molecules with which they interact to develop drugs to treat insomnia. At least three other genes are associated with insomnia. The most severe condition is "fatal familial insomnia," which is caused by infectious protein misfolding (see section 10.4).

Familial Advanced Sleep Phase Syndrome

Daily rhythms such as the sleep-wake cycle are set by cells that form a "circadian pacemaker" in two clusters of neurons in the brain called the suprachiasmatic nuclei. In these cells, certain "clock" genes are expressed in response to light or dark in the environment.

The function of clock genes is startling in families that have a mutation. For example, five generations of a family in Utah have familial advanced sleep phase syndrome. Affected people fall asleep at 7:30 P.M. and awaken at 4:30 A.M. They have a single DNA base mutation in a gene on chromosome

2, *period 2,* that interferes with synchronization of the sleep-wake cycle with daily sunrise and sunset, which is an example of a circadian rhythm. The condition is autosomal dominant, and each generation has several affected individuals. A young woman in another family goes to sleep at 9 A.M. and awakens at 5 P.M., as do her father and paternal grandmother. She wrote a newspaper column called "The Night Girl Finds a Day Boy" about living with a partner who has a normal sleep-wake cycle.

Mutation at a different part of the *period 2* gene affects how early we tend to awaken. People with two adenines at this site (genotype *AA*) awaken an hour earlier than people with two guanines (genotype *GG*). Heterozygotes, genotype *AG,* awaken at times between those of the two homozygotes.

Key Concepts Questions 8.2

1. What is the evidence that sleep habits are inherited?
2. Describe two mutations that alter sleep.

8.3 Intelligence and Intellectual Disability

Intelligence is a complex and variable trait that is subject to many genetic and environmental influences, and whose definition is highly subjective. Definitions of intelligence therefore vary. In general, intelligence refers to the ability to reason, learn, remember, connect ideas, deduce, and create.

The first intelligence tests, developed in the late nineteenth century, assessed sensory perception and reaction times to various stimuli. In 1904, Alfred Binet in France developed a test with questions based on language, numbers, and pictures to predict the success of developmentally disabled youngsters in school. The test was later modified at Stanford University to assess white, middle-class Americans. An average score on this "intelligence quotient," or IQ test, is 100, with two-thirds of all people scoring between 85 and 115 in a bell-shaped curve,

---

A GLIMPSE OF HISTORY

Sir Francis Galton, a half–first cousin of Charles Darwin, investigated the characteristic of genius, which he defined as "a man endowed with superior faculties." He identified successful and prominent people in Victorian-era English society and then assessed success among their relatives. In his 1869 book, *Hereditary Genius,* Galton wrote that relatives of eminent people were more likely to also be successful than people in the general population. The closer the blood relationship, he concluded, the more likely the person was to succeed. This, he claimed, established a hereditary basis for intelligence.

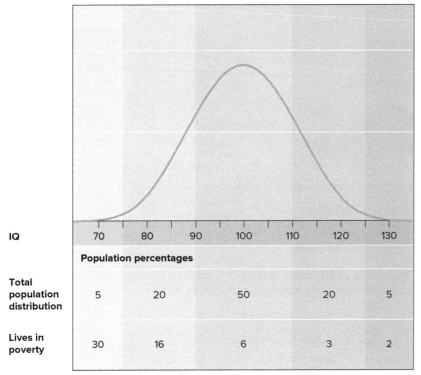

IQ	70	80	90	100	110	120	130
Population percentages							
Total population distribution	5	20		50		20	5
Lives in poverty	30	16		6		3	2

Figure 8.3 **Success and IQ.** IQ scores predict success in school and the workplace in U.S. society. The bell-shaped curve for IQ indicates that most people fall between 85 and 115. However, those living in poverty tend to have lower IQs.

due to genes or the environment? The researchers discovered a "substantial genetic correlation" of 0.62 for intelligence at both ends of the life span. That is, if someone was a smart 11-year-old, chances are excellent that he or she will be a smart senior citizen.

Genetics can explain some cases of intellectual disability, which affects 3 in 100 individuals in the United States. (Intellectual disability is a broad term that includes mental retardation.) Causes are many: abnormal genes and chromosomes, but also environmental exposure to toxins, infections, malnutrition, and noninherited birth defects. Probably at least 1,000 of our 20,325 or so genes affect the brain and their variants can therefore impact intelligence.

Among the best-studied genetic causes of intellectual disability are Down syndrome (see Bioethics in chapter 13) and fragile X syndrome (see Clinical Connection 12.1), and one of the most familiar environmental causes is fetal alcohol syndrome (see figure 3.22). Exome and genome sequencing have been helpful in assigning causative genes in children with intellectual disability who have not been diagnosed with these or other more common conditions.

or normal distribution (**figure 8.3**). An IQ between 50 and 70 is considered mild intellectual disability and below 50, severe intellectual disability. IQ has been a fairly accurate predictor of success in school and work, but it does not solely represent biological influences on intelligence. Low IQ correlates with societal situations, such as poverty, not graduating from high school, incarceration (males), and having a child out of wedlock (females).

Psychologists use a general intelligence ability, called "*g*," to represent the four basic skills that IQ encompasses. These are verbal fluency, mathematical reasoning, memory, and spatial visualization skills. The general intelligence measure eliminates the effect of unequal opportunities on assessment of intelligence.

Traditional methods of assessing multifactorial traits, such as adoption and twin studies, show a heritability of 0.5 to 0.8 for intelligence, indicating a high contribution from genetics. New types of studies confirm the heritability established with older techniques, and also identify specific genes that contribute to or affect intelligence.

Researchers used a variation of a genome-wide association study (see section 7.3) to evaluate change in intelligence over a lifetime. They considered general intelligence measured with a standard test given to nearly 2,000 people in Scotland when they were 11 years old and again when they were older (age 65, 70, or 79). They asked the question: *To what extent are differences in intelligence between childhood and old age*

Key Concepts Questions 8.3

1. What is intelligence?
2. How does using measurement of general intelligence improve upon using IQ score to assess the inheritance of intelligence?
3. How are newer genetic technologies being used to discover gene variants that cause or contribute to intellectual disability?

8.4 Drug Addiction

One person sees a loved one battling lung cancer and never smokes again. Another person actually has lung cancer, yet takes frequent breaks from using her oxygen tank to smoke. Evidence is mounting that genes play a large role in making some individuals prone to dependency on certain substances, and others not.

Substance Use Disorders

Our ancient ancestors must have discovered that ingesting certain natural substances, particularly from plants, provided a feeling of well-being. That tendency persists today. Substance use disorders can develop in individuals who repeatedly ingest certain psychoactive chemical compounds, even if the dangers are clear. The most severe type of substance use disorder is drug addiction. It is defined as compulsively seeking and taking a drug despite knowing its dangers.

The two identifying characteristics of drug addiction are tolerance and dependence. Tolerance is the need to take more of the drug to achieve the same effects as time goes on. Dependence is the onset of withdrawal symptoms upon stopping use of the drug. Both tolerance and dependence contribute to the biological and psychological components of craving the drug. The behavior associated with drug addiction can be extremely difficult to break.

Whether or not a particular chemical compound is addictive or not depends on which receptors on brain cells it binds to and where in the brain those cells are located. Drug addiction produces long-lasting changes in brain structure and function. Craving and high risk of relapse remain even after a person has abstained for years. Heritability of drug addiction is 0.4 to 0.6, with a two- to threefold increase in risk among adopted individuals who have one affected biological parent. Twin studies also indicate an inherited component to drug addiction.

Brain imaging techniques localize the "seat" of drug addiction in the brain by highlighting the cell surface receptors that bind neurotransmitters when a person craves the drug. The brain changes that contribute to addiction occur in parts called the nucleus accumbens, the prefrontal cortex, and the ventral tegmental area, which are part of a larger set of brain structures called the limbic system (**figure 8.4**). The effects of cocaine seem to be largely confined to the nucleus accumbens, whereas alcohol affects the prefrontal cortex.

The specific genes and proteins that are implicated in addiction to different substances may vary, but several general routes of interference in brain function are at play. Proteins involved in drug addiction are those that

- produce neurotransmitters, such as enzymes required for neurotransmitter synthesis;
- remove excess neurotransmitters from the synapse (called reuptake transporters);
- form receptors on the postsynaptic (receiving) neuron that are activated or inactivated when specific neurotransmitters bind; and
- convey chemical signals in the postsynaptic neuron.

Addictive Drugs

Many addictive drugs are derived from plants, including cocaine, opium, and tetrahydrocannabinol (THC), the main psychoactive component of cannabis (marijuana). These substances bind to

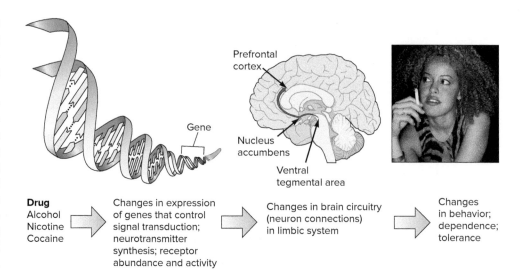

Drug Alcohol Nicotine Cocaine → Changes in expression of genes that control signal transduction; neurotransmitter synthesis; receptor abundance and activity → Changes in brain circuitry (neuron connections) in limbic system → Changes in behavior; dependence; tolerance

Figure 8.4 **The events of addiction.** Addiction is manifest at several levels: at the molecular level, in neuron-neuron interactions in the brain, and in behavioral responses. © Ryan McVay/Getty Images RF

receptors on human neurons, which indicates that our bodies have versions of these plant-derived compounds. The human equivalents of the opiates are the endorphins and enkephalins, and the equivalent of THC is anandamide. The endorphins and enkephalins relieve pain. Anandamide modulates how brain cells respond to stimulation by binding to neurotransmitter receptors on presynaptic (sending) neurons, and are involved in response to stress and in social behavior. In contrast, neurotransmitters bind to receptors on postsynaptic neurons. **Clinical Connection 8.1** discusses a second component of cannabis that is being tested to treat seizures and other conditions.

Amphetamines and LSD (lysergic acid diethylamide) produce their effects by binding to receptors on neurons that normally bind neurotransmitters called trace amines, which are found throughout the brain at low levels. LSD causes effects similar to the symptoms of schizophrenia (see section 8.6), suggesting that the trace amine receptors, which are proteins, may be implicated in the illness. LSD produces hallucinations, distorted perception, and heightened states of awareness, but is not addictive.

A GLIMPSE OF HISTORY

Industrial chemist Albert Hofmann had to go home one afternoon in 1943 when he accidentally ingested the chemical compound he was working with—LSD—as he investigated ways to treat behavioral disorders. He discovered the hallucinogenic results that would make the drug recreational in the 1960s: a 2-hour bombardment of kaleidoscope-like colors and shapes in his vision, while he remained in a happy, dreamlike state.

People addicted to various drugs share certain gene variants that must be paired with environmental stimuli for addiction to occur. For example, people who are homozygous for the

Cannabis: Addictive Drug and Seizure Treatment

The plant *Cannabis sativa* includes more than 500 chemical compounds, 113 of which are classified as cannabinoids because they bind the same family of receptors (**figure 8A**). The cannabinoid that provides the psychoactive and addictive effects of marijuana (the flowers of the plant) is tetrahydrocannabinol (THC). Genome-wide association studies (see figure 7.11) identify three regions in the human genome that include genes that control dependency on THC.

Another abundant cannabinoid is cannabidiol (CBD). Plants that are very high in THC tend to be low in CBD, and vice versa. CBD binds to several types of receptors and does not produce dependency or psychoactive effects. However, CBD, given as an oil, alleviates the intractable seizures of a genetic condition called Dravet syndrome. CBD is currently in dozens of clinical trials, to test treatments for other seizure disorders as well as anxiety, schizophrenia, irritable bowel syndrome, attention deficit hyperactivity disorder, and pain from multiple sclerosis.

Plant-based pharmaceuticals often have long histories. In the nineteenth century, marijuana was used to treat epilepsy, and small studies have tested it over the years. CBD came to public attention in 2013, with the story of a 5-year-old who has Dravet syndrome. She was having three to four short seizures per hour, interspersed with long ones that would last 2 to 4 hours. Her heart had stopped many times, she couldn't eat or walk or talk, and she'd often been close to death. After no standard treatments for seizures were effective, the family turned to medical marijuana, which required support from two physicians in the state where they live.

The parents found growers who had cultivated a strain of cannabis that wouldn't sell as a recreational drug because it had very low THC, but it had high CBD. When the child drank oil from the plant, her seizures soon ceased—for a week! Over time, with continuing treatment with the oil, her seizure frequency plummeted from 300 episodes a week to 2 to 3 per month.

About 80 percent of people with Dravet syndrome have a mutation in the sodium channel gene *SCN1A,* as does the girl. The mutation is autosomal dominant, and 95 percent of cases result from a *de novo* mutation. If a child survives the first year, prolonged seizures continue, with involuntary twitches and impaired intellectual and physical development.

Today the child takes the oil twice a day mixed with food, and feels so well that she can ride a bike. Hundreds of children are now being treated, as physicians carefully monitor their responses and long-term effects. Researchers continue to explore the

Figure 8A *Cannabis sativa* © travelab India/Alamy

therapeutic possibilities from a component of a plant once known only for its recreational use.

Questions for Discussion

1. Possession of *Cannabis sativa* has been illegal in the United States until recently, when a few states have allowed use of small amounts of recreational marijuana and have expanded uses for medical marijuana. Yet *Cannabis sativa* contains an addictive substance. How can products derived from the plant be allowed, yet the risk of addiction minimized?

2. Some of the gene variants linked to cannabis addiction are also implicated in elevating risk of developing major depressive disorder and schizophrenia, discussed later in the chapter. How might these connections explain cannabis dependency?

3. Discuss possible reasoning behind the fact that possession of *Cannabis sativa* still is a crime in many states, and its addictive nature only recently understood, while tobacco, known since the 1960s to cause lung cancer and to be highly addictive, has been legal.

A1 allele of the dopamine D(2) receptor gene variant are over-represented among those with alcoholism and other addictions. Genome-wide association studies have found more than 50 chromosomal regions that may include genes that contribute to drug craving. Studies of gene expression flesh out this picture by providing a real-time view of biochemical changes that happen when a person seeks a drug, and then takes it.

Discovering the genetic underpinnings of nicotine addiction is increasing our knowledge of addiction in general and may have practical consequences. Genetics may explain how the nicotine in tobacco products causes lung cancer and why nicotine is so highly addictive.

Each year, 35 million people try to quit smoking, yet only 7 percent of them succeed. It is easy to see on a whole-body level how this occurs: Nicotine levels peak 10 seconds after an inhalation and the resulting pleasurable release of the neurotransmitter dopamine in brain cells fades away in minutes. To keep the good feeling, smoking must continue.

On a molecular level, nicotine does damage through a five-part assembly of proteins called a nicotinic receptor. The receptor normally binds the neurotransmitter acetylcholine, but it also binds the similarly-shaped nicotine molecule (**figure 8.5**). Certain variants of the receptor bind nicotine very strongly, which triggers a nerve impulse that, in turn, stimulates the pleasurable dopamine release into the synapse. That may explain the addiction. These receptors are also located on several types of lung cells, where they bind carcinogens. So the nicotine in tobacco causes addiction and susceptibility to lung cancer, and the smoke delivers the carcinogens right to the sensitive lung cells.

A different gene encodes each of the five parts of the nicotinic receptor. If two of the five parts are certain variants, then a person experiences desire to continue smoking after the first cigarette.

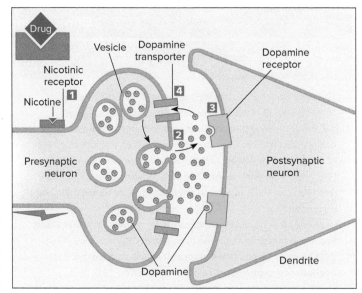

Figure 8.5 **Nicotine's effects at the cellular level. (1)** Binding of nicotine to nicotinic receptors, which also bind the neurotransmitter acetylcholine, triggers release of dopamine **(2)** from vesicles into the synapse. Some dopamine binds receptors on the postsynaptic (receiving) neuron **(3)** and some dopamine reenters the presynaptic neuron through a dopamine transporter protein **(4)**. Uptake of dopamine into the postsynaptic cell triggers pleasurable feelings. The inset illustrates how a smoking cessation drug blocks nicotinic receptors.

Key Concepts Questions 8.4

1. What is drug addiction?
2. How do genes control addiction?
3. How can a human become addicted to a chemical from plants?
4. Explain the basis of nicotine addiction.

8.5 Mood Disorders

We all have moods, but mood disorders, which affect millions of people, impair the ability to function on a day-to-day basis. Context is important in evaluating extreme moods. For example, symptoms that may lead to a diagnosis of depression if they occur for no apparent reason are normal in the context of experiencing profound grief. The two most prevalent mood disorders are major depressive disorder and bipolar disorder (also called manic depression).

Major Depressive Disorder

Major depressive disorder is more than being "down in the dumps," nor can a person simply "snap out of it." Clinical depression is a disabling collection of symptoms that go beyond sadness. A person with depression no longer enjoys favorite activities. Fatigue may be great, and the person has trouble making decisions, concentrating, and recalling details. Motivation lags, anxiety and irritability rise, and crying may start unexpectedly, often several times a day. People with depression have difficulty falling and staying asleep, which intensifies the daytime fatigue. Routine, simple tasks may become overwhelming, and working or studying may become difficult. Weight may drop as interest in food, as in most everything else, diminishes. Fifteen percent of people hospitalized for severe, recurrent depression ultimately end their lives.

Genes and the environment contribute about equally to depression. Risk is greater for people with an affected first-degree relative (parent or sibling). Researchers initially discovered genes that increase risk of depression by focusing on disturbed sleep. They looked for genes known to control circadian rhythms (sleep, hormone levels, and body temperature), and discovered eight that show altered expression in the brains of people who had depression compared to brains from people who hadn't been depressed. About twice as many women as men suffer from depression.

A likely cause of depression is deficiency of the neurotransmitter serotonin in synapses. Serotonin affects mood, emotion, appetite, and sleep. Antidepressant drugs called selective serotonin reuptake inhibitors (SSRIs) prevent presynaptic neurons from taking up serotonin from the synapse, leaving more of it available to stimulate the postsynaptic cell (**figure 8.6**). This action apparently offsets the neurotransmitter deficit. The genetic connection is that a protein in the plasma membrane of the presynaptic neuron, called the serotonin transporter, may function too efficiently in depressed individuals, removing too much serotonin from the synapse.

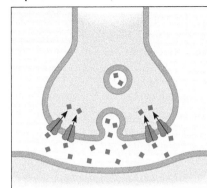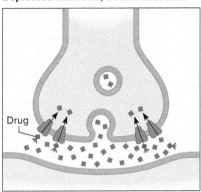

Nondepressed individual

Presynaptic neuron

Receptors for serotonin reuptake

Serotonin

Synapse

Postsynaptic neuron

Depressed individual, untreated

Depressed individual, treated with SSRI

Drug

Figure 8.6 **How an antidepressant works.** Selective serotonin reuptake inhibitors (SSRIs) block the reuptake of serotonin, leaving more of the neurotransmitter in synapses. This corrects a neurotransmitter deficit that presumably causes the symptoms. Overactive or overabundant reuptake receptors can cause the deficit.

SSRIs lock the transporter in an "outward-only" position that prevents serotonin from binding to the presynaptic plasma membrane receptor. The drugs may begin to produce effects after 1 week, often enabling a person with moderate or severe depression to return to some activities, but full response can take up to 8 weeks. Other drugs similarly target the neurotransmitters dopamine or noradrenaline.

In the past, to treat patients who have depression, physicians would try one antidepressant drug after another, based on experience with other patients. This trial-and-error approach would often take months. As the genes in the human genome are identified and their variants described, researchers are establishing genetic profiles that are associated with response to a particular drug. For example, variants of a gene that encodes cytochrome P450 protein control the rate at which an individual metabolizes a drug. This information may be helpful in selecting a dosage that will be high enough to be effective but not so high that it causes adverse effects. Selecting antidepressants based on genotype is an example of a form of genetic testing called **pharmacogenetics**, discussed in section 20.2.

Bipolar Disorder

Bipolar disorder is rarer than depression. It affects 2.6 percent of the population and has a general population lifetime risk of 0.5 to 1 percent. Weeks or months of depression alternate with periods of mania, when the person is hyperactive and restless and may experience a rush of ideas and excitement. A person who is normally frugal might, when manic, spend lavishly. In one subtype, the "up" times are termed hypomania, and they seem more a temporary reprieve from depression than the bizarre behavior of full mania. Bipolar disorder with hypomania may appear to be depression. Distinguishing between major depressive disorder and bipolar disorder with hypomania in a clinical situation is important because different drugs are used to treat the conditions.

Many gene variants contribute to developing bipolar disorder. Early genetic studies looked at large Amish families, in whom the manic phase was quite obvious amid their restrained

lifestyle. But studies in other families implicated different genes. Studies disagree because many gene variant combinations cause or contribute to the distinctive phenotype of bipolar disorder, but only a few such variants are seen in any one family. A systematic scrutiny of all studies of the genes behind bipolar disorder suggests as much as 10 percent of the genome—that is, hundreds of genes—are part of the clinical picture.

Key Concepts Questions 8.5

1. What are the symptoms of major depressive disorder?
2. To what extent is depression inherited?
3. What types of genes may affect the risk of developing depression?
4. What are the symptoms of bipolar disorder?
5. How is the role of genes in bipolar disorder complex?

8.6 Schizophrenia

Schizophrenia is a debilitating loss of the ability to organize thoughts and perceptions, which leads to a withdrawal from reality. It is a form of psychosis, which is a disease of thought and sense of self. In contrast, the mood disorders are emotional, and the dementias are cognitive. Various forms of schizophrenia together affect about 1 percent of the world's population. Ten percent of affected individuals commit suicide.

Identifying genetic contributions to schizophrenia illustrates the difficulties in analyzing a behavioral condition:

- Some of the symptoms are also associated with other illnesses.
- Variants of many genes cause or contribute to schizophrenia.
- Several environmental factors mimic schizophrenia.

Signs and Symptoms

The first signs of schizophrenia often affect thinking. In late childhood or early adolescence, a person might develop trouble paying attention in school. Learning may become difficult as memory falters and information-processing skills diminish.

Symptoms of psychosis begin between ages 17 and 27 for males and 20 and 37 for females. These include delusions and hallucinations, sometimes heard, sometimes seen. A person with schizophrenia may hear a voice giving instructions. What others recognize as irrational fears, such as being followed by monsters, are real to a person with schizophrenia. Speech reflects the garbled thought process, skipping from topic to topic with no obvious thread of logic. Responses may be inappropriate, such as laughing at sad news.

Artwork by a person with schizophrenia can display the characteristic fragmentation of the mind (**figure 8.7**). Schizophrenia means "split mind," but it does not cause a split or multiple personality. In many patients the course of schizophrenia plateaus (evens out) or becomes episodic. It is not a continuous decline, as is the case for dementia.

A Disorder of Synaptic Pruning

A heritability of 0.9 and empiric risk values indicate a strong role for genes in causing schizophrenia (**table 8.2**). However, it is possible to develop some of the symptoms, such as disordered thinking, from living with and imitating people who have schizophrenia. Although concordance is high, a person who has an identical twin with schizophrenia has a 52 percent chance of *not* developing it. Therefore, the condition has a significant environmental component, too.

Several environmental factors may increase the risk of developing schizophrenia. These include birth complications, fetal oxygen deprivation, herpesvirus infection at birth, and traumatic brain injury or malnutrition in the mother (see the

Table 8.2	Empiric Risks for Schizophrenia
Relationship	**Risk**
MZ twin	48%
DZ twin	17%
Child	13%
Sibling	9%
Parent	6%
Half sibling	6%
Grandchild	5%
Niece/nephew	4%
Aunt/uncle	2%
First cousin	2%
General population	1%

opening essay to chapter 11). Infection during pregnancy is a well-studied environmental factor. When a pregnant woman is infected, her immune system bathes the brain of the embryo or fetus with cytokines, which are proteins that cause inflammation and can subtly alter brain development. It isn't the infection that sets the stage for later schizophrenia, but the maternal immune response to it. Investigators have long noted a tendency of people with schizophrenia to have been born during the winter, especially during years of flu pandemics. Research supports this connection between schizophrenia and response to maternal infection. Blood stored from 12,000 pregnant women during the 1950s and 1960s shows an association between high levels of the cytokine interleukin-8 and having a child who developed schizophrenia.

Evidence from many studies has strengthened the role of an overactive or inappropriate immune response operating at key times in development in causing schizophrenia. Such critical periods include the time of prenatal development and again during late adolescence, when symptoms typically become noticeable. The immune response introduces both an environmental component—infection—and a genetic component in the deployment of immune system proteins.

Genetic research has gradually focused in on a narrower set of candidate genes behind schizophrenia, with one emerging as a strong factor. At first, genome-wide screens of families with schizophrenia revealed at least twenty-four sites where affected siblings share alleles much more often than the 50 percent of the time that Mendel's first law (segregation) predicts. Then genome-wide association studies (see figure 7.11) implicated 108 sites in the genome that could include schizophrenia genes. One part stood out: the major histocompatibility complex, an extensive, gene-packed part of chromosome 6 (see section 17.1). Researchers traced the "signal" to a pair of

Figure 8.7 Schizophrenia alters thinking. People with schizophrenia communicate the disarray of their thoughts in characteristically disjointed drawings. © Robert Gilliam

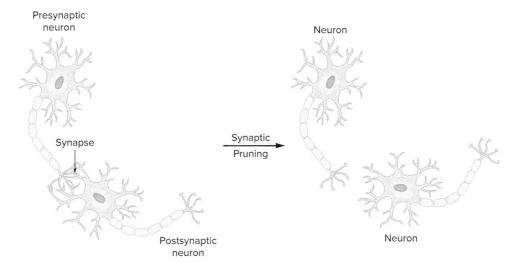

Figure 8.8 **Synaptic pruning.** Microglia bind to *C4A* protein "tags" (not shown) on dendrites, breaking the connections between neurons. The result: fewer synapses and some loss of brain function.

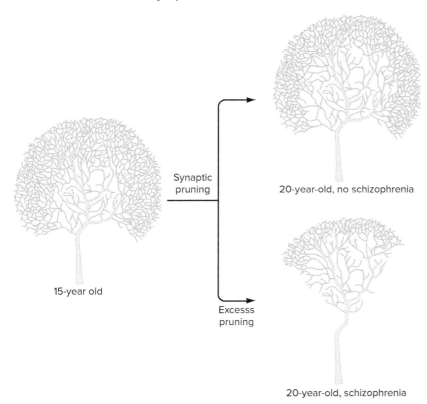

15-year old

Synaptic pruning

20-year-old, no schizophrenia

Excesss pruning

20-year-old, schizophrenia

Figure 8.9 **Schizophrenia may result from overactive synaptic pruning.** Losing up to 2 billion brain synapses a day is normal as adulthood begins, but an acceleration in the process may lie behind schizophrenia.

them. Certain variants of *C4A* are much more common among individuals with schizophrenia. Consideration of what the encoded protein does makes sense: It destroys synapses.

In late adolescence and in the early twenties, in all human brains, the dense bushes of dendrites emanating from brain neurons are trimmed, as part of normal development (**figure 8.8**). In people who have the *C4A* gene variant associated with schizophrenia, this "pruning" is too extensive, leaving too few connections in the parts of the brain that control thinking, planning, and the ability to link thoughts into accurate perceptions of the world (**figure 8.9**). Cells called microglia do the pruning, in response to *C4A* protein "tags" on the targeted synapses. Microglia are a type of neuroglia that had been thought to function mainly in cleaning up cellular debris. When they are too active, perhaps in response to infection, a person loses too many synapses, neuron-to-neuron communication in the brain suffers, and the symptoms of schizophrenia appear.

nearly identical genes called *C4 complement component 4*. The genes are called *C4A* and *C4B*. "Complement" is a series of interacting proteins that are part of the immune response (see section 17.2).

The *C4* genes are highly variable—that is, people have different sizes of these genes, different numbers of copies of the genes, and single nucleotide polymorphisms (SNPs) within

---

Key Concepts Questions 8.6

1. What are symptoms of schizophrenia?

2. How does schizophrenia differ from the mood disorders?

3. Explain how a role of the immune system in raising risk of or causing schizophrenia accounts for environmental and genetic factors.

4. Explain how excessive synaptic pruning may set the stage for development of schizophrenia.

8.7 Autism

Autism spectrum disorders impair socialization and communication skills, with symptoms appearing before age 3. One in 68 children in the United States has autism, and boys outnumber girls four to one. About 25 percent of affected children have seizures as they grow older. Although 70 percent of people with autism have intellectual disability, others may be very intelligent, and many are talented in science or the arts (**table 8.3**).

In severe autism, a toddler may lose vocabulary words and cease starting conversations. He or she stops playing with other children, preferring to be alone. Rocking back and forth or clutching a treasured object and using repetitive motions are soothing behaviors for some children. The child may refer to herself by her name, rather than "me," not make eye contact, and appear oblivious to nonverbal cues such as facial expressions or tone of voice. Special education programs that provide a rigid routine can help an individual with autism to connect with the world. People who have one type of autism (Asperger syndrome) may have the socialization effects but not the language deficits.

In most cases of autism, many genes contribute to different degrees against a backdrop of environmental influences. Environmental triggers include prenatal exposures to rubella (German measles) and to the drug valproate. Evidence is accumulating that exposure to folic acid supplements late in gestation can cause epigenetic changes that increase the risk of autism. A better way to identify environmental risk factors may be to consider different genetic subtypes, so that studies can compare individuals with the same underlying problem.

Heritability of autism is high: about 0.90 (90% of variability due to genetics). Although siblings of affected children are at a 15 percent risk of being affected, compared to the less than 1.5 percent for the general population, in some MZ (identical) twin pairs, one twin has autism and one does not. More than thirty susceptibility or causative genes have been identified.

In about 15 percent of people with autism, the condition is part of a syndrome, including single-gene disorders such as Rett syndrome (see the chapter 2 opener), mitochondrial diseases (see section 5.2), and chromosomal conditions such as fragile X syndrome (see Clinical Connection 12.1) and Down syndrome (see Bioethics in chapter 13). A diagnostic workup for autism includes tests for these conditions as well as chromosomal microarray analysis, which detects copy number variants (repetitions of DNA sequences that vary in number among individuals).

Our understanding of the genetics of autism parallels development of genetic technology. At first, the rare individuals who had autism as well as an abnormal chromosome pointed the way to a place in the genome—the abnormal chromosome—that likely harbors a predisposing or causative mutation. Then genome-wide association studies highlighted genome regions that include genes that increase risk of developing autism. **Table 8.4** lists other genetic syndromes that include autism.

Table 8.3	Famous People Who Had Autistic Behaviors
Person	**Characteristics**
Albert Einstein	Did not like to be touched
	Difficulty learning in school
	Awkward in social situations
Charles Darwin	Preferred letter writing to face-to-face interactions
	Fixated on certain objects and topics
Emily Dickinson	Heightened sense of smell
	Preferred white clothing
	Reclusive
	Fascinated with flowers
Michelangelo	Limited interests but obsessed with work
	Poor communication and social skills
	Relied on rigid routines

Table 8.4	Genetic Syndromes That Include Autism	
Syndrome	**Gene**	**Phenotype (in addition to autism)**
Joubert syndrome	*INPP5E*	Missing brain structure ("molar tooth sign"), abnormal breathing, abnormal eye movement, cognitive impairment
Phenylketonuria	*PAH*	Severe intellectual disability if remedial diet is not followed
PTEN macrocephaly syndrome	*PTEN*	Large head, cancer predisposition
Smith-Lemli-Opitz syndrome	*DHCR7*	Intellectual disability, self-injury, poor temperature regulation, disturbed sleep
Timothy syndrome	*CACNA1C*	Webbed fingers and toes, heart defects, unusual facial features, developmental delay
Tuberous sclerosis 1	*TSC1*	Seizures, tumors, skin lesions

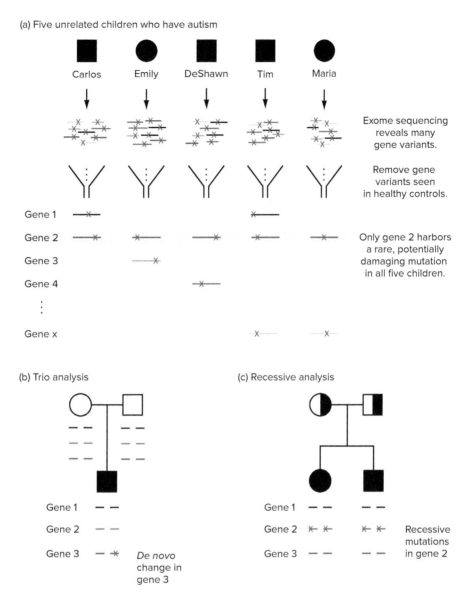

(a) Five unrelated children who have autism

Carlos   Emily   DeShawn   Tim   Maria

Exome sequencing reveals many gene variants.

Remove gene variants seen in healthy controls.

Gene 1

Gene 2   Only gene 2 harbors a rare, potentially damaging mutation in all five children.

Gene 3

Gene 4

Gene x

(b) Trio analysis

(c) Recessive analysis

Gene 1

Gene 2

Gene 3   *De novo* change in gene 3

Gene 1

Gene 2   Recessive mutations in gene 2

Gene 3

Figure 8.10 **Three approaches to using exome sequencing to discover genes that contribute to autism.** **(a)** Comparison to a reference genome identifies genes that unrelated people with autism share. **(b)** Trio (parents-child) analysis detects *de novo* mutations in a child. **(c)** Recessive analysis detects mutations in families with more than one affected child.

The fact that autism is part of so many genetic syndromes suggests that it might be a common response to variants of multiple genes. Autism tends to occur in syndromes that also include unusual facial features, developmental delay, and/or intellectual disability. Each genetic syndrome accounts for less than 1 percent of autism cases, but a significant percentage of individuals with each condition has autistic behaviors.

Today exome and whole-genome sequencing projects are enabling researchers to identify many more genes behind autism. Experiments focus on candidate genes within the huge datasets that these technologies generate. **Figure 8.10** illustrates three general strategies for isolating autism genes (or other genes for a shared phenotype) from exome or genome sequencing data.

The first approach, "multiple unrelated affected subjects," compares the exome or genome sequences of many unrelated individuals with autism to "reference" genomes from healthy people compiled in a population database. In which genes do affected children share variants that are not in unaffected children? In one study, researchers compared genome sequences from affected members of 55 families that each had several people with autism to reference genomes from nonaffected families. They identified 153 parts of the genome. Then the researchers made probes—copies of those identified DNA sequences that incorporate a fluorescent label—and applied them to the DNA of 2,175 children with autism and 5,801 children without autism. In those 153 genome regions, the investigators narrowed down 15 sites, with a total of 24 gene variants, seen in the affected children and associated with a twofold increase in the risk of autism.

The second approach is "trio analysis," which compares the DNA in parents-child trios. Figure 4.18 and Clinical Connection 1.1 discuss the trio analysis of a girl whose genetic

Finding Fault for Autism

The behaviors that we call autistic have been around for centuries and likely longer, unnamed but perhaps not unnoticed. An early reference is to Victor, the "wild boy of Avalon," who lived alone in the woods. In 1910 Swiss psychiatrist Paul Bleuler coined the term "autismus" (in German), from the Greek *autos* for "self" and *ismos* for "state"—the definition was "morbid self-absorption."

American child psychiatrist Leo Kanner is often credited with the first modern report on autism, published in 1943 on 11 children with what he termed "infantile autism." The children shared obsession with certain objects, inability to interact with others, and a "need for sameness." In 1944 pediatrician Hans Asperger, at the University of Vienna, published a report on a similar group of young patients, but his work was unnoticed for many years.

Before autism was recognized as a medical condition, children with more severe manifestations tended to be ignored or abused, some even institutionalized or subjected to barbaric treatments. In 1959, psychologist Bruno Bettelheim published a controversial article, "Joey, a Mechanical Boy," in *Scientific American,* in which he blamed a deficiency of maternal affection for the child's behavior. Dr. Kanner at first agreed, using the term "refrigerator mother," but eventually recognized the accumulating evidence of a biological explanation of autistic behaviors. The maternal connection could not explain why families with multiple children had only one child with autism.

Once the term "autism" became well recognized, some desperate parents were drawn in by promises of miracle cures. Yet in some individuals with autism, astonishing talents tended to overshadow the unusual behaviors.

In the 1980s British psychiatrist Lorna Wing rediscovered Dr. Asperger's contribution and described the part of the spectrum that takes his name, to denote individuals who have language skills and who may have remarkable skills or gifts. Autism was added to the *Diagnostic and Statistical Manual* in 1980. Since the Individuals with Disability Education Act of 1990, public schools in the United States have provided services to children with autism spectrum disorders.

Questions for Discussion

1. Some people argue that autism is an extreme of normal behavior that has been "medicalized," or is a manifestation of "neurodiversity." What type of evidence supports the view that autism is a medical condition, versus an extreme of normal behavior?

2. What are the risks and benefits of autism being considered a disease?

3. Research famous people who were or are "on the spectrum." What are possible benefits of studying these individuals?

4. Why might Dr. Bettelheim have blamed mothers, and what type of evidence from everyday life disproved his hypothesis?

---

syndrome includes autism. These studies either find that parents are carriers of autosomal recessive or X-linked conditions, or that the autism in the child arose *de novo.*

The third approach, "recessive analysis," compares DNA sequences in families in which more than one child is affected but the parents are not, seeking genes that are mutant in two copies in the children with autism. A study that used this approach looked at three very large families from the Middle East that have many cousin-cousin marriages, which are at higher risk for passing along autosomal recessive conditions due to inheritance of mutations from shared ancestors. When the researchers discovered mutations in three genes already known to cause other diseases, they hypothesized that some cases of autism are actually mild or atypical versions of recognized syndromes. The researchers found seven other known autosomal recessive conditions that also cause autism. These diseases are pleiotropic—they have several manifestations, and in some children, autism is the only symptom.

After identifying mutations that might cause or contribute to autism based on their occurrence in people with the condition, the next step is to consider what these genes do. How might malfunction of the gene impair socialization and communication skills? For example, abnormalities in either of two types of cell adhesion proteins found at synapses, called **neurexins** and **neuroligins**, are responsible for a small percentage of cases of autism.

Three types of neurexins emanate from the "sending" presynaptic neuron, and a neuroligin extends toward the neurexin from the "receiving" postsynaptic neuron. When the two protein types meet, they draw the neurons closer together so that one can signal the other, using the neurotransmitter glutamate. Mutations misfold the two proteins so that they cannot interact as they normally do, which disrupts the sending of signals from one nerve cell to another (**figure 8.11**). If this interference happens in early childhood, when synapses naturally form as a child experiences the environment, the behaviors of autism could emerge from fewer or impaired synapses in brain parts that oversee learning and memory. **Bioethics** looks at the evolution of treatments for and attitudes toward people with autism spectrum disorders.

Identifying the genes behind autism and the other behavioral (also known as neuropsychiatric) disorders discussed in this chapter can have clinical benefits. Genetic information can add precision to a traditional diagnosis based on

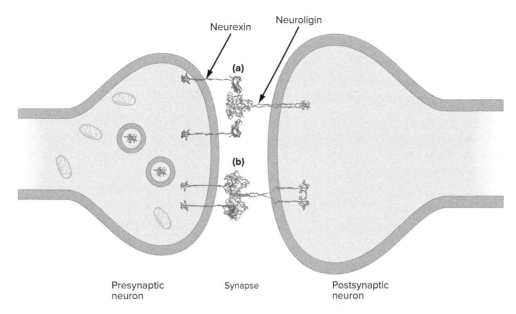

Neurexin   Neuroligin

(a)

(b)

Presynaptic   Synapse   Postsynaptic
neuron   neuron

Figure 8.11 **Understanding autism.** In some cases of autism, neurexins and neuroligins cannot interact strongly enough at synapses **(a).** Normally the two kinds of proteins can interact as shown in **(b).**

symptoms, define clinical subtypes of a condition detectable with existing diagnostic tests, and provide a way to predict risk that relatives will develop the condition, which is greater than the general population risk. Finally, identifying genes behind behavioral disorders is important in drug discovery research. One day we may have genetically based personalized treatments for subtypes of these multifactorial conditions that we have traditionally considered together based on their phenotypes.

Key Concepts Questions 8.7

1. What is autism?
2. What is the evidence that genes contribute to autism risk?
3. What are some environmental factors that contribute to the development of autism?
4. How have exome and genome sequencing led to the discovery of genes implicated in autism?
5. How do interactions between neurexins and neuroligins explain some cases of autism?

Summary

8.1 Genes and Behavior

1. Genes affect how the brain responds to environmental stimuli. **Neurons** and **neuroglia** are the two major types of brain cells.
2. Candidate genes for behavioral traits and disorders affect **neurotransmitters** crossing **synapses** (neurotransmission).
3. Old and new genetic tools are being used to describe the biological causes of various behavioral disorders.
4. Variants of hundreds of genes contribute in different degrees to behavioral disorders.

8.2 Sleep

5. Twin studies and single-gene disorders that affect the sleep-wake cycle reveal a large inherited component to sleep-wake behavior.
6. Clock genes such as *period 2* enable a person to respond to day and night environmental cues.

8.3 Intelligence and Intellectual Disability

7. The general intelligence ability (*g*) measures the inherited portion of IQ that may underlie population variance in IQ test performance.
8. Studies indicate that general intelligence is maintained throughout life. The fact that many syndromes that result from abnormal chromosomes affect intelligence suggests high heritability.
9. Exome and genome sequencing are identifying mutations that cause or contribute to intellectual disability, which has many causes.

8.4 Drug Addiction

10. Drug addiction is a severe substance use disorder that arises from developing tolerance to and dependence on a drug. Addiction produces stable changes in certain brain parts.

11. Proteins involved in drug addiction affect neurotransmitter synthesis and neurotransmitter removal from synapses, and receptor structure and function.

12. Candidate genes for drug addiction encode the dopamine D(2) receptor and nicotinic receptor parts.

8.5 Mood Disorders

13. Major depressive disorder is common and associated with mutations in genes that regulate circadian rhythms and deficits of serotonin.

14. Bipolar disorder includes depressive periods and periods of mania or hypomania. Variants of hundreds of genes may raise the risk of developing bipolar disorder. Combinations of such predisposing gene variants in the presence of certain environmental influences can lead to the disorder.

8.6 Schizophrenia

15. Schizophrenia greatly disrupts the ability to think and perceive the world. Onset is typically in early adulthood, and the course is episodic or it plateaus.

16. Empiric risk values and a heritability of 0.9 indicate a large genetic component to schizophrenia.

17. An overactive or inappropriate immune response during prenatal development or in late adolescence may increase the risk of developing schizophrenia.

18. Inheriting a variant of a gene (*C4A*) that encodes an immune system protein increases schizophrenia risk by increasing the rate of synaptic pruning during late adolescence or young adulthood.

8.7 Autism

19. Autism is a loss of communication and social skills beginning in early childhood.

20. Exposure to certain pathogens and chemicals can increase the risk of autism.

21. Autism is part of many genetic syndromes, but these account for a small percentage of autism cases.

22. Three approaches are used to analyze exome or genome sequences to identify gene variants that contribute to developing autism. The "multiple unrelated affected subjects" approach seeks gene variants in individuals with autism that are not in population databases of unaffected children. "Trio analysis" distinguishes carrier parents from *de novo* mutations in the child. "Recessive analysis" searches for gene variants that two or more affected siblings share.

23. **Neurexins** and **neuroligins** are proteins embedded in the plasma membranes of the presynaptic and postsynaptic neurons, respectively, that cannot interact normally at synapses in some cases of autism.

Review Questions

1. What are the two major types of cells in the brain, and what do they do?

2. Why are behavioral traits multifactorial?

3. List the pathways or mechanisms involving proteins that, when absent or atypical, affect behavior.

4. Describe how variants of the following genes affect the risk of developing specific behavioral disorders.

   a. the gene that encodes the orexin receptor
   b. the *period 2* gene
   c. the gene for the dopamine D(2) receptor
   d. the gene for nicotinic receptor parts
   e. the gene for the serotonin transporter
   f. *C4 complement component 4*
   g. genes that encode neurexins and neuroligins

5. What evidence would indicate that a family with familial advanced sleep phase syndrome has a genetic condition rather than just all the members becoming used to keeping unusual hours?

6. What were some of the prejudices that were part of studying the inheritance of intelligence and the causes of autism?

7. Why is it so difficult to identify factors that cause intellectual disability?

8. What are the two defining characteristics of addiction?

9. What is the evidence that our bodies have their own uses for cocaine, THC, and opium?

10. Explain how both environmental and genetic factors can cause schizophrenia.

11. Distinguish among the three approaches to autism gene discovery that figure 8.10 shows.

12. Explain how abnormal interactions between neurexins and neuroligins at synapses can cause autism.

Applied Questions

1. Identify a gene variant and an environmental factor that together might significantly increase the risk of an individual developing posttraumatic stress disorder. (See Forensics Focus question 1.)

2. The drug Belsomna inhibits orexin. What condition does it treat?

3. Many older individuals experience advanced sleep phase syndrome. Even though this condition is probably a normal part of aging, how might research on the Utah family with an inherited form of the condition help researchers develop a drug to assist the elderly in sleeping through the night and awakening later in the morning?

4. Some technical terms are being replaced so that they do not make people uncomfortable. "Mental retardation" is now considered one condition of "intellectual disability," and "drug addiction" is one manifestation of "drug use disease." Do you think that these redefinitions are helpful, harmful, or just confusing? Cite a reason for your answer.

5. How does the subunit structure of the nicotinic receptor provide a mechanism for epistasis? (Epistasis is an interaction in which one gene masks or affects the expression of another; see chapter 5.)

6. What might be the advantages and disadvantages of a genetic test done at birth that indicates whether a person is at high risk for developing a drug addiction?

7. What aspect of bipolar disorder may account for the fact that the condition is overrepresented among very creative individuals?

8. How can discovery of the role of the *C4A* gene in schizophrenia be used to develop a treatment?

9. In the United States, the incidence of autism has dramatically increased since 2012.
    a. Does this finding better support a genetic cause or an environmental cause for autism?
    b. What is a nongenetic factor that might explain the increased incidence of autism?

Forensics Focus

1. When 43-year-old Francis discovered his soon-to-be ex-wife in bed with her boyfriend, he shot and killed them both. The defense ordered a pretrial forensic psychiatric workup that included genotyping for the gene for the enzyme monoamine oxidase *A* (*MAOA*), which breaks down the neurotransmitters serotonin, dopamine, and noradrenaline, and for the gene for the serotonin transporter (*SLC6A4*). A "high-activity" allele for *MAOA* is associated with violence in people who also suffered child abuse. Inheriting one or two "short" alleles of *SLC6A4* is associated with depression and suicidal thoughts, in people who have experienced great stress. Francis had the high-activity *MAOA* genotype, two short *SLC6A4* alleles, and a lifetime of abuse and stress. However, the judge ruled that the science was not far enough along to admit the genotyping results as evidence.

    a. Under what conditions or situations do you think it is valid to include genotyping results in cases like this?
    b. In a Dutch family, a mutation disables *MAOA*, causing a syndrome of intellectual disability and abnormal behavior. Family members had committed arson, attempted rape, and shown exhibitionism. How can both high and low levels of an enzyme cause behavioral problems?
    c. What is a limitation to use of behavioral genotyping in a criminal trial?

Case Studies and Research Results

1. The drug MDMA, known on the street as ecstasy, X, E, and disco biscuit, is in clinical trials for evaluation to treat PTSD. It calms patients who do not respond well to psychotherapy. MDMA interacts with the serotonin transporter in a way that increases levels of the neurotransmitter in brain synapses. Which behavioral disorder mentioned in the chapter responds to other drugs that increase the availability of serotonin?

2. Variants of a gene called *CYP2A6* are associated with the speed at which a person metabolizes nicotine. A "fast-metabolizer" smokes more cigarettes than a "slow-metabolizer." Design a clinical trial using this information to test whether the *CYP2A6* genotype can be used to predict which individuals will respond to specific smoking cessation medications.

3. In a short e-book by Karen Russell called "Sleep Donation," overuse of lit screens leads to a pandemic of an infectious, deadly insomnia due to abnormal expression of the gene encoding a protein mentioned in the chapter. Name the implicated gene, and state whether it would have to be overexpressed or underexpressed to cause insomnia. (The book is fiction.)

4. Many studies have shown that the relatives of people who have bipolar disorder or schizophrenia are overrepresented among the creative professions, including writers, actors, dancers, musicians, and visual artists. Researchers in Iceland analyzed genetic information on hundreds of thousands of people and found that gene variants associated with the two conditions were also found in family members that did not have any psychiatric diagnoses, but who were members of Iceland's artistic societies.

    a. How might the same gene variants cause bipolar disorder or schizophrenia in some family members, but creativity in others?
    b. Do you think that in light of these findings, either of these disorders, or both, should be considered extremes of "normal" behavior? How might this designation be advantageous or dangerous?

CHAPTER 9

DNA Structure and Replication

Source: U.S. Fish & Wildlife Service/Miriam Westervelt

Comparing DNA sequences is a way to solve problems, including exposing crimes such as poaching wildlife.

Learning Outcomes

9.1 Experiments Identify and Describe the Genetic Material

1. Describe the experiments showing that DNA is the genetic material and that protein is not.
2. Explain how Watson and Crick deduced the three-dimensional structure of DNA.

9.2 DNA Structure

3. List the components of a DNA nucleotide building block.
4. Explain how nucleotides joined into two chains form the strands of a DNA molecule.
5. Describe how the great length of a DNA molecule is folded and looped.

9.3 DNA Replication—Maintaining Genetic Information

6. Explain the semiconservative mechanism of DNA replication.
7. List the steps of DNA replication.
8. Explain how the polymerase chain reaction amplifies DNA outside cells.

9.4 Sequencing DNA

9. Explain the basic strategy used to determine the base sequence of a DNA molecule.
10. Explain how next-generation sequencing improves upon Sanger sequencing.

The BIG Picture

DNA is the basis of life because of three qualities: It holds information, it copies itself, and it changes. DNA sequences can be used to identify individuals and as the basis of diagnostic tests for genetic diseases.

Elephant Forensics

Each year in Africa, poachers take more than 50 tons of ivory from the tusks of elephants. The animals, which now number about 500,000, will soon become extinct.

To identify the geographical sources of poached ivory, researchers collected dung from 1,350 elephants in 71 locations in 29 African nations—1,001 animals from the savanna and 349 from the forest. They genotyped the DNA using short repeats from 16 genome regions, much like human DNA is profiled at 13 sites (see figure 14.9). Then they genotyped ivory from 28 large seizures—nearly half a ton, representing 20 percent of all large seizures from 1996 through 2014.

The DNA component of tusks is in the outer cementum layer. The researchers pulverized samples, then used chemicals to extract the mineral and protein components. Enough DNA remained to determine genotypes.

When the researchers compared the geographic map of elephant habitats to the map of ivory seizures, they could deduce where the crimes occurred. The savanna elephant tusks came from Mozambique and Tanzania, and the forest elephant tusks came from Gabon, the Republic of Congo, and the Central African Republic. DNA analysis also revealed that ivory is moved from storage sites to shipping sites.

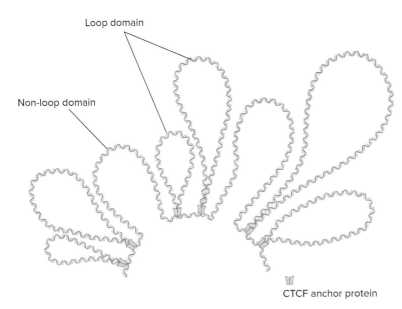

Loop domain

Non-loop domain

CTCF anchor protein

Figure 9.14 DNA looping. Proteins called CTCF bring together parts of a long DNA molecule that include genes whose expression is coordinated.

attachment to the nuclear membrane are not random, and the placement may control which genes a cell is using to make proteins. Progeria, the genetic disorder that resembles rapid aging, described in the chapter 3 opener, disrupts chromatin binding to the nuclear envelope.

Key Concepts Questions 9.2

1. What are the components of the DNA double helix?
2. Explain the basis of antiparallelism.
3. Explain the basis of complementary base pairing.
4. How are very long DNA molecules wound so that they fit inside cell nuclei?
5. What is the significance of DNA loops?

9.3 DNA Replication— Maintaining Genetic Information

DNA must be copied, or replicated, so that the information it holds can be maintained and passed to future cell generations, even as that information is accessed to guide the manufacture of proteins. As soon as Watson and Crick deciphered the structure of DNA, its mechanism for replication became obvious. They ended their report on the structure of DNA with the statement, *"It has not escaped our notice that the specific pairing we have postulated immediately suggests a possible copying mechanism for the genetic material."*

DNA Replication Is Semiconservative

Watson and Crick envisioned the two strands of the DNA double helix unwinding and separating. Then the exposed unpaired bases would attract their complements from free, unattached nucleotides available in the cell from nutrients. In this way, two identical double helices would form from one original, parental double helix. This route to replication is called **semiconservative** because each new DNA double helix conserves half of the original. However, separating the long strands poses a great physical challenge.

In 1957, two young researchers, Matthew Meselson and Franklin Stahl, demonstrated the semiconservative mechanism of DNA replication with a series of "density shift" experiments. They labeled replicating DNA from bacteria with a dense, heavy form of nitrogen, and traced the pattern of distribution of the nitrogen. The higher-density nitrogen was incorporated into one strand of each daughter double helix (**figure 9.15**). Alternative mechanisms that the experiments ruled out were replicating a daughter DNA double helix built of entirely "heavy" labeled nucleotides (a conservative mechanism) or a daughter double helix in which both strands were composed of joined pieces of "light" and "heavy" nucleotides (a dispersive mechanism).

Steps of DNA Replication

DNA replication occurs during S phase of the cell cycle (see figure 2.12). When DNA is replicated, it first unwinds and then locally separates. The hydrogen bonds holding the base pairs together break, and two identical nucleotide chains are built from one, as the bases form new complementary pairs (**figure 9.16**). A site where DNA is locally opened, resembling a fork, is called a **replication fork**.

DNA replication begins when an unwinding protein called a helicase breaks the hydrogen bonds that connect

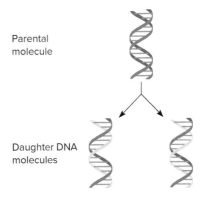

Parental molecule

Daughter DNA molecules

Figure 9.15 DNA replication is semiconservative. The two daughter double helices are identical to the original parental double helix in DNA sequence. However, the light blue helix halves of the daughter DNA molecules indicate that each consists of one strand of parental DNA and one strand of newly replicated DNA.

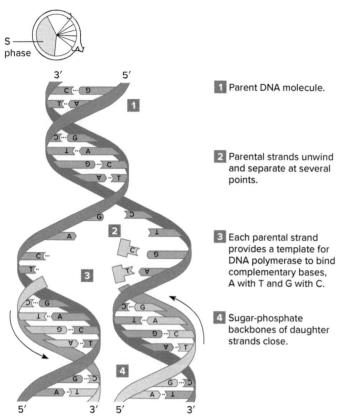

S phase

1 Parent DNA molecule.

2 Parental strands unwind and separate at several points.

3 Each parental strand provides a template for DNA polymerase to bind complementary bases, A with T and G with C.

4 Sugar-phosphate backbones of daughter strands close.

Figure 9.16 Overview of DNA replication. After experiments demonstrated the semiconservative nature of DNA replication, the next challenge was to decipher the steps of the process.

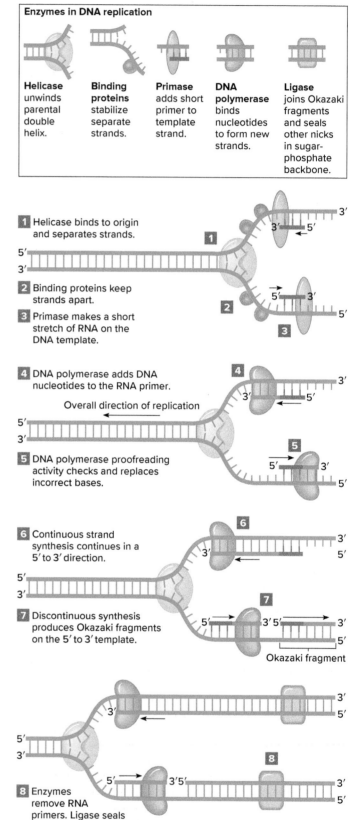

Enzymes in DNA replication

Helicase unwinds parental double helix.

Binding proteins stabilize separate strands.

Primase adds short primer to template strand.

DNA polymerase binds nucleotides to form new strands.

Ligase joins Okazaki fragments and seals other nicks in sugar-phosphate backbone.

1 Helicase binds to origin and separates strands.

2 Binding proteins keep strands apart.

3 Primase makes a short stretch of RNA on the DNA template.

4 DNA polymerase adds DNA nucleotides to the RNA primer.

Overall direction of replication

5 DNA polymerase proofreading activity checks and replaces incorrect bases.

6 Continuous strand synthesis continues in a 5′ to 3′ direction.

7 Discontinuous synthesis produces Okazaki fragments on the 5′ to 3′ template.

Okazaki fragment

8 Enzymes remove RNA primers. Ligase seals sugar-phosphate backbone.

Figure 9.17 Activities at the replication fork. DNA replication takes many steps. (The continuous strand is also called the leading strand, and the discontinuous strand is the lagging strand.)

a base pair (**figure 9.17**). The helicase opens up a localized area, enabling other enzymes to guide assembly of a new DNA strand. Binding proteins hold the two single strands apart. Another enzyme, primase, then attracts complementary RNA nucleotides to build a short piece of RNA, called an RNA primer, at the start of each segment of DNA to be replicated. The RNA primer is required because the major replication enzyme, **DNA polymerase (DNAP)**, can only add bases to an existing nucleic acid strand. (A polymerase is an enzyme that builds a polymer, which is a chain of chemical building blocks.)

Next, the RNA primer attracts DNAP, which brings in DNA nucleotides complementary to the exposed bases on the parental strand; the parental strand serves as a mold, or template. New bases are added one at a time, starting at the RNA primer. The new DNA strand grows as hydrogen bonds form between the complementary bases and the sugar-phosphate backbone links the newly incorporated nucleotides into a strong chain. DNA polymerase carries out both of these activities, and also removes the RNA primer once replication is under way and replaces it with the correct DNA bases. Nucleotides are abundant in cells, and are synthesized from dietary nutrients. The base-pairing rules ensure that each base on the parental strand pulls in its complement.

Because of the antiparallel configuration of the DNA molecule, DNAP works directionally. It adds nucleotides to the exposed 3′ end of the sugar in the growing strand. Replication adds nucleotides in a 5′ to 3′ direction, because this is the only chemical configuration in which DNAP can function. How can the growing fork proceed in one direction, when both parental strands must be replicated and run in opposite directions? The answer is that on one strand, replication is discontinuous. It is accomplished in small pieces from the inner part of the fork outward, in a pattern similar to backstitching. Next, an enzyme called a **ligase** seals the sugar-phosphate backbones of the pieces, building the new strand. These pieces, up to 150 nucleotides long, are called Okazaki fragments, after their discoverer (see figure 9.17).

DNA polymerase also "proofreads" as it goes, excising mismatched bases and inserting correct ones. Yet another enzyme, called an annealing helicase, rewinds any sections of the DNA molecule that remain unwound. Finally, ligases seal the entire sugar-phosphate backbone. *Ligase* comes from a Latin word meaning "to tie."

Human DNA replicates at a rate of about 50 bases per second. To accomplish the job in about an hour rather than the month it would take if starting from only one point, a human chromosome replicates simultaneously at hundreds of points, called replication bubbles. Then the pieces are joined (**figure 9.18**). As a human body grows to trillions of cells, DNA replication occurs about 100 quadrillion times. However, telomeres (chromosome tips) do not replicate, and the chromosomes shrink with each cell division.

Polymerase Chain Reaction

DNA replication is necessary in a cell to perpetuate genetic information. Researchers use DNA replication conducted outside cells in a biotechnology called DNA amplification. It has diverse applications (**table 9.2**). The first and best-known DNA amplification technique is the **polymerase chain reaction** (**PCR**). PCR uses DNA polymerase to rapidly replicate a specific DNA sequence in a test tube.

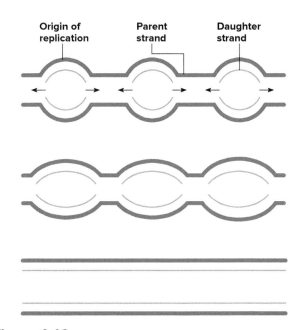

Figure 9.18 **DNA replication bubbles.** DNA replication occurs simultaneously along the 150 million nucleotides of the average human chromosome. The sites of replication resemble bubbles that coalesce as the daughter double helices form.
© James Cavallini/Science Source

Figure 9.19 presents the steps in amplifying DNA using PCR to identify a specific DNA sequence in a virus causing respiratory illness. The parenthetical numbers in the next paragraph correspond to the numbers in the figure.

PCR requires the following: (1) knowing a target DNA sequence from the suspected pathogen; (2) two types of lab-made, single-stranded short pieces of DNA called primers—these are complementary in sequence to opposite ends of the target sequence; (3) many copies of the four types of DNA nucleotides; (4) Taq1, which is a DNA polymerase from a bacterium that lives in hot springs. The enzyme eases PCR because it withstands heat, which is necessary to separate the DNA strands.

Table 9.2	Uses of the Polymerase Chain Reaction

PCR has been used to amplify DNA from:

- A cremated man, from skin cells left in his electric shaver, to diagnose an inherited disease in his children.
- A preserved quagga (a relative of the zebra) and a marsupial wolf, both extinct.
- Microorganisms that cannot be cultured for study.
- The brain of a 7,000-year-old human mummy.
- The digestive tracts of carnivores, to reveal food web interactions.
- Roadkills and carcasses washed ashore, to identify locally threatened species.
- Products illegally made from endangered species.
- Genetically altered bacteria that are released in field tests, to follow their dispersion.
- One cell of an 8-celled human embryo to detect a disease-related genotype.
- Poached moose meat in hamburger.
- Remains in Jesse James's grave, to make a positive identification.
- The guts of genital crab lice on a rape victim, which matched the DNA of the suspect.
- Fur from Snowball, a cat that linked a murder suspect to a crime.

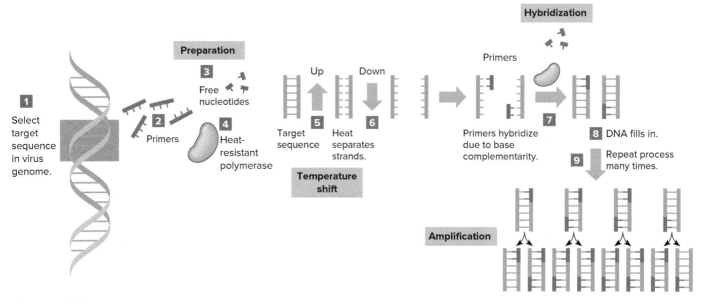

Figure 9.19 **PCR amplifies a specific nucleic acid sequence.** Many types of viruses cause upper respiratory infections. This figure illustrates how PCR is used on secretions to identify an adenovirus, which has double-stranded DNA as its genetic material. Influenza viruses have RNA as the genetic material, and a different form of PCR, called reverse transcription PCR, is used to analyze them.

PCR was born in the mind of Kary Mullis on a moonlit night in northern California in 1983. As he drove through the hills, Mullis was thinking about the precision of DNA replication, and a way to tap into it popped into his mind. He excitedly explained his idea to his girlfriend and then went home to think it through. "It was difficult for me to sleep with deoxyribonuclear bombs exploding in my brain," he wrote much later.

The idea behind PCR was so simple that Mullis had trouble convincing his superiors at Cetus Corporation that he was onto something. Over the next year, he used the technique to amplify a well-studied gene. Mullis published a landmark paper in 1985 and filed patent applications, launching the field of DNA amplification. He received a $10,000 bonus for his invention, which the company sold to another company for $300 million. Mullis did, however, win a Nobel Prize in 1993.

In the first step of PCR, the temperature is raised (5) and then lowered (6), which separates the two strands of the target DNA. Next (7), many copies of the two short DNA primers and Taq1 DNA polymerase are added. Primers bind by complementary base pairing to the separated target strands. The polymerase then fills in the bases opposite their complements, creating two daughter strands from each separated target double helix (8). The four double helices resulting from the first round of amplification then serve as the target sequences for the next round (9), and the process continues by again raising the temperature.

Pieces of identical DNA accumulate exponentially. The number of amplified pieces of DNA equals $2n$, where n is the number of temperature cycles. After 30 cycles, PCR yields more than 10 billion copies of the target DNA sequence. PCR is useful in forensic investigations to amplify small DNA samples (see **Bioethics**).

Key Concepts Questions 9.3

1. What does DNA replication accomplish?
2. What does semiconservative replication mean?
3. Explain how the steps of semiconservative DNA replication differ from the steps of the two other possible mechanisms.
4. Explain how DNA can be replicated fast enough to sustain a cell.
5. Discuss how the polymerase chain reaction is based on DNA replication.

9.4 Sequencing DNA

In 1977, Frederick Sanger, a British biochemist and two-time Nobel Prize winner, invented a way to determine the base sequence of a small piece of DNA. "Sanger sequencing" remains the conceptual basis for techniques that today can sequence an entire human genome in a day. Sanger's method is still used to sequence individual genes or to check the accuracy

Infidelity Testing

Bridgette came home a day early from a business trip to find her husband Roy drinking coffee in their kitchen with Tiffany, his business associate. They were laughing so hard that it took a few moments for them to notice Bridgette standing there. When they did, Tiffany blushed and Roy quickly withdrew his hand from her arm, knocking over her coffee mug.

Bridgette went upstairs to unpack. Flinging her purse on the bed, she noticed several strands of red hair on her pillow. Bridgette's hair was dark brown. She also noticed a crumpled tissue on the floor, partway under the bed.

Bridgette had read an article on the plane about companies that test "abandoned DNA." She went back downstairs for some plastic bags and picked up Tiffany's coffee mug, carrying it all back upstairs. In the bedroom, she quickly collected evidence—the telltale hairs, the discarded used tissue, and a cotton swab rubbed along the inside rim of the lipstick-stained mug. Bridgette emailed gotchaDNA.com and received a cheek swab collection kit a few days later, which she used to send in her own DNA for comparison, plus the $600 fee. Then she waited.

The technicians at gotchaDNA.com extracted the DNA from the samples and amplified it. First they checked for Y chromosome markers, which they found on the crumpled tissue. Then they looked for several short tandem repeats (STRs), which are short DNA sequences that are found in certain places in the genome but in different numbers of repeats in different individuals. STRs are used as markers to identify individuals in forensic investigations (see

figures 14.9 and 14.10). The STRs confirmed Bridgette's suspicion—the DNA on the mug that Tiffany had used and in the hair cells matched each other and not Bridgette's DNA. Tiffany, or at least her red hair, had somehow found its way onto Bridgette's pillow.

Cells use DNA to manufacture protein. People use DNA to identify people. Dozens of companies offer "infidelity DNA testing." Although a few websites provide documents for attesting that the samples are given willingly, many do not—and even list suggested sources of DNA for "adultery tracing." These sources include underwear, toothbrushes, dental floss, nail clippings, gum, cigarette butts, and razor clippings.

Questions for Discussion

1. In the United Kingdom, a law was enacted to prohibit sampling of a celebrity's DNA after someone tried to steal hair from Prince Harry to determine whether or not Prince Charles is his biological father. The United States has no such law. Do you think that one is warranted?

2. Do you think that DNA data obtained without consent should be admissible in a court of law? State a reason for your answer.

3. Discuss one reason in support of infidelity testing with DNA and one reason against it.

4. Identify the individuals in the scenario whom you believe behaved unethically.

---

of a selected sequenced part of a genome. The ability to sequence millions of small DNA pieces at once is the basis of newer methods, called "next-generation sequencing," that can handle much larger DNA molecules much faster.

Sanger sequencing generates a series of DNA fragments of identical sequence that are complementary to the DNA sequence of interest, which serves as a template strand. The fragments differ in length from each other by one end base, as follows:

Sequence of interest:	T A C G C A G T A C
Complementary sequence:	**A** T G C G T C A T G
Series of fragments:	**T** G C G T C A T G
	G C G T C A T G
	C G T C A T G
	G T C A T G
	T C A T G
	C A T G
	A T G
	T G
	G

DNA sequencing technologies are based on clever applications of chemistry. Sanger sequencing uses an approach called "chain termination." In today's version, PCR replicates a sequence of DNA many times, incorporating chemically altered bases that bear fluorescent tags, a different color for A, T, C, and G. When a tagged base is incorporated into a growing DNA chain, the DNA replication halts because another base cannot be added. It is a little like someone breaking into a line of people waiting to get into an event and then turning around and preventing others from extending the line. The result, in PCR, is a partial DNA molecule.

When chain termination happens many times, at different points in the replication of many copies of the same sequence, a collection of partial molecules results. The end bases reveal the sequence. The pieces are separated by size. A flash of laser light excites the four fluorescent tags and produces signals captured in a readout (**figures 9.20** and **9.21**). An algorithm deduces and displays the DNA sequence as a series of wavelengths and as the letters A, T, C, and G.

In 2007, new DNA sequencing "platforms" began to become available to researchers, and these are improved regularly, with storage of megadata in clouds. The sequencing technologies follow the basic approach of Sanger sequencing,

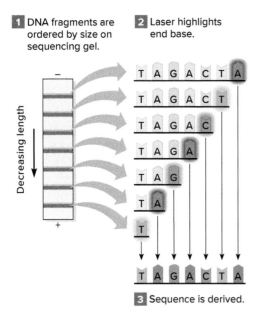

1 DNA fragments are ordered by size on sequencing gel.

2 Laser highlights end base.

Decreasing length

T A G A C T **A**

T A G A C **T**

T A G A **C**

T A G **A**

T A **G**

T **A**

T

3 Sequence is derived.

T A G A C T A

Figure 9.20 Reading a DNA sequence. A computer algorithm detects and records the end base from a series of size-ordered DNA fragments.

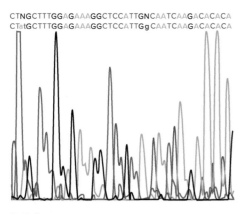

CTNGCTTTGGAGAAAGGCTCCATTGNCAATCAAGACACACA
CTatGCTTTGGAGAAAGGCTCCATTGgCAATCAAGACACACA

Figure 9.21 DNA sequence data. In automated first-generation DNA sequencing, a readout of sequenced DNA is a series of wavelengths that represent the terminal DNA base labeled with a fluorescent molecule.

but use different chemistries and materials as well as different ways to display the DNA pieces. The new techniques still fragment the human genome into millions of DNA strands, but then capture the pieces onto a solid surface such as glass or beads, and then make identical copies of each fragment on its unique bead or position on the glass. A reiterative process called "sequencing by synthesis" then reads one base at a time on the tens of millions of clusters or beads, yielding 40 billion to 600 billion bases of sequence information per experiment.

Several companies provide variations of next-generation sequencing. One product attaches short DNA pieces in a flow cell, which resembles a microscope slide. Another device uses tiny DNA-fringed beads in a water-oil mixture. A laser reads off the bases corresponding to the fluorescently tagged bases that are added and stream past the strands, binding their complements.

Another approach uses nanomaterials to distinguish how each of the four nucleotide bases disrupts an electrical field as a DNA molecule passes through nanopores at about 1,000 bases per second. One such material is a one-atom-thick sheet of carbon called grapheme that is strong, very thin, and conducts electricity. DNA is threaded through nanopores in it. Each of the four DNA base types disrupts an electrical field in a slightly different way. An algorithm converts the voltage changes into the DNA sequence that they represent. The voltage approach is also used on semiconductor chips.

The new ways to sequence DNA, some of which are company secrets, build on Sanger's invention, but differ in scale. Next-generation sequencing reads millions of sequences at once, and so the general approach is called "massively parallel DNA sequencing." The sequenced short pieces are called "reads." A sequencing experiment will yield many copies of each and every subset of contiguous bases that makes up a target sequence or genome. The more times any particular site in the target is represented among the pieces, the more accurate the deduced sequence—like learning more from reading this book over and over, from different starting points. For example, 30-fold coverage of a genome sequence is more accurate than 20-fold coverage. Thanks to next-generation sequencing, a human genome can be sequenced in hours—the first sequencing took 10 years!

Chapter 22 continues the discussion of how to overlap short DNA sequences to derive a genome sequence. The chapters in between address what is really important—what our genes do, and what all those sequences of A, T, C, and G can tell us about who we are, where we came from, and even where we may be headed.

Key Concepts Questions 9.4

1. Explain how Sanger sequencing works.
2. How are many copies of DNA pieces generated to obtain the sequence?
3. Why must several copies of a genome be cut up to sequence it?
4. What are next-generation sequencing technologies?

Summary

9.1 Experiments Identify and Describe the Genetic Material

1. DNA encodes information that the cell uses to synthesize protein. DNA can also be replicated, passing on its information during cell division.

2. Many scientists contributed to the discovery of DNA as the hereditary material. Miescher identified DNA in white blood cell nuclei. Garrod connected heredity to enzyme abnormalities. Griffith identified a "transforming principle" that transmitted infectiousness in pneumonia-causing bacteria; Avery, MacLeod, and McCarty discovered that the transforming principle is DNA; and Hershey and Chase confirmed that the genetic material is DNA and not protein.

3. Levene described the three components of a DNA building block (**nucleotide**) and found that they appear in DNA in equal amounts, and that nucleic acids include the sugar **ribose** or **deoxyribose**. Chargaff discovered that the amount of **adenine** (A) equals the amount of **thymine** (T), and the amount of **guanine** (G) equals that of **cytosine** (C). Franklin showed that the molecule is a certain type of helix. Watson and Crick deduced DNA's double helix structure.

9.2 DNA Structure

4. A **gene** encodes a protein. A nucleotide building block of a gene consists of a deoxyribose, a phosphate, and a nitrogenous base. A and G are **purines**; C and T are **pyrimidines**.

5. The rungs of the DNA double helix consist of hydrogen-bonded **complementary base pairs** (A with T, and C with G). The rails are chains of alternating sugars and phosphates (the **sugar-phosphate backbone**) that run **antiparallel** to each other. DNA is highly coiled, and complexed with protein to form **chromatin**. DNA winds around **histone** proteins and forms beadlike structures called **nucleosomes**. Loops enable different regions of the same DNA molecule to interact.

9.3 DNA Replication—Maintaining Genetic Information

6. Meselson and Stahl demonstrated the **semiconservative** nature of DNA replication with density shift experiments.

7. During replication, the DNA unwinds locally at several sites. **Replication forks** form as hydrogen bonds break between base pairs. Primase builds short RNA primers, which DNA sequences eventually replace. Next, **DNA polymerase** fills in DNA bases, and **ligase** seals remaining gaps, filling in the sugar-phosphate backbone.

8. Replication proceeds in a 5′ to 3′ direction, so the process must be discontinuous over short stretches on one strand.

9. Nucleic acid amplification (often by the **polymerase chain reaction**, or **PCR**) uses the power and precision of DNA replication enzymes to selectively mass produce selected DNA sequences. In PCR, primers corresponding to a DNA sequence of interest direct polymerization of supplied nucleotides to make many copies of that sequence.

9.4 Sequencing DNA

10. Sanger sequencing deduces a DNA sequence by aligning pieces of different sizes that differ from each other at the end base. Variations on this theme label, cut, and/or immobilize the DNA pieces in different ways, greatly speeding the process.

11. Next-generation sequencing uses a massively parallel approach, and different types of materials on which to immobilize DNA pieces, to read and overlap millions of pieces at once. It greatly speeds DNA sequencing.

Review Questions

1. List the components of a nucleotide.

2. How does a purine differ from a pyrimidine?

3. Distinguish between the locations of the phosphodiester bonds and the hydrogen bonds in a DNA molecule.

4. Why must DNA be replicated?

5. Why would a DNA structure in which each base type could form hydrogen bonds with any of the other three base types not produce a molecule that is easily replicated?

6. What part of the DNA molecule encodes information?

7. Explain how DNA is a directional molecule in a chemical sense.

8. What characteristic of the DNA structure does the following statement refer to? "New DNA forms in the 5′ to 3′ direction, but the template strand is read in the 3′ to 5′ direction."

9. Match each experiment in 1–5 to the appropriate description or conclusion from a–e.

1. Density shift experiments	a. First experiments in identifying hereditary material.
2. Discovery on dirty bandages of an acidic substance that includes nitrogen and phosphorus	b. Complementary base pairing is part of DNA structure and maintains a symmetrical double helix.
3. "Blender experiments" showing that the part of a virus that infects bacteria contains phosphorus, but not sulfur	c. Identification of nuclein.
4. Determination that DNA contains equal amounts of guanine and cytosine, and of adenine and thymine	d. DNA, not protein, is the hereditary material.
5. Discovery that bacteria can transfer a "factor" that transforms a harmless strain into a lethal one	e. DNA replication is semiconservative.

10. Place the following enzymes in the order in which they function in DNA replication: ligase, primase, helicase, and DNA polymerase.

11. How can very long DNA molecules fit into a cell's nucleus?

12. Place in increasing size order: nucleosome, histone protein, and chromatin.

13. Explain how loop formation enables gene-gene interactions.

14. How are very long strands of DNA replicated without becoming twisted into a huge tangle?

15. List the steps in DNA replication.

16. Why must DNA be replicated continuously as well as discontinuously?

17. Explain how RNA participates in DNA replication.

18. Is downloading a document from the Internet analogous to replicating DNA? Cite a reason for your answer.

Applied Questions

1. The Rhino DNA Indexing System (RhODIS) was established in South Africa to track rhinos, which are hunted for their horns. In some Asian countries, rhino horn is believed to cure many illnesses and is collected as a status symbol. The black rhino became extinct in 2011, and three of the remaining five species are critically endangered. Explain how law enforcement officials can use RhODIS to capture poachers.

2. Bloom syndrome causes short stature, sun sensitivity, and susceptibility to certain types of cancer. An autosomal recessive mutation in the gene that encodes ligase causes the condition. Which step in DNA replication is affected?

3. DNA contains the information that a cell uses to synthesize a particular protein. How do proteins assist in DNA replication?

4. A person with deficient or abnormal ligase may have an increased cancer risk and chromosome breaks that cannot heal. The person is, nevertheless, alive. Why are there no people who lack DNA polymerase?

5. Write the sequence of a strand of DNA replicated using each of the following base sequences as a template:
   a. T C G A G A A T C T C G A T T
   b. C C G T A T A G C C G G T A C
   c. A T C G G A T C G C T A C T G

6. Cite an example of how knowing a DNA sequence could be abused and an example of how knowing a DNA sequence could be helpful.

7. People often use the phrase "the gene for" to describe traits that do not necessarily or directly arise from a protein's actions, such as "the gene for jealousy" or "the gene for acting." How would you explain to them what a gene actually is?

8. The first several DNA bases in the gene that encodes serum albumin, which is a protein that is abundant in blood as well as in egg white, are as follows:

   A G C T T T T C T C T T C T G T C A A C

   This is the strand that holds the information the cell uses to construct the protein. Write the sequence of the complementary DNA strand.

9. To diagnose a rare form of encephalitis (brain inflammation), a researcher needs a million copies of a viral gene. She decides to use PCR on a sample of the patient's cerebrospinal fluid. If one cycle takes 2 minutes, how long will it take to perform a millionfold amplification?

Case Studies and Research Results

1. Which do you think was the more far-reaching accomplishment, determining the structure of DNA or sequencing human genomes? State a reason for your answer.

2. Find further information on one of the inventors or researchers mentioned in the chapter, and describe his or her inspiration or thought process.

3. To identify the "loop-ome," researchers had to chop the human genome into pieces about 1,000 bases long and then observe interactions of the genes that are part of the pieces. Explain how this was computationally a huge challenge.

4. The heath hen is an extinct bird that resembled a prairie chicken. It lived along the east coast of the United States, but by the mid-nineteenth century, was found only on Martha's Vineyard, an island off Cape Cod, Massachusetts. The last heath hen succumbed in 1932. A non-profit organization is proposing to use genome editing (see figure 19.11) to "bring back" the heath hen using "genetic rescue." They plan to identify genes unique to the heath hen in samples of the birds from the Martha's Vineyard museum, introduce those genes into the genomes of modern prairie chickens, and then breed heath hens.

   a. Explain how the plan to bring back the heath hen differs from bringing back the Przewalskii's horse. This "wild ass" was thought to have become extinct in the 1960s but was brought back using conventional breeding after a few of them were discovered roaming the plains of Mongolia.

   b. Do you think "de-extinction" by manipulating DNA is a good or a bad idea?

   c. How would you recreate a woolly mammoth?

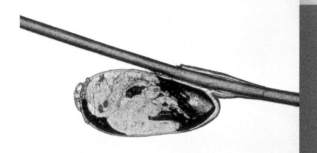

CHAPTER 10

Gene Action: From DNA to Protein

The egg of a scalp louse clings to a human hair. When humans began to wear clothing 170,000 years ago, lice that were better able to mix protein-encoding parts of certain genes, enabling them to lay eggs in the seams and folds of garments, led to the evolution of the subspecies of "body lice."

Source: Centers for Disease Control and Prevention (CDC)

Learning Outcomes

10.1 The Importance of Proteins
1. Dietary proteins are digested into amino acids.
2. Genes provide instructions for linking amino acids into proteins.

10.2 Transcription Copies the Information in DNA
3. List the major types of RNA molecules and their functions.
4. Explain the importance of transcription factors.
5. List the steps of transcription.

10.3 Translation of a Protein
6. Discuss how researchers deduced the genetic code.
7. List the steps of protein synthesis.

10.4 Processing a Protein
8. Define the four components of a protein's shape.
9. Explain the importance of protein folding.

 ## The BIG Picture

DNA sequences are the blueprints of life. Cells must maintain this information, yet also access it to manufacture proteins. RNA acts as the go-between, linking DNA sequences to the amino acid sequences of proteins.

Gene Splicing in Lice and the Challenge of Clothing

The louse *Pediculus humanus* deposits its eggs ("nits") onto scalp hairs, much to the distress of parents whose children bring these unwelcome guests home from school. Until the invention of clothing about 170,000 years ago, lice enjoyed the vast landscapes of hairy humans. How did the blood-sucking insects cope with their shrinking turf? A few were able to lay their eggs into garment seams and folds, thanks to a flexibility in their genomes. Over time, the species began to diverge into the scalp variety and the newer, slightly larger "body lice."

Lice adapted to the new "human clothing environment" because certain of their genes are in "pieces." These genes can be spliced: transcribed into RNA but then certain pieces removed and the remainder reattached, creating variants of the encoded protein. Over time, insects that could splice certain genes in ways that enabled them to lay their eggs in more places left more offspring, and the species began to diverge.

Researchers discovered "alternate splicing" to create different versions of proteins in about a third of the louse genome. A closer look at the proteins that differ between scalp and body lice made sense. In order to move from hairy heads to fabric folds, body lice:

- had different behaviors,
- had metabolic changes that enabled them to survive longer without a blood meal, and
- had changes to their salivary glands.

Unfortunately for us, the genomically retooled lice can give us, in addition to nasty nits, diseases such as typhus, trench fever, and relapsing fever.

10.1 The Importance of Proteins

Proteins are familiar as one of the three major types of nutrients, abundant in meats, eggs, and legumes. We eat protein molecules and our digestive systems break them down into **amino acids**, which are molecules small enough to enter the bloodstream at the small intestine. From there, amino acids can enter our cells, where RNA molecules transcribed from genes guide their assembly into new proteins. The muscle protein myosin from a hamburger does not become the myosin in a muscle of the person who eats it. Instead, the cow myosin is broken down into its constituent amino acids, which are absorbed and then linked to form a protein that is part of the human body, perhaps even the human version of myosin.

More than 500 types of amino acids are chemically possible, but only 20 are required in our diets for the body to produce its own protein molecules. An adult can synthesize 12 of the 20 amino acids; the other 8 are termed "essential" because they must come from the diet. Children can synthesize 10 of the 20 types of amino acids.

The 20 biological amino acids have similar frameworks. Each has a central carbon atom that bonds to an amino group (NH_2), an acid group (COOH), a hydrogen atom (H), and an "R" group (**figure 10.1**). It is the R group that distinguishes the 20 types of amino acids the human body uses. In a protein molecule, the many linked amino acids fold into a three-dimensional structure that is essential to the protein's function. A protein consists of one or more long chains of amino acids called **polypeptides**. Shorter chains of amino acids are called peptides. We return to protein structure in section 10.4.

It is impossible to imagine a living human body that cannot manufacture protein! Without proteins, blood couldn't clot, muscles couldn't contract, antibodies couldn't fight infection, and we'd have no hair, skin, or connective tissue, nor would any biochemical reactions that require enzymes proceed fast enough to keep us alive.

The inability to make just one of the 20,325 or so types of proteins that the human exome encodes is devastating. A child with giant axonal neuropathy (see Clinical Connection 2.2) can no longer walk because her motor neurons cannot manufacture gigaxonin protein. People who have cystic fibrosis (see figures 1.3

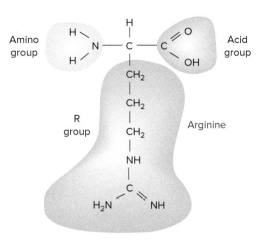

Figure 10.1 Amino acid structure. An amino acid consists of a central carbon atom bonded to a hydrogen atom, an amino group, an acid group, and a distinctive R group. This amino acid is arginine.

and 1.4 and Clinical Connection 4.1) have respiratory and digestive problems because ion channel proteins cannot fold properly. The boy described in the opener to chapter 5) cannot see because cells that form a very thin layer at the backs of his eyes cannot synthesize a protein needed to utilize vitamin A.

Chapter 9 detailed how researchers narrowed down the candidates for the genetic material to DNA. Protein is equally important because the way that genotype becomes phenotype is through protein actions and interactions.

Key Concepts Questions 10.1

1. Describe the building blocks of proteins.
2. Explain the role of genes in manufacturing proteins.
3. List some of the ways that proteins are important.

10.2 Transcription Copies the Information in DNA

A cell uses two processes to manufacture proteins using genetic instructions. **Transcription** first synthesizes an RNA molecule that is complementary to one strand of the DNA double helix for a particular gene. The RNA copy is taken out of the nucleus and into the cytoplasm. There, the process of **translation** uses the information in the RNA to manufacture a protein by aligning and joining specified amino acids. Finally, the protein folds into a specific three-dimensional form necessary for its function.

Accessing the genome to manufacture protein is a huge, constant task. Cells replicate their DNA only during S phase of the cell cycle, but transcription and translation occur continuously, except during M phase. The two processes supply the proteins essential to be alive, as well as those that give a cell its specialized characteristics.

Figure 10.2

DNA to RNA to protein. Some of the information stored in DNA is copied to RNA (transcription), some of which is used to assemble amino acids into proteins (translation). DNA replication perpetuates genetic information. This figure repeats throughout the chapter, with the part under discussion highlighted.

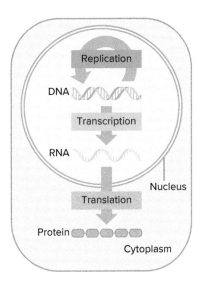

Shortly after Watson and Crick published their structure of DNA in 1953, they described the relationship between nucleic acids and proteins as a directional flow of information called the "central dogma" (**figure 10.2**). As Francis Crick explained in 1957, *"The specificity of a piece of nucleic acid is expressed solely by the sequence of its bases, and this sequence is a code for the amino acid sequence of a particular protein."* This statement inspired more than a decade of intense research to discover exactly how cells make proteins. The process centers around RNA.

RNA Structure and Types

RNA bridges gene and protein, as **figure 10.3** depicts. The bases of an RNA sequence are complementary to those of one strand of the double helix, which is called the **template strand**. An enzyme, **RNA polymerase**, builds an RNA molecule. The other, nontemplate strand of the DNA double helix is called the **coding strand**.

RNA and DNA have similarities and differences (**figure 10.4** and **table 10.1**). Both are nucleic acids, consisting

DNA
Stores RNA- and protein-encoding information, and transfers information to daughter cells

(a)

RNA
Carries protein-encoding information, and helps to make proteins

Double-stranded

Generally single-stranded

(b)

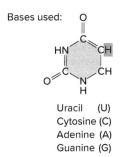

Deoxyribose as the sugar

Ribose as the sugar

(c)

Bases used:

Bases used:

Thymine (T)
Cytosine (C)
Adenine (A)
Guanine (G)

Uracil (U)
Cytosine (C)
Adenine (A)
Guanine (G)

(d)

Figure 10.4 DNA and RNA differences. (a) DNA and RNA have different functions. **(b)** DNA is double-stranded; RNA is usually single-stranded. **(c)** DNA nucleotides include deoxyribose; RNA nucleotides have ribose. **(d)** Finally, DNA nucleotides include the pyrimidine thymine, whereas RNA nucleotides have uracil.

of sequences of nitrogen-containing bases joined by sugar-phosphate backbones. However, their structures are different. RNA is usually single-stranded, whereas DNA is double-stranded.

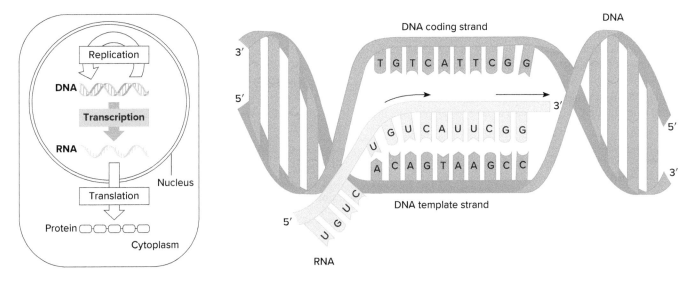

Figure 10.3 The relationship among RNA, the DNA template strand, and the DNA coding strand. The RNA sequence is complementary to the DNA template strand and is the same sequence as the DNA coding strand, with uracil (U) in place of thymine (T).

Table 10.1	How DNA and RNA Differ
DNA	**RNA**
1. Usually double-stranded	1. Usually single-stranded
2. Thymine as a base	2. Uracil as a base
3. Deoxyribose as the sugar	3. Ribose as the sugar
4. Maintains protein-encoding information	4. Carries protein-encoding information and controls how information is used
5. Cannot function as an enzyme	5. Can function as an enzyme
6. Persists	6. Short-lived

Also, RNA has the pyrimidine base **uracil** where DNA has thymine. RNA (*ribo*nucleic acid) nucleotides include the sugar ribose. DNA (*deoxyribo*nucleic acid) nucleotides include the sugar deoxyribose. DNA and RNA differ in function, too. DNA stores genetic information, whereas RNA controls how that information is used. The presence of the OH group at the 2′ position of ribose makes RNA much less stable than DNA. This distinction is critical in its function as a short-lived carrier of genetic information.

As RNA is synthesized along DNA, it folds into a three-dimensional shape, or **conformation**, that arises from complementary base pairing within the same RNA molecule. For example, a sequence of AAUUCC might hydrogen bond to a sequence of UUAAGG—its complement—elsewhere in the same molecule, a little like touching elbows to knees. Conformation is very important for RNA's functioning. The three major types of RNA, which have distinctive conformations, are messenger RNA, ribosomal RNA, and transfer RNA (**table 10.2**). Other classes of RNA control which genes are expressed (transcribed and translated) under specific circumstances (see table 11.2 for their descriptions).

Messenger RNA (mRNA) carries the information that specifies a particular protein. Each set of three consecutive mRNA bases forms a genetic code word, or **codon**, that specifies a certain amino acid. Because genes vary in length, so do mature mRNA molecules. Most mRNAs are 500 to 4,500 bases long. Differentiated cells can carry out specialized functions

because they express certain subsets of genes—that is, they produce certain mRNA molecules, which are also called transcripts. The information in the transcripts is then used to manufacture the encoded proteins. A muscle cell, for example, has many mRNAs that correspond to the contractile proteins actin and myosin, whereas a skin cell contains many mRNAs that specify scaly keratin proteins.

To use the information in an mRNA sequence, a cell requires the two other major classes of RNA. **Ribosomal RNA (rRNA)** molecules range from 100 to nearly 3,000 nucleotides long. Ribosomal RNAs associate with certain proteins to form a ribosome. Recall from chapter 2 that a ribosome is an organelle made up of many different protein and RNA subunits. Overall, a ribosome functions like a machine to assemble and link amino acids to form proteins (**figure 10.5**).

A ribosome has two subunits that are separate in the cytoplasm but join at the site of initiation of protein synthesis. The larger ribosomal subunit has three types of rRNA molecules, and the small subunit has one. Ribosomal RNA, however, is more than a structural support. Certain rRNAs catalyze the formation of the peptide bonds between amino acids. Such an RNA with enzymatic function is called a ribozyme. Other rRNAs help to align the ribosome and mRNA.

The third major type of RNA molecule, **transfer RNA (tRNA)**, binds an mRNA codon at one end and a specific amino acid at the other. A tRNA molecule is only 75 to 80 nucleotides long. Some of its bases form weak chemical bonds with each other, folding the tRNA into loops in a characteristic cloverleaf shape (**figure 10.6**). One loop of the tRNA has three bases in a row that form the **anticodon**, which is complementary to an mRNA codon. The end of the tRNA opposite the anticodon

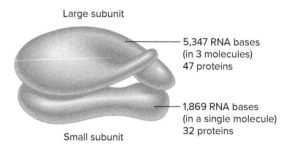

Large subunit

5,347 RNA bases (in 3 molecules) 47 proteins

1,869 RNA bases (in a single molecule) 32 proteins

Small subunit

Figure 10.5 **The ribosome.** A ribosome from a eukaryotic cell has two subunits; together, they consist of 79 proteins and 4 rRNA molecules.

Table 10.2	Major Types of RNA	
Type of RNA	**Size (number of nucleotides)**	**Function**
Messenger RNA (mRNA)	500 to 4,500+	Encodes amino acid sequence
Ribosomal RNA (rRNA)	100 to 3,000	Associates with proteins to form ribosomes, which structurally support and catalyze protein synthesis
Transfer RNA (tRNA)	75 to 80	Transports specific amino acids to the ribosome for protein synthesis

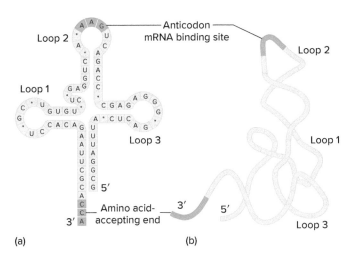

(a)   (b)

Figure 10.6 **Transfer RNA.** **(a)** Certain nucleotide bases within a tRNA hydrogen bond with each other, giving the molecule a "cloverleaf" conformation that can be represented in two dimensions. The darker bases at the top form the anticodon, the sequence that binds a complementary mRNA codon. Each tRNA terminates with the sequence CCA, where a particular amino acid covalently bonds. **(b)** A three-dimensional representation of a tRNA depicts the loops that interact with the ribosome.

strongly bonds to a specific type of amino acid, and a tRNA with a particular anticodon sequence always carries the same type of amino acid. For example, a tRNA with the anticodon sequence GAA always picks up the amino acid phenylalanine. Enzymes attach amino acids to tRNAs that bear the appropriate anticodons (**figure 10.7**).

Transcription Factors

If all of the genes in the human genome were being transcribed and translated at the same time, chaos would result. It would be like trying to open all of the files and programs on a computer at once. Instead, accessing genetic information is selective and efficient.

Different cell types express different subsets of genes. To manage this, proteins called **transcription factors** come together and interact, forming an apparatus that binds DNA at certain sequences and initiates transcription at specific sites

Figure 10.7 **A tRNA** with a particular anticodon sequence always binds the same type of amino acid.

on chromosomes. The transcription factors respond to signals from outside the cell, such as hormones and growth factors, and form a pocket for RNA polymerase to bind and begin building RNA molecules at specific genes. Transcription factors include regions called binding domains that guide them to the genes they control. The DNA binding domains have colorful names, such as "helix-turn-helix," "zinc fingers," and "leucine zippers," that reflect their distinctive shapes. Harnessing these binding domains underlies genome editing technologies such as CRISPR-Cas9 and zinc finger nucleases (see section 19.4).

Types of transcription factors are few, but they work in combinations, providing great specificity in controlling gene expression. Overall, transcription factors link the genome to the environment. For example, lack of oxygen, such as from choking or smoking, sends signals that activate transcription factors to turn on dozens of genes that enable cells to handle the stress of low-oxygen conditions.

Mutations in transcription factor genes can be devastating. Rett syndrome (see the chapter 2 opener) and the homeotic mutations described in Clinical Connection 3.1 result from mutations in transcription factor genes. Overexpressed transcription factors can cause cancer. Transcription factors are themselves controlled by each other and by other classes of molecules.

Steps of Transcription

Transcription is described in three steps: initiation, elongation, and termination. The process is called "transcription" because it copies the DNA information into RNA, but keeps the information in the genetic language of DNA nucleotide bases.

Transcription factors and RNA polymerase use clues in the DNA sequence to "know" where to bind to DNA to begin transcribing a specific gene. In transcription initiation, transcription factors and RNA polymerase are attracted to a **promoter**, which is a special sequence that signals the start of the gene, like a capital letter at the start of a sentence.

Figure 10.8 illustrates transcription factor binding, which sets up a site to receive RNA polymerase. The first transcription factor to bind, called a TATA binding protein, is chemically attracted to a DNA sequence called a TATA box—the base sequence TATA surrounded by long stretches of G and C. Once the first transcription factor binds, it attracts others in groups. Finally RNA polymerase joins the complex, binding just in front of the start of the gene sequence. The assembly of these components is transcription initiation.

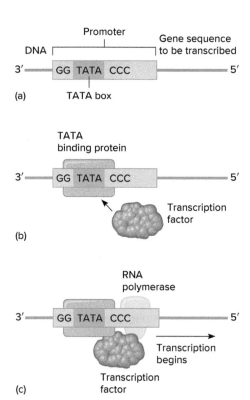

(a)

Promoter
DNA | Gene sequence to be transcribed
3′ — GG TATA CCC — 5′
TATA box

(b)

TATA binding protein
3′ — GG TATA CCC — 5′
Transcription factor

(c)

RNA polymerase
3′ — GG TATA CCC — 5′
Transcription begins
Transcription factor

Figure 10.8 **Setting the stage for transcription to begin. (a)** Proteins that initiate transcription recognize specific sequences in the promoter region of a gene. **(b)** A binding protein recognizes the TATA region and binds to the DNA. This allows other transcription factors to bind. **(c)** The bound transcription factors form a pocket that allows RNA polymerase to bind and begin making RNA.

In the next stage, transcription elongation, enzymes unwind the DNA double helix locally, and free RNA nucleotides bond with exposed complementary bases on the DNA template strand (see figure 10.3). RNA polymerase adds the RNA nucleotides in the sequence the DNA specifies, moving along the DNA strand in a 3′ to 5′ direction, synthesizing the RNA molecule in a 5′ to 3′ direction. A terminator sequence in the DNA indicates where the gene's RNA-encoding region ends, like the period at the end of a sentence. When this spot is reached, the third stage, transcription termination, occurs (**figure 10.9**). A typical rate of transcription in humans is 20 bases per second.

RNA is transcribed using only the gene sequence on the template strand.

However, different genes on the same chromosome may be transcribed from different strands of the double helix. The coding strand of the DNA is so-called because its sequence is identical to that of the RNA, except with thymine (T) in place of uracil (U).

Several RNAs may be transcribed from the same DNA template strand simultaneously (**figure 10.10**). Because mRNA is short-lived, with about half of it degraded every 10 minutes, a cell must constantly transcribe certain genes to maintain supplies of essential proteins.

To determine the sequence of RNA bases transcribed from a gene, write the RNA bases that are complementary to the template DNA strand, using uracil opposite adenine. For example, a DNA template strand that has the sequence

$$3′\ C\ C\ T\ A\ G\ C\ T\ A\ C\ 5′$$

is transcribed into RNA with the sequence

$$5′\ G\ G\ A\ U\ C\ G\ A\ U\ G\ 3′$$

and the coding DNA sequence is

$$5′\ G\ G\ A\ T\ C\ G\ A\ T\ G\ 3′$$

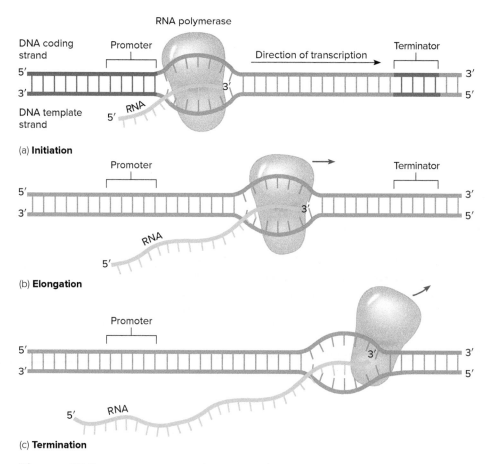

(a) **Initiation**

(b) **Elongation**

(c) **Termination**

Figure 10.9 **Transcription of RNA from DNA.** Transcription occurs in three stages: initiation, elongation, and termination. Initiation is the control point that determines which genes are transcribed. RNA nucleotides are added during elongation. A terminator sequence signals the end of transcription.

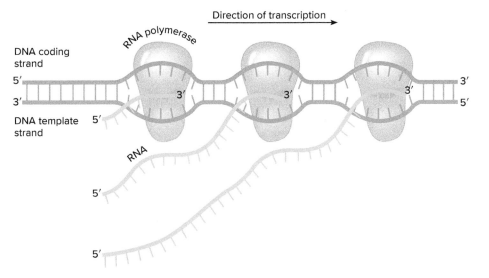

RNA polymerase

DNA coding strand

5′

3′

DNA template strand 5′

3′

3′

3′

3′

5′

RNA

5′

5′

5′

Figure 10.10 **Many identical copies of RNA are transcribed simultaneously.** Usually 100 or more DNA bases lie between RNA polymerases.

RNA Processing

In bacteria, RNA is translated into protein as soon as it is transcribed from DNA because a nucleus does not physically separate the two processes. In eukaryotic cells like ours, mRNA must first exit the nucleus and enter the cytoplasm, where ribosomes are located. Messenger RNA is altered in several steps before it is translated in these more complex cells.

First, after mRNA is transcribed, a short sequence of modified nucleotides, called a cap, is added to the 5′ end of the molecule. The cap consists of a backwardly inserted guanine (G), which attracts an enzyme that adds methyl groups (CH₃) to the G and one or two adjacent nucleotides. This methylated cap is a recognition site for protein synthesis. At the 3′ end, a special polymerase adds about 200 adenines, forming a "poly A tail." The poly A tail is necessary for protein synthesis to begin, and may also stabilize the mRNA so that it stays intact longer.

Further changes occur to the capped, poly A tailed mRNA before it is translated into protein. Parts of the mRNA called **introns** (short for "intervening sequences") that were transcribed are removed. The ends of the remaining molecule are spliced together before the mRNA is translated. The parts of mRNA that remain and are translated into amino acid sequences (protein) are called **exons** (**figure 10.11**).

Once introns are spliced out, enzymes check, or proofread, the remaining mRNA. Messenger RNAs that are too short or too long may be held in the nucleus. Proofreading also monitors tRNAs, ensuring that they assume the correct cloverleaf shape.

Prior to intron removal, the mRNA is called premRNA. Introns control their own removal. They associate with certain proteins to form small nuclear ribonucleoproteins (snRNPs), or "snurps." Four snurps form a structure called a spliceosome that cuts introns out and attaches exons

to each other to form the mature mRNA that exits the nucleus.

Introns range in size from 65 to 10,000 or more bases; the average intron is 3,365 bases. The average exon, in contrast, is only 145 bases long. The number, size, and organization of introns vary from gene to gene. The coding portion of the average human gene is 1,340 bases, whereas the average total size of a gene is 27,000 bases. The dystrophin gene is 2,500,000 bases, but its corresponding mRNA sequence is only 14,000 bases! The gene contains 80 introns.

The discovery of introns in the 1970s surprised geneticists, who had thought genes were like sentences in which all of the information has meaning. At first, some geneticists called introns "junk DNA"—a term that has unfortunately persisted even as researchers have discovered the functions of many introns. Some introns encode RNAs that control gene expression, whereas others are actually exons on the complementary strand of DNA. Introns may also be vestiges of ancient genes that have lost their original function, or are remnants of the DNA of viruses that once integrated into chromosomes. Chapter 11 explores the functions of introns.

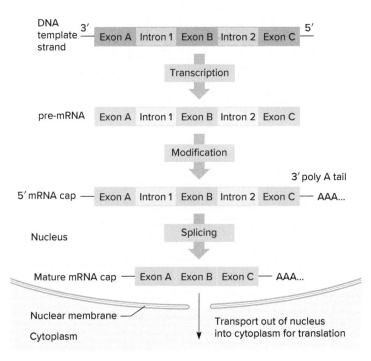

Figure 10.11 **Messenger RNA processing—the maturing of the message.** Several steps process pre-mRNA into mature mRNA. First, a large region of DNA containing the gene is transcribed. Then a modified nucleotide cap and poly A tail are added and introns are spliced out. Finally, the intact, mature mRNA is sent out of the nucleus.

The intron/exon organization of genes maximizes genetic information. Different combinations of exons of a gene encode different versions of the protein product, termed isoforms. From 40 to 60 percent of human genes encode isoforms, and the mechanism of combining exons of a gene in different ways is called **alternate splicing**. In this way, cell types can use versions of the same protein in slightly different ways in different tissues. For example, a protein that transports fats is shorter in the small intestine, where it carries dietary fats, than it is in the liver, where it carries fats made in the body. The chapter opener describes how alternate splicing enabled lice to expand their ecological niche to encompass both our scalps and our clothed parts.

Key Concepts Questions 10.2

1. Distinguish the structures and functions of DNA and RNA.
2. Explain how messenger RNA transmits instructions to build proteins.
3. What is the function of transfer RNAs in synthesizing proteins?
4. List the steps of transcription.
5. What is alternate splicing?

10.3 Translation of a Protein

Translation assembles a protein using the information in the mRNA sequence. Particular mRNA codons correspond to particular amino acids (**figure 10.12**). This correspondence between the chemical languages of mRNA and protein is the **genetic code** (**table 10.3**). Translation takes place on free ribosomes in the cytoplasm as well as on ribosomes that are in the endoplasmic reticulum (ER).

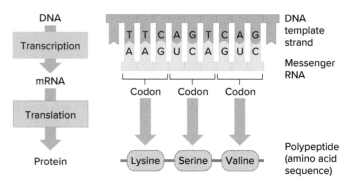

Figure 10.12 **From DNA to RNA to protein.** Messenger RNA is transcribed from a locally unwound portion of DNA. In translation, transfer RNA matches mRNA codons with amino acids.

Table 10.3   The Genetic Code*

		Second Letter				
First Letter		**U**	**C**	**A**	**G**	**Third Letter**
U	UUU	Phenylalanine (Phe)	UCU	Serine (Ser)	UAU	Tyrosine (Tyr)	UGU	Cysteine (Cys)	U

Second Letter

		U	C	A	G	
U	UUU — Phenylalanine (Phe)	UCU — Serine (Ser)	UAU — Tyrosine (Tyr)	UGU — Cysteine (Cys)	U
	UUC — Phenylalanine (Phe)	UCC — Serine (Ser)	UAC — Tyrosine (Tyr)	UGC — Cysteine (Cys)	C
	UUA — Leucine (Leu)	UCA — Serine (Ser)	UAA — "stop"	UGA — "stop"	A
	UUG — Leucine (Leu)	UCG — Serine (Ser)	UAG — "stop"	UGG — Tryptophan (Trp)	G
C	CUU — Leucine (Leu)	CCU — Proline (Pro)	CAU — Histidine (His)	CGU — Arginine (Arg)	U
	CUC — Leucine (Leu)	CCC — Proline (Pro)	CAC — Histidine (His)	CGC — Arginine (Arg)	C
	CUA — Leucine (Leu)	CCA — Proline (Pro)	CAA — Glutamine (Gln)	CGA — Arginine (Arg)	A
	CUG — Leucine (Leu)	CCG — Proline (Pro)	CAG — Glutamine (Gln)	CGG — Arginine (Arg)	G
A	AUU — Isoleucine (Ile)	ACU — Threonine (Thr)	AAU — Asparagine (Asn)	AGU — Serine (Ser)	U
	AUC — Isoleucine (Ile)	ACC — Threonine (Thr)	AAC — Asparagine (Asn)	AGC — Serine (Ser)	C
	AUA — Isoleucine (Ile)	ACA — Threonine (Thr)	AAA — Lysine (Lys)	AGA — Arginine (Arg)	A
	AUG — Methionine (Met) and "start"	ACG — Threonine (Thr)	AAG — Lysine (Lys)	AGG — Arginine (Arg)	G
G	GUU — Valine (Val)	GCU — Alanine (Ala)	GAU — Aspartic acid (Asp)	GGU — Glycine (Gly)	U
	GUC — Valine (Val)	GCC — Alanine (Ala)	GAC — Aspartic acid (Asp)	GGC — Glycine (Gly)	C
	GUA — Valine (Val)	GCA — Alanine (Ala)	GAA — Glutamic acid (Glu)	GGA — Glycine (Gly)	A
	GUG — Valine (Val)	GCG — Alanine (Ala)	GAG — Glutamic acid (Glu)	GGG — Glycine (Gly)	G

First Letter (left column), Third Letter (right column)

*The genetic code consists of mRNA triplets and the amino acids that they specify.

In the 1960s, researchers described and deciphered the genetic code. They used logic and clever experiments on simple genetic systems such as viruses, and synthetic DNA molecules, to discover which mRNA codons correspond to which amino acids.

The Genetic Code

Before researchers could match mRNA codons to the amino acids they encode, they had to establish certain requirements for such a code.

1. The Code Is Triplet.

The number of different protein building blocks (20 amino acids) exceeds the number of different mRNA building blocks (4 bases). Therefore, each codon must include more than one mRNA base. If a codon consisted of only one mRNA base, then codons could specify only four different amino acids, one corresponding to each of the four bases: A, C, G, and U. If each codon consisted of two bases, then only 16 (4^2) different amino acids could be specified, one corresponding to each of the 16 possible combinations of two RNA bases. If a codon consisted of three bases, then the genetic code could specify as many as 64 (4^3) different amino acids, sufficient to encode the 20 different amino acids that make up biological proteins. Therefore, the minimum number of bases in a codon is three.

Francis Crick and his coworkers showed that the code is triplet by adding or removing one, two, or three bases to or from a viral gene with a well-known sequence and protein product. Altering the DNA sequence by one or two bases produced a different amino acid sequence. This happened because the change disrupted the **reading frame**, which is the sequence of amino acids encoded from a certain starting point in a DNA sequence. However, adding or deleting three contiguous bases added or deleted only one amino acid in the protein without disrupting the reading frame. The rest of the amino acid sequence was retained. The code, the researchers deduced, is triplet (**figure 10.13**).

Further experiments confirmed the triplet nature of the genetic code. Adding a base at one point in the gene and deleting a base at another point disrupted the reading frame only between these sites. The result was a protein with a stretch of the wrong amino acids, like a sentence with a few misspelled words in the middle after one letter is added or removed.

2. The Code Does Not Overlap.

Consider a hypothetical mRNA sequence:

AUGCCCAAG

If the genetic code is triplet and a DNA sequence is "read" in a nonoverlapping manner, then this sequence has only three codons and specifies three amino acids:

AUGCCCAAG

AUG (methionine)

CCC (proline)

AAG (lysine)

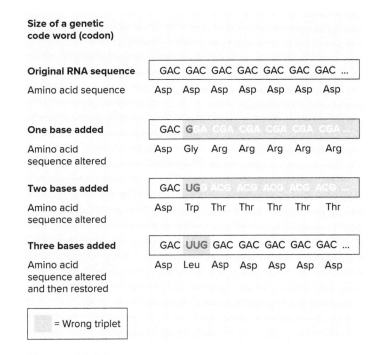

Size of a genetic code word (codon)

Original RNA sequence	GAC GAC GAC GAC GAC GAC GAC ...
Amino acid sequence	Asp Asp Asp Asp Asp Asp Asp

One base added	GAC G GA CGA CGA CGA CGA CGA ...
Amino acid sequence altered	Asp Gly Arg Arg Arg Arg Arg

Two bases added	GAC UG ACG ACG ACG ACG ACG ...
Amino acid sequence altered	Asp Trp Thr Thr Thr Thr Thr

Three bases added	GAC UUG GAC GAC GAC GAC GAC ...
Amino acid sequence altered and then restored	Asp Leu Asp Asp Asp Asp Asp

☐ = Wrong triplet

Figure 10.13 Three at a time. Adding or deleting one or two nucleotides in a DNA sequence results in a frameshift that disrupts the encoded amino acid sequence. Adding or deleting three bases does not disrupt the reading frame because the code is triplet. This is a simplified representation of the Crick experiment.

If the DNA sequence is overlapping, however, the sequence specifies seven codons:

AUGCCCAAG

AUG (methionine)

UGC (cysteine)

GCC (alanine)

CCC (proline)

CCA (proline)

CAA (glutamine)

AAG (lysine)

In an overlapping DNA sequence, certain amino acids would always follow certain others, constraining protein structure. For example, AUG would always be followed by an amino acid whose codon begins with UG. This does not happen in nature. Therefore, the protein-encoding DNA sequence is not overlapping.

Even though the genetic code is nonoverlapping, any DNA or RNA sequence can be read in three different reading frames, depending upon the "start" base. **Figure 10.14** depicts the three reading frames for the sequence just discussed, slightly extended. It encodes three different trios of amino acids.

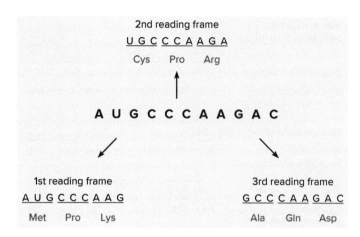

Figure 10.14 Reading frames—where the sequence begins. A sequence of DNA has three reading frames.

3. The Code Includes Controls.

The genetic code includes directions for starting and stopping translation. The codon AUG signals "start," and the codons UGA, UAA, and UAG signify "stop." A sequence of DNA that does not include a stop codon is called an **open reading frame**, and is a sign of a possible protein-encoding gene. If a DNA sequence was just random, assuming equal numbers of each base, a stop codon would arise about every 21 codons.

Another form of "punctuation" in the genetic code is a short sequence of bases at the start of each mRNA that enables the mRNA to form hydrogen bonds with rRNA in a ribosome. It is called a leader sequence.

4. The Code Is the Same in All Species.

All species use the same mRNA codons to specify the same amino acids, and therefore the same genetic code. References to the "human genetic code" usually mean a human genome sequence. The simplest explanation for the "universality" of the genetic code is that all life evolved from a common ancestor. No other mechanism as efficient at directing cellular activities has emerged and persisted. The only known exceptions to the universality of the genetic code are a few codons in mitochondria and in certain single-celled eukaryotes (ciliated protozoa).

Once researchers knew that a codon consists of three bases and the code is nonoverlapping, the next step was to match codons to amino acids. **A Glimpse of History** explains the clever experiments that revealed the genetic code. Sixty of the possible 64 codons specify amino acids, three codons indicate "stop," and one encodes both the amino acid methionine and "start." Some amino acids are specified by more than one codon. For example, UUU and UUC encode phenylalanine.

Different codons that specify the same amino acid are termed **synonymous codons**, just as synonyms are words with the same meaning. The genetic code is termed "degenerate" because most amino acids are not uniquely specified. Several synonymous codons differ by the base in the third position. The corresponding base of a tRNA's anticodon is called the "wobble" position because it can bind to more than one type of base in synonymous

codons. The degeneracy of the genetic code protects against mutation, because changes in the DNA that substitute a synonymous codon do not alter the protein's amino acid sequence. Codons that encode different amino acids are called **nonsynonymous codons**.

Sequencing human genomes added to the genetic code experiments of the 1960s by identifying the DNA sequences that are transcribed into tRNAs. Sixty-one different tRNAs could exist, one for each codon that specifies an amino acid (the 64 triplets minus 3 stop codons). However, only 49 different genes encode tRNAs. This is because the same type of tRNA can detect synonymous codons that differ only in whether the wobble (third) position has U or C, such as the codons UUU and UUC that specify phenylalanine.

Building a Protein

Protein synthesis requires several participants: mRNA, tRNA molecules carrying amino acids, ribosomes, energy-storing molecules such as adenosine triphosphate (ATP) and guanosine triphosphate (GTP), and protein factors. These pieces meet during translation initiation (**figure 10.15**). Chemical bonds hold the components together.

First, the mRNA leader sequence forms hydrogen bonds with a short sequence of rRNA in a small ribosomal subunit.

Translation initiation

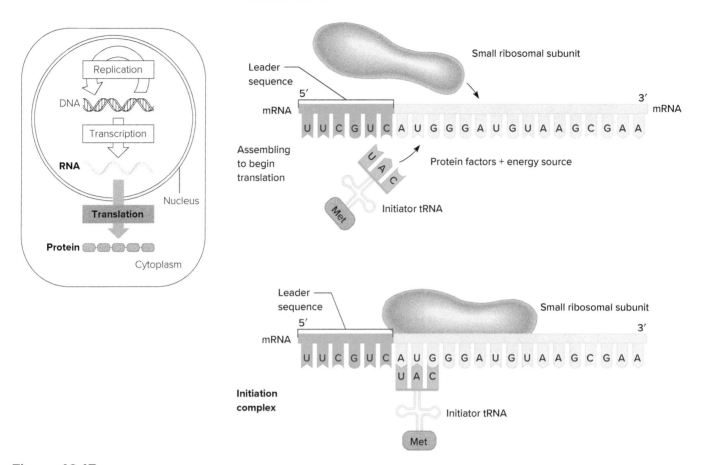

Figure 10.15 **Translation begins as the initiation complex forms.** Initiation of translation brings together a small ribosomal subunit, mRNA, and an initiator tRNA, and aligns them in the proper orientation to begin translation.

The first mRNA codon to specify an amino acid is always AUG, which attracts an initiator tRNA that carries the amino acid methionine (abbreviated Met). This methionine signifies the start of a polypeptide. The small ribosomal subunit, the mRNA bonded to it, and the initiator tRNA with its attached methionine, form the initiation complex at the appropriate AUG codon of the mRNA.

To begin elongation, a large ribosomal subunit bonds to the initiation complex. The codon adjacent to the initiation codon (AUG), which is GGA in **figure 10.16**, then bonds to its complementary anticodon, which is part of a free tRNA that carries the amino acid glycine. The two amino acids (Met and Gly in the example), still attached to their tRNAs, align.

The part of the ribosome that holds the mRNA and tRNAs together can be described as having two sites. The positions of the sites on the ribosome remain the same with respect to each other as translation proceeds, but they cover different parts of the mRNA as the ribosome moves. The P ("peptide") site holds the growing amino acid chain, and the A ("acceptor") site next to it holds the next amino acid to be added to the chain. In figure 10.16*a*, when the forming protein consists of only the first two amino acids, Met occupies the P site and Gly the A site.

The amino acids link by a type of chemical bond called a peptide bond, with the help of rRNA that functions as a

ribozyme (an RNA with enzymatic activity). Then the first tRNA is released to pick up another amino acid of the same type and be used again, a little like a shopping cart ferrying bags of the same type of dog food over and over. Special enzymes ensure that tRNAs always bind the correct amino acids. The ribosome moves down the mRNA by one codon, so that the former A site is now the P site (figure 10.16*b*). The attached mRNA is now bound to a single tRNA, with two amino acids extending from it at the P site. This is the start of a polypeptide.

Next, the ribosome moves down the mRNA by another codon, and a third tRNA brings in its amino acid (Cys in figure 10.16*b*). This third amino acid aligns with the other two and forms a peptide bond to the second amino acid in the growing chain. The tRNA attached to the second amino acid is released and recycled. The polypeptide builds one amino acid at a time, each piece brought in by a tRNA whose anticodon corresponds to a consecutive mRNA codon as the ribosome moves down the mRNA (figure 10.16*c*).

Elongation halts when the A site of the ribosome has a "stop" codon (UGA, UAG, or UAA), because no tRNA molecules correspond to it. A protein release factor starts to free the polypeptide. The last tRNA leaves the ribosome, the ribosomal subunits separate and are recycled, and the new polypeptide is released (**figure 10.17**).

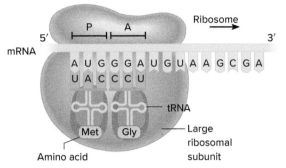

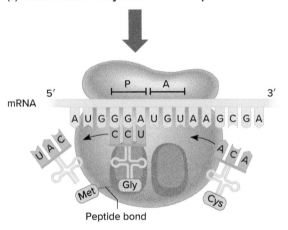

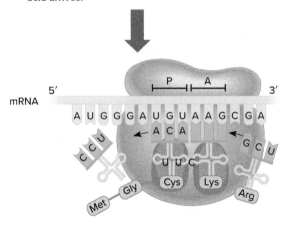

(a) **Second amino acid joins initiation complex.**

(b) **First peptide bond forms as new amino acid arrives.**

(c) **Amino acid chain extends.**

Figure 10.16 **Building a polypeptide.** **(a)** A large ribosomal subunit binds to the initiation complex, and a tRNA bearing a second amino acid (glycine, in this example) forms hydrogen bonds between its anticodon and the mRNA's second codon at the A site. The first amino acid, methionine, occupies the P site. **(b)** The methionine brought in by the first tRNA forms a peptide bond with the amino acid brought in by the second tRNA, and a third tRNA arrives, in this example carrying the amino acid cysteine, at the temporarily vacant A site. **(c)** A fourth and then fifth amino acid link to the growing polypeptide chain. The process continues until reaching a stop codon.

Protein synthesis is efficient. A cell can produce large numbers of a particular protein molecule from just one or two copies of a gene. A plasma cell in the immune system, for example, manufactures 2,000 identical antibody molecules per second. To mass produce proteins at this rate, RNA, ribosomes, enzymes, and other proteins are continually recycled. In addition, transcription produces many copies of a particular mRNA, and each mRNA may bind dozens of ribosomes, as **figure 10.18** shows. Many protein molecules are made from the same mRNA.

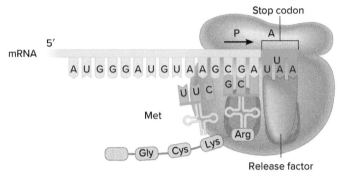

(a) Ribosome reaches stop codon.

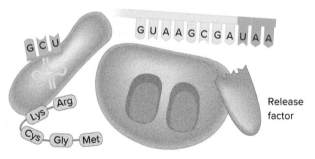

(b) Components disassemble.

Figure 10.17 **Terminating translation.** **(a)** A protein release factor binds to the stop codon, releasing the completed polypeptide from the tRNA and **(b)** freeing all of the components of the translation complex.

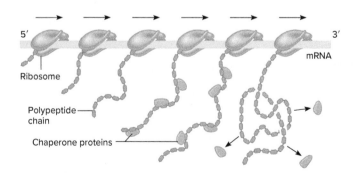

Figure 10.18 **Making many copies of a protein.** Several ribosomes can simultaneously translate a protein from a single mRNA. These ribosomes hold different-sized polypeptides—the closer to the end of a gene, the longer the polypeptide. Proteins called chaperones help fold the polypeptide.

Some proteins undergo further alterations, called post-translational modifications, before they can function. For example, insulin is initially translated as the 80-amino-acid-long polypeptide proinsulin, but enzymes cut it to 51 amino acids. Some proteins must have sugars attached or sometimes polypeptides must aggregate to become active.

Key Concepts Questions 10.3

1. What are the general characteristics of the genetic code?
2. What are the steps of translation?
3. How is translation efficient?

10.4 Processing a Protein

Proteins fold into one or more conformations. This folding arises from chemistry: attraction and repulsion between atoms of the proteins as well as interactions of proteins with chemicals in the immediate environment. For example, thousands of water molecules surround a growing chain of amino acids. Because some amino acids are attracted to water and some are repelled by it, the protein contorts in water. Sulfur atoms also affect protein conformation by bridging the two types of amino acids that contain them.

The conformation of a protein is described at several levels (**figure 10.19**). The amino acid sequence of a polypeptide chain is its **primary (1°) structure**. Chemical attractions between amino acids that are close together in the 1° structure fold the

H₂N— Ala — Thr — Cys — Tyr — Glu — Gly —COOH

(a) **Primary structure**—the sequence of amino acids in a polypeptide chain

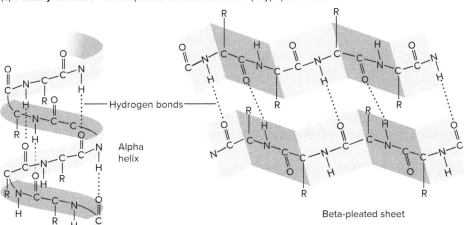

(b) **Secondary structure**—loops, coils, sheets, or other shapes formed by hydrogen bonds between neighboring carboxyl and amino groups

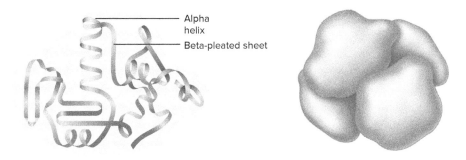

(c) **Tertiary structure**—three-dimensional forms shaped by bonds between R groups, interactions between R groups and water

(d) **Quaternary structure**—protein complexes formed by bonds between separate polypeptides

Figure 10.19 Four levels of protein structure. (a) The amino acid sequence of a polypeptide forms the primary structure. Each amino acid has an amino end (H₂N) and a carboxyl end (COOH), and an "R" group distinguishes each of the 20 types of amino acids. **(b)** Hydrogen bonds between non–R groups create secondary structures such as helices and sheets. **(c)** The tertiary structure arises when R groups interact with other R groups or with water, folding the polypeptide into a unique shape. **(d)** A protein consisting of more than one polypeptide has a quaternary structure.

polypeptide chain into its **secondary (2°) structure**, which may form loops, coils, barrels, helices, or sheets. Two common secondary structures are an alpha helix and a beta-pleated sheet. Secondary structures wind into larger **tertiary (3°) structures** as more widely separated amino acids attract or repel in response to water molecules. Finally, proteins consisting of more than one polypeptide form a **quaternary (4°) structure**. Hemoglobin, the blood protein that carries oxygen, has four polypeptide chains (see figure 11.1). The liver protein ferritin has 20 identical polypeptides of 200 amino acids each. In contrast, the muscle protein myoglobin is a single polypeptide chain.

Mutations may alter the primary structure of a protein if the genetic change is nonsynonymous, which means that it changes the amino acid. In contrast, more than one tertiary or quaternary structure may be possible for a protein if an amino acid chain can fold in different ways.

Protein Folding and Misfolding

Proteins begin to fold within a minute after the amino acid chain winds away from the ribosome. A small protein might contort into its final, functional form in one step, taking microseconds. Larger proteins may initially fold into a series of short-lived intermediates.

Proteins start to move toward their destinations as they are being synthesized. In some proteins, part of the start of the amino acid chain forms a tag that helps direct the protein in the cell. The first few amino acids in a protein that will be secreted or lodge in a membrane form a "signal sequence" that leads it and the ribosome to which it binds into a pore in the ER membrane. Once in the ER, the protein enters the secretory network (see figure 2.5). Proteins destined for the mitochondria bear a different signal sequence.

Signal sequences are not found on proteins synthesized on free ribosomes in the cytoplasm. These proteins may function right where they are made; examples include the protein tubules and filaments of the cytoskeleton (see figure 2.9) or enzymes that take part in metabolism. Proteins destined for the nucleus, such as transcription factors, are synthesized on free ribosomes.

Various proteins assist in precise folding, whatever the destination. **Chaperone proteins** stabilize partially folded regions in their correct form, and prevent a protein from getting "stuck" in a useless intermediate form. Chaperone proteins are being developed as drugs to treat diseases that result from misfolded proteins. Other proteins help new chemical bonds to form as the final shape arises, and yet others monitor the accuracy of folding. If a protein misfolds, an "unfolded protein response" slows or even stops protein synthesis while accelerating transcription of genes that encode chaperone proteins and the other folding proteins. Proper protein folding is quickly restored.

If a protein misfolds despite these protections, cells have ways to either refold the protein correctly, or get rid of it. Misfolded proteins are sent out of the ER back into the cytoplasm, where they are "tagged" with yet another protein, called ubiquitin. A misfolded protein bearing just one ubiquitin tag may straighten and refold correctly, but a protein with more than one tag is taken to another cellular machine called a **proteasome** (**figure 10.20**). A proteasome is a tunnel-like multiprotein structure. As a misfolded protein moves through the proteasome opening, it is stretched out, chopped up, and its peptide pieces degraded into amino acids, a little like a wood chipper. The amino acids may be recycled to build new proteins.

Proteasomes also destroy properly folded proteins that are in excess or no longer needed. For example, a cell must dismantle excess transcription factors, or the genes that they control may remain activated or repressed for too long. Proteasomes also destroy proteins from pathogens, such as viruses.

Diseases of Protein Misfolding

Proteins misfold in two ways: either from a mutation or by having more than one conformation. A mutation may change the amino acid sequence in a way that alters the attractions and repulsions between parts of the protein, contorting it. In many cases of cystic

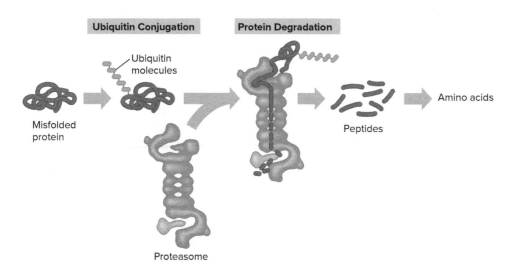

Figure 10.20 Proteasomes provide quality control. Ubiquitin binds to a misfolded protein and escorts it to a proteasome. The proteasome, which is composed of several proteins, encases the misfolded protein, straightening and dismantling it.

fibrosis, CFTR protein misfolds in the endoplasmic reticulum, attracts too many chaperones, and is sent to the cytoplasm, never reaching the plasma membrane where normally it would form a chloride channel (see figures 1.3 and 1.4 and Clinical Connection 4.1).

Some proteins have more than one conformation, and one conformation may become "infectious," converting molecules with the other conformation to more copies of itself. The two forms of the same protein have identical amino acid sequences, but fold differently. An infectious protein is called a **prion** (pronounced "pree-on").

In several diseases that affect the brain, misfolded proteins aggregate, forming masses that clog the proteasomes and block them from processing any malformed proteins. Different proteins are affected in different diseases. In Huntington disease, for example, extra glutamines in the protein huntingtin cause it to obstruct proteasomes. Misfolded proteins that clog proteasomes also form in the diseases listed in **table 10.4**, but it isn't always clear whether the accumulated proteins cause the disease or are a response to it. Chapter 12 discusses some of these diseases further.

Prion Diseases

Prion diseases were first described in sheep, which develop a disease called scrapie when they eat prion-infected brains of other sheep. The name denotes the way the sick sheep rub against things to scratch themselves. Their brains became riddled with holes. More than 85 animal species develop similar diseases. **Table 10.5** lists prion diseases of humans.

The first prion disease recognized in humans was kuru, which affected the native Fore people who lived in the remote mountains of Papua New Guinea (**figure 10.21**). *Kuru* means "to shake." The disease began with wobbly legs, then trembling

Table 10.4	Diseases Associated with Protein Misfolding
Disease	**Misfolded Protein(s)**
Alzheimer disease	Amyloid beta precursor protein, tau proteins
Familial amyotrophic lateral sclerosis	Superoxide dismutase, TDP-43
Frontotemporal dementia	Tau proteins, TDP-43
Huntington disease	Huntingtin
Parkinson disease	Alpha synuclein
Lewy body dementia	Alpha synuclein
PKU	Phenylalanine hydroxylase
Prion diseases	Prion protein

(All but Huntington disease are genetically heterogeneic; that is, abnormalities in different proteins cause similar syndromes.)

Table 10.5	Prion Diseases of Humans
Disease	
Creutzfeldt-Jakob disease	
Fatal familial insomnia	
Gerstmann-Sträussler-Scheinker disease	

and whole-body shaking. Uncontrollable laughter led to the name "laughing disease." Speech slurred, thinking slowed, and the person became unable to walk or eat. Death came within a year. The disease, which affected mostly women and children, was traced to a ritual in which the people ate their war heroes. When the women and children prepared the brains, prions entered cuts and they became infected. At first, the disease seemed to be inherited because it affected families.

Not many people knew about the plight of the Fore in the 1950s, but in the mid-1990s, prion diseases dominated the headlines when "mad cow disease" in the United Kingdom led to a human version, called variant Creutzfeldt-Jakob disease. More than 120 people ate infectious prions in beef and became ill.

Prions cause disease both by spreading the alternate form (infectious or mutant) and by aggregation of the protein. These aggregates are also seen in more familiar diseases, such as beta amyloid plaques and tau protein neurofibrillary tangles in Alzheimer disease, and alpha synuclein deposits in Parkinson disease (see table 10.4).

Figure 10.21 Kuru. Kuru affected the Fore people of New Guinea until they gave up a cannibalism ritual that spread infectious prion protein. © *The Nobel Foundation, 1976.*

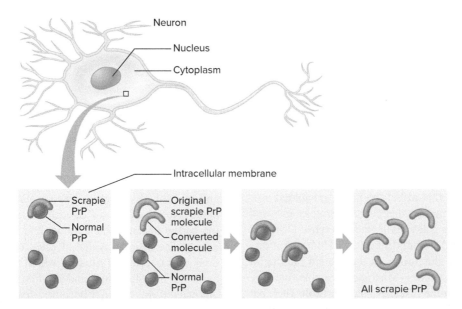

Figure 10.22 Prions change shape. Scrapie is a disease of sheep that was the first recognized prion disease. A single scrapie prion protein (PrP) contacts a normal PrP and changes it into the scrapie conformation. As the change spreads, disease results. Accumulated scrapie prion proteins clog brain tissue, eventually causing symptoms.

The rare prion diseases like kuru, scrapie, and mad cow disease, as well as the more common ones, are all diseases of protein folding. The infectious forms have beta-pleated sheets in places where their normal counterparts have alpha helices. **Figure 10.22** depicts schematically how normal proteins become prions, which can directly destroy brain areas and aggregate. In rare cases of these conditions, a mutation misfolds the protein. Drug discovery efforts are being directed toward refolding misfolded proteins.

Key Concepts Questions 10.4

1. How are proteins folded?
2. How does a cell handle misfolded proteins?
3. How can a protein that has more than one conformation cause a disease?

Summary

10.1 The Importance of Proteins

1. Dietary proteins are digested into **amino acids**. Genes control the linking of amino acids into proteins.
2. Biological proteins consist of 20 types of amino acids. Each amino acid has an amino group, an acid group, a hydrogen atom, and an R group, bonded to a central carbon atom. Amino acids are linked into long chains called **polypeptides**. A protein molecule consists of one or more polypeptides.

10.2 Transcription Copies the Information in DNA

3. **Transcription** copies DNA information into RNA. **Translation** uses the information in RNA to connect amino acids into proteins.
4. RNA is transcribed from the **template strand** of DNA. The other DNA strand is called the **coding strand**.
5. RNA is a single-stranded nucleic acid similar to DNA but with uracil and ribose rather than thymine and deoxyribose.

6. **Messenger RNA (mRNA)** carries a protein-encoding gene's information in the form of three-base units called **codons**. **Ribosomal RNA (rRNA)** associates with certain proteins to form ribosomes, on which proteins are synthesized. **Transfer RNA (tRNA)** is cloverleaf-shaped, with a three-base anticodon that is complementary to an mRNA on one end and bonds to a specific amino acid on the other end.
7. **Transcription factors** regulate which genes are transcribed in a particular cell type under particular conditions.
8. Transcription begins when transcription factors help **RNA polymerase** bind to a gene's **promoter** and then adds RNA nucleotides to a growing chain, in a sequence complementary to the DNA template strand.
9. After a gene is transcribed, the mRNA receives a "cap" of modified nucleotides at the 5′ end and a poly A tail at the 3′ end.
10. For many mRNAs, after transcription **introns** are removed and **exons** are then translated into protein. Introns may outnumber and outsize exons. **Alternate splicing** increases protein diversity.

10.3 Translation of a Protein

11. Each three consecutive mRNA bases form a **codon** that specifies a certain amino acid. The **genetic code** is the correspondence between each codon and the amino acid it specifies. Of the 64 possible codon types, 60 specify amino acids, one specifies the amino acid methionine and "start," and three signal "stop." More than one type of codon may encode a single amino acid. The genetic code is triplet and nonoverlapping, includes controls, and is universal. **Synonymous codons** encode the same amino acid, and **nonsynonymous codons** encode different amino acids.

12. The **reading frame** is the starting point of a DNA sequence that encodes a protein. An **open reading frame** is a stretch of DNA without a stop codon that might indicate a protein-encoding gene.

13. In translation, an initiation complex forms when mRNA, a small ribosomal subunit, and a tRNA carrying methionine join. The amino acid chain elongates when a large ribosomal subunit joins the small one. Next, a second tRNA binds by its anticodon to the next mRNA codon, and its amino acid bonds with the first amino acid. Transfer RNAs add more amino acids. The ribosome moves down the mRNA as the chain grows. The P site bears the growing amino acid chain, and the A site holds the next amino acid to be added. When the ribosome reaches a "stop" codon, it falls apart into its two subunits and is released. The new polypeptide breaks free.

14. After translation, some polypeptides are cleaved, have sugars added, or aggregate. The cell uses or secretes the protein.

10.4 Processing a Protein

15. A protein must fold into a specific **conformation** to function. A protein's **primary (1°) structure** is its amino acid sequence. Its **secondary (2°) structure** forms as amino acids close in the primary structure attract. **Tertiary (3°) structure** appears as more widely separated amino acids attract or repel in response to water molecules. **Quaternary (4°) structure** forms when a protein consists of more than one polypeptide.

16. **Chaperone proteins** help the correct conformation arise. Other proteins help new bonds form and oversee folding accuracy. Ubiquitin attaches to misfolded proteins and enables them to refold or escorts them to **proteasomes** for dismantling. Protein misfolding causes diseases. This can happen by mutation or if a protein can fold into several conformations, some of which are "infectious" and can cause disease.

17. At least one conformation of a **prion** protein is infectious.

Review Questions

1. Explain how complementary base pairing is responsible for
   1. the structure of the DNA double helix.
   b. DNA replication.
   c. transcription of RNA from DNA.
   d. the attachment of mRNA to a ribosome.
   e. codon-anticodon pairing.
   f. tRNA conformation.

2. State the functions of these proteins.
   a. RNA polymerase
   b. ubiquitin
   c. a chaperone protein
   d. a transcription factor

3. Explain where a hydrogen bond forms and where a peptide bond forms in the transmission of genetic information.

4. List the differences between RNA and DNA.

5. Identify where in a cell DNA replication, transcription, and translation occur.

6. Explain how transcription controls cell specialization.

7. How can the same mRNA codon be at an A site on a ribosome at one time but at a P site at another time?

8. Describe the events of transcription initiation.

9. List the three major types of RNA and their functions.

10. Describe three ways RNA is altered after it is transcribed.

11. What are the components of a ribosome?

12. Why would an overlapping genetic code be restrictive?

13. Why would two-nucleotide codons be insufficient to encode the number of amino acids in biological proteins?

14. How are the processes of transcription and translation efficient?

15. What factors determine how a protein folds into its characteristic conformation?

16. How do a protein's primary, secondary, and tertiary structures affect conformation? Which is the most important determinant of conformation?

17. Explain how a protein can be infectious.

Applied Questions

1. Explain the genetic basis of the ability of lice to adapt to humans wearing clothing.

2. List the RNA sequence transcribed from the DNA template sequence TTACACTTGCTTGAGAGTC.

3. Reconstruct the corresponding DNA template sequence from the partial mRNA sequence GCUAUCUGUCAUAAAAGAGGA.

4. List three different mRNA sequences that could encode the amino acid sequence histidine-alanine-arginine-serine-leucine-valine-cysteine.

5. Write a DNA sequence that would encode the amino acid sequence valine-tryptophan-lysine-proline-phenylalanine-threonine.

6. In the film *Jurassic Park,* which is about cloned dinosaurs, a cartoon character named Mr. DNA talks about the billions of genetic codes in DNA. Why is this statement incorrect?

7. Titin is a muscle protein named for its size—its gene has the largest known coding sequence of 80,781 DNA bases. How many amino acids long is it?

8. An extraterrestrial life form has a triplet genetic code with five different bases. How many different amino acids can this code specify, assuming no degeneracy?

9. In malignant hyperthermia, a person develops a life-threateningly high fever after taking certain types of anesthetic drugs. In one family, mutation deletes three contiguous DNA bases in exon 44. How many amino acids are missing from the protein?

10. The protein that serves as a receptor that allows insulin to enter cells has a different number of amino acids in a fetus and in an adult. Explain how this may happen.

11. In "hypomyelination with atrophy of the basal ganglia and cerebellum," a mutation in a gene that encodes a form of the cytoskeletal protein tubulin, *TUBB4A,* changes an aspartic acid (Asp) to an asparagine (Asn). The resulting brain shrinkage causes seizures, developmental delay, and a spastic gait. Only 22 cases are known.

    a. Does the mutation involve synonymous or nonsynonymous codons?

    b. Consult the genetic code table (table 10.3) and determine two mutations that could account for the amino acid changes.

    c. What might be the transcription pattern in the body for this gene?

Case Studies and Research Results

1. Five patients meet at a clinic for families with early-onset Parkinson disease. This condition causes rigidity, tremors, and other motor symptoms. Only 2 percent of cases of Parkinson disease are inherited. The five patients all have mutations in a gene that encodes the protein parkin, which has 12 exons. For each patient, indicate whether the mutation shortens, lengthens, or does not change the size of the protein, or if it isn't possible to tell what the effect might be.

    1. Manny's *parkin* gene is missing exon 3.
    2. Frank's *parkin* gene has a duplication in intron 4.
    3. Theresa's *parkin* gene lacks six contiguous nucleotides in exon 1.
    4. Elyse's *parkin* gene has an altered splice site between exon 8 and intron 8.
    5. Scott's *parkin* gene is deleted.

2. The human genome sequence encodes many more mRNA transcripts than there are genes. Why isn't the number the same?

3. Francis Crick envisioned "20 different kinds of adaptor molecule, one for each amino acid, and 20 different enzymes to join the amino acid to their adaptors." What type of molecule was Dr. Crick describing?

4. Before 1985, approximately 30,000 people were injected with human growth hormone. Many of these patients were children being treated for severe growth deficiency. The growth hormone was extracted and pooled from thousands of pituitary glands from cadavers. Many years later, some of these individuals developed Creutzfeldt-Jakob disease. What might have caused the condition?

Courtesy Bomber Command Museum of Canada

CHAPTER 11

Gene Expression and Epigenetics

The Allies dropped food over the Netherlands, in 1945, stopping the Dutch Hunger Winter in just 2 days. By altering gene expression, starvation before birth may have led to schizophrenia years later in many individuals.

Learning Outcomes

11.1 Gene Expression Through Time and Tissue

1. Define *epigenetic*.
2. Explain how globin chain switching, development of organs, and the types of proteins cells make over time illustrate gene expression.

11.2 Control of Gene Expression

3. Explain how small molecules binding to histone proteins control gene expression by remodeling chromatin.
4. Explain how microRNAs control transcription.

11.3 Maximizing Genetic Information

5. Explain how division of genes into exons and introns maximizes the number of encoded proteins.

11.4 Most of the Human Genome Does *Not* Encode Protein

6. Discuss how viral DNA, noncoding RNAs, and repeated DNA sequences account for large proportions of the human genome.

 ## The BIG Picture

Discovering the nature of the genetic material, determining the structure of DNA, deciphering the genetic code, and sequencing the human genome led to today's challenge: learning how genes are expressed and interact through tissue and time.

The Dutch Hunger Winter

When a fetus is starving, control of gene expression changes in ways that enable immediate survival, but may harm health decades later. An ongoing investigation called the Dutch Famine Study is tracing the effects of prenatal starvation during World War II on health decades later. The study has been publishing findings since 1976.

The "Dutch Hunger Winter" lasted from November 1944 through May 1945, when the Nazis blocked all food supplies from entering six large cities in western Holland. As severe malnutrition weakened and killed people, a cruel experiment took place, in retrospect. By consulting food ration records that showed exact daily caloric intakes, time during pregnancy for specific women, and diagnoses from psychiatric registries and military induction records for the people exposed before birth, researchers at Columbia University and Leiden University in the Netherlands discovered clear links between prenatal malnutrition and health problems years later. They compared the health of people born during the Dutch Hunger Winter to their same-sex siblings born during better times.

Prenatal nutrition affects an adult phenotype because starvation alters the pattern in which methyl groups (CH_3) bind to DNA, selectively silencing genes. The DNA of the "Dutch Hunger

Winter children," when studied 60 years later, had different methylation patterns in genes that control metabolism than their siblings. The exposed children, when grown, had poor glucose control, high body mass index, and elevated serum cholesterol—clearly adaptations to prenatal starvation. They also have a strikingly increased risk of developing schizophrenia that perhaps is related to altered expression of the gene that encodes insulin-like growth factor 2 (IGF-2), which controls genes that affect thinking. Researchers hypothesize that schizophrenia may develop in some people born into famine when *IGF-2* has too few methyl groups and is overexpressed in the brain.

11.1 Gene Expression Through Time and Tissue

A genome is like an orchestra. Just as not all of the musical instruments play with the same intensity at every moment, not all genes are expressed continually at the same levels. Some genes are always transcribed and translated, in all cells. Because they keep cells running, these genes are called "housekeeping genes." Other genes have more specialized roles, and become active as a cell differentiates.

Before the field of genomics began in the 1990s, the study of genetics proceeded one gene at a time, like hearing the separate contributions of a violin, a viola, and a flute. Many genetic investigations today, in contrast, track the crescendos of gene activity that parallel events in an organism's life. This new view introduced the element of time to genetic analysis. Unlike the gene maps of old, which ordered genes linearly on chromosomes, new types of maps are more like networks that depict the timing of gene expression in unfolding programs of development and response to the environment.

The discoveries of the 1950s and 1960s about DNA structure and function answered some questions about the control of gene expression, while raising many more: How does a bone cell "know" to transcribe the genes that control the synthesis of collagen and not to transcribe genes that specify muscle proteins? What shifts the proportions of blood cell types into leukemia? How do chemical groups "know" to shield a specific DNA sequence from transcription in one circumstance, yet expose it in others?

Changes to the chemical groups that associate with DNA greatly affect which parts of the genome are accessible to transcription factors and under which conditions. Such changes to the molecules that bind to DNA and are transmitted to daughter cells when the cell divides are termed **epigenetic**, which means "outside the gene." Epigenetic changes do not alter the DNA base sequence and are reversible. For a few sites in the genome, an epigenetic change may persist through meiosis to a third generation, but this appears to be rare. Figure 6.12 shows how methyl groups bind to DNA, causing epigenetic changes.

This chapter looks at how cells access the information in DNA. We begin with two examples of gene expression at the molecular and organ levels: (1) hemoglobin switching during development and (2) specialization of the two major parts of the pancreas.

Globin Chain Switching

The globin proteins transport oxygen in the blood. These proteins vividly illustrate control of gene expression because they assemble into different hemoglobin molecules depending upon stage of development. This process is called globin chain switching.

A hemoglobin molecule in the blood of an adult consists of four polypeptide chains, each wound into a globular conformation (**figure 11.1**). Two of the chains are 146 amino acids long and are called "beta" (β). The other two chains are 141 amino acids long and are termed "alpha" (α). Genes on different chromosomes encode the alpha and beta globin polypeptide chains.

As a human develops, different globin polypeptide chains are used to make molecules of hemoglobin. The different forms of hemoglobin are necessary because of changes in blood oxygen levels that happen when a newborn begins breathing and no longer receives oxygen through the placenta. The promoters of the globin genes include binding sites for transcription factors, which orchestrate the changes in hemoglobin molecules through prenatal development. Other DNA sequences in the globin gene clusters turn off expression of genes no longer needed. **Figure 11.2** tracks the changes to the globin proteins before and after birth.

The chemical basis for globin chain switching is that different globin polypeptide chains attract oxygen molecules to different degrees. In the embryo, as the placenta forms, hemoglobin consists first of two epsilon (ε) chains, which are in the beta globin group, and two zeta (ζ) chains, which are in the alpha globin group. About 4 percent of the hemoglobin in the embryo includes "adult" beta chains. This percentage gradually increases. Globin

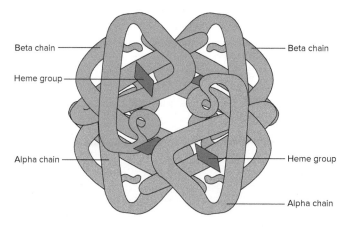

Figure 11.1 The structure of hemoglobin. A hemoglobin molecule is made up of two globular protein chains from the beta (β) globin group and two from the alpha (α) globin group. Each globin surrounds an iron-containing chemical group called a heme.

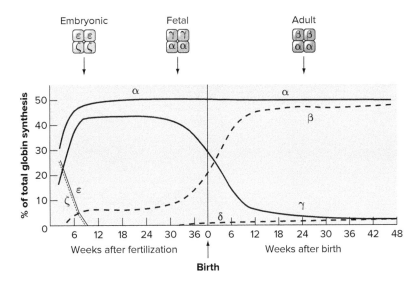

Figure 11.2 Globin chain switching. The subunit composition of human hemoglobin changes as the concentration of oxygen in the environment changes. Each globin quartet has two polypeptide chains encoded by genes in the alpha (α) globin cluster (chromosome 16) and two polypeptide chains from the beta (β) globin cluster (chromosome 11). When the oxygen source switches from the placenta to the newborn's lungs, beta globin begins to replace gamma (γ) globin.

chains are manufactured first in the yolk sac in the embryo, then in the liver and spleen in the fetus, and finally primarily in the bone marrow after birth.

As the embryo develops into a fetus, the epsilon and zeta globin polypeptide chains decrease in number as gamma (γ) and alpha chains accumulate. Hemoglobin consisting of two gamma and two alpha chains is called fetal hemoglobin. Because the gamma globin subunits bind very strongly to oxygen released from maternal red blood cells into the placenta, fetal blood carries 20 to 30 percent more oxygen than an adult's blood. As the fetus matures, beta chains gradually replace the gamma chains. At birth, however, the hemoglobin is not fully of the adult type—fetal hemoglobin (two gamma and two alpha chains) comprises from 50 to 85 percent of the blood. By 4 months of age, the proportion drops to 10 to 15 percent, and by 4 years, less than 1 percent of the child's hemoglobin is the fetal form.

Building Tissues and Organs

Blood is a structurally simple tissue that is easy to obtain and study because its components are easily separated. A solid gland or organ is much more complex. It has a distinctive three-dimensional form and is constructed from specialized types of cells and commonly more than one type of tissue. The specific, solid organization of an organ must persist throughout a lifetime of growth, repair, and changing external conditions, while its cells must maintain their specializations.

In all tissues and organs, genes are turned on and off during development, as stem cells self-renew and yield more specialized daughter cells. The organization of the pancreas is simple enough to reveal how its cells specialize. It is a dual gland, with two types of cell clusters. The exocrine part releases digestive enzymes into ducts. The endocrine part

secretes directly into the bloodstream polypeptide hormones that control nutrient use. The endocrine cell clusters are called pancreatic islets.

The dual nature of the pancreas unfolds in the embryo, when ducts form. Within duct walls reside rare stem cells and progenitor cells (see figure 2.18). A transcription factor is activated and controls expression of other genes in a way that stimulates some progenitor cells to divide. Certain daughter cells follow an exocrine pathway and will produce digestive enzymes. Other progenitor cells respond to different signals and divide to yield daughter cells that follow the endocrine pathway. **Figure 11.3** shows the differentiated cell types that form from the two cell lineages. The most familiar pancreatic hormone is insulin, which the beta cells of the pancreas secrete. The absence of insulin, or the inability of cells to recognize it, causes diabetes mellitus.

Shared Gene Expression Connects Diseases

A more complete portrait of gene expression emerges through **transcriptomics**, which looks at all of the mRNA molecules made in a specific cell under specific circumstances, and **proteomics**, which identifies all the proteins a cell manufactures under specific conditions. Other "ome" words narrow the focus of examining mRNAs and/or proteins in a cell. Genes whose encoded proteins control lipid synthesis, for example, constitute the "lipidome," and proteins that monitor carbohydrates form the "glycome."

Identifying and grouping the mRNAs and proteins in a cell is only a first step in understanding how cells specialize. The next challenge is to determine how proteins with related functions interact, which can reveal links between medical conditions. **Figure 11.4** shows part of a huge disease map

Figure 11.3 Building a pancreas. A single type of stem cell theoretically divides to give rise to an exocrine/endocrine progenitor cell that in turn divides to yield more restricted progenitor cells that give rise to both mature exocrine and endocrine cells. The endocrine progenitor cell in turn divides, yielding specialized daughter cells that produce specific hormones.

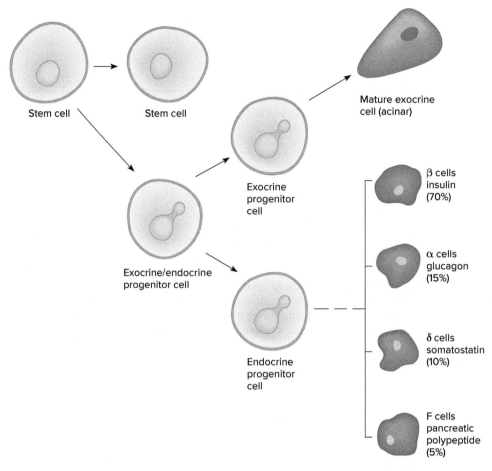

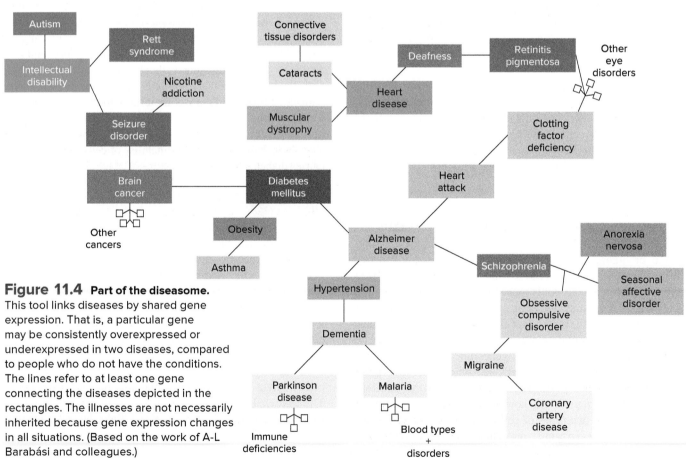

Figure 11.4 Part of the diseasome. This tool links diseases by shared gene expression. That is, a particular gene may be consistently overexpressed or underexpressed in two diseases, compared to people who do not have the conditions. The lines refer to at least one gene connecting the diseases depicted in the rectangles. The illnesses are not necessarily inherited because gene expression changes in all situations. (Based on the work of A-L Barabási and colleagues.)

called the "diseasome." It connects diseases that share genes that have altered expression. Some of the links and clusters are well known, such as obesity, hypertension, and diabetes. Others are surprises, such as Duchenne muscular dystrophy (DMD; see figure 2.1) and heart attacks. Researchers are now testing cardiac drugs on boys with DMD. In other cases, the association of a disease with genes whose expression goes up or down can suggest new drugs to try or develop.

Transcriptomics and proteomics are also useful in better understanding basic biology. One proteomic experiment, for example, compared the relative contributions of major categories of proteins from conception through birth and from conception through old age. The differences in the levels of proteins made at these times make sense. Transcription factors are more abundant before birth because of the extensive cell differentiation of this period, as organs form. During the prenatal period, enzymes are less abundant, perhaps because the fetus receives some enzymes through the placenta. Immunoglobulins appear after birth, when the immune system begins to function by responding to environmental stimuli.

Gene expression profiles for different cell types under various conditions are the basis for many new medical tests that assess risk, diagnose disease, or monitor response to treatment. For example, 55 genes are overexpressed and 480 are underexpressed in cells of a prostate cancer that has a high likelihood of spreading, but not in a prostate cancer that will probably not spread. A test based on such findings can assist physicians in deciding which patients can safely delay or avoid invasive and risky treatment.

Key Concepts Questions 11.1

1. How does gene expression change over time for the polypeptide components of hemoglobin?

2. How does gene expression change as a pancreas develops?

3. What are transcriptomics and proteomics?

11.2 Control of Gene Expression

In a human genome, the blueprints for building proteins take up significantly less DNA than the instructions for when and where to do so. A protein-encoding gene contains controls of its own expression, and is also controlled by the configuration of the chromatin of which it is a part.

An internal control of a gene's expression is the promoter sequence. Recall from chapter 10 that the promoter is the part of the gene where RNA polymerase and transcription factors bind, marking the start point of transcription. Variations in the promoter sequence of a gene can affect how quickly the encoded protein is synthesized. In a form of early-onset Alzheimer disease, for example, a mutation in the promoter for the gene that encodes amyloid precursor protein (see Clinical

Connection 5.1) causes the sticky protein to accumulate in the brain twice as fast as normal because it isn't cleared quickly enough. A second way that expression of a gene can exceed the normal pace is if a person has more than one copy of it.

Stepping back from a single gene perspective, cells specialize as different combinations of transcription factors bind to the many genes that provide the cell's characteristics. Transcription factors can bind at a gene's promoter or to a DNA sequence away from the gene, providing long-distance regulation, too.

Control of gene expression happens in steps. In **chromatin remodeling**, the histone proteins associated with DNA interact with other chemical groups in ways that expose some sections of DNA to transcription factors and shield other sections, blocking their expression. Later in the protein production process, small RNAs called, appropriately, **microRNAs**, bind to certain mRNAs, preventing their translation into protein. Overall, these two processes—chromatin remodeling and blocking of translation by microRNAs—determine the ebb and flow of different proteins, enabling cells to adapt to changing conditions.

Chromatin Remodeling

Recall from figure 9.13 that DNA associates with proteins and RNA to form chromatin. For many years, biologists thought that the histone proteins that wind long DNA molecules into nucleosomes were little more than tiny spools. However, histones do much more. Enzymes add or delete small organic chemical groups to histones, affecting expression of the protein-encoding genes that these proteins bind.

The three major types of small molecules that bind to histones are acetyl groups, methyl groups, and phosphate groups (**figure 11.5**). The key to the role histones play in controlling gene expression lies in acetyl groups (CH_3CO_2). They bind to specific sites on certain histones, particularly to the amino acid lysine.

Figure 11.6 shows how acetyl binding shifts histone interactions in a way that allows transcription to begin. A series of proteins moves the histone complex away from a DNA sequence called a TATA box, exposing it enough for RNA polymerase to bind and transcription to begin (see figure 10.8). First, a group of proteins called an enhanceosome attracts the enzyme (acetylase) that adds acetyl groups to specific lysines on specific histones, which neutralizes the histones' positive charge. Because DNA carries a negative charge, this change moves the

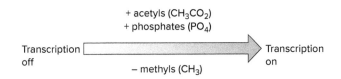

Figure 11.5 Chromatin remodeling. Chromatin remodeling adds or removes certain organic chemical groups to or from histones. The pattern of binding controls whether the DNA wrapped around the histones is transcribed or not. A gene is expressed when acetyl groups and phosphate groups are added and when methyl groups are removed.

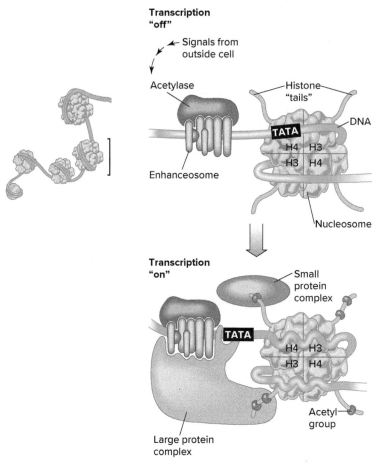

Transcription "off"

Signals from outside cell

Acetylase

Histone "tails"

DNA

TATA

H4 H3
H3 H4

Enhanceosome

Nucleosome

Transcription "on"

Small protein complex

TATA

H4 H3
H3 H4

Acetyl group

Large protein complex

Figure 11.6 **Acetylated histones allow transcription to begin.** Once acetyl groups are added to particular amino acids in the tails of certain histones, the TATA box becomes accessible to transcription factors. H3 and H4 are histone types.

histones away from the DNA, making room for transcription factors to bind and begin transcription. Enzymes called deacetylases remove acetyl groups, which shuts off gene expression.

In most cell types, only about 2 percent of the chromatin is "open" and "easy to read" (**figure 11.7**). Researchers can separate these regions from the rest of the DNA and identify the genes being transcribed in a particular cell type under particular conditions.

Methyl groups (CH_3) are also added to or taken away from histones. When CH_3 binds to a specific amino acid in a specific histone type, a protein is attracted that prevents transcription. As CH_3 groups are added, methylation spreads from the tail of one histone to the adjacent histone, propagating the gene silencing. Methyl groups also turn off transcription by binding to DNA directly, to cytosines at about 16,000 places in the genome. The "methylome" is the collection of all the methylated sites in the genome. Different cell types have different genes that are methylated and turned off.

The modified state of chromatin can be passed on when DNA replicates. These changes in gene expression

are heritable from cell generation to cell generation, but they do not alter the DNA base sequence—that is, they are epigenetic. Addition and removal of acetyl groups, methyl groups, or phosphate groups are all epigenetic changes. Effects of methylation can sometimes be seen when MZ (identical) twins inherit the same disease-causing genotype, but only one twin is sick. The reason for the discordance may be different patterns of methylation of one or more genes.

Enzymes that add or delete acetyl, methyl, and phosphate groups must be in a balance that controls which genes are expressed and which are silenced. Upset this balance, and disease can result. In a blood cancer called mixed lineage leukemia (see figure 18.16), for example, a single abnormal protein binds to more than 150 genes and alters their associated chromatin. Among these genes are several that normally stimulate frequent cell division in the stem cells that give rise to blood cells. In the leukemia, these overexpressed genes send the affected white blood cells back in developmental time, to a state in which the rapid cell division causes cancer. In addition to these types of cancer, at least 44 Mendelian conditions arise from abnormal functioning of the "epigenetic machinery." Nearly all of them include intellectual disability, which makes sense considering that many genes control brain function. One limitation to altering chromatin remodeling to treat inherited disease is that this action could affect the expression of many genes—not just the gene (or genes) implicated in the disease.

MicroRNAs

Chromatin remodeling determines which genes are transcribed. MicroRNAs act later in gene expression, preventing the translation of mRNA transcripts into protein. If chromatin remodeling is considered as an on/off switch to transcription, then the control of microRNAs is more like that of a dimmer switch—fine-tuning gene expression at a later stage. **Figure 11.8** schematically compares chromatin remodeling and the actions of

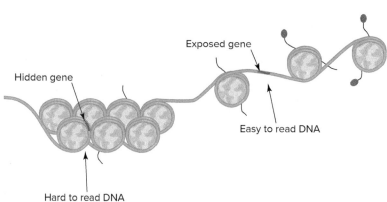

Exposed gene

Hidden gene

Easy to read DNA

Hard to read DNA

Figure 11.7 **Chromatin may be closed and not accessible or open and accessible.** The red section of DNA represents a gene that is either hidden and hard to read or exposed and easy to read.

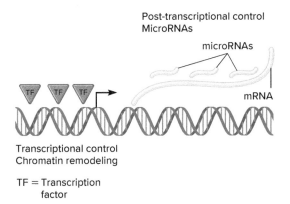

Post-transcriptional control
MicroRNAs

microRNAs

mRNA

TF TF TF

Transcriptional control
Chromatin remodeling

TF = Transcription
       factor

Figure 11.8 **Control of gene expression.** Chromatin opens to allow transcription factors to bind, whereas microRNAs bind to specific mRNAs, blocking their translation into protein.

microRNAs, and **figure 11.9** places these events in the overall flow of gene expression.

MicroRNAs are so-named because they are small—just 21 or 22 bases long. They are cut from precursor RNA molecules. MicroRNAs are a type of noncoding RNA, which means that they do not encode an amino acid sequence. The human genome encodes at least 2,555 distinct sequences of microRNAs that regulate at least one-third of the protein-encoding genes. A typical human cell has from 1,000 to 200,000 microRNAs.

Each type of microRNA binds to complementary sequences within a particular set of mRNAs. When a microRNA binds a "target" mRNA, it prevents translation. Because a single type of microRNA has many targets, it controls the expression of sets of genes. In turn, a single type of mRNA can bind several different microRNAs.

Within the patterns of microRNA function may lie clues to new ways to fight disease, because discovering the mRNAs that a certain microRNA targets may reveal genes that interact. Discovering these relationships can suggest new uses for and ways to combine existing drugs. The first practical applications will likely be in cancer treatment because certain microRNAs are more or less abundant in cancer cells than in healthy cells of the same type from which the cancer cells formed. Restoring the levels of microRNAs that normally suppress the too-rapid cell cycling of cancer, or blocking production of microRNAs that are too abundant in cancer, could help to restore a normal cell division rate.

Key Concepts Questions 11.2

1. What is the role of histone proteins in controlling gene expression?

2. How do acetyl, phosphate, and methyl groups control transcription?

3. How do microRNAs control gene expression?

11.3 Maximizing Genetic Information

A human genome maximizes information in the 20,325 genes that encode about 100,000 mRNAs, which in turn specify more than a million proteins. Figure 11.9 depicts this increase in information from gene to RNA to protein on the left and lists the mechanisms that maximize the information on the right.

Several events account for the fact that proteins outnumber genes. Alternate splicing of the exons and introns of a gene enable it to encode several versions of a protein, termed **isoforms** (**figure 11.10**). Which isoform a cell makes is a response to circumstance. For example, when an infection begins, an immune system cell first secretes a short version of an antibody molecule that is displayed on the cell's surface, where it alerts other cells. As the infection progresses, the cell transcribes an additional exon that extends the antibody in a way that enables it to be secreted into the bloodstream, where it attacks the pathogen. Rarely, transcripts of exons from different genes can combine into a single mRNA.

Alternate splicing explains how a long sequence of DNA can specify more mRNAs than genes.

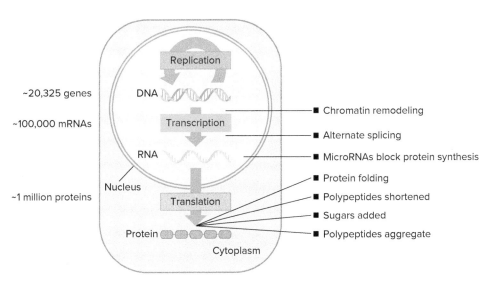

~20,325 genes

~100,000 mRNAs

~1 million proteins

DNA

Replication

Transcription

RNA

Nucleus

Translation

Protein

Cytoplasm

■ Chromatin remodeling
■ Alternate splicing
■ MicroRNAs block protein synthesis
■ Protein folding
■ Polypeptides shortened
■ Sugars added
■ Polypeptides aggregate

Figure 11.9 **A summary of the events of gene expression.** Chromatin remodeling determines which genes are transcribed. Alternate splicing creates different forms of a protein. MicroRNAs bind to mRNAs by complementary base pairing, blocking translation. After translation, a protein must fold a certain way. Certain polypeptides are shortened, attached to sugars, and/or aggregated.

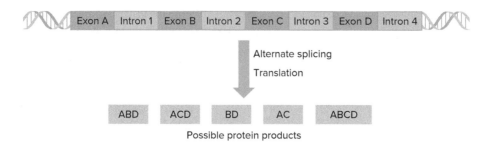

Possible protein products

Figure 11.10 **Exons provide flexibility.** Alternate splicing enables a cell to manufacture different versions (isoforms) of a protein by adding or deleting parts. Introns are removed and exons retained.

On a part of chromosome 22, for example, 245 genes yield 642 mRNA transcripts. About 90 percent of all human genes are alternately spliced.

Another way that the genome holds more information than it may appear to is that a DNA sequence that is an intron in one context may encode protein (function as an exon) in another. Consider prostate specific antigen (PSA). It is a protein on certain cell surfaces in the prostate gland that liquifies semen and is overproduced when the gland is enlarged, which could indicate a benign growth or cancer. PSA level in the blood is regularly measured in men over age 50, as a biomarker of possible cancer.

The gene encoding PSA has five exons and four introns. It is alternately spliced to encode seven isoforms. One of them, called PSA-linked molecule (PSA-LM), is encoded by the first exon and the fourth intron (**figure 11.11**). The two proteins (PSA and PSA-LM) work against one another. When the level of one is high, the other is low. Blood tests that measure levels of both proteins may more accurately assess the risk of developing prostate cancer than testing PSA alone.

Yet another way that introns may increase the number of proteins over the number of genes is that a DNA sequence that is an intron in a gene on the template strand may encode protein on the coding strand. This is so for the gene that encodes

neurofibromin (NF). Mutations in the gene cause neurofibromatosis type 1, an autosomal dominant condition that causes benign tumors beneath the skin and spots on the skin surface (**figure 11.12**). Within an intron of the gene, but on the coding strand, are instructions for three other genes. Finding such dual meaning in a gene is a little like reading a novel backwards and discovering a second story!

Genome information is maximized after translation when a protein is modified into different forms by adding sugars or lipids to create glycoproteins and lipoproteins, respectively. Many proteins on cell surfaces have these additions, marking them as part of a particular tissue, organ, and individual.

Another way that one gene can encode more than one protein is if the protein is cut in two. This happens in dentinogenesis imperfecta, which is an autosomal dominant condition that causes discolored, misshapen teeth with peeling enamel. The dentin, which is the bonelike substance beneath the enamel that forms the bulk of the tooth, is abnormal in this condition.

Dentin is a complex mixture of extracellular matrix proteins, 90 percent of which are forms of collagen. However, two proteins are unique to dentin: the abundant dentin phosphoprotein (DPP) and the rare dentin sialoprotein (DSP). DSP regulates mineral deposition into dentin, and DPP controls maturation of the mineralized dentin. A single gene encodes

Figure 11.11 **Two genes from one.** Embedded in the PSA gene are two protein-encoding sequences. The PSA portion consists of five exons, and the PSA-LM part consists of two exons, one of which lies within an intron of the gene. (Not drawn to scale; introns are much larger than exons.)

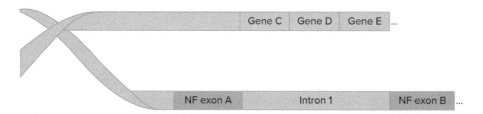

Figure 11.12 **Genes in introns.** An intron of the neurofibromin gene harbors three genes on the opposite DNA strand.

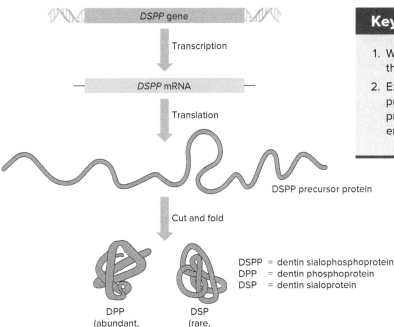

DSPP gene

Transcription

DSPP mRNA

Translation

DSPP precursor protein

Cut and fold

DPP (abundant, degraded slowly)

DSP (rare, degraded rapidly)

DSPP = dentin sialophosphoprotein
DPP = dentin phosphoprotein
DSP = dentin sialoprotein

Figure 11.13 **Sculpting two genes from one, in a different way than in figure 11.11.** In dentinogenesis imperfecta, a large precursor protein is cut in two, but one of the products is degraded much faster than the other.

both proteins. DPP and DSP are translated from a single mRNA molecule as the precursor protein dentin sialophosphoprotein (DSPP) (**figure 11.13**). Then they are separated. The two proteins are present in differing amounts because one (DSP) is degraded much faster than the other (DPP). The disease results from abnormal comparative levels of the two final proteins, DSP and DPP.

Table 11.1 summarizes mechanisms that maximize genetic information.

Table 11.1	Mechanisms That Maximize Genetic Information
Mechanism	**Example**
Alternate splicing	Cell surface and secreted forms of antibodies
An intron encoding one isoform is an exon in another	Prostate specific antigen (PSA) and PSA-linked molecule (PSA-LM)
An intron on one DNA strand is an exon on the other	Neurofibromin and three other genes
Adding sugars to form glycoproteins or lipids to form lipoproteins	Cell surface molecules important in cell-cell recognition
Precursor protein is cut to yield two proteins	Dentinogenesis imperfecta

11.4 Most of the Human Genome Does *Not* Encode Protein

Only about 1.5 percent of human DNA encodes protein. The rest of the genome includes viral sequences, sequences that encode RNAs other than mRNA, introns, promoters and other control sequences, and repeated sequences. In fact, most of the genome is transcribed—a DNA sequence is not "junk" if it does not encode protein.

Viral DNA

Our genomes include DNA sequences that represent viruses. Viruses are nonliving infectious particles that consist of a nucleic acid (DNA or RNA) encased in a protein coat (see Clinical Connection 17.1). A virus mass produces itself using a cell's transcriptional and translational machinery. New viruses may exit the host cell, or the viral nucleic acid may remain in the cell. A DNA virus may insert its DNA into a host chromosome or remain outside the nucleus in a circle of DNA called an episome. An RNA virus first uses an enzyme called **reverse transcriptase** to copy its genetic material into DNA, which then inserts into a host chromosome.

About 100,000 sequences in our DNA, of varying lengths and comprising about 8 percent of the genome, were once a type of RNA virus called a **retrovirus**. The name refers to a retrovirus's direction of genetic information transfer, which is opposite DNA to RNA to protein. Retroviral sequences in our chromosomes are termed "endogenous" because they are carried from generation to generation of the host, rather than acquired as an acute infection. The retroviruses whose genetic material is in our chromosomes are called human endogenous retroviruses, or HERVs.

HERVs likely infected our ancestors' genomes about 5 million years ago and were retained in our own. Researchers deduced this by comparing HERV sequences to similar viruses in other primates. Most HERV sequences have exchanged parts (recombined) and mutated to the extent that they no longer make us sick. Harmless HERVs silently pass from human generation to generation as parts of our chromosomes. They increase in number with time. However, several small studies show that people with certain diseases (forms of amyotrophic lateral sclerosis, multiple sclerosis, and melanoma) have overexpressed HERVs.

Noncoding RNAs

Much more of a human genome is transcribed than would be predicted based on the number and diversity of proteins in a human body. The two general classes of RNAs are coding (the mRNAs) and noncoding (ncRNAs), which include everything else. The best-studied noncoding RNAs are the tRNAs and rRNAs. **Table 11.2** lists noncoding RNAs.

The rate of transcription of a cell's tRNA genes is attuned to cell specialization. The proteins of a skeletal muscle cell, for example, require different numbers of certain amino acids than the proteins of a white blood cell, and therefore different numbers of the corresponding tRNAs. Human tRNA genes are dispersed among the chromosomes in clusters. Altogether, our 500 or so types of tRNA genes account for 0.1 percent of the genome.

The 243 types of rRNA genes are grouped on six chromosomes, each cluster harboring 150 to 200 copies of a 44,000-base repeat sequence. Once transcribed from these clustered genes, the rRNAs go to the nucleolus, where another type of ncRNA called small nucleolar RNA (snoRNA) cuts them into their final forms.

Hundreds of thousands of noncoding RNAs are not tRNA, rRNA, snoRNAs, microRNAs, nor the other less abundant types described in table 11.2. Some noncoding RNAs correspond to DNA sequences called **pseudogenes**. A pseudogene is similar in sequence to a protein-encoding gene that may be transcribed, but it is not translated into protein, because it is altered in sequence from an ancestral gene in a way that may impair its translation or folding. Pseudogenes may be remnants of genes that functioned in the past, variants that diverged from the normal sequence too greatly to encode a working protein. The alpha and beta globin gene complexes (see figure 11.1) include pseudogenes.

Another class of RNA molecules are the 12,000 **long noncoding RNAs**, which are each 200 or more nucleotides long. They are transcribed from exons, introns, and regions between genes, and they associate with chromatin and therefore control gene expression. A third of the long noncoding RNAs are found only in the genomes of primates, and many are transcribed only in the brain. These findings suggest that the long noncoding RNAs may hold clues to what makes us human—a topic discussed in chapter 16. More practically, low levels of a noncoding RNA called lnc13 have been correlated to the overactive immune response that lies behind gluten intolerance (celiac disease).

Repeats

Human genomes are riddled with highly repetitive sequences that hold a different type of information than a protein's amino acid sequence. The most abundant type of repeat is a sequence of DNA that can move about the genome. It is called a transposable element, or **transposon** for short. Geneticist Barbara McClintock originally identified transposons in corn in the 1940s, and they were rediscovered in bacteria in the 1960s.

Transposons comprise about 45 percent of the human genome sequence, and typically are present in many copies. Some transposons include parts that encode enzymes that cut them out of one chromosomal site and integrate them into another. Unstable transposons may lie behind inherited diseases that have

Table 11.2	Some Non-Protein-Encoding Parts of the Human Genome
Type of Sequence	**Function or Characteristic**
Viral DNA	Evidence of past infection
Noncoding RNA genes	
tRNA genes	Connect mRNA codons to amino acids
rRNA genes	Parts of ribosomes
Long noncoding RNAs	Control of gene expression
Pseudogenes	DNA sequences very similar to known genes that are not translated
Piwi-interacting RNA (piRNA)	Keeps transposons out of germline cells
Large intergenic noncoding RNAs	Between genes
Small nucleolar RNAs (snoRNAs)	Process rRNAs in nucleolus
Small nuclear RNAs (snRNAs)	Parts of spliceosomes
Telomerase RNA	Adds bases to chromosome tips
Xist RNA	Inactivates one X chromosome in cells of females
Introns	Parts of genes that are cut out of mRNA
Promoters and other control sequences	Guide enzymes that carry out DNA replication, transcription, or translation
Small interfering RNAs (siRNAs)	Control translation
MicroRNAs (miRNAs)	Control translation of many genes
Repeats	
Transposons	Repeats that move around the genome
Telomeres	Protect chromosome tips
Centromeres	Largest constrictions in chromosomes, providing attachment points for spindle fibers

several symptoms, because they insert into different genes. This is the case for Rett syndrome (see the chapter 2 opener).

An example of a specific type of repeat is an Alu sequence. Each Alu repeat is about 300 bases long, and a human genome may contain 300,000 to 500,000 of them. Alu repeats comprise 2 to 3 percent of the genome, and they have been increasing in number over time because they can copy themselves. We don't know exactly what these common repeats do, if anything. They may serve as attachment points for proteins that bind newly replicated DNA to parental strands before anaphase of mitosis, when replicated chromosomes pull apart.

Rarer classes of repeats comprise telomeres, centromeres, duplications of 10,000 to 300,000 bases, copies of pseudogenes, and simple repeats of one, two, or three bases. In fact, the entire human genome may have duplicated once or even twice.

Repeats may make sense in light of evolution, past and future. Pseudogenes are likely vestiges of genes that functioned in our nonhuman ancestors. Perhaps the repeats that seem to have no obvious function today will serve as raw material from which future genes may arise by mutation.

Key Concepts Questions 11.4

1. What can RNA do in addition to encoding protein?
2. What are some types of noncoding RNAs?
3. What type of noncoding RNA might reflect our past?

Summary

11.1 Gene Expression Through Time and Tissue

1. Changes in gene expression occur over time at the molecular and organ levels. **Epigenetic** changes to DNA alter gene expression, but do not change the DNA sequence.
2. **Transcriptomics** and **proteomics** catalog the types of mRNAs and proteins, respectively, made in particular cells. Each provides a global view of gene expression.

11.2 Control of Gene Expression

3. In **chromatin remodeling**, acetylation and phosphorylation of certain histone proteins enables the transcription of associated genes, whereas methylation selectively prevents transcription.
4. **MicroRNAs** bind to certain mRNAs, blocking translation.

11.3 Maximizing Genetic Information

5. A small part of the genome encodes protein, but the number of proteins is much greater than the number of genes.
6. Alternate splicing, use of introns, protein modification, and cutting of a precursor protein into two proteins contribute to protein diversity.

11.4 Most of the Human Genome Does *Not* Encode Protein

7. The non-protein-encoding part of the genome includes viral sequences, sequences that encode noncoding RNAs, **pseudogenes**, introns, promoters and other controls, and repeats such as **transposons**.
8. **Reverse transcriptase** copies DNA from RNA, enabling viruses to insert their genetic material into human chromosomes.
9. **Long noncoding RNAs** control gene expression.

Review Questions

1. Why is control of gene expression necessary?
2. Define *epigenetic*.
3. Distinguish between the information in the DNA sequence of a protein-encoding gene and epigenetic information.
4. Describe three types of cells mentioned anywhere in the book and explain how they likely differ in gene expression from each other.
5. What is the environmental signal that stimulates globin chain switching?
6. How does development of the pancreas illustrate differential gene expression?
7. Explain how a mutation in a promoter can affect gene expression.
8. How do histones control gene expression, yet genes also control histones?
9. What controls whether histones enable the DNA wrapped around them to be transcribed?
10. State two ways that methyl groups control gene expression.
11. Name a mechanism that silences transcription of a gene and a mechanism that blocks translation of an mRNA.
12. Explain why evaluating microRNA functions in human genomes is computationally complex.
13. Provide three reasons the number of proteins exceeds the number of protein-encoding genes in human genomes.
14. How can alternate splicing generate more than one type of protein from the information in a gene?

15. In the 1960s, a gene was defined as a continuous sequence of DNA that is located permanently at one place on a chromosome and specifies a sequence of amino acids from one strand. List three ways this definition has changed.

16. What is the evidence that some long noncoding RNAs may hold clues to human evolution?

Applied Questions

1. Possible causes of schizophrenia are discussed in section 8.6 and in this chapter's opener. Suggest how the two types of causes described might represent the same phenomenon.

2. The World Anti-Doping Agency warns against gene doping, which it defines as "the non-therapeutic use of cells, genes, genetic elements, or of the modulation of gene expression, having the capacity to improve athletic performance." The organization lists the following genes as candidates for gene doping when overexpressed:

   > Insulin-like growth factor (*IGF-1*)
   > Growth hormone (*GH*)
   > Erythropoietin (*EPO*)
   > Vascular endothelial growth factor (*VEGF*)
   > Fibroblast growth factor (*FGF*)

   Select one of these genes, look up information about it, and explain how its overexpression might improve athletic performance.

3. What might be the effect of a mutation in the part of the gamma globin gene that normally binds a transcription factor in an adult?

4. Researchers compared the pattern of microRNAs in breast cancer tumors from women whose disease did not spread to the brain to the pattern in tumors that did spread to the brain. Suggest a clinical application of this finding.

5. Several new drugs inhibit the enzymes that either put acetyl groups on histones or take them off. Would a drug that combats a cancer caused by underexpression of a gene that normally suppresses cell division add or remove acetyl groups?

6. Chromosome 7 has 863 protein-encoding genes, but encodes many more proteins. The average gene is 69,877 bases long, but the average mRNA is 2,639 bases long. Explain both of these observations.

7. How many different proteins encoded by two exons can be produced from a gene that has three exons?

Forensics Focus

1. Establishing time of death is critical information in a murder investigation. Forensic entomologists can estimate the "postmortem interval" (PMI), or the time at which insects began to deposit eggs on the corpse, by sampling larvae of specific insect species and consulting developmental charts to determine the stage. The investigators then count the hours backwards to estimate the PMI. Blowflies are often used for this purpose, but their three larval stages look remarkably alike in shape and color, and development rate varies with environmental conditions. With luck, researchers can count back 6 hours from the developmental time for the largest larvae to estimate the time of death.

   In many cases, a window of 6 hours is not precise enough to narrow down suspects when the victim visited several places and interacted with many people in the hours before death. Suggest a way that gene expression profiling might be used to more precisely define the PMI and extrapolate a probable time of death.

Case Studies and Research Results

1. Kabuki syndrome is named for the resemblance of an affected individual to a performer wearing the dramatic makeup used in traditional Japanese theater called Kabuki. The face has long lashes, arched eyebrows, flared eyelids, a flat nose tip, and large earlobes. The syndrome is associated with many symptoms, including developmental delay and intellectual disability, seizures, a small head (microcephaly), weak muscle tone, fleshy fingertips, cleft palate, short stature, hearing loss, and heart problems. Both genes associated with the condition result in too many regions of closed chromatin. Drugs that inhibit histone deacetylases (enzymes that remove acetyl groups from histone proteins) are effective. Explain how the drugs work.

2. To make a "reprogrammed" induced pluripotent stem (iPS) cell (see figure 2.20), researchers expose fibroblasts taken from skin to "cocktails" that include transcription factors. The fibroblasts divide and give rise to iPS cells, which, when exposed to other transcription factors, divide and yield daughter cells that specialize in distinctive ways that make them different from the original fibroblasts. How do transcription factors orchestrate these changes in cell type?

3. Using an enzyme called DNAse 1, researchers can determine which parts of the genome are in the "open chromatin" configuration in a particular cell. How could this technique be used to develop a new cancer treatment?

Courtesy of Marc Pieterse

CHAPTER 12

Gene Mutation

Vincent's long eyelashes, asymmetrical ears, autistic behaviors, learning disability, and other traits and symptoms arise from a mutation in a ribosomal protein, revealed through exome sequencing. For years he was the only one in the world known to have the mutation.

Learning Outcomes

12.1 The Nature of Gene Variants

1. Distinguish between mutation and mutant.
2. Distinguish between mutation and polymorphism.
3. Distinguish between germline and somatic mutations.

12.2 A Closer Look at Two Mutations

4. Describe mutations in the genes that encode beta globin and collagen.

12.3 Allelic Diseases

5. Provide examples of how mutations in a single gene can cause more than one illness.

12.4 Causes of Mutation

6. Explain how mutations arise spontaneously and how they may be induced.

12.5 Types of Mutations

7. Describe the two types of single-base mutations.
8. Explain the consequences of a splice-site mutation.
9. Discuss mutations that add, remove, or move DNA nucleotides.
10. Describe how pseudogenes and transposons can cause mutations.

12.6 The Importance of Position

11. Give examples of how the location of a mutation in a gene affects the phenotype.
12. Describe a conditional mutation.

12.7 DNA Repair

13. List the types of damage that DNA repair mechanisms fix.
14. Describe the types of DNA repair.

 ## The BIG Picture

The informational nature of DNA means that many variants of each gene are possible. Some variant combinations (genotypes) cause disease or increase susceptibility to a disease, yet some may be protective. The nature of the genetic code as well as natural DNA repair mechanisms minimize the effects of mutation and DNA damage.

Vincent's Diagnostic Odyssey

For years, Marc P. thought his son Vincent was the only person in the world with his unusual combination of characteristics: arched eyebrows, extra-long lashes, bluish eye whites, offset ears, additional teeth, a webbed neck, a small lower jaw, elastic skin, brittle hair, and a few other oddities. Vincent also has developmental delay, autistic behaviors, and a learning disability. But he's happy and active and loves watching and sketching the birds that he passes on his way to school.

Marc has an interest in genetics and recognized what appeared to be a syndrome in his young son. A few years ago, after many doctors had failed to put a name to Vincent's condition, Marc, his wife, and their son had their exomes sequenced as part of a research project at the University Medical Center Nijmegen, in the Netherlands, where the family lives. Exome sequencing found that Vincent has a *de novo* mutation in the gene *RPS23*, which encodes one of the 79 types of proteins that, with RNA, make up ribosomes. A *de novo* mutation is new, originating in the person who has it, rather than being inherited. Millie McWilliams, whose story is told in Clinical Connection 1.1, also has a *de novo* mutation, hers discovered through genome sequencing.

Marc assembled an international group of researchers to explore his son's mutation in a variety of model organisms that have versions of the gene, such as zebrafish and mice. Model organisms revealed aspects of the gene's function, but to connect the mutation to Vincent's phenotype required finding other individuals like him. That didn't happen for years, until in September 2015, Marc entered Vincent's information in GeneMatcher, an online tool. GeneMatcher is part of an international effort called the Matchmaker Exchange that is connecting parents, researchers, and health care professionals to work together to identify the mutations behind rare diseases.

Just 3 months after entering Vincent's mutation, Marc learned of a second child who had the same phenotype and genotype! The child had the same facial features, including the gorgeous eyelashes, as well as the offset ears and autistic behaviors. With more cases identified and the condition better understood, investigators can begin testing existing drugs and therapies and perhaps develop new ones to help these children. In the meantime, cell biologists can learn more about ribosomes, and Marc's efforts have revealed a new possible explanation for autism.

12.1 The Nature of Gene Variants

A **mutation** is a change in a DNA sequence that is rare in a population and typically affects the phenotype. "Mutate" refers to the process of altering a DNA sequence. Mutations range in magnitude from substitution of a single DNA base; to deletion or duplication of tens, hundreds, thousands, or even millions of bases; to deletion or duplication of entire chromosomes. This chapter discusses smaller-scale mutations, and chapter 13 considers mutations at the chromosomal level. However, the extent of mutation is a continuum.

Mutation can affect any part of the genome: sequences that encode proteins or control transcription, introns, repeats, and sites critical to intron removal and exon splicing. Not all DNA sequences are equally likely to mutate.

The effects of mutation vary. Mutations may impair a function, have no effect, or even be beneficial. A deleterious (harmful) mutation can stop or slow production of a protein, overproduce it, or impair the protein's function—such as altering its secretion, location, or interaction with another protein. The effect of a mutation is called a "loss-of-function" when the gene's product is reduced or absent, or a "gain-of-function" when the gene's activity changes. Most mutations are recessive and cause a loss-of-function (see figure 4.8). Gain-of-function mutations tend to be dominant and are also called "toxic."

Mutation, Polymorphism, or Variant?

The terms *mutation* and *polymorphism* each denote a genetic change from wild type (the most common form). **Polymorphism** is a general term meaning, literally, "many forms." Recall from chapter 7 that a single nucleotide polymorphism, or SNP, is a single

base change. So are many mutations. The distinction between mutation and polymorphism before many human genomes were sequenced was largely artificial, reflecting frequency in a particular population. According to that older definition, a mutation was much rarer than a polymorphism. The reasoning for the distinction was that if a genetic change greatly impairs health, individuals with it are unlikely to reproduce, and the mutant allele remains uncommon. A polymorphism that does not harm health, elevates risk of illness only slightly, or is even beneficial will remain prevalent in a population or even increase in frequency because it does not hamper reproduction.

Sequencing many genomes has shown that a DNA base change that is a mutation in one population may be a harmless polymorphism in another, due to the effects of other genes and different environments. Therefore, the distinction between polymorphism and mutation reflects gene function as well as frequency. Because of this confusion, the term "variant" is increasingly being used to mean both polymorphism and mutation. The **Bioethics** box at the end of this chapter discusses the clinical challenges of identifying and reporting to patients "variants of uncertain significance" that arise with genetic testing.

A mutation may be helpful. For example, about 1 percent of the general population is homozygous for a recessive allele that encodes a cell surface protein called CCR5 (see figure 17.11). To infect an immune system cell, HIV must bind CCR5 and another protein. Because the mutation prevents CCR5 from moving to the cell surface from inside the cell, HIV cannot bind. Homozygotes for a *CCR5* mutation cannot be infected with HIV and heterozygotes are partially protected. The opener to chapter 17 describes how mimicking the *CCR5* mutation treats HIV infection.

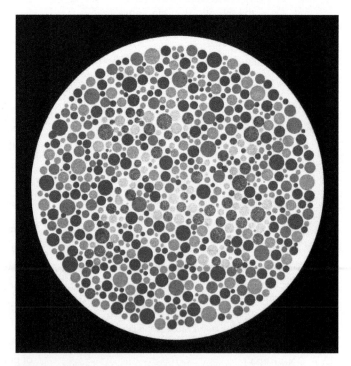

Figure 12.1 **The concept of a mosaic.** A test used to detect colorblindness illustrates the idea of a mosaic. In somatic mosaicism, only some cells in an individual have the mutation. © Steve Allen/Getty Images

The term *mutation* refers to genotype—that is, a change at the DNA or chromosome level. The familiar term **mutant** refers to phenotype and also describes an allele. The nature of a mutant phenotype depends upon how the mutation affects the gene's product or activity, and usually connotes an abnormal or unusual characteristic. However, a mutant phenotype may also be a rare variant that is nevertheless "normal," such as red hair.

Germline and Somatic Mutation

A mutation may be present in all the cells of an individual or only in some cells. In a **germline mutation**, the change occurs during the DNA replication that precedes *meiosis*. The resulting gamete and all the cells that descend from it after fertilization have the mutation—that is, every cell in the body. Germline mutations are transmitted to the next generation of individuals.

A **somatic mutation** happens during DNA replication before *mitosis*, and is passed to the next generation of cells, but not to all the cells in the individual's body. All the cells that descend from the original changed cell are altered, but they might only comprise a small part of the body. A person with a somatic mutation has **somatic mosaicism**. Somatic mutations are more likely to occur in cells that divide often, such as skin and blood cells, because there are more opportunities for DNA replication errors. **Figure 12.1** shows the concept of a mosaic—subunits of one color scheme (blue, green, gray, and yellow) are obvious against another color scheme (reds and oranges). (This is actually a test used to detect colorblindness and is not meant to represent a mosaic eye.)

Some genetic diseases are so severe that they are only seen in mosaics. That is, having the mutation in every cell is incompatible with life. A mutation occurs in one cell of an early embryo, and leads to an individual with patches of tissue that descend from the cell in which the mutation originated (**figure 12.2a**). Some cases of neurofibromatosis, for example, are so mild, causing only a few darkened areas on the skin, that people can survive with the mutation in every cell. This is

(a)

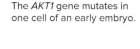

The *AKT1* gene mutates in one cell of an early embryo.

The cells of the embryo divide, including the cell with the *AKT1* mutation. Cells migrate.

The fetus has two cell populations: A larger one with cells that have non-mutant *AKT1* genes, and a smaller population of cells bearing the mutation.

After birth, cells with the mutation grow faster, leading to body asymmetry.

(b)

Figure 12.2
Somatic mosaicism.
(a) In an early embryo, this mutation leads to patches of the body in which cells have the mutation in *AKT1*.
(b) Note the overgrown fingers and hands of this young man who has Proteus syndrome. His neck and spine are also enlarged. Only some of his cells have the mutation. *(a): Source: The Proteus Syndrome Foundation, UK (b): Courtesy of The Proteus Syndrome Foundation, UK*

not the case for Proteus syndrome, which occurs *de novo* and is known to affect only a few hundred people worldwide. They have overgrowth of skin and bone, but only in the body parts that bear a mutation in the gene *AKT1*. The different growth rates in different parts of the body over time lead to disfigurement, apparent by 18 months of age (**figure 12.2b**). The disfigurement worsens over time.

In general, somatic mutations occur as errors during DNA replication at about every 300 mitotic cell divisions. Researchers confirmed the fact that somatic mutations are more likely to happen in cell types that divide frequently when they autopsied a woman named Henne Holstege, who died in 2005 at the age of 115. Whole genome sequencing of Henne's white blood cells revealed 450 mutations, but that of her brain neurons, none. White blood cells descend from many divisions of stem cells, providing many opportunities for replication error, but neurons never divide. Chapter 18 considers how series of somatic mutations cause and "drive" the progression of most cancers.

Sequencing the genomes of individual cells from different organs from the same individual has shown that somatic mosaicism is not as rare as had been thought. In addition to somatic mutation, mosaicism can result from having a twin that died before birth or, for women, from fetal cells retained from a pregnancy. Most of us may be, in some way, genomic mosaics.

Key Concepts Questions 12.1

1. Where do mutations occur in the genome?
2. How are mutations and polymorphisms alike and how do they differ?
3. Distinguish between the consequences of a germline and a somatic mutation.
4. Explain why some conditions are only seen as somatic mosaics.

12.2 A Closer Look at Two Mutations

Identifying how a mutation causes symptoms has clinical applications, and also reveals the workings of biology. Following are two examples of well-studied mutations that cause disease.

The Beta Globin Gene Revisited

The first genetic illness understood at the molecular level was sickle cell disease. The tiny mutation that causes it substitutes the amino acid valine for the glutamic acid that is normally the sixth amino acid in the beta globin polypeptide chain (**figure 12.3**). At the DNA level, the change is even smaller—a CTC is changed to a CAC, corresponding to RNA codons GAG and GUG. Valine at this position changes the surfaces of hemoglobin molecules so that in low-oxygen conditions they attach at many more points than they would if the wild type glutamic acid were at the site.

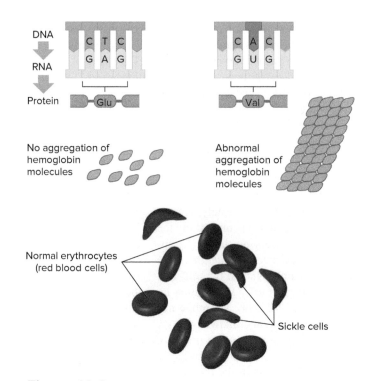

Figure 12.3 **Sickle cell disease results from a single DNA base change that substitutes one amino acid in the protein (valine replaces glutamic acid).** This changes the surfaces of the hemoglobin molecules, and they aggregate into long, curved rods that deform the red blood cell. The illustration shows the appearance of sickle cells.

The aggregated hemoglobin molecules form ropelike cables that make red blood cells sticky and able to deform, and then bend the cells into rigid, fragile sickle shapes. The cells lodge in narrow blood vessels, cutting off local blood supplies. Once a blockage occurs, sickling speeds up and spreads, as the oxygen level falls. The result is great pain in the blocked body parts, particularly the hands, feet, and intestines. The bones ache, and depletion of normal red blood cells causes the great fatigue of anemia.

Other conditions result from mutations in the beta globin gene. In 1925, Thomas Cooley and Pearl Lee described severe anemia in Italian children, and in the decade following, others described a milder version of "Cooley's anemia," also in Italian children. The disease was named thalassemia, from the Greek for "sea," in light of its high prevalence in the Mediterranean area. The two anemias turned out to be the same. The severe form, sometimes called thalassemia major, results from a homozygous mutation in the beta globin gene at a site other than the one that causes sickle cell disease. The milder form, called thalassemia minor, affects some individuals who are heterozygous for the mutation.

Once researchers had worked out the structure of hemoglobin, and learned that different globins function in the embryo and fetus (see figure 11.2), the molecular basis of thalassemia became clear. The disease that is common in the Mediterranean is more accurately called beta thalassemia, because the symptoms result from too few beta globin chains. Without them, not enough hemoglobin molecules are assembled to effectively deliver oxygen to tissues. Fatigue and bone pain arise during the first year of life as the child depletes fetal hemoglobin, and the "adult" beta globin genes are not transcribed and translated on schedule.

As severe beta thalassemia progresses, red blood cells die because the excess of alpha globin chains prevents formation of hemoglobin molecules. Liberated iron slowly destroys the heart, liver, and endocrine glands. Periodic blood transfusions can control the anemia, but they hasten iron buildup and organ damage. Drugs called chelators that entrap the iron can extend life past early adulthood, but they are costly and not available in some nations.

Disorders of Orderly Collagen

The protein collagen is a major component of connective tissue. Collagen accounts for more than 60 percent of the protein in bone and cartilage and provides 50 to 90 percent of the dry weight of skin, ligaments, tendons, and the dentin of teeth.

Genetic control of collagen synthesis and distribution is complex; more than 35 collagen genes encode more than 20 types of collagen molecules. Other genes affect collagen, too. Mutations in the genes that encode collagen, not surprisingly, lead to a variety of medical conditions (**table 12.1**). These disorders are particularly devastating, not only because collagen is nearly everywhere, but because collagen has an extremely precise conformation that is easily disrupted, even by slight alterations that might have little effect in proteins with other shapes (**figure 12.4**).

Collagen is trimmed from a longer precursor molecule called procollagen, which consists of many repeats of the amino acid sequence *glycine-proline-modified proline*. Three procollagen chains entwine. Two of the chains are identical and are encoded by one gene, and the other is encoded by a second gene

In 1904, young medical intern Ernest Irons noted "many pear-shaped and elongated forms" in a blood sample from a dental student in Chicago who had anemia. Irons sketched this first view of sickle cell disease at the cellular level, and reported his findings to his supervisor, physician James Herrick. Alas, Herrick published the work without including Irons and has been credited with the discovery ever since.

In 1949, Linus Pauling found that hemoglobin from healthy people and from people with the anemia, when placed in a solution in an electrically charged field, moved to different positions. Hemoglobin molecules from the parents of people with the anemia, who were carriers, moved to both positions.

The difference between the two types of hemoglobin lay in beta globin. Recall from figure 11.1 that adult hemoglobin consists of two alpha polypeptide subunits and two beta subunits. Protein chemist V. M. Ingram took a shortcut to localize the sickle cell mutation in the 146-amino-acid-long beta subunit. He cut normal and sickle hemoglobin with a protein-digesting enzyme, separated the pieces, stained them, and displayed them on filter paper. The patterns of fragments—known as peptide fingerprints—were different for the two types of beta globin. This meant, Ingram deduced, that the two molecules differ in amino acid sequence. One piece of the molecule in the fingerprint, fragment four, occupied a different position for each type of beta globin. Because this peptide was only eight amino acids long, Ingram needed to decipher only that short sequence to find the site of the mutation. It was a little like finding which sentence on a page contains a miskeyed word.

Table 12.1	Some Collagen Disorders	
Disorder	**Mutations (Genotype)**	**Signs and Symptoms (Phenotype)**
Alport syndrome	Mutations in any of three genes (*COL4A3, COL4A4, COL4A5*) affect type IV collagen, which disrupts tissue boundaries.	Deafness and inflamed kidneys
Chondrodysplasia	Deletion, insertion, or missense mutation replaces Gly with bulky amino acids in *COL2A1* type II collagen gene.	Stunted growth, deformed joints
Dystrophic epidermolysis bullosa	Mutation in *COL7A1* gene that encodes type VII collagen breaks down fibrils that attach epidermis to dermis.	Skin blisters upon any touch
Ehlers-Danlos syndrome	Diverse mutations in at least a dozen genes affect collagens or the molecules to which they bind.	Stretchy, easily scarred skin, lax joints
Osteoarthritis	Missense mutation in α1 collagen gene (*COL1A1*) substitutes Cys for Arg.	Painful joints
Osteogenesis imperfecta type I	Inactivation of α1 collagen gene (*COL1A1 or COL1A2*) reduces number of collagen triple helices by 50%.	Easily broken bones; blue eye whites; deafness
Stickler syndrome	Nonsense mutations in type II procollagen gene (*COL2A1 or COL11A1*) reduce number of collagen molecules.	Joint pain, degeneration of vitreous gel and retina

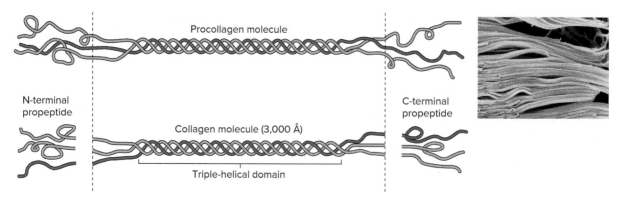

Procollagen molecule

N-terminal
propeptide

C-terminal
propeptide

Collagen molecule (3,000 Å)

Triple-helical domain

Figure 12.4 Collagen has a precise conformation. The α1 collagen gene encodes the two blue polypeptide chains, and the α2 procollagen gene encodes the third (red) chain. The procollagen triple helix is shortened before it becomes functional, forming the fibrils and networks that comprise much of the human body. The inset shows aligned collagen fibrils. © *Science Photo Library RF/Getty Images RF*

and has a different amino acid sequence. The electrical charges and interactions of the amino acids with water coil the procollagen chains into a very regular triple helix, with space in the middle only for tiny glycine. Enzymes snip off the ragged ends of the polypeptides, forming mature collagen. The collagen fibrils continue to associate with each other outside the cell, building the fibrils and networks that hold the body together. **Figure 12.5** shows the characteristic stretchy skin that results from unassembled collagen molecules.

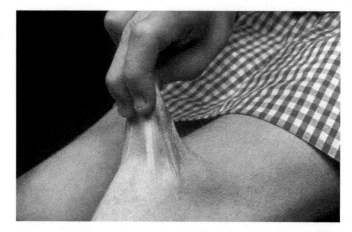

Figure 12.5 Collagen mutations can affect the skin. Ehlers-Danlos syndrome type I causes highly extensible joints and stretchy skin, due to inability of collagen molecules to assemble properly. © *Biophoto Associates/Science Source*

Table 12.2	**How Mutations Cause Disease**		
Disease	**Protein**	**Mutations (Genotype)**	**Signs and Symptoms (Phenotype)**
Cystic fibrosis	Cystic fibrosis transmembrane regulator (CFTR)	Missing amino acid or other variant alters conformation of chloride channels in certain epithelial cell plasma membranes. Water enters cells, drying out secretions.	Frequent lung infection, pancreatic insufficiency
Duchenne muscular dystrophy	Dystrophin	Deletion eliminates dystrophin, which normally binds inner face of muscle cell to plasma membrane. Muscles weaken.	Gradual loss of muscle function
Familial hypercholesterolemia	LDL receptor	Deficient LDL receptors cause cholesterol to accumulate in blood.	High blood cholesterol, early heart disease
Hemophilia B	Factor IX	Absent or deficient clotting factor causes hard-to-control bleeding.	Slow or absent blood clotting
Huntington disease	Huntingtin	Extra bases add amino acids to the protein, which impairs certain transcription factors and proteasomes.	Uncontrollable movements, personality changes
Marfan syndrome	Fibrillin or transforming growth factor β receptor	Deficient proteins in lenses cause cataracts and in the wall of the aorta cause aneurysm (bursting).	Long limbs, weakened aorta, spindly fingers, sunken chest, lens dislocation
Neurofibromatosis type 1	Neurofibromin	Defect in protein that normally suppresses activity of a gene that causes cell division, leading to abnormal growths.	Pigmented skin patches and benign tumors of nervous tissue beneath skin

So important is the precision of collagen formation that a mutation that controls placement of a single hydroxyl chemical group (—OH⁻) on collagen causes a form of osteogenesis imperfecta ("brittle bone disease"). Other collagen mutations remove procollagen chains, kink the triple helix, and disrupt aggregation outside the cell.

Aortic aneurysm is a serious connective tissue abnormality that can occur by itself or as part of Marfan syndrome (see figure 5.5). Detection of mutations that cause Marfan syndrome before symptoms arise can be lifesaving, because frequent ultrasound exams can detect aortic weakening early enough to patch the vessel before it bursts. **Table 12.2** describes how other mutations mentioned in this and other chapters impair health.

Key Concepts Questions 12.2

1. Describe how mutations affect the beta globin gene.
2. Explain why collagen genes are unusually prone to mutation.

12.3 Allelic Diseases

Analysis of human genomes is changing the way we describe single-gene diseases. In the past, geneticists were inconsistent when assigning disease names to mutations. For certain genes, mutations cause the same disease, yet for other genes, this isn't the case. For example, mutations in the *CFTR* gene cause cystic fibrosis, which may include the full spectrum of impaired breathing and digestion, or just male infertility or frequent bronchitis. Yet different mutations in the beta globin gene cause sickle cell disease and beta thalassemia.

Mutations in some genes correspond to many diseases. This is the case for the gene *lamin A,* in which different mutations cause different diseases in different tissues. The diseases include the rapid aging condition Hutchinson-Gilford progeria syndrome (see the chapter 3 opener), muscular dystrophies, and a heart disease. Lamin A proteins form a network beneath the inner nuclear membrane that interacts with chromatin. Different mutations affect lamin A's interactions with chromatin in ways that cause the diverse associated diseases.

As researchers discovered more cases of different diseases arising from mutations in the same gene, it became clear that the phenomenon is not unusual, and is not merely a matter of how we name diseases. Different clinical phenotypes caused by mutations in the same gene are termed **allelic diseases**.

The same gene can underlie different diseases in different ways (**table 12.3**). A pair of allelic diseases may result from mutations in different parts of the gene, mutations that are localized (a single base change) or catastrophic (a missing gene), or mutations that alter the protein in ways that affect its interactions with other proteins.

Allelic diseases may arise from a mutation that affects a protein that is used in different tissues. Some researchers are reclassifying cystic fibrosis as two allelic diseases, based on whether the lungs are affected. At least nine CF genotypes cause male infertility, pancreatitis, sinusitis, or a combination of these, but not respiratory impairment.

Key Concepts Questions 12.3

1. How are the ways that cystic fibrosis and sickle cell disease are named inconsistent?
2. Explain how mutations in *lamin A* cause wide-ranging effects on the phenotype.
3. Name a pair of allelic diseases and the gene from which they arise.

Table 12.3	Allelic Diseases	
Gene	**Function**	**Associated Diseases**
ATP7A	Copper transport	Menkes ("kinky hair") disease; peripheral neuropathy
DMD	Dystrophin muscle protein	Duchenne and Becker muscular dystrophy
FBN1	Encodes fibrillin-1, which forms tiny fibrils outside cells; a connective tissue protein	Marfan syndrome; systemic sclerosis (scleroderma; "stiff skin syndrome")
FGFR3	Fibroblast growth factor	Two types of dwarfism
GBA	Glucocerebrosidase	Gaucher disease; Parkinson disease
PSEN1	Presenilin 1 (enzyme part that trims membrane proteins)	Acne inversa; Alzheimer disease
RET	Oncogene (causes cancer)	Multiple endocrine neoplasia; Hirschsprung disease
TRPV4	Calcium channel	Peripheral neuropathy; spinal muscular atrophy

12.4 Causes of Mutation

A mutation can occur spontaneously or be induced by exposure to a chemical or radiation. An agent that causes mutation is called a **mutagen**.

Spontaneous Mutation

A spontaneous mutation can be a surprise. For example, two healthy people of normal height have a child with achondroplasia due to a *de novo* (new) autosomal dominant mutation (see figure 5.1a). Siblings have no higher risk of inheriting the dwarfism than anyone in the general population (unless the parents are mosaics), but each of the affected person's children faces a 50 percent chance of inheriting it.

A spontaneous mutation usually originates as an error in DNA replication. One trigger stems from the tendency of free DNA bases to exist in two slightly different chemical structures, called tautomers. For extremely brief times, each base is in an unstable tautomeric form. If, by chance, such an unstable base is inserted into newly forming DNA, an error will be generated and perpetuated when that strand replicates. **Figure 12.6** shows how this can happen.

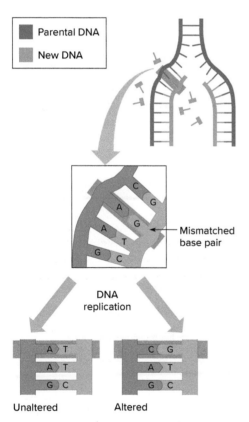

Figure 12.6 Spontaneous mutation. DNA bases exist fleetingly in alternate forms. A replication fork encountering a base in its unstable form can insert a noncomplementary base. After another round of replication, one of the daughter cells has a different base pair than the one in the corresponding position in the original DNA. (This figure depicts two rounds of DNA replication.)

Spontaneous Mutation Rate

Different genes have different spontaneous mutation rates. The gene that, when mutant, causes neurofibromatosis type 1 (NF1; see figure 11.12), for example, has a very high mutation rate, arising in 40 to 100 of every million gametes. NF1 affects 1 in 3,000 births, about half in families with no prior cases. The gene's large size may contribute to its high mutability—there are more ways for its sequence to change, just as there are more opportunities for a misspelling to occur in a long sentence than in a short one. In contrast, the mutation that causes the clotting disorder hemophilia B happens in only 1 to 10 of every million gametes formed.

Spontaneous mutation lies behind many instances of somatic mosaicism, such as Proteus syndrome (see figure 12.2). Spontaneous mutation also manifests as **gonadal mosaicism**. In this situation, a parent has a mutation in some sperm or oocytes, because a spontaneous mutation occurred in the developing testis or ovary and was transmitted only to the cells descended from the original cell bearing the mutation. Gonadal mosaicism is suspected when more than one child in a family has a genetic condition but both parents do not have the mutation in the cells sampled for genetic testing.

Sequencing of human genomes has revealed that spontaneous mutations are fairly common. For example, at least 10 percent of heart defects in newborns are due to mutations not present in the parents. Each of us has about 175 spontaneously mutated alleles. Mitochondrial genes mutate at a higher rate than genes in the nucleus because mitochondria cannot repair their DNA (see section 12.7).

Estimates of the spontaneous mutation rate for a particular gene are usually derived from observations of new, dominant conditions, such as achondroplasia. This is possible because a new dominant mutation is detectable simply by observing the phenotype. In contrast, a new recessive mutation is not obvious until two heterozygotes have a homozygous recessive child with a noticeable phenotype.

The spontaneous mutation rate for autosomal genes can be estimated using this formula: number of *de novo* cases/2X, where X is the number of individuals examined. The denominator has a factor of 2 to account for the nonmutated homologous chromosome.

Spontaneous mutation rates in human genes are difficult to assess because our generation time is long—usually 20 to 30 years. In bacteria, a new generation arises every half hour or so, and mutation is therefore much more frequent. The genetic material of viruses also spontaneously mutates rapidly, because they reproduce quickly and do not repair DNA errors.

Mutational Hot Spots

In some genes, mutations are more likely to occur in regions called hot spots, where sequences are repetitive. It is as if the molecules that guide and carry out replication become "confused" by short repeated sequences, much as an editor scanning a manuscript might miss the spelling errors in the words "hipppopotamus" and "bananana" (**figure 12.7**).

Repeat of a nucleotide   A A A A A A A A A

Direct repeat of a dinucleotide   G C G C G C G C

Direct repeat of a trinucleotide   T A C T A C T A C

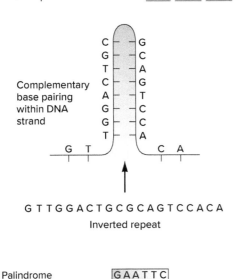

Complementary base pairing within DNA strand

```
    C ── G
    G ── C
    T ── A
    C ── G
    A ── T
    G ── C
    G ── C
    T ── A
G  T       C  A
```

G T T G G A C T G C G C A G T C C A C A

Inverted repeat

Palindrome   G A A T T C
                        C T T A A G

Figure 12.7 **DNA symmetry may increase the likelihood of mutation.** These examples show repetitive and symmetrical DNA sequences that may "confuse" replication enzymes, causing errors.

The increased incidence of mutations in repeats has a physical basis. Within a gene, when DNA strands locally unwind to replicate in symmetrical or repeated sequences, bases located on the same strand may pair. A stretch of ATATAT might pair with TATATA elsewhere on the same strand, creating a loop that interferes with replication and repair enzymes. Errors may result. This is the case for mutations in the gene for clotting factor IX, which causes hemophilia B. Mutations occur 10 to 100 times as often at any of 11 sites in the gene that have extensive direct repeats of CG than they do elsewhere in the gene.

Small additions and deletions of DNA bases are more likely to occur near sequences called palindromes (figure 12.7). These sequences read the same, in a 5′ to 3′ direction, on complementary strands. Put another way, the sequence on one strand is the reverse of the sequence on the complementary strand. Palindromes probably increase the spontaneous mutation rate by disturbing replication.

The blood disease alpha thalassemia illustrates the confusing effect of direct (as opposed to inverted) repeats of an entire gene. A person who does not have the disease has four genes that specify alpha globin chains, two next to each other on each chromosome 16. Homologs with repeated genes can misalign during meiosis when the first sequence on one chromosome lies opposite the second sequence on the homolog. If such a misalignment happens, crossing over can result in a sperm or oocyte that has one or three alpha globin genes instead of the normal two (**figure 12.8**). Fertilization with a normal gamete then results in a zygote with one extra or one missing alpha globin gene. At least three dozen conditions

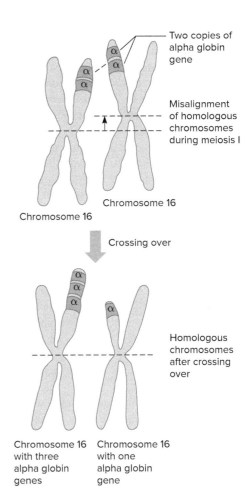

Figure 12.8 **Gene duplication and deletion.** The repeated alpha globin genes are prone to mutation by mispairing during meiosis.

result from this unequal crossing over, including colorblindness (see Clinical Connection 6.1).

The number of alpha globin genes affects health. A person with only three alpha globin genes produces enough hemoglobin, and is a healthy carrier. Individuals with only two copies of the gene are mildly anemic and tire easily, and a person with a single alpha globin gene is severely anemic. A fetus lacking all alpha globin genes does not survive.

Induced Mutation

Researchers can infer a gene's normal function by observing what happens when mutation alters it. Because the spontaneous mutation rate is much too low to be a practical source of genetic variants for experiments, researchers make mutants, using mutagens on "model" organisms such as mice and fruit flies. These types of experiments can yield insights into human health.

Intentional Use of Mutagens

Chemicals or radiation are used to induce mutation. Alkylating agents, for example, are chemicals that remove a DNA base, which is replaced with any of the four bases—three of which are a mismatch against the complementary strand. Dyes called acridines add or remove a single DNA base. Because the DNA sequence is read three bases in a row, adding or deleting a single base can destroy a gene's information, altering the amino acid sequence of the encoded protein. Several other mutagenic chemicals alter base pairs, so that an A-T replaces a G-C, or vice versa. X rays and other forms of radiation delete a few bases or break chromosomes.

Researchers have several ways to test the mutagenicity of a substance. The best known, the Ames test, assesses how likely a substance is to harm the DNA of rapidly reproducing bacteria. One version of the test uses a strain of *Salmonella* that cannot grow when the amino acid histidine is absent from its medium. If exposure to a substance enables bacteria to grow on the deficient medium, then a gene has mutated that allows it to do so.

In a variation of the Ames test, researchers exposed human connective tissue cells growing in culture to liquefied cigarette smoke. The chemicals from the smoke cut chromosomes through both DNA strands. Broken chromosomes can join with each other in different ways that can activate cancer-causing genes. Hence, the experiment may have modeled one way that cigarettes cause cancer. Because many mutagens are also carcinogens (cancer-causing agents), the substances that the Ames test identifies as mutagens may also cause cancer. Common products that contain mutagens are hair dye, smoked meats, certain flame retardants used in children's sleepwear, and food additives.

A limitation of using a mutagen is that it cannot cause a specific mutation. In contrast, a technique called site-directed mutagenesis changes a gene in a desired way. A gene is mass produced (see section 9.3), but the copies include an intentionally substituted base, just as an error in a manuscript is printed in every copy of a book. Site-directed mutagenesis is faster and more precise than waiting for nature or a mutagen to produce a useful gene variant. The gene editing techniques described in section 19.4 and section 20.4 can also rapidly introduce changes to a DNA sequence.

Accidental Exposures to Mutagens

Some mutagen exposure is unintentional. Such exposure occurs from workplace contact before the danger is known, from industrial accidents, from medical treatments such as chemotherapy and radiation, from exposure to weapons that emit radiation, and from natural disasters that damage radiation-emitting equipment. For example, on April 25, 1986, between 1:23 and 1:24 A.M., Reactor 4 at the Chernobyl Nuclear Power Station in Ukraine exploded, sending a great plume of radioactive isotopes into the air that spread for thousands of miles. The reactor had been undergoing a test, its safety systems temporarily disabled, when it overloaded and rapidly flared out of control. Twenty-eight people died of acute radiation exposure in the days following the explosion. Today the level of radioactivity has diminished by 75 percent. Tourists visit limited areas of the site, and vegetation has returned.

Evidence of a mutagenic effect of acute radiation poisoning is the 10-fold increased rate of thyroid cancer among children who were living near the Chernobyl plant in Belarus when the disaster happened. The thyroid glands of young people soaked up iodine that, in a radioactive form, bathed Belarus in the days after the explosion.

One way that researchers tracked mutation rates after the Chernobyl explosion was to compare the lengths of short DNA repeats, called minisatellite sequences, in children born in 1994 and in their parents, who lived in the exposed district at the time of the accident and have remained there. Minisatellites are the same length in all cells of an individual. A minisatellite size in a child that does not match the size of either parent indicates that a mutation occurred in a parent's gamete. Such a mutation was twice as likely to occur in exposed families as in families living elsewhere. Because mutation rates of nonrepeated DNA sequences are too low to provide useful information on the effects of radiation exposure, investigators track minisatellites as a sensitive test of genetic change.

Natural Exposure to Mutagens

Simply being alive exposes us to radiation that can cause mutation. Natural environmental sources of radiation account for 81 percent of our exposure, including cosmic rays, sunlight, and radioactive substances in the Earth's crust, such as radium, which produces radon. Medical X rays and occupational radiation hazards add risk. Job sites with increased radiation exposure include weapons facilities, research laboratories, health care facilities, nuclear power plants, and certain manufacturing plants.

Most of the potentially mutagenic radiation we are exposed to is ionizing, which means that it has sufficient energy to remove electrons from atoms. Unstable atoms that emit ionizing radiation exist naturally, and we make them. Ionizing radiation breaks the DNA sugar-phosphate backbone.

Ionizing radiation is of three major types. Alpha radiation is the least energetic and most short lived, and the skin absorbs most of it. Uranium and radium emit alpha radiation. Beta radiation can penetrate the body farther, and emitters include tritium (a form of hydrogen), carbon-14, and strontium-70. Both alpha and beta rays tend not to harm health, although they can do damage if inhaled or eaten. In contrast is the third type of ionizing radiation, gamma rays. These can penetrate the body, damaging tissues. Plutonium and cesium isotopes used in weapons emit gamma rays, and this form of radiation is used to kill cancer cells.

X rays are the major source of exposure to human-made radiation. They have less energy and do not penetrate the body to the extent that gamma rays do.

The effects of radiation damage to DNA depend upon the functions of the mutated genes. Mutations in oncogenes or tumor suppressor genes, discussed in chapter 18, can cause cancer. Radiation damage can be widespread, too. Exposing cells to radiation and then culturing them causes a genome-wide destabilization, so that mutations may occur even after the cell has divided a few times. Cell culture studies have also identified a "bystander effect," when radiation harms cells not directly exposed.

Chemical mutagens are in the environment, too. Evaluating the risk that a specific chemical exposure will cause a mutation is difficult, largely because people vary greatly in inherited susceptibilities, and are exposed to many chemicals. The risk that exposure to a certain chemical will cause a mutation is often less than the natural variability in susceptibility within a population, making it nearly impossible to track the true source and mechanism of any mutational event.

Key Concepts Questions 12.4

1. How do spontaneous mutations occur?
2. How does the DNA sequence affect the likelihood of mutations?
3. What are some mutagens?
4. Why are the effects of mutagens encountered in the environment difficult to assess?

12.5 Types of Mutations

Mutations are classified by whether they remove, alter, or add a function, or by how they structurally alter DNA. The same single-gene disease can result from different types of mutations. **Table 12.4** summarizes types of mutations using an analogy to an English sentence.

Point Mutations

A **point mutation** is a change in a single DNA base. A **transition** is a point mutation that replaces a purine with a purine (A to G or G to A) or a pyrimidine with a pyrimidine (C to T or T to C).

Table 12.4	Types of Mutations

A sentence comprised of three-letter words provides analogies to the effects of mutations on a gene's DNA sequence:

Normal	THE ONE BIG FLY HAD ONE RED EYE
Missense	THQ ONE BIG FLY HAD ONE RED EYE
Nonsense	THE ONE BIG _____
Frameshift	THE ONE QBI GFL YHA DON ERE DEY
Deletion	THE ONE BIG ___ HAD ONE RED EYE
Insertion	THE ONE BIG FLY WET HAD ONE RED EYE
Duplication	THE ONE BIG FLY FLY HAD ONE RED EYE
Expanding mutation	
Generation 1	THE ONE BIG FLY HAD ONE RED EYE
Generation 2	THE ONE BIG FLY FLY FLY HAD ONE RED EYE
Generation 3	THE ONE BIG FLY FLY FLY FLY FLY HAD ONE RED EYE

A **transversion** replaces a purine with a pyrimidine, or vice versa (A or G to T or C). Addition or deletion of a single DNA base is also considered to be a point mutation. A point mutation can have any of several consequences—or it may have no obvious effect at all on the phenotype, acting as a silent mutation.

Missense and Nonsense Mutations

A point mutation that changes a codon that normally specifies a particular amino acid into one that codes for a different amino acid is called a **missense mutation**. If the substituted amino acid alters the protein's conformation significantly or occurs at a site critical to its function, signs or symptoms of disease or an observable variant of a trait may result. About a third of missense mutations harm health. **Figure 12.9a** shows a missense mutation in the gene encoding the low-density lipoprotein (LDL) receptor that is abnormal in familial hypercholesterolemia (see figure 5.2); also, figure 12.3 depicts the missense mutation behind sickle cell disease.

A point mutation that changes a codon specifying an amino acid into a "stop" codon—UAA, UAG, or UGA in mRNA—is a **nonsense mutation**, shown in **Figure 12.9b**. A premature stop codon is one that occurs before the natural end of the gene. It shortens the protein product, which can greatly influence the phenotype. For example, in factor XI deficiency, which is a blood clotting disorder, a GAA codon specifying glutamic acid is changed to UAA, signifying "stop." The shortened clotting factor cannot halt the profuse bleeding that occurs during surgery or from an injury. Nonsense mutations are predictable because we can identify which codons can mutate to become a "stop" codon. Cells have a response to shortened proteins called **nonsense-mediated decay** that destroys mRNAs with premature stop codons. This response is protective because some shortened proteins can cause a "gain-of-function" and damage the cell beyond the absence of the normal function.

In the opposite of a nonsense mutation, a normal stop codon mutates into a codon that specifies an amino acid. The resulting protein is too long, because translation continues through what is typically a stop codon.

Point mutations can control transcription, affecting the quantity rather than the quality of a protein. For example, in 15 percent of people who have Becker muscular dystrophy, the muscle protein dystrophin is normal, but its levels are reduced. The mutation is in the promoter for the dystrophin gene. This slows transcription, and dystrophin protein is scarce. The other 85 percent of individuals who have Becker muscular dystrophy have shortened proteins, rather than too few normal-length proteins.

Splice-Site Mutations

A point mutation can greatly affect a gene's product if it alters a site where introns are normally removed from the mRNA. This is

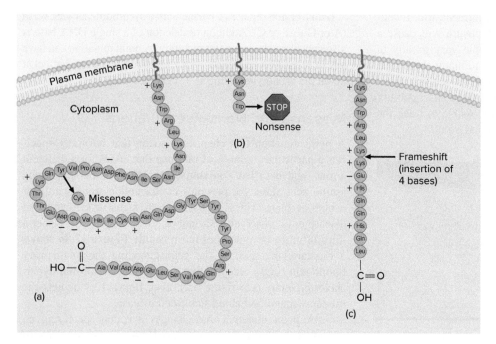

Figure 12.9 Different mutations in a gene can cause the same disease. In familial hypercholesterolemia, several types of mutations alter the LDL receptor normally anchored in the plasma membrane. LDL receptor **(a)** bears a missense mutation—a cysteine substitutes for a tyrosine, bending the receptor enough to impair its function. The short LDL receptor in **(b)** results from a nonsense mutation, in which a stop codon replaces a tryptophan codon. In **(c)**, a 4-base insertion adds an amino acid and then alters the reading frame; + and − indicate electrical charges on the amino acids.

called a **splice-site mutation**. It can affect the phenotype if an intron is translated into amino acids, or if an exon is skipped instead of being translated, shortening the protein.

Retaining an intron is unusual because most introns have stop codons in all reading frames. However, if a stop codon is not encountered, a retained intron adds bases to the protein-coding part of an mRNA. For example, in a family with severe cystic fibrosis, a mutation alters an intron site so that it is not removed from the mRNA. The extra amino acids make the encoded ion channel protein too bulky to move to its normal position in the plasma membrane (see figure 1.3).

The exome includes only exons. Because of this, whole exome sequencing would not detect a mutation that disrupts an intron. This was the case for a child who had a single-base mutation in an intron of a gene called *RTTN,* which encodes a protein called rotatin. The child had a very small head, large face, dwarfism, and reduced brain folds. Tests for 11 known conditions with these symptoms were negative, and exome sequencing did not find a mutation that could explain the condition. The child had a point mutation in a splice site. This led to inclusion of the amino acids encoded in part of an intron until a stop codon ended translation of the extended protein.

A missense mutation that creates an intron splicing site where there should not be one can cause **exon skipping**, which removes a few contiguous amino acids from the protein product. An entire exon is "skipped" when the mRNA is translated into protein, as if it were an intron. It is a little like leaving out

a word when cutting and pasting a sentence in a document. An exon-skipping mutation is a deletion at the mRNA level, but it is a point (single-base) mutation at the DNA level. For example, a disease called familial dysautonomia (FD) can result from exon skipping in the gene encoding an enzyme necessary for the survival of certain neurons that control sensation and involuntary responses. Symptoms include pneumonia, vomiting and retching, extremely high fevers, chills, rapid heartbeat, rashes, and seizures. FD also reduces sensations of pain, heat, and cold, and impairs motor skills, affecting eating, swallowing, and breathing.

A peculiarity of some diseases caused by exon skipping, including FD, is that some cells ignore the mutation and manufacture a normal protein from the affected gene—possible because the amino acid sequence information is still there. Depending upon which cells actually make the full encoded protein, the phenotype may be less severe than in individuals with the same disease but with a different type of mutation in an exon. In FD, cells in which the exon is skipped are the cells that contribute to symptoms. That is, many cells from the brain and spinal cord skip the exon, but cells from muscle, lung, liver, white blood cells, and glands produce normal-length proteins. For FD, clinical trials are testing compounds that restore retention of the skipped exon in mRNAs in neurons of affected children.

The natural exon skipping seen in FD inspired development of drugs to treat Duchenne muscular dystrophy (DMD). One drug consists of synthetic DNA-like molecules that are complementary, or "antisense," to part of the huge dystrophin gene near a mutation that causes the disease by altering the reading frame (a frameshift mutation, see figure 10.13). A frameshift mutation results in a shortened protein, as shown in **Figure 12.9c**. The drug causes transcription to skip the exon containing the mutation in a way that restores the reading frame, so that skeletal muscle cells produce some of the protein. The drug was approved because it leads to dystrophin production, but research continues to investigate effects on mobility in patients over time and how to deliver the drug to enough muscle cells to make a difference in quality of life and survival.

Deletions and Insertions

In genes, the number 3 is very important, because triplets of DNA bases specify amino acids. Adding or deleting a number of bases that is not a multiple of 3 devastates a gene's function because it disrupts the gene's reading frame. Most exons are "readable" (have no stop codons) in only one of the three possible reading frames. Recall from chapter 10 that a reading frame that is readable because it is translatable into protein is called an open reading frame.

A **deletion mutation** removes DNA. A deletion that removes three bases or a multiple of three bases will not cause a frameshift, but can still alter the phenotype. Deletions range from a single DNA nucleotide to thousands of bases to large parts of chromosomes. An **insertion mutation** adds DNA, which can offset the reading frame. In one form of Gaucher disease, for example, an inserted single DNA base prevents production of an enzyme that normally breaks down glycolipids in lysosomes (see figure 2.6). The resulting buildup of glycolipid enlarges the liver and spleen and causes easily fractured bones and neurological impairment.

One type of insertion mutation repeats part of a gene's sequence. The insertion is usually adjacent or close to the original sequence, like a keying error repeating a word word. Two complete copies of a gene next to each other is a type of mutation called a **tandem duplication**. A form of Charcot-Marie-Tooth disease, which causes numb hands and feet, results from a 1.5-million-base-long tandem duplication.

Pseudogenes and Transposons Revisited

Recall from chapter 11 that a pseudogene is a DNA sequence that is very similar to the sequence of a protein-encoding gene. A pseudogene is not translated into protein, although it may be transcribed. The pseudogene may have descended from a gene sequence that was duplicated when DNA strands misaligned during meiosis (similar to the situation depicted in figure 12.8 for the alpha globin gene). When this happens, a gene and its copy end up next to each other on the chromosome. The original gene or the copy then mutates to such an extent that it is no longer functional and becomes a pseudogene. Its partner lives on as the functional gene.

Even though a pseudogene is not translated, it can interfere with the expression of the functional gene and cause a mutation. For example, some cases of Gaucher disease result from a crossover between the working gene and its pseudogene, which has 96 percent of the same sequence and is located 16,000 bases away. The result is a fusion gene, which is part functional gene and part pseudogene. The fusion gene does not retain enough of the normal gene sequence to enable the cell to synthesize the encoded enzyme, and Gaucher disease results.

Chapter 11 also discussed transposons. These "jumping genes" can alter gene function in several ways. They can disrupt the site they jump from, shut off transcription of the gene they jump into, or alter the reading frame if they are not a multiple of three bases. For example, a boy with X-linked hemophilia A had a transposon in his factor VIII gene—a sequence that was also in his carrier mother's genome, but on her chromosome 22. Apparently, in the oocyte, the transposon jumped from chromosome 22 into the factor VIII gene on the X chromosome, eventually causing the boy's hemophilia.

Expanding Repeats

In a type of mutation called an **expanding repeat**, a gene actually grows as a small part of the DNA sequence is copied and added. The family described in the opener to chapter 4 offers an extreme example of the devastation that an expanding repeat mutation can cause.

One of the first expanding repeat diseases studied was myotonic dystrophy, which begins earlier and causes more severe symptoms from one generation to the next. A grandfather might experience mild weakness in his forearms, but his daughter may have more noticeable arm and leg weakness and a flat facial expression. Her children might have severe muscle weakness.

For many years, clinicians thought that the "anticipation"—the worsening of symptoms over generations—was psychological. Then, with the ability to sequence DNA, researchers found that the gene expands! The gene that causes the first recognized type of myotonic dystrophy, on chromosome 19, has an area rich in CTG repeats (GAC mRNA repeats). The wild type repeat number is 5 to 37; a person with the disease has from 50 to thousands of copies (**figure 12.10**).

Expanding triplet (also called trinucleotide) repeats have been discovered to underlie more than 15 diseases. Usually, a repeat number of fewer than 40 copies is stably transmitted to the next generation and doesn't produce symptoms. Larger repeats are unstable, growing with each generation and causing symptoms that are more severe and begin sooner. **Clinical Connection 12.1** describes the triplet repeat condition fragile X syndrome.

The mechanism behind triplet repeat diseases lies in the DNA sequence. The bases of the repeated triplets implicated in the expansion diseases, unlike others, bond to each other in ways that bend the DNA strand into shapes, such as

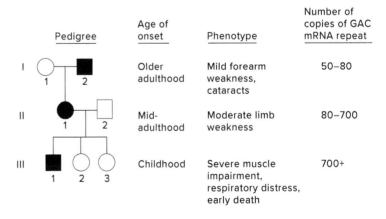

Myotonic Dystrophy

Pedigree	Age of onset	Phenotype	Number of copies of GAC mRNA repeat
I 1 2	Older adulthood	Mild forearm weakness, cataracts	50–80
II 1 2	Mid-adulthood	Moderate limb weakness	80–700
III 1 2 3	Childhood	Severe muscle impairment, respiratory distress, early death	700+

Figure 12.10 Expanding genes explain anticipation. In some diseases, symptoms that worsen from one generation to the next—termed *anticipation*—have a physical basis: The gene is expanding as the number of repeats increases.

Fragile X Mutations Affect Boys and Their Grandfathers

Fragile X syndrome is the most common inherited form of intellectual disability and accounts for 3 percent of autism cases. In the 1940s, geneticists thought that a gene on the X chromosome caused intellectual disability, which was then termed mental retardation, because more affected individuals were male. In 1969, a clue emerged to the genetic basis of X-linked intellectual disability. Two brothers with the condition and their mother had an unusual X chromosome. The tips at one chromosome end dangled by a thin thread (**figure 12A**). When grown in culture medium lacking folic acid, this part of the X chromosome was prone to breaking—hence, the name fragile X syndrome. Worldwide, it affects 1 in 2,000 males, accounting for 4 to 8 percent of all males with intellectual disability. One in 4,000 females is affected. They usually have milder cases because their cells have normal X chromosomes, too.

In fragile X syndrome, MRI scans show brain differences as early as age 2. By young adulthood, the face becomes long and narrow, with a pronounced jaw and protruding ears; and the testicles are oversized. Affected individuals may have learning disabilities, repetitive speech, hyperactivity, shyness, social anxiety, a short attention span, language delays, and temper outbursts.

Fragile X syndrome is inherited in an unusual pattern. The syndrome should be transmitted like any X-linked trait, from carrier mother to affected son. However, penetrance is incomplete. One-fifth of males who inherit the chromosomal abnormality have no symptoms. Because they transmit the affected chromosome to all their daughters—half of whom have some mental impairment—they are called "transmitting males." A transmitting male's grandchildren may inherit fragile X syndrome.

A triplet repeat mutation causes fragile X syndrome. In unaffected individuals, the fragile X area contains 29 or 30 repeats of the sequence CGG, in a gene called the fragile X mental retardation gene (*FMR1*). In people who have the fragile chromosome and show its effects, this region is expanded to 200 to 2,000 CGG repeats. Transmitting males, as well as females who have mild symptoms or have affected sons, may have a "premutation" of 55 to 200 repeats. People with the premutation may develop mild neurological problems, such as tremors and poor balance. About a fifth of women with the premutation have infertility due to ovarian failure.

The *FMR1* gene encodes fragile X mental retardation protein (FMRP). This protein, when abnormal, binds to and disables several types of mRNA molecules whose encoded proteins are crucial for brain neuron function.

Mysteries remain about fragile X syndrome. A distinct type of disease is seen in the maternal grandfathers of boys who have fragile X syndrome. Clinicians noticed that mothers of boys with fragile X syndrome often reported the same symptoms in their fathers—tremors, balance problems, and then cognitive or psychiatric difficulties. The grandfathers were sometimes misdiagnosed with Parkinson disease due to the tremors and their age. However, Parkinson patients can walk a straight line, but the grandfathers could not. The grandfathers' symptoms worsened with time and sometimes led to premature death.

Further investigation of the grandfathers led to the description of fragile X–associated tremor/ataxia syndrome (FXTAS) (**table 12A**). (Ataxia is poor balance and coordination.) The discovery of FXTAS has genetic counseling implications. As neurologists learn to distinguish this disease from others, such as Parkinson disease, daughters can be counseled that they might pass the condition on to sons and be offered testing.

Table 12A	Prevalence of FXTAS in Grandfathers of Fragile X Syndrome Grandsons
Age	**Prevalence**
50s	17%
60s	38%
80+	75%

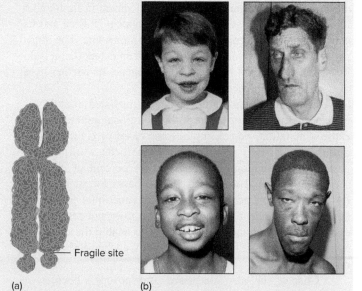

Fragile site

(a)   (b)

Figure 12A **Fragile X syndrome.** A fragile site on the tip of the long arm of the X chromosome **(a)** is associated with intellectual disability and a characteristic long face that becomes pronounced with age **(b).** *(b): From R. Simensen, R. Curtis Rogers, "Fragile X Syndrome," American Family Physician, 39:186 May 1989. @ American Academy of Family Physicians.*

hairpins. These shapes then interfere with DNA replication, which extends the expansion. Once the repeats are translated, the extra-long proteins harm cells in several ways:

- They bind to parts of transcription factors that have stretches of amino acid repeats similar to or matching the expanded repeat.
- They block proteasomes, enabling misfolded proteins to persist.
- They trigger apoptosis.

Triplet repeat proteins may also enter the nucleus, even though their wild type versions function only in the cytoplasm, or vice versa.

The triplet repeat diseases cause a "dominant toxic gain-of-function," illustrated in figure 4.8. They cause something novel to happen, rather than removing a function, such as occurs with a recessive enzyme deficiency. The idea of a gain-of-function arose from the observation that people who have deletions of these genes do not have associated symptoms. Several triplet repeat conditions are "polyglutamine diseases" that are due to repeats of the mRNA codon CAG, which encodes the amino acid glutamine.

For some triplet repeat diseases, the mutation blocks gene expression before a protein is even made. In myotonic dystrophy type 1, the expansion is in the initial untranslated region of the gene on chromosome 19, resulting in a huge mRNA. When genetic testing became available for the disease, researchers discovered a second form of the illness in patients who had wild type alleles of the gene on chromosome 19. They have myotonic dystrophy type 2, caused by an expanding *quadruple* repeat of CCTG in a gene on chromosome 3. Affected individuals have more than a hundred copies of the repeat, compared to the normal maximum of 10 copies.

Copy Number Variants

Recall from chapter 7 that a copy number variant (CNV) is a different form of information than a DNA sequence variant. A CNV is a specific DNA sequence that varies in number of copies from person to person. Our genomes have hundreds to thousands of them. Sequencing the first human genomes missed CNVs because the technology could then detect a DNA sequence only once. It was a little like searching this book for the word "variant," and not the number of times it is used.

A language analogy is useful to distinguish point mutations and single nucleotide polymorphisms (SNPs) from CNVs. If a wild type short sequence and a variant with two SNPs are written as:

> *The sad rat sat on a red cat*   (wild type)
>
> *The sad rat sat in a red hat*   (two SNPs)

then the sequence with two CNVs might be:

> *The sad sad rat sat on a red red red cat*

CNVs may contribute significantly to the differences among us. A CNV can range in size from a few DNA bases to millions, and copies may lie next to each other on a chromosome ("tandem") or might be far away—even parts of other chromosomes. Duplications, deletions, and triplet repeats are types of copy number variants.

CNVs may have no effect on the phenotype, or they can disrupt a gene's function and harm health. A CNV may have a direct effect by inserting into a protein-encoding gene and offsetting its reading frame or have an indirect effect by destabilizing surrounding sequences. CNVs are particularly common among people who have behavioral disorders, such as attention deficit hyperactivity disorder (ADHD), autism, and schizophrenia.

Key Concepts Questions 12.5

1. What is a point mutation?
2. Distinguish between a transversion and a transition.
3. Distinguish between a missense mutation and a nonsense mutation.
4. Explain how mutations outside the coding region of a gene can affect its function.
5. How do insertions and deletions affect gene function?
6. How do pseudogenes and transposons affect gene function?
7. Describe expanding repeat mutations and their effects.
8. What are copy number variants?

12.6 The Importance of Position

The degree to which a mutation alters a phenotype depends upon where in the gene the change occurs, and how the mutation affects the folding, conformation, activity, or abundance of an encoded protein. A mutation that replaces an amino acid with a similar one would probably not affect the phenotype greatly, because it wouldn't substantially change the conformation of the protein. Even substituting a very different amino acid would not have much effect if the change is in part of the protein not crucial to its function. The effects of specific mutations are well studied in hemoglobin.

Globin Gene Variants

Hundreds of globin gene mutations are known. Mutations in these genes can cause anemia with or without sickling, or cause cyanosis (a blue pallor due to poor oxygen binding). Rarely, a mutation boosts the molecule's affinity for oxygen. Some globin gene variants exert no effect at all and are considered "clinically silent" (**table 12.5**).

Different mutations at the same site in a gene can have different effects. For example, hemoglobin S and hemoglobin C result from mutations that change the sixth amino acid in the beta globin polypeptide, but in different ways. Homozygotes for hemoglobin S have sickle cell disease, yet homozygotes for hemoglobin C are healthy. Both types of homozygotes are resistant to malaria because the unusual hemoglobin alters the shapes and surfaces of red blood cells in ways that keep out the parasite that causes the infectious illness (see figure 15.13).

A rare but very noticeable condition of abnormal hemoglobin affects the "blue people of Troublesome Creek" (**figure 12.11**). Seven generations ago, in 1820, a French orphan named Martin Fugate who settled in this area of Kentucky brought in an autosomal recessive mutation that causes methemoglobinemia type I, also known as "blue person disease." Martin's mutation was in the *CYP5R3* gene, which encodes an enzyme (cytochrome b_5 methemoglobin reductase) that normally catalyzes a reaction that converts a type of hemoglobin with poor oxygen affinity, called methemoglobin, back into normal hemoglobin by adding an electron. Martin was a heterozygote but still slightly bluish. His wife, Elizabeth Smith, was also a carrier for this rare disease, and four of their seven

Figure 12.11 **The Fugate family of Troublesome Creek, Kentucky, have a mutation that affects hemoglobin.** Their skin is blue. Artist Walt Spitzmiller painted this haunting image based on photographs. *"The Blue People," copyright © 1982–2016 by Walt Spitzmiller. All Rights Reserved*

children were blue. After extensive inbreeding in the isolated community—their son married his aunt, for example—a large pedigree of "blue people" of both sexes arose. (Exposure to certain drugs causes a noninherited form of the condition.)

In "blue person disease," excess oxygen-poor hemoglobin causes a dark blue complexion. Carriers may have bluish lips and fingernails at birth, which usually lighten. Treatment is simple: A tablet of methylene blue, a commonly used dye, or of ascorbic acid (vitamin C) for mild cases, adds the electron back to methemoglobin, converting it to normal hemoglobin.

In most members of the Fugate family, blueness was the only symptom. Normally, less than 1 percent of hemoglobin molecules are the methemoglobin form, which binds less oxygen. The Fugates had 10 to 20 percent in this form. People with the inherited condition who have more than 20 percent

Table 12.5	Globin Mutations	
Associated Phenotype	**Name**	**Mutation**
Clinically silent	Hemoglobin (Hb) Wayne	Single-base deletion in alpha globin gene causes frameshift, changing amino acids 139–141 and adding amino acids
	Hb Grady	Nine extra bases add three amino acids between amino acids 118 and 119 of alpha chain
Oxygen binding	Hb Chesapeake	Change from arginine to leucine at amino acid 92 of beta chain
	Hb McKees Rocks	Change from tyrosine to STOP codon at amino acid 145 in beta chain
Anemia	Hb Constant Spring	Change from STOP codon to glutamine elongates alpha chain
	Hb S	Change from glutamic acid to valine at amino acid 6 in beta chain causes sickling
	Hb Leiden	Amino acid 6 deleted from beta chain
Protection against malaria	Hb C	Change from glutamic acid to lysine at amino acid 6 in beta chain causes sickling

methemoglobin may suffer seizures, heart failure, and even death. The disease remains rare. It is still seen in the Kentucky family, among certain families in Alaska and Algeria, and among Navajo Indians.

Factors That Lessen the Effects of Mutation

Mutation is a natural consequence of DNA's ability to change. This flexibility is essential for evolution because it generates new gene variants, some of which may resist environmental change and enable a population or even a species to survive. However, many factors minimize the negative effects of mutations on phenotypes.

The genetic code protects against mutation to an extent. Recall from chapter 10 that synonymous codons specify the same amino acid. Mutation in the third position in a codon is called "silent" when the mutated and original codons are synonymous. For example, a change from RNA codon CAA to CAG does not change the encoded amino acid, glutamine, so a protein whose gene contains the change would not change. However, synonymous codons can affect splicing out of introns and mRNA stability differently. Therefore, expression of genes that have mutations that result in synonymous codons can differ.

Other genetic code nuances prevent synthesis of highly altered proteins. For example, mutations in the second position in a codon sometimes replace one amino acid with another that has a similar conformation, minimizing disruption of the protein's shape. GCC mutated to GGC, for instance, replaces alanine with equally small glycine.

A **conditional mutation** affects the phenotype only under certain circumstances. This can be protective if an individual avoids the exposures that trigger symptoms. Consider a common variant of the X-linked gene that encodes glucose-6-phosphate dehydrogenase (G6PD), an enzyme that immature red blood cells use to extract energy from glucose. One hundred million people worldwide have G6PD deficiency. It can cause life-threatening hemolytic anemia, but only under rather unusual conditions—eating fava beans or taking certain antimalarial drugs (**figure 12.12**).

Another protection against mutation is in stem cells. When a stem cell divides to yield another stem cell and a progenitor

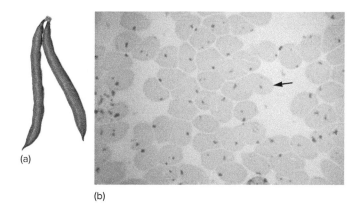

Figure 12.12 **Sickness and circumstance.** A conditional mutation causes some cases of G6PD deficiency hemolytic anemia. Exposure to two biochemicals in fava beans **(a)** unfolds hemoglobin molecules, causing spots called Heinz bodies **(b)** and bending red blood cells out of shape until they burst. *(a): © Burke Triolo Productions/Getty Images RF; (b): © CNRI/SPL/Custom Medical Stock Photo*

or differentiated cell, the oldest DNA strands segregate with the stem cell, and the most recently replicated DNA strands go to the more specialized daughter cells (see figure 2.18). This makes sense in organs where stem cells actively yield specialized daughter cells, such as the skin and small intestine. Because mutations occur when DNA replicates, this skewed distribution of chromosomes sends the DNA most likely to harbor mutations into cells that will soon be shed (from a towel rubbed on skin or in a bowel movement) while keeping mutations away from the stem cells that must continually regenerate the tissues.

Key Concepts Questions 12.6

1. What characteristics of a mutation determine its effects?
2. Describe a mutation in a globin gene.
3. Name three phenomena that lessen the effects of mutations.

12.7 DNA Repair

Any manufacturing facility tests a product in several ways to see whether it has been assembled correctly. Mistakes in production are rectified before the item goes on the market—most of the time. The same is true for a cell's manufacture of DNA.

Damage to DNA becomes important when the genetic material is replicated, because the error is passed on to daughter cells. In response to damage, the cell may die by apoptosis or it may repair the error. If the cell doesn't die or the error is not repaired, cancer may result. Fortunately, DNA replication is very accurate—only 1 in 100 million or so bases is incorrectly incorporated. This is quite an accomplishment, because DNA replicates approximately 10^{16} times during an average human lifetime. However, most such mutations occur in somatic cells, and do not affect the phenotype.

DNA polymerase as well as "DNA damage response" genes oversee the accuracy of replication. In DNA repair, a cell detects damage and then signaling systems in the cell respond by repairing the damage or signaling apoptosis to kill the cell. More than 50 DNA damage response genes have been identified. Mitochondrial DNA cannot repair itself, which contributes to its higher mutation rate.

Many types of organisms repair their DNA, some more efficiently than others. The master at DNA repair is a large, reddish microbe, *Deinococcus radiodurans*. It tolerates 1,000 times the radiation level that a person can, and it can even live amidst the intense radiation of a nuclear reactor. The bacterium realigns its radiation-shattered pieces of DNA. Then enzymes bring in new nucleotides and assemble the pieces.

Types of DNA Repair

Exposure to radiation is a fact of life. The Earth, since its beginning, has been periodically bathed in UV radiation. Volcanoes, comets, meteorites, and supernovas all depleted ozone in the atmosphere, which allowed ultraviolet wavelengths of light to reach organisms. The longer UV wavelengths—UVA—are not dangerous, but the shorter UVB wavelengths damage DNA by forming an extra covalent bond between adjacent (same-strand) pyrimidines, particularly thymines. The linked thymines are called thymine dimers. Their extra bonds kink the double helix enough to disrupt replication and permit insertion of a noncomplementary base. For example, an A might be inserted opposite a G or C, instead of opposite a T. Thymine dimers also disrupt transcription.

Early in the evolution of life, organisms that could survive UV damage had an advantage. Enzymes enabled them to do this, and because enzymes are gene-encoded, DNA repair came to persist.

In many modern species, three types of DNA repair mechanisms check the genetic material for mismatched base pairs. In the first type, enzymes called photolyases absorb energy from visible light and use it to detect and bind to pyrimidine dimers, then break the extra bonds. This type of repair, called photoreactivation, enables UV-damaged fungi to recover with exposure to sunlight. Humans do not have this type of DNA repair.

In the early 1960s, researchers discovered a second type of DNA self-mending, called **excision repair**, in mutant *E. coli* that were unable to repair UV-induced DNA damage. Enzymes cut the bond between the DNA sugar and base and excise the pyrimidine dimer and surrounding bases. Then, a DNA polymerase fills in the correct nucleotides, using the exposed template as a guide. DNA polymerase also detects and corrects mismatched bases in newly replicated DNA.

Humans have two types of excision repair. **Nucleotide excision repair** replaces up to 30 nucleotides and removes errors that result from several types of causes, including exposure to chemical carcinogens, UVB in sunlight, and oxidative damage (**figure 12.13**). Thirty different proteins carry out nucleotide excision repair.

The second type of excision repair, **base excision repair**, replaces one to five nucleotides at a time, but specifically corrects errors that result from oxidative damage. Oxygen free radicals are highly reactive forms of oxygen that arise during chemical reactions such as those of metabolism and transcription. Free radicals damage DNA. Genes that are actively transcribed face greater oxidative damage from free radicals; base excision repair targets this type of damage.

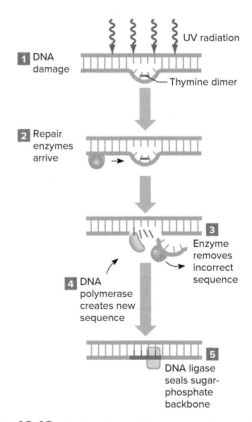

Figure 12.13 **Nucleotide excision repair.** Human DNA damaged by UV light is corrected by nucleotide excision repair, which removes and replaces the pyrimidine dimer and a few surrounding bases.

A third mechanism of DNA sequence correction is **mismatch repair**. Enzymes "proofread" newly replicated DNA for small loops that emerge from the double helix. The enzymes excise the mismatched base and replace it with the correct one (**figure 12.14**). The small loops form where the two strands do not precisely align, but instead slip and misalign. This happens where very short DNA sequences repeat. These sequences, called microsatellites, are scattered throughout the genome. Like the minisatellites mentioned in section 12.4, microsatellite lengths can vary from person to person, but within an individual, they are usually the same length. Excision and mismatch repair differ in the cause of the error—UV-induced pyrimidine dimers versus replication errors—and in the types of enzymes involved.

Excision repair and mismatch repair in human cells relieve the strain on thymine dimers or replace incorrectly inserted bases. Another form of repair can heal a broken sugar-phosphate backbone in both DNA strands, which can result from exposure to ionizing radiation or oxidative damage. Such a double-stranded break is especially damaging because it breaks a chromosome, which can cause cancer. At least two

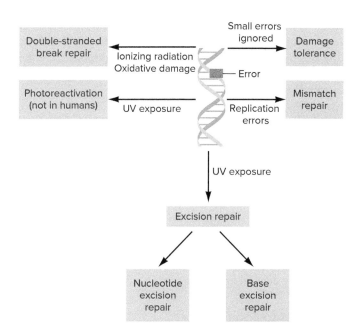

Figure 12.15 **DNA repair mechanisms.**

types of multiprotein complexes reseal the sugar-phosphate backbone, either by rejoining the broken ends or recombining with DNA on the unaffected homolog. Double-stranded breaks and their natural repair form the basis of genome editing technologies, discussed in section 19.4 and section 20.4.

In yet another type of DNA repair called damage tolerance, a "wrong" DNA base is left in place, but replication and transcription proceed. "Sloppy" DNA polymerases, with looser adherence to the base-pairing rules, read past the error, randomly inserting any other base. It is a little like retaining a misspelled word in a sentence—usually the meaning remains clear.

Figure 12.15 summarizes DNA repair mechanisms.

DNA Repair Disorders

The ability to repair DNA is crucial to health. If both copies of a repair gene are mutant, a disorder can result. Heterozygotes who have one mutant repair gene may be more sensitive to damage from environmental factors, such as toxins and radiation.

A well-studied DNA repair gene encodes a protein called p53. It controls whether DNA is repaired and the cell salvaged, or the cell dies by apoptosis (see figure 18.3). Signals from outside the cell activate p53 proteins to aggregate into complexes of four molecules. These quartets bind certain genes that slow the cell cycle, enabling repair to take place. If the damage is too severe, the p53 protein quartets instead increase the rate of transcription of genes that promote apoptosis, and the cell dies.

In DNA repair disorders, chromosome breakage persists. Mutations in repair genes greatly increase susceptibility to certain types of cancer following exposure to ionizing radiation or chemicals that affect cell division. Cancers develop as errors in the DNA sequence accumulate and are perpetuated to a much greater extent than they are in people with functioning

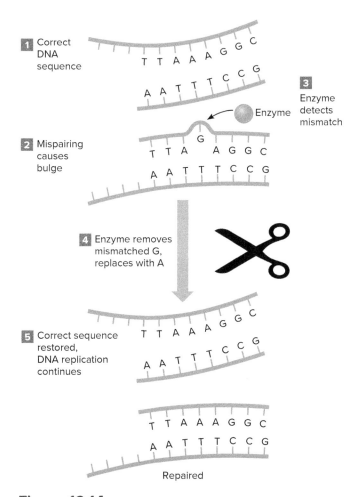

Figure 12.14 **Mismatch repair.** In this form of DNA repair, enzymes detect loops and bulges in newly replicated DNA that indicate mispairing. The enzymes correct the error. Highly repeated sequences are more prone to this type of error.

repair systems. We conclude this chapter with a closer look at repair disorders (they are generally termed disorders and not diseases because the symptoms do not arise directly from the mutations). Chapter 18 discusses other cancers that arise from disrupted DNA repair.

Trichothiodystrophy

At least five genes can cause trichothiodystrophy. At its worst, this condition causes dwarfism, intellectual disability, and brittle hair and scaly skin, both with low sulfur content. Although the child may appear to be normal for a year or two, growth soon slows dramatically, signs of premature aging begin, and life ends early. Hearing and vision may fail. Symptoms reflect accumulating oxidative damage. Individuals have faulty nucleotide excision repair, base excision repair, or both. The trichothiodystrophies are unusual in that they do not increase cancer risk.

Inherited Colon Cancer

Hereditary nonpolyposis colon cancer (HNPCC), a group of seven disorders also known as Lynch syndrome, was linked to a DNA repair defect when researchers discovered different-length short repeated sequences of DNA (microsatellites) within an individual. People with this type of colon cancer have a breakdown of mismatch repair, which normally keeps a person's microsatellites all the same length. HNPCC affects 1 in 200 people and accounts for 3 percent of newly diagnosed colorectal cancers. Genetic testing in all people newly diagnosed with colon cancer is advised because if they have a germline mutation (that is, in all of their cells), their relatives can be tested to see if they inherited the mutation, too. If healthy relatives test positive, frequent colonoscopies can detect disease early, at a more treatable stage. Penetrance of HNPCC is about 45 percent by age 70—considered a high cancer risk.

Xeroderma Pigmentosum (XP)

A child with XP must stay indoors in artificial light, because even the briefest exposure to sunlight causes painful blisters. Failing to cover up and use sunblock can cause skin cancer (**figure 12.16**), which more than half of all children with XP develop before adolescence. People with XP have a 1,000-fold increased risk of developing skin cancer compared to others, and a 10-fold increased risk of developing tumors inside the body.

XP is autosomal recessive, and results from mutations in any of seven genes. It can reflect malfunction of nucleotide excision repair or deficient "sloppy" DNA polymerase, both of which allow thymine dimers to stay and block replication. One of the genes that causes XP, when mutant, also causes trichothiodystrophy and another disease, Cockayne syndrome. The different symptoms arise from the different ways that mutations disrupt the encoded protein, which is a helicase that helps unwind replicating DNA.

Only about 250 people in the world are known to have XP. A family living in upstate New York runs a special

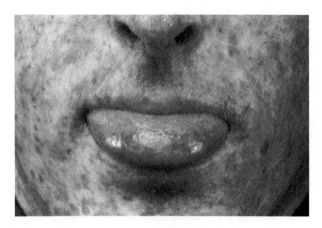

Figure 12.16 A DNA repair disorder. The marks on this person's face result from sun exposure. Xeroderma pigmentosum, an impairment of excision repair, makes the skin highly sensitive to the sun. © *Wellcome Image Library/Custom Medical Stock Photo*

summer camp for children with XP, where night is turned into day. Activities take place at night, or in special areas where the windows are covered and light comes from low-ultraviolet incandescent lightbulbs.

Ataxia Telangiectasia (AT)

This multisymptom condition is the result of a defect in a kinase that functions as a cell cycle checkpoint (see figure 2.15). In AT, cells proceed through the cell cycle without pausing just after replication to inspect the new DNA and to repair any mispaired bases. Some cells die through apoptosis if the damage is too great to repair. Because of the malfunctioning cell cycle, individuals who have this autosomal recessive disorder have 50 times the risk of developing cancer, particularly of the blood. About 40 percent of individuals with AT have cancer by age 30. Additional symptoms include poor balance and coordination (ataxia), red marks on the face (telangiectasia), delayed sexual maturation, and high risk of infection and diabetes mellitus.

AT is rare, but heterozygotes are not. They make up from 0.5 to 1.4 percent of various populations. Carriers may have mild radiation sensitivity, which causes a two- to sixfold increase in cancer risk over that of the general population. For people who are AT carriers, dental or medical X rays may cause cancer.

DNA's changeability, so vital for evolution of a species, comes at the cost of occasional harm to individuals. Each of us harbors about 175 new mutations, many old ones, and many polymorphisms. The **Bioethics** box considers a situation in which learning about gene variants can be confusing.

Key Concepts Questions 12.7

1. Describe the different types of DNA repair mechanisms.
2. Describe disorders that result from faulty DNA repair.

A Diagnostic Dilemma: What to Do with a VUS?

"I have a *what*?" Ellie asked her genetic counselor, tearing up.

"You have a variant of uncertain significance. We call it a VUS. It's a bit confusing, but please let me explain."

Ellie had taken a genetic test to find out if she had a mutation in either the *BRCA1* or *BRCA2* cancer susceptibility genes, because she has an aunt with ovarian cancer and two cousins who had breast cancer. Although the family is in a high-risk group (Ashkenazi Jewish), none of Ellie's relatives has had genetic testing. She had prepared for either a "*yes you have a mutation*" or "*no you don't*" answer—but not this.

"One of your two *BRCA1* genes has an unusual DNA sequence. It doesn't match any of the mutations that we know confers cancer susceptibility, but it is not the 'normal,' or wild type sequence. We just don't know," said the counselor.

"And what am I supposed to do with that information?" Ellie asked, a bit angrily. She had been considering surgery to remove her breasts and ovaries if she had a mutation. Now she didn't know what to do.

All genes vary in many ways, and some variants affect the encoded protein and the phenotype, and others do not. As a result, some gene variants are much more dangerous than others. The *BRCA1* and *BRCA2* mutations that are more common among the Ashkenazim delete one or two DNA bases, which offsets the reading frame and introduces inappropriate "stop" codons. The encoded proteins are too short to function as they normally would in the mismatch repair reactions that prevent cancer. It was good news that Ellie didn't have those mutations—but what *did* she have?

The genetic counselor explained that as researchers identify specific gene variants and associate them with diseases because they are found exclusively in people with the disease, databases are built and tools such as the Matchmaker Exchange (see chapter opener) are consulted to interpret what gene variants actually mean. Until all gene variants are known and described, clinicians are using five categories of risk **(figure 12B)**. A gene variant is one of the following:

- pathogenic
- likely pathogenic
- a variant of uncertain significance
- likely benign
- benign

Figure 12B **Gene variants are classified as pathogenic, likely pathogenic, of uncertain significance, likely benign, or benign.** © *mediacolor's/Alamy*

More than 2,000 variants are known for the *BRCA1* and *BRCA2* genes. About 15 percent of people having the genes sequenced receive the same news as Ellie did—the middle category, the mysterious VUS.

The VUS issue will resolve with time. It is a bioinformatics challenge that requires identifying all gene variants that nearly always track with a suspected clinical condition, and almost never with people who don't have the condition. But until then, a "VUS" situation can be upsetting.

Questions for Discussion

1. At which of the five levels of variant interpretation should a patient be informed about the test results? Cite a reason for your answer.

2. Explain how the number of variants of uncertain significance might increase for a time and then decrease.

3. Should a nurse, physician, or genetic counselor deliver news of a VUS? Cite a reason for your answer.

4. What would you want to know before considering surgery or another type of procedure to prevent a genetic condition that has not yet produced symptoms?

Summary

12.1 The Nature of Gene Variants

1. A **mutation** is a change in a gene's base sequence that is rare in a population and can cause a **mutant** phenotype.

2. A mutation disrupts the function or abundance of a protein or introduces a new function. Most loss-of-function mutations are recessive, and most gain-of-function mutations are dominant.

3. A **polymorphism** is a more common and typically less harmful genetic change. Mutations and polymorphisms are types of gene variants (alleles).

4. A **germline mutation** occurs during DNA replication before meiosis, affects all cells of an individual, and can be transmitted to the next generation in gametes. A **somatic mutation** occurs before mitosis and affects a subset of cells.

5. A severe genetic disease may be seen only in individuals with **somatic mosaicism**.

12.2 A Closer Look at Two Mutations

6. Sickle cell disease results from a single base mutation in the beta globin gene, and beta thalassemia results from too few beta globin chains.

7. Many medical conditions result from mutations in collagen genes because of the precise conformation of the protein and its pervasiveness in the body.

12.3 Allelic Diseases

8. Whether different mutations in a gene cause the same or distinct illnesses varies; the naming of diseases associated with mutations was inconsistent in the past.

9. **Allelic diseases** have different phenotypes but result from mutations in the same gene.

12.4 Causes of Mutation

10. A spontaneous mutation arises due to chemical damage or to an error in DNA replication. The spontaneous mutation rate is characteristic of a gene, and mutation is more likely to occur in regions of DNA sequence repeats. In **gonadal mosaicism**, only some gametes have a spontaneous mutation.

11. **Mutagens** are agents that delete, substitute, or add DNA bases, causing mutations. An organism may be exposed to a mutagen intentionally, accidentally, or naturally.

12.5 Types of Mutations

12. A **point mutation** alters a single DNA base. It may be a **transition** (purine to purine or pyrimidine to pyrimidine) or a **transversion** (purine to pyrimidine, or vice versa).

13. A **missense mutation** substitutes one amino acid for another. A **nonsense mutation** substitutes a "stop" codon for a codon that specifies an amino acid, shortening the encoded protein.

14. **Nonsense-mediated decay** destroys mRNAs that have premature stop codons. Splice-site mutations add or delete amino acids and can shift the reading frame. A missense mutation that creates an intron splicing site can cause **exon skipping**.

15. A **deletion mutation** removes genetic material and an **insertion mutation** adds it. A **frameshift mutation** alters the sequence of amino acids (disrupting the reading frame). A **tandem duplication** is a copy of a gene next to the original.

16. A pseudogene results when a duplicate of a gene mutates. It may disrupt chromosome pairing, causing mutation. Transposons may disrupt the functions of genes they jump into.

17. **Expanding repeats** are mutations that add stretches of the same amino acid to a protein. The DNA repeats expand because they attract each other, which causes replication errors.

18. Copy number variants are DNA sequences that are repeated a different number of times among individuals and include deletions, duplications, and repeats.

12.6 The Importance of Position

19. Mutations in the globin genes may affect the ability of the blood to transport oxygen, or they may have no effect.

20. Synonymous codons limit the effects of mutation. Changes in the second position in a codon may substitute a similarly shaped amino acid.

21. **Conditional mutations** are expressed only in response to certain environmental triggers.

22. Sending the most recently replicated DNA into cells headed for differentiation, while sending older strands into stem cells, protects against the effects of a new mutation.

12.7 DNA Repair

23. DNA polymerase proofreads DNA, but repair enzymes correct errors in other ways.

24. Photoreactivation repair uses light energy to split pyrimidine dimers, but humans do not have it.

25. **Nucleotide excision repair** replaces up to 30 nucleotides to fix errors from various sources of mutation, including UV damage. **Base excision repair** replaces up to five nucleotides to correct errors due to oxidative damage.

26. **Mismatch repair** proofreads newly replicated DNA for loops that indicate noncomplementary base pairing.

27. Another type of DNA repair fixes the sugar-phosphate backbone. Damage tolerance enables replication to continue beyond a mismatch.

28. Mutations in DNA repair genes break chromosomes and increase cancer risk.

29. DNA repair disorders include trichothiodystrophy, a form of inherited colon cancer, xeroderma pigmentosum, and ataxia telangiectasia.

Review Questions

1. Distinguish between a loss-of-function and a gain-of-function mutation.

2. How do a mutation and a polymorphism differ?

3. Give an example of a gene variant that is not harmful.

4. Distinguish between a germline and a somatic mutation.

5. How is a human body a genomic mosaic?

6. Explain how different people with Proteus syndrome have deformities in different parts of the body.

7. Explain why Henne Holstege's white blood cells had hundreds of mutations, but her brain neurons had none.

8. Explain how a mutation causes sickle cell disease.

9. Why are mutations in collagen genes especially likely to affect health?

10. Distinguish among any three of the mutations described in tables 12.1 and 12.2.

11. Describe a pair of allelic diseases.

12. How can DNA spontaneously mutate?

13. Distinguish between
    a. a transition and a transversion.
    b. a missense mutation and a nonsense mutation.

14. List two types of mutations that can alter the reading frame.

15. List four ways that DNA can mutate without affecting the phenotype.

16. Cite two ways a transposon can disrupt gene function.

17. What is a molecular explanation for the worsening of an inherited illness over generations?

18. How can short repeats within a gene, long triplet repeats within a gene, and repeated genes cause disease?

19. Explain how a copy number variant differs from a missense mutation.

20. How can a mutation that retains an intron's sequence and a triplet repeat mutation have a similar effect on a gene's encoded protein?

21. Explain how a single-base mutation can encode a protein that is missing many amino acids.

22. Cite three ways in which the genetic code protects against the effects of mutation.

23. What is a conditional mutation?

24. How do excision repair and mismatch repair differ?

25. Explain how semiconservative DNA replication makes it possible for stem cells to receive the DNA least likely to bear mutations.

Applied Questions

1. The chapter opener describes a genetic syndrome in a boy named Vincent. Why are his parents not concerned about passing the condition to other children?

2. In McCune-Albright syndrome, fibrous connective tissue replaces bone, tan patches (café-au-lait spots) dot the skin, and hormone abnormalities cause early puberty and malfunction of the thyroid, pituitary, and adrenal glands. The phenotype is highly variable, and all patients are somatic mosaics for the mutation, which is in the gene *GNAS1*. Why is the condition seen only in mosaics?

3. Explain why designating allelic diseases or terming a mutation a "variant of uncertain significance" is based more on the state of our knowledge than on biological phenomena.

4. Which of the following DNA sequences is most likely to mutate? Cite a reason for your answer.
   a. CTGAGTCGTGACTCGTACGTCAT
   b. ATGCTCAGCTACTCAGCTACGGA
   c. ATCCTAGTCGTCGTCGTCGTCATC

5. Retinitis pigmentosa refers to a group of conditions that cause night blindness and loss of peripheral vision before age 20. A form of X-linked retinitis pigmentosa is caused by a frameshift mutation that deletes 199 amino acids. How can a simple mutation have such a great effect?

6. A mutation that changes a C to a T causes a type of Ehlers-Danlos syndrome, forming a "stop" codon and resulting in shortened procollagen. Consult the genetic code (see table 10.3) and suggest one way that this can happen.

7. Part of the mRNA sequence of an exon of a gene that encodes a blood protein is: AUGACUCAUCGCUGUAGUUUACGA. Consult the genetic code to answer the following questions.
   a. What is the sequence of amino acids that this mRNA encodes?
   b. What is the sequence if a point mutation changes the 10th base from a C to an A?
   c. What is the effect of a point mutation that changes the 15th base from a U to an A?
   d. How does the encoded amino acid sequence change if a C is inserted between the fourth and fifth bases?
   e. Which would be more devastating to the encoded amino acid sequence, insertion of three bases in a row or insertion of two bases in a row?

8. Susceptibility to developing prion diseases (see section 10.4) arises from a mutation that changes aspartic acid (Asp) to asparagine (Asn). Which nucleotide base changes could make this happen?

9. Two teenage boys meet at a clinic for treatment of muscular dystrophy. The boy who is more severely affected has a two-base insertion at the start of his dystrophin gene. The other boy has the same two-base insertion but also has a third base inserted a few bases away. Why is the second boy's illness milder?

10. Two missense mutations in the gene that encodes an enzyme called superoxide dismutase cause a form of amyotrophic lateral sclerosis (ALS, or Lou Gehrig disease). This disease causes loss of neurological function over a 5-year period. One mutation alters the amino acid asparagine (Asn) to lysine (Lys). The other changes an isoleucine (Ile) to a threonine (Thr). List the codons involved and describe how single-base mutations can alter the specified amino acids.

11. A man develops Huntington disease at age 48. He is tested and his mutant gene has 54 repeats. His 20-year-old daughter is tested, and she has a gene with 68 repeats. Explain how the gene likely expanded.

12. In one family, Tay-Sachs disease stems from a four-base insertion, which changes an amino-acid-encoding codon into a "stop" codon. What type of mutation is this?

13. A biotechnology company has encapsulated DNA repair enzymes in fatty bubbles called liposomes. Why would these be a valuable addition to a suntanning lotion?

Forensics Focus

1. Late one night, a man broke into Colleen's apartment and raped her. He wore a mask and it was dark, so she couldn't see his face, but she yanked out some of his long, greasy hair. The forensic investigator, after examining the hair, asked Colleen if her boyfriend had been in the bed or with her earlier in the evening, because the hairs were of two genotypes for one of the repeated sequences that was analyzed. What is another explanation for finding two genotypes in the hair?

Case Studies and Research Results

1. The parents of the second child identified with the genotype that Vincent has in the *RPS23* gene did not want to participate in any research (see chapter opener). How can a child's genetic information be used in a way that protects the child, but contributes to research?

2. Researchers scrutinized the genome sequences from nearly 600,000 adults for 874 genes known to cause specific single-gene diseases that have severe symptoms in early childhood. The study identified 13 adults who had eight of these mutations, but were healthy. Headlines trumpeted the existence of these "genetic superheroes." Explain how a person can inherit a genotype associated with a severe illness but not be affected.

3. Robinow syndrome causes a face that resembles that of a fetus, with a high prominent forehead, wide eyes, a flat nose, and large head and face. Other symptoms include heart defects, hearing loss, and a dense skull. The gene that is mutant, *DVL1*, consists of 2,941 nucleotides. It has 15 exons and encodes a 670-amino-acid-long protein. A study of eight affected individuals found that they have different mutations, but all the mutations deleted a number of nucleotides from exon 14 that is not a multiple of 3, which results in a shortened protein.

    a. What is the significance of the fact that the small deletions are not a multiple of three nucleotides?

    b. Researchers were surprised to identify mRNAs encoding 14 exons of the *DVL1* gene from the eight people with Robinow syndrome. Why were they surprised?

    c. How many nucleotides are in the introns of this gene?

4. Latika and Keyshauna meet at a clinic for college students who have cystic fibrosis. Latika's mutation results in exon skipping. Keyshauna's mutation is a nonsense mutation. Which young woman probably has more severe symptoms? Cite a reason for your answer.

5. Marshall and Angela have skin cancer resulting from xeroderma pigmentosum. They meet at an event for teenagers with cancer. However, their mutations affect different genes. They decide to marry but not to have children because they believe that each child would have a 25 percent chance of inheriting XP because it is autosomal recessive. Are they correct? Why or why not?

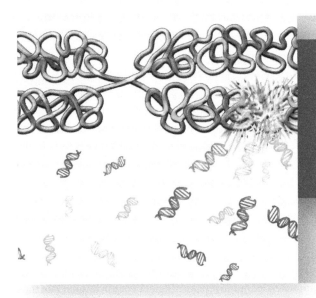

CHAPTER 13

Chromosomes

Chromothripsis is a rare event that spontaneously shatters a chromosome—sometimes more than one. Discovered in 2011 in cancer cells, chromothripsis may also account for some cases of severe birth defects.

 ## The BIG Picture

A human genome has 20,000-plus protein-encoding genes dispersed among 24 chromosome types. Chromosome-level illnesses reflect disruption of individual genes or their regulation. Because chromosomes determine which sets of genes are transmitted together to the next generation, detecting and analyzing chromosomes can provide clinically useful information.

Shattered!

In 1965, *The New England Journal of Medicine* published a case report about a 9-year-old who was the only person in the world known to have an immune condition that causes warts and susceptibility to bacterial infections. Over the years, a few other cases were identified, and the condition was attributed to a mutation in an immune system gene, *CXCR4*. A mutation in one copy of the gene causes the encoded protein, which is a receptor on certain white blood cells, to be overactive, keeping the cells in the bone marrow rather than releasing them to the circulation to fight bacteria. The warts come from susceptibility to human papillomavirus.

The disease is now called WHIM, named for the first patient. A few years ago, then in her late fifties, she brought her two grown daughters to the National Institutes of Health Clinical Center, reporting that they had inherited her WHIM. Strangely though, the woman was well: no infections, no warts. What had happened? A hematopoietic (blood) stem cell had undergone an extremely rare event, previously known only to occur in cancer cells, called **chromothripsis**. Double-stranded breaks had occurred in many places in one of her chromosomes, as if the chromosome had exploded. Chromothripsis is Greek for "shattered colored bodies."

When the researchers sequenced the woman's genome in some of her white blood cells, they found that one copy of her chromosome 2 was shorter than the other copy—it was missing 164 genes, including the copy of the *CXCR4* gene that had caused her disease. The chromosome had shattered and sent pieces integrating into other chromosomes, but they had healed, thanks to DNA repair. The white blood cells that had descended from the altered stem cell had begun to leave the bone marrow and function, and her infections ceased. However, half of her oocytes still had the mutation, which explains how her daughters inherited WHIM.

A few cases are now known of children with severe birth defects whose mothers show evidence of past chromothripsis in all of their cells. DNA repair knit together the chromosome pieces in the mothers, and they are healthy because all of their genetic material is present, just rearranged. However, during meiosis, the patchwork chromosomes can mispair as pieces from homologous chromosomes attract from odd places, resulting in offspring that have extra or missing genetic material, which can cause severe birth defects. It is as if the mothers' genomes are like large jigsaw puzzles, cut up and pasted back together, but the offspring inherit too few or too many puzzle pieces.

13.1 Portrait of a Chromosome

Mutations range from single-base changes to entire extra sets of chromosomes. A mutation is considered a chromosomal aberration if it is large enough to be seen with a light microscope using stains and/or fluorescent probes to highlight missing, extra, or moved genetic material.

In general, too little genetic material has more severe effects on health than too much. Extensive chromosome abnormalities that are present in all cells of an embryo or fetus disrupt or halt prenatal development. As a result, only 0.65 percent of all newborns have chromosomal abnormalities that produce symptoms. An additional 0.20 percent of newborns have chromosomal rearrangements in which chromosome parts have flipped or swapped, but the rearrangements do not produce symptoms unless they disrupt the structures or functions of genes that affect health.

Cytogenetics is the classical area of genetics that links chromosome variations to specific traits, including illnesses. This chapter explores several ways that chromosomes can be atypical (used synonymously with abnormal) and affect health. Actual cases introduce some of them.

Required Parts: Telomeres and Centromeres

A chromosome consists primarily of DNA and proteins with a small amount of RNA. Chromosomes are duplicated and transmitted, via cell division (mitosis or meiosis), to the next cell generation. The different chromosome types have long been described and distinguished by size and shape, using stains and dyes to contrast dark **heterochromatin** with the lighter **euchromatin** (**figure 13.1**). Heterochromatin consists mostly of highly repetitive DNA sequences, whereas euchromatin has many protein-encoding sequences.

A chromosome must include structures that enable it to replicate and remain intact. Everything else is informational (protein-encoding genes and their controls). The essential parts of a chromosome are

- telomeres;
- origin of replication sites, where replication forks begin to form; and
- the centromere.

Recall from figure 2.16 that **telomeres** are chromosome tips. In humans, each telomere repeats the sequence TTAGGG. In most cell types, telomeres shorten with each mitotic cell division.

The **centromere** is the largest constriction of a chromosome. It is where spindle fibers attach when the cell divides. A chromosome without a centromere is no longer a chromosome. It vanishes from the cell as soon as division begins because there is no way to attach to the spindle.

Centromeres, like chromosomes, consist mostly of DNA and protein. Many of the hundreds of thousands of DNA bases that form the centromere are copies of a specific 171-base DNA sequence. The size and number of repeats are similar in many species, although the sequence differs. The similarity among species suggests that these repeats have a structural role in maintaining chromosomes rather than an informational one from their sequence. Certain centromere-associated proteins are synthesized only when mitosis is imminent, forming

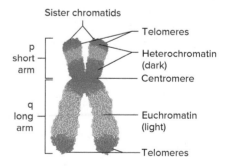

Figure 13.1 Portrait of a chromosome. Tightly wound, highly repetitive heterochromatin forms the centromere (the largest constriction) and the telomeres (the tips) of chromosomes. Elsewhere, lighter-staining euchromatin includes many protein-encoding genes. The centromere divides this chromosome into a short arm (p) and a long arm (q). This chromosome is in the replicated form.

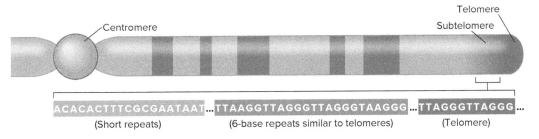

ACACACTTTCGCGAATAAT ... TTAAGGTTAGGGTTAGGGTAAGGG ... TTAGGGTTAGGG ...
(Short repeats)   (6-base repeats similar to telomeres)   (Telomere)

Figure 13.2 **Subtelomeres.** The repetitive sequence of a telomere gradually diversifies toward the centromere. The centromere is depicted as a buttonlike structure to more easily distinguish it, but it is composed of DNA like the rest of the chromosome.

a structure called a kinetochore that contacts the spindle fibers, enabling the cell to divide.

Centromeres replicate toward the end of S phase of the cell cycle. A protein that may control the process is centromere protein A, or CENP-A. Molecules of CENP-A stay with centromeres as chromosomes replicate, covering about half a million DNA base pairs. When the replicated (sister) chromatids separate at anaphase, each member of the pair retains some CENP-A. The protein therefore passes to the next cell generation, but it is *not* DNA. This is an epigenetic change.

Centromeres lie within vast stretches of heterochromatin. The arms of the chromosome extend outward from the centromere. Gradually, with increasing distance from the centromere, the DNA includes more protein-encoding sequences. Gene density varies greatly among chromosomes. Chromosome 21 is a gene "desert," harboring a million-base stretch with no protein-encoding genes at all. Chromosome 22, in contrast, is a gene "jungle." These two tiniest chromosomes are remarkably similar in size, which is why they are out of place in karyotypes. Chromosome 22 contains 545 genes to chromosome 21's 225!

The chromosome parts that lie between protein-rich areas and the telomeres are termed subtelomeres (**figure 13.2**). These areas extend from 8,000 to 300,000 bases inward toward the centromere from the telomeres. Subtelomeres include some protein-encoding genes and therefore bridge the gene-rich regions and the telomere repeats. The transition is gradual. Areas of 50 to 250 bases, right next to the telomeres, consist of 6-base repeats, many of them very similar to the TTAGGG of the telomeres. Then, moving inward from the 6-base zone are many shorter repeats. Their function isn't known. Finally, the sequence diversifies and protein-encoding genes appear.

At least 500 protein-encoding genes lie in the subtelomere regions. About half are members of multigene families (groups of genes of very similar sequence next to each other) that include pseudogenes. These multigene families may reflect recent evolution: Apes and chimps have only one or two genes for many of the large gene families in humans. Such gene organization may explain why our genome sequence is so similar to that of our primate cousins—but we are clearly different animals. Our genomes differ from those of other primates more in gene copy number and chromosomal organization than in DNA base sequence.

Karyotypes Chart Chromosomes

Even in this age of genomics, the standard chromosome chart, or **karyotype**, remains a major clinical tool. By showing which genes are transmitted together because they are part of the same chromosome, a karyotype can reveal certain conditions that DNA sequencing can miss. **Clinical Connection 13.1** in section 13.4 returns to this point.

A karyotype displays chromosomes in pairs by size and by physical landmarks that appear during mitotic metaphase, when DNA coils tightly, enabling it to be visualized. **Figure 13.3** shows a karyotype with one extra chromosome, which is called a **trisomy**.

The 24 human chromosome types are numbered from largest to smallest—1 to 22. The other two chromosomes are

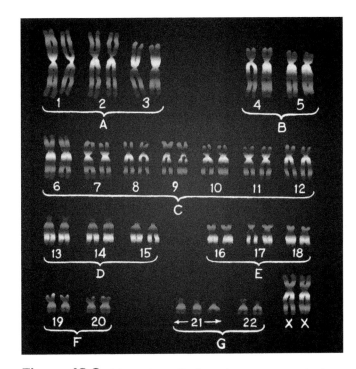

Figure 13.3 **A karyotype displays chromosome pairs in size order.** Note the extra chromosome 21 that causes trisomy 21 Down syndrome. (Chromosomes are color-enhanced; A-G denote generalized groups.) © *Biophoto Associates/Science Source. Colorization by Mary Martin*

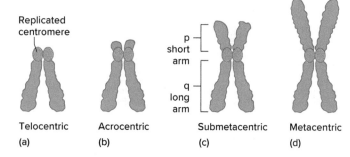

Replicated
centromere

p
short
arm

q
long
arm

Telocentric   Acrocentric   Submetacentric   Metacentric
(a)   (b)   (c)   (d)

Figure 13.4 **Centromere position distinguishes chromosomes.** **(a)** A telocentric chromosome has the centromere near one end, although telomere DNA sequences are at the tip. Humans do not have telocentric chromosomes. **(b)** An acrocentric chromosome has the centromere near an end. **(c)** A submetacentric chromosome's centromere creates a long arm (q) and a short arm (p). **(d)** A metacentric chromosome's centromere establishes more equal-size arms.

the X and the Y. Early attempts to size-order chromosomes resulted in generalized groupings because many of the chromosomes are of similar size. Use of dyes and stains made it easier to distinguish chromosomes by creating patterns of bands.

Centromere position is one physical feature of chromosomes. A chromosome is **metacentric** if the centromere divides it into two arms of approximately equal length. It is **submetacentric** if the centromere establishes one long arm and one short arm, and **acrocentric** if it pinches off only a small amount of material toward one end (**figure 13.4**). Some species have telocentric chromosomes that have only one arm, but humans do not. The long arm of a chromosome is designated *q*, and the short arm *p* (for "petite").

Five human chromosomes (13, 14, 15, 21, and 22) have bloblike ends, called satellites, that extend from a thinner, stalklike bridge from the rest of the chromosome. The stalk regions do not bind stains well. The stalks carry many copies of genes encoding ribosomal RNA and ribosomal proteins. These areas coalesce to form the nucleolus, a structure in the nucleus where ribosomal building blocks are produced and assembled (see figure 2.3).

Karyotypes are useful at several levels. When a baby is born with the distinctive facial features of Down syndrome, a karyotype confirms the clinical diagnosis. Within families, karyotypes are used to identify relatives with a particular chromosome aberration that can affect health. In one family, several adults died from a rare form of kidney cancer. Karyotypes revealed that the affected individuals all had an exchange of genetic material between chromosomes 3 and 8. When karyotypes showed that two healthy young family members had the unusual chromosomes, physicians examined and monitored their kidneys. Cancer was found very early and successfully treated.

Karyotypes of individuals from different populations can reveal the effects of environmental toxins, if abnormalities appear only in a group exposed to a contaminant. Because chemicals and radiation that can cause cancer and birth defects often break chromosomes into fragments or rings, detecting this genetic damage can alert researchers to the possibility that certain cancers may appear in the population.

Key Concepts Questions 13.1

1. What are the basic parts of a chromosome?
2. What characteristics of chromosomes do karyotypes display?

13.2 Detecting Chromosomes

Laboratory tests to detect or analyze chromosomes can be done on cells sampled at any stage of prenatal development or at any age. Many such tests are conducted on cells from a fetus or the structures that support it. Direct methods of prenatal testing collect cells and separate the chromosomes, then add a stain or a DNA probe (a labeled piece of DNA that binds its complement) to distinguish the different chromosomes, or collect small pieces of DNA from the maternal bloodstream that are from the placenta and sequence it. Indirect methods detect changes in levels of biochemicals or rely on clinical findings (such as symptoms or ultrasound scans) associated with a particular chromosomal condition.

Direct Visualization of Chromosomes

The two traditional technologies that provide images of chromosomes from a fetus are **amniocentesis** and **chorionic villus sampling (CVS)**. The images are organized into karyotypes that show the normal 46 chromosomes or abnormal conditions, such as extra or missing chromosomes, inverted chromosomes, or chromosomes that have exchanged parts.

Postnatally, any cell other than a mature red blood cell (which lacks a nucleus) can be used to examine chromosomes, but some cells are easier to obtain and culture than others. A person collects saliva in a tube or swirls a brush, called a buccal swab, in the mouth to collect cheek lining cells. Saliva and cheek scrapings include white blood cells and epithelial cells, which are the sources of the DNA. Blood tests yield DNA from the white blood cells. A person might require a chromosome test if he or she has a family history of a chromosome abnormality or has a history of unexplained infertility.

Couples who receive a prenatal diagnosis of a chromosome abnormality in a fetus can arrange for treatment of the newborn, if possible; learn more about the condition and contact support groups and plan care; or terminate the pregnancy. These choices are best made after a genetic counselor or physician provides information on the medical condition and treatment options.

Amniocentesis

The first fetal karyotype was constructed in 1966 using amniocentesis. In this procedure, a doctor removes a small sample of amniotic fluid from the uterus with a needle passed through

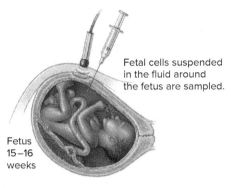

Fetal cells suspended in the fluid around the fetus are sampled.

Fetus 15–16 weeks

(a) Amniocentesis

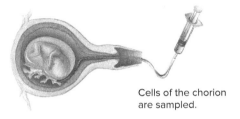

Cells of the chorion are sampled.

(b) Chorionic villus sampling

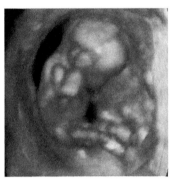

(c) Ultrasound image

Figure 13.5 **Checking fetal chromosomes, directly and indirectly. (a)** Amniocentesis shows chromosomes from fetal cells in amniotic fluid. **(b)** Chorionic villus sampling examines chromosomes from cells of structures that develop into the placenta. **(c)** Ultrasound images the fetus and may reveal an anomaly that is associated with a specific chromosome abnormality. © Dr. Najeeb Layyous/Science Source

the woman's abdominal wall (**figure 13.5a**). The fluid contains a few shed fetal cells, which are cultured for 7 to 10 days, and then 20 cells are karyotyped. The sampled amniotic fluid may also be examined for deficient, excess, or abnormal biochemicals that could indicate an inborn error of metabolism. Tests for specific single-gene diseases may be done on cells in the amniotic fluid and are tailored to family history. Ultrasound is used to follow the needle's movement and to visualize the fetus or its parts (**figure 13.5c**).

Amniocentesis can detect approximately 1,000 of the more than 5,000 known chromosomal and biochemical problems. The most common chromosomal abnormality detected a

trisomy. Amniocentesis is usually performed between 14 and 16 weeks of gestation, when the fetus isn't yet very large but amniotic fluid is plentiful. Amniocentesis can be carried out anytime after this point.

Health care providers recommend amniocentesis if the risk that the fetus has a detectable condition exceeds the risk that the procedure will cause a miscarriage. Until recently, this cutoff was age 35, when the risk to the fetus of a detectable chromosome problem about equals the risk of amniocentesis causing pregnancy loss—1 in 350. While it is still true that the risk of a chromosomal problem increases steeply after maternal age 35, amniocentesis has become much safer since the statistics were obtained that have been used for most risk estimates (**figure 13.6**). A more recent risk estimate for amniocentesis causing miscarriage of about 1 in 1,600 pregnancies has led some physicians to offer amniocentesis to women under 35. The procedure is also warranted if a couple has had several spontaneous abortions or children with birth defects who have a chromosome abnormality, irrespective of maternal age. However, tests of placental DNA in the woman's circulation are becoming so accurate at detecting fetal chromosome imbalances that the more invasive techniques (amniocentesis and CVS) are increasingly being used to follow-up such detection, rather than being used based solely on age.

Chorionic Villus Sampling

During the 10th through 12th weeks of pregnancy, chorionic villus sampling obtains cells from the chorionic villi, which are fingerlike structures that develop into the placenta (see **figure 13.5b**). A karyotype is prepared directly from the collected cells, rather than first culturing them, as in amniocentesis. Results are ready in days.

The basis of CVS is that chorionic villus cells descend from the fertilized ovum, so their chromosomes should be identical to those of the embryo and fetus. Occasionally, an atypical chromosome appears only in a cell of the embryo, or

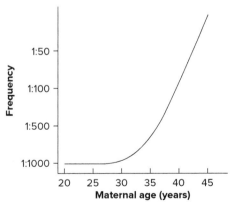

Trisomy 21 in liveborn infants

Figure 13.6 **The maternal age effect.** The risk of conceiving an offspring with trisomy 21 rises dramatically with maternal age. *Source: Color Atlas of Genetics by Eberhard Passage, p. 401. Thieme Medical Publishers, Inc.*

only in a chorionic villus cell. Chromosomal mosaicism results—the karyotype of a villus cell differs from that of an embryo cell. Clinical consequences are great. If CVS indicates an abnormality in villus cells that is not also in the fetus, then a couple may elect to terminate the pregnancy when the fetus is actually chromosomally normal. In the opposite situation, the results of the CVS may be normal, but the fetus has atypical chromosomes.

CVS is slightly less accurate than amniocentesis, and in about 1 in 1,000 to 3,000 procedures, it halts development of the feet and/or hands and may be lethal. Also, CVS does not sample amniotic fluid, so tests for inborn errors of metabolism are not possible. The advantage of CVS is earlier results, but the disadvantage is a greater risk of spontaneous abortion. However, CVS has become much safer in recent years.

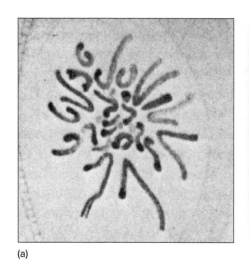
(a)

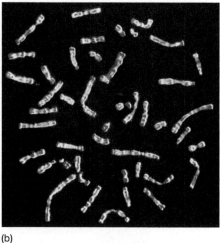
(b)

Figure 13.7 **Viewing chromosomes, then and now. (a)** The earliest drawings of chromosomes, by German biologist Walther Flemming, date from 1882. His depiction captures the random distribution of chromosomes as they splash down on a slide. **(b)** Highlighted bands distinguish the chromosomes in this micrograph. *(a): Drawing by Walther Flemming, 1882; (b): © CNRI/Science Source*

Preparing Cells and Chromosomes

The difficulty in distinguishing chromosomes is physical—it is challenging to prepare a cell in which chromosomes do not overlap (**figure 13.7**). To count chromosomes, scientists had to find a way to capture them when they are most condensed—during cell division—and also spread them apart. Since the 1950s, cytogeneticists have used colchicine, an extract of the crocus plant, to arrest cells during division.

In 1951, an accident led to discovery of how to untangle chromosomes. A technician mistakenly washed white blood cells being prepared for chromosome analysis in a salt solution that was less concentrated than the interiors of the cells. Water rushed into the cells, swelling them and separating the chromosomes. Then cell biologists found that drawing cell-rich fluid into a pipette and dropping it onto a microscope slide prepared with stain burst the cells and freed the mass of chromosomes. Adding a glass coverslip spread the chromosomes enough to be counted. Researchers finally could see that the number of chromosomes in a diploid human cell is 46, and that the number in gametes is 23.

Karyotypes were once constructed using a microscope to locate a cell where the chromosomes were not touching, photographing the cell, developing a print, cutting out the individual chromosomes, and arranging them into a size-ordered chart. It was a literal cut-and-paste! Today, a computer scans ruptured cells in a drop of stain and selects one in which the chromosomes are the most discernible. Image analysis software recognizes the band patterns of each stained chromosome pair, sorts the structures into a size-ordered chart, and sends the karyotype to the electronic medical record. If the software recognizes an atypical band pattern, a database pulls out identical or similar karyotypes from records of other patients.

The first karyotypes used dyes to stain chromosomes a uniform color. Chromosomes were grouped into size classes, designated A through G, in decreasing size order. In 1959, scientists described the first chromosomal abnormalities—Down syndrome (an extra chromosome 21), Turner syndrome (also called XO syndrome, a female with only one X chromosome), and Klinefelter syndrome (also called XXY syndrome, a male with an extra X chromosome). These first chromosome stains could highlight only large deletions and duplications.

Describing smaller chromosomal aberrations required better ways to distinguish chromosomes. By the 1970s, new stains created banding patterns unique to each chromosome. These stains are specific for AT-rich or GC-rich stretches of DNA, or for heterochromatin, which is dark staining. A band represents at least 5 to 10 million DNA bases. Synchronizing the cell cycle revealed even more bands per chromosome.

A technique called fluorescence *in situ* hybridization (FISH) introduced the ability to highlight individual genes. FISH is more targeted than conventional chromosome staining because it uses DNA probes. The probes are attached to molecules that fluoresce when illuminated, producing a flash of color precisely where the probe binds to a complementary DNA sequence among the chromosomes in a patient's sample. Using a FISH probe is a little like a search engine finding the word "hippopotamus" in a book compared to pulling out all words that have the letters *h, p,* and *o*.

FISH can "paint" entire karyotypes by probing each chromosome with several different fluorescent molecules. A computer integrates the images and creates a unique false color for each chromosome. Many laboratories that perform amniocentesis or CVS use FISH probes for chromosomes 13, 18, 21, and X and Y to quickly identify the most common problems. In **figure 13.8**, FISH reveals the extra chromosome 21 in cells from a fetus with trisomy 21 Down syndrome—three dots.

Once amniocentesis or CVS produces a karyotype, the pertinent information is abbreviated by listing chromosome number, sex chromosome constitution, and atypical autosomes. Symbols describe the type of aberration, such as a deletion or translocation; numbers correspond to specific bands. A chromosomally normal male is 46,XY; a normal female is 46,XX. Band notations describe gene locations. For example, the gene encoding the β-globin subunit of hemoglobin is located at 11p15.5 —the short arm of chromosome 11. **Table 13.1** gives examples of chromosomal shorthand.

Graphical representations are increasingly being used to display chromosome information. An ideogram (**figure 13.9**) shows chromosome arms (p is short and q is long) and numbered regions, called bands, and subbands. The numbers assigned to

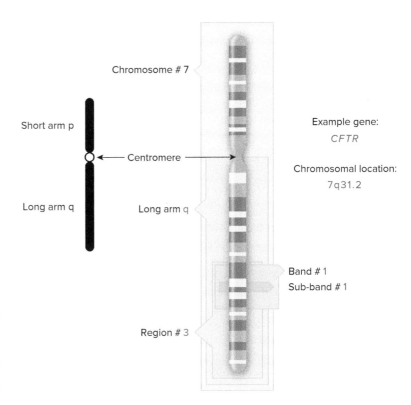

Figure 13.9 Ideogram. An ideogram is a schematic chromosome map. It indicates chromosome arm (p or q) and major regions delineated by banding patterns. This is chromosome 7, which includes the *CFTR* gene that causes cystic fibrosis when mutant. *Source: U.S. National Library of Medicine*

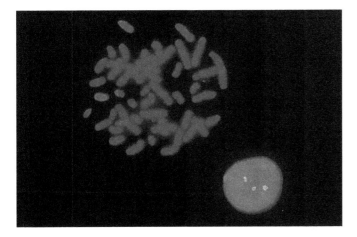

Figure 13.8 FISHing for genes and chromosomes. FISH shows three fluorescent dots that correspond to three copies of chromosome 21. Each dot represents a specific DNA sequence with which the fluorescently labeled probe forms complementary base pairs. © *James King-Holmes/Science Source*

Table 13.1	Chromosomal Shorthand
Abbreviation	**Meaning**
46,XY	Normal male
46,XX	Normal female
45,X	Turner syndrome (female)
47,XXY	Klinefelter syndrome (male)
47,XYY	Jacobs syndrome (male)
46,XY,del (7q)	A male missing part of the long arm of chromosome 7
47,XX,+21	A female with trisomy 21 Down syndrome
46,XY,t(7;9)(p21.1; q34.1)	A male with a translocation between the short arm of chromosome 7 at band 21.1 and the long arm of chromosome 9 at band 34.1
48,XXYY	A male with an extra X and an extra Y chromosome

bands and subbands increase with distance from the centromere. Specific gene loci are sometimes listed on the right side.

A digital karyotype is another kind of illustration that shows copy number variants (repeats) and is color-coded to represent chromosomes that have exchanged parts. Digital karyotypes are used to trace the fate of chromosomes that have undergone chromothripsis, described in the chaper opener. As more human genomes are sequenced and annotated, ways to summarize the information on illustrations of chromosomes will evolve.

Indirect Detection of Extra Chromosomes

Amniocentesis and chorionic villus sampling provide a prenatal diagnosis, because they directly detect chromosome abnormalities known to cause specific clinical conditions. Indirect methods that measure or detect a characteristic that is associated with having a chromosomal abnormality are used as screening tests to identify individuals at elevated risk—not to provide a diagnosis. Use of these indirect approaches will likely decline in coming years as more tests will be based on analyzing cell-free DNA from the placenta in the maternal bloodstream, including complete sequencing of fetal genomes.

Maternal serum markers (**table 13.2**) are biochemicals whose levels in the blood are within a certain range in a pregnant woman carrying a fetus with the normal number of chromosomes, but lie outside that range in fetuses whose cells have an extra copy of a certain chromosome. The more markers tested, the more predictive the information. What these biomarkers actually do is not as important for testing purposes as their concentrations.

Testing maternal serum markers didn't begin as a way to detect chromosomal conditions, but to identify fetuses that had neural tube defects (NTDs). Recall from chapter 3 that in an NTD, part of the brain or spinal cord protrudes. In the 1980s, a researcher who had a child with an NTD developed a test based on the finding that the level of alpha fetoprotein (AFP) is higher in fetuses with an open neural tube defect. AFP is made in the yolk sac and leaves the fetal circulation and enters the maternal bloodstream at a certain rate. Too much AFP in the maternal bloodstream indicates an open neural tube in the fetus.

As data accumulated to detect elevated AFP, researchers noted that the level of AFP is lower in fetuses that have trisomy 21 Down syndrome. With further discoveries, the markers listed in table 13.2 were incorporated into testing. Altogether, combined with ultrasound findings and considering maternal age, maternal serum markers offer a greater than 90 percent probability of detecting trisomies 13, 18, or 21 if they are present. For trisomy 21 Down syndrome, ultrasound findings include short limbs, flattened noses, and excess fluid at the back of the neck (called nuchal translucency).

Maternal serum marker test results are based on cutoff levels adjusted for personal statistics (such as race, other medical conditions like diabetes, and a twin pregnancy). A value called "MoM," which stands for "multiples of the median," is used. The MoM value indicates how far individual results deviate from the normal range of concentration for each marker. A MoM above 2.0 means that the level of the biomarker is twice as high or twice as low as the average in a normal pregnancy, and this is considered elevated risk. The woman can then have the more invasive amniocentesis or chorionic villus sampling procedure to confirm or rule out the trisomy.

Cell-Free Fetal DNA Testing Infers Chromosomes

Testing fetal DNA in the woman's bloodstream is replacing measurement of maternal serum markers, and is done at 10 weeks into the pregnancy or later. The tests are marketed under a variety of names, and the technology is also called noninvasive prenatal diagnosis (or testing). Testing maternal blood for pieces of DNA from the fetus is based on the fact that small pieces of DNA are normally in a pregnant woman's bloodstream, and up to 20 percent of those pieces actually come from the placenta, and therefore represent the fetus in that the genome should be identical. These pieces are outside cells ("cell-free") and can be separated because they are shorter than pieces of DNA not from the fetus (**figure 13.10**).

The first versions of cell-free fetal DNA tests detected trisomies 13, 18, and 21 (discussed in section 13.3) by finding proportionately more copies of DNA pieces from these chromosomes than from the other chromosomes, indicating an extra "dose." For example, blood from a woman whose fetus has trisomy 21 Down syndrome has about 50 percent more DNA pieces from chromosome 21 than it does from the other chromosomes, representing the third copy. Chromosomes 13, 18, and 21 account for most trisomies, which is why the first tests looked for them. Some test brands also check the sex chromosomes. An abnormal finding on a cell-free fetal DNA test is typically followed up with amniocentesis to construct a karyotype of the fetal DNA.

Testing cell-free fetal DNA removes the focus on maternal age. That is, it can identify women over 35 whose fetuses are *not* at elevated risk of trisomy, avoiding the more invasive amniocentesis or CVS, while also identifying women under age 35 who would not have had either invasive procedure based on age alone. Testing cell-free fetal DNA is so sensitive that it

Table 13.2	Maternal Serum Markers for Trisomy 21
Marker	**Normal Function**
High Level	
Beta human chorionic gonadotropin (hCG)	Implantation of the embryo
Inhibin A	Lowers levels of follicle-stimulating hormone
Low Level	
Estradiol (UE3)	Female hormone that binds AFP
Alpha fetoprotein (AFP)	Binds estradiol
Pregnancy-associated plasma protein (PAPP-A)	Cell division

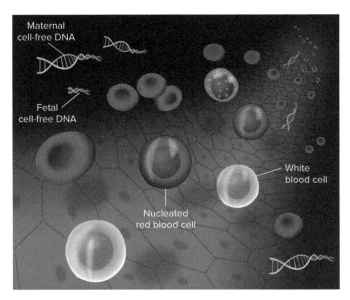

Figure 13.10 **Testing cell-free fetal DNA.** The fact that pieces of cell-free fetal DNA in the maternal circulation are much shorter than pieces of maternal DNA provides a way to collect the fetal material and sequence the genome by overlapping pieces. Comparison of a fetal genome sequence to that of the parents can reveal inherited diseases and new mutations.

has even picked up DNA sequences from cancer cells in some pregnant women who did not have any symptoms, but did in fact have cancer. Complete fetal exomes and genomes can also be sequenced from cell-free fetal DNA. These more complete analyses are used to distinguish *de novo* from inherited mutations, such as in parents-child trios in which a child has a syndrome but the parents do not have any mutations that could have transmitted the condition (see Clinical Connection 1.1).

Key Concepts Questions 13.2

1. From which cells are chromosomes typically obtained to be visualized?
2. Describe how amniocentesis and chorionic villus sampling directly reveal fetal chromosomes.
3. Describe indirect ways to detect abnormal chromosome numbers.
4. Explain how presence of an extra chromosome is deduced from testing cell-free fetal DNA.

13.3 Atypical Chromosome Number

A human karyotype is atypical (abnormal) if the number of chromosomes in a somatic cell is not 46, or if individual chromosomes have extra, missing, or rearranged genetic material.

More discriminating technologies can detect very small numbers of extra or missing nucleotides. As a result, more people are being diagnosed with chromosomal abnormalities than in the days when stains made all chromosomes look alike.

Atypical chromosomes account for at least 50 percent of spontaneous abortions, yet only 0.65 percent of newborns have them. Therefore, most embryos and fetuses with atypical chromosomes stop developing before birth. **Table 13.3** summarizes the types of chromosome variants in the order in which they are discussed.

Polyploidy

The most extreme upset in chromosome number is an entire extra set. A cell with extra sets of chromosomes is **polyploid**. An individual whose cells have three copies of each chromosome is a triploid (designated 3N, for three sets of chromosomes). Two-thirds of all triploids result from fertilization of an oocyte by two sperm. The other cases arise from formation of a diploid gamete, such as when a normal haploid sperm fertilizes a diploid oocyte. Triploids account for 17 percent of spontaneous abortions (**figure 13.11**). Very rarely, an infant survives a few days, with defects in nearly all organs. However, certain human cells may be polyploid. The liver, for example, has some tetraploid (4N) and even octaploid (8N) cells.

Polyploids are common among flowering plants, including roses, cotton, barley, and wheat, and in some insects. Fish farmers raise triploid salmon, which cannot breed because their gametes contain different numbers of chromosomes.

Table 13.3	Chromosome Abnormalities
Type of Abnormality	**Definition**
Polyploidy	Extra chromosome sets
Aneuploidy	An extra or missing chromosome
Monosomy	One chromosome absent
Trisomy	One chromosome extra
Deletion	Part of a chromosome missing
Duplication	Part of a chromosome present twice
Translocation	Two chromosomes join long arms or exchange parts
Inversion	Segment of chromosome reversed
Isochromosome	A chromosome with identical arms
Ring chromosome	A chromosome that forms a ring due to deletions in telomeres, which cause ends to adhere
Chromothripsis	One or more chromosomes shatters

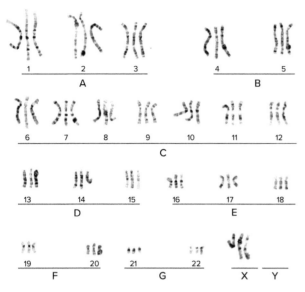

Figure 13.11 **Polyploids in humans are nearly always lethal.** Individuals with three copies of each chromosome (triploids) in every cell account for 17 percent of all spontaneous abortions and 3 percent of stillbirths and newborn deaths. © CNRI/Science Source

Aneuploidy

Cells missing a single chromosome or having an extra chromosome are **aneuploid**, which means "not good set." Rarely, aneuploids can have more than one missing or extra chromosome, indicating abnormal meiosis in a parent. A normal chromosome number is **euploid**, which means "good set."

Most autosomal aneuploids (with a missing or extra nonsex chromosome) are spontaneously aborted. Those that survive have specific syndromes, with symptoms depending upon which chromosomes are missing or extra. Intellectual disability is common in aneuploidy because development of the brain is so complex and of such long duration that nearly any chromosome-scale disruption affects genes whose protein products affect the brain. Sex chromosome aneuploidy usually produces milder symptoms.

Most children born with a chromosome number other than 46 have an extra chromosome (a trisomy) rather than a missing one (a **monosomy**). Most monosomies are so severe that an affected embryo ceases developing. Trisomies and monosomies are named for the chromosomes involved, and in the past the associated syndromes were named for the discoverers. Today, cytogenetic terminology is used because it is more precise. For example, Down syndrome can result from a trisomy or a translocation. The distinction is important in genetic counseling. Translocation Down syndrome, although accounting for only 4 percent of cases, has a much higher recurrence risk within a family than does trisomy 21 Down syndrome, a point we return to later in the chapter.

The meiotic error that causes aneuploidy is called **nondisjunction**. Recall that in normal meiosis, homologs separate and each of the resulting gametes receives only one member of each chromosome pair. In nondisjunction, a chromosome pair does not separate at anaphase of either the first or second meiotic division. This unequal division produces a sperm or oocyte that has two copies of a particular chromosome, or none, rather than one copy (**figure 13.12**). When such a gamete meets its partner at fertilization, the resulting zygote has either 45 or 47 chromosomes, instead of the normal 46. Different trisomies tend to be caused by nondisjunction in the male or female, at meiosis I or II.

A cell can have a missing or extra chromosome in 49 ways—an extra or missing copy of each of the 22 autosomes, plus the sex chromosome combinations of Y, X, XXX, XXY, and XYY. (Some individuals have four or even five sex chromosomes.) However, only nine types of aneuploids are recognized in newborns. Others are seen in spontaneous abortions or fertilized ova intended for *in vitro* fertilization.

Most of the 50 percent of spontaneous abortions that result from extra or missing chromosomes are 45,X individuals (missing an X chromosome), triploids, or trisomy 16. About 9 percent of spontaneous abortions are trisomy 13, 18, or 21. More than 95 percent of newborns with atypical chromosome numbers have an extra chromosome 13, 18, or 21, or an extra or missing X or Y chromosome. These conditions are all rare at birth—together they affect only 0.1 percent of all children; however, nondisjunction occurs in 5 percent of recognized pregnancies.

Types of chromosome abnormalities differ between the sexes. Atypical oocytes mostly have extra or missing chromosomes, whereas atypical sperm more often have structural variants, such as inversions or translocations, discussed later in the chapter.

Aneuploidy and polyploidy also arise during mitosis, producing groups of somatic cells with the extra or missing chromosome (see figure 12.2). A mitotic abnormality that occurs early in development, so that many cells descend from the unusual one, can affect health. For example, a chromosomal mosaic for a trisomy may have a mild version of the associated condition. This is usually the case for 1 to 2 percent of people with Down syndrome; they are mosaics. The phenotype depends upon which cells have the extra chromosome. Unfortunately, prenatal testing cannot reveal which cells are affected.

Autosomal Aneuploids

Most autosomal aneuploids cease developing long before birth. Following are cases and descriptions of the most common autosomal aneuploids among liveborns. The most frequently seen extra autosomes in newborns are chromosomes 21, 18, and 13, because these chromosomes carry many fewer protein-encoding genes than the other autosomes, compared to their total amount of DNA. Therefore, extra copies of these chromosomes are tolerated well enough for some fetuses with them to survive to be born (**table 13.4**).

Trisomy 21

When David was born in 1994, doctors told his 19-year-old mother, Toni, to put him into an institution, because he would never walk or talk. David had other ideas. He grew up to graduate high school.

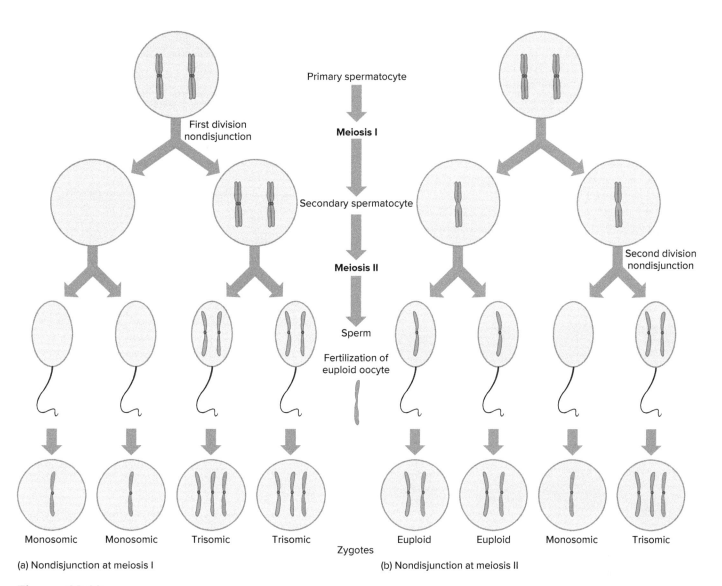

(a) Nondisjunction at meiosis I

(b) Nondisjunction at meiosis II

Figure 13.12 Extra and missing chromosomes—aneuploidy. Unequal division of chromosome pairs can occur at either the first or second meiotic division. **(a)** A single pair of chromosomes is unevenly partitioned into the two cells arising from meiosis I in a male. The result: Two sperm cells have two copies of the chromosome, and two sperm cells have no copies. When a sperm cell with two copies of the chromosome fertilizes a normal oocyte, the zygote is trisomic; when a sperm cell lacking the chromosome fertilizes a normal oocyte, the zygote is monosomic. **(b)** This nondisjunction occurs at meiosis II. Because the two products of the first division are unaffected, two of the mature sperm are normal and two are aneuploid. Oocytes can undergo nondisjunction as well, leading to zygotes with extra or missing chromosomes when normal sperm cells fertilize them.

Table 13.4	Comparing and Contrasting Trisomies 13, 18, and 21	
Type of Trisomy	Incidence at Birth	Percentage of Conceptions That Survive 1 Year After Birth
13 (Patau)	1/12,500–1/21,700	<5%
18 (Edwards)	1/6,000–1/10,000	<5%
21 (Down)	1/800–1/826	85%

Like other teens, David has held part-time jobs, gone to dances, and uses a computer. But he is unlike most other teens in that his cells have an extra chromosome 21, which limits his intellectual abilities. "Maybe he's not book smart, but when you look around at what he can do, he's smart," Toni says. His speech is difficult to understand, and he has facial features characteristic of Down syndrome, but he has a winning personality and close friends.

Sometimes David gets into unusual situations because he thinks literally. He once dialed 911 when he stubbed his toe, because he'd been told to do just that when he was hurt. Today David lives in a group home and attends community college.

The most common autosomal aneuploid among liveborns is trisomy 21, because this chromosome has the fewest genes. A person with Down syndrome is usually short and has straight, sparse hair and a tongue protruding through thick lips. The hands have an atypical pattern of creases, the joints are loose, and poor reflexes and muscle tone give a "floppy" appearance. Developmental milestones (such as sitting, standing, and walking) come slowly, and toilet training may take several years. Intelligence varies greatly. Parents of a child with Down syndrome can help their child reach maximal potential by providing a stimulating environment (**figure 13.13**).

Many people with Down syndrome have physical problems, including heart and kidney defects and hearing and visual loss. A suppressed immune system can make influenza deadly. Digestive system blockages are common and may require surgical correction. A child with Down syndrome is 15 times more likely to develop leukemia than a child who does not have the

Figure 13.13 Trisomy 21. Many years ago, people with Down syndrome were institutionalized. Today, thanks to tremendous strides in both medical care and special education, people with the condition can hold jobs and attend college. This young lady enjoys painting. © *Angela Hampton/ Bubbles Photolibrary/Alamy*

syndrome, but this is still only a 1 percent risk. However, people with Down syndrome are somewhat protected against developing solid tumors. Many of the medical problems associated with Down syndrome are treatable, so that average life expectancy is now 60 years. In 1910, life expectancy was only 9 years.

Some people with Down syndrome older than 40 develop the black fibers and tangles of amyloid beta protein in their brains characteristic of Alzheimer disease, although they usually do not become severely demented (see Clinical Connection 5.1). The chance of a person with trisomy 21 developing Alzheimer disease is 25 percent, compared to 6 percent for the general population. A gene on chromosome 21 causes one inherited form of Alzheimer disease. Perhaps the extra copy of the gene in trisomy 21 has a similar effect to a mutation in the gene that causes Alzheimer disease, such as causing amyloid beta buildup.

Before human genome sequencing began, researchers studied people who have a third copy of only part of chromosome 21 to deduce which genes cause which symptoms. This approach led to the discovery that genes near the tip of the long arm of the chromosome contribute most of the abnormalities. Two specific genes control many aspects of Down syndrome by controlling a third gene, which encodes a transcription factor and therefore affects expression of many other genes.

A newer approach, **genome editing**, is being used to study Down syndrome at the cellular level. Section 19.4 discusses how genome editing cuts double-stranded DNA at a specific site on a chromosome and allows insertion of a specific DNA sequence. For trisomy 21 Down syndrome, the technique "borrows" the mechanism that shuts off one X chromosome in the cells of females. Researchers insert the DNA that encodes the long noncoding RNA sequence called XIST, which normally shuts off one X chromosome (see figure 6.10), into one chromosome 21 of induced pluripotent stem cells (see Table 2.3) made from skin cells of a boy with trisomy 21. The shut-off chromosome 21 forms a Barr body, which normally happens when XIST silences one X chromosome in a female cell. (Male cells do not normally have Barr bodies because they have only one X chromosome.) The treated male cells have Barr bodies (**figure 13.14**), meaning one of the three copies of chromosome 21 is turned off. A challenge will be learning how to adapt this approach to help people with trisomy 21.

The likelihood of giving birth to a child with trisomy 21 Down syndrome increases dramatically with the age of the

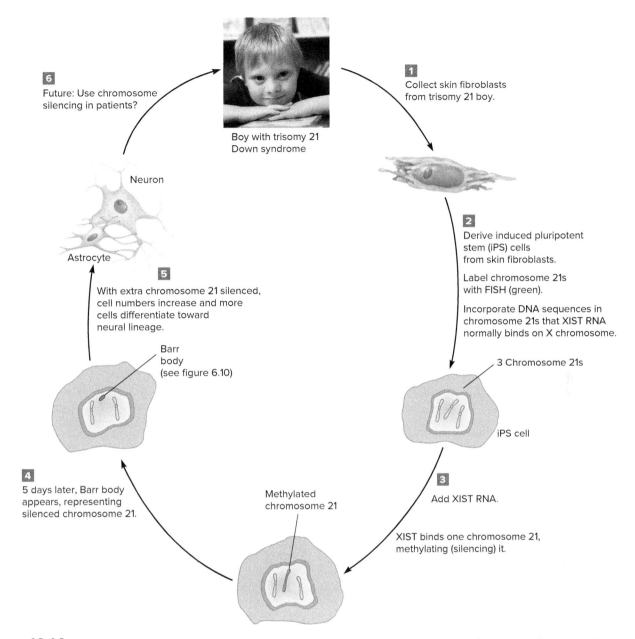

Figure 13.14 **Turning off the extra chromosome 21 of Down syndrome.** One application of genome editing uses the mechanism of X inactivation to "turn off" the extra chromosome of trisomy 21, silencing the extra genetic information—in induced pluripotent stem cells from an affected individual. Observing what happens as the stem cells are stimulated to differentiate as neurons and comparing this to what happens to trisomic cells may reveal new ways to treat the syndrome. © Stockbyte/Veer RF

mother (see figure 13.6). However, 80 percent of children with trisomy 21 are born to women under age 35, because younger women are more fertile and, until noninvasive prenatal testing on cell-free fetal DNA became available, less likely to have prenatal tests than older women. About 90 percent of trisomy 21 conceptions are due to nondisjunction during meiosis I in the female. The 10 percent of cases due to the male result from nondisjunction during meiosis I or II. The chance that trisomy 21 will recur in a family, based on empirical data (how often it actually does recur in families), is 1 percent.

The age factor in trisomy 21 Down syndrome and other trisomies may reflect the fact that the older a woman is, the

longer her oocytes have been arrested on the brink of completing meiosis. During this 15 to 45 years, oocytes may have been exposed to toxins, viruses, and radiation. A second explanation for the maternal age effect is that females have a pool of immature aneuploid oocytes resulting from spindle abnormalities that cause nondisjunction. As a woman ages, selectively releasing normal oocytes each month, the atypical ones remain.

The association between maternal age and Down syndrome has been recognized for a long time, because affected individuals were often the youngest children in large families. Before the chromosome connection was made in 1959, the syndrome was attributed to syphilis, tuberculosis, thyroid

Will Trisomy 21 Down Syndrome Disappear?

In the 1960s, a newborn with Down syndrome was usually sent to an institution. In the 1970s, testing for trisomy 21 Down syndrome became available but was riskier than it is today because the only technologies available—amniocentesis and chorionic villus sampling—can cause spontaneous abortion. Only women over age 35 or with a family history of chromosome abnormalities were tested, because their risk of carrying a fetus with a detectable condition exceeded the risk of the test causing spontaneous abortion. Maternal serum marker testing in the 1980s and then cell-free fetal DNA testing starting in 2011 made it safer and easier for any pregnant woman to learn if the fetus had trisomy 21 Down syndrome.

The ability to detect trisomy 21 prenatally has led to a decrease in the population of people who have this condition. In the United Kingdom, about 90 percent of couples learning that their child will have trisomy 21 end the pregnancy. In the United States, it is about 60 percent. At the same time, some organizations representing people with Down syndrome, distressed that affected individuals will have fewer peers as time goes on and prenatal testing becomes more widespread, began to pressure pregnant women not to avoid having children with the condition. Laws have been passed in some states that ban pregnancy termination if test results indicate Down syndrome.

Questions for Discussion

1. Should state laws control the circumstances under which women can end pregnancies?
2. Look up the definition of eugenics, or read about it in chapter 15. Do you think that use of prenatal testing for trisomy 21 is a eugenic measure?
3. A researcher who has pioneered cell-free fetal DNA testing to detect trisomies has received threats from people accusing her of killing babies. She counters that the test will save lives. How can cell-free fetal DNA testing save lives?
4. What impact do you think the wider availability of early, safe, and accurate testing for trisomy 21 Down syndrome will have on society? Discuss views that might come from a pregnant woman and her partner, a parent of an adult child with Down syndrome, and a physician who offers prenatal testing.

---

malfunction, alcoholism, emotional trauma, or "maternal reproductive exhaustion." The increased risk of Down syndrome correlates to maternal age, not to the number of children in the family. **Bioethics** examines the effect of prenatal diagnosis on the prevalence of people with trisomy 21 Down syndrome.

Trisomy 18

When an ultrasound scan early in pregnancy revealed a small fetus with low-set ears, a small jaw, a pocket of fluid in the brain, and a clenched fist, the parents-to-be were offered amniocentesis to confirm the extra chromosome 18 that the scan indicated might be there. Further ultrasound scans showed that only one kidney worked, the heart had holes between the chambers, and part of the intestine lay outside the stomach in a sac. The boy was delivered at 36 weeks of gestation, after his heart rate became erratic during a routine prenatal visit. He lived only 22 days.

Most individuals with trisomy 18 (**figure 13.15**) (Edwards syndrome) or trisomy 13 (Patau syndrome) are not born or die in infancy, but a few have lived into young adulthood. Most children who have trisomy 18 have great physical and intellectual disabilities, with developmental skills stalled at the 6-month level. Major abnormalities include heart defects, a displaced liver, growth retardation, and oddly clenched fists. Milder signs include overlapping fingers, a narrow and flat skull, low-set ears, a small mouth, unusual fingerprints, and

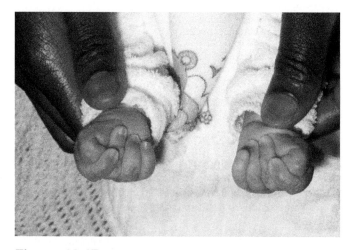

Figure 13.15 Trisomy 18. An infant with trisomy 18 clenches the fists in a characteristic way, with fingers overlapping. © Wellcome Image Library/Custom Medical Stock Photo

"rocker-bottom" feet. Most cases of trisomy 18 arise from nondisjunction in meiosis II of the oocyte.

Trisomy 13

The 15-month-old is a "long-term survivor" of trisomy 13. She is small for her age, at the 5th percentile for weight, but she is happy, curious, and playful. Her physical skills, however, lag.

She can finally, with great effort, sit up, but cannot yet crawl. She has about 20 minor seizures a day, which look like startles, and has difficulty eating because of persistent acid reflux. She is also missing a rib. Early surgeries corrected a cleft lip and palate, removed an extra finger and toe, and repaired a hernia. The girl is intellectually disabled, but her parents hope she will live long enough to attend preschool. Unlike many others with trisomy 13, she can see and hear.

Trisomy 13 has different signs and symptoms than trisomy 18. Most striking is fusion of the developing eyes into one large eyelike structure in the center of the face, or a small or absent eye. Major abnormalities affect the heart, kidneys, brain, face, and limbs. The nose is often malformed, and cleft lip and/or palate is present in a small head. There may be extra fingers and toes. Ultrasound examination of an affected newborn may reveal an extra spleen, atypical liver, rotated intestines, and an abnormal pancreas.

Sex Chromosome Aneuploids: Female

People with sex chromosome aneuploidy have extra or missing sex chromosomes. **Table 13.5** indicates how these aneuploids can arise. Some conditions can result from nondisjunction in meiosis in the male *or* female. Sex chromosome aneuploids are generally associated with much less severe symptoms and characteristics than autosomal aneuploids.

XO Syndrome

At age 17, the young woman still looked about 12. She was short, her breasts had never developed, and she had never menstruated. When she turned 16, her physician suggested that she have her chromosomes checked. The karyotype revealed absence of an X chromosome (XO, or Turner syndrome). The diagnosis explained other problems, such as poor hearing, high blood pressure, low thyroid function, and the "beauty marks" that dotted her skin.

Like the autosomal aneuploids, XO syndrome is more frequent among spontaneously aborted fetuses than among newborns—99 percent of XO fetuses are not born. The syndrome affects 1 in 2,500 female births. However, if amniocentesis or CVS was not done, a person with XO syndrome would likely not know she has a chromosome abnormality until she lags in sexual development and has her chromosomes checked. Two X chromosomes are necessary for normal sexual development in females.

At birth, a girl with XO syndrome looks normal, except for puffy hands and feet caused by impaired lymph flow. In childhood, signs of XO syndrome include wide-set nipples, soft nails that turn up at the tips, slight webbing at the back of the

Table 13.5	How Nondisjunction Leads to Sex Chromosome Aneuploids		
Situation	**Oocyte**	**Sperm**	**Consequence**
Normal	X	Y	46,XY normal male
	X	X	46,XX normal female
Female nondisjunction	XX	Y	47,XXY Klinefelter syndrome
	XX	X	47,XXX triplo-X
		Y	45,Y nonviable
		X	45,X Turner syndrome
Male nondisjunction (meiosis I)	X		45,X Turner syndrome
	X	XY	47,XXY Klinefelter syndrome
Male nondisjunction (meiosis II)	X	XX	47,XXX triplo-X
	X	YY	47,XYY Jacobs syndrome
	X		45,X Turner syndrome
Male and female nondisjunction	XX	YY	48,XXYY syndrome

neck, short stature, coarse facial features, and a low hairline at the back of the head. About half of people with XO syndrome have impaired hearing and frequent ear infections due to a small defect in the shape of the coiled part of the inner ear. They cannot hear certain frequencies of sound.

At sexual maturity, sparse body hair develops, but the girls do not ovulate or menstruate, and their breasts do not develop. The uterus is very small, but the vagina and cervix are normal size. In the ovaries, oocytes mature too fast, depleting the supply during infancy. XO syndrome may impair the ability to solve math problems that entail envisioning objects in three-dimensional space, and may cause memory deficits, but intelligence is normal. Low doses of hormones (estrogen and progesterone) can stimulate development of secondary sexual structures for girls diagnosed before puberty, and growth hormone can maximize height.

Individuals who are mosaics (only some cells are XO) may have children, but their offspring are at high risk of having too many or too few chromosomes. XO syndrome is unrelated to the age of the mother. The effects of XO syndrome continue past the reproductive years. Adults with XO syndrome are more likely to develop certain diseases than the general population, including osteoporosis, types 1 and 2 diabetes, and colon cancer. The many signs and symptoms of XO syndrome result from loss of specific genes. Life span is shortened slightly.

Triplo-X

About 1 in every 1,000 females has an extra X chromosome in each of her cells, a condition called triplo-X. The only symptoms are tall stature and menstrual irregularities. Although triplo-X females are rarely intellectually disabled, they tend to be less intelligent than their siblings who have the normal number of chromosomes. The lack of symptoms reflects the protective effect of X inactivation—all but one of the X chromosomes is inactivated.

Sex Chromosome Aneuploids: Male

Any individual with a Y chromosome is a male. A man can have one or more extra X or Y chromosomes.

XXY Syndrome

Looking back, the man's only indication of XXY syndrome was small testes. When his extra X chromosome was detected when he was 25, suddenly his personality quirks made sense. As a child and teen, he had been very shy, had trouble making friends, and angered easily. His parents had consulted several clinicians. Psychologists, psychiatrists, and therapists diagnosed learning disabilities. One said that the man would never be able to attend college. Yet he earned two bachelor's degrees and became a successful software engineer. The man heads a support group for men with XXY syndrome.

About 1 in 500 males has the extra X chromosome that causes XXY (Klinefelter) syndrome. Severely affected men are underdeveloped sexually, with rudimentary testes and prostate glands and sparse pubic and facial hair. They have very long arms and legs, large hands and feet, and may develop breast tissue.

Testosterone injections during adolescence can limit limb lengthening and stimulate development of secondary sexual characteristics. Boys and men with XXY syndrome may be slow to learn, but they are usually not intellectually disabled unless they have more than two X chromosomes, which is rare.

XXY syndrome is the most common genetic or chromosomal cause of male infertility. Doctors can select sperm that contain only one sex chromosome and use the sperm to fertilize oocytes *in vitro*. However, sperm from men with XXY syndrome are more likely to have extra chromosomes—usually X or Y, but also autosomes—than sperm from men who do not have XXY syndrome.

XXYY Syndrome

The boy's problems were at first so common that it was years before a chromosome check revealed he had an extra X and an extra Y chromosome. He was late to sit, crawl, walk, and talk. In preschool, he had frequent outbursts and made inappropriate comments. He was tall and clumsy and drooled and choked easily. The boy would run about flapping his arms, then hide under a chair. Severe ulcers formed on his legs.

By the second grade, the boy's difficulties alarmed his special education teacher, who suggested to his parents that they and their son have their chromosomes checked. The parents' chromosomes were normal. He must have been conceived from a very unusual oocyte meeting a very unusual sperm, both arising from nondisjunction. The extra sex chromosomes explained many of the boy's problems, and even a few that hadn't been recognized, such as curved pinkie fingers, flat feet, and scoliosis. He began receiving testosterone injections so that his teen years would be more normal than his difficult childhood had been.

About 1 in 17,000 newborn boys have an extra X chromosome *and* an extra Y chromosome, and, until recently, were diagnosed with Klinefelter syndrome. These XXYY males have more severe behavioral problems than males with XXY syndrome and tend to develop foot and leg ulcers, resulting from poor venous circulation. Attention deficit disorder, obsessive compulsive disorder, autism, and learning disabilities typically develop by adolescence. In the teen years, testosterone level is low, development of secondary sexual characteristics is delayed, and the testes are undescended. A man with XXYY syndrome is infertile.

XYY Syndrome

In 1965, cytogeneticist Patricia Jacobs surveyed 197 inmates at Carstairs, a high-security prison in Scotland. Of 12 men with unusual chromosomes, seven had an extra Y. Jacobs's findings were repeated for mental institutions, and soon after, Newsweek magazine ran a cover story on "congenital criminals," blaming such behavior on the extra Y chromosome. XYY syndrome, also known as Jacobs syndrome, became a legal defense for committing a violent crime.

In the early 1970s, newborn screens began in hospital nurseries in England, Canada, Denmark, and Boston. Social workers and psychologists visited XYY boys to offer "anticipatory guidance" to the parents on how to deal with their toddling future criminals. By 1974, geneticists and others halted the programs, pointing out that singling out these boys on the basis of a few statistical studies was inviting self-fulfilling prophecy.

One male in 1,000 has an extra Y chromosome. Today, we know that 96 percent of XYY males are *not* destined to become criminals, but instead may be very tall and have acne. Problems with speech and understanding language are subtle. The mother of one young man who has an extra Y chromosome and is in middle school, but is intellectually at a third-grade level, explains his difficulties: *He did not talk until speech therapy at age 5. He cannot tell a joke—he does not have the ability to relate information and put it together on his own, so a lot of his language is adopted from what he sees and hears. He is emotionally age 5 to 7, and he is almost 12 and just started to dress himself fully.*

The continued prevalence of XYY among mental and penal institution populations may have indirect roots more in psychology than biology. Large body size may lead teachers, employers, parents, and others to expect more of these people, and a few XYY individuals may deal with this stress aggressively.

XYY syndrome arises from nondisjunction in the male, producing a sperm with two Y chromosomes that fertilizes a normal oocyte. Geneticists have never observed a sex chromosome constitution of one Y and no X because the X chromosome carries so many genes that a YO embryo stops developing very early.

Key Concepts Questions 13.3

1. How many chromosomes are in a normal human karyotype?
2. What is a polyploid?
3. Explain how nondisjunction generates aneuploids.
4. Compare the severities of monosomies and trisomies, and of sex chromosome and autosomal aneuploidies.
5. Describe the most common aneuploid conditions.

13.4 Atypical Chromosome Structure

A chromosome can be structurally atypical in several ways. It may have too much or too little genetic material, or a stretch of DNA that is inverted or moved and inserted into a different type of chromosome (**figure 13.16**). Atypical chromosomes are balanced if they have the normal amount of genetic material and unbalanced if they have extra or missing DNA sequences.

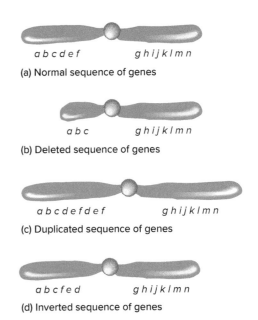

a b c d e f   g h i j k l m n

(a) Normal sequence of genes

a b c   g h i j k l m n

(b) Deleted sequence of genes

a b c d e f d e f   g h i j k l m n

(c) Duplicated sequence of genes

a b c f e d   g h i j k l m n

(d) Inverted sequence of genes

Figure 13.16 **Chromosome abnormalities.** If a hypothetical normal gene sequence appears as shown in **(a)**, then **(b)** represents a deletion, **(c)** a duplication, and **(d)** an inversion.

Deletions and Duplications

The beautiful baby had a strange cry. When the child was a few weeks old, her mother was talking on the phone to her sister, a physician. The sister, overhearing the baby's cries, asked when they had gotten a cat. They hadn't. Alarmed, the sister advised telling the child's pediatrician about the cry. The pediatrician heard the mewling and agreed. The baby's karyotype revealed what both doctors had suspected: 5p⁻ (5p minus or deletion) syndrome, also called cri-du-chat syndrome. One of the baby's copies of chromosome 5 is missing part of the short arm. The predictions were dire, but the child has defied expectations and can walk, eat, see, and hear, although she has autistic behaviors, is small, cannot use a toilet, and spikes fevers. She can't speak, but communicates quite well in other ways and goes to school.

Deletion and **duplication** mutations are missing and extra DNA sequences, respectively. They are types of copy number variants (CNVs). The more genes involved, the more severe the associated syndrome. **Figure 13.17** depicts a common duplication, of part of chromosome 15. Many deletions and duplications are *de novo* (neither parent has the abnormality).

A technique called **comparative genomic hybridization (CGH)** is used to detect very small CNVs, which are also termed **microdeletions** and **microduplications**. The technique compares the abundance of copies of a particular CNV in the same amount of DNA from two people—one with a medical condition, one healthy. CGH is used to help diagnose autism, intellectual disability, learning disabilities, and other behavioral conditions. For example, the technique showed that a young boy who had difficulty concentrating and sleeping and would often scream for no apparent reason had a small

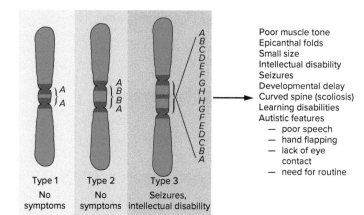

Figure 13.17 A duplication. A study of duplications of parts of chromosome 15 revealed that small duplications do not affect the phenotype, but larger ones may. The letters indicate specific DNA sequences, which serve as markers to compare chromosome regions. Note that the duplication is also inverted.

duplication in chromosome 7. A young girl plagued with head-banging behavior, digestive difficulties, severe constipation, and great sensitivity to sound had a microdeletion in chromosome 16. Other microdeletions cause male infertility.

Deletions and duplications can arise from chromosome rearrangements. These include translocations, inversions, and ring chromosomes.

Translocation Down Syndrome

When the child was born, while her parents marveled at her beauty, the obstetrician noted that the broad, tilted eyes and sunken nose looked like the face of a child with Down syndrome. The doctor knew that the mother had experienced two spontaneous abortions, a family history suggesting a chromosome problem. So the doctor looked for the telltale single crease in the palms of people with Down syndrome, and found it. Gently, she told the parents that she'd like to check their daughter's chromosomes.

A few days later, the new parents learned that their daughter has an unusual form of Down syndrome that she inherited from one of them, rather than the more common "extra chromosome" (trisomy) form. Because the father's mother and sister had also had several pregnancy losses, the exchanged chromosomes likely came from his side. Karyotypes of the parents confirmed that the father was a translocation carrier: One chromosome 14 had attached to one chromosome 21, and distribution of this unusual chromosome in meiosis had led to various imbalances, depicted in **figure 13.18**.

In a translocation, different (nonhomologous) chromosomes exchange or combine parts. Translocations can be inherited because they can be present in carriers, who have the normal amount of genetic material, but it is rearranged. A translocation can affect the phenotype if it breaks a gene or leads to duplications or deletions in the chromosomes of offspring.

There are two major types of translocations, as well as a few rarer types. In a **Robertsonian translocation**, the short arms of two different acrocentric chromosomes break, leaving sticky ends on the two long arms that join, forming a single, large chromosome with two long arms (see chromosome 14/21 in figure 13.18). The tiny short arms are lost, but their DNA sequences are repeated elsewhere in the genome, so the loss does not cause symptoms. The person with the large, trans-located chromosome, called a **translocation carrier**, has 45 chromosomes instead of 46, but may not have symptoms if no crucial genes have been deleted or damaged. Even so, he or she may produce unbalanced gametes—sperm or oocytes with too many or too few genes. This can lead to spontaneous abortion or birth defects.

In 1 out of 20 cases of Down syndrome, a parent has a Robertsonian translocation between chromosome 21 and another, usually chromosome 14. That parent produces some gametes that do not have either of the involved chromosomes and some gametes that have extra material from one of the translocated chromosomes. In such a case, each fertilized ovum has a 1 in 2 chance of being spontaneously aborted, and a 1 in 6 chance of developing into an individual with Down syndrome. The risk of giving birth to a child with Down syndrome is theoretically 1 in 3, because the spontaneous abortions are not births. However, because some Down syndrome fetuses spontaneously abort, the empiric risk of a couple in this situation having a child with Down syndrome is about 15 percent. The other two outcomes—a fetus with normal chromosomes or a translocation carrier like the parent—have normal phenotypes. Either a male or a female can be a translocation carrier, and the condition is not related to age. About 1 in 1,000 individuals in the general population carries a Robertsonian translocation.

Much rarer than being a heterozygote for a Robertsonian translocation is being a homozygote, because such individuals can only arise from inheriting one copy of the unusual chromosome from each parent—which typically means the parents are related and inherited the translocation from a shared ancestor, like a common great-grandparent. Robertsonian homozygotes have 44 chromosomes rather than the normal 46. A case report describes a healthy 25-year-old man from China who has 44 chromosomes because each chromosome 14 joins a chromosome 15. His parents, both translocation carriers, were first cousins. The Chinese man's sperm carry 21 autosomes and an X or Y, and he can only be fertile with a woman whose cells have the same 44 chromosome types.

Robertsonian translocation homozygotes bring up an interesting scenario. If several people with the same

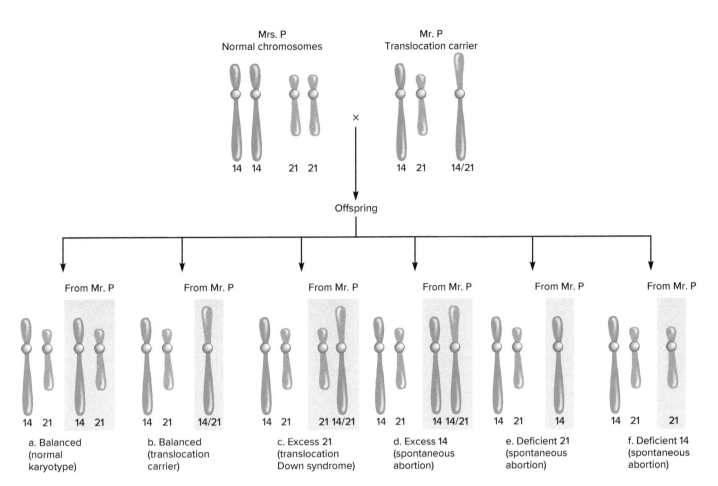

Figure 13.18 **A Robertsonian translocation.** Mr. P. has only 45 chromosomes because the long arm of one chromosome 14 has joined the long arm of one chromosome 21. He has no symptoms. Mr. P. makes six types of sperm. **(a)** A sperm with one normal chromosome 14 and one normal 21 yields a normal child when fertilized. **(b)** A sperm carrying the translocated chromosome produces a child who is a translocation carrier, like Mr. P. **(c)** If a sperm contains Mr. P.'s normal 21 and his translocated chromosome, the child receives too much chromosome 21 material and has Down syndrome. **(d)** A sperm containing the translocated chromosome and a normal 14 leads to excess chromosome 14 material, which is lethal in the embryo or fetus. If a sperm does not have chromosome 21 **(e)** or 14 **(f),** a monosomy results, which is lethal prenatally. (Chromosome arm lengths are not precisely accurate.)

44 chromosome types breed among themselves, they could, theoretically, start a new human subspecies. The probability of two carriers of the same Robertsonian translocation randomly meeting and having children together is about 1 in 4 million. **Clinical Connection 13.1** describes how genomewide SNP analysis of sequence variants can miss a Robertsonian translocation if all of the parts of the affected chromosomes are present.

In a **reciprocal translocation**, the second major type of translocation, two different chromosomes exchange parts (**figure 13.19**). About 1 in 600 people is a carrier for a reciprocal translocation. FISH can be used to highlight the involved chromosomes. If the chromosome exchange does not break any genes, then a person who has both translocated chromosomes is healthy and a translocation carrier. He or she has the normal amount of genetic material, but it is rearranged. A reciprocal translocation carrier can have symptoms if one of the two breakpoints lies in a gene, disrupting its function. For example,

a translocation between chromosomes 11 and 22 causes infertility in males and recurrent pregnancy loss in females. However, a reciprocal translocation that arises *de novo* in a sperm or oocyte can affect health if fertilization occurs and development proceeds, because the resulting individual would have extra or missing genetic material.

Reciprocal translocations usually occur in specific chromosomes that have unstable parts. Vulnerable parts of chromosomes are located where the DNA is so symmetrical in sequence that complementary base pairing occurs within the same DNA strand, folding it into loops during DNA replication. These contortions can break both DNA strands, which can enable parts of two different chromosomes to switch places.

Unbalanced gametes can result when a reciprocal translocation is inherited or *de novo,* just as they can from a Robertsonian translocation. The four possibilities are (1) transmitting two normal copies of the two involved chromosomes; (2) transmitting two abnormal copies, with no effect on the phenotype

Distinguishing a Robertsonian Translocation from a Trisomy

In this age of genome sequencing, we can lose sight of the importance of how our genomes are distributed over 23 pairs of chromosomes. Rearrangements of the chromosome pairs are invisible to sequencing if two complete genomes are present, or may lead to diagnosis of the wrong type of chromosome abnormality. However, the distribution of genes on chromosomes, and the fact that genes on the same chromosome are transmitted together unless separated during meiosis by a crossover, is important in genetic counseling of parents-to-be. Consider a case in which DNA sequencing masked a Robertsonian translocation, leading to an error in deducing the type of chromosomal abnormality.

The young couple had suffered two early pregnancy losses. The second time, the woman collected some of the tissue she had passed and gave it to her obstetrician, who sent it for genetic testing. The report came back indicating extra material from chromosome 22. Although 22 is a tiny chromosome, it is packed with genes and so the excess is enough to halt development just as an embryo is becoming a fetus. The lab report profiled single nucleotide polymorphisms (SNPs), landmarks among the chromosomes, and detected overrepresentation of those on chromosome 22. The report did *not* provide a photograph of the actual chromosomes in the fetus, an old-fashioned karyotype. The obstetrician assumed that extra chromosome 22 material meant a trisomy—an anomaly that was only likely to recur in future pregnancies at the population empiric risk of 1 in 100. The doctor reassured the patient and her partner.

Then the couple had a third pregnancy loss. The SNP test again came back with extra chromosome 22 material. The genetic counselor was quick to note the significance. If two fetuses in the same family had the same extra chromosome 22 material, then it was much more likely due to a translocation, and not a trisomy. That distinction would mean that every conception for this couple had a 2 in 3 risk of having an imbalance of genetic material (see figure 13.18). On the counselor's advice and interpretation, the physician ordered a karyotype of the products of conception. As the counselor suspected, the lab identified a Robertsonian

Figure 13A **Muntjacs are especially prone to Robertsonian translocations.** The ancestral species *Muntiacus reevesi* has 46 chromosomes, which coalesced via Robertsonian translocations into the six female and seven male chromosomes of the Indian muntjac *Muntiacus muntjac*, shown here. © *Louise Heusinkveld/Getty*

translocation in which one large chromosome 14 had some chromosome 22 material—in addition to two normal copies of chromosome 22. The overload stopped prenatal development.

Robertsonian translocations are very rare in humans, but more common in certain other species (**figure 13A**).

Questions for Discussion

1. Explain why the risk of recurrence of extra chromosome 22 material in this family is much higher for a Robertsonian translocation than for trisomy 22.
2. Describe the karyotype of the parent who passed on the translocated chromosome.
3. What other types of chromosomal problems might arise in future pregnancies for this couple?

of an offspring unless a vital gene is disrupted; or (3 and 4) transmitting either translocated chromosome, which introduces extra or missing genetic material and likely would affect the phenotype.

A rare type of translocation is an insertional translocation, in which part of one chromosome inserts into a nonhomologous chromosome. Symptoms may result if the inserted DNA disrupts a vital gene or if crucial DNA sequences are lost or present in excess.

A carrier of any type of translocation can produce unbalanced gametes—sperm or oocytes that have deletions or duplications of some of the genes in the translocated chromosomes. The phenotype depends upon the genes that the rearrangement disrupts and whether genes are extra or missing. A translocation and a deletion can cause the same syndrome if they affect the same part of a chromosome.

A genetic counselor suspects a translocation when a family has a history of birth defects, pregnancy loss, and/or stillbirths. Prenatal testing may reveal a translocation in a fetus, which can then be traced back to a parent who is a translocation carrier. If neither parent has the translocation in cells typically used to check chromosomes, then the translocation is *de novo*,

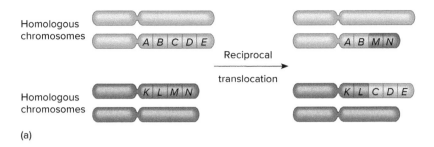

Homologous chromosomes

Homologous chromosomes

Reciprocal translocation

A B C D E

K L M N

A B M N

K L C D E

(a)

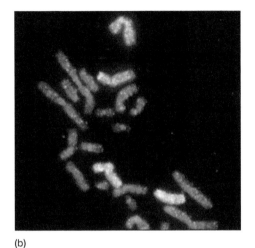

(b)

Figure 13.19 **A reciprocal translocation.** In a reciprocal translocation, two nonhomologous chromosomes exchange parts. In **(a)**, genes *C, D,* and *E* on the blue chromosome exchange positions with genes *M* and *N* on the red chromosome. In **(b)**, a reciprocal translocation has been highlighted using FISH. The pink chromosome with the dab of blue, and the blue chromosome with a small section of pink, are the translocated chromosomes. *Courtesy of Lawrence Livermore National Laboratory*

or a parent is a somatic mosaic and only some gametes have the atypical chromosome.

Inversions

The couple was excited about getting the results of the amniocentesis because a pregnancy had not gone this far before. Then they were called to come in for the results. Expecting bad news, they were surprised to learn that the fetus had an inverted chromosome. Some of the bands that normally appear on chromosome 11 were flipped. The genetic counselor advised them to have their own chromosomes checked. Karyotyping revealed that the woman had the same inversion, so it likely wouldn't harm their daughter, although someday she, too, might experience pregnancy loss.

An inverted sequence of chromosome bands disrupts important genes and harms health in only 5 to 10 percent of cases. If neither parent has the inversion, then it arose in a gamete *de novo* or because a parental ovary or testis is mosaic. The specific effects of an inverted chromosome may depend upon which genes the flip disrupts. Consulting the "reference" human genome (a consensus of common sequences compiled

from many genomes) can help to identify genes that might be implicated in a particular inversion.

Like a translocation carrier, an adult who is heterozygous for an inversion can be healthy, but have reproductive problems. One woman had an inversion in the long arm of chromosome 15 and had two spontaneous abortions, two stillbirths, and two children who died within days of birth. She did eventually give birth to a healthy child. How did the inversion cause these problems?

Inversions with such devastating effects can be traced to meiosis, when a crossover occurs between the inverted chromosome segment and the noninverted homolog. To allow the genes to align, the inverted chromosome forms a loop. When crossovers occur within the loop, some areas are duplicated and some deleted in the resulting recombinant chromosomes. In inversions, the atypical chromosomes result from the chromatids that crossed over.

Two types of inversions are distinguished by the position of the centromere relative to the inverted section. **Figure 13.20** shows a **paracentric inversion**, in which the inverted section does not include the centromere. A single crossover within the inverted segment gives rise to one normal, one

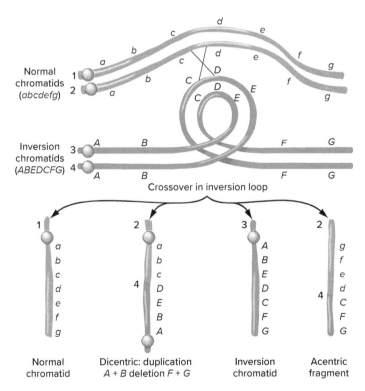

Figure 13.20 **Paracentric inversion.** A paracentric inversion in one chromosome leads to one normal chromatid, one inverted chromatid, one with two centromeres (dicentric), and one with no centromere (an acentric fragment) if a crossover occurs with the normal homolog. The letters *a* through *g* denote genes.

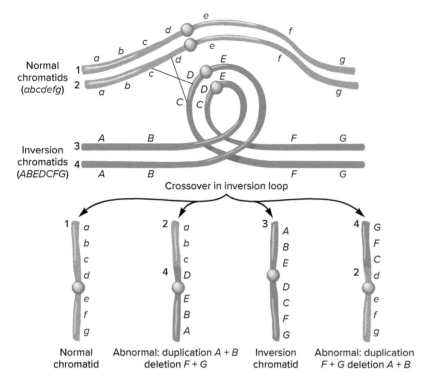

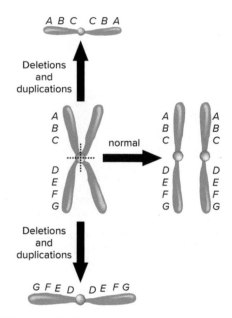

Figure 13.21 **Pericentric inversion.** A pericentric inversion in one chromosome leads to two chromatids with duplications and deletions, one normal chromatid, and one inverted chromatid if a crossover occurs with the normal homolog.

Figure 13.22 **Isochromosomes have identical arms.** They form when chromatids divide along the wrong plane (in this depiction, horizontally rather than vertically).

inversion, and two highly atypical chromatids. One abnormal chromatid retains both centromeres and is termed dicentric. When the cell divides, the two centromeres are pulled to opposite sides of the cell, and the chromatid breaks, leaving pieces with extra or missing segments. The second type of atypical chromatid resulting from a crossover within an inversion loop is a small piece that lacks a centromere, called an acentric fragment. When the cell divides, the fragment is lost because a centromere is required for cell division.

A **pericentric inversion** includes the centromere within the loop. A crossover in the inversion loop produces two chromatids that have duplications and deletions, but one centromere each, plus one normal and one inversion chromatid (**figure 13.21**).

Isochromosomes and Ring Chromosomes

An **isochromosome** is the result of another meiotic error that leads to unbalanced genetic material. It is a chromosome that has identical arms. An isochromosome forms when, during division, the centromeres part in the wrong plane (**figure 13.22**). Isochromosomes are known for chromosomes 12 and 21 and for the long arms of the X and the Y. Some women with Turner syndrome do not have the more common XO abnormality, but have an isochromosome with the long arm of the X chromosome duplicated but the short arm absent.

Chromosomes shaped like rings form in 1 of 25,000 conceptions. Ring chromosomes may arise when telomeres are lost,

leaving sticky ends that adhere (**figure 13.23**). Exposure to radiation can form rings. Any chromosome can circularize. Most ring chromosomes consist of DNA repeats and do not affect health. The few that do, listed in **table 13.6**, arise *de novo*. They all cause a small head (microcephaly), learning disabilities or intellectual disabilities, slow growth, and unusual facial features.

Table 13.7 summarizes causes of types of abnormal chromosomes.

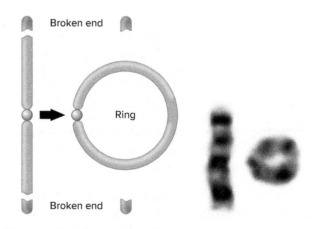

Figure 13.23 **A ring chromosome.** A ring chromosome may form if the chromosome's tips (telomeres) break, forming sticky ends. Genes can be lost or disrupted, possibly causing symptoms. *Courtesy of Ring Chromosome 20 Foundation*

Table 13.6 — Ring Chromosome Syndromes

Chromosome	Symptoms
9	Microcephaly; triangular forehead; closely spaced, slanted, protruding eyes; arched brows; small jaw; congenital heart defects; abnormal male genitalia; intellectual disability
13	Microcephaly, small eyes, developmental delay, underdevelopment of certain brain parts, difficulty feeding, unusual hands and feet, small or closed anus, abnormal genitals, heart defects, abnormal kidneys
14	Microcephaly, puffy hands and feet, seizures, delayed speech and motor skills, recurrent infections, intellectual disability
20	Microcephaly, short stature, seizures, intellectual disability
22	Microcephaly, wide nose, large ears and eyes, floppiness (hypotonia), delayed speech and language, autistic behaviors, learning disabilities

Table 13.7 — Causes of Abnormal Chromosomes

Abnormalities	Causes
Numerical Abnormalities	
Polyploidy	Multiple fertilization or diploid gamete leads to extra sets of chromosomes
Aneuploidy	Nondisjunction (in meiosis or mitosis) leads to lost or extra chromosomes when not all chromatid pairs separate in anaphase
Structural Abnormalities	
Deletions and duplications	Translocation
	Crossover between a chromosome that has a pericentric inversion and its noninverted homolog
Translocation	Exchange between nonhomologous chromosomes at parts that are unstable due to symmetrical DNA sequences.
Inversion	Breakage and reunion of fragment in same chromosome, but with wrong orientation
Dicentric and acentric chromatids	Crossover between a chromosome with a paracentric inversion and its noninverted homolog
Ring chromosome	A chromosome loses telomeres and the ends fuse, forming a circle
Chromothripsis	Up to three chromosomes spontaneously shatter

Key Concepts Questions 13.4

1. Which chromosome rearrangements can cause deletions and duplications?
2. Distinguish between the two major types of translocations.
3. What must occur for a translocation or inversion to cause symptoms?
4. How do isochromosomes and ring chromosomes arise?

13.5 Uniparental Disomy—A Double Dose from One Parent

If nondisjunction (unequal chromosome division) occurs in both a sperm and an oocyte that join, a pair of chromosomes (or their parts) can come solely from one parent, rather than one from each parent, as Mendel's law of segregation predicts. For example, if a sperm that does not have a chromosome 14 fertilizes an ovum that has two of that chromosome, then an individual with the normal 46 chromosomes results, but both copies of chromosome 14 come from the female.

Inheriting two chromosomes or chromosome segments from one parent is called **uniparental disomy (UPD)** ("two bodies from one parent"). UPD can also arise from a trisomic embryo in which some cells lose the extra chromosome, leaving two homologs from one parent. For example, an embryo may have trisomy 21, with the extra chromosome 21 coming from the father. If in some cells the chromosome 21 from the mother is lost, then both remaining copies of the chromosome are from the father.

Because UPD requires the simultaneous occurrence of two rare events—either nondisjunction of the same chromosome in sperm and oocyte or trisomy followed by chromosome loss—it is indeed rare. In addition, many cases of UPD are probably never seen, because bringing together identical homologs inherited from one parent could give the fertilized ovum homozygous lethal alleles. Development would halt. Other cases of UPD may go undetected if they cause known recessive conditions and both parents are assumed to be carriers, when only one parent contributed to the offspring's illness. This situation was how UPD was discovered.

In 1988, Arthur Beaudet of the Baylor College of Medicine saw an unusual patient with cystic fibrosis (see Clinical Connection 4.1). In comparing *CFTR* alleles of the patient to those of her parents, Beaudet found that only the mother was a carrier—the father had two normal alleles. Beaudet constructed haplotypes for each parent's chromosome 7, which includes the *CFTR* gene, and he found that the daughter had two copies from her mother, and none from her father (**figure 13.24**). How did this happen?

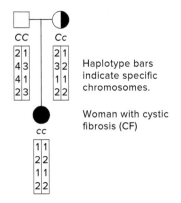

Figure 13.24 **Uniparental disomy.** Uniparental disomy doubles part of one parent's genetic contribution. In this family, the woman with CF inherited two copies of her mother's chromosome 7, and neither of her father's. Unfortunately, it was the chromosome with the disease-causing allele that she inherited in a double dose.

Apparently, in the patient's mother, nondisjunction of chromosome 7 in meiosis II led to formation of an oocyte bearing two identical copies of the chromosome, instead of the usual one copy. A sperm that had also undergone nondisjunction and did not have a chromosome 7 then fertilized the abnormal oocyte. The mother's extra genetic material compensated for the father's deficit, but unfortunately, the child inherited a double dose of the mother's chromosome that carried the mutant *CFTR* allele. In effect, inheriting two of the same chromosome from one parent shatters the protection that combining genetic material from two individuals offers. This protection is the defining characteristic of sexual reproduction.

UPD may also cause disease if it removes the contribution of the important parent for an imprinted gene. Recall from chapter 6 that an imprinted gene is expressed if it comes from one parent, but silenced if it comes from the other (see figure 6.13). If UPD removes the parental genetic material that must be present for a critical gene to be expressed, a mutant phenotype results. The classic example of UPD disrupting imprinting is Prader-Willi syndrome and Angelman syndrome, for which UPD causes 20 to 30 percent of cases (see figure 6.15). These diseases arise from mutations in different genes that are closely linked in a region of the long arm of chromosome 15, where imprinting occurs. They both cause intellectual disability and a variety of other symptoms, but are quite distinct.

Some children with Prader-Willi syndrome have two parts of the long arm of chromosome 15 from their mothers. The disease results because the father's Prader-Willi gene must be expressed for the child to not have the associated illness. For Angelman syndrome, the situation is reversed. Children have a double dose of their father's DNA in the same chromosomal region implicated in Prader-Willi syndrome, with no maternal contribution. The mother's gene must be present for health.

People usually learn their chromosomal makeup only when something goes wrong—when they have a family history of reproductive problems, exposure to a toxin, cancer, or symptoms of a known chromosomal disease. While researchers analyze human genome sequences, chromosome studies will continue to be part of health care.

Key Concepts Questions 13.5

1. What is uniparental disomy?
2. What are two ways that uniparental disomy can arise?
3. What are two ways that uniparental disomy can cause disease?

Summary

13.1 Portrait of a Chromosome

1. **Cytogenetics** is the study of chromosome variations and their effects on phenotypes.

2. **Heterochromatin** stains darkly and harbors many DNA repeats. **Euchromatin** is light staining and contains many protein-encoding genes.

3. A chromosome consists of DNA and proteins. Essential parts are the **telomeres**, **centromeres**, and origin of replication sites.

4. Centromeres include DNA repeats and proteins that enable the cell to divide.

5. Subtelomeres have telomere-like repeats that gradually change inward toward the centromere, as protein-encoding genes predominate.

6. Chromosomes are distinguishable by size, centromere position, satellites, and staining patterns. They are imaged and displayed in charts called **karyotypes**.

7. A **metacentric** chromosome has two fairly equal arms. A **submetacentric** chromosome has a long arm and a short arm. An **acrocentric** chromosome's centromere is near a tip, so that it has one long arm and one very short arm.

13.2 Detecting Chromosomes

8. **Amniocentesis** and **chorionic villus sampling (CVS)** directly obtain and analyze DNA from fetal cells or their supporting structures.

9. Fluorescence *in situ* hybridization (FISH) provides more specific information about chromosome bands than dyes. Ideograms display chromosome bands and digital karyotypes show chromosome parts.

10. Chromosomal shorthand indicates chromosome number, sex chromosome constitution, and type of abnormality.

11. Maternal serum markers and ultrasound findings are used to screen for elevated risk of a trisomy in a fetus.

12. Cell-free fetal DNA analysis is used for noninvasive prenatal diagnosis of trisomies by detecting overrepresentation of DNA pieces from the implicated chromosome, and for fetal exome and genome sequencing.

13.3 Atypical Chromosome Number

13. A **euploid** somatic human cell has twenty-two pairs of autosomes and one pair of sex chromosomes.

14. **Polyploid** cells have extra chromosome sets.

15. **Aneuploids** have extra or missing chromosomes. **Trisomies** (an extra chromosome) are less harmful than **monosomies** (absence of a chromosome), and sex chromosome aneuploidy is less severe than autosomal aneuploidy. **Nondisjunction** is uneven distribution of chromosomes in meiosis. It causes aneuploidy. Most autosomal aneuploids cease developing as embryos.

13.4 Atypical Chromosome Structure

16. **Deletions** and **duplications** result from crossing over after pairing errors in synapsis. Crossing over in an inversion heterozygote can also generate deletions and duplications. **Microdeletions** and **microduplications** explain many diseases.

17. In a **Robertsonian translocation**, the short arms of two acrocentric chromosomes break, leaving sticky ends on the long arms that join to form an unusual, large chromosome.

18. In a **reciprocal translocation**, two nonhomologous chromosomes exchange parts.

19. An insertional translocation places a DNA sequence from one chromosome into a nonhomologous chromosome.

20. A **translocation carrier** may have an associated phenotype and produces some unbalanced gametes.

21. A heterozygote for an inversion may have reproductive problems if a crossover occurs between the inverted region and the noninverted homolog, generating deletions and duplications. A **paracentric inversion** does not include the centromere; a **pericentric inversion** does.

22. **Isochromosomes** repeat one chromosome arm but delete the other. They form when the centromere divides in the wrong plane in meiosis. Ring chromosomes form when telomeres are removed, leaving sticky ends that adhere.

13.5 Uniparental Disomy—A Double Dose from One Parent

23. In **uniparental disomy (UPD)**, two copies of a chromosome or part of one are inherited from one parent, and not from the other. It can result from nondisjunction in both gametes, or from a trisomic cell that loses a chromosome.

24. Uniparental disomy causes symptoms if it creates a homozygous recessive state associated with an illness, or if it affects an imprinted gene.

Review Questions

1. Identify the structures and/or DNA sequences that must be present for a chromosome to carry information and withstand the forces of cell division.

2. How does the DNA sequence change with distance from the telomere?

3. Distinguish among a euploid, aneuploid, and polyploid cell.

4. Explain how analysis of fetal DNA in the maternal circulation can provide information about a trisomy.

5. How are amniocentesis and chorionic villus sampling (CVS) alike yet different?

6. Distinguish between a prenatal screening test and diagnostic test for trisomy 21 in a fetus.

7. What happens during meiosis to produce
   a. an aneuploid fertilized ovum?
   b. a polyploid fertilized ovum?
   c. increased risk of trisomy 21 in the offspring of a woman over age 40 at the time of conception?
   d. recurrent spontaneous abortions for a couple in which the man has a pericentric inversion?
   e. several children with Down syndrome in a family where one parent is a translocation carrier?

8. A human liver has patches of octaploid cells—they have eight sets of chromosomes. Explain how this condition might arise.

9. List the gender and possible phenotype of individuals with the following chromosomes:
   a. 47,XXX
   b. 45,X
   c. 47,XX, trisomy 21
   d. XXYY

10. Which chromosomal anomaly might be more prevalent among members of the National Basketball Association than in the general public? Cite a reason for your answer.

11. Explain the difference between a triploid and a trisomic embryo.

12. About 80 percent of cases of Edwards syndrome are caused by trisomy 18, another 10 percent are caused by mosaic trisomy 18, and 10 percent are attributed to translocation. Distinguish among these three chromosome abnormalities.

13. List three types of abnormal chromosomes that illustrate that the amount of genetic material that is missing or extra affects the severity of the associated phenotype.

14. List three types of atypical chromosomes that can cause duplications and/or deletions, and explain how they do so.

15. Distinguish among three types of translocations.

16. Why would having the same inversion on both members of a homologous chromosome pair not lead to unbalanced gametes, as having only one inverted chromosome would?

17. Define or describe the following technologies.
    a. FISH
    b. amniocentesis
    c. chorionic villus sampling (CVS)
    d. cell-free fetal DNA analysis

18. How many chromosomes does a person have who has Klinefelter syndrome and also trisomy 21?

19. Explain why a female cannot have XXY syndrome and a male cannot have XO syndrome.

20. List three causes of XO/Turner syndrome.

21. Explain how the sequence of genes on an isochromosome differs from normal.

22. Describe how a ring chromosome forms.

23. Explain how uniparental disomy does not appear to follow Mendel's first law (segregation).

Applied Questions

1. The chapter opener describes chromothripsis, in which one or a few chromosomes shatter. Why might whole genome sequencing of white blood cells not reveal chromothripsis?

2. Describe the evidence that a fetus has trisomy 21 provided by the following procedures.
    a. amniocentesis
    b. chorionic villus sampling (CVS)
    c. maternal serum markers
    d. cell-free fetal DNA analysis

3. Amniocentesis indicates that a fetus has the chromosomal constitution 46,XX,del(5)(p15). What does this mean? What might be the child's phenotype?

4. In a college genetics laboratory course, a healthy student constructs a karyotype from a cell from inside her cheek. She finds only one chromosome 3 and one chromosome 21, plus two unusual chromosomes that do not seem to have matching partners.
    a. What type of chromosomal abnormality does she have?
    b. Why doesn't she have any symptoms?
    c. Would you expect any of her relatives to have particular medical problems?

5. A fetus ceases developing in the uterus. Several of its cells are karyotyped. Approximately 75 percent of the cells are diploid and 25 percent are tetraploid (four copies of each chromosome). What happened, and when in development did it probably occur?

6. Distinguish between translocation and trisomy 21 Down syndrome.

7. A couple has a son diagnosed with XXY syndrome. Explain how the son's chromosome constitution could have arisen from either parent.

8. 22q11.2 deletion syndrome, also called DiGeorge syndrome, causes atypical parathyroid glands, a heart defect, and an underdeveloped thymus gland. About 85 percent of patients have a microdeletion of part of chromosome 22. A girl, her mother, and a maternal aunt have very mild DiGeorge syndrome. They all have a reciprocal translocation of chromosomes 22 and 2.
    a. How can a microdeletion and a translocation cause the same symptoms?
    b. Why were the people with the translocation less severely affected than the people with the microdeletion?
    c. What other problems might arise in the family with the translocation?

9. From 2 to 6 percent of people with autism have an extra chromosome that consists of two long arms of chromosome 15. It includes two copies of the chromosome 15 centromere. Two normal copies of the chromosome are also present. What type of chromosome abnormality in a gamete can lead to this karyotype, which is called isodicentric 15?

Case Studies and Research Results

1. An ultrasound of a pregnant woman detects a fetus and a similarly sized and shaped structure that has disorganized remnants of facial features at one end. Amniocentesis reveals that the fetus is 46,XX, but cells of the other structure are 47,XX,trisomy 2. No cases of trisomy 2 infants have ever been reported. However, individuals who are mosaics for trisomy 2 have a collection of defects, including a rotated and underdeveloped small intestine, a small head, a hole in the diaphragm, and seizures.
    a. How do the chromosomes of cells from the fetus and the other structure differ?
    b. What is the process that occurred during meiosis to yield the abnormal structure?

    c. Why are people with a complete extra chromosome 2 not seen, even though people with an extra chromosome 21 can live many years?
    d. List two factors that affect the type and severity of abnormalities in an individual who is an aneuploid mosaic.

2. Two sets of parents who have children with Down syndrome meet at a clinic. The Phelpses know that their son has trisomy 21. The Watkinses have two affected children, and Mrs. Watkins has had two spontaneous abortions. Why should the Watkinses be more concerned about future reproductive problems than the Phelpses? How are the offspring of the two families different, even though they have the same symptoms?

3. The genomes of 4 of 291 people with intellectual disability have a microdeletion in chromosome 17q21.3. The children have large noses, delayed speech, and mild intellectual disability. Each has a parent with an inversion in the same part of chromosome 17.

    a. Which arm of chromosome 17 is implicated in this syndrome?
    b. How can an inversion in a parent's chromosome cause a deletion in a child's chromosome?
    c. What other type of chromosome abnormality might occur in these children's siblings?

4. A 38-year-old woman, Dasheen, has amniocentesis. She learns that the fetus she is carrying has an inversion in chromosome 9 and a duplication in chromosome 18. She and her husband Franco have their chromosomes tested, and they learn that she has the duplication and Franco has the inversion. Both of the parents are healthy. Should they be concerned about the health of the fetus? Cite a reason for your answer.

5. What is a difficulty of applying the research technique used to silence the extra chromosome 21 that causes trisomy 21 Down syndrome to treating the condition? (See figure 13.14).

6. A child has severe intellectual disability, developmental delay, autism, and unusual facial features. His mother had a learning disability and slightly delayed development. Genome sequencing as well as karyotyping reveal that the child has excess material from chromosomes 9 and 14 and too little from chromosome 10. The mother has all of the genetic material, but she has four "derivative" chromosomes—they have seemingly mixed and matched their parts. These chromosomes consist of pieces of 9, 10, and 14; 9, 10, and 16; 10 and 16; and 9 and 4.

    a. What type of chromosomal event has likely occurred?
    b. How can the nature of the chromosomal event explain the mother's learning disability and developmental delay, but good health otherwise and a full genome?

7. In a family, three adult siblings had 44 chromosomes. Each had two chromosomes consisting of chromosomes 13 and 14.

    a. What type of chromosome abnormality do the siblings have?
    b. Describe the chromosomes of the parents of the three siblings.
    c. Explain why the siblings all have fertility problems.

8. Clinical Connection 2.2 describes the diagnostic odyssey of young Hannah Sames, who has the extremely rare, autosomal recessive condition giant axonal neuropathy. For many years her physicians assumed that her parents were each carriers. When researchers were planning to publish findings, they sequenced the gigaxonin gene in both parents, and were surprised to discover that Hannah's father is wild type. Hannah's deletions on both of her copies of chromosome 16 match the deletion on one of her mother's chromosomes. What type of chromosome abnormality discussed in the chapter does Hannah have?

PART 4 Population Genetics

© Tom Grill/Corbis RF

CHAPTER 14

Constant Allele Frequencies and DNA Forensics

A forensic scientist consults a DNA profile. The black bars represent short tandem repeats that form patterns used to exclude suspects in a crime.

Learning Outcomes

14.1 Population Genetics Underlies Evolution
1. State the unit of information of genetics at the population level.
2. Define *gene pool*.
3. List the five processes that cause microevolutionary change.
4. State the consequence of macroevolutionary change.

14.2 Constant Allele Frequencies
5. State the genotypes represented in each part of the Hardy-Weinberg equation.
6. Explain the conditions necessary for Hardy-Weinberg equilibrium.

14.3 Applying Hardy-Weinberg Equilibrium
7. Explain how Hardy-Weinberg equilibrium uses population incidence statistics to predict the probability of a particular phenotype.

14.4 DNA Profiling Uses Hardy-Weinberg Assumptions
8. Explain how parts of the genome that are in Hardy-Weinberg equilibrium can be used to identify individuals.

 ## The BIG Picture

Human genetics at the population level considers allele frequencies. Some parts of the genome that have changed over time enable us to trace our origins, migrations, and relationships. Allele frequencies that do not change in response to environmental factors provide a way to distinguish individuals.

Postconviction DNA Testing

Josiah Sutton had served 4.5 years of a 25-year sentence for rape when he was exonerated, thanks to the Innocence Project. This nonprofit legal clinic and public policy organization, created in 1992, has used DNA retesting to free hundreds of wrongfully convicted prisoners, most of whom were "poor, forgotten, and have used up all legal avenues for relief." Sutton became a suspect after a woman in Houston identified him and a friend 5 days after she had been raped, threatened with a gun, and left in a field. The two young men supplied saliva and blood samples, from which DNA profiles were done and compared to DNA profiles from semen found in the victim and in her car. At the trial, a crime lab employee testified that the probability that Sutton's DNA matched that of the evidence by chance was 1 in 694,000, leading to a conviction. Jurors ignored the fact that Sutton's physical description did not match the victim's description of her assailant.

The DNA evidence came from more than one individual, yielded different results when the testing was repeated, and most importantly, looked at only seven of the parts of the genome that are typically compared in a DNA profile, or fingerprint. Doing the test correctly and considering more forensic "alleles" revised the statistics dramatically: Sutton's pattern was shared not with 1 in 694,000 black men, as had originally been claimed, but with 1 in 16.

While in jail, Sutton read about DNA profiling and requested independent testing, but was refused. Then journalists investigating the Houston crime laboratory learned of his case and alerted the Innocence Project. Retesting the DNA evidence led to freeing Sutton. DNA profiling is based on how common a suspect's gene variants are in the appropriate population. The forensic (criminal) technology is a direct application of population genetics.

14.1 Population Genetics Underlies Evolution

The language of genetics at the family and individual levels is written in DNA sequences. However, at the population level, the language of genetics is allele (gene variant) frequencies. Genetics at the population level goes beyond science, embracing information from history, anthropology, human behavior, and sociology. Population genetics enables us to trace our beginnings, understand our diversity today, and imagine the future.

A biological **population** is any group of members of the same species in a given geographical area that can mate and produce fertile offspring (**figure 14.1**). A population in a sociological sense may be more restrictive, such as an ethnic group or economic stratum. **Population genetics** is a branch of genetics that considers all the alleles of all the genes in a biological

Figure 14.1 **A biological population is a group of interbreeding organisms living in the same place.** Populations of sexually reproducing organisms include many genetic variants that provide flexibility to enhance species survival. To us, these hippos look alike, but they can undoubtedly recognize phenotypic differences in each other. © Comstock/Punchstock RF

population, which constitute the **gene pool**. The "pool" refers to the collection of gametes in the population; an offspring represents two gametes from the pool. Alleles can move between populations when individuals migrate and mate. This movement, termed **gene flow**, underlies evolution, which is explored in the next two chapters.

Thinking about genes at the population level begins by considering frequencies—that is, how often a particular gene variant occurs in a particular population. Such frequencies can be calculated for alleles, genotypes, or phenotypes, and may include single-base mutations or numbers of short, repeated DNA sequences. For example, an allele frequency for the cystic fibrosis gene (*CFTR*) might be the number of $\Delta F508$ alleles among the residents of San Francisco. $\Delta F508$ is the most common allele that, when homozygous, causes the disease. The allele frequency derives from the two $\Delta F508$ alleles in each person with CF, plus alleles carried in heterozygotes, considered as a proportion of all alleles for that gene in the gene pool of San Francisco. The genotype frequencies are the proportions of heterozygotes and the two types of homozygotes in the population. Finally, a phenotypic frequency is simply the percentage of people in the population who have CF (or who do not). The situation becomes more complex with multiple alleles for a single gene, because many more phenotypes and genotypes are possible.

Phenotypic frequencies are determined empirically—that is, by observing how common a condition or trait is in a population. Genetic counselors use phenotype frequency to estimate the risk that a particular inherited disease will occur in an individual when there is no family history of the illness. **Table 14.1** shows disease incidence for phenylketonuria (PKU), an inborn error of metabolism that causes intellectual disability unless the person follows a special, low-protein diet from birth. Note how the frequency differs in different populations. A person from Turkey without a family history of PKU would have a higher risk of having an affected child than a person from Japan.

On a broader level, shifting allele frequencies in populations reflect small steps of genetic change, called **microevolution**. These small, incremental changes alter genotype

Table 14.1	Frequency of PKU in Various Populations	
Population	**Frequency of PKU**
Chinese	1/16,000
Irish, Scottish, Yemenite Jews	1/5,000
Japanese	1/119,000
Swedes	1/30,000
Turks	1/2,600
U.S. Caucasians	1/10,000

frequencies and underlie evolution. Genotype frequencies rarely stay constant. They can change under any of the following conditions:

1. Individuals of one genotype are more likely to reproduce with each other than with individuals of other genotypes (*nonrandom mating*).
2. Individuals move between populations (*migration*).
3. Random sampling of gametes alters allele frequencies (*genetic drift*).
4. New alleles arise (*mutation*).
5. People with a particular genotype are more likely to produce viable, fertile offspring under a specific environmental condition than individuals with other genotypes (*natural selection*).

All of these conditions, except mutation, are quite common. Therefore, genetic equilibrium—when allele frequencies are *not* changing—is rare. Given our tendency to pick our partners and move about, microevolution is not only possible, but also nearly unavoidable. (Chapter 15 considers these factors in depth.)

When sufficient microevolutionary changes accumulate to keep two fertile organisms of opposite sexes from producing fertile offspring together, a new species forms. Changes that are great enough to result in speciation are termed **macroevolution**. Speciation can occur through many small changes over time, and/or a few changes that greatly affect the phenotype.

Key Concepts Questions 14.1

1. What is a biological population?
2. Define *gene pool*.
3. Distinguish microevolution from macroevolution.
4. List the five factors that can change genotype frequencies in populations.

14.2 Constant Allele Frequencies

Before examining the pervasive genetic evidence for evolution, consider the interesting, but unusual, situation in which frequencies for certain alleles stay constant. This is a condition called **Hardy-Weinberg equilibrium**.

Hardy-Weinberg Equilibrium

In 1908, Cambridge University mathematician Godfrey Harold Hardy (1877–1947) and Wilhelm Weinberg (1862–1937), a German physician interested in genetics, independently used algebra to explain how allele frequencies can be used to predict phenotypic and genotypic frequencies in populations of diploid, sexually reproducing organisms.

Hardy unintentionally cofounded the field of population genetics with a simple letter published in the journal *Science*. The letter began with a curious mix of modesty and condescension:

> *I am reluctant to intrude in a discussion concerning matters of which I have no expert knowledge, and I should have expected the very simple point which I wish to make to have been familiar to biologists.*

Hardy-Weinberg Equilibrium

Hardy continued to explain how mathematically inept biologists had incorrectly deduced from Mendel's work that dominant traits would increase in populations while recessive traits would become rarer. At first glance, this seems logical. However, it is untrue because recessive alleles enter populations by mutation or migration and are maintained in heterozygotes. Recessive alleles also become more common when they confer a reproductive advantage, thanks to natural selection.

Hardy and Weinberg disproved the assumption that dominant traits increase while recessive traits decrease using algebra. The expression of population genetics in algebraic terms begins with the simple equation

$$p + q = 1.0$$

where p represents the frequency of all dominant alleles for a gene and q represents the frequency of all recessive alleles. The expression "$p + q = 1.0$" means that all the dominant alleles and all the recessive alleles comprise all the alleles for that gene in a population.

Next, Hardy and Weinberg described the possible genotypes for a gene with two alleles using the binomial expansion

$$p^2 + 2pq + q^2 = 1.0$$

In this equation, p^2 represents the proportion of homozygous dominant individuals, q^2 represents the proportion of homozygous recessive individuals, and $2pq$ represents the proportion of heterozygotes (**figure 14.2**). The letter p designates the frequency of a dominant allele, and q is the frequency of a recessive allele. **Figure 14.3** shows how the binomial expansion is derived from allele frequencies. Note that the derivation is conceptually the same as tracing alleles in a monohybrid cross, in which the heterozygote forms in two ways: recessive allele from the mother and dominant allele from the father, or vice versa (see figure 4.4).

The binomial expansion used to describe genes in populations became known as the Hardy-Weinberg equation. It can reveal the changes in allele frequency that underlie evolution. If the proportion of genotypes remains the same from generation to generation, as the equation indicates, then that gene is not

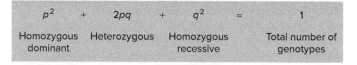

p^2	+	$2pq$	+	q^2	=	1
Homozygous dominant		Heterozygous		Homozygous recessive		Total number of genotypes

Figure 14.2 The Hardy-Weinberg equation in English.

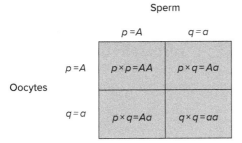

Sperm

	$p = A$	$q = a$
$p = A$	$p \times p = AA$	$p \times q = Aa$
$q = a$	$p \times q = Aa$	$q \times q = aa$

Oocytes

Figure 14.3 Source of the Hardy-Weinberg equation.
A variation on a Punnett square reveals how random mating in a population in which gene *A* has two alleles—*A* and *a*—generates genotypes *aa*, *AA*, and *Aa*, in the relationship $p^2 + 2pq + q^2$.

evolving (changing). This situation, Hardy-Weinberg equilibrium, is theoretical. It happens only if the population is large and undisturbed. That is, its members mate at random, do not migrate, and there is no genetic drift, mutation, or natural selection.

Hardy-Weinberg equilibrium is rare for protein-encoding genes that affect the phenotype, because an organism's appearance and health affect its ability to reproduce. Harmful allele combinations are weeded out of the population while helpful ones are passed on. Hardy-Weinberg equilibrium *is* seen in many DNA sequences that do not affect the phenotype. However, if a gene in Hardy-Weinberg equilibrium is closely linked on its chromosome to a gene that is subject to natural selection, it may be pulled from equilibrium by being inherited along with the selected gene. This bystander effect is called **linkage disequilibrium (LD)**.

Solving a Problem Using the Hardy-Weinberg Equation

To understand Hardy-Weinberg equilibrium, it helps to follow the frequency of two alleles of a gene from one generation to the next. Mendel's laws underlie such calculations.

Consider an autosomal recessive trait: a middle finger that is shorter than the second and fourth fingers. If we know the frequencies of the dominant and recessive alleles, then we can calculate the frequencies of the genotypes and phenotypes and trace the trait through the next generation. The dominant allele *D* confers normal-length fingers; the recessive allele *d* confers a short middle finger (**figure 14.4**). We can deduce the frequencies of the dominant and recessive alleles by observing the frequency of homozygous recessives, because this phenotype—short finger—reflects only one genotype. If 9 of 100 individuals in a population have short fingers—genotype *dd*—the frequency is 9/100, or 0.09. Since *dd* equals q^2, then *q* equals 0.3. Since $p + q = 1.0$, knowing that *q* is 0.3 tells us that *p* is 0.7.

Next, we can calculate the proportions of the three genotypes that arise when gametes combine at random:

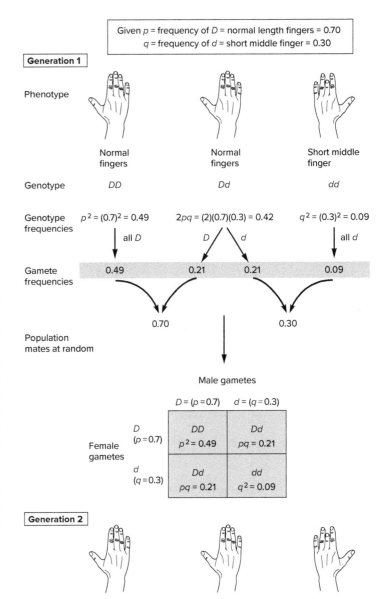

Figure 14.4 Hardy-Weinberg equilibrium. In Hardy-Weinberg equilibrium, allele frequencies remain constant from one generation to the next.

Homozygous dominant $= DD$
$= 0.7 \times 0.7 = 0.49$
$= 49$ percent of individuals in generation 1

Homozygous recessive $= dd$
$= 0.3 \times 0.3 = 0.09$
$= 9$ percent of individuals in generation 1

Heterozygous $= Dd + dD$
$= 2pq = (0.7)(0.3) + (0.3)(0.7) = 0.42$
$= 42$ percent of individuals in generation 1

POSSIBLE MATINGS			FREQUENCY OF OFFSPRING GENOTYPES		
Male	**Female**	**Proportion in Population**	**DD**	**Dd**	**dd**
0.49 *DD*	0.49 *DD*	0.2401 (*DD* × *DD*)	0.2401		
0.49 *DD*	0.42 *Dd*	0.2058 (*DD* × *Dd*)	0.1029	0.1029	
0.49 *DD*	0.09 *dd*	0.0441 (*DD* × *dd*)		0.0441	
0.42 *Dd*	0.49 *DD*	0.2058 (*Dd* × *DD*)	0.1029	0.1029	
0.42 *Dd*	0.42 *Dd*	0.1764 (*Dd* × *Dd*)	0.0441	0.0882	0.0441
0.42 *Dd*	0.09 *dd*	0.0378 (*Dd* × *dd*)		0.0189	0.0189
0.09 *dd*	0.49 *DD*	0.0441 (*dd* × *DD*)		0.0441	
0.09 *dd*	0.42 *Dd*	0.0378 (*dd* × *Dd*)		0.0189	0.0189
0.09 *dd*	0.09 *dd*	0.0081 (*dd* × *dd*)			0.0081
		Resulting offspring frequencies:	0.49	0.42	0.09
			DD	*Dd*	*dd*

The proportion of homozygous individuals is calculated by multiplying the allele frequency for the recessive or dominant allele by itself. The heterozygous calculation is $2pq$ because D can combine with d in two ways—a D sperm with a d egg, and a d sperm with a D egg.

In this population, 9 percent of the individuals have a short middle finger. Now jump ahead a few generations, and assume that people choose mates irrespective of finger length. This means that each genotype of a female (*DD, Dd,* or *dd*) is equally likely to mate with each of the three types of males (*DD, Dd,* or *dd*), and vice versa. **Table 14.2** multiplies the genotype frequencies for each possible mating, which leads to offspring in the familiar proportions of 49 percent *DD,* 42 percent *Dd,* and 9 percent *dd.* This gene, therefore, is in Hardy-Weinberg equilibrium—the allele and genotype frequencies do not change from one generation to the next.

Key Concepts Questions 14.2

1. What does the Hardy-Weinberg equation describe?
2. Explain the components of the Hardy-Weinberg equation.
3. Why aren't all genes under Hardy-Weinberg equilibrium?

14.3 Applying Hardy-Weinberg Equilibrium

A young woman pregnant for the first time learns that her health care providers offer all patients a blood test to see if they are carriers for cystic fibrosis. The test detects more than 100 of the most common CF mutations. The woman had never thought about CF. She and her partner are from the same population (Caucasians of European descent) and neither has a family history of the disease. What is her risk of being a carrier? If it is not high, perhaps she will not take the test. The Hardy-Weinberg equation can answer the patient's question by determining the probability that she and her partner are carriers for CF. If they are carriers, then Mendel's first law predicts that the risk that an offspring inherits the disease is 1 in 4, or 25 percent. The couple might then opt for prenatal testing.

To derive carrier risks, the Hardy-Weinberg equation is applied to population statistics on genetic disease incidence. To determine allele frequencies for autosomal recessively inherited characteristics, we need to know the frequency of one genotype in the population. This is typically the homozygous recessive class, because its phenotype indicates its genotype.

The incidence (frequency) of an autosomal recessive disease in a population is used to help calculate the risk that a particular person is a heterozygote. Returning to the example of CF, the incidence of the disease, and therefore also of carriers, may vary greatly in different populations (**table 14.3**).

Table 14.3	Carrier Frequency for Cystic Fibrosis
Population Group	**Carrier Frequency**
African Americans	1 in 66
Asian Americans	1 in 150
Caucasians of European descent	1 in 23
Hispanic Americans	1 in 46

CF affects 1 in 2,000 Caucasian newborns. Therefore, the homozygous recessive frequency—*cc* if *c* represents the disease-causing allele—is 1/2,000, or 0.0005 in the population. This equals q^2. The square root of q^2 is about 0.022, which equals the frequency of the *c* allele. If *q* equals 0.022, then *p*, or $1 - q$, equals 0.978. Carrier frequency is equal to $2pq$, which equals (2)(0.978)(0.022), or 0.043—about 1 in 23. **Figure 14.5** summarizes these calculations.

Because there is no CF in the woman's family, her risk of having an affected child, based on population statistics, is low. The chance of *each* potential parent being a carrier is about 4.3 percent, or 1 in 23. The chance that *both* are carriers is 1/23 multiplied by 1/23—or 1 in 529—because the probability that two independent events will occur equals the product of the probability that each event will happen alone (the product rule). However, if they *are* both carriers, each of their children would face a 1 in 4 chance of inheriting the illness, based on Mendel's first law (gene segregation). Therefore, the risk that these two unrelated Caucasian individuals with no family history of CF will have an affected child is $1/4 \times 1/23 \times 1/23$, or 1 in 2,116. The woman takes the test, but is much less worried than she was when she first learned it was an option.

For X-linked traits, different predictions of allele frequencies apply to males and females. For a female, who can be homozygous recessive, homozygous dominant, or a heterozygote, the standard Hardy-Weinberg equation, $p^2 + 2pq + q^2 = 1.0$, applies. However, in males, the allele frequency is the phenotypic frequency, because a male who inherits an X-linked recessive allele exhibits it in his phenotype.

The incidence of X-linked hemophilia A, for example, is 1 in 10,000 male (X^hY) births. Therefore, *q* (the frequency of the *h* allele) equals 0.0001. Using the formula $p + q = 1$, the

Cystic Fibrosis

incidence (autosomal recessive class) = 1/2,000 = 0.0005

$\therefore q^2 = 0.0005$

$\therefore q = \sqrt{0.0005} = 0.022$

$\therefore p = 1 - q = 1 - 0.022 = 0.978$

$\therefore$ carrier frequency = $2pq$ = (2) (0.978) (0.022) = 0.043 = 1/23

Figure 14.5 Calculating the carrier frequency given population incidence: autosomal recessive.

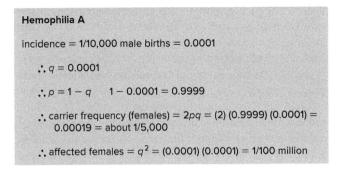

Hemophilia A

incidence = 1/10,000 male births = 0.0001

$\therefore q = 0.0001$

$\therefore p = 1 - q \quad 1 - 0.0001 = 0.9999$

$\therefore$ carrier frequency (females) = $2pq$ = (2) (0.9999) (0.0001) = 0.00019 = about 1/5,000

$\therefore$ affected females = q^2 = (0.0001) (0.0001) = 1/100 million

Figure 14.6 Calculating the carrier frequency given population incidence: X-linked recessive.

frequency of the wild type allele is 0.9999. The incidence of carriers (X^HX^h), who are all female, equals $2pq$, or (2)(0.0001) (0.9999), which equals 0.00019; this is 0.0002, or 0.02 percent, which equals about 1 in 5,000. The incidence of a female having hemophilia A (X^hX^h) is q^2, or $(0.0001)^2$. This is about 1 in 100 million. **Figure 14.6** summarizes these calculations.

Neat allele frequencies such as 0.6 and 0.4, or 0.7 and 0.3, are unusual. In actuality, single-gene diseases are rare, and so the *q* component of the Hardy-Weinberg equation contributes little. Because this means that the value of *p* approaches 1, the carrier frequency, $2pq$, is close to $2q$. Thus, the carrier frequency is approximately twice the frequency of the rare, disease-causing allele.

Consider Tay-Sachs disease, which occurs in 1 in 3,600 Ashkenazim (Jewish people of Eastern European descent). This means that q^2 equals 1/3,600, or about 0.0003. The square root, *q*, equals 0.017. The frequency of the dominant allele (*p*) is then $1 - 0.017$, or 0.983. What is the likelihood that an Ashkenazim carries Tay-Sachs disease? It is $2pq$, or (2)(0.983)(0.017), or 0.033. This is close to double the frequency of the mutant allele (*q*), 0.017. Modifications of the Hardy-Weinberg equation are used to analyze genes that have more than two alleles.

Key Concepts Questions 14.3

1. How are allele frequencies in populations inferred?
2. How does the Hardy-Weinberg equation apply to X-linked traits?
3. What is the approximate carrier frequency for rare diseases?

14.4 DNA Profiling Uses Hardy-Weinberg Assumptions

Hardy-Weinberg equilibrium is useful in understanding the conditions necessary for evolution to occur, but in a practical sense it provides the foundation for **DNA profiling**, described in the chapter opener. The connection between the theoretical and the practical is that many parts of the genome that do not affect the phenotype, such as some short repeated sequences that do not encode amino acids and are not closely linked to genes that do, are in Hardy-Weinberg equilibrium. Variability in these sequences can be used to identify individuals if the frequencies are known in particular populations, several sites in the genome are considered at the same time, and the product rule is applied.

In DNA profiling, the number of copies of a repeated sequence is considered to be an allele because it is information—just not protein-encoding information. A person is classified as a heterozygote or a homozygote based on the number of copies of the same repeat at the same chromosomal locus on the two homologs. A homozygote has the same number of copies of a repeat on both homologs, such as individual 2 in **figure 14.7**. A heterozygote has two different numbers of copies of a repeat, such as the other two individuals in the figure. The copy numbers are distributed in the next generation according to Mendel's law of segregation. A child of individual 1 and individual 2 in figure 14.7, for example, could have any of the two possible combinations of the parental copy numbers, one from each parent: two repeats and three repeats, or four repeats and three repeats.

DNA profiling was pioneered on detecting copy number variants of very short repeats and using them to identify or distinguish individuals. In general, the technique calculates the probability that certain combinations of repeat numbers will be in two DNA sources by chance. For example, if a DNA profile of skin cells taken from under the fingernails of an assault victim matches the profile from a suspect's hair, and the likelihood is very low that those two samples would match by chance, that is strong evidence of guilt rather than a coincidental similarity. DNA evidence is more often valuable in excluding a suspect, and should be considered along with other types of evidence.

An Evolving Technology

Obtaining a DNA profile is a molecular technique, but interpretation requires statistical analysis of population genetic data. The methods for generating a DNA profile grew out of genetic marker testing that identified DNA sequences closely linked to disease-causing genes. Genetic markers for disease were used in the 1980s, before we knew the sequences of human disease-causing genes and could detect them directly. Sir Alec Jeffreys at Leicester University in the United Kingdom was the first investigator to use DNA typing in forensic cases. He followed DNA sequences called variable number of tandem repeats—VNTRs—that are each 10 to 80 bases long (**table 14.4**). **Clinical Connection 14.1** describes one of the earliest legal cases using DNA profiling.

A GLIMPSE OF HISTORY

The criminal case that made Sir Alec Jeffreys—and what was then called DNA fingerprinting—famous was that of Colin Pitchfork, a 27-year-old baker and father who raped and strangled two 15-year-old girls in the Leicester countryside. Other early cases proved that a boy was the son of a British citizen so that he could enter England and freed a man unjustly imprisoned for raping two schoolgirls.

The first exoneration thanks to DNA profiling came in 1993, in the United States. Kirk Bloodsworth, while on death row for the 1985 conviction of raping and murdering a 9-year-old, read about Colin Pitchfork and requested DNA testing of a stain on the victim's underwear. Bloodsworth's DNA and the DNA on the evidence did not match. The real criminal had lived a floor below Mr. Bloodsworth in prison!

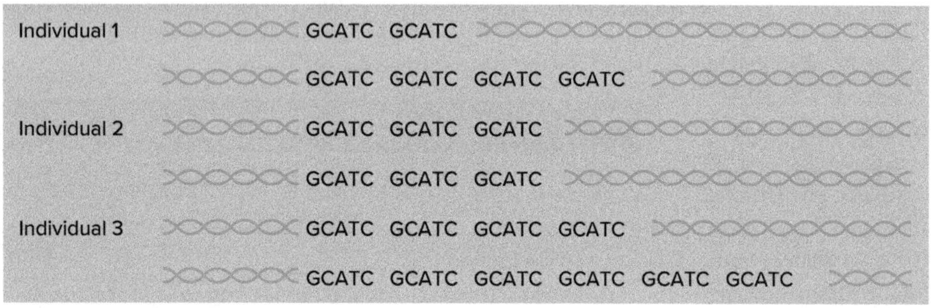

Figure 14.7 **DNA profiling detects differing numbers of repeats at specific chromosomal loci.** Individuals 1 and 3 are heterozygotes for the number of copies of a 5-base sequence at a particular chromosomal locus. Individual 2 is a homozygote, with the same number of repeats on the two copies of the chromosome. (Repeat number is considered an allele.)

Table 14.4	Characteristics of Repeats Used in DNA Profiling			
Type	**Repeat Length**	**Distribution**	**Example**	**Fragment Sizes**
VNTRs (minisatellites)	10–80 bases	Not uniform	TTCGGGTTGT	50–1,500 bases
STRs (microsatellites)	2–10 bases	More uniform	ACTT	50–500 bases

DNA for profiling can come from any cell with a nucleus, including hair, skin, secretions, white blood cells, and cells scraped from the inside of the cheek. Sir Jeffreys cut DNA with naturally occurring, scissorlike enzymes, called restriction enzymes, that cut at specific sequences. He then separated the DNA pieces using a technique called agarose gel electrophoresis. The DNA fragments carry a negative electrical charge because of the phosphate groups. When placed in an electrical field on a gel strip, the DNA pieces move toward the positively charged end by size. The shorter the piece, the faster it travels. When the different-sized fragments stop moving, a pattern of smears remains. A person heterozygous

DNA Profiling: Molecular Genetics Meets Population Genetics

DNA profiling is a powerful tool in forensic investigations, agriculture, paternity testing, and historical research. Until 1986, it was unheard of outside of scientific circles. A dramatic rape case changed that.

Tommie Lee Andrews watched his victims months before he attacked so that he knew when they would be home alone. On a balmy Sunday night in May 1986, Andrews awaited Nancy Hodge, a young computer operator at Disney World in Orlando, Florida. The burly man surprised her when she was in her bathroom removing her contact lenses. He covered her face, then raped and brutalized her repeatedly.

Andrews was very careful not to leave fingerprints, threads, hairs, or any other indication that he had ever been in Hodge's home. But he left DNA. Thanks to a clear-thinking crime victim and scientifically savvy lawyers, Andrews was soon at the center of a trial that would judge the technology that helped to convict him.

After the attack, Hodge went to the hospital, where she provided a vaginal secretion sample containing sperm. Two district attorneys who had read about DNA testing sent the sperm to a biotechnology company that extracted DNA and cut it with restriction enzymes. The sperm's DNA pieces were then mixed with labeled DNA probes that bound to complementary sequences.

The same extracting, cutting, and probing of DNA was done on white blood cells from Hodge and from Andrews, who had been held as a suspect in several assaults. When the radioactive DNA pieces from each sample, which were the sequences where the probes had bound, were separated and displayed by size, the resulting pattern of bands—the DNA profile—matched exactly for the sperm sample and Andrews's blood, differing from Hodge's DNA (**figure 14A**). (This differs from the STR approach used today.)

Andrews's allele frequencies were compared to those for a representative African American population. At the first trial the

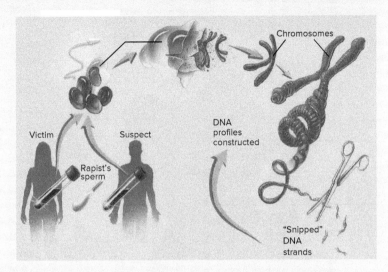

Figure 14A DNA profiling. The 1986 Andrews rape case used radioactive DNA probes on sperm from the victim's body, her white blood cells, and the suspect's white blood cells.

judge, fearful too much technical information would overwhelm the jury, did not allow the prosecution to cite population-based statistics. Without the appropriate allele frequencies, DNA profiling was reduced to just a comparison of smeary lines on test papers to see whether the patterns of DNA pieces in the forensic sperm sample looked like those for Andrews's white blood cells. Although population statistics indicated that the possibility that Andrews's DNA would match the evidence by chance was 1 in 10 billion, the prosecution was not allowed to mention this critical interpretation.

After a mistrial was declared, the prosecution cited the precedent of using population statistics to derive databases on standard blood types. When Andrews stood trial just 3 months later for raping a different woman, the judge permitted population analysis. This time, Andrews was convicted.

for a particular VNTR—meaning a different number of copies of the repeat on each of the two homologs—would have a DNA profile with two bands at that site. A homozygote would have just one band, because his or her DNA would have only one size piece corresponding to that part of the genome. **Figure 14.8** shows one of these older DNA profiles, done to test whether Dolly the cloned sheep actually developed from the claimed breast cell of a ewe. She did.

Researchers began to use a different type of repeat, called a **short tandem repeat (STR)**, because shorter DNA molecules were more likely to persist in a violent situation, such as an explosion, fire, or natural disaster. Today the Federal

Bureau of Investigation (FBI) in the United States tracks 13 STRs for DNA profiling, which generates 26 data points, because a person has two copies of each STR. The FBI system, called CODIS for Combined DNA Index System, shares DNA profiles among local, state, and federal crime laboratories. The power of DNA profiling is that the number of repeats of each of the 13 STRs occurs with certain probabilities in any particular population. The probability that any two unrelated individuals have the same 26 CODIS markers (13 pairs) by chance is 1 in 250 trillion.

STRs range in size from two to 10 bases, but most used for forensic applications are four bases long. **Figure 14.9** shows

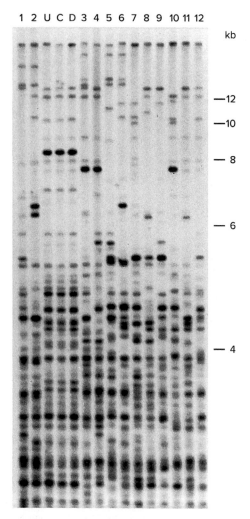

Figure 14.8 Comparing DNA profiles. These DNA profiles compare VNTRs from the DNA of Dolly the cloned sheep (lane D), fresh donor udder tissue (U), and cultured donor udder tissue (C). The other 12 lanes represent other sheep. The match between Dolly and the two versions of her nucleus donor is obvious. Dolly was born in 1996 and died in 2003—young for an ewe, perhaps due to her first cell's earlier existence. *Courtesy of Esther N. Signer*

the 13 STRs used in DNA profiling. The technique also includes the genotype for a gene called amelogenin to determine the gender of the person who left the DNA sample. Amelogenin is present on both the X and Y chromosomes but is six bases shorter on the X chromosome. Therefore, a long amelogenin allele indicates a male.

Today a technique called capillary electrophoresis is used to separate DNA pieces (**figure 14.10**). Fluorescently labeled short pieces of DNA that correspond to parts of the genome including the CODIS STRs are applied to sample DNA, where they bind to their complements. Then the polymerase chain reaction (PCR; see figure 9.19) copies the selected sequences, amplifying all 13 CODIS markers simultaneously. Instead of the slab of gel used when the technology was developed, the DNA pieces travel through hair-thin tubes, called capillaries. A laser shone through a window excites the fluorescent dyes that tag each DNA piece, and the strength of the fluorescence is measured and displayed as a peak on a readout. A single tall peak for a particular STR indicates that the person is a homozygote—for example, six copies of the repeat on each homolog. Two peaks of lesser intensity indicate that the person is a heterozygote, such as having six copies on one homolog but eight on its mate.

The FBI maintains a National DNA Index (NDIS). It has DNA from more than 13 million convicted offenders, 3 million arrested individuals, and 0.8 million samples collected from crime scenes but never associated with an individual. The FBI claims that NDIS has aided more than 380,000 investigations.

If DNA is so degraded that even STRs are destroyed, mitochondrial DNA (mtDNA) is often used instead, particularly two regions of repeats, called hypervariable regions I and II, that are distinct in different populations. Because a single cell can yield hundreds or thousands of copies of the mitochondrial genome, even extremely small forensic samples contain this DNA. MtDNA analysis was critical in analyzing evidence from the September 11 terrorist attacks, most of which was extremely degraded.

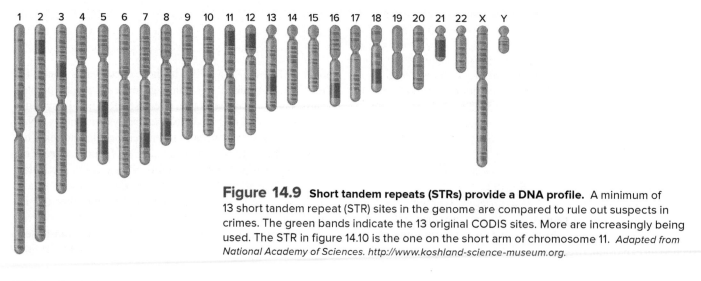

Figure 14.9 Short tandem repeats (STRs) provide a DNA profile. A minimum of 13 short tandem repeat (STR) sites in the genome are compared to rule out suspects in crimes. The green bands indicate the 13 original CODIS sites. More are increasingly being used. The STR in figure 14.10 is the one on the short arm of chromosome 11. *Adapted from National Academy of Sciences. http://www.koshland-science-museum.org.*

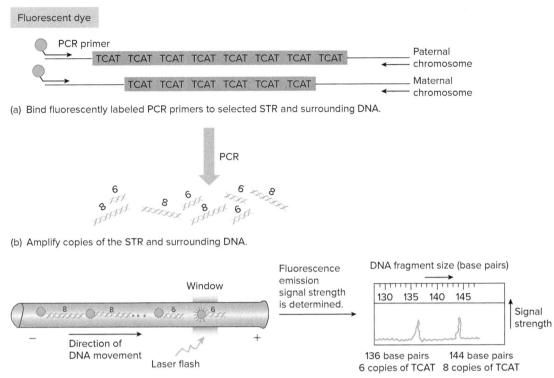

(a) Bind fluorescently labeled PCR primers to selected STR and surrounding DNA.

PCR

(b) Amplify copies of the STR and surrounding DNA.

(c) Run amplified copies of labeled STR-containing DNA pieces through capillary electrophoresis device. Electropherogram shows two peaks for heterozygotes (two fragment sizes) and one larger peak for homozygotes (same fragment size from both homologs). This person is heterozygous for the TCAT STR on chromosome 11.

Figure 14.10 Several steps identify STRs. In this example, the STR is called TH01, consisting of TCAT repeats on chromosome 11. **(a)** Short pieces of DNA called primers that are complementary to regions near the STR of interest are added to the DNA sample. Each primer is bound to a fluorescent dye. **(b)** The primers direct the polymerase chain reaction to make many copies of the specific STR. **(c)** The amplified DNA pieces—in this case either six or eight copies of the four-base repeat TCAT—enter a gel electrophoresis device, where the smaller pieces move toward the positive charge faster than the larger pieces. Laser excitation of the dye bound to the primers emits a signal that is converted into an image called an electropherogram. Two peaks indicate a heterozygote and a single, larger peak a homozygote—for this STR.

Using Population Statistics to Interpret DNA Profiles

In forensics in general, the more clues, the better. The power of DNA profiling comes from tracking repeats on several chromosomes. The numbers of copies of a repeat are assigned probabilities (likelihood of being present) based on their observed frequencies in the population from which the source of the DNA comes. Considering repeats on different chromosomes makes it possible to use the product rule to calculate the probabilities of particular combinations of repeat numbers occurring in a population, based on Mendel's law of independent assortment.

The Hardy-Weinberg equation and the product rule are applied to derive the statistics that back up a DNA profile, using the following steps:

1. The pattern of peaks on the electropherograms indicates whether an individual is a homozygote or a heterozygote for each repeat.
2. Calculate genotype frequencies using parts of the Hardy-Weinberg equation: p^2 and q^2 denote each of

the two homozygotes for a two-allele repeat, and $2pq$ represents the heterozygote.
3. Multiply the frequencies to reveal the probability that this combination of repeat copy numbers would occur in a particular population.
4. If the genotype combination is rare in the population the suspect comes from, and the genotype is in the suspect's DNA and in crime scene evidence, such as a rape victim's body or stolen property, guilt is highly likely.

Figure 14.11 summarizes the procedure for considering multiple STRs.

For the sequences used in DNA profiling, Hardy-Weinberg equilibrium is assumed. When it doesn't apply, problems can arise. For example, the requirement of nonrandom mating for Hardy-Weinberg equilibrium wouldn't be met in a community with a few very large families where distant relatives might inadvertently marry each other—a situation in many small towns. A particular DNA profile for one person might be shared with his or her cousins. In one case, a young man was convicted of rape based on a DNA profile—which he shared with his father, the actual rapist. Considering a larger

Figure 14.11 **To solve a crime.** A man was found brutally murdered, with bits of skin and blood beneath a fingernail. The evidence was sent to a forensics lab, where the patterns of five STRs in the blood from beneath the fingernail was compared to patterns for the same STRs in blood from the victim and blood from a suspect. The pattern for the crime scene evidence matched that for the suspect. Allele frequencies from the man's ethnic group used in the Hardy-Weinberg equation yielded the probability that his DNA matched that of the skin and blood under the murdered man's fingernail by chance. This is a hypothetical case. Using the full set of 13 STRs yields much more significant probabilities. The numbers under the peaks are the numbers of copies of the STR. The decimals are allele frequencies. © Lisa Zador/Getty Images RF

DNA collected from evidence at crime scene (blood, skin under victim's fingernail)

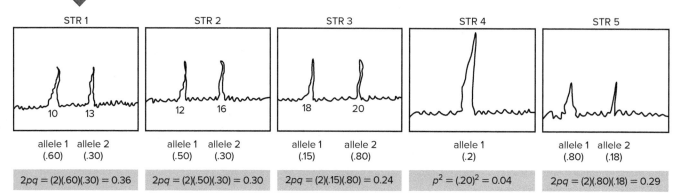

STR 1	STR 2	STR 3	STR 4	STR 5
10 13	12 16	18 20		
allele 1 allele 2	allele 1 allele 2	allele 1 allele 2	allele 1	allele 1 allele 2
(.60) (.30)	(.50) (.30)	(.15) (.80)	(.2)	(.80) (.18)
$2pq = (2)(.60)(.30) = 0.36$	$2pq = (2)(.50)(.30) = 0.30$	$2pq = (2)(.15)(.80) = 0.24$	$p^2 = (.20)^2 = 0.04$	$2pq = (2)(.80)(.18) = 0.29$

$$0.36 \times 0.30 \times 0.24 \times 0.04 \times 0.29 \approx 0.00031 \approx 1/3,226$$

Conclusion: The probability that another person in the suspect's population group has the same pattern of these alleles is approximately 1 in 3,226.

number of repeat sites can minimize such complications. If more repeat sites had been considered in the rape case, chances are that they would have revealed a polymorphism that the son had inherited from his mother, but that the guilty father lacked. Presence of the polymorphism would have indicated that the son was not guilty, but a close male relative might be.

The accuracy and meaning of a DNA profile depend upon the population that is the source of the allele frequencies. If populations are too broadly defined, then allele frequencies are typically low, leading to very low estimates of the likelihood that a suspect matches evidence based on chance. In one oft-quoted trial, the prosecutor concluded: *The chance of the DNA fingerprint of the cells in the evidence matching blood of the defendant by chance is 1 in 738 trillion.* The numbers were accurate, but did they really reflect the gene pool compositions of actual populations?

The first DNA profiling databases relied on populations that weren't realistic enough to yield valid statistics. They placed different groups into just three categories: Caucasian, black, or Hispanic. People from Poland, Greece, or Sweden were all considered Caucasian, and a dark-skinned person from Jamaica and one from Somalia would be considered blacks. Perhaps the most incongruous of all were the Hispanics. Cubans and Puerto Ricans are part African, whereas people from Mexico and Guatemala have mostly Native American gene variants. Spanish and Argentinians have neither black African nor Native American genetic backgrounds. Native Americans and Asians were left out altogether. Analysis of just three databases—Caucasian, black, and Hispanic—revealed significantly more homozygous recessives for certain polymorphic genes than the Hardy-Weinberg equation would predict, confirming that allele frequencies were not in equilibrium.

More restrictive ethnic databases are required to apply allele frequencies to interpret DNA profiles. A frequency of 1 in 1,000 for an allele in all Caucasians may actually be much higher or lower in a subgroup whose members intermarry. However, even narrowly defined ethnic databases may be insufficient to interpret DNA profiles from people of mixed heritage, such as someone whose mother was Scottish/French and whose father was Greek/German. As it becomes easier for people from different parts of the world to meet, human population substructure will grow more complex, and DNA profiling more challenging.

Using DNA Profiling to Identify Victims

DNA profiling was first used in criminal cases and to identify human remains from plane crashes. Then terrorist attacks and natural disaster applications brought the technology to a new level.

Identifying World Trade Center Victims

In late September 2001, a company that provides breast cancer tests received three unusual types of DNA samples:

- evidence from the World Trade Center in New York City;
- cheek brush scrapings from relatives of people missing from the site; and
- "reference samples" from the victims' toothbrushes, razors, and hairbrushes.

Technologists analyzed the DNA for copy numbers of the 13 standard STRs and the sex chromosomes (the amelogenin gene). STR analysis yielded results on pieces of soft tissue, but bone bits that persisted despite the ongoing fire at the site required hardier mtDNA analysis. If the DNA pattern in crime scene evidence matched DNA from a victim's toothbrush, identification was fairly certain. Forensic investigators used the DNA results to match family members to victims. DNA profiling provides much more reliable information on identity than traditional forensic identifiers such as dental patterns, scars, and fingerprints, and objects found with the evidence, such as jewelry.

Identifying Natural Disaster Victims

Different types of disasters present different challenges for DNA profiling (**table 14.5**). Whereas New York City workers searched rubble in a limited area for remains, the approximately 250,000 bodies strewn about by the Indian Ocean tsunami in 2004 were everywhere. Disaster workers had to exhume bodies that had been buried quickly to stem the spread of infectious disease. Remains that were accessible after the waves hit quickly decayed in the hot, wet climate. These conditions, combined with the lack of roads and labs, led to 75 percent of the bodies being identified by standard dental record analysis, and 10 percent from fingerprints. Fewer than half of 1 percent of the victims were identified by their DNA.

Forensic scientists had learned from 9/11 the importance of matching victim DNA to that of relatives, to avoid errors

Table 14.5	Challenges to DNA Profiling in Mass Disasters

- Climate that hastens decay
- Inability to reach remains
- No laboratory facilities
- Number of casualties
- Lack of relatives
- Destruction of personal item evidence
- Poor DNA quality (too fragmented, scarce, degraded)
- Lack of availability of DNA probes and statistics for a population

when two people matched at several genome sites by chance. In New York City, many of those relatives were from nearby neighborhoods; in the 2004 tsunami, 12 countries were directly affected and victims came from 30 countries. Entire families were washed away, leaving few and many times no relatives to provide DNA, even if everyday evidence such as toothbrushes had remained.

To compensate for the barriers to implementing DNA profiling in mass disasters, Sir Alec Jeffreys advises assessing 15 to 20 STRs, and some investigators recommend using 50. Tragic as these disasters are, they have spurred forensic scientists to develop ways to better integrate many types of evidence, including that found in DNA sequences.

Confirming Genocide

DNA profiling can be used to provide evidence of mass murder. During four days in July 1995, Bosnian Serbs killed 8,100 Muslim men and boys in the Srebrenica genocide of Bosnia and Herzegovina. The dead were initially dumped into mass graves but then were dug up, broken apart, and widely dispersed in an attempt to cover up what had happened. Many of 32,000 other bodies had been buried deeply in clay-rich soil, where the lack of oxygen turned the soft tissue clinging to bones into a soaplike substance called "grave wax."

An investigation by the International Commission on Missing Persons (ICMP), which formed in 2000, used DNA profiling to identify more than 7,000 of the victims of the July 1995 Srebrenica massacre. Researchers collected blood samples from more than 85,000 relatives in search of loved ones and matched the samples to the DNA profiles of the remains, using STRs. At the ICMP laboratory in northern Sarajevo, investigators cleaned and pulverized bone samples, then extracted the DNA and tested it for the STRs commonly used in forensic investigations. The DNA profiles provided a great deal more information than had classical evidence, such as clothing found with bodies and dental patterns—enough to convict war criminal Radovan Karadzic in 2016, after a six-year trial. DNA profiling has identified more than 70 percent of the 40,000 people missing in total after the conflict. In addition, by noting where each bone fragment was found, researchers could reconstruct the movements to hide the bodies. DNA profiling is also being used on war crime victims from Syria, Iraq, and Libya.

Genetic Privacy

Before the information age, population genetics was an academic discipline that was more theoretical than practical. Today, with the combination of information technology, genome-wide association studies, genome sequencing, and shortcuts to identify people by SNP or copy number patterns, population genetics presents a powerful way to identify individuals. **Bioethics** discusses how the Supreme Court dealt with a matter involving use of genetic information. The Bioethics feature in chapter 16 returns to the challenge of maintaining genetic privacy, in a research setting.

The human genome is 3.2 billion bits of information, each of which can be one of four possibilities: A, C, T, or G. Because of this huge capacity for diversity, our genomes can vary many more ways than there are people—about 7.4 billion worldwide. Only 30 to 80 genome sites need be considered to uniquely describe each person. This is why forensic tests can compare only 13 STRs to rule out or establish identity.

The ease of assigning highly individualized genetic name tags may be helpful in forensics, but it poses privacy issues. Consider a "DNA dragnet," which is a forensic approach that compiles DNA profiles of all residents of a town where a violent crime is unsolved. Sir Jeffreys conducted some of the first DNA dragnets in the late 1980s. The largest to date occurred in 1998 in Germany, where more than 16,000 men had their DNA profiled in a search for the man who raped and murdered an 11-year-old. The dragnet indeed caught the killer.

A controversial application of DNA profiling is a familial DNA search, which is based on the fact that close relatives share large portions of their genomes. DNA from a crime scene is compared to DNA in databases from convicted felons. If nearly half of the CODIS sites match, then a first-degree relative of the convict becomes a suspect. This might be a son, brother, or the father of a man in prison, for example.

In the case of California's "Grim Sleeper," who killed nine women and one girl over more than two decades, the suspect was finally arrested when DNA found on a victim closely matched that of a young man in prison for trafficking weapons.

Bioethics

Should DNA Collected Today Be Used to Solve a Past Crime?

In 2003, a masked man broke into a woman's home in Salisbury, Maryland, and raped her. She couldn't describe her assailant, but forensic investigators collected and stored the man's DNA. The crime remained unsolved—until 2009.

Alonzo Jay King was arrested for assault in 2009 in Wicomico County, Maryland, for "menacing a group of people with a shotgun." State law required that he provide a DNA sample on a cheek swab, because the crime was considered violent. Maryland law also holds that taking a DNA sample and entering the information in a database is legal as long as the suspect has been arraigned, but the sample must be destroyed if the suspect is not convicted.

When FBI forensic scientists ran King's DNA profile against a database of half a million unidentified DNA samples taken from crime scenes (part of CODIS), the sample matched DNA on the underwear of the victim from the 2003 unsolved rape case. When a second sample from King matched, too, he was brought before a grand jury, indicted, then tried and convicted of the older crime and sentenced to life in prison without parole. However, the conviction was based on the DNA from the scene of the more recent crime. A court of appeals ruled that taking the DNA in 2009 had been an unlawful seizure and unreasonable search and set the conviction aside.

The case went to the Supreme Court, which ruled in 2013 that taking DNA from King was like taking fingerprints or a photograph—a noninvasive way to establish or confirm identity. The court deemed DNA profiling a "legitimate police booking procedure." The five judges in the assenting opinion argued for the accuracy of DNA profiling compared to other forensic methods and the fact that the parts of the genome that are compared have nothing to do with either traits or health, so privacy was not at issue. The dissenting opinion of four judges focused more on common sense—DNA wasn't needed to identify King in 2009. Everyone knew who he was. What wasn't known, and the DNA revealed, was that he'd committed the *earlier* crime. Whether or not DNA collected at one time for one crime can be used to convict an individual for a *different* crime is the bioethical issue.

Questions for Discussion

1. What should be criteria for requiring a suspect to provide a DNA sample to law enforcement officials?
2. Why would a DNA profile not reveal anything about the arrested individual's health?
3. Do you agree with the majority opinion or the dissenting opinion of the Court? Cite a reason for your answer.
4. How is a DNA profile like a fingerprint and unlike one?
5. Why is destroying DNA evidence futile?
6. How do you think the rape victim might have felt about the use of DNA testing in this case?

The young man did not have brothers, was too young to have a son old enough to be the Grim Sleeper, but did have a father who could have committed the crimes. After an intense investigation, Los Angeles police collected DNA from the father, a garbage truck driver named Lonnie David Franklin Jr., from a discarded pizza slice. A search of his home turned up 180 photographs of potential victims, which were posted on the police department's website so that the public could help in identifying them. He was convicted in 2016. Yet for every criminal that familial DNA searches identify, some innocent people are accused, based on the chance of sharing CODIS markers with convicted felons.

DNA profiling is based on population genetics, but it requires logic to avoid false accusations. Investigators must consider how DNA came to be at a crime scene, which is called DNA transfer, or "touch DNA." Primary transfer occurs when the suspect's DNA is on an object, such as a glove or a weapon. Secondary transfer occurs when the person who touches the object had earlier contacted another person's DNA, such as a rapist having first shaken hands with someone. In one complicated case, a professor at an Ivy League school was accused of his wife's murder. When he and his wife had shared a towel earlier that day, his skin cells had been transferred to her face. Later, the murderer, wearing gloves, touched her face, picked up the husband's cells, and unknowingly transferred them to his weapon. Reconstructing this scenario helped to exonerate the husband.

The Hardy-Weinberg equilibrium that makes DNA profiling possible is extremely rare in the real world, for most genes. Chapter 15 considers the familiar and common circumstances that change allele frequencies.

Key Concepts Questions 14.4

1. How does the concept of Hardy-Weinberg equilibrium enable interpretation of DNA profiles?
2. What type of DNA sequence is used to construct DNA profiles?
3. How do the terms *heterozygote* and *homozygote* apply to repeated DNA sequences?
4. How is probability used to interpret DNA profiles?
5. Describe how capillary electrophoresis is used to detect STRs to provide a DNA profile.
6. What is the source of the power behind DNA profiling?
7. How can researchers improve the accuracy of the interpretation of DNA profiles?

Summary

14.1 Population Genetics Underlies Evolution

1. A **population** is a group of interbreeding members of the same species in a particular area. All the alleles of all their genes constitute the population's **gene pool**.
2. **Population genetics** considers allele, genotype, and phenotype frequencies to understand **microevolution**. Phenotypic frequencies can be determined empirically, then used in algebraic expressions to derive other frequencies. Changes great enough to cause speciation are termed **macroevolution**.
3. Genotype frequencies change if migration, nonrandom mating, genetic drift, mutations, or natural selection operate.

14.2 Constant Allele Frequencies

4. In **Hardy-Weinberg equilibrium**, allele frequencies are not changing. Hardy and Weinberg proposed an algebraic equation to explain the constancy of allele frequencies. This shows why dominant traits do not increase and recessive traits do not decrease in populations. A gene in Hardy-Weinberg equilibrium tightly linked to a gene subject to natural selection, called **linkage disequilibrium (LD)**, may disturb the equilibrium.
5. The Hardy-Weinberg equation is a binomial expansion used to represent genotypes in a population. In Hardy-Weinberg equilibrium, gamete frequencies do not change as they recombine in the next generation.

Evolution is not occurring. When the equation $p^2 + 2pq + q^2 = 1.0$ represents a gene with one dominant and one recessive allele, p^2 corresponds to the frequency of homozygous dominant individuals; $2pq$ is the frequency of heterozygotes; and q^2 represents the frequency of the homozygous recessive class. The frequency of the dominant allele is p, and of the recessive allele, q.

14.3 Applying Hardy-Weinberg Equilibrium

6. If we know either p or q, we can calculate genotype frequencies, such as carrier risks. Often such information comes from knowing the value of q^2, which represents the frequency of homozygous recessive individuals in a population.
7. For X-linked recessive traits, the mutant allele frequency for males equals the trait frequency. For rare diseases or traits, the value of p approaches 1, so the carrier frequency ($2pq$) is approximately twice the frequency of the rare trait (q).

14.4 DNA Profiling Uses Hardy-Weinberg Assumptions

8. The numbers of copies of repeated DNA sequences that do not encode protein are presumably in Hardy-Weinberg equilibrium and can be compared to establish individual DNA profiles.

9. To obtain a **DNA profile**, technicians determine numbers of **short tandem repeats (STRs)** and multiply population-based allele frequencies to derive the probability that profiles from two sources match by chance.
10. Forensic DNA profiles are constructed using 13 STRs, mtDNA sequences, and amelogenin gene size to determine sex.
11. DNA profiling is used to identify victims of natural disasters, terrorist attacks, and genocide. It is used to solve crimes, but may compromise individual privacy.

Review Questions

1. Define *gene pool.*

2. Why are Hardy-Weinberg calculations more complicated if a gene has many alleles that affect the phenotype?

3. How can evolution occur at both microscopic and macroscopic levels?

4. Explain the differences among an allele frequency, a phenotypic frequency, and a genotypic frequency.

5. What are the conditions under which Hardy-Weinberg equilibrium cannot be met?

6. Why is knowing the incidence of a homozygous recessive condition in a population important in deriving allele frequencies?

7. How is the Hardy-Weinberg equation used to predict the recurrence of X-linked recessive traits?

8. Why were VNTR sequences replaced with STRs for obtaining a DNA profile?

9. How would a heterozygote and homozygote for the same STR be represented differently on an electropherogram?

10. Why are specific population databases needed to interpret DNA profiles?

11. Why is a probability value assigned to a particular copy number variant? Where do the probabilities come from?

12. Under what circumstances is analysis of repeats in mtDNA valuable?

13. Explain and provide an example of how a familial DNA search can lead to a false accusation.

Applied Questions

1. Discuss how DNA profiling can lead to conviction and exoneration.

2. "We like him, he seems to have a terrific gene pool," say the parents upon meeting their daughter's boyfriend. Why doesn't their statement make sense?

3. Two couples want to know their risk of conceiving a child with cystic fibrosis. In one couple, neither partner has a family history of the disease; in the other, one partner knows he is a carrier. How do their risks differ?

4. How does calculation of allele frequencies differ for an X-linked trait or disease compared to one that is autosomal recessive?

5. Profiling of Y chromosome DNA implicated Thomas Jefferson in fathering a child of his slave, discussed in chapter 1. What might have been a problem with the conclusion?

6. Torsion dystonia is a movement disease that affects 1 in 1,000 Jewish people of Eastern European descent (Ashkenazim). What is the carrier frequency in this population?

7. Maple syrup urine disease (MSUD) (see Clinical Connection 2.1) is autosomal recessive and causes intellectual and physical disability, difficulty feeding, and a sweet odor to urine. In Costa Rica, 1 in 8,000 newborns inherits the condition. What is the carrier frequency of MSUD in this population?

8. Ability to taste phenylthiocarbamide (PTC) is mostly determined by the gene *PTC*. The letters *T* and *t* are used here to simplify analysis. *TT* individuals taste a strong, bitter taste; *Tt* people experience a slightly bitter taste; *tt* individuals taste nothing.

   A fifth-grade class of 20 students tastes PTC that has been applied to small pieces of paper, rating the experience as "very yucky" (*TT*), "I can taste it" (*Tt*), and "I can't taste it" (*tt*). For homework, the students test their parents, with these results:

   - Of six *TT* students, four have two *TT* parents and two have one parent who is *TT* and one parent who is *Tt*.
   - Of four students who are *Tt*, two have two parents who are *Tt*, and two have one parent who is *TT* and one parent who is *tt*.
   - Of the 10 students who can't taste PTC, four have two parents who also are *tt*, but four students have one parent who is *Tt* and one who is *tt*. The remaining two students have two *Tt* parents.

   Calculate the frequencies of the *T* and *t* alleles in the two generations. Is Hardy-Weinberg equilibrium maintained?

9. The examples of DNA profiling in the chapter concern criminals, war crimes, or natural disasters. Explain how the 13 STRs can be used to rule out paternity.

10. MeowPlex is a test that detects 11 four-base STRs on nine autosomes plus the *SRY* gene of domestic cats, pumas, and ocelots. Describe a scenario where this test might be useful.

Forensics Focus

1. A DNA profiling method that is simplified so that it can be used in the field uses only three STRs. How should the three be selected to best distinguish individuals?

2. The Indian Ocean tsunami of 2004 gave forensic investigators the idea to form and implement a Global DNA Response Team to collect samples after a natural disaster. What rules should such a program have?

3. Irene is 80 years old and lives alone with her black cat Moe. One day when she is taking out the garbage, a man jumps out from behind a garbage can and shoves her. As she struggles to get up, he runs inside, pushes open her door, enters, and grabs her purse, which is next to a slumbering Moe on the kitchen table. Moe, sensing something is wrong or perhaps just upset that his nap has been interrupted, scratches the man, who yelps and forcefully flings the cat against a wall, yanking out one of the animal's claws in the process.

   Meanwhile, outside, Irene is back on her feet, trying to get her bearings when the fleeing thief knocks her down again. This time she blacks out, and the man escapes. A few minutes later a neighbor finds her and calls the police. At the crime scene, they collect a drop of blood on the table, a curly blond hair, a few straight black hairs, several gray hairs, and a cat's claw with human skin under it. The police send Irene to the hospital, Moe to the veterinary clinic, and the claw, blood, and hairs to the state forensics lab. The police then thoroughly search the immediate area and neighborhood, but do not find any suspects. When Irene regains consciousness, she remembers nothing about her attacker.

   a. List DNA tests that could help identify the perpetrator.

   b. A familial DNA search closely matches two of the forensic samples to a 58-year-old man in prison for murder. One sample matches at 12 CODIS sites and the other at 10 sites. The investigators find that the convict has four sons and four nephews. What should the police do with this information?

4. DNA dragnets have been so successful that some people have suggested storing DNA samples of everyone at birth, so that a DNA profile could be obtained from anyone at any time. Do you think that this is a good idea or not? Cite reasons for your answer.

5. Rufus the cat was discovered in a trash can by his owners, his body covered in cuts and bite marks and bits of gray fur clinging to his claws—gray fur that looked a lot like the coat of Killer, the huge hound next door. Fearful that Killer might attack their other felines, Rufus's distraught owners brought his body to a vet, demanding forensic analysis. The vet suggested that the hair might have come from a squirrel, but agreed to send samples to a genetic testing lab. Identify the samples that the vet might have sent, and what information each could contribute to the case.

6. In a crime in Israel, a man knocked a woman unconscious and raped her. He didn't leave any hairs at the crime scene, but he left eyeglasses with unusual frames, and an optician helped police locate him. The man also left a half-eaten lollipop at the scene. DNA from blood taken from the suspect matched DNA from cheek-lining cells collected from the base of the telltale lollipop at four STRs on different chromosomes. Allele frequencies from the man's ethnic group in Israel are listed beside the profile pattern shown here:

STR1	Homozygote	Allele frequency 0.2
STR2	Allele 1 = 0.3	
	Allele 2 = 0.7	
STR3	Homozygote	Allele frequency = 0.1
STR4	Allele 1 = 0.4	
	Allele 2 = 0.2	

   a. How would the electropherogram for STRs 1 and 3 look different than that for STRs 2 and 4?

   b. What is the probability that the suspect's DNA matches that of the lollipop rapist by chance? (Do the calculation.)

   c. The man's population group is highly inbred. How does this affect the accuracy or reliability of the DNA profile? (P.S.—He was so frightened by the DNA analysis that he confessed!)

Case Studies and Research Results

1. An extra row of eyelashes is an autosomal recessive trait seen in 900 of the 10,000 residents of an island in the South Pacific. Greta knows that she is a heterozygote for this gene because her eyelashes are normal but she has an affected parent. She wants to have children with a homozygous dominant man so that the trait will not affect her offspring. What is the probability that a person with normal eyelashes in this population is a homozygote for the wild-type allele of this gene?

2. Glutaric aciduria type 1 was the first disease investigated at the Clinic for Special Children, which was founded in 1989 in the heart of the Old Order Amish and Mennonite community in Lancaster, Pennsylvania (see Clinical Connection 15.1). The disease causes severe movement problems that lead to paralysis, brain damage, and early death. In this population, 0.25 percent of newborns have the disease. Researchers at the clinic developed a special formula with the levels of amino acids tailored to counter the metabolic abnormality. Children who use the formula can avoid the symptoms.

   a. What percentage of this population are carriers for glutaric aciduria type 1?

   b. What effect might the treatment have on the mutant allele frequency in this population in the future?

CHAPTER 15

Changing Allele Frequencies

The ability to digest lactose (milk sugar) became more prevalent in human populations after agriculture introduced dairy foods, providing an example of evolution in action.

© Digital Vision/Getty RF

Learning Outcomes

15.1 Population Matters: Steel Syndrome in East Harlem

1. Explain how DNA variants that are inherited together in members of a family reflect genetic ancestry.
2. Discuss how understanding population substructure can aid in health care.

15.2 Nonrandom Mating

3. Explain how nonrandom mating changes allele frequencies in populations.

15.3 Migration

4. Explain how migration changes allele frequencies in populations.

15.4 Genetic Drift

5. Explain how the random fluctuations of genetic drift affect genetic diversity.
6. Discuss how founder effects and population bottlenecks amplify genetic drift.

15.5 Mutation

7. Discuss how mutation affects population genetic structure.

15.6 Natural Selection

8. Provide examples of negative, positive, and artificial selection.
9. Explain how balanced polymorphism maintains certain diseases in populations.

15.7 Eugenics

10. Explain how eugenics attempts to alter allele frequencies.

The BIG Picture

Several forces mold the genetic composition of populations and, over time, drive evolution. The forces of evolutionary change are nonrandom mating, migration, genetic drift, mutation, and natural selection.

The Evolution of Lactose Tolerance

For millions of people who have lactose (milk sugar) intolerance, dairy food causes cramps, bloating, gas, and diarrhea. They no longer produce lactase, which is an enzyme made in early childhood that breaks down the milk sugar lactose into more easily digested sugars. But people who have lactose intolerance may represent the "normal," or wild type condition. Only 35 percent of people in the world can digest lactose into adulthood. People who *can* digest dairy foods have lactase persistence. One gene controls the ability to digest milk sugar.

Clues in DNA suggest that agriculture drove the differences in our abilities to digest lactose. As dairy farming spread around the world, from 5,000 to 10,000 years ago, people who had gene variants enabling them to digest milk into adulthood had an advantage. They could eat a greater variety of the now more plentiful foods, were healthier, and had more children. Over time, populations that consumed dairy foods had more people with lactase persistence. In contrast, in populations with few or no dairy foods, lactose intolerance was not a problem, and so gene variants impairing the ability to digest lactose remained prevalent because they caused no harm.

The link between lactose intolerance and agriculture is why today, the European American population only has 10 percent lactose intolerance. Among Asian Americans, who eat far less dairy, 90 percent have lactose intolerance. That is, the inability to digest lactose doesn't bother them. Seventy-five percent of African Americans and Native Americans have lactose intolerance.

15.1 Population Matters: Steel Syndrome in East Harlem

Many aspects of modern life alter allele (gene variant) frequencies in populations. As a result, many genes are not in Hardy-Weinberg equilibrium. Recall from Chapter 14 that this equilibrium means unchanging allele frequencies from generation to generation.

Several factors alter allele frequencies. Human behavior can shape genetic "substructures" within populations in terms of the genotypes of people who have children together. Religious restrictions and personal preferences guide our choices of mates. Wars and persecution diminish certain populations. Economic and political systems enable some groups to have more children than others. We travel, shuttling genes in and out of populations. Natural disasters and new diseases shrink populations, which then resurge, at the expense of genetic diversity. These influences, plus mutation and a reshuffling of genes at each generation, make gene pools quite fluid.

Researchers can trace the genetic ancestry of the individuals who make up a population through linked alleles and SNPs that are passed, as haplotypes, from generation to generation (see figure 5.19 and section 5.3). These DNA sequences, linked on the same copy of a chromosome and transmitted together through generations, are said to be "**identical by descent**." When both chromosome copies have the same identical by descent DNA sequence, the phenomenon is termed a "**run of homozygosity**." It indicates that two individuals shared a recent ancestor, such as a common great-great-grandmother. The more recent the shared ancestor, the longer the run of homozygosity. Because of the tendency of some stretches of DNA to remain together during meiosis (linkage disequilibrium), even if two people haven't shared an ancestor for 10 generations, runs of homozygosity may reach 4 million DNA bases. These areas of sameness within genomes are what one researcher calls "hidden patterns of relatedness" that can be used to reconstruct demography and ancestry, extending family trees back in time, perhaps farther than diaries and memories can. A person who knows nothing about her or his ancestors can find a matching population group using DNA data. In a practical sense, merging such genetic information with genealogical records and electronic health records can lead to more precise diagnoses.

Understanding the genetic substructure of populations, especially in cities where people are from many diverse backgrounds, can enhance health care. This is the case for Steel syndrome. It is an inherited bone disease that is rare in the general population, but clustered among the 8,000 people of Puerto Rican ancestry living in the New York City community of East Harlem (**figure 15.1**). Without knowing that a patient is from this group, a physician might misdiagnose the condition and possibly perform hip surgery that can worsen symptoms.

Steel syndrome causes joint pain, hip dislocation, pinching of the spinal cord in the neck, short stature, and a large head with a long, oval face, broad nose, small low ears rotated slightly backward, and a prominent forehead. Researchers at the Icahn School of Medicine at Mt. Sinai in New York City analyzed 600,000 SNPs in 11,000 DNA samples from a biobank. They identified a genetic signature of Puerto Rican descent so distinctive that they could even tell people who knew only that they were Hispanic/Latino that their ancestors came from the island.

The researchers noticed that several individuals of Puerto Rican ancestry in East Harlem who were short also had joint pain, had undergone joint replacements at young ages, and had pinched necks. Plus, the unusual symptoms clearly ran in families as an autosomal recessive trait. At about the same time, another group of researchers reported a disease of cartilage and bone stemming from a specific point mutation in a collagen gene (*COL27A1*): Steel syndrome, named for the orthopedist who first described it in 1993 in 23 children from Puerto Rico. (See Table 12.1 for other medical conditions resulting from abnormal collagen.) Other symptoms include soft bone tips, dislocated elbows and hips, fused finger and toe bones, and scoliosis. Most importantly, Dr. Steel advised against hip surgery because the abnormality differs from common reasons for the surgery, such as a sports injury or accident. Steel syndrome, extremely rare among Hispanics as a whole but more common among the Puerto Rican population, provides a compelling example of the clinical importance of recognizing where a person lives and his or her genetic ancestry.

Key Concepts Questions 15.1

1. Define "identical by descent" and "run of homozygosity."
2. Explain how considering population genetic substructure can help in providing health care.

Figure 15.1 Considering population subgroups can aid disease diagnosis. Steel syndrome is ultra-rare among the general Hispanic population but clustered among people of Puerto Rican ancestry living in East Harlem, in New York City. © *blvdone/Shutterstock*

15.2 Nonrandom Mating

The ever-present and interacting forces of nonrandom mating, migration, genetic drift, mutation, and natural selection shape populations at the allele level. Changing allele frequencies can change genotype frequencies, which in turn can change phenotype frequencies. In a series of illustrations throughout this chapter, colored shapes represent alleles. Figure 15.14 combines the illustrations to summarize the chapter. We begin our look at the forces that change allele frequencies in populations with nonrandom mating (**figure 15.2**).

In the theoretical state of Hardy-Weinberg equilibrium, individuals of all genotypes are equally likely to successfully mate and to choose partners at random. In reality we may choose partners for many reasons: physical appearance, ethnic background, intelligence, earning potential, and shared interests, to name a few. Nonrandom mating is a major factor in changing allele frequencies in human populations.

Nonrandom mating affects a population when certain individuals contribute more to the next generation than others. This is common in agriculture when semen from one prize bull is used to inseminate thousands of cows. A similar situation in people arose when many families used the same sperm donor to conceive 150 children.

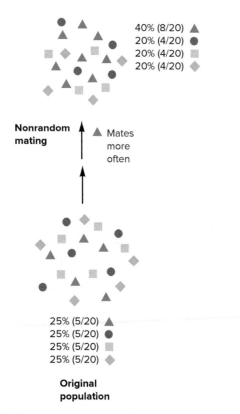

Nonrandom mating

40% (8/20) ▲
20% (4/20) ●
20% (4/20) ▪
20% (4/20) ◆

▲ Mates more often

25% (5/20) ▲
25% (5/20) ●
25% (5/20) ▪
25% (5/20) ◆

Original population

Figure 15.2 Nonrandom mating alters allele frequencies. More successful mating among individuals with the blue triangle allele will skew allele frequencies in the next generation.

High prevalence of an otherwise rare inherited condition can be due to nonrandom mating. For example, a form of albinism is uncommon in the general U.S. population, but it affects 1 in 200 Hopi Indians who live in Arizona. The reason for the trait's prevalence is cultural—men with albinism often stay back and help the women, rather than risk severe sunburn in the fields with the other men. They contribute more children to the population because they have more contact with the women.

The events of history echo in nonrandom mating patterns. When a group of people is subservient to another, genes tend to "flow" from one group to the other as the males of the ruling class have children with females of the underclass—often forcibly. Historical records and DNA sequences show such directional gene flow. A Glimpse of History in section 16.3 offers the compelling example of slave owners in the Deep South of the United States prior to the Civil War marking the genomes of the descendants of many African American women through rape.

Traits may mix randomly in the next generation if we are unaware of them or do not consider them in choosing partners. In populations where AIDS is extremely rare or nonexistent, for example, the two mutations that render a person resistant to HIV infection are in Hardy-Weinberg equilibrium (see figure 17.11). This situation would change, over time, if HIV began to infect the population, because the people with these mutations would become more likely to survive to produce offspring—and some of them would perpetuate the protective mutation. Natural selection would intervene, ultimately altering allele frequencies.

Many blood types are in Hardy-Weinberg equilibrium because we do not choose partners by blood type. Yet sometimes the opposite situation occurs. People with mutations in the same gene meet when their families participate in programs for people with the associated disease. Consider that more than two-thirds of relatives visiting a camp for children with cystic fibrosis are likely to be carriers, compared to the 1 in 23 or fewer in more diverse population groups.

People can avoid genetic disease with controlled mate choice and reproduction. In a program that began in New York City called Dor Yeshorim, for example, young people take tests for selected severe recessive diseases that are much more common among Jewish people of eastern European descent (Ashkenazim) or Spanish or Portuguese descent (Sephardim). Results are stored in a confidential database. Two people wishing to have children together can find out if they are carriers for the same disease. If so, they may elect not to have children. Thousands of people have been tested, and the program led to the near-disappearance of Tay-Sachs disease among Ashkenazim. The very few cases each year are usually in non-Jews who have not taken carrier tests.

A population that practices consanguinity has highly nonrandom mating. Recall from chapter 4 that in a consanguineous relationship, "blood" relatives have children together. On the family level, this practice increases the likelihood that harmful recessive alleles from shared ancestors will combine and pass to offspring, causing disease. The birth defect rate in offspring of blood relatives is 2.5 times the normal rate of about 3 percent. On a population level, consanguinity decreases genetic diversity. The proportion of homozygotes rises as the proportion of heterozygotes falls, and runs of homozygosity gather in genomes.

Some populations encourage marriage between cousins, which increases the incidence of certain recessive diseases. In parts of the Middle East, Africa, and India, 20 to 60 percent of marriages are between cousins or between uncles and nieces. The tools of molecular genetics can reveal these relationships. Researchers traced DNA sequences on the Y chromosome and in mitochondria among residents of an ancient, geographically isolated "micropopulation" on the island of Sardinia, near Italy. They consulted archival records dating from the village's founding by 200 settlers around the year 1000 to determine familial relationships. Between 1640 and 1870, the population doubled, reaching 1,200 by 1990. Fifty percent of the present population descends from just two paternal and four maternal lines, and 86 percent of the people have the same X chromosome. Researchers are analyzing health conditions that are especially prevalent in this population, which include hypertension and a kidney disease.

Worldwide, about 960 million married couples are related, and know of their relationship. Also contributing to nonrandom mating is endogamy, which is marriage within a community. In an endogamous society, spouses may be distantly related and be unaware of the connection.

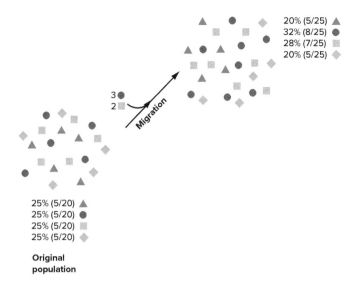

Figure 15.3 **Migration alters allele frequencies.** If the population travels and picks up new individuals, allele frequencies can change. Individuals leaving the traveling group can also alter allele frequencies.

Key Concepts Questions 15.2

1. Explain why human mating is usually not random.
2. Discuss the effects of consanguinity and endogamy on population genetic structure.

15.3 Migration

Large cities, with their pockets of ethnicity, defy Hardy-Weinberg equilibrium by their very existence. New York City exemplifies the contributions that waves of immigration bring to a population's complex and dynamic substructure. The original Dutch settlers of the 1600s had different alleles than those in today's metropolis of English, Irish, Slavs, Africans, Hispanics, Italians, Asians, and many others. Today the New York City neighborhood of Elmhurst, Queens, is the most genetically diverse community in the world!

Figure 15.3 presents a simplified view of the effect on allele frequencies of individuals joining a migrating population. Clues to past migrations lie in historical documents as well as in differing allele frequencies in regions defined by

geographical or language barriers. A migration caused by a group of people forced to leave their homeland is a common feature of history and is called a "diaspora," which means "to scatter."

The frequency of the allele that causes galactokinase deficiency in several European populations reveals how people with this autosomal recessive disease migrated (**figure 15.4**). Galactokinase deficiency causes cataracts (clouding of the lens) in infants. It is common among a population of 800,000 gypsies, called the Vlax Roma, who live in Bulgaria. The disease affects 1 in 1,600 to 2,500 people among them, and 5 percent of the population are carriers. But among all gypsies in Bulgaria as a whole, the incidence drops to 1 in 52,000. As the map in figure 15.4 shows, the disease becomes rarer to the west. This pattern may have arisen when people with the allele settled in Bulgaria, with only a few individuals or families moving westward.

Allele frequencies often reflect who ruled whom. For example, the frequency of ABO blood types in certain parts of the world today mirrors past Arab rule. The distribution of ABO blood types is very similar in northern Africa, the Near East, and southern Spain, all areas where Arabs ruled until 1492. Uneven distribution of allele frequencies can reveal when and where nomadic peoples stopped. For instance, in the eighteenth century, European Caucasians called trekboers migrated to the Cape area of South Africa. The men stayed and had children with the native women of the Nama tribe. The mixed society remained fairly isolated, leading to the distinctive allele frequencies found in the present-day people of color of the area.

In a situation termed a **cline**, allele frequencies change from one neighboring population to another. A cline develops as immigrants introduce alleles and emigrants remove them.

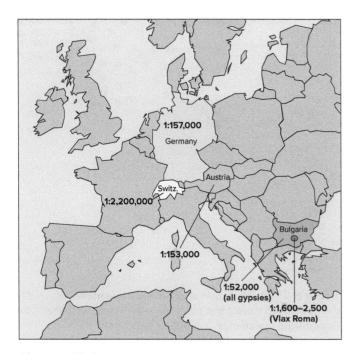

Figure 15.4 **Galactokinase deficiency in Europe illustrates a cline.** This autosomal recessive disease that causes blindness varies in prevalence across Europe. It is most common among the Vlax Roma gypsies in Bulgaria. The condition becomes much rarer to the west, as the shading from dark to light green indicates.

Clines may be gradual, if nothing blocks migration paths, or abrupt if barriers block gene flow. Barriers include geography and language.

Geographical formations such as mountains and bodies of water may block migration, maintaining population differences in allele frequencies on either side of the barrier. Consider agriculture, which arose in several places at different times. Agriculture spread faster from east to west in Eurasia because latitude did not change much, and there is less evidence of gene flow in this relatively steady environment. Even if people migrated, they tended to have the same gene variants as the people in their new homes. In the Americas, the situation was different, due to geography. Here, agriculture spread north to south. Hours of daylight and climate that differed with longitude presented different environmental challenges, which are reflected in greater genetic change from north to south than from east to west.

Language differences may isolate alleles if people who cannot communicate do not have children together. For example, in Italy, a nation rich in family history records with several language variants, blood types tend to cluster with linguistic group more than with geographical area. Perhaps differences in language prevent people from socializing, keeping alleles within groups.

Allele frequencies up and down the lush strip of fertile land that hugs the Nile River illustrate the concept of clines. Researchers found a gradual change in mitochondrial DNA sequences in 224 people who live

on either side of the Nile, an area settled 15,000 years ago. The farther apart two individuals live along the Nile, the less alike their mtDNA sequence. The story told in mDNA is consistent with evidence from mummies and historical records that indicate the area was once home to kingdoms separated by wars and language differences. If the region had been one large interacting settlement, then the mtDNA sequences would have been more evenly distributed.

Key Concepts Questions 15.3

1. How does migration alter allele frequencies?
2. What are clines?
3. How do geography and language affect clines?

15.4 Genetic Drift

Genetic drift refers to fluctuations in allele frequencies from generation to generation that happen by chance, to gametes. It is a characteristic of all populations. The changes in allele frequencies that cause genetic drift occur at random and are unpredictable.

The effects of genetic drift are accelerated when the population becomes very small, and sampling changes allele frequencies. Populations may shrink, amplifying genetic drift, in several circumstances: because of migration, due to a natural disaster or geographical barrier that isolates small groups, or as the consequences of human behavior (**figure 15.5**). Members of a small ethnic community within a larger population might have children only among themselves, keeping certain alleles more prevalent within the smaller group. For example, the skin-lightening condition vitiligo is much more common in a small community isolated in the mountains of northern Romania than

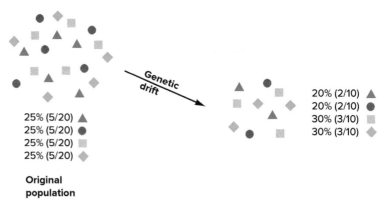

Figure 15.5 **Genetic drift alters allele frequencies.** If members of a population leave or do not reproduce, allele frequencies can change by chance sampling of a small population. When half of this population does not contribute to the next generation, two alleles increase in frequency and two decrease.

elsewhere in the nation. **Clinical Connection 15.1** discusses genetic diseases among the Old Order Amish and Mennonite populations of North America, in which genetic drift has isolated and amplified disease-causing alleles brought from Europe more than 300 years ago.

Two factors that can accelerate genetic drift are a **founder effect** and a **population bottleneck**, which both decrease population size. A founder effect results when some individuals leave a larger group or become reproductively isolated from them. In contrast, a population bottleneck is a large decrease in the size of an original population.

The Founder Effect

In a founder effect, a small group leaves a population to start a new settlement, and the new colony has different allele frequencies than the original population (**table 15.1**). Geneticists recognize a founder effect in a community known from local history to have descended from a few founders who brought in inherited traits and illnesses that are rare elsewhere.

A founder effect is easiest to trace when historical or genealogical records are available. This is the case for the 2.5 million Afrikaners of South Africa, who descend from a small group of Dutch, French, and German immigrants who had very large families. In the nineteenth century, some Afrikaners migrated northeast to the Transvaal Province, where they lived in isolation until the Boer War in 1902 introduced better transportation. Today, 30,000 Afrikaners who have an abnormal hemoglobin condition called porphyria variegata descended from one couple who came from the Netherlands in 1688! Their many children also had large families, passing on and amplifying the dominant mutation.

Another type of evidence for a founder effect is when all individuals in a population with a certain illness have the same mutation, which present-day patients inherited from shared ancestors. The people of Puerto Rican ancestry living in East Harlem, described in section 15.1, all have the same point mutation causing Steel syndrome. In contrast, a population

with several mutations that cause the same disease is more likely to have picked up those variants from unrelated people joining the group. This is the situation for breast cancer caused by mutation in the *BRCA1* gene. Affected individuals of Ashkenazi Jewish ancestry share a specific 3-base-pair deletion, whereas families from the Ivory Coast in Africa, the Bahamas, and the southeastern United States share a 10-base-pair deletion, probably inherited from West Africans ancestral to all three modern groups. Slaves brought the disease to the United States and the Bahamas between 1619 and 1808, while relatives remaining in Africa have perpetuated the mutant allele there.

Population Bottlenecks

A population bottleneck occurs when many members of a group die and only a few are left to replenish the numbers. The new population has only those alleles in the small group that survived the catastrophe. An allele in the remnant population might become more common in the replenished population than it was in the original larger group. Therefore, the new population has a much more restricted gene pool than the larger ancestral population, with some variants amplified, others diminished.

Population bottlenecks can occur when people (or other animals) colonize islands. An extreme example is seen among the Pingelapese people of the eastern Caroline Islands in Micronesia. Four to 10 percent of the people are born with "Pingelapese blindness." This is an autosomal recessive condition that includes colorblindness, nearsightedness, and cataracts. It is also called achromatopsia. Elsewhere, only 1 in 20,000 to 50,000 people inherits the condition. Nearly 30 percent of the Pingelapese are carriers. The prevalence of the blindness among the Pingelapese stems from a typhoon in 1780 that killed all but 9 males and 10 females who founded the present population. This severe population bottleneck, plus geographic and cultural isolation, increased the frequency of the blindness mutation as the population resurged.

A more widespread series of population bottlenecks occurred as a consequence of the early human expansion from Africa, discussed in chapter 16. As numbers dwindled during the journeys and then were replenished as people settled down, mating among relatives led, over time, to an increase in homozygous recessive genotypes compared to the larger and more genetically diverse ancestral populations that remained in Africa. These bottlenecks contrast today with the persistence of the greatest genetic diversity among African populations.

Figure 15.6 illustrates schematically the dwindling genetic diversity that results from a population bottleneck. Today's cheetahs live in just two isolated populations of a few thousand animals in South and East Africa. Their numbers once exceeded 10,000. The South African cheetahs are so alike genetically that even unrelated animals can accept skin grafts from each

Table 15.1	Founder Populations		
Population	**Number of Founders**	**Number of Generations**	**Population Size Today**
Costa Rica	4,000	12	2,500,000
Finland	500	80–100	5,000,000
Hutterites	80	14	36,000
Japan	1,000	80–100	120,000,000
Iceland	25,000	40	300,000
Newfoundland	25,000	16	500,000
Quebec	2,500	12–16	6,000,000
Sardinia	500	400	1,660,000

The Founder Effect and "Plain" Populations

The Old Order Amish and Mennonite people, called the Plain Populations, arrived in North America from Switzerland in the early eighteenth century, to escape religious persecution. They were descendants of a group called the Swiss Anabaptists. The earliest immigrants settled in Pennsylvania, and additional migrations from Europe established small farming communities in Ohio. The Plain people spread across the Midwest and into Canada, isolating themselves and maintaining simpler ways of living. Today more than a million of their descendants live in hundreds of communities in South, Central, and North America. The genetic traits that are overrepresented in these populations originated in Europe and are still there today.

Genetic drift plus a powerful founder effect, nonrandom mating, and migration contributed to the higher prevalence of certain inherited diseases among these people. Some families went to Children's Hospital of Philadelphia for diagnosis and treatment. Geneticists helped the families and investigated the causes of their inherited diseases, and so the Plain People have contributed greatly to our modern knowledge of genetic disease. Now several nonprofit health care centers are bringing genetic testing, exome sequencing, and new therapies to the Plain People. Earlier and more accurate diagnoses are making more treatments possible.

Statistics reveal founder effects in the Plain Populations. In Lancaster County, Pennsylvania, for example, maple syrup urine disease (MSUD; see Clinical Connection 2.1) affects 1 in 400 newborns, but affects only 1 in 225,000 newborns in the general population. Some conditions are rare variants of more common diseases, such as "Amish cerebral palsy." A physician from Philadelphia, Holmes Morton, discovered that the disease was an inborn error of metabolism called glutaric acidemia type 1, and was not due to lack of oxygen at birth, as others had thought. He went from farm to farm, tracking cases against genealogical records and drawing pedigrees. He found that nearly every family with MSUD descended from a couple who transmitted the recessive alleles. They had come to the settlement in 1730.

Dr. Morton founded the Clinic for Special Children in 1989, constructed at the site of a cornfield in Strasburg, Pennsylvania. The goal was to use genetic tools to diagnose children early, when they were still healthy enough to treat. Since the late 1990s, the clinic has used genomic technology. For example, for an Amish infant born with very thick skin, no hair, and a quickly developing life-threatening bacterial infection, comparing genetic markers to those of her seven healthy siblings revealed that she had a mutation in a gene, RAG1, that caused a form of severe combined immune deficiency. A stem cell transplant from a sister saved her life.

Research on the Plain People helps the wider community, too. For example, several related Amish children who have autism and seizures led researchers to mutations in a gene, CNTNAP2, that causes some cases of unexplained autism, seizures, schizophrenia, and language problems in the broader population (**figure 15A**). Depression, bipolar disorder, and attention deficit hyperactivity disorder also affect many populations but may stand out among the Plain People.

The Clinic for Special Children's approach of catching genetic disease early and treating symptoms as they arise has worked. Physicians at the clinic can now treat nearly half of the 110 genetic diseases that they detect. They can also recognize conditions that are lethal, saving children from pointless and painful treatments that physicians less familiar with the diseases unique to the Plain communities might provide.

The Amish and Mennonites not only have at least 100 rare diseases, but also have their own variants of the more common single-gene diseases, including cystic fibrosis, phenylketonuria, fragile X syndrome, and clotting and immune disorders. **Table 15A** lists some of these conditions.

Figure 15A **The Clinic for Special Children.** Genetic drift amplifies mutations from Europe among Plain Populations. The child seated with Dr. Kevin Strauss, Medical Director at the Clinic for Special Children, inherited a homozygous single-base deletion in a gene called CNTNAP2, which causes seizures and autism (cortical dysplasia–focal epilepsy syndrome). The family is Beachy Amish, a sect. *All rights reserved, Clinic for Special Children, 2013*

(Continued)

Table 15A	Selected Genetic Diseases More Common in Plain Populations	
Disease	**Gene**	**Symptoms**
Achromatopsia 2	CNGA3	Colorblindness
Crigler-Najjar 1 syndrome	UGT1A1	Jaundice in newborn
Ellis-van Creveld syndrome	LBN, EVC	Dwarfism, heart disease, polydactyly, fused wrist bones
Liebenberg syndrome	LBNBG	Hands develop as feet (see Clinical Connection 3.1)
Malignant hyperthermia	APOA4	Death on exposure to certain anesthetics
Periodic fever, familial	TNFRSF1A	Fever, abdominal pain, skin lesions, muscle pain
Pierson syndrome	LAMB2	Scarred kidneys, weak muscles, narrow pupils
Salla disease	SLC17A5	Poor muscle tone, uncoordinated movement, intellectual disability, coarse facial features

Questions for Discussion

1. Public perception is that diseases of the Amish are unique to them. Explain why this is not true.

2. If natural selection removes deleterious alleles from a population, how have certain single-gene diseases become much more prevalent among the Plain People than in other groups?

3. Because the Amish do not permit prenatal diagnosis, inborn errors of metabolism are detected with tests on cord blood that midwives collect. Explain how such analysis can lead to diagnosis of a single-gene disease, even if the child does not have symptoms.

4. Research a condition listed in Table 15A and describe treatments that may alter the course of a child's life if the condition is detected early.

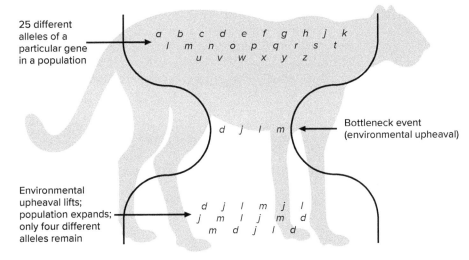

Figure 15.6 Population bottlenecks. A population bottleneck occurs when the size of a genetically diverse population falls, remains at this level for a time, and then expands again. The new population loses some genetic diversity if alleles are lost.

other. Researchers attribute the cheetahs' genetic uniformity to two bottlenecks—one at the end of the most recent ice age, when habitats changed, and another following mass slaughter by humans in the nineteenth century. However, the good health of the animals today indicates that the alleles that have persisted enable the cheetahs to thrive in their environment.

Human-wrought disasters that kill many people can cause population bottlenecks that greatly alter gene pools because aggression is typically directed against particular groups, whereas a typhoon indiscriminately kills anyone in its path. For example, after the many waves of killings, called pogroms, of Jewish people, only a few thousand remained in eastern Europe by the end of the eighteenth century. Then their numbers grew again, and from 1800 to 1939, the Jewish population in eastern Europe swelled to several million, only to be decimated again by the Holocaust.

Among Jewish people, nonrandom mating and population bottlenecks changed allele frequencies and contributed to the nearly 10 times higher incidence of certain inherited diseases compared to other populations. "Jewish genetic disease" tests are not meant to stereotype, but are based on a genetic fact of life—some illnesses are more common in

Table 15.2	Autosomal Recessive Genetic Diseases More Prevalent Among Ashkenazi Jewish Populations		
Disease	**Gene**	**Signs and Symptoms (Phenotype)**	**Carrier Frequency**
Bloom syndrome	*RECQL3*	Sun sensitivity, short stature, poor immunity, impaired fertility, increased cancer risk	1/110
Breast cancer	*BRCA1 BRCA2*	Malignant breast tumor	3/100
Canavan disease	*ASPA*	Brain degeneration, seizures, developmental delay, early death	1/40
Familial dysautonomia	*IKAKAP*	No tears, cold hands and feet, skin blotching, drooling, difficulty swallowing, excess sweating	1/32
Gaucher disease	*GBA*	Enlarged liver and spleen, bone degeneration, nervous system impairment	1/12
Niemann-Pick disease type A	*SMPD1*	Lipid accumulation in cells, particularly in the brain; intellectual and physical disability, death by age 3	1/90
Tay-Sachs disease	*HEXA*	Brain degeneration causing intellectual disability, paralysis, blindness, death by age 4	1/26
Fanconi anemia type C	*FANCA*	Deficiencies of all blood cell types, poor growth, increased cancer risk	1/89

certain populations, due to human behavior. However, DNA itself does not discriminate. The "Jewish" mutations can arise anew in anyone. **Table 15.2** describes some inherited diseases that are more common among Ashkenazi Jewish populations.

Key Concepts Questions 15.4

1. What is genetic drift?
2. Explain how founder effects and population bottlenecks amplify the effects of genetic drift.

15.5 Mutation

A continual source of genetic variation in populations is mutation, which changes one allele into another (**figure 15.7**). Genetic variability also arises from crossing over and independent assortment during meiosis, but these events recombine existing traits rather than introduce new ones.

If a DNA base change occurs in the part of a gene that encodes part of a protein necessary for the protein's function, then an altered trait may result. Another way that genetic change can occur from generation to generation is in the numbers of repeats of copy number variants (CNVs), which function as alleles.

Natural selection, discussed in section 15.6, eliminates alleles that adversely affect reproduction. Yet harmful recessive alleles are maintained in heterozygotes and are reintroduced by new mutation. Therefore, all populations have some alleles that would be harmful if homozygous. The collection of deleterious recessive alleles in a population is called its **genetic load**.

Human behavior and the events of history can influence the diversity of mutations in a population. Consider phenylketonuria (PKU), an autosomal recessive condition that causes intellectual disability unless a special diet restricts one amino acid type. PKU mutations worldwide are highly diverse, indicating that the disease has arisen more than once. Conversely, a distinct mutation found in many groups of people is probably more ancient, having occurred before those groups separated and perpetuated it.

Mutations found only in a small geographical region are more likely to be of recent origin, perhaps set apart due

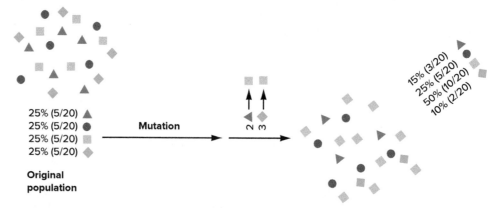

25% (5/20) ▲
25% (5/20) ●
25% (5/20) ■
25% (5/20) ◆

Original population

Mutation

15% (3/20)
25% (5/20)
50% (10/20)
10% (2/20)

Figure 15.7 **Mutation alters allele frequencies.** If one allele changes into another from one generation to the next, genotype frequencies can change.

to genetic drift. These mutations have had less time to spread. For example, people from Turkey, Norway, Jewish people from Yemen living in Israel, and French-Canadians have their "own" PKU alleles. The Yemeni mutation is so distinctive—a large deletion, compared to point mutations in other populations—that researchers used court and religious records to trace the mutation from families leaving the capital of Yemen north to three towns, and then to Israel.

The contribution of mutation in countering Hardy-Weinberg equilibrium is small compared to the influence of migration and nonrandom mating, because mutation is rare. The spontaneous mutation rate is about 170 bases per haploid genome in each gamete. Each of us probably has at least five "lethal equivalents"—alleles or allele combinations that if homozygous would kill us or make us too sick to have children. Natural selection has the greatest influence of all the factors on allele frequencies.

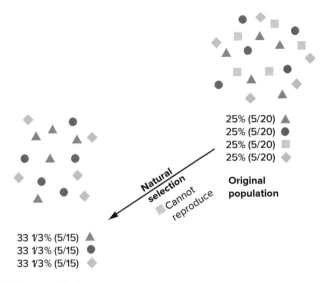

Figure 15.8 Natural selection alters allele frequencies. If health conditions impair the ability of individuals who have certain alleles to reproduce, allele frequencies in a population can change.

Key Concepts Questions 15.5

1. Explain how mutation increases genetic variation in a way that crossing over and independent assortment do not.
2. How does mutation alter allele frequencies?
3. Define *genetic load*.

15.6 Natural Selection

Environmental change can alter allele frequencies when individuals with certain phenotypes are more likely to survive and reproduce than others. This differential survival to reproduce guided by environmental change is **natural selection (figure 15.8)**. The chapter opener chronicles natural selection acting on gene variants that enable people to digest the sugar lactose.

Another illustration of natural selection related to a dietary pattern involves copy number variants (CNVs), in contrast to the point mutations that cause most instances of lactase persistence. In populations that follow high-carbohydrate diets, most people have more copies of the gene that encodes salivary amylase, the digestive enzyme that begins to break down carbohydrates in the mouth. Members of populations that follow low-carbohydrate diets have fewer copies of the gene and presumably less of the enzyme.

In natural selection, reproductive success is all important, because this is what transmits favorable alleles and weeds out the unfavorable ones, ultimately impacting population structure and therefore driving microevolution. In the common phrase used synonymously with natural selection—"survival of the fittest"—"fit" refers to reproductive success, not to physical prowess or intelligence (unless those traits lead directly to reproductive success). In a Darwinian sense, an unattractive and out-of-shape parent of 10 is more "fit" than a gorgeous triathlete with one child.

Negative and Positive Selection

Natural selection acts negatively if a trait diminishes in a population because it adversely affects reproduction. Natural selection acts positively if a trait becomes more common in a population because it enhances the chances of reproductive success. However, "negative" and "positive" selection are not value judgments—they simply refer to a trait's decreasing in a population or increasing. In Darwin's time, natural selection was thought to be primarily negative. Today, the ability to sequence DNA has revealed "signatures" of positive selection in the human genome.

One sign of positive selection is a gene in humans that has a counterpart in other primates, but in humans has at least one distinctive difference in the amino acid sequence, not just the DNA sequence. A change in the DNA sequence that does *not* substitute an amino acid does not change the protein, and therefore has no effect on the phenotype. Such a change isn't subject to natural selection, which acts on phenotypes.

The allele for lactase persistence illustrates positive selection. The allele is part of a large haplotype (genes linked on a chromosome) that shows very little sequence diversity among individuals. The persistence of the DNA sequence of an allele that has greatly increased in frequency in the population over a very short time—since agriculture began—indicates positive selection. Section 16.4 describes traits that were positively selected during recent human evolution.

A dramatic example of positive selection in humans is seen among native people ("highlanders") who live more than two miles above sea level on the Tibetan plateau (**figure 15.9**). Thanks to a "high-altitude adaptation," they thrive in the oxygen-poor air that makes others very ill and not likely to reproduce. Even the Han Chinese, who are a related population that lives in the lowlands, cannot survive up in the mountains for long without developing the headache, dizziness, ringing in the ears, heart palpitations, breathlessness, fatigue, sleep disturbance, lack of appetite, and

Figure 15.9 **Positive selection enables the Sherpa to thrive at elevations that make others very sick.** The Sherpa have migrated from their native Tibet to Nepal over the past 300 to 400 years. Many of them work as mountaineering guides, especially for climbers ascending Mt. Everest. They are short, strong, and hardy. A variant of the hypoxia inducible factor 2 (*EPAS1*) gene is one of several alleles responsible for their astonishing adaptation to low oxygen conditions. © *Goodshoot/ Alamy RF*

confusion of chronic mountain sickness. If they ascend to great heights, their bloodstreams become crammed with extra red blood cells that extract maximal oxygen from the thin air, raising the risks of heart attack and stroke.

In people other than the native Tibetans exposed to low oxygen conditions, a protein called hypoxia inducible factor 2 signals the kidneys to release the hormone EPO (erythropoietin). EPO stimulates the bone marrow to make more red blood cells, increasing hemoglobin levels in the blood. About 90 percent of native Tibetans have a deletion that removes part of the gene *EPAS1* that encodes hypoxia inducible factor 2. The deletion is extremely rare or not present at all in other populations. The Tibetans who have the deletion also share a set of five SNPs in *EPAS1*. The highlanders' version of the protein enables them to thrive in the thin air. Whole-genome sequencing has revealed two other genes that affect adaptation to thin air. The Tibetans have resided in the area only about 10,000 years. Their ability to survive with less oxygen is considered an example of rapid and recent evolution.

Dog Breeding Illustrates Artificial Selection

Natural selection acts on preexisting genetic variants. It is uncontrolled and largely unpredictable. In contrast, artificial selection is controlled breeding to perpetuate individuals with a particular phenotype, such as a crop plant or purebred dog. Darwin's idea of natural selection grew from his observations of artificially selected pigeons.

We have created our pets by controlling their evolution (**figure 15.10**). Pets arose from initial domestication of wild species, followed by artificial selection. Cats were domesticated in the Near East when agriculture began, about 10,000 years ago. They descended from one of five subspecies of wildcats. Dogs

were thought to have been domesticated one time in Southeast Asia about 30,000 years ago, from gray wolves. However, genome sequencing of several dozen ancient dogs supports a hypothesis of two original ancestral wolf populations in Eastern and Western Eurasia giving rise to two dog populations, with the eastern dogs following people to Europe between 6,400 and 14,000 years ago. Most of today's dog breeds were artificially selected in the nineteenth century, but some, such as the Australian bulldog and the silken windhound, were bred in the 1990s. The intense inbreeding required to fashion breeds cuts their genetic diversity, resulting in more than 300 inherited diseases, from bladder stones in Dalmatians to hip problems in St. Bernards. Yet the differences among dog breeds for some traits stem from only a few genes. For example, variants in only three genes account for 95 percent of the variability in the texture and length of canine coats.

The recent history of dog domestication explains why only a few genes define breeds. We select oddities and quirks in dogs, and these are often the consequence of a mutation in a single gene, rather than the collection of variants in many genes that may underlie a multifactorial trait that hasn't been artificially selected. In nature, that mutant may not have survived; we intentionally breed to select it. For example, ear floppiness and dwarfism, both seen in basset hounds, are each the result of a mutation in a single gene.

Analysis of the genomes of domesticated dogs and their wild relatives confirmed what we know from history—that our pets are no longer like wolves. Blocks of neighboring SNPs on chromosomes distinguish modern breeds. The fact that breeds share few SNP blocks indicates that we have selected away many ancestral DNA sequences. In addition, runs of homozygosity—long DNA sequences identical on both chromosomes—reflect intense inbreeding. The fact that individual domesticated breeds differ from each other less than wild dogs differ from each other indicates population bottlenecks that accompanied dog breeding, severely narrowing the breeds' gene pools.

Figure 15.10 **Dogs small and large.** It is hard to believe that the diminutive Chihuahua and the Great Dane are members of the same species. Thanks to artificial selection and intense inbreeding, body size in domesticated dogs varies more than it does in any other terrestrial mammal. © *Brand X Pictures/Getty Images RF*

Antibiotic Resistance Illustrates Natural Selection

Antibiotic drugs are used to treat bacterial infections, but many are no longer effective because of bacteria that are genetically equipped to survive in the presence of a particular drug. More than 2 million antibiotic-resistant infections occur in the United States each year, causing 23,000 deaths. Antibiotic resistance arises from the interplay between mutation and natural selection.

Genetic variants arise in populations of pathogenic bacteria in our bodies, some of which enable the microorganisms to survive in the presence of a particular antibiotic drug. When an infected person takes the drug, symptoms abate as sensitive bacteria die. However, resistant mutants reproduce, taking over the niche the antibiotic-sensitive bacteria vacated (**figure 15.11**). Soon, enough antibiotic-resistant bacteria accumulate to once again cause symptoms as the immune system responds to the returning infection. Usually antibiotic resistance genes already exist in the bacterial populations, and exposure to the drug selects the resistant bacteria. However, some antibiotics actually induce mutation.

Resistant bacteria circumvent antibiotic actions in several ways. Penicillin kills bacteria by tearing apart their cell walls. Resistant bacteria produce enzyme variants that dismantle penicillin, or have altered cell walls that the drug cannot bind. Erythromycin, streptomycin, tetracycline, and gentamicin are antibiotic drugs that kill bacteria by attacking their ribosomes, which are different from ribosomes in a human. The drugs cannot bind the ribosomes of resistant bacteria.

Antibiotic resistance can begin in either of two ways. Bacterial DNA can mutate, passing the resistance from one bacterial generation to the next by cell division. Or, groups of resistance genes are transmitted on mobile pieces of DNA called transposons, which move from bacterium to bacterium as part of DNA circles called plasmids. Bacteria usually pass transposons to similar bacteria, but in the unnatural environment of a hospital, genes may flit to any bacterium, quickly passing drug resistance among bacterial species. This is what has happened with infection by the bacterium *Staphylococcus aureus*.

S. aureus is normally present in low numbers in the nose and on the skin, but in high numbers it causes boils, food poisoning,

toxic shock syndrome, pneumonia, and wound infections. *S. aureus* became resistant to penicillin soon after the drug was introduced in the 1940s. A related penicillin, methicillin, worked for a time, but in 2000 resistant bacterial strains appeared, called MRSA, for methicillin-resistant *S. aureus*. By 2014, MRSA had become a global concern, accounting for 90 percent of *S. aureus* infections in some nations. Another antibiotic, vancomycin, works against some cases of MRSA infection, but *S. aureus* is becoming resistant to it, too. Some plasmids harbor resistance genes to both drugs.

In the rise of MRSA infection, natural selection benefits the pathogen, not us. That is, bacteria that can resist the drugs that we use to fight them will survive and reproduce, ensuring that *S. aureus* infection continues.

Balanced Polymorphism

If natural selection eliminates individuals with detrimental phenotypes from a population, then how do harmful mutant alleles remain in a gene pool? Harmful recessive alleles are replaced in two ways: by new mutation and by persistence in heterozygotes.

A recessive condition can remain prevalent if the heterozygote enjoys a health advantage that affects reproduction, such as resisting an infectious disease or surviving an environmental threat. This "heterozygous advantage" that maintains a recessive, disease-causing allele in a population is called **balanced polymorphism**. Recall that *polymorphism* means "variant"; the effect is *balanced* because the protective effect of the noninherited condition counters the negative effect of the deleterious allele in two copies, maintaining its frequency in the population. Balanced polymorphism is a type of balancing selection, which more generally refers to maintaining heterozygotes in a population. A few examples follow, and these and others are summarized in **table 15.3**.

Sickle Cell Disease and Malaria

Sickle cell disease is autosomal recessive and causes anemia, joint pain, a swollen spleen, and frequent, severe infections. It is the classic example of balanced polymorphism: carriers are resistant to malaria or develop very mild cases.

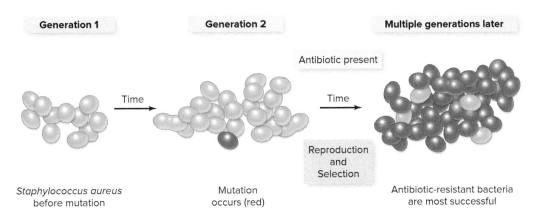

Generation 1 · **Generation 2** · **Multiple generations later**

Antibiotic present

Time → Time →

Reproduction and Selection

Staphylococcus aureus before mutation

Mutation occurs (red)

Antibiotic-resistant bacteria are most successful

Figure 15.11 **Antibiotic resistance.** Resistance to an antibiotic drug develops when a mutation in a bacterial cell gives the cell the ability to survive and reproduce in the presence of the drug. Over time, bacteria harboring the mutation dominate the population. This means return or resurgence of the infection.

Table 15.3	Balanced Polymorphism				
Disease 1 (inherited, carrier)	Protects against →	Disease 2	Because →	Mechanism	
Sickle cell disease		Malaria		Atypical red blood cells cannot retain parasites	
G6PD deficiency		Malaria		Parasite cannot reproduce in atypical red blood cells	
Phenylketonuria		Fungal infection in fetuses		Elevated phenylalanine inactivates fungal toxin	
Prion protein mutation		Transmissible spongiform encephalopathy		Prion protein cannot misfold in presence of infectious prion protein	
Cystic fibrosis		Diarrheal disease (cholera, typhoid fever)		Fewer chloride channels in intestinal cells prevent water loss	
Smith-Lemli-Opitz syndrome		Cardiovascular disease		Lowered serum cholesterol	

Malaria is an infection by the parasite *Plasmodium falciparum* or related species that causes cycles of chills and fever. The parasite spends the first stage of its life cycle in the salivary glands of the mosquito *Anopheles gambiae*. When an infected female mosquito draws blood from a human, malaria parasites enter red blood cells, which transport the parasites to the liver (**figure 15.12a**). The red blood cells burst, releasing parasites throughout the body.

In sickle cell disease, the blood becomes unwelcoming to the malaria parasite in several ways. The blood is thicker than normal, which may hamper the parasite's ability to infect. The bent shape of the cells has a dual effect (**figure 15.12b**). It prevents parasites from growing, and also blocks them from producing a protein required to go to the red blood cell surface and enable the cell to bind to other types of host cells, spreading the infection. In addition, many red blood cells burst

too soon, expelling the parasites. A sickle cell disease carrier's blood has enough atypical cells to hamper the actions of the parasite, but usually not enough of those cells to block circulation. Two other globin abnormalities that block malaria parasites are alpha thalassemia and hemoglobin C.

A clue to the protective effect of being a carrier for sickle cell disease came from striking differences in the incidence of the two diseases in different parts of the world (**figure 15.13**). In the United States, 8 percent of African Americans are sickle cell carriers, whereas in parts of Africa, up to 45 to 50 percent are carriers. Although Africans had known about a painful disease that shortened life, the sickled cells weren't reported in a medical journal until 1910 (see section 12.2). In 1949, British geneticist Anthony Allison found that the frequency of sickle cell carriers in tropical Africa was higher in regions where malaria rages all year long. Blood tests from children hospitalized with malaria

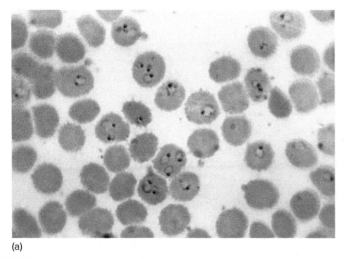
(a)

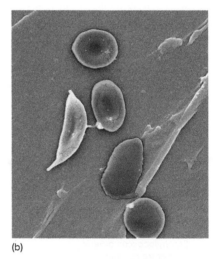
(b)

Figure 15.12 **Sickle cell disease protects against malaria.** **(a)** For part of the life cycle, one parasite that causes malaria, *Plasmodium falciparum*, lives inside red blood cells (dark areas). **(b)** Carriers of sickle cell disease do not contract malaria, or have very mild cases, because the misshapen red blood cells are inhospitable to the parasite. *(a): © CDC/ Steven Glenn, Laboratory & Consultation Division; (b): Source: CDC/Sickle Cell Foundation of Georgia: Jackie George, Beverly Sinclair/photo by Janice Haney Carr*

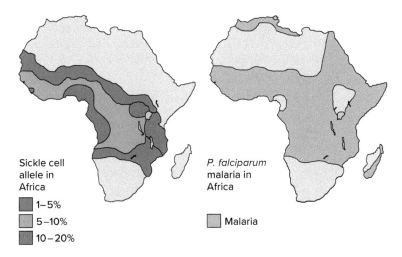

Sickle cell
allele in
Africa

■ 1–5%
☐ 5–10%
■ 10–20%

P. falciparum
malaria in
Africa

☐ Malaria

Figure 15.13 **Geography reveals balanced polymorphism.**
Comparing the distribution of people with malaria and people with
sickle cell disease in Africa reveals balanced polymorphism. Carriers
for sickle cell disease are resistant to malaria because changes in
their blood do not impair health, but inhibit the malaria parasite.

showed that nearly all were homozygous for the wild type sickle
cell allele. The few sickle cell carriers among them had the
mildest cases of malaria. Was malaria enabling the sickle cell
allele to persist by felling people who did not inherit it? The fact
that sickle cell disease is rarer where malaria is rare supports the
idea that sickle cell heterozygosity protects against the infection.

Further evidence of a sickle cell carrier's advantage in
a malaria-ridden environment is the fact that the rise of sickle

cell disease parallels the cultivation of crops that provide
breeding grounds for *Anopheles* mosquitoes. About 1000
B.C.E., Malayo-Polynesian sailors from Southeast Asia
traveled in canoes to East Africa, bringing new crops
of bananas, yams, taros, and coconuts. When the jungle
was cleared to grow these crops, mosquitoes occupied the
open space. The insects offered a habitat for part of the
life cycle of the malaria parasite.

The sickle cell allele may have been brought to
Africa by people migrating from southern Arabia and
India, or it may have arisen directly by mutation in East
Africa. However it happened, people who inherited one
copy of the sickle cell allele survived or never contracted
malaria—the essence of natural selection. These healthy
carriers had more children and passed the protective
allele to approximately half of them. Gradually, the frequency of the sickle cell allele in East Africa rose from
0.1 percent to 45 percent in 35 generations. Carriers paid
the price for this genetic protection, however, whenever
two of them produced a child with sickle cell disease.

A cycle set in. Settlements with large numbers of
sickle cell carriers escaped malaria. They were strong enough
to clear even more land to grow food, and support the disease-
bearing mosquitoes. Today, however, in African nations that
control malaria well, the frequency of the sickle cell allele has
decreased. The selective pressure is off. A Glimpse of History
discusses a time when malaria was common in parts of the
United States.

A GLIMPSE OF HISTORY

For three centuries, malaria plagued the United States, as
human activities opened niches for the mosquitoes that carried
the parasites that cause the disease. Starting with Christopher
Columbus, European explorers brought malaria, but in mild
forms that did not spread. It wasn't until parasite-carrying
mosquitoes arrived with slaves from western Africa that malaria
became a deadly infectious disease in the United States.

By 1776, swarms of mosquitoes carried malaria from
Georgia up through Pennsylvania, as pioneers took the
disease westward. Clearing forests and grasslands and
digging canals opened up vast new environments for
the mosquitoes. Some Africans were protected because
they were heterozygous (carriers) for sickle cell disease,
but Caucasians were vulnerable, and Native Americans
the most vulnerable of all. Settlers from France and Spain
brought malaria to Louisiana and Texas, respectively. By
1850, summertime epidemics of malaria stretched from
Florida up through the middle of New England and even into
the valleys of California. By the 1920s, public health workers
began to use dynamite to dig drainage ditches to remove
the mosquitoes' habitat (**figure 15B**).

By the beginning of World War II, malaria in the United States
was largely confined to the South, but watery conditions
around military training grounds again spread the disease.

In 1942, the Office of
Malaria Control in War
Areas began a huge
effort to eradicate the
disease. Nearly 5 million
homes were sprayed with
DDT, and residents used
mosquito netting at night,
as is common today in
parts of Africa. By 1951, the
disease was gone. The
Office of Malaria Control
in War Areas eventually
became the Centers
for Disease Control and
Prevention. Today a
thousand or so cases are
reported each year in the
country, nearly all from
travelers to areas where
malaria is endemic. Had
malaria not been stopped
in the United States,
natural selection might
have favored carriers of
sickle cell disease.

Figure 15B **Stopping
the spread of malaria.** In
the 1920s, public
health workers blew
up tree stumps to drain
the standing water in
which mosquito larvae
thrive. *Source: Centers
for Disease Control and
Prevention (CDC)*

Cystic Fibrosis and Diarrheal Disease

Balanced polymorphism may explain why CF is so common—its cellular defect protects against certain diarrheal illnesses. Diarrheal disease epidemics have left their mark on many human populations, and continue to be a major killer in the developing world.

Severe diarrhea rapidly dehydrates the body and leads to shock, kidney and heart failure, and death in days. In cholera, bacteria produce a toxin that opens chloride channels in cells of the small intestine. As salt (NaCl) leaves the intestinal cells, water rushes out, producing diarrhea. The CFTR protein does just the opposite, closing chloride channels and trapping salt and water in cells, which dries out mucus and other secretions. A person with CF is very unlikely to contract cholera, because the toxin cannot open the chloride channels in the small intestine cells.

CF carriers enjoy the mixed blessing of balanced polymorphism. They do not have enough abnormal chloride channels to cause the labored breathing and clogged pancreas of CF, but they have enough of a defect to block the cholera toxin. During cholera epidemics throughout history, individuals carrying mutant CF alleles had a selective advantage, and they disproportionately transmitted those alleles to future generations.

Because CF arose in western Europe and cholera originated in Africa, an initial increase in CF heterozygosity may have been a response to a different diarrheal infection—typhoid fever. The causative bacterium, *Salmonella typhi*, rather than producing a toxin, enters cells lining the small intestine—but only if CFTR channels are present. The cells of people with severe CF manufacture CFTR proteins that never reach the cell surface, and therefore bacteria cannot get in. Cells of CF carriers admit some bacteria. Protection against infections that produce diarrhea may therefore have kept CF in populations.

Human societies are highly complex, and so the forces of evolutionary change—nonrandom mating, migration, genetic drift, mutation, and natural selection—interact all the time. **Figure 15.14** summarizes the effects of these five forces, and **table 15.4** lists examples in the chapter with the mechanisms that they illustrate.

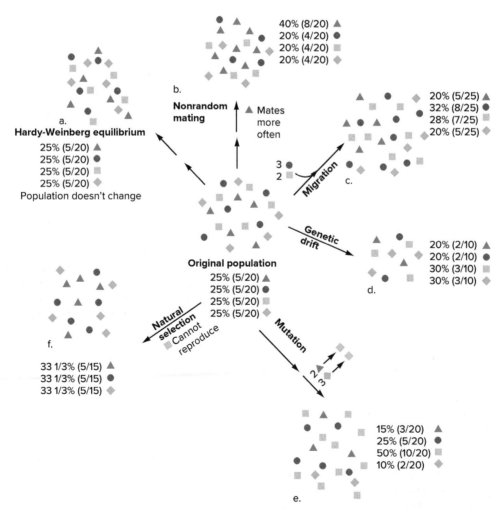

Figure 15.14 Forces that change allele frequencies.

Table 15.4	Forces That Change Allele Frequencies	

Mechanism of Allele Frequency Change	Examples
Nonrandom mating	Agriculture
	Hopi Indians with albinism
	U.S. slave owners and slaves
Migration	Galactokinase deficiency in Europe
	ABO blood type distribution
	Clines along the Nile
	Blood type and linguistic group in Italy
Genetic drift	
Founder effect	Steel syndrome in East Harlem
	Old Order Amish and Mennonites
	Afrikaners and porphyria variegata
Population bottleneck	Pingelapese blindness
	Cheetahs
	Pogroms against Ashkenazi Jews
Mutation	See Chapters 12 and 13
Natural selection	Lactose intolerance
	Tibetan adaptation to high altitude
	Antibiotic resistance in bacteria
	Sickle cell disease and malaria
	CF and diarrheal disease

Key Concepts Questions 15.6

1. How do cycles of infectious disease prevalence and virulence reflect natural selection?

2. Explain why disease-causing alleles do not vanish from populations.

3. Give an example of how balanced polymorphism maintains an inherited disease in a population.

15.7 Eugenics

We usually think of artificial selection in the context of Darwin's pet pigeons or purebred cats, dogs, and horses. We also practice artificial selection through control of our own reproduction, through individual choices as well as at the societal level.

Some people attempt to control the genes in their offspring. They do this by seeking mates with certain characteristics, by choosing egg or sperm donors with particular traits for use in assisted reproductive technologies (see chapter 21), or by ending pregnancies after a test reveals or predicts an untreatable, severe disease. On a societal level, **eugenics** refers

to programs or policies that control human reproduction with the intent of changing the genetic structure of the population. Eugenics works in two directions. Positive eugenics creates incentives for reproduction among those considered superior; negative eugenics interferes with reproduction of those judged inferior. Obviously, eugenic measures are highly subjective. **Table 15.5** includes some famous examples of eugenics.

The word *eugenics* was coined in 1883 by Sir Francis Galton to mean "good in birth." He defined eugenics as "the science of improvement of the human race germplasm through better breeding." One vocal supporter of the eugenics movement was Sir Ronald Aylmer Fisher. In 1930, he tried to apply the principles of population genetics to human society, writing that those at the top of a society tend to be "genetically infertile," producing fewer children than the less-affluent classes. This difference in family size, he claimed, was the reason why civilizations ultimately topple. He offered several practical suggestions to remedy this situation, including state monetary gifts to high-income families for each child born to them.

Early in the twentieth century, eugenics focused on maintaining purity. One prominent geneticist, Luther Burbank, realized the value of genetic diversity at the beginning of a eugenic effort. Known for selecting interesting plants and crossing them to breed plants with useful characteristics, Burbank in 1906 applied his ideas to people. In a book called *The Training of the Human Plant,* he encouraged immigration to the United States so that advantageous combinations of traits would appear as the new Americans interbred. Burbank's plan ran into problems, however, at the selection stage, which allowed only those with "desirable" traits to reproduce. Who decides which traits are desirable?

On the East Coast of the United States, Charles Davenport led the eugenics movement. In 1910, he established the Eugenics Record Office at Cold Spring Harbor, New York. There he headed a massive effort to compile data from institutions, prisons, circuses, and general society. He attributed nearly every trait rather simplistically to a single gene, such as "feeblemindedness," "criminality," "promiscuity," and "social dependency." A famous case was that of 17-year-old Carrie Buck. In 1927, she was tried in her hometown of Charlottesville for having a mother who lived in an asylum for the feebleminded and having a daughter out-of-wedlock (following rape) who was also deemed feebleminded, as was Carrie herself, although she was a B student in school. Ruled Sir Oliver Wendell Holmes, Jr., *"three generations of imbeciles are enough."* Carrie Buck was the first person sterilized to prevent having another "socially inadequate offspring." A crude pedigree drawn at the time showed Carrie Buck and her "inherited trait" of feeblemindedness.

Other nations practiced eugenics. From 1934 until 1976, the Swedish government forced certain individuals to be sterilized as part of a "scientific and modern way of changing society for the better," according to one historian. At first, only people with mental illness were sterilized, but poor, single mothers were later included. The women's movement in the 1970s pushed for an end to forced sterilizations.

Table 15.5	A Chronology of Eugenics-Related Events
1883	Sir Francis Galton coins the term *eugenics*.
1889	Sir Francis Galton's writings are published in the book *Natural Inheritance*.
1896	Connecticut enacts law forbidding sex with a person who has epilepsy or is "feebleminded" or an "imbecile."
1904	Galton establishes the Eugenics Record Office at the University of London to keep family records.
1907	First eugenic law in the United States orders sterilization of institutionalized intellectually disabled males and criminal males when experts recommend it.
1910	Eugenics Record Office founded in Cold Spring Harbor, New York, to collect family and institutional data.
1924	Immigration Act limits entry into the United States of "idiots, imbeciles, feebleminded, epileptics, insane persons," and restricts immigration to 7 percent of the U.S. population from a particular country according to the 1890 census—keeping out those from southern and eastern Europe.
1927	Supreme Court (*Buck v. Bell*) upholds compulsory sterilization of the intellectually disabled by a vote of 8 to 1, leading to many state laws.
1934	Eugenic sterilization law of Nazi Germany orders sterilization of individuals with conditions thought to be inherited, including epilepsy, schizophrenia, and blindness, depending upon rulings in Genetic Health Courts.
1939	Nazis begin killing 5,000 children with birth defects or intellectual disability, then 70,000 "unfit" adults.
1956	U.S. state eugenic sterilization laws are repealed, but 58,000 people have already been sterilized.
1965	U.S. immigration laws reformed, lifting many restrictions.
1980s	California's Center for Germinal Choice established, where Nobel Prize winners can deposit sperm to inseminate selected women.
1990s	U.S. state laws passed to prevent health insurance or employment discrimination based on genotype.
2003	Many governments recommend certain genetic tests and enact legislation to prevent genetic discrimination.
2004	Genocide of black Africans in Sudan.
2009	U.S. Genetic Information Nondiscrimination Act enacted but is limited in scope.
Future	Will widespread genome sequencing before birth for health reasons or curiosity have eugenic effects?

In 1994, China passed the Maternal and Infant Health Care Law. It proposed "ensuring the quality of the newborn population" and forbade procreation between two people if physical exams showed "genetic disease of a serious nature" that included intellectual disability, mental illness, and seizures, conditions that are ill-defined in the law and, if inherited, are typically multifactorial. Ironically, physical exams would miss carriers of the same genetic disease.

Another guise of eugenics is war, if the fighting groups differ genetically. Throughout history, war and conflict have altered gene pools, sometimes dramatically. These effects are eugenic when they take the form of rape of women of one group by men from another, with the intent of "diluting" the genes of the rape victims. In recent years in Rwanda, Congo, and Darfur, the conquerors claimed that their intent was to diminish the genetic contributions of their victims and spread their own genes. That statement is a working definition of eugenics.

Humanity arose in Africa some 200,000 years ago. As pockets of peoples spread across the globe, our behaviors and intelligence introduced culture. Chapter 16 explores some of our journeys, through clues to the past found in the sequences of our DNA.

Key Concepts Questions 15.6

1. What is eugenics?
2. Distinguish between positive and negative eugenics.
3. How do eugenics and medical genetics differ?

Designer Babies: Is Prenatal Genetic Testing Eugenic?

Modern genetics is sometimes compared to eugenics because genetic technologies may affect reproductive choices and can influence which alleles are passed to the next generation. However, medical genetics and eugenics differ in their intent. Eugenics aims to allow only people with certain "valuable" genotypes to reproduce, for the supposed benefit of the population as a whole. The goal of medical genetics, in contrast, is to prevent and alleviate suffering in individuals and families. But the once-clear line between eugenics and genetics is starting to blur as access to prenatal DNA testing widens, the scope of testing broadens to exomes and genomes, and sequencing cost plummets.

For decades prenatal genetic testing has focused on detecting the most common aneuploids—extra or missing chromosomes—or single-gene diseases known to occur in a family. As chapter 13 describes, chorionic villus sampling and amniocentesis have been used to visualize fetal chromosomes. In 2011 it became possible to collect, sequence, and overlap small pieces of placenta-derived cell-free DNA in the circulation of a pregnant woman and reconstruct a full genome sequence (see figure 13.10). Then, the sequence can be analyzed for genotypes that cause or increase the risk of developing known diseases. This information would theoretically allow a quality control of sorts in terms of which fetuses, with which characteristics, complete development. Such extensive analysis is not (yet) commercially available, but is done as part of research protocols.

Preimplantation genetic diagnosis (see figure 21.6) may provide a form of selection because it checks the genes of early embryos and chooses those with certain genotypes to continue development. Finally, use of technologies such as genome editing (see section 20.4) enable manipulation of a fertilized ovum's DNA, although such germline intervention in humans is generally prohibited.

The ability to sequence genomes has the potential to extend prenatal investigation from the more common chromosomal conditions to many aspects of a future individual's health and perhaps other characteristics, such as personality traits, appearance, and intelligence. **Figure 15C** takes a simple view of a complex idea—altering the frequency of inherited traits in a future human population.

Figure 15C **Designer babies.** Will widespread use of genetic technologies to create or select perfect children have eugenic effects? © Finn Brandt/Getty Images

Questions for Discussion

1. Is the lower birth rate of people with trisomy 21 Down syndrome a sign of eugenics (see Bioethics in chapter 13)? Cite a reason for your answer.

2. Is genetic manipulation to enhance an individual a eugenic measure?

3. Do you think that eugenics should be distinguished from medical genetics based on intent, or can widespread genetic testing to prevent disease have an effect on the population that is essentially eugenic?

4. Tens of thousands of years ago, humans with very poor eyesight were likely not to have survived to reproductive age. Is wearing corrective lenses a eugenic measure? Why or why not?

Summary

15.1 Population Matters: Steel Syndrome in East Harlem

1. The conditions that counter Hardy-Weinberg equilibrium—nonrandom mating, migration, genetic drift, mutation, and natural selection—create population substructures that can be valuable to consider in health care.

2. Researchers use haplotypes of **identical by descent** gene variants or SNPs to trace ancestry.

3. **Runs of homozygosity** (homologs with the same identical by descent haplotypes) indicate that two individuals shared a recent ancestor.

4. The higher prevalence of Steel syndrome among people of Puerto Rican descent living in East Harlem in New York City compared to the broader Hispanic population illustrates the value of recognizing a population genetic substructure.

15.2 Nonrandom Mating

5. Hardy-Weinberg equilibrium assumes all individuals mate with the same frequency and choose mates without regard to phenotype. This rarely happens. We choose mates based on certain characteristics, and some people have many more children than others.

6. DNA sequences that do not cause a phenotype important in mate selection or reproduction may be in Hardy-Weinberg equilibrium.

7. Consanguinity increases the proportion of homozygotes in a population, which may lead to increased incidence of recessive illnesses or traits.

15.3 Migration

8. **Clines** are changes in allele frequencies from one area to another. They reflect geographical barriers or linguistic differences and may be abrupt or gradual.

9. Human migration patterns through history explain many cline boundaries. Forces behind migration include escape from persecution and a nomadic lifestyle.

15.4 Genetic Drift

10. **Genetic drift** is the random fluctuation of allele frequencies from generation to generation. New allele frequencies may result from chance sampling.

11. A **founder effect** occurs when a few individuals found a settlement and their alleles form a new gene pool, amplifying their alleles and eliminating others.

12. A **population bottleneck** is a narrowing of genetic diversity that occurs after many members of a population die and the few survivors rebuild the gene pool.

13. Founder effects and population bottlenecks can amplify the impact of genetic drift.

15.5 Mutation

14. Mutation continually introduces new alleles into populations. It can occur as a consequence of DNA replication errors.

15. The **genetic load** is the collection of deleterious recessive alleles in a population.

15.6 Natural Selection

16. Environmental conditions influence allele frequencies via **natural selection.** Alleles that do not enable an individual to reproduce in a particular environment are selected against and diminish in the population, unless conditions change. Beneficial alleles are retained.

17. In **balanced polymorphism,** the frequencies of some deleterious alleles are maintained when heterozygotes have a reproductive advantage under certain conditions.

15.7 Eugenics

18. **Eugenics** is the control of individual reproduction to serve a societal goal.

19. Positive eugenics encourages those deemed superior to reproduce. Negative eugenics restricts reproduction of those considered inferior.

20. Some aspects of genetic technology affect reproductive choices and allele frequencies, but the goal is to alleviate or prevent suffering, not to change society.

Review Questions

1. What does a "run of homozygosity" indicate?

2. Give examples of how nonrandom mating, migration, a population bottleneck, and a mutation can each alter allele frequencies from Hardy-Weinberg equilibrium.

3. How might a mutant allele that causes an inherited illness in homozygotes persist in a population?

4. Why can increasing homozygosity in a population be detrimental?

5. Describe two scenarios for human populations, one of which accounts for a gradual cline and one for an abrupt cline.

6. How does a knowledge of history, sociology, and anthropology help geneticists interpret allele frequency data?

7. Define *genetic drift.*

8. How does a founder effect or a population bottleneck amplify the effect of genetic drift?

9. Explain the influence of natural selection on bacterial resistance to antibiotics.

10. Explain how negative and positive selection can simultaneously affect a population.

11. What do artificial selection and natural selection have in common? How do they differ?

12. Name an inherited disease allele that protects against an infectious illness.

13. How do genetic drift, nonrandom mating, and natural selection interact?

14. How have human activities contributed to balanced polymorphism between sickle cell disease and malaria? What effect, if any, might climate change have on the carrier frequency for sickle cell disease?

15. Cite three examples of eugenic actions or policies.

16. Distinguish between positive and negative selection, and between positive and negative eugenics. How do selection and eugenics differ?

17. Distinguish eugenics from China's past one-child policy, described in the opener to chapter 6.

Applied Questions

1. Why do more Asian Americans have lactose intolerance than European Americans?

2. Explain the value of considering the population subgroup to which a patient belongs, such as the people of Puerto Rican descent in the Hispanic community of East Harlem.

3. Begin with the original population represented at the center of figure 15.14, and deduce the overall, final effect of the following changes:
   - Two yellow square individuals join the population when they visit and decide to stay.
   - Four red circle individuals are asked to leave as punishment for criminal behavior.
   - A blue triangle man has sex with many females, adding five blue triangles to the next generation.
   - A green diamond female produces an oocyte with a mutation that adds a yellow square to the next generation.
   - A new infectious disease affects only blue triangles and yellow squares, removing two of each from the next generation.

4. What effect might dating apps have on the nonrandom mating aspect of population genetic change?

5. Before the year 1500, medieval Gaelic society in Ireland isolated itself from the rest of Europe, physically and culturally. Men in the group are called "descendants of Niall," and they have Y chromosomes inherited from a single ancestor. In the society, men took several partners, and sons born out of wedlock were fully accepted. Today, in a corner of northwest Ireland, one in five men has the "descendant of Niall" Y chromosome. In all of Ireland, the percentage of Y chromosomes with the Niall signature is 8.2 percent. In western Scotland, where the Celtic language is similar to Gaelic, 7.3 percent of the males have the Niall Y. In the United States, among those of European descent, it is 2 percent. Worldwide, the Niall Y chromosome makes up only 0.13 percent of the total. What concept from the chapter do these findings illustrate?

6. Indicate whether each of the following situations, examples, or findings illustrates nonrandom mating, migration, genetic drift, mutation, natural selection, or a combination of these factors.

   a. The "granny hypothesis" holds that grandparents who do not develop dementia contribute to the intellectual development of their grandchildren, increasing the likelihood of the grandchildren surviving to reproduce. The evidence for this apparent protective effect is that variants of 11 genes that lower the risk of vascular (blood-vessel-associated) dementia are more prevalent among the aged.

   b. The Orinoco crocodile of Venezuela is highly endangered. Researchers collected 20 egg clutches, raised the babies, and compared the 335 hatchlings for 17 genetic markers (short DNA repeats) to identify the parents. Although 16 females and 14 males were identified, six of the males had fathered more than

   90 percent of the offspring, and three of those six had fathered more than 70 percent.

   c. Ten castaways are shipwrecked on an island. The first mate has blue eyes, the others brown. A coconut falls by chance on the first mate, killing him.

   d. The Old Order Amish of Lancaster, Pennsylvania, have more cases of polydactyly (extra fingers and toes) than the rest of the world combined. All of the affected individuals descend from the same person, in whom the dominant mutation originated.

   e. The CCR5 mutation discussed in the chapter 17 opener that keeps HIV out of human cells also blocks infection by the bacterium that causes plague. Seven centuries ago, in Europe, the "Black Death" plague epidemic increased the protective allele in the population. Today the mutation makes 3 million people in the United States and the United Kingdom resistant to HIV infection.

   f. African Americans develop a form of end-stage kidney disease associated with elevated blood pressure that Europeans do not. Two variants in a gene on chromosome 22, called ApoL1, cause the kidney disease. The encoded protein is secreted into the blood, but only the forms in the African Americans who have the kidney disease also kill the parasites that cause African sleeping sickness. The mutations persist because they protect against African sleeping sickness.

7. Use the information in chapters 14 and 15 to explain why

   a. porphyria variegata is more prevalent among Afrikaners than other South African populations.
   b. cystic fibrosis and sickle cell disease remain common.
   c. the Sherpa from Tibet can tolerate thin air.
   d. the Amish in Lancaster County have a high incidence of genetic diseases that are ultra-rare elsewhere.
   e. the frequency of the allele that causes galactokinase deficiency varies across Europe.

8. Which principles discussed in this chapter do the following science fiction plots illustrate? Some plots may illustrate more than one concept.

   a. In the novel and film *SevenEves,* the moon shatters into seven pieces. The earth's human population has two years to prepare for the pieces to smash into the surface and make the planet uninhabitable for centuries. Some people already living on huge space stations survive, and others are selected to join them. Everyone else dies under the barrage of moon junk and intense heat. Above, on the "Cloud Ark," the human species dwindles, but eventually resurges from seven surviving women, with help from assisted reproductive technologies to make babies. Five thousand years after the moon blows up, the human population, ready to inhabit a healed earth, has resurged to 3 billion.

   b. In *The Time Machine,* set in the distant future on Earth, one group of people is forced to live on the surface while another group is forced to live in caves. Over many generations, the groups diverge. The Morlocks that live below ground have dark skin, dark hair, and behave aggressively. The Eloi that live above ground are blond, fair-skinned, and meek.

c. In *Children of the Damned,* all of the women in a small town are suddenly impregnated by genetically identical beings from another planet.

d. In Dean Koontz's novel *The Taking,* giant mutant fungi kill nearly everyone on Earth, sparing only young children and the few adults who protect them. The human race must reestablish itself from the survivors.

e. In the television series *Wayward Pines,* most present-day humans die in the aftermath of environmental collapse triggered by climate change. The survivors, over the ensuing centuries, become increasingly ferocious, although they maintain human family structures.

9. Syndrome X consists of obesity, type 2 diabetes, hypertension, and heart disease. Researchers sampled blood from nearly all of the 2,188 residents of the Pacific Island of Kosrae, and found that 1,709 of them are part of the same pedigree. The incidence of syndrome X is much higher in this population than for other populations. Suggest a reason for this finding, and indicate why it would be difficult to study these traits, even in an isolated population.

10. By which mechanisms discussed in this chapter do the following situations alter Hardy-Weinberg equilibrium?

a. In ovalocytosis, a protein that anchors the red blood cell plasma membrane to the cytoplasm is abnormal, making the membrane so rigid that parasites that cause malaria cannot enter.

b. In the mid-eighteenth century, a multi-toed male cat from England crossed the sea and settled in Boston, where he left behind many kittens, about half of whom also had extra digits. People loved the odd felines and bred them. Today, in Boston and nearby regions, multi-toed cats are much more common than in other parts of the United States.

c. Many slaves in the United States arrived in groups from Nigeria, which is an area in Africa with many ethnic subgroups. They landed at a few sites and settled on widely dispersed plantations. Once emancipated, former slaves in the South were free to travel and disperse.

d. About 300,000 people in the United States have Alzheimer disease caused by a mutation in the presenilin-2 gene. They all belong to five families that came from two small villages in Germany and migrated to Russia in the 1760s and then to the United States from 1870 through 1920.

Forensics Focus

1. In the 1870s, prison inspector and self-described sociologist Richard Dugdale noticed that many inmates at his facility in Ulster County, New York, were related. He began studying them, calling the family "Jukes," although he kept records of their real names. Dugdale traced the family back seven generations to a son of Dutch settlers, named Max, who was a pioneer and lived off the land. Margaret, "the mother of criminals," as Dugdale wrote in his 1877 book *The Jukes: A Study in Crime, Pauperism, Disease and Heredity,* married one of Max's sons, and the couple supposedly were ancestors of 540 of the 709 criminals on Dugdale's watch. Dugdale attributed the Jukes's less desirable traits to heredity.

   The Jukes study influenced social scientists to probe other families with misfits who were all Caucasian, descended from colonial settlers, and poor. Poverty was not seen as an economic problem, but due to inborn degeneracy that if left unchecked would cost society greatly.

   Dugdale's book fed the fledgling eugenics movement. In 1911, researchers at the Eugenics

Record Office in Cold Spring Harbor described the Jukes' phenotype as "feeblemindedness, indolence, licentiousness, and dishonesty." The Jukes story and others were used to support compulsory sterilization of those deemed unfit. But the original research on the Jukes family was flawed, and its accuracy never questioned. Less notorious Jukes family members served in respected professions, some even holding public office. The Jukes were vindicated in 2003, when archives at the State University of New York at Albany revealed the original names of the people in Dugdale's account; most were not even related. The Jukes family curse was more legend than fact.

a. Was the original jailing of the people called Jukes eugenics or not? Cite a reason for your answer.

b. How could studies on one family harm others?

c. Cite an example of an idea based on eugenics that is current or from the recent past.

d. If you were a contemporary of Dugdale's, what type of evidence would you have sought to counter his ideas?

Case Studies and Research Results

1. Lana seemed to be a healthy newborn. She reached developmental milestones ahead of schedule, trying to lift her head up at only 3 weeks. But then she rapidly lost skills. Her head flopped, she stopped trying to turn over, and her arms and legs became spastic. When she no longer made eye contact, her anxious parents took her to the pediatrician. The doctor referred the family to a pediatric neurologist, who was puzzled. "She has all the symptoms of Canavan disease, but that can't be. She's not Jewish." Explain how the neurologist was incorrect.

2. The population of India is divided into many castes, and the people follow strict rules governing who can marry whom. Researchers compared several genes among 265 Indians of different castes and 750 people from Africa, Europe, and Asia. The study found that the genes of higher Indian castes most closely resembled those of Europeans and that the genes of the lowest castes most closely resembled those of Asians. In addition, maternally inherited genes (mitochondrial DNA) more closely resembled Asian versions of those genes, but paternally inherited genes (on the Y chromosome) more closely resembled European DNA sequences. Construct a historical scenario to account for these observations.

© Peter Johnson/Corbis/VCG

CHAPTER 16

Human Ancestry and Evolution

The genomes of the Khoisan, the modern people whose roots go back the farthest, are diverse. Their gene variants reflect their long adaptation to a hunter-gatherer lifestyle. They live in the Kalahari Desert in Botswana and Namibia.

Learning Outcomes

16.1 Human Origins

1. State two ways that DNA sequences provide information about our ancestry.
2. Explain why we do not have DNA evidence from *Australopithecus* and *Homo* species that dwelled in Africa.
3. Describe what might have happened to Neanderthals and Denisovans and how evidence of their existence persists in some modern human genomes.
4. Explain how DNA analysis of human remains left since agriculture began reveals ancient migration patterns.

16.2 Methods to Study Molecular Evolution

5. Explain how DNA and protein sequences and chromosome banding patterns hold clues to evolution.
6. Describe how a "molecular clock" estimates the passage of long time periods.
7. Explain how mitochondrial DNA and Y chromosome sequences are used to track human ancestry.
8. Explain how haplotypes provide clues to ancient migrations.

16.3 The Peopling of the Planet

9. Explain the meaning of mitochondrial Eve.
10. Describe the expansion of groups of people out of Africa.

16.4 What Makes Us Human?

11. Compare and contrast the human genome to genomes of other primate species.
12. List traits that are unique to humans.
13. List genes that distinguish us from our closest relatives.

The BIG Picture

Our genes and genomes hold clues to our deep past and our present diversity. Ancestors of anatomically modern humans (us) interbred with archaic humans, who died out. How will our species continue to evolve?

Indigenous Peoples

It is difficult to imagine a group of people today who do not have cell phones, Internet access, or other signs of modern human existence, but about 5,000 such groups are scattered over the planet. An indigenous group of people can trace its ancestry back farther in a particular geographical region than any other group and has retained its uniqueness in cultural practices, social organization, and/or language. Physical and/or cultural isolation from colonists also keeps the gene pool of an indigenous people separate. In those gene pools lie signs of adaptations to past ways of life and, by comparison to other modern genomes, hold clues to how we are continuing to evolve. Indigenous groups are also called "native" or "first" peoples.

Today less than 5 percent of the world population is indigenous, accounting for about 370 million individuals. They live in 90 nations and the groups range from just a few dozen people to sizable portions of a country's population. Some live in distinct tribes yet go to school and work and dress just like anyone else. But some indigenous tribes are not very different in lifestyle from their hunter-gatherer ancestors.

Indigenous peoples living today include the Khoisan and Pygmies of Africa, the Etas of Japan, the Hill People of New Guinea, and the Brazilian Arawete, who number only 130 individuals. A fascinating look into the past comes from the genomes of the Khoisan, the modern people whose roots go back the farthest. These hunter-gatherers live today, as they have for millennia, in the Kalahari Desert in southern Africa. They are also known as "San" or "Bushmen," and their language uses "click" sounds.

Researchers compared the complete genome sequences of a Khoisan man named !Gubi to that of Archbishop Desmond Tutu, a well-known South African civil rights activist and a member of the majority Bantu group, as well as to partial genome sequences of three other Khoisan who live near each other. The results indicate that the great genetic diversity from which humanity sprung in Africa persists in the Khoisan today, whose genomes are as different from each other as a modern European genome is from that of a modern Asian genome. The four Khoisan genomes and Desmond Tutu's differ from each other at more than a million places. Comparing the Khoisan genomes to those of later arriving groups may indicate adaptations to an agricultural way of life among more recent Africans.

Khoisan gene variants reflect their lifestyle in the desert. A variant of the actinin-3 muscle gene promotes sprinting over distance running. A gene variant that encodes a chloride channel conserves water. The "bitter taste" gene variant would have enabled a hunter-gatherer to avoid poison and perhaps locate medicinal plants. The Khoisan are also defined by what is not in their genomes. They lack a gene variant that in other populations protects against malaria. Selection would have ignored it in the dry climate where malaria-bearing mosquitoes cannot live.

16.1 Human Origins

Who are we and where did we come from? We have sparse evidence of our beginnings—pieces of a puzzle in time, some out of sequence, many missing. Traditionally, paleontologists (scientists who study evidence of ancient life) have consulted the record in the Earth's rocks—fossils—to glimpse the ancestors

of *Homo sapiens sapiens*, our species. Researchers assign approximate ages to fossils by observing which rock layers fossils are in, and by extrapolating the passage of time from the ratios of certain radioactive chemicals in surrounding rock.

Harsh environments destroy most remains of past life. However, fossils aren't the only way to peek into species' origins and relationships, which is why the subject of human origins is part of a genetics textbook. Modern organisms provide clues to the past in their DNA. Sequences of DNA change over time, as mutations occur, at a rate that provides a "molecular clock." Another source of genetic information over time is found in the frequencies of gene variants (alleles) that change in populations due to the forces of nonrandom mating, migration, genetic drift, and natural selection, the topics of chapter 15.

The premise behind DNA sequence comparisons is that the more closely related two species are, the more similar their sequences. Similar DNA sequences are more likely to come from individuals or species sharing ancestors than from the exact same set of spontaneous mutations occurring by chance. Rarely, DNA from ancient specimens can add to what we know from DNA sequences of modern organisms. However, the heat of equatorial areas in Africa destroyed much DNA from the past.

Treelike diagrams are used to depict evolutionary relationships, based on fossil evidence and/or inferred from DNA sequence similarities. Typically the oldest species form the trunk of the tree, with more recent species diverging like branches. The branchpoints represent the most recent common (shared) ancestor. Humans and chimps diverged from a shared ancestor; humans didn't evolve directly from chimps. Similarly, two second cousins share great-grandparents, but one cousin did not descend from the other.

This chapter explores human origins and considers how genetic and genomic evidence adds to our view of our evolution. It concludes with a look at more recent events in our ancestry and a consideration of which characteristics distinguish our species.

Our Place in the Primate Family Tree

A species includes organisms that can successfully produce healthy offspring only among themselves. *Homo sapiens sapiens* ("the wise human") probably first appeared about 200,000 years ago.

We and animals ancestral only to us are members of a taxonomic (biological classification) "tribe" called the Hominini that diverged from an ancestor we shared with other African apes (gorillas, chimpanzees, and bonobos) about 6 million years ago, in Africa. **Figure 16.1** shows our place among our closest primate relatives.

We know of at least three types of Hominini (also called hominins) who lived shortly (in geological time) after the split from our ancestor shared with chimpanzees: *Ardipithecus kadabba* from Ethiopia, *Sahelanthropus tchadensis* from Chad, and *Orrorin tugenensis* from Kenya. Fossil evidence for them is scant. Another hominin, *Ardipithecus ramidus*, lived more recently, about 4.4 million years ago. "Ardi" was discovered in the Afar rift valley of Ethiopia in 1994, an area whose rocks hold fossilized remains of human ancestors going back nearly

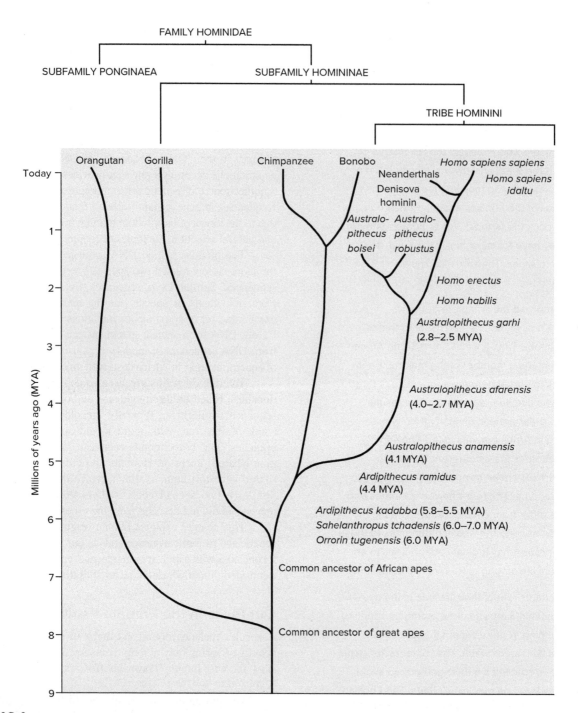

Figure 16.1 Our place on the human family tree. Our most recent ancestors, the archaic humans, are just below us at the upper right of the tree.

6 million years. Ardi was taller than more ancient hominins and was partly able to walk on two legs (bipedal), perhaps while on the forest floor, but could also use all four limbs to climb trees. Ardi likely ate plants.

Fossil evidence is more complete for our ancestors who lived 2 to 4 million years ago. They walked fully upright and conquered vast new habitats on the plains. Several species of a hominin called *Australopithecus* lived at this time, probably following a hunter-gatherer lifestyle. More than one species could coexist because the animals lived in small, widely separated groups that probably never came into contact. The

australopithecines were gradually replaced with members of our own genus, *Homo*. The following sections introduce a few of these ancestors known from rare fossil remains and what our computer modeling and imaginations can fill in. DNA evidence becomes important when we reach the archaic humans.

Australopithecus

The australopithecines had a mix of apelike and humanlike characteristics. They had flat skull bases, as do all modern primates except humans. Their legs were shorter than ours but their arms were longer. The angle of preserved bones from the pelvis, and discovery of *Australopithecus* fossils with those of grazing animals, suggest that this ape-human had left the forest for grasslands. Paleontologists have pieced together what little we know about the australopithecines from fossil finds. These hominins lived too long ago to have left DNA evidence.

The oldest species of australopithecine known, *Australopithecus anamensis*, lived about 4.1 million years ago. Another species, *Australopithecus afarensis*, is represented by a partial skeleton, named Lucy, who lived for 20 years about 3.6 million years ago in the Afar river basin of Ethiopia. Lucy was about 4 feet tall, but partial skeletons from slightly older males of her species stood about 5 feet 3 inches tall and were more human-like than chimplike. Stone tools found with other members of *A. afarensis* from 3.4 million years ago show distinctive cut marks that indicate these hominins sliced meat from bones and removed, and presumably ate, the marrow. Their jaws were strong enough to crack nuts.

Other fossils offer additional clues to australopithecine life. Two parallel paths of footprints, preserved in volcanic ash in the Laetoli area of Tanzania, are contemporary with Lucy. A family may have left the prints, which are from a large and small individual walking close together, with a third following in the steps of the larger animal in front. A fossilized 3-year-old girl from 3.3 million years ago, the "Dikika infant," apparently died in a flash flood, perhaps while trying to catch fish in rivulets of the Awash river delta near where Lucy lived 300,000 years earlier.

Toward the end of the australopithecine reign, *Australopithecus garhi* may have coexisted with the earliest members of *Homo*. *A. garhi* fossils from the Afar region date from about 2.5 million years ago. Remains of an antelope found near the australopithecine fossils suggest butchering. The ends of the long antelope bones had been cleanly cut with tools, the marrow removed, meat stripped, and the tongue cleanly sliced off. *A. garhi* stood about 4.5 feet tall, and like the Dikika infant and Lucy, the long legs were like those of a human, but the long arms were more like those of an ape. The small cranium and large teeth hinted at apelike ancestors.

Early *Homo*

By 2.3 million years ago, *Australopithecus* coexisted with hominins of genus *Homo*. We do not know how *Homo* replaced *Australopithecus*, or even whether there was just one or several species of early *Homo*. Some australopithecines were "dead ends" that died off. Based on older fossil evidence paleontologists have described several *Homo* species, but more recent fossil finds that reveal variability in bone structure may lead to revision in the classification to a single but variable type of *Homo*.

An early possible species was *Homo habilis*, who cared intensively for its young in caves. *Habilis* means handy, and this primate was the first to use tools for tasks more challenging than stripping meat from bones. *H. habilis* may have descended from hominins who ate a more varied diet than other ape-humans, enabling them to live in a wider range of habitats.

H. habilis coexisted with and was followed by *Homo erectus*. One of the first *H. erectus* individuals described,

"Nariokotome Boy," influenced descriptions of the species as tall and thin for two decades, until discovery of more specimens revealed that *H. erectus* matured faster than we do, but their heights and builds varied. Another famed *H. erectus* fossil, named "Daka" for the place where he was found in the Afar region, is from an individual who lived about a million years ago. He had a shallow forehead, massive brow ridges, a brain about a third smaller than ours, and strong, thick legs. Daka lived on a grassland, with elephants, wildebeests, hippos, antelopes, many types of pigs, and giant hyenas. **Figure 16.2** depicts a partial skeleton of an *H. erectus* individual who lived about 1.8 million years ago in the country of Georgia.

H. erectus left fossil evidence of cooperation, social organization, tools, and use of fire. Fossilized teeth and jaws suggest that they ate meat. The distribution of fossils indicates that they lived in families of male-female pairs (most primates have harems). The male hunted, and the female nurtured the young. They were the first to have an angled skull base that enabled them to produce a greater range of sounds, making

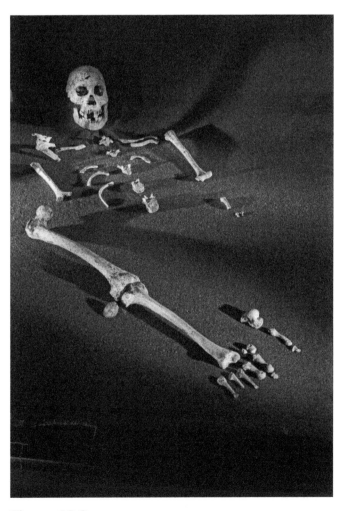

Figure 16.2 Reconstructing *Homo erectus*. Paleontologists assembled fossilized bones from several *Homo erectus* individuals from the Dmanisi site in the country of Georgia to try to understand what this hominin looked like and how variable it was. *H. erectus* lived about 1.8 million years ago.
© *Volker Steger/Science Source*

speech possible. *H. erectus* fossils have been found in China, Java, Africa, Eurasia, and Southeast Asia, indicating that these animals could migrate farther than earlier primates.

A dwarfed variant of *H. erectus* is known from jaw and tooth fragments from three individuals discovered on the island of Flores in Indonesia. The trio were dated back to about 700,000 years ago. Remains of another individual from the island made headlines when they were discovered in 2003. The remains were initially given too recent a date based on erroneously analyzing charcoal, but were then dated to between 100,000 and 50,000 years ago. She was initially given her own species designation, *H. floresiensis*, or "the hobbit" after the fictional character, but many researchers consider her an offshoot of *H. erectus* with a small body.

Archaic Humans

DNA evidence can flesh out fossil evidence for more recent members of *Homo*. *Homo erectus* may have lived until as recently as 35,000 years ago. Meanwhile, the branching of evolution led to hominins that were more like us and that we call the "archaic humans." We know of two types of archaic humans: the Neanderthals (*Homo sapiens neanderthalensis*) and the Denisovans (*Homo sapiens denisova*) (**figure 16.3**). In contrast to archaic humans, we are "anatomically modern humans."

As some of our ancestors left Africa and expanded beyond that continent, reproductively successful populations would eventually have begun to overlap geographically. If populations interbred, then gene flow would occur, leaving evidence in preserved ancient DNA as well as in the genomes of modern peoples. Both have happened. We probably do not yet know anything about some of the peoples who migrated out of Africa

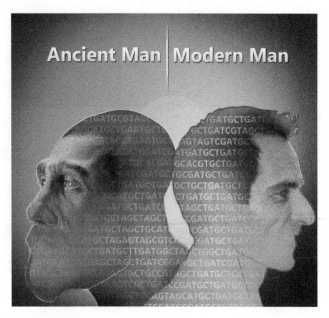

Figure 16.3 **Archaic humans.** Several types of archaic humans likely coexisted. They probably looked more similar to than different from us. *Source: National Human Genome Research Institute*

over the past tens of thousands of years. The following sections describe the Neanderthals and the Denisovans. These archaic humans may have branched from a shared ancestor with us as far back as 600,000 years ago or even earlier.

The Neanderthals

The ancestors of the Neanderthals left Africa in separate migrations, beginning about 400,000 years ago. They headed toward Europe and west Asia, but did not establish themselves in these places long enough to have left evidence (as far as we know) until about 230,000 years ago. Their gene pools had time to diverge from the ancestral African genomes, perhaps as natural selection favored adaptations to cooler environments. Anatomically modern humans first appeared about 200,000 years ago, as a patchwork of peoples in Africa, Europe, and Asia. The earlier migrations from Africa of Neanderthal ancestors explains why modern Africans do not have uniquely Neanderthal DNA sequence variants in their genomes, but Europeans and Asians, who met up with the Neanderthals much later, have up to about 2 percent.

Between 100,000 and 50,000 years ago, our ancestors left Africa and encountered Neanderthals in the Middle East en route to Eurasia. The two types of hominins interbred—a discovery deduced from shared DNA sequence variants. When researchers first began to sequence mitochondrial DNA from Neanderthal bones, they concluded that interbreeding was highly unlikely, but sequencing entire Neanderthal genomes changed the long-held idea of our separateness.

Neanderthals may have lived as recently as 30,000 years ago, in warm caves in Gibraltar when northern Europe was under ice. They might have inhabited these caves since 100,000 years ago. The fossil record indicates that the Neanderthals vanished, but their gene variants persist in modern humans, suggesting strongly that they may have been assimilated as different groups interbred and Neanderthal fertility declined. Without a time machine, we may never know exactly what happened (see **A Glimpse of History**).

The genome sequences of modern humans and Neanderthals are 99.7 percent identical, but the Neanderthal contribution to our ancestry is small. Researchers describe the contribution as the proportion of a person's ancestors who were Neanderthal, going back a certain number of generations or years. For example, 50 percent of a person's ancestry is from the mother and 50 percent from the father. Each grandparent contributes 25 percent, and each great-grandparent contributes 12.5 percent, and so on. Going back 100,000 years ago for Asians, approximately 1.8 percent of ancestors were Neanderthals, and for Europeans the proportion is about 1.5 percent. Modern humans who lived longer ago have greater Neanderthal ancestry. DNA sequencing of a jaw bone from a modern human who lived about 40,000 years ago in Romania, named "Oase1," revealed 6 to 9 percent Neanderthal ancestors who lived about 100,000 years ago, considerably greater than for us today. The long length of the Neanderthal portion of Oase1's genome suggests that the sexual encounter with an archaic partner may have happened just four to six generations earlier!

The first bones discovered from Neanderthals came from a limestone cave in Neander Valley, Germany, when quarry workers blasted them out on a summer day in 1856. Assembling the scattered skeletal remains revealed large, brutish individuals with jutting faces and brows and giant jaws and teeth above small chins. Half a century later, discovery of bones from the "Old Man" of La Chapelle-aux-Saints in France led to the depiction of Neanderthals as primitive and slow-witted, stooped perhaps due to arthritis. Another find, of 10 Neanderthals from Shanidar Cave in Iraq whose bones were excavated from 1957 through 1961, included tools and remains of flowers. An old man from that discovery, who researchers named "Nandy," died of what was then very old age—40 to 50 years. Preserved dental plaque on the teeth held starch grains that had been cooked.

The first samples of Neanderthal DNA came from the accidental discovery by cave explorers of bone fragments about 700 feet in from the entrance of a cave called El Sidrón in northwestern Spain, on the westernmost part of the Iberian peninsula. The environment had been warm enough to sustain a pleasant existence for the Neanderthals, but with caves cool enough to preserve their remains, including their DNA. Excavation of the El Sidrón cave began in 2000, and by 2005, researchers had sequenced the first Neanderthal genes from mitochondrial DNA, which is more likely to survive harsh environmental conditions than DNA from cell nuclei.

Neanderthal DNA from a hip bone of a 38,000-year-old female found in Vindija Cave in Croatia was used to reconstruct the first full genome, published in 2010 from researchers at the Max Planck Institute for Evolutionary Anthropology in Leipzig, Germany. They used dental drills to delicately remove pulverized "dust" from bones from three individuals to keep the bones intact for future studies. DNA sequencing in a "clean room" prevented the investigators' DNA contaminating the samples, and other techniques avoided sequencing bacterial or fungal DNA. In 2014 researchers published a second Neanderthal genome sequence from a female who lived 50,000 to 100,000 years ago in the Altai mountains of Siberia. This must have been a highly isolated population, and the high levels of homozygous gene variants in the woman's genome indicate intensive inbreeding—her parents were siblings.

Computer simulations that use mutation rates and population data suggest that in a population living at about Oase1's time or slightly earlier, modern humans outnumbered Neanderthals 10 to 1, and they lived as one interbreeding group. Genome sequencing reveals that Neanderthal gene variants in modern human genomes aren't evenly dispersed among the chromosomes. The apparent nonrandomness suggests that natural selection was at play once breeding with modern humans began. That is, when Neanderthals lived in tiny, isolated groups, inbreeding was common, and genetic diversity as well as fertility fell.

In small populations, genetic drift would have had a large effect. But once Neanderthals became part of a larger population, natural selection would have acted both negatively to remove harmful gene variants and positively to retain helpful ones.

The term **introgression** refers to entry of a specific gene variant into a genome from an individual of a different species or subspecies. **Admixture** is sometimes used as a synonym for introgression, but some researchers use it to mean introgression on a genome-wide scale, such as when individuals from previously separate populations interbreed. Figure 16.11 depicts admixture.

Which genes from Neanderthals persist in modern human genomes? Some Neanderthals had variants of the *MCR1* gene that confers freckles, pale skin, and reddish hair (**figure 16.4**). Before genome sequencing, they were thought to have all had dark hair and eyes. Researchers can propose hypotheses based on genome differences. For example, the Neanderthal Y chromosome is about twice as old as the oldest known modern Y chromosome (from the Mbo people of the Bantu tribe in Cameroon) and has distinct variants in several genes

Figure 16.4 A Neanderthal phenotype. Mutations found in Neanderthal DNA suggest that some of them may have had pale skin and red hair. Others had darker hair and skin. © *Knut Finstermeier, Max Planck Institute for Evolutionary Anthropology*

that affect the immune system. Perhaps the immune systems of modern women impregnated with Neanderthal sperm attacked the embryos, leading to miscarriage, which might explain why so few of our gene variants are from Neanderthals. Births of healthy children from matings between Neanderthals and our ancestors must have been rare events.

Neanderthal gene variants persisting in some modern human genomes form about 12,000 haplotypes (sets of linked genes). Identifying the differences in the modern human and Neanderthal genomes enables researchers to reconstruct scenes from prehistory and to understand some modern diseases. That is, variants of several genes that were likely adaptive in Neanderthals may cause disease in modern humans. To investigate this hypothesis, researchers turned to information technology, consulting a database called eMERGE (Electronic Medical Records and Genomics) that lists disease diagnostic billing codes and DNA sequence data. They looked for 6,000 Neanderthal haplotypes among the 28,416 people of European descent represented in the database, to see if any diagnoses tracked with specific Neanderthal gene variants. **Table 16.1** lists Neanderthal gene variants identified from this clever study that might affect our health.

Gene flow apparently also occurred from anatomically modern humans to Neanderthals. Genome comparisons indicate that modern humans other than the group that led to us left Africa about 125,000 to 100,000 years ago, and mated with the isolated Altai Neanderthals in Siberia—but not with the European Neanderthals. Introgression of modern human DNA into Neanderthal genomes might also have occurred in Cyprus, Egypt, Iraq, Israel, Jordan, Lebanon, Syria, and Turkey and the Persian gulf area as Neanderthals moved east to Europe during the interglacial period about 100,000 years ago. Pure Neanderthals may have become extinct because they were about 40 percent less reproductively fit as modern humans.

The Denisovans

The Denisovan story begins in 2010, with a description based on DNA extracted from a pinky finger bone discovered in Denisova Cave in the Altai Mountains of southern Siberia (**figure 16.5**). Researchers at first thought that the bone was from a Neanderthal, because Neanderthals had occupied the cave as well. However, sequencing of the mitochondrial DNA in finger bone cells showed that the digit was neither Neanderthal nor modern human. By 2012, researchers had sequenced the genomes of cells from two molars from other Denisovans discovered in different layers of rock in the cave wall, indicating that the teeth had been left at different times. These Denisovan genomes were quite different from each other, suggesting that the cave residents might have lived long enough and left enough offspring to foster some genetic diversity.

Table 16.1	Neanderthal Genes in Modern European Genomes	
Phenotype	**Genes**	**Hypothesis**
Faster blood clotting	SELP (P-selectin)	Benefited Neanderthals in healing from injuries and surviving childbirth. In longer-living modern humans, faster clotting might cause stroke and heart attack.
Innate immunity	TLR1, 6, 10 (Toll-like receptors)	Receptor variants alert immune system to respond to microbes and parasites present 100,000–550,000 years ago. Today they may increase risk of allergies, autoimmunity, and inflammatory diseases.
Tobacco use disease	SLC6A11 (solute carrier family 6 member 11)	Impaired ability of neurons to take up gamma-aminobutyric acid, a neurotransmitter, from synapses in brain's addiction center.
Thiamine (vitamin B1) transport	SLC35F3 (solute carrier family 35 member F3)	Protein transports vitamin to intestines to metabolize complex carbohydrates that were common in Neanderthal diet, but comprise far less of modern refined-carbohydrate diets.
Urinary problems (infection, incontinence, bladder pain	STIM1 (stromal interaction molecule 1)	Neanderthal variant not expressed well in brain parts that control bladder function.
Actinic keratosis	Circadian clock genes	Neanderthal adaptations to natural patterns of daylight at higher latitudes predispose modern humans to these precancerous skin lesions.
Depression	Circadian clock genes	Gene variants that cause depression in modern humans are closely linked to circadian rhythm genes that had different variants in Neanderthals.

Figure 16.5 **The Denisovan cave.** A Siberian cave holds abundant evidence of archaic humans. At different times, Neanderthals and Denisovans were residents. © *Max Planck Institute for Evolutionary Anthropology*

The Denisovan girl or young woman who left the pinky bone, nicknamed "Denise," lived between 32,000 to 50,000 years ago, according to fossil as well as DNA evidence. The genome sequence reveals that Denise had dark skin and brown eyes and hair. These archaic humans may have lived throughout Asia. Today the people most closely related to the Denisovans are from Papau New Guinea, Melanesia, Oceania, and Polynesia. Some aboriginal Australians have about 3.5 percent of their ancestors who were Denisovans going back about 100,000 years. People who live on Northern Island in Melanesia, in Papua New Guinea, are the closest living relatives of the Denisovans.

The Denisovan cave was also home to the Altai Neanderthals, and likely others. The extensive runs of homozygosity that pepper the Altai Neanderthal genomes indicate siblings having children together, perhaps a consequence of their isolation in an extreme environment. Comparing genomes of the archaic humans to our own indicates that the Denisovans and Neanderthals were more closely related to each other than either was to us and that Neanderthal genes flowed into older Denisovan genomes. As complicated as this emerging picture is, we do not yet have all of the puzzle pieces—the Denisovan genome clearly indicates admixture with at least one other, yet unknown, type of archaic human. Genes also entered Denisovan and Neanderthal genomes from our very distant ancestors. Because we have such scant evidence, we cannot really know what happened as humanity sorted itself out. Several small populations of archaic humans probably coexisted and eventually mixed for at least 100,000 years before we modern humans, retaining some archaic genes, emerged and persisted. A few ancient gene variants remain in some of us.

Figure 16.6 attempts to capture some of this complex back-and-forth gene flow. A caveat: what we know of the Neanderthals and Denisovans is subject to change with new discoveries and DNA sequences!

Modern Humans

Cave art from about 14,000 years ago indicates that by that time, our ancestors had developed fine-hand coordination and could use symbols. These were milestones in cultural evolution. By 10,000 years ago, people had expanded from the Middle East across Europe, bringing agricultural practices.

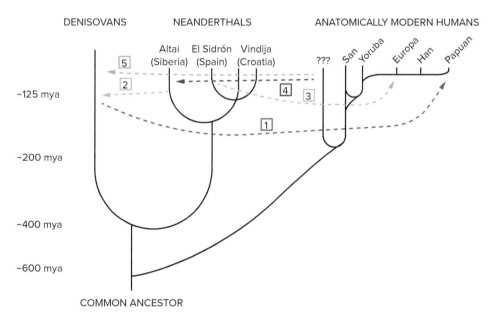

Figure 16.6 **Gene flow among archaic and anatomically modern humans.** DNA sequence similarities indicate that genes flowed (individuals mated) from Denisovans to some groups of anatomically modern humans (1), from Siberian Neanderthals to Denisovans (2), from Spanish Neanderthals to Europeans (3), and from unknown anatomically modern humans to Siberian Neanderthals (4) and to Denisovans (5). (The abbreviation mya stands for "millions of years ago.") *Adapted from the US National Library of Medicine National Institutes of Health.*

Early Farmers

DNA evidence has sketched one scene from the spread of agriculture. Researchers sequenced mitochondrial DNA haplotypes from 21 bodies found in a graveyard in Germany, about 100 miles south of Berlin, and compared them to DNA from 36 modern Eurasian populations. The bodies were from about 7,100 years ago. The comparisons yielded a clear cline, with genetic similarities indicating a long-ago migration of early farmers from Turkey, Syria, Iraq, and other Near Eastern cultures westward from the Balkans, north along the Danube into central Europe—not just the spread of their agricultural techniques by word of mouth. The migration took centuries. According to the DNA, the farmers arrived in Europe and encountered hunter-gatherers descended from the original population from 40,000 B.C.E., and the two groups of people interbred. **Figure 16.7** shows what the ancient farmers from 7,100 years ago might have looked like.

Ötzi the Ice Man

In 1991, hikers in the Ötztaler Alps of northern Italy discovered an ancient man frozen in the ice (**figure 16.8**). Named "Ötzi the Ice Man," he was on a mountain more than 10,000 feet high 5,200 years ago when he perished. He wore furry leggings, leather suspenders, a loincloth, fanny pack, bearskin cap and cape, and sandal-like snowshoes. He had stained his skin to fashion tattoos, and indentations in his ears suggest that he might have worn earrings. He carried mushrooms

Figure 16.8 A 5,200-year-old man. Hikers discovered Ötzi the Ice Man in the Ötztaler Alps of northern Italy in 1991. © Andrea Solero/AFP/Getty Images

Figure 16.7 **Farmers from 7,100 years ago probably looked like modern people.** © Karol Schauer/State Museum of Prehistory, Saxony-Anhalt, Germany

that had antibiotic properties. Berries found with him place the season as late summer or early fall. His last meal was ibex and venison.

Ötzi died following a fight. He had a knife in one hand, cuts and bruises, and an arrowhead embedded in his left shoulder that nicked a vital artery. The wound bore blood from two other individuals and his cape had the blood of a third person. Moss found on his body may have been used as wound dressings. When X rays indicated swelling on the right side of Ötzi's brain, researchers delicately removed two tiny samples of the tissue and analyzed the proteins present. Several were related to blood clotting. Whether Ötzi fell and hit his head or was attacked remains a mystery. However he perished, Ötzi landed in a ditch, where snow covered him and preserved his body, including his DNA.

Ötzi's genome held clues to his health and ancestry. He was lactose intolerant and had type O blood, and had he not been buried in snow immediately following a fight, he might have eventually died of heart disease. Ötzi shared ancestors with men living today in Sardinia and Corsica in the Mediterranean area, according to his Y chromosome haplogroup. However, Ötzi's mitochondrial genome apparently died out about 5,000 years ago. One interpretation is that some of his male ancestors had left the isolated Alps, migrating toward the Mediterranean sea, while the women who had his mitochondrial genome perished in the local, very harsh environment.

Key Concepts Questions 16.1

1. What types of animals are the hominins?
2. Describe the australopithecines and early *Homo* species.
3. Describe how researchers compare archaic to modern humans.
4. What has DNA evidence revealed about early farmers and the 5,200-year-old ice man?

16.2 Methods to Study Molecular Evolution

Molecules of DNA and protein change in sequence over time as mutations occur and are perpetuated. The more alike a gene or protein sequence is between two species or two individuals of the same species, the more closely related the two are presumed to be—that is, the more recently they shared an ancestor. The field of **molecular evolution** compares genomes, DNA or protein sequences, and chromosome banding patterns. Knowing the mutation rates for specific genes provides a way to measure the passage of time using a sequence-based "molecular clock."

Comparing Genes and Genomes

We can assess similarities in DNA sequences between two species for a piece of DNA, a single gene, a chromosome segment, a chromosome, mitochondrial DNA, or an entire exome or genome. In general, DNA sequences that encode protein are often very similar among closely related species, which presumably inherited the gene from a shared ancestor. A change in that gene would not persist in a population unless it provided a selective advantage (positive natural selection). At the same time, negative natural selection would have weeded out proteins that did not promote survival to reproduce. DNA sequences that do not affect the phenotype can vary because natural selection does not act on them.

Similar DNA or amino acid sequences found in different species are said to be "highly conserved." Such similarities are the basis of animal models of human disease. **Figure 16.9** compares part of the dystrophin gene, which is mutant in Duchenne muscular dystrophy (see figure 2.1), among wild type humans, mice, and dogs. The mouse does not become very ill so is not a good model, but "golden retriever muscular dystrophy" (GRMD) is strikingly like the disease in boys. The dogs lose weight rapidly from when they are born and usually do not survive beyond 1 year because their breathing muscles fail. The GRMD mutation is in a splice site that removes an exon in the giant gene, depicted as the blue G in the figure, and as a result the dogs cannot make dystrophin protein. German shorthair pointers with the disease have a full gene deletion, cocker spaniels have a 4-base deletion, Tibetan terriers have a several-exon-long deletion, and Labrador retrievers have an insertion into an intron.

Sequences that are similar in closely related species but that do not encode protein often control transcription or translation, and so are also vital and therefore subject to natural selection. In contrast, some genome regions that vary widely among species do not affect the phenotype, and are therefore not subject to natural selection. Within a protein-encoding gene, the exons tend to be highly conserved, but the introns, which are removed from the corresponding RNA, are not.

Comparing Chromosomes

Before gene and genome sequencing became possible, researchers considered similarities in chromosome banding patterns to assess evolutionary relatedness. Human chromosome banding patterns most closely match those of chimpanzees, then gorillas, and then orangutans (**table 16.2**). The karyotypes of humans, chimpanzees, and gorillas differ from each other by nine inversions, one translocation, and one chromosome fusion.

Chromosome banding patterns are like puzzle pieces. If both copies of human chromosome 2 were broken in half, we would have 48 chromosomes, as the three species of apes do,

Species	Part of dystrophin gene	# differences
Human wild type	T A T G T G T A T G T G T T T A G G C C A G A C C T A T T T G A C T G	
Mouse wild type	T G T A T G T G T T T G T T T C A G G C C C G A C C T G T T T G A T T G G	8
Dog wild type	T A T G T G T G T G T G T T T C A G G C C A G A C C T G T T T G A T T G G	4
Golden retriever	T A T G T G T G T G T G T T T C G G G C C A G A C C T G T T T G A T T G G	5

Figure 16.9 The dystrophin gene. The gene that, when mutant, causes Duchenne muscular dystrophy is highly conserved among 48 mammalian species. Different dog breeds have specific mutations in the gene. Golden retrievers suffer severe illness. *Adapted from Current Genomics, Aug 2013, 14(5):330-342, PubMed Central.*

Table 16.2	Percent of Common Chromosome Bands Between Humans and Other Species
Chimpanzees	99⁺%
Gorillas	99⁺%
Orangutans	99⁺%
African green monkeys	95%
Domestic cats	35%
Mice	7%

Table 16.3	Cytochrome c Evolution	
	Organism	Number of Amino Acid Differences from Humans
	Chimpanzee	0
	Rhesus monkey	1
	Rabbit	9
	Cow	10
	Pigeon	12
	Bullfrog	20
	Fruit fly	24
	Wheat germ	37
	Yeast	42

instead of 46. Human chromosome 2 therefore arose from a fusion event. The banding pattern of chromosome 1 in humans, chimps, gorillas, and orangutans matches that of two small chromosomes in the African green monkey, suggesting that this monkey was ancestral to the other primates. Most of the karyotype differences among these three primates and more primitive primates are translocations.

We can also compare chromosome patterns between species that are not as closely related as we are to other primates. All mammals, for example, have identically banded X chromosomes. One section of human chromosome 1 that is alike in humans, apes, and monkeys is remarkably similar to parts of chromosomes in cats and mice. A human even shares chromosomal segments with a horse, but our karyotype is much less like that of the aardvark, the most primitive placental mammal.

The information gained from comparing chromosome bands using stains is broad and imprecise compared to information from DNA sequences. This is because a chromosome band can contain many genes that differ from those within a band at a corresponding locus in another species' genome. In contrast, DNA probes used in a FISH analysis highlight specific genes (see figure 13.8). FISH can indicate direct correspondence of gene order, or **synteny**, between species, which is solid evidence of close evolutionary relationships. For example, versions of eleven genes are closely linked on human chromosome 21, mouse chromosome 16, and chromosome U10 in cows.

Comparing Proteins

Many different types of organisms use the same proteins, with only slight variations in amino acid sequence. The similarity of protein sequences is compelling evidence for descent from shared ancestors—that is, evolution. Many proteins in humans and chimps are alike in 99 percent of their amino acids, and several are identical. When analyzing a human gene's function, researchers routinely consult databases of known genes in many other types of organisms.

Cytochrome c is one of the most ancient and well-studied proteins. It helps to extract energy from nutrients, in the mitochondria. The more closely related two species are, the more alike their cytochrome c amino acid sequence (**table 16.3**).

Human cytochrome c differs from the horse version by twelve amino acids and from kangaroo cytochrome c by eight amino acids. The human protein is identical to chimpanzee cytochrome c. The homeotic genes, discussed in Clinical Connection 3.1, are another class of genes that has changed little across evolutionary time.

How alike proteins are among different species has medical consequences. Consider insulin, which enables cells to take up glucose from the blood. Inability to make insulin causes type 1 diabetes mellitus.

Insulin is two linked peptide chains, one consisting of 21 amino acids and the other of 30. Human insulin differs from pig and dog insulin by just one amino acid, from cow insulin by three, and from cat insulin by four. The source of insulin is important, because even though the amino acid sequences are very similar among species, the molecules differ in how quickly the insulin is absorbed into cells, how soon after absorption the level peaks, and the duration of the effect—all important factors when trying to stabilize blood glucose levels. Cats, for example, metabolize insulin twice as fast as dogs or people because their metabolisms are much faster. Even though insulin is highly conserved, the source is important when it is used as a drug to treat diabetes, as **A Glimpse of History** explains.

Molecular Clocks

A clock measures the passage of time as its hands move through a certain degree of a circle in a specific and constant interval—a second, a minute, or an hour. In the same way, an informational molecule can be used as a "molecular clock" if its building blocks are replaced at a known and constant rate.

The similarity of nuclear DNA sequences in different species can be used to estimate the time when the organisms diverged from a common ancestor, if the rate of base substitution mutation is known. For example, many nuclear genes studied in humans and chimpanzees differ in 5 percent of their bases, and substitutions occur at a rate of 1 percent per million

The first use of insulin to treat diabetes was in 1922 in Toronto, Canada. Sir Frederick Banting, a surgeon, his medical student Charles Best, and biochemist Bertram Collip conducted experiments on 10 beagles, and then used a bovine (cow) insulin extract first on themselves and then to treat 14-year-old Leonard Thompson, who was near death from diabetes. He fully recovered rapidly.

Bovine insulin differs from the human peptide by three amino acids. In 1946, diabetes treatment began to use pig insulin, which differs by only one amino acid. Then in 1978, researchers used recombinant DNA technology to engineer bacterial cells to produce human insulin (see section 19.2), which became available in 1982. Since then several brands of these artificial human insulins have been in widespread use. They are called analog insulins.

Figure 16A Drs. Banting and Best and one of their experimental subjects. *Source: Library and Archives Canada*

years. Therefore, 5 million years have presumably passed since the two species diverged. Mitochondrial DNA (mtDNA) is also tracked to estimate time of divergence from shared ancestors.

Time scales based on fossil evidence and molecular clocks can be superimposed on evolutionary tree diagrams constructed from DNA or protein sequence data. However, evolutionary trees can become complex when data can be arranged into different tree configurations. A tree for 17 mammalian species, for example, can be constructed in 10,395 different ways! With new sequence information, tree possibilities change.

Researchers use statistical methods to select out of many possibilities an evolutionary tree that is likely to represent what really happened. An algorithm connects all evolutionary tree sequence data using the fewest possible number of mutational events to account for observed DNA base sequence differences. Because mutations are rare events, the tree that requires the fewest mutations is most likely to reflect reality. Statistical methods can account for different mutation rates for different genes and for different positions of mutations within genes, and for incomplete data.

Using Mitochondrial and Y Chromosome DNA Sequences

To track ancient human migration patterns, researchers use the types of genetic markers that are used to track traits in modern families. This approach, called genetic genealogy or genetic ancestry, uses as markers:

- single nucleotide polymorphisms (SNPs)
- short tandem repeats (STRs) or microsatellites
- other copy number variants (CNVs)

Following markers in mitochondrial DNA provides information on the female lineage, and markers on the Y chromosome, the male lineage, but as **figure 16.10** illustrates, these ancestors are only a few of many. Markers on the autosomes flesh out the others, and are called "ancestry informative markers." It is easy to see that the contribution of a particular ancestral DNA sequence decreases as the number of generations increases. This is why Neanderthal DNA is only a small percentage of the DNA from ancestors of modern Europeans—it has been many generations since the two types of humans mated.

MtDNA is ideal for monitoring recent events because it mutates faster than DNA in the nucleus. Its sequences change by 2 to 3 percent per million years. Mutations accumulate faster because mtDNA has no DNA repair. Another advantage of typing mtDNA is that it is more abundant than nuclear DNA because mitochondria have several copies of it, and a cell has many mitochondria. It is also more able to withstand harsh environments than nuclear DNA. These are the reasons why the first studies of DNA from archaic humans examined mtDNA.

For many years, researchers compared certain rapidly mutating DNA sequences in mitochondria, but today they compare complete mitochondrial genomes, or "mitogenomes." A study of mitogenomes of the Ashkenazi Jews, for example, showed that many maternal lineages came, not from the Near East, as had long been thought from the more limited mtDNA analysis, but from many places in Europe. A picture is emerging of Ashkenazi men traveling in Europe, where they married local women, who converted to Judaism.

Most of the Y chromosome DNA sequence offers the advantage of not recombining. Crossing over, which it could only do with an X chromosome because there is no second Y, would break the linkage from the past generation and therefore make tracing relationships impossible.

Sets of SNPs along mitochondrial and Y chromosome DNA define long DNA stretches termed **haplogroups**.

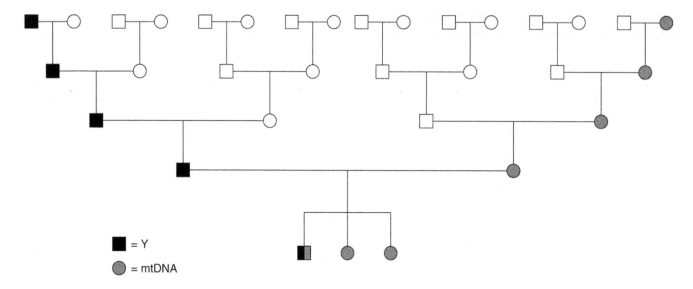

= Y

= mtDNA

Figure 16.10 **Genetic genealogy.** Y chromosome and mtDNA sequences represent only some of a person's ancestors. Considering autosomal sequences can capture the contributions of other relatives, represented as the unfilled symbols in the interior of this pedigree. The symbol for the man in the most recent generation has two shades because his Y chromosome DNA (black) and mitochondrial DNA (gray) were tested.

(The haplotypes used to describe linkage in chapter 5 are shorter DNA sequences.) Haplogroups track with geography, rather than with racial or ethnic groups, which are social designations and not biological ones.

Y haplogroups are classified from "A" through "T," with several subgroups, called subclades, indicated by alternating letters and numbers. Haplogroups and their subgroups also describe mtDNA. Populations can be classified into both mtDNA and Y chromosome groups, indicating the sources of female and male lineages, respectively. Sub-Saharan Africans, for example, have Y haplogroups E1, E2, and E3a and mtDNA haplogroup L3. Europeans, however, have Y haplogroups R, I, E3b, and J, and their mtDNA haplogroup is R, which includes three subgroups.

Admixture

The first molecular evolution studies used Y chromosome and mitochondrial DNA sequences because their gender specificity made them easiest to track. Because we now know of millions of SNPs across the genome, and can sequence genomes, we can fill in more of the blanks in the history of human migrations, by tracing transmission of haplogroups and haplotypes to reveal admixture. Nearly all modern human populations are admixed, to differing extents.

We can trace the occurrence of admixture by following how the pattern of haplotypes changes from one generation to the next if a crossover occurs. The crossover breaks up the linkage of a haplotype or haplogroup, and the admixed generation as a result has shorter chunks of ancestral DNA sequences. If we know the crossover rate for a particular gene that is involved in the exchange of genetic material, then we can extrapolate

back to when admixture occurred, assuming a generation time, such as 25 years. **Figure 16.11** illustrates schematically the break up of haplotypes over generations.

Computational tools use changing haplotype data to reconstruct what one researcher calls "snapshots in time" as far back as 2 million years ago, by tracking admixture. One study, for example, looked at 2,400 35,000-base segments of the genome, with a generation time of 25 years, to estimate that an ethnic group called the Yoruba left their native southwestern Nigeria in 1550 and in 1790, times already known from history to correspond to the start of the slave trade and its peak.

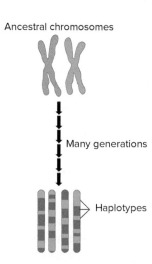

Ancestral chromosomes

Many generations

Haplotypes

Figure 16.11 **In admixture, crossing over recombines segments of linked genes representing ancestral populations.** The shorter the haplotype blocks, the more generations have passed.

Key Concepts Questions 16.2

1. What is the basis of comparing DNA sequences, protein sequences, chromosome banding patterns, and genome sequences to learn about human origins and evolution?

2. What is the basis of a molecular clock?

3. How are mitochondrial DNA and Y chromosome sequences used to trace human lineages?

4. How does analyzing admixture provide information about the past?

16.3 The Peopling of the Planet

Fossil evidence and extrapolating and inferring relationships from DNA sequence data provide clues to the major movements that peopled the planet (**figure 16.12**). The evidence so far has been like reading chapters from different parts of a novel and trying to understand the whole story. Three such chapters in the saga of modern human origins stand out: our beginnings 200,000 years ago, expansion from Africa, and populating the New World.

Mitochondrial Eve

Theoretically, if a particular sequence of mtDNA could have mutated to yield the mtDNA sequences in modern humans, then that ancestral sequence may represent a very early human or humanlike female—a mitochondrial "Eve," or metaphorical first woman. **Figure 16.13** shows how one maternal line may have persisted.

When might this theoretical "first" woman, the most recent female ancestor common to us all, have lived? Researchers in the mid-1980s compared mtDNA sequences for protein-encoding as well as noncoding DNA regions in a variety of people, including Africans, African Americans, Europeans, New Guineans, and Australians. They deduced that the hypothesized ancestral woman lived about 200,000 years ago, in Africa. More recent analysis of mtDNA from 600 living East Africans estimated 170,000 years ago for the beginning of the modern human line. One way to reach this time estimate is by comparing how much the mtDNA sequence differs among modern humans to how much it differs between humans and chimps. The differences in mtDNA sequences among contemporary humans are 1/25 as many as the differences between humans and chimps. The two species diverged about 5 million years ago, according to extrapolation from fossil and molecular evidence. Multiplying 1/25 by 5 million gives a value of 200,000 years ago, assuming that the mtDNA mutation rate is constant over time.

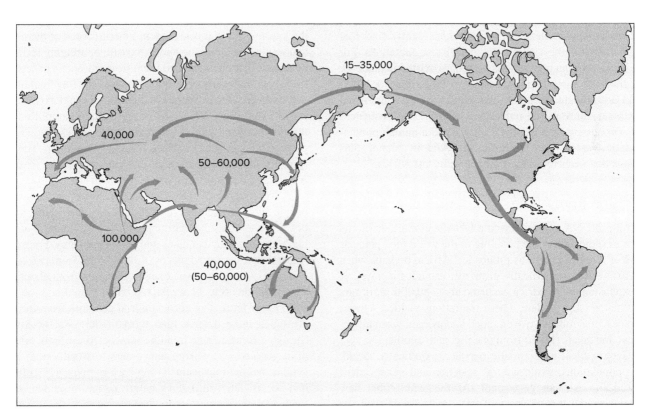

Figure 16.12 **The peopling of the planet.** This map depicts major migratory paths based on DNA haplogroup information, mostly from mtDNA and the Y chromosome. Fossil and DNA evidence suggest that humanity arose in East Africa, but a southern African origin is possible, too. Other routes to the Americas may have been possible too. These routes are hypotheses, based on available evidence. *Adapted from Hua Liu, Frank Prugnolle, Andrea Manica and Francois Balloux, "A Geographically Explicit Genetic Model of Worldwide Human-Settlement History," The American Journal of Human Genetics, Vol. 79: 2, 2006.*

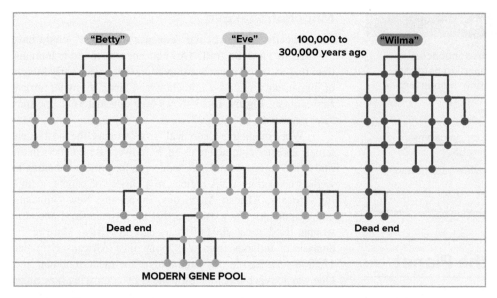

Figure 16.13 Mitochondrial Eve. According to the mitochondrial Eve hypothesis, modern mtDNA retains some sequences from a figurative first woman, "Eve," who lived in Africa 300,000 to 100,000 years ago. In this schematic, the lines represent generations, and the circles, females. Lineages cease whenever a woman does not have a daughter to pass on the mtDNA.

Where did Eve live? The locations of fossil evidence support an African origin, and Charles Darwin suggested it, too. In addition, studies comparing mitochondrial and nuclear DNA sequences among modern populations consistently find that Africans have the most numerous and diverse mutations. For this to be so, African populations must have existed in place for longer than other populations, because it takes time for mutations to accumulate. In many evolutionary trees constructed by computer analysis, the individuals whose DNA sequences form the bases are from Africa. Other modern human populations all have at least part of an ancestral African genome, plus mutations that occurred after their ancestors left Africa.

Expansion from Africa

Data from mtDNA, Y chromosome DNA, and markers on the autosomes indicate that the peopling of the world took place as a series of founder effects as groups left Africa, perhaps when the Sahara desert periodically grew wet enough for travel. A large expansion out of Africa occurred about 50,000 years ago, but estimates range widely. These migrations yielded "chains of colonies" that may have overlapped and merged when neighbors met and genes flowed from one region to another.

Geographical and climatic barriers periodically shrank human populations, while natural selection and genetic drift narrowed the African gene pool. At the same time, new mutations among the groups that left Africa established the haplogroups that are consulted to trace non-African ancestry today. Human evolution continued in Africa as groups left and populated other parts of the world. The descendants of some people who left Africa returned. Global climate change, such as ice ages, forced people into habitable areas, called refugia.

Today gene flow continues, both out of and back into Africa.

Populating the New World

People spread across Eurasia and elsewhere by 40,000 years ago, lastly through southern Siberia and Mongolia. By 20,000 years ago, humanity was everywhere except the Americas and Antarctica (**figure 16.14**). From Siberia, people could cross the Bering Land Bridge, which emerged between Siberia and Alaska during times when the glaciers had retreated. The land bridge (also called an ice-free corridor) stretched for about 1,000 miles from north to south, surfacing as winds from the southwest blew snow away. The areas for several hundred miles on either side of the bridge, and the bridge itself, are called Beringia. The land bridge was present from 25,000 to 15,000 years ago, according to most evidence.

Mutations occurred in the population of Beringia. Then between 23,000 and 19,000 years ago, a severe population bottleneck killed off all but only about 1,000 people who survived the journey over the bridge from Siberia. Some of them continued southward along the Pacific coastline, bringing some of the mutations that distinguished them from European ancestors. As the ice age ended about 18,000 years ago, the tiny founding population in the Americas began a period of rapid expansion that lasted 3,000 years. The burst of reproduction amplified alleles that had survived while at the same time alleles unique to people who had perished vanished from the population. The people spread south and east through the Americas as the first Native Americans, a process that may have taken about 2,000 years.

Mitochondrial DNA evidence from descendants of people who crossed the land bridge is particularly valuable because the coastal migratory path is now under water, hiding both archeological clues and preserved DNA. For example, researchers discovered pieces of skull bones from two babies in a part of Alaska that was once in Beringia from about 11,500 years ago, one 6 to 12 weeks old when it died and the other a stillborn or fetus. The mitochondrial genomes from the infants are much more diverse than most modern Native American lineages. Researchers hypothesize that these two mitochondrial genomes were from latecomers to North America. The genomes of their ancestors going back as many as 10,000 years had accumulated mutations before people migrated. Perhaps the harsh environment or the chance sampling of genetic drift whittled away at the original human genetic diversity that came across the land bridge.

Today, Native American genomes reflect the long-ago trek across the land bridge: five major mtDNA haplogroups (A, B, C, D and X) and two Y chromosome haplogroups (C and Q),

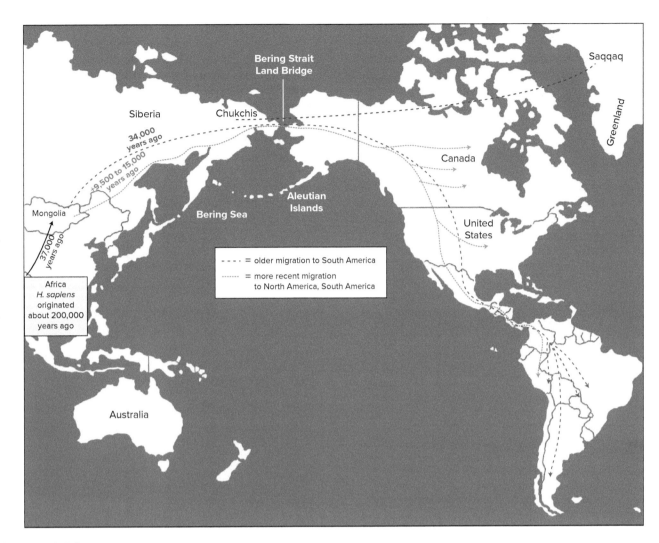

Figure 16.14 **Tracing Native American origins.** Analyses of mitochondrial DNA and Y chromosome DNA sequences reveal that the ancestors of Native Americans came from Mongolia and Siberia.

plus a few rare haplogroups. These markers are seen in all present-day indigenous populations in southern Siberia and some in northern Siberia, indicating a single gene pool traveling in a single migration. However, a complete genome sequence from a boy who lived 24,000 years ago in Siberia had gene variants from the European edge of Asia. Today about a third of Native American genomes come from western Eurasia and two-thirds from eastern Asia. Some ancestors of Native Americans did not survive to contribute to modern gene pools. That is, genetic diversity among Native American populations has decreased over time. More recently, human behavior has had much to do with that.

DNA sequence information extrapolated back from present-day Native American populations is consistent with molecular clock data from ancient DNA. In addition, Native American populations have an STR marker and some mtDNA haplogroups that are not seen in eastern Siberian peoples, indicating mutations that happened after crossing the Bering Strait.

A comparison of 678 autosomal STR markers from 29 Native American populations and 49 other indigenous groups worldwide found that Native Americans are very different from other populations, yet are very much like each other—as might be expected from a multigenerational journey southward by a small but hardy group, along the coastline of the Americas.

By 14,000 years ago, Native Americans had arrived inland, indicated in fossilized excrement. DNA from the skeleton of an adolescent girl discovered underwater in the Yucatan peninsula dates from 12,000 years ago and includes sequences from Asia and from the Americas.

It seems that with every new DNA discovery, we rewrite prehistory or add details. This was the case for the discovery in Greenland of a 4,000-year-old tuft of hair from a male Paleo-Eskimo from the first group of settlers, the Saqqaq. The DNA from the hair indicated that the man had coarse, dark hair and skin, brown eyes, shovel-shaped teeth, dry earwax, and, if he had lived longer, would have become bald. He had type A-positive blood, he was at increased risk for high blood pressure and diabetes, he had a high tolerance for alcohol, and he may have become addicted to nicotine had it been possible to cultivate tobacco in the frozen wasteland. The most revealing information concerned the Saqqaq Eskimo's origins. His mitochondrial DNA was distinct from that of either Native Americans or modern Eskimos, yet was very similar to that of the Chukchis people of Siberia. Apparently, the ancestors of the Greenland Eskimos crossed a land bridge about 5,400 years ago, separate from the crossings that founded the Native Americans and modern Eskimos, and left no present-day descendants.

Ancestry Testing

Today, people can learn about their ancestry by mailing a DNA sample to a genetic genealogy company that tests mitochondrial DNA to trace maternal lineages, Y chromosome DNA sequences to trace paternal lineages, and some autosomal sequences. Markers are compared to extensive databases to identify genome regions that individuals share. Companies provide information on both "deep ancestry" and possible contemporary relatives. "Deep ancestry" assigns likely origin to a major part of the world—sub-Saharan Africa, for example—or to a major population group or subgroup, such as "African American" or "African American with roots in Nigeria or Senegal." **A Glimpse of History** describes much more recent influences on African American genomes.

At a contemporary level, DNA testing can reveal whether any two individuals living today share an ancestor, and how many generations back they did so. Such tests assign an approximate generation to the "most recent common ancestor" (MRCA). The more markers tested, the more meaningful the results. If two people share all 37 of 37 tested markers, there is a 50 percent chance that their MRCA was no more than two generations ago. The people are so alike because not enough generations have passed for their genomes to have diverged. Sharing 25 markers

The transatlantic slave trade from many parts of Africa to the United States was declared illegal in 1808, when 90 percent of the 360,000 or so slaves lived in the South. Yet slaves continued to arrive until 1865, and the 1870 U.S. census listed 4.88 million "colored" people. African American genomes came to include significant gene variants from Europeans, with a directional gene flow from male to female reflecting widespread rape of slaves. Today about 40 million people in the United States identify as black or African American.

During the "Great Migration" from 1910 through 1970, millions of African Americans left the South in search of economic opportunity. They traveled, mostly by rail, to cities in the northeastern, midwestern, and western parts of the United States. Genetic evidence supports historical evidence of their journeys.

Researchers analyzed more than half a million SNPs among 3,726 African Americans living today in all 50 states, tracking the proportions of gene variants from African, European, and Native American ancestors, and extrapolating backwards using haplotype data. They inferred that:

- Overall 82.1 percent of the ancestors of the average modern African American were African, 16.7 percent were European, and 1.2 percent were Native American. The proportions vary slightly geographically, with average African ancestry 83 percent in the South, 80 percent in the North, and 79 percent in the West.
- Native American DNA sequences entered in the seventeenth century, mostly from females.
- European admixture occurred in two phases, around 1740 and then 1863. Admixture declined as the Civil War ended.
- Similar levels of African ancestry for all modern age groups suggest little genomic input from Europeans from 1930 through 1960.
- African Americans with a greater proportion of European genes were more likely to have left the South during

Figure 16B **SNP genotyping reconstructed migration routes from the southern United States to cities in the North and West.** The genetic data directly parallel the railroad routes that the people used to escape. This family traveled from the South to Chicago in 1919. © *Chicago History Museum/Getty Images*

the first wave of migration, perhaps because some could pass for white, or because owners helped them. Genetic data are consistent with historical data that indicate earlier migrants had lighter skin.
- When the researchers compared all possible pairs of the 3,726 individuals for identical by descent genome regions and geographical census data, the map of the United States that emerged nearly exactly parallels the rail lines that freed slaves and their descendants used during the Great Migration, including lines up the Atlantic coast and the Illinois Central line to Chicago.

gives a 50 percent chance that the MRCA 25 was not more than three generations ago. If two people share 12 markers, there is a 50 percent chance that the MRCA was no longer ago than seven generations.

For some people, such as Ashkenazi Jews or Caucasian Europeans, deep ancestry testing may reveal something that they already know from family history. In contrast, for African American families who cannot trace their history back before slavery using documents and oral histories, ancestry testing can tell them if their forebears came from any of more than two dozen places along the coast of West Africa and inland.

Companies that offer ancestry testing match new customers' DNA to sequences in their databases, and report back possible distant cousins. A person might then receive email from dozens of fifth cousins—we each have 4,688 of them! Ancestry companies can apply an algorithm to the DNA data and family tree information that includes birth locations to identify shared ancestors. As genetic genealogy databases grow with sequenced genomes, ancestry testing will provide more detailed information.

Ancestry testing has limitations. Testing of mtDNA and Y chromosome DNA considers much less than 1 percent of the genome, and traces only some lineages. Another limitation is that a haplogroup may come from more than one geographic region, due to gene flow. In addition, not all human haplogroups have been discovered, so a person may be erroneously placed into one group to which he or she partially matches, because the true haplogroup has not yet been described. **Bioethics** discusses privacy issues that may arise with ancestry testing.

Key Concepts Questions 16.3

1. Who does "mitochondrial Eve" represent?
2. Describe the main events of how we think people populated the world.
3. How might people have gotten to the Americas?
4. What can we learn from ancestry testing and what are its limitations?

Bioethics

Genetic Privacy: A Compromised Genealogy Database

The intent of the 1000 Genomes Project was to collect a large number of human genome sequences to study inherited variation. The database was to be anonymous. Part of the informed consent form read, ". . . *it will be hard for anyone to find out anything about you personally from any of this research.*" However, online searches shattered the premature promise of privacy.

Yaniv Erlich, a researcher at the Whitehead Institute who had worked with databases at financial banks, and his student Melissa Gymrek, tried to identify people who'd anonymously donated DNA to the 1000 Genomes Project—just to see if they could. If so, they could warn government officials that the supposedly anonymous database was not private.

The researchers looked at short tandem repeats (STRs). Recall from chapter 14 that STRs are the non-protein-encoding short repeated DNA sequences used in forensics investigations. They are also used in genealogy research. Erlich and Gymrek considered STRs on the Y chromosome and consulted public genealogy databases to find surnames corresponding to specific Y haplotypes. Further clues came from basic public information such as state of residence and birth year. Often an Internet search on some of the data led to family websites that instantly confirmed that the researchers had correctly identified an individual from his DNA. Cross-referencing to the DNA sequences of cells that the 1000 Genomes Project had deposited in the Coriell Cell Repositories in New Jersey identified women.

When the researchers had identified 50 people and realized how easy it was, they contacted officials at the National Institutes of Health. The agency took swift measures to hide some of the data, such as year of birth. The researchers and *Science* magazine published the findings to begin a discussion and possibly prevent further exposures of personal genetic information. The problem of maintaining genetic privacy for individuals who participate in research projects that may benefit many people will continue, and likely grow.

Questions for Discussion

1. How can researchers collect enough genome data to better understand human inherited variation, yet protect identities?

2. Ten years from now, storing the information in our genomes may be routine, perhaps even required. It is already possible to store a genome sequence on a smartphone. What measures can researchers take now to protect genetic privacy?

3. How does genome data differ from other biomedical information, such as cholesterol level?

4. Discuss how use of direct-to-consumer ancestry testing can raise issues of genetic privacy.

5. Children with rare diseases whose mutations are listed in mutation databases have been identified from information on hometown, age, and other nongenetic characteristics. Suggest a way to better protect their identities.

6. Is complete de-identification in a database possible?

7. Personal genetic information, as well as any other type of information, can be inadvertently shared via social media. What can be done to avoid this problem?

16.4 What Makes Us Human?

We can investigate the traits and abilities that distinguish us from chimpanzees and bonobos, our closest relatives, because they live today. For the archaic humans that we know about only from bits of bones, genome sequencing has provided clues to some of their traits, suggesting abilities gained as the archaic humans interbred with our ancestors.

We assess similarity between ourselves and chimps and bonobos by DNA sequence, numbers of copies of sequences, or sequences not in our genomes. We share about 98.7 percent sequence similarity with our closest primate relatives, but we also differ in the number of copies of certain DNA sequences. These include insertions and deletions, which are collectively called **indels**. Considering indels, our degree of genome similarity to chimps and bonobos is only about 96.6 percent. The degree of similarity may even be as low as 94 percent if sequences not in the human genome are considered. That is, what *isn't* present defines us as well as what *is* present.

Evolution of Human Genomes

Comparisons of human genome sequences to those of other species are interesting and humbling. For example, our genome is not that different from that of a pufferfish, with added introns and repeated sequences. Our exomes are remarkably similar for animals that live in such different environments.

Overall, the human genome has a more complex organization of the same basic parts that make up the fruit fly and roundworm genomes. For example, the human genome has 30 copies of the gene that encodes fibroblast growth factor, compared to two copies in the fly and worm genomes. This growth factor is important for the development of highly complex organs.

Genome studies indicate that over a great span of evolutionary time, genes and gene pieces provided vertebrates, including humans, with certain defining characteristics:

- complex neural networks
- blood clotting pathways
- adaptive immunity (see section 17.2)
- refined apoptosis
- greater control of transcription
- complex development
- more intricate signaling within and among cells

Comparing the human genome to itself provides clues to evolution, too. The many duplicated genes and chromosome segments in the human genome suggest that it doubled, at least once, since diverging from the genome of a vertebrate ancestor about 500 million years ago. Either the human genome doubled twice, followed by loss of some genes, or one doubling was followed by additional duplication of certain DNA sequences.

The extensive duplication within the human genome is what distinguishes our genome from those of other primates. Some of the doublings are vast. Half of chromosome 20 repeats, rearranged, on chromosome 18. Much of chromosome 2's short arm reappears as almost three-quarters of chromosome 14, and a block on its long arm is echoed on chromosome 12. The gene-packed yet tiny chromosome 22 includes eight huge duplications and several gene families. Repeated DNA sequences in the human genome may provide raw material for future evolution. A second copy of a DNA sequence can mutate, allowing a cell to experiment with a new function while the old one carries on, a little like trying a new car before selling the old one. More often, the twin gene mutates into a silenced pseudogene, leaving a ghost of the gene behind as a similar but untranslated DNA sequence.

A duplication can be located near the original DNA sequence it was copied from, or away from it. A sequence repeated right next to itself is called a tandem duplication, and it usually results from mispairing during DNA replication. A copy of a gene on a different chromosome may arise when messenger RNA is copied (reverse transcribed) into DNA, which then inserts elsewhere among the chromosomes.

Duplication of an entire genome results in polyploidy, discussed in section 13.3. It is common in plants and some insects, but not vertebrates. If a polyploid event was followed by loss of some genes and duplication of others, the result would look much like the modern human genome (**figure 16.15**). The remnants of such an ancient whole-genome duplication would have become further muddled with time, as inversions and translocations altered the ancestral DNA landscape.

Genes That Help to Define Us

Comparison of the chimpanzee and human genomes has revealed "human accelerated regions." These are highly conserved sequences that show signs of positive selection in humans, such as an amino acid change seen in all human groups but not in the chimp or orangutan versions of the same gene. (Recall from section 15.6 that positive selection maintains a beneficial trait.) Signs of positive selection in the human genome can complement views of our ancestry from fossils.

Relatively few DNA sequence differences distinguish us from the Denisovans and Neanderthals: 31,000 single nucleotide changes (SNPs and mutations), and fewer than 125 small losses or gains of DNA bases (indels). Only 96 of the DNA

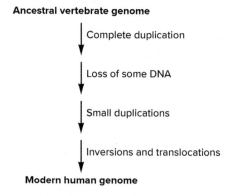

Figure 16.15 Evolution of the human genome. The many duplicated DNA sequences in the human genome suggest a complete duplication followed by other chromosome-level events.

differences in the human genome alter amino acids in proteins (are nonsynonymous) and can therefore be subject to natural selection. Subtracting Denisovan DNA sequences that are conserved in ape and monkey genomes, indicating their antiquity, reduces the list of distinctive human mutations to 23. Eight of them have to do with brain function and development, neural connectivity and synapses, and two of these mutations are implicated in autism. About 3,000 sequence differences are in parts of the human genome that regulate gene expression, so just a few distinctions can nonetheless have large effects.

Traits that may be uniquely human include spoken language, abstract reasoning ability, highly opposable thumbs, and larger frontal lobes of the brain. We can't know whether the Denisovans and Neanderthals were able to speak and the extent of their abilities to think. Following are examples of traits determined by single genes that help to distinguish us from our closest relatives (**figure 16.16**).

Speech

A family living in London whose members have unintelligible speech led to the discovery of a single gene (*FOXP2*) that controls speaking ability. Modern humans and Neanderthals have the same two *FOXP2* variants that are not in the chimpanzee genome.

Different Developmental Timetables

More primitive primates lack or have very little fetal hemoglobin (see figure 11.2). In more recently evolved and more complex primates, fetal hemoglobin extends the fetal period, allowing time for substantial brain growth and development before birth. With larger brains came greater skills. Other single genes make possible the longer childhood and adolescence of humans compared to chimps.

Bare Skin

Nearly all mammals have thick, abundant hairs that insulate, protect, and are important in social displays and species recognition. Chimpanzees and gorillas express a keratin gene whose counterpart in humans is not expressed.

Our naked skins make sense in terms of evolution in a few ways. Relatively hairless skin enables us to sweat. When our forebears moved onto the plains as the forests shrank, about 2.5 to 3 million years ago, individuals with less hair and more abundant sweat glands could travel farther in search of food, or to avoid becoming food. Natural selection favored them. Today, our skin is peppered with sweat glands that pump out quarts of sweat a day, compared to our furred fellow mammals, who have sparser and less efficient sweat glands. Lack of hair may have also enabled our ancestors to shed skin parasites such as lice. Human skin is thin yet strong, thanks to a unique combination of keratin proteins that fill the flattened skin cells.

Walking

Walking requires the ability to place one foot in front of the other. People who have Joubert syndrome can't do this. Nerve cell fibers cannot cross from their origin on one side of the brain to the other, so moving just one arm or leg is impossible; both move at once. The part of the brain that controls posture, balance, and coordination is compromised. The gene that causes Joubert syndrome, *AHI1*, is identical in all modern human groups examined, but has different alleles in chimps, gorillas, and orangutans. Perhaps in the lineage leading to humans, the gene came to control walking by making it possible to place one foot in front of the other.

Running

Homo erectus was the first hominin to be able to run long distances, thanks to specific anatomical adaptations. The nuchal ligament that connects the skull to the neck became more highly developed in *H. erectus*, keeping the head in place while running. Leg muscles were more highly developed than those of chimps or australopithecines, acting as springs. *H. erectus* originated large buttocks, whose muscles contract during running. All three of these structures are not merely the result of being able to walk, but enabled early *Homo*, and us, to run.

Big brain
Single genes, copy number variants, and gene expression patterns contribute to human cognition.

Speech
FOXP2 gene

Abundant sweat glands

Bare skin
Silenced keratin genes

Walking + running
Ability to put one foot in front of the other; *AHI1* gene

Figure 16.16 What makes us human? A collection of traits distinguishes us from our closest primate relatives.
© Imgram Publishing RF

This skill would have helped our ancestors to escape predators, find food, and locate new homes.

A Big Brain

Only a few genes may control the difference in brain size between us and chimps. About 2.4 million years ago, a gene called *MYH16* underwent a nonsense mutation, which prevented production of a type of muscle protein called a myosin. The mutation is seen in all modern human populations, but not in other primates. Without this type of myosin, jaws developed less, which allowed expansion of the bony plates of the skull, permitting greater brain growth. Researchers nicknamed the mutation "room for thought." Fossil evidence indicates that the switch from "big jaw, small brain" to "small jaw, big brain" happened when *Homo* gradually replaced *Australopithecus*, about 2 million years ago. The genetic analysis is new, but the idea isn't. Charles Darwin wrote in 1871 that different-sized jaw muscles were at the root of the distinction between apes and humans.

The major reason humans and chimps look and behave differently but are genetically so similar seems to have more to do with gene expression than genome sequence. For example, a study comparing gene expression in the liver and brain found many more differences between the two species in the brain than in the liver, suggesting that our mental capabilities have extended beyond those of chimps more than have the functions of our livers. The brain, therefore, is a big part of what makes us human.

A larger brain presumably led to greater cognitive skills. In this area, gene copy number may be more important than the nature of the genes. At least 134 genes are present in more copies in the human genome than in genomes of apes. Many of these genes affect brain structure or function. Some of the genes foster long-term memory, and others, when mutant, cause intellectual disability or impair language skills. Single genes implicated in fueling human brain growth control the migration of nerve cells in the front of the fetal brain.

Recent mutations may have propelled our brain power, too. A mutation in a gene that encodes an enzyme called a fatty acid desaturase enables the digestive system to use long-chain fatty acids, which are found in fish and shellfish and are useful in brain development. The mutation occurred in Africa from 180,000 to 80,000 years ago, when some of our ancestors lived near the lakes of Central Africa. Positive selection greatly increased the prevalence of the mutation, but at a cost—the encoded enzyme also causes inflammation and may be responsible for the increased prevalence of hypertension, stroke, type 2 diabetes, and coronary heart disease among African Americans.

Sense of Smell

Our chemical senses—smell and taste—have decreased as our reliance on them has decreased. The sense of smell derives from a 1-inch-square patch of tissue high in the nose that consists of 12 million cells that bear odorant receptor (OR) proteins. (In contrast, a bloodhound has 4 billion such cells!) Molecules from a smelly substance bind to combinations of these receptors, which then signal the brain in a way that creates the perception of the associated odor.

The 906 human odorant receptor genes are clustered on chromosome 17. More than half of them are pseudogenes, similar to active "smell" genes but riddled with nonsense mutations. Perhaps these genes are remnants of a distant past, when we depended more upon our chemical senses for survival, and natural selection has silenced them. At the same time, natural selection has acted positively in a way that has retained functioning OR genes. While the pseudogenes contain many diverse SNPs, the functional OR genes are remarkably alike in sequence among individuals, indicating a successful function. In addition, the nucleotide differences that persist among the retained OR genes alter the encoded amino acids, suggesting that natural selection favored these sequences.

The Future of Humanity?

How will evolution mold humanity in the future? Perhaps the most profound characteristic that distinguishes us from archaic humans is our ability to alter the environment on a large scale, which has led to our domination of the planet. We exist in varieties of skin color, body size, intellectual ability, and hundreds of other observable and measurable traits. Yet at the same time, people from different populations are meeting and mixing their genomes at unprecedented levels, thanks to transportation and communication over vast distances. The planet is far different from the days of the australopithecines, who dwelled in geographical areas so dispersed that they existed as an overlapping series of distinct species, or even the Neanderthals and Denisovans who might never have ventured far from their caves. Today, phenotypes persist that in millennia past would have disappeared due to negative natural selection, thanks to health care and technology.

What will humans be like 10,000 years from now? Will admixture increase so that we all have genome parts that were once associated with specific population groups? Will new technologies and medical care improve our genomes, or allow gene variants that would have been lethal in the past to persist and be passed to future generations? Will we manipulate our genomes to direct our own future evolution? Understanding our origins poses many questions about our futures.

Key Concepts Questions 16.4

1. How does the human genome differ from the genomes of other animals, in a general sense?
2. What are some genes that provide traits that are uniquely human?
3. How do we influence our own future evolution?

Summary

16.1 Human Origins

1. DNA provides information on evolution because sequence similarities indicate descent from common (shared) ancestors. Changing allele frequencies indicate events at a population level.

2. Evolutionary tree diagrams depict descent from shared ancestors.

3. Humans are hominins, a taxonomic group that includes our ancestors who lived after the split from other African apes about 6 million years ago.

4. The australopithecines preceded and then coexisted with early *Homo* species, who lived in caves, had strong family units, and used tools, then lived in societies and used fire.

5. Neanderthals and Denisovans were archaic humans. A small percentage of their genomes persists in the genomes of some modern human populations, indicating **introgression** (introduction of gene variants from one species or subspecies to another) and **admixture** (presence of DNA variants from more than one species or subspecies).

6. The spread of agriculture is reflected in changing allele frequencies that resulted from ancient migrations.

7. The DNA of Ötzi the Ice Man compared to DNA sequences from other groups indicates that his maternal ancestors died out but his paternal ancestors escaped the harsh Alps and survived.

16.2 Methods to Study Molecular Evolution

8. **Molecular evolution** considers differences in genomes, chromosome banding patterns, protein sequences, and DNA base sequences, using mutation rates of individual genes to estimate species relatedness.

9. Genes in the same order on chromosomes in different species show **synteny**.

10. For a highly conserved gene, the DNA sequence is similar or identical in different species, indicating importance and shared ancestry.

11. Evolutionary tree diagrams represent gene sequence information from several species, using molecular clocks based on mutation rates.

12. Molecular clocks based on mtDNA date recent events through the maternal line because this DNA mutates faster than nuclear DNA. Y chromosomes are used to trace paternal lineage. Markers (SNPs, STRs, and CNVs) in mtDNA, Y chromosome DNA, and autosomal DNA are used to study human origins and expansions. Many linked markers inherited together form **haplogroups**. Haplotypes are shorter than haplogroups.

13. Changing haplotype patterns indicate admixture, which, using crossover rates and generation times, estimates when populations mixed.

16.3 The Peopling of the Planet

14. The rate of mtDNA mutation and current mtDNA diversity can be extrapolated to hypothesize that a theoretical first woman lived in Africa about 200,000 years ago. *Homo sapiens* began to leave Africa about 56,000 years ago.

15. A series of migrations, creating founder effects, peopled the planet, with genetic diversity decreasing from that of the ancestral African population, but with new mutations occurring.

16. After the last ice age, people crossed the Bering Strait from Siberia, occupying the Americas.

17. Ancestry testing considers mitochondrial DNA to trace maternal lineages, Y chromosome sequences to trace paternal lineages, and selected autosomal sequences to fill in ancestors, providing information on deep ancestry (geographic origins) and contemporary relatives.

16.4 What Makes Us Human?

18. Humans and chimps share 98.7 percent of their protein-encoding gene sequences. **Indels**, introns, and repeats create genome differences between humans and chimps. The two species also differ in gene expression.

19. The human genome shows many signs of past duplication.

20. Humans differ from other apes in adaptations that allowed greater brain growth, less hairy skins and the advantages that they provide, and the abilities to speak, walk, and run.

21. We affect our own evolution by treating disease and altering the environment.

Review Questions

1. What types of evidence are used to construct evolutionary tree diagrams?

2. Explain how the evolutionary tree diagram in figure 16.1 indicates that we did not descend directly from chimpanzees.

3. Why were there several species of *Australopithecus* but possibly only one of early *Homo* in Africa?

4. What is a limitation of relying on fossil evidence in describing extinct species?

5. Cite the evidence that Neanderthals and our ancestors interbred.

6. Why are Neanderthal DNA sequences in the genomes of modern Europeans but not in those of modern Africans?

7. Discuss how a specific gene variant might have been adaptive in Neanderthals but harms health in modern humans.

8. How can DNA sequences indicate that parents were also siblings?

9. Why is the gene flow among Neanderthals, Denisovans, and our ancestors so complicated?

10. What do we know about ancient farmers and Ötzi the Ice Man from their DNA?

11. State two assumptions behind molecular evolution studies.

12. Which of the forces discussed in chapter 15 explains why some DNA sequences vary greatly among species, while others are highly conserved?

13. Why are exons highly conserved but introns not?

14. Why are active odorant receptor genes highly conserved but pseudogene versions of them not highly conserved?

15. What is a limitation of an evolutionary tree diagram constructed using DNA or protein sequence data?

16. Describe the type of information that Y chromosome and mitochondrial DNA sequences provide, and explain why it is incomplete.

17. How can haplotypes provide information on genetic admixture and ancient migrations?

18. Why is the statement that 4 percent of some modern human genomes are from Neanderthals incorrect?

19. Name a human population that has not admixed with others.

20. Describe how people expanded out of Africa, while evolution continued in Africa.

21. Provide an example of how changing geography or climate affected human evolution.

22. How can we affect our own future evolution?

Applied Questions

1. Explain what comparing genomes from indigenous peoples to other modern peoples can reveal about evolution.

2. Select an example from this chapter and explain how it illustrates one of the forces of evolutionary change discussed in chapter 15 (natural selection, nonrandom mating, migration, genetic drift, or mutation).

3. A geneticist aboard a federation starship must deduce how closely related humans, Klingons, Romulans, and Betazoids are. Each organism walks on two legs, lives in complex societies, uses tools and technologies, looks similar, and reproduces in the same manner. Each can interbreed with any of the others. The geneticist finds the following data:

   - Klingons and Romulans each have 44 chromosomes. Humans and Betazoids have 46 chromosomes. Human chromosomes 15 and 17 resemble part of the same large chromosome found in Klingons and Romulans.
   - Humans and Klingons have 97 percent of their chromosome bands in common. Humans and Romulans have 98 percent of their chromosome bands in common, and humans and Betazoids show 100 percent correspondence. Humans and Betazoids differ only by an extra segment on chromosome 11, which appears to be a duplication.
   - The cytochrome *c* amino acid sequence is identical in humans and Betazoids, differs by one amino acid between humans and Romulans, and differs by two amino acids between humans and Klingons.
   - The gene for collagen contains 50 introns in humans, 50 introns in Betazoids, 62 introns in Romulans, and 74 introns in Klingons.
   - Mitochondrial DNA analysis reveals many more individual differences between Klingons and Romulans than between humans and Betazoids.

   a. Suggest a series of chromosomal abnormalities or variants that might explain the karyotypic differences among these four types of organisms.
   b. Which are our closest relatives among the Klingons, Romulans, and Betazoids? What is the evidence for this?

   c. Are Klingons, Romulans, humans, and Betazoids distinct species? What information reveals this?
   d. Which of the evolutionary tree diagrams is consistent with the data?

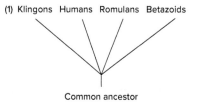

(1) Klingons Humans Romulans Betazoids

Common ancestor

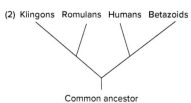

(2) Klingons Romulans Humans Betazoids

Common ancestor

(3) Klingons Romulans Humans Betazoids

Common ancestor

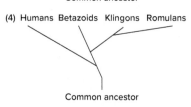

(4) Humans Betazoids Klingons Romulans

Common ancestor

4. A man with white skin checks the box on forms about personal information for "African American," claiming that we are all, if we go back far enough in our family trees, from Africa. Is he correct?

5. What do drug developers need to consider to create a product based on or using a protein from a species other than humans?

6. In Central Africa, the Mbuti Pygmies are hunter-gatherers who live amid agricultural communities of peoples called the Alur, Hema, and Nande. Researchers compared autosomal STRs, mitochondrial DNA, and Y chromosome DNA haplogroups among these four types of people. The pygmies had the most diverse Y chromosomes, including about a third of the sequence that was the same as those among the agricultural groups, who had greater mtDNA diversity than the pygmies. None of the agricultural males had Y chromosome DNA sequences that matched those of the pygmies. Create a narrative of gene flow to explain these findings.

7. Explain why analyzing mtDNA or Y chromosome DNA cannot provide a complete picture of a person's ancestry.

8. Explain how a female can trace her paternal lineage if she doesn't have a Y chromosome.

9. Would you want your ancestry information and identity posted in an ancestry database so that cousins can contact you?

10. Name a technology that is contributing to human admixture.

11. Humans have chins, but other primates, including *Australopithecus* and Neanderthals, do or did not. Propose a function that might explain the positive selection of chins in anatomically modern humans.

Forensics Focus

1. How can the technology used to describe the 4,000-year-old Saqqaq Paleo-Eskimo from Greenland be applied to forensics techniques used on crime scene evidence?

2. The Millers raise chickens in their backyard. On their block are six pet dogs, all mixed breeds. One day, Mr. Miller discovers that his chickens are dead, and amid the feathers and signs of struggle are clumps of dog hair. He collects the hair and sends it to a dog ancestry testing company, for clues to which of his canine neighbors is the culprit. What will the company test?

3. In many African cultures, "family" is not dictated by genetics, but by who cares for whom. Any adult can be "mother" or "father" to any child. Since the early 1990s, many parts of East Africa have been under civil war. Thousands of Africans have asked to be admitted to the United States to join relatives. The "family reunification resettlement program" enables parents, siblings, and children of U.S. citizens to come to the United States.

In 2008, addressing rumors that many people were lying that they were related to people in the United States, the State Department began to ask refugees from Kenya to voluntarily provide a DNA sample, to be compared to that of the U.S. citizen claimed to be a relative. When the DNA testing turned up many cases of people claiming to be family who were not blood relatives, the resettlement program was stopped.

a. Do you think that DNA testing should have been imposed on people seeking asylum in the United States?

b. Should the people have been compelled to have their DNA tested?

c. How should cultural definitions, such as that of "family," be handled?

d. How should the situation be resolved?

e. How might DNA testing be used to either help refugees enter the US, or to hinder their efforts, based on recent events and government policies?

Case Studies and Research Results

1. The Australian aborigines are one of the oldest peoples outside of Africa, tracing their ancestry back 55,000 years. When geneticists first approached them in the 1990s about studying their DNA, the people became upset. They regarded ancestry genetic research as a threat to their traditional belief systems, as challenging their identity and claims to land, and feared stigma from genetic test results. They also objected strenuously to one research project referring to Aborigines as a "vanishing people."

An Aboriginal genome was sequenced in 2011, from a hair sample of remains of a young man from the 1920s donated to a museum. The researchers went into the Aborigine community and obtained consent. Now the researchers would like to sequence the genomes of additional members of the community whose DNA was sampled years ago, before the first human genomes had been sequenced.

Should the researchers request new informed consent to sequence the genomes in the samples, and if so, what information should they provide? Should they sequence genomes from samples taken from people who have died or cannot be located?

2. "Neanderthals are not totally extinct; they live on in some of us," said Svante Paabo, the leader of the Neanderthal Genome Project. What does he mean?

3. The 4,000-year-old Saqqaq Paleo-Eskimo from Greenland is known only from a tuft of hair. How did researchers learn about other characteristics?

4. A Y chromosome haplotype has mutations for the *SRY* gene and genes called *M96* and *P29*. Modern Africans have three variants of this haplotype. Two variants occur only in Africans, but the third variant, E3, is also seen in western Asia and parts of Europe. Researchers examined specific subhaplotypes (variations of the variations) and

found that one type, E-M81, accounts for 80 percent of the Y chromosomes sampled in northwest Africa, falling sharply in incidence to the east, and not present in sub-Saharan Africa. That same haplotype is found in a small percentage of the Y chromosomes in Spain and Portugal. Consult a map, and propose a scenario for this gene flow. What further information would be useful in reconstructing a migration pattern?

5. Roland has always considered himself African American, but ancestry testing of his Y chromosome indicates a Chinese background. He is very upset, concluding that he is not African American after all. Explain how he has misinterpreted the test results or the test results could be in error.

6. "Glossogenetics" is the field of study that associates language with patterns of genetic variation. Explain the finding that languages in a particular geographic area track with Y chromosome sequences, but not with mtDNA sequences.

7. The Beringia hypothesis has been accepted for many years, but newer DNA evidence from sagebrush, trees, woolly mammoths, bison, rabbits, and other species found in sediments of ancient lakes that once separated the retreating glaciers suggest that the most recent land bridge was actually 12,600 years ago and not 25,000 to 15,000 years ago. How would these new findings alter the explanation for evidence of humans in North America dating to before 12,600 years ago?

8. For more than 20 million years, lice have lived on the skins of primates (see the opener to chapter 10). Researchers compared a 1,525-base-pair sequence of mtDNA among modern varieties of lice and, applying the mutation rate, derived an evolutionary tree. It depicts a split in the louse lineage, with one group of head and body lice living throughout the world, and another group of only head lice living in the Americas.

   a. What events in human evolution roughly correspond to the branch points in the louse evolutionary tree?

   b. What might be the significance of the similarity between the evolutionary trees for lice and humans?

   c. What is the evidence that lice moved from archaic humans to modern humans?

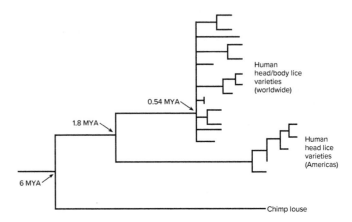

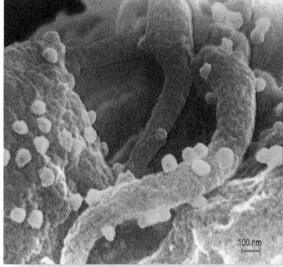

Source: CDC/C. Goldsmith, P. Feorino, E.L. Palmer, W.R. McManus

CHAPTER 17

Genetics of Immunity

These HIV particles (green) are attached to a white blood cell (pink). In people with a certain mutation, cells lack one of two types of receptors that HIV requires to enter.

Learning Outcomes

17.1 The Importance of Cell Surfaces
1. List the components of the immune system.
2. Describe the basis of blood groups.
3. Explain what human leukocyte antigens are and what they indicate about health.

17.2 The Human Immune System
4. Distinguish among physical barriers, innate immunity, and adaptive immunity.
5. Distinguish between the humoral and cellular immune responses.

17.3 Abnormal Immunity
6. Discuss conditions that result when the immune system is underactive, overactive, and misdirected.

17.4 Altering Immunity
7. Describe how medical technologies boost or suppress immunity to prevent or treat disease.
8. Explain the requirements for the body to accept an organ from another person.

17.5 Using Genomics to Fight Infection
9. Discuss how we can use knowledge of the genome sequences of pathogens.

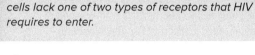

The BIG Picture

The immune system enables us to share the planet with other organisms, while it ignores the members of the microbiome that normally live in and on our bodies. Genes control the immune response. We can alter immunity to prevent and treat several types of diseases.

Mimicking a Mutation to Protect Against HIV

In 2008, a 40-year-old man received a stem cell transplant to treat leukemia at a hospital in Berlin. He had been HIV positive for at least a decade, and had taken anti-HIV drugs for 4 years. Leukemia was his first HIV-related illness. Known at first as "the Berlin patient," Timothy Brown became the star of a groundbreaking experiment.

Brown's stem cell donor was both a tissue match for Brown and one of the 0.5 percent of Caucasians who are genetically resistant to HIV infection. As a homozygote for the gene *CCR5 delta 32*, the donor did not have a type of receptor, called CCR5, that HIV must bind to enter a cell (figure 17.11 shows this receptor). Could a transplant of the man's stem cells into Brown fight his leukemia *and* his HIV infection?

So far, the answer is yes! The transplanted stem cells eventually replaced Brown's blood and bone marrow with HIV-resistant cells, and all signs of the infection vanished. As Brown made headlines, researchers were already thinking about how to mimic the natural mutation that protects against HIV. One approach is genome editing, discussed in chapters 19 and 20.

In genome editing to treat HIV infection, researchers take cells most disabled with virus (CD4 T cells) from patients, culture the cells in the laboratory, and use enzymes (zinc finger nucleases) to remove the *CCR5* gene. The manipulated cells are infused back into the patients, who stop taking their HIV medications to see if the altered cells keep viral levels undetectable. When patients discontinue their anti-HIV drugs after the cell infusion, the infection is still present but definitely dampened, and gradually the fixed cells take over.

17.1 The Importance of Cell Surfaces

We share the planet with plants, microbes, fungi, and other animals, but we can become ill when some of them, or their parts, enter our bodies, causing an infection. The immune system protects us against infectious illnesses. It also detects and destroys cancer cells, the topic of chapter 18. The human immune system is a mobile army of about 2 trillion cells, the biochemicals they release, and the organs where they are produced and stored. Genes control many aspects of immune system function, including susceptibility to infection.

Protection against infection is based on the ability of the immune system to recognize "foreign" or "nonself" cell surfaces that are not part of the body. These include surfaces of certain microorganisms such as bacteria and yeasts; nonliving "infectious agents" such as viruses and prions; and tumor cells and transplanted cells. If stimulated, the immune system launches a highly coordinated attack that includes both general and highly specific responses. Organisms or infectious agents that cause disease are called pathogens. **Clinical Connection 17.1** highlights one commonly encountered type of pathogen—viruses. **Figure 17.1** shows another—bacteria. A bacterium is a cell; a virus is simpler than a cell.

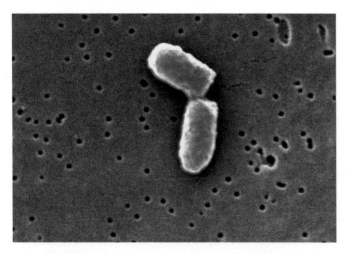

Figure 17.1 **A bacterial pathogen.** *Escherichia coli* is a normal resident of the human small intestine, but under certain conditions can produce a toxin that causes severe diarrhea ("food poisoning") and can damage the kidneys. *Source: CDC/Janice Haney Carr*

Understanding how genes control immunity makes it possible to enhance or redirect the system's ability to fight disease. Mutations can impair immune function, causing immune deficiencies, autoimmune diseases, allergies, and cancer. Genes affect immunity by conferring susceptibilities or resistances to certain infectious diseases. Like other inherited characteristics, degree of immune protection varies from person to person. One individual may suffer frequent respiratory infections, yet another is rarely ill. A study of the immune systems of people who survived the 1918 flu pandemic revealed that many decades later, they can still rapidly destroy a flu virus. Yet 50 million people died of that flu, many in just days.

A few types of genes encode proteins that powerfully affect immunity. **Antibodies** are proteins that directly attack foreign **antigens**, which are any molecules that can elicit an immune response. Most antigens are proteins or carbohydrates. Genes also specify the cell surface antigens that mark the body's cells as "self."

Blood Groups

Some of the antigens that dot our cell surfaces determine blood types. Figures 5.3 and 5.4 describe the familiar ABO blood types. Humans have 33 major blood types based on protein and carbohydrate antigens on the surfaces of red blood cells. More may be recognized as human genome sequences are annotated. Each blood type includes many subtypes, generating hundreds of ways that the topographies of our red blood cells differ from individual to individual. **Table 17.1** lists a few blood groups.

For blood transfusions, blood is typed and matched from donor to recipient. For more than a century, an approach called serology typed blood according to red blood cell antigens. A newer way to type blood is to identify the *instructions* for the cell-surface antigens—that is, the genes that encode these proteins. This approach, termed genotyping, uses a tiny device that detects 100 distinct DNA "signatures" for blood types. Genotyping is especially useful for people who have a chronic disease that requires multiple transfusions, such as leukemia or

Table 17.1	Blood Groups
Blood Group	**Description**
MN	Codominant alleles *M, N,* and *S* determine six genotypes and phenotypes. The antigens bind two glycoproteins.
Lewis	Allele *Le* encodes fucosyltransferase (FUT3), which adds an antigen to the sugar fucose, which the product of the *H* gene places on red blood cells. *H* gene expression is necessary for the ABO phenotype (see figure 5.3). People with *LeLe* or *Lele* have the Lewis antigen on red blood cells and in saliva. People of genotype *lele* do not.
Secretor	People with the *Se* allele secrete A, B, and H antigens into body fluids.

Viruses

A viral infection can bring much misery, but many of the aches and pains are actions of the human immune system.

A virus is a single or double strand of RNA or DNA wrapped in protein, and in some types, an outer envelope. A virus can reproduce only if it enters and uses a host cell's energy resources, protein synthetic machinery, and secretion pathway. It is very streamlined. A virus may have only a few protein-encoding genes, but many copies of the same protein can assemble to form an intricate covering, like the panes of glass in a greenhouse. Ebola viruses, for example, produce only seven types of proteins, but these form a structure that can reduce a human body to little more than a bag of blood and decomposed tissue in days. In contrast, the smallpox virus codes for more than 100 different types of proteins.

Many DNA viruses reproduce by inserting their DNA into the host cell's DNA, but an RNA virus must copy its RNA into DNA before it can insert into a human chromosome. A viral enzyme called reverse transcriptase copies viral RNA into DNA. Certain RNA viruses are called retroviruses because they transmit genetic information opposite the usual direction—instead of DNA to RNA to protein, viral RNA is copied into DNA, which may then be copied back into RNA to guide the synthesis of viral proteins. HIV is a retrovirus.

Once viral DNA integrates into the host cell's DNA, it can either remain and replicate along with the host's DNA without causing harm, or it can take over and kill the cell. Activated viral genes direct the host cell to replicate viral DNA and then transcribe and translate it, producing viral proteins. The cell bursts, releasing many newly-assembled viruses. Human chromosomes include viral DNA sequences that are vestiges of past infections, passed on, silently, from ancestors.

Diverse viruses infect all types of organisms. Their genetic material cannot repair itself, so the mutation rate may be high. This is one reason we cannot develop an effective vaccine against the common cold, or norovirus, and why new influenza vaccines must be developed each year.

Influenza viruses have RNA as their genetic material and are of three strains: A, B, and C. Influenza A is the most common and comes from birds, sometimes passing through pigs. Each viral strain has subtypes, based on two types of glycoproteins on their surfaces. Specifically, influenza A has many copies of any of 16 variants of a large surface glycoprotein called a hemagglutinin (HA), and many copies of any of nine variants of another surface glycoprotein called a neuraminidase (NA). The 1918 Spanish flu and the swine flu of 2009 were "H1N1," whereas the "bird flu" of 2004 was "H5N1." Vaccines typically consist of two types of influenza A and one type of the less common influenza B, but efforts are ongoing to develop a "universal" vaccine that targets parts that all flu viruses share.

Some viral diseases have been around for centuries, such as influenza, but others have emerged more recently, such as AIDS and SARS (severe acute respiratory syndrome). It can be difficult to understand how and why a viral pathogen suddenly affects many people. This is what happened with Zika virus (**figure 17A**). It was discovered in a monkey in a forest in Uganda in 1947, with the first human case recorded in 1952. The first outbreak among humans was in the Pacific island of Yap in 2007, but then in 2015 and 2016 the infection appeared and rapidly spread in several countries, mostly in South America. Zika virus has a single strand of RNA as its genetic material and is most closely related to dengue virus. It causes mild symptoms in some adults, but is devastating to a child exposed prenatally (see figure 3.23).

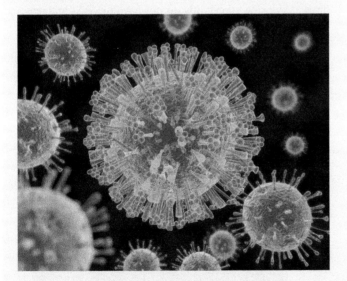

Figure 17A Zika virus. The teratogenic effects of Zika virus infection appeared suddenly in South America in 2015 and 2016. Zika is a flavivirus and is related to the virus that causes dengue fever. Both are transmitted by mosquitoes.
© AuntSpray/Shutterstock

Questions for Discussion

1. How does the structure of a virus differ from that of a cell?

2. How does an RNA virus insert its genetic material into a human chromosome?

3. Why is it difficult to create a vaccine that protects against HIV?

4. What part of an influenza virus is used to manufacture a vaccine?

sickle cell disease. They produce so many antibodies against so many types of donor blood that it is often difficult to determine their blood types by serology.

The Major Histocompatibility Complex

Many proteins on our cell surfaces are encoded by genes that are part of a 6-million-base-long DNA sequence on the short arm of chromosome 6 called the major histocompatibility complex (MHC). This region includes more than 200 genes, and confers about 50 percent of the genetic influence on immunity. MHC genes are classified into three groups based on their functions.

MHC class III genes encode proteins that are in plasma (the liquid portion of blood) and provide nonspecific immune functions. MHC classes I and II genes encode the **human leukocyte antigens (HLAs),** so-named because they were first studied in leukocytes (white blood cells). The HLA proteins link to sugars, forming branchlike glycoproteins that extend from cell surfaces.

The proteins that the class I and II HLA genes encode differ in the types of immune system cells they alert. Some HLA glycoproteins bind bacterial and viral proteins, displaying them like badges to alert other immune system cells. This action, called antigen processing, is often the first step in an immune response. The cell that displays the foreign antigen is called an **antigen-presenting cell. Figure 17.2** shows how a large cell called a **macrophage** displays bacterial antigens. Certain white blood cells called T cells (or T lymphocytes) are also antigen-presenting cells. **Dendritic cells** are antigen-presenting cells found in places where the body contacts the environment, such as in the skin and in the linings of the respiratory and digestive tracts. Dendritic cells signal T cells, initiating an immune response. Dendritic cells and macrophages are known as "sentinel cells" because they alert other components of the immune system to the presence of a pathogen.

A person's HLA "type" identifies all of his or her cells as "self," or as belonging to that individual. The three class I HLA genes encode proteins that vary greatly and are found on all types of cells that have nuclei. These proteins display peptides from within the cells that may come from pathogens. The six class II genes encode proteins found mostly on antigen-presenting cells. In addition to the common HLA markers are more specific markers that distinguish particular tissue types.

An individual's overall HLA type is based on six major HLA genes. So variable are these genes that only 2 in every 20,000 unrelated people match for the six major HLA genes by chance. When transplant physicians attempt to match donor tissue to a potential recipient, they determine how alike the two individuals are in terms of these six genes. Usually at least four of the genes must match for a transplant to have a reasonable chance of success. Before DNA profiling, HLA genotyping was the predominant type of blood test used in forensic and paternity cases to rule out involvement of certain individuals. However, HLA genotyping has become very complex because hundreds of alleles are now known.

More than 100 diseases are strongly associated with inheriting particular HLA types. This is the case for ankylosing spondylitis, which inflames and deforms vertebrae. A person with either of two subtypes of an HLA called B27 is 100 times as likely to develop the condition as someone who lacks either form of the antigen. HLA-associated risks are not absolute. More than 90 percent of people who suffer from ankylosing spondylitis have the B27 antigen, which occurs in only 5 percent of the general population. However, 10 percent of people who have ankylosing spondylitis do *not* have the B27 antigen, and some people who have the antigen never develop the disease.

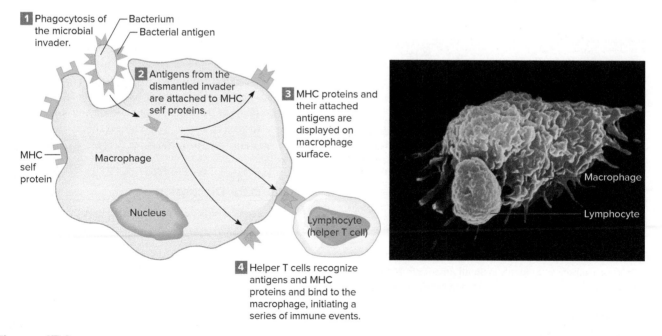

Figure 17.2 Macrophages are antigen-presenting cells. A macrophage engulfs a bacterium, then displays foreign antigens on its surface, which are held in place by major histocompatibility complex (MHC) self proteins. This "antigen presentation" sets into motion many immune reactions, including binding to a lymphocyte. © *CNRI/Science Source*

17.2 The Human Immune System

The human immune system is a network of vessels called lymphatics that transport lymph fluid to bean-shaped structures throughout the body called lymph nodes. Lymph fluid carries white blood cells called lymphocytes and the wandering, scavenging macrophages that capture and degrade bacteria, viruses, and cellular debris. Figure 2.2 shows a macrophage engulfing bacteria. **B cells** and **T cells** are the two major types of lymphocytes. The spleen and thymus gland are also part of the immune system (**figure 17.3**).

The genetic connection to immunity is the proteins required to carry out an immune response. The immune response attacks pathogens, cancer cells, and transplanted cells with two lines of defense—an immediate generalized **innate immune response** and a more specific, slower **adaptive immune response**. These defenses act after various physical barriers block pathogens. **Figure 17.4** summarizes the basic components of the immune system, discussed in the following sections. ("Immune response" is used synonymously with "immunity.")

Physical Barriers and Innate Immunity

Several familiar structures and fluids keep pathogens from entering the body in the innate immune response: unbroken skin, mucous membranes such as the lining inside the mouth, earwax, and waving cilia that push debris and pathogens up and out of the respiratory tract. Most microbes that reach the stomach perish in a vat of churning acid or are flushed out in diarrhea. These physical barriers are nonspecific. That is, they keep out anything foreign, not just particular pathogens.

If a pathogen breaches physical barriers, then innate immunity provides a rapid, broad defense. The term *innate* means that these general defenses are ready to fight infection all the time, without requiring specific stimulation. Many of the aches and pains we experience from an infection are actually due to the innate immune response, not directly to the actions of the pathogens.

Once inside the body, the pathogen encounters the sentinel cells (macrophages and dendritic cells) that are festooned with proteins, called **toll-like receptors**, that span the plasma membrane. Toll-like receptors bind highly conserved proteins, which means that they respond to broad classes of pathogens that have some of the same molecules on their surfaces. Binding to these receptors activates both the innate and adaptive immune responses, through cascades of signal transduction. As a result, macrophages degrade bacteria while displaying some bacterial antigens like flags to further signal the immune response. Virally infected cells may produce virus-fighting biochemicals (interferons) or die by apoptosis, which limits spread of the infection. Interestingly, modern Europeans share with Neanderthals genes that encode toll-like receptors that fight infection by West Nile virus, dengue, and influenza A.

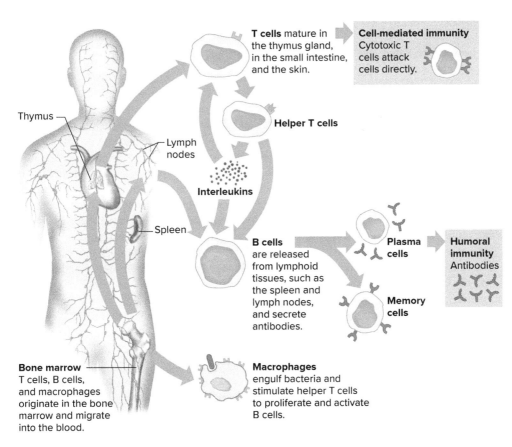

T cells mature in the thymus gland, in the small intestine, and the skin.

Cell-mediated immunity
Cytotoxic T cells attack cells directly.

Helper T cells

Interleukins

Thymus

Lymph nodes

Spleen

B cells are released from lymphoid tissues, such as the spleen and lymph nodes, and secrete antibodies.

Plasma cells

Humoral immunity Antibodies

Memory cells

Bone marrow T cells, B cells, and macrophages originate in the bone marrow and migrate into the blood.

Macrophages engulf bacteria and stimulate helper T cells to proliferate and activate B cells.

Figure 17.3 Cells of the immune system. T cells, B cells, and macrophages build an overall immune response. All three types of cells originate in the bone marrow and circulate in the blood.

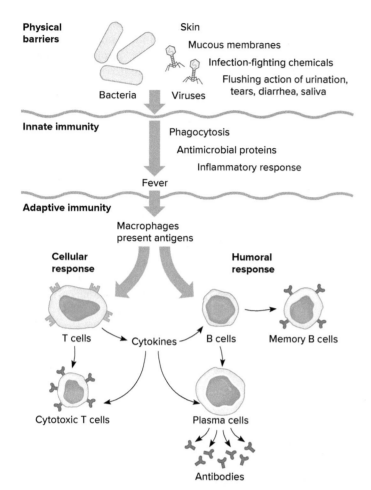

Physical barriers
- Skin
- Mucous membranes
- Infection-fighting chemicals
- Flushing action of urination, tears, diarrhea, saliva

Bacteria   Viruses

Innate immunity
- Phagocytosis
- Antimicrobial proteins
- Inflammatory response
- Fever

Adaptive immunity

Macrophages present antigens

Cellular response — Humoral response

T cells — Cytokines — B cells — Memory B cells

Cytotoxic T cells   Plasma cells

Antibodies

Figure 17.4 **Levels of immune protection.** Pathogens must breach physical barriers, then encounter nonspecific cells and molecules of the innate immune response. If this is ineffective, the adaptive immune response begins: Antigen-presenting cells stimulate T cells to produce cytokines, which activate B cells to divide and differentiate into plasma cells, which secrete antibodies. Once activated, some B cells "remember" the pathogen, allowing faster responses to subsequent encounters.

Table 17.2	Types of Proteins in Innate Immunity		
Protein Class	**Protein**		**Function**
Complement			Puncture bacteria, dismantle viruses, activate mast cells to release histamine, attract phagocytes
Collectins			Recognize how surfaces of bacteria, yeasts, and some viruses differ from the surfaces of human cells
Cytokines	Colony-stimulating factors		Stimulate bone marrow cells to produce lymphocytes
	Interferons		Block viral replication, stimulate macrophages to engulf viruses, stimulate B cells to produce antibodies, attack cancer cells
	Interleukins		Control lymphocyte differentiation and growth, cause fever that accompanies bacterial infection
	Tumor necrosis factor		Stops tumor growth, releases growth factors, stimulates lymphocyte differentiation, dismantles bacterial toxins

A central part of the innate immune response is **inflammation**, which creates a hostile environment for certain types of pathogens at an injury site. Inflammation sends in cells that engulf and destroy pathogens. Such cells are called phagocytes, and their engulfing action is phagocytosis. Certain types of white blood cells and the large, wandering macrophages are phagocytes. Also at the infection site, plasma accumulates, which dilutes toxins and brings in antimicrobial chemicals. Increased blood flow with inflammation warms the area, turning it swollen and red.

In addition to inflammation, three classes of proteins participate in innate immunity (**table 17.2**). These are the complement system, collectins, and cytokines. Mutations in the genes that encode these proteins lower resistance to infection.

The **complement** system consists of plasma proteins that assist, or complement, several other defenses. Some complement proteins puncture bacterial plasma membranes, bursting the cells. Other complement proteins dismantle viruses or trigger release of histamine from **mast cells**, another type of immune system cell that is involved in allergies. Histamine dilates blood vessels, increasing fluid flow to the infected or injured area. Still other complement proteins attract phagocytes to an injury site.

Collectins broadly protect against bacteria, yeasts, and some viruses by detecting slight differences in their surfaces from the surfaces of human cells. Groups of human collectins respond to the surfaces of different pathogens, such as the distinctive sugars on yeast, the linked sugars and lipids of certain bacteria, and the surface features of some viruses.

Cytokines play roles in both innate and adaptive immunity. As part of the innate immune response, cytokines called **interferons** alert other components of the immune system to the presence of cells infected with viruses. These cells are then destroyed, which limits the spread of the viral infection. **Interleukins** are cytokines that cause fever, temporarily triggering a higher body temperature that directly kills some infecting bacteria and viruses. Fever also counters microbial growth indirectly, because higher body temperature reduces the iron level in the blood. Because bacteria and fungi require more iron as the body temperature rises, they cannot survive in a fever-ridden body. Phagocytes also attack more vigorously when the temperature rises. Tumor necrosis factor is another type of cytokine that activates other protective biochemicals, destroys certain bacterial toxins, and attacks cancer cells.

Adaptive Immunity

Adaptive immunity must be stimulated into action. It may take days to respond, compared to minutes for innate immunity. Adaptive immunity is highly specific and directed.

B cells and T cells carry out the adaptive immune response, which has two components. In the **humoral immune response**, B cells produce antibodies after activation by T cells. ("Humor" means fluid; antibodies are carried in fluids.) In the **cellular immune response**, T cells produce cytokines and activate other cells. B and T cells differentiate in the bone marrow and migrate to the lymph nodes, spleen, and thymus gland, as well as circulate in the blood and tissue fluid.

The adaptive arm of the immune system has three basic characteristics. It is *diverse,* vanquishing many types of pathogens. It is *specific,* distinguishing the cells and molecules that cause disease from those that are harmless. The immune system also *remembers,* responding faster to a subsequent encounter with a foreign antigen than it did the first time. The first assault initiates a **primary immune response**. The second assault, based on the system's "memory," is a **secondary immune response**. The immune system's memory is why we get some infections, such as mononucleosis, only once. However, upper respiratory infections and influenza recur because the causative viruses mutate, presenting a slightly different face to our immune systems each season.

The Humoral Immune Response—B Cells and Antibodies

An antibody response begins when an antigen-presenting macrophage activates a T cell. The T cell in turn contacts a B cell that has surface receptors that can bind the type of foreign antigen the macrophage presents. The immune system has so many B cells, each with different combinations of surface antigens, that one or more B cells corresponding to a particular foreign antigen is nearly always available. Turnover of B cells is high. Each day, millions of B cells perish in the lymph nodes and spleen, while millions more form in the bone marrow, each with a unique combination of surface molecules.

Once an activated T cell finds a B cell match, it releases cytokines that stimulate the B cell to divide. When the B cell divides, it yields two types of cells (**figure 17.5**). The first, **plasma cells**, are antibody factories, each secreting 1,000 to 2,000 identical antibodies per second into the bloodstream. Plasma cells live only days. These cells provide the primary immune response. Plasma cells derived from different B cells secrete different antibodies. Each type of antibody corresponds to a specific part of the pathogen, like touching a person on different parts of the body. This multipronged attack is called a polyclonal antibody response (**figure 17.6**). The second type

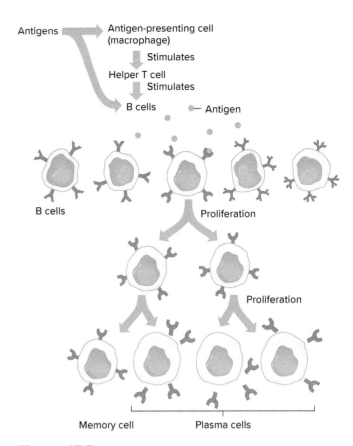

Figure 17.5 **Production of antibodies.** In the humoral immune response, B cells proliferate and mature into antibody-secreting plasma cells. Only the B cell that binds the antigen proliferates; its descendants may develop into memory cells or plasma cells. Plasma cells greatly outnumber memory cells.

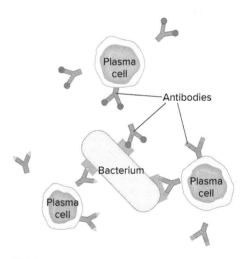

Figure 17.6 **An immune response recognizes many targets.** A humoral immune response is polyclonal, which means that different plasma cells produce antibody proteins that recognize and bind to different features of a foreign cell's surface.

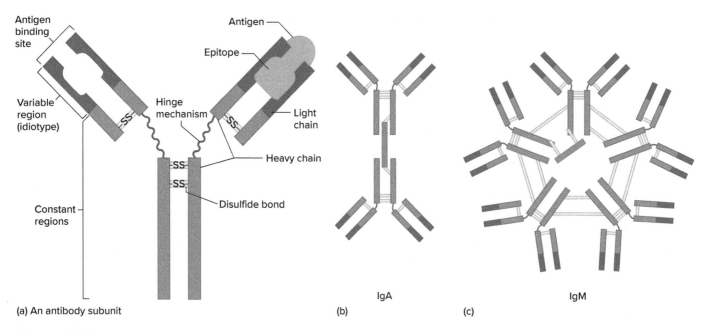

Antigen binding site

Variable region (idiotype)

Constant regions

Antigen

Epitope

Hinge mechanism

Light chain

Heavy chain

Disulfide bond

(a) An antibody subunit

IgA

(b)

IgM

(c)

Figure 17.7 **Antibody structure.** The simplest antibody molecule **(a)** consists of four polypeptide chains, two heavy and two light, joined by disulfide bonds between pairs of sulfur atoms. Part of each polypeptide chain has a constant sequence of amino acids, and the remainder varies. The tops of the Y-shaped molecules form antigen binding sites. **(b)** IgA consists of two Y-shaped subunits, and **(c)** IgM consists of five subunits.

of B cell descendants, **memory cells**, are far fewer and usually dormant. They respond to the foreign antigen faster and with more force should it appear again. This is a secondary immune response.

An antibody molecule is built of several polypeptides and is therefore encoded by several genes. The simplest type of antibody molecule is four polypeptide chains connected by disulfide (sulfur-sulfur) bonds, forming a shape like the letter Y (**figure 17.7**). A large antibody molecule might consist of three, four, or five such Y-shaped subunits.

In a Y-shaped antibody subunit, the two longer polypeptides are called **heavy chains** and the other two **light chains**. The lower portion of each chain is an amino acid sequence that is similar in all antibody molecules, even in different species. These areas are called constant regions, and they provide the *activity* of the antibody. The amino acid sequences of the upper portions of each polypeptide chain, the variable regions, can differ greatly among antibodies. These parts provide the *specificities* of particular antibodies to particular antigens.

Antibodies can bind certain antigens because of the three-dimensional shapes of the tips of the variable regions. These specialized ends are **antigen binding sites**, and the parts that actually contact the antigen are called **idiotypes**. The parts of the antigens that idiotypes bind are **epitopes**. An antibody contorts to form a pocket around the antigen.

Antibodies have several functions. Antibody-antigen binding may inactivate a pathogen or neutralize the toxin it produces. Antibodies can clump pathogens, making them more visible to macrophages, which then destroy them. Antibodies also activate complement, extending the innate immune

response. In some situations, the antibody response can be harmful.

Antibodies are of five major types, distinguished by where they act and what they do (**table 17.3**). (Antibodies are also called immunoglobulins, abbreviated *Ig*.) Different antibody types predominate in different stages of an infection.

Table 17.3	Types of Antibodies	
Type*	**Location**	**Functions**
IgA	Milk, saliva, urine, and tears; respiratory and digestive secretions	Protects against pathogens at points of entry into body
IgD	On B cells in blood	Stimulates B cells to make other types of antibodies, particularly in infants
IgE	In secretions with IgA and in mast cells in tissues	Acts as receptor for antigens that cause mast cells to secrete allergy mediators
IgG	Blood plasma and tissue fluid; passes to fetus	Protects against bacteria, viruses, and toxins, especially in secondary immune response
IgM	Blood plasma	Fights bacteria in primary immune response; includes anti-A and anti-B antibodies of ABO blood groups

*The letters A, D, E, G, and M refer to the specific conformation of heavy chains characteristic of each class of antibody.

How can a human body manufacture seemingly limitless varieties of antibodies from the information in a limited number of antibody genes? This great diversity is possible because parts of different antibody genes combine. During the early development of B cells, sections of their antibody genes move to other chromosomal locations, creating new genetic instructions for antibodies.

The assembly of antibody molecules is like putting together many different outfits from the contents of a closet containing 200 pairs of pants, a drawer containing 15 different shirts, and four belts. Specifically, each variable region of a heavy chain and a light chain consists of three sections, called V (for variable), D (for diversity), and J (for joining). The V, D, and J genes—several of each—for the heavy chains are on chromosome 14, and the corresponding genes for the light chains are on chromosomes 2 and 22. C (constant) genes encode the constant regions of each heavy and light chain. A promoter sequence precedes the V genes and an enhancer sequence precedes the C genes. These control sequences oversee the mixing and matching of the V, D, and J genes. **Figure 17.8** shows how the genetic instructions for the antibody parts are combined in different ways to encode the heavy and light polypeptide chains.

Enzymes cut and paste the pieces of antibody gene parts. The number of combinations of parts that can be used to build antibodies is so great that virtually any antigen that a person with a healthy immune system might encounter will elicit an immune response.

The Cellular Immune Response—T Cells and Cytokines

T cells provide the cellular immune response. It is called "cellular" because the T cells themselves travel to where they act, unlike B cells, which secrete antibodies into the bloodstream. T cells descend from stem cells in the bone marrow, then travel to the thymus gland ("T" refers to thymus). As the immature T cells, called thymocytes, migrate toward the interior of the thymus, they display diverse cell surface receptors. Then selection happens. As the wandering thymocytes touch lining cells in the gland that are studded with "self" antigens, thymocytes that do not attack the lining cells begin maturing into T cells, whereas those that harm the lining cells die by apoptosis—in great numbers. Gradually, T cells-to-be that recognize self cells persist, while those that harm body cells are destroyed.

Several types of T cells are distinguished by the types and patterns of receptors on their surfaces, and by their functions. Helper T cells have many functions: They recognize foreign antigens on macrophages, stimulate B cells to produce antibodies, secrete cytokines, and activate another type of T cell called a cytotoxic T cell (also called a killer T cell). Regulatory T cells help to suppress an immune response when it is no longer required. The cytokines that helper T cells secrete include interleukins, interferons, tumor necrosis factor, and colony-stimulating factors, which stimulate white blood cells in bone marrow to mature. Cytokines interact with and signal each other, sometimes in complex cascades.

Distinctive surfaces distinguish subsets of helper T cells. Certain antigens called cluster-of-differentiation antigens, or CD antigens, enable T cells to recognize foreign antigens displayed on macrophages. One such cell type, called a CD4 helper T cell, is an early target of HIV. Considering the critical role helper T cells play in coordinating immunity, it is little wonder that HIV infection ultimately topples the entire system.

Cytotoxic T cells do not have CD4 receptors but have CD8 receptors. These cells attack virally infected and cancerous cells by attaching to them and releasing chemicals. They do this by linking two surface peptides to form structures called T cell receptors that bind foreign antigens. When a cytotoxic T cell encounters a nonself cell—a cancer cell, for example—the T cell receptors draw the two cells into physical contact. The T cell then releases a protein called perforin, which pierces the cancer cell's plasma membrane, killing it (**figure 17.9**). Cytotoxic T cell receptors also attract cells that are covered with certain viruses, destroying the cells before the

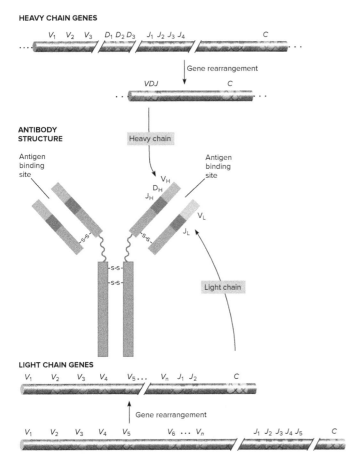

Figure 17.8 **Antibody diversity.** The human immune system can produce antibodies to millions of possible antigens because each polypeptide is encoded by more than one gene. That is, the many components of antibodies can combine in many ways.

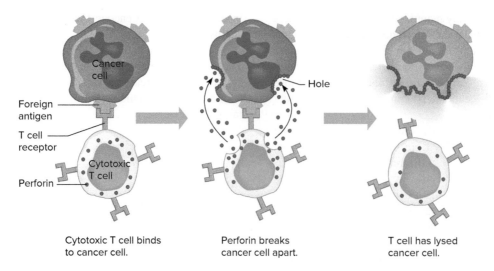

Cytotoxic T cell binds to cancer cell.

Perforin breaks cancer cell apart.

T cell has lysed cancer cell.

Figure 17.9 **Death of a cancer cell.** A cytotoxic T cell binds to a cancer cell and injects perforin, a protein that pierces (lyses) the cancer cell's plasma membrane. The cancer cell dies, leaving debris that macrophages clear away.

viruses on them can enter, replicate, and spread the infection. Cytotoxic T cells continually monitor body cells, recognizing and eliminating virally infected cells and tumor cells. **Table 17.4** lists and defines types of immune system cells.

Table 17.4	Types of Immune System Cells
Cell Type	**Function**
Macrophage	Presents antigens
	Performs phagocytosis
Dendritic cell	Presents antigens
Mast cell	Releases histamine in inflammation
	Releases allergy mediators
B cell	Matures into antibody-producing plasma cell or into memory cell
T cell	
Helper	Recognizes nonself antigens presented on macrophages
	Stimulates B cells to produce antibodies
	Secretes cytokines
	Activates cytotoxic T cells
Cytotoxic	Attacks cancer cells and cells infected with viruses upon recognizing antigens
Neutrophil	Attacks bacteria
Natural killer	Attacks cancer cells and cells infected with viruses without recognizing antigens; activates other white blood cells
Suppressor	Inhibits antibody production

Key Concepts Questions 17.2

1. What are the three major parts of the immune response?
2. How does innate immunity differ from adaptive immunity?
3. Describe antibody structure and function.
4. What do helper T cells do?
5. What do cytotoxic T cells do?

17.3 Abnormal Immunity

The immune system continually adapts to environmental change. Because the immune response is so diverse, its breakdown affects health in many ways. Immune system malfunction may be inherited or acquired, and immunity may be too weak, too strong, or misdirected. Abnormal immune responses may be multifactorial, with several genes contributing to susceptibility to infection, or caused by mutation in a single gene.

Inherited Immune Deficiencies

The more than 20 types of inherited immune deficiencies affect innate and/or adaptive immunity (**table 17.5**). These conditions can arise in several ways.

In chronic granulomatous disease, white blood cells called neutrophils engulf bacteria, but, due to deficiency of an enzyme complex (NADPH oxidase), the neutrophils cannot produce the activated oxygen compounds that would kill the bacteria. Mutations in any of five genes disrupt the enzyme complex. One gene is on the X chromosome and causes an X-linked recessive form of the condition, and the other four genes are on autosomes, causing autosomal recessive forms. Antibiotics and gamma interferon are used to prevent bacterial

Table 17.5 — Inherited Immune Deficiencies

Table 17.5 — Inherited Immune Deficiencies

Disease	Inheritance	Defect
Chronic granulomatous disease	ar, AD, XIr	Abnormal neutrophils cannot kill engulfed bacteria
Immune defect due to absence of thymus	ar	No thymus, no T cells
Neutrophil immunodeficiency syndrome	ar	Deficiencies of T cells, B cells, and neutrophils
Severe combined immune deficiency (SCID)		
Adenosine deaminase deficiency	ar	No T or B cells
Adenosine deaminase deficiency with sensitivity to ionizing radiation	ar	No T, B, or natural killer cells
IL-2 receptor mutation	XIr	No T, B, or natural killer cells
X-linked lymphoproliferative disease	XIr	Absence of protein that enables T cells to bind B cells
X1	XIr	Abnormal interleukin-2

Note: ar = autosomal recessive; AD = autosomal dominant; XIr = X-linked recessive.

infections in these patients, and a bone marrow or an umbilical cord stem cell transplant can cure the disease.

Mutations in genes that encode cytokines or T cell receptors impair cellular immunity, which primarily targets viruses and cancer cells. Because T cells activate the B cells that manufacture antibodies, abnormal cellular immunity (T cell function) disrupts humoral immunity (B cell function). Mutations in the genes that encode antibody segments, that control how the segments join, or that direct maturation of B cells mostly impair immunity against bacterial infection. Inherited immune deficiency can also result from defective B cells, which usually increases vulnerability to certain bacterial infections.

Severe combined immune deficiencies (SCIDs) are called "combined" because they affect both humoral and cellular immunity. About half of SCID cases are X-linked. In a less severe form, the individual lacks B cells but has some T cells. Before antibiotic drugs became available, children with this form of SCID died before age 10 of overwhelming bacterial infection. In a more severe form of X-linked SCID, lack of B and T cells causes death by 18 months of age, usually of severe and diverse infections.

A young man named David Vetter taught the world about the difficulty of life without immunity years before AIDS arrived. David had an X-linked recessive form of SCID, called SCID-X1, that caused him to be born without a thymus gland. His T cells could not mature and activate B cells, leaving him defenseless in a germ-filled world. Born in Texas in 1971, David spent his short life in a large vinyl "bubble" enclosure, awaiting a treatment that never came (**figure 17.10**). As he reached adolescence, David wanted to leave his bubble. He did, and received a bone marrow transplant from his sister. Sadly, her bone marrow contained Epstein-Barr virus. Her healthy immune system could fight the virus so she had no symptoms, but the virus caused lymphoma, a cancer of the immune system, in David. He

Figure 17.10 David Vetter, the original "bubble boy," had severe combined immune deficiency (SCID) X1. Because his T cells could not mature, he was virtually defenseless against infection. Gene therapy can now treat this disease.
© Bettmann/Corbis

died in just a few weeks. Today gene therapy can treat SCID-X1 as well as another form of SCID called adenosine deaminase deficiency (see figure 20.7).

Acquired Immune Deficiency Syndrome

AIDS is not inherited, but acquired by infection with HIV, a virus that gradually shuts down the immune system. The effect of HIV on a human body is especially astounding because the virus is so simple. Its genome is a millionth the size of ours, and its nine genes, consisting of about 9,000 RNA bases, encode only 15 proteins! But HIV affects more than 200 human proteins as it invades the immune system.

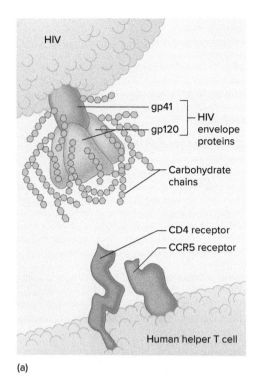

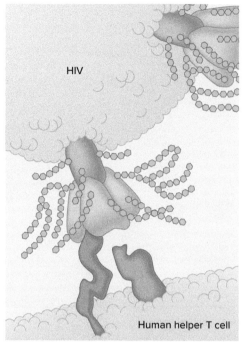

HIV

gp41 ⎤ HIV
gp120 ⎦ envelope
proteins

Carbohydrate
chains

CD4 receptor
CCR5 receptor

Human helper T cell

(a)

HIV

Human helper T cell

(b)

Figure 17.11 **HIV binds to a helper T cell. (a)** The part
of HIV that binds to helper T cells is gp120 (gp stands for
glycoprotein). **(b)** The carbohydrate chains that shield the
protein part of gp120 move aside as they approach the cell
surface, allowing the viral molecule to bind to a CD4 receptor.
After also binding to the CCR5 receptor, the viral envelope fuses
with the plasma membrane and HIV enters. (The size of HIV is
greatly exaggerated.) The man described in the chapter opener
received stem cells from a donor whose cells lack CCR5.

HIV infection begins as the virus enters macrophages,
impairing this first line of defense. In these cells, and later in
helper T cells, the virus adheres with its surface protein, called
gp120, to two coreceptors on the host cell surface, CD4 and
CCR5 (**figure 17.11**). (CCR5 is the glycoprotein altered by
mutation that was discussed in the chapter opener.) Another
glycoprotein, gp41, anchors gp120 molecules into the viral
envelope. When the virus binds both coreceptors, virus and cell
surface contort in a way that enables the virus to enter the cell.
Once in the cell, reverse transcriptase copies the viral RNA
into DNA, which replicates to form a DNA double helix. The
viral DNA then enters the nucleus and inserts into a chromosome. As the viral genes are transcribed and translated, the cell
fills with viral pieces, which are assembled into complete new
viral particles that eventually bud from the cell (**figure 17.12**).

Once helper T cells start to die at a high rate, bacterial
infections begin, because B cells aren't activated to produce
antibodies. Much later in infection, HIV variants arise that can
bind to a receptor called CXCR4 on cytotoxic T cells, killing
them. Loss of these cells renders the body vulnerable to viral
infections and cancer.

HIV replicates quickly, changes quickly, and can hide.
The virus mutates easily because it cannot repair replication

errors, which are frequent—1 per
every 5,000 or so bases—because
of the "sloppiness" in copying viral
RNA into DNA. The immune system cannot keep up; antibodies
against one viral variant are useless
against the next. For several years,
the bone marrow produces 2 billion
new T and B cells a day. A million
to a billion new HIV particles bud
daily from infected cells.

So genetically diverse is the
population of HIV in a human host
that, within days of the initial infection, variants arise that resist the
drugs used to treat AIDS. HIV's
changeable nature is why combining drugs that work in different
ways is the most effective strategy

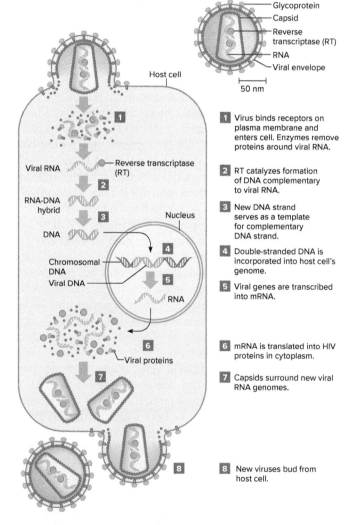

Glycoprotein
Capsid
Reverse
transcriptase (RT)
RNA
Viral envelope

50 nm

Host cell

Viral RNA

Reverse transcriptase
(RT)

RNA-DNA
hybrid

DNA

Nucleus

Chromosomal
DNA
Viral DNA

RNA

Viral proteins

1 Virus binds receptors on
plasma membrane and
enters cell. Enzymes remove
proteins around viral RNA.

2 RT catalyzes formation
of DNA complementary
to viral RNA.

3 New DNA strand
serves as a template
for complementary
DNA strand.

4 Double-stranded DNA is
incorporated into host cell's
genome.

5 Viral genes are transcribed
into mRNA.

6 mRNA is translated into HIV
proteins in cytoplasm.

7 Capsids surround new viral
RNA genomes.

8 New viruses bud from
host cell.

Figure 17.12 **How HIV infects.** HIV integrates into the
host chromosome, then commandeers transcription and
translation, ultimately producing more virus particles.

Table 17.6	Anti-HIV Drugs
Drug Type	**Mechanism**
Reverse transcriptase inhibitor	Blocks copying of viral RNA into DNA
Protease inhibitor	Blocks shortening of certain viral proteins
Fusion inhibitor	Blocks ability of HIV to bind a cell
Entry inhibitor	Blocks ability of HIV to enter a cell

to slow the disease into a chronic, lifelong, but treatable illness, instead of a killer (**table 17.6**). The several classes of anti-HIV drugs work at different points of infection: blocking binding or entry of the virus into T cells, replicating viral genetic material, and processing viral proteins.

Clues to developing new drugs to treat HIV infection come from people at high risk who resist infection. Researchers identified variants of four receptors or the molecules that bind to them that block HIV from entering cells by looking at the DNA of people who had unprotected sex with many partners but who never became infected. Some of them were homozygous recessive for the 32-base deletion in the *CCR5* gene described in the chapter opener. The abnormal CCR5 coreceptors are too stunted to reach the infected cell's surface, so HIV has nowhere to dock. Heterozygotes, with one copy of the deletion, can become infected, but they remain healthy for several years longer than people who do not have the deletion. Curiously, the same mutation may have enabled people to survive various plagues in Europe during the Middle Ages. Apparently, more than one pathogen uses the CCR5 entryway into human cells.

Autoimmunity

In **autoimmunity**, the immune system produces antibodies that attack the body's own cells. These antibodies are called **autoantibodies**. About 5 percent of the population has an autoimmune disorder. The signs and symptoms resulting from autoimmune disorders correspond to the cell types under attack (**table 17.7**).

Most autoimmune conditions are not single-gene diseases. However, the fact that different autoimmune disorders affect members of the same family, and may respond to the same drugs, suggests that these conditions stem from shared susceptibilities. For example, in autoimmune polyendocrinopathy syndrome type I, caused by a mutation in a single gene on chromosome 21, autoantibodies attack endocrine glands in a sequence, so that different members of a family may have very different symptoms. However, autoimmunity is usually polygenic. Dozens of genes are each associated with more than one autoimmune disorder. An autoimmune disorder may result from the actions of variants of several genes that each contributes to susceptibility, perhaps in the presence of a specific environmental trigger such as a food. This is the case for the digestive condition Crohn's disease, which has been associated with 32 genome regions and a few genes that have variants that increase risk.

Some of the more common autoimmune disorders may arise in several ways when parts of the immune response overact. This is the case for systemic lupus erythematosus, better known as "lupus." A butterfly-shaped rash on the cheeks is characteristic, but the condition also produces autoantibodies that affect the connective tissue of many organs. These are the kidneys, joints, lungs, brain, spinal cord, and the heart and blood vessels. A person may need dialysis when the kidneys are involved, blood pressure medication to counter increasing pressure in the lungs, and drugs to minimize buildup of fatty deposits on interior artery walls. Lupus can also cause strokes, memory loss, fever, seizures, headache, and psychosis.

Table 17.7	Autoimmune Disorders	
Disorder	**Symptoms**	**Autoantibodies Against**
Diabetes mellitus (type 1)	Thirst, hunger, weakness, weight loss	Pancreatic beta cells
Graves disease	Restlessness, weight loss, irritability, increased heart rate and blood pressure	Thyroid gland cells
Hemolytic anemia	Fatigue, weakness	Red blood cells
Multiple sclerosis	Weakness, poor coordination, failing vision, disturbed speech	Myelin in the white matter of the central nervous system
Myasthenia gravis	Muscle weakness	Neurotransmitter receptors on skeletal muscle cells
Rheumatic fever	Weakness, shortness of breath	Heart valve cells
Rheumatoid arthritis	Joint pain and deformity	Cells lining joints
Systemic lupus erythematosus	Red facial rash, fever, weakness, joint pain	Connective tissue
Ulcerative colitis	Lower abdominal pain	Colon cells

Lupus involves several aspects of the immune response, including cell surface characteristics, secretion of interferons, production of autoantibodies, activation of B and T cells, antigen presentation, adhesion of immune system cells to blood vessel linings, inflammation, removal of complexes of immune cells and foreign antigens, and cytokine production. Variants of at least 10 genes can predispose a person to develop lupus.

Autoimmunity may arise in several ways:

- A virus replicating in a cell incorporates proteins from the cell's surface onto its own. When the immune system "learns" the surface of the virus to destroy it, it also learns to attack human cells that normally bear the protein.
- Some cells that should have died in the thymus escape the massive die-off, persisting to attack "self" tissue later on.
- A nonself antigen coincidentally resembles a self antigen, and the immune system attacks both. In rheumatic fever, for example, antigens on heart valve cells resemble those on *Streptococcus* bacteria; antibodies produced to fight a strep throat also attack the heart valve cells.

- If X inactivation is skewed, a female may have a few cells that express the X chromosome genes of one parent (see figure 6.10). The immune system may respond to these cells as foreign if they have surface antigens that are not also on the majority of cells. Skewed X inactivation may explain why some autoimmune diseases are much more common in females.

Clinical Connection 17.2 highlights a special situation in which two immune systems must coexist—pregnancy.

Allergies

An **allergy** is an immune system response to a substance, called an **allergen**, that does not actually present a threat. Many allergens are particles small enough to be carried in the air and enter a person's respiratory tract. The size of the allergen may determine the type of allergy. For example, grass pollen is large and remains in the upper respiratory tract, where it causes hay fever. But allergens from house dust mites, cat dander, and cockroaches are small enough to infiltrate the lungs, triggering asthma. Asthma is a chronic disease in which contractions of the airways, inflammation, and accumulation of mucus block air flow.

Clinical Connection 17.2

A Special Immunological Relationship: Mother-to-Be and Fetus

The immune system recognizes "self" cell surfaces and protects the body from foreign, "nonself" cells and molecules. This is quite helpful when the nonself triggers are parts of infecting bacteria, but what tempers the immune system of a pregnant woman to accept cells from her fetus? Half of a fetal genome comes from the father, and so fetal cell surfaces likely include some antigens from him that would be "foreign" to the mother-to-be. Similarly, some of her antigens might be foreign to the fetus. Yet pregnant woman and fetus routinely swap cells. Most women retain fetal cells after pregnancy. The presence of cell populations from more than one individual in one body is called microchimerism ("little mosaic"). We do not completely understand how immune tolerance evolved between pregnant woman and fetus, but following are three examples of the immunological "crosstalk" between the two.

T Regs

Samples of lymph nodes from fetuses indicate that up to 1 percent of the cells are maternal. The woman's cells stimulate the fetal immune system to produce "regulatory T cells," called "T regs," which dampen the fetal immune response. The maternal immune system similarly produces T regs that inhibit response to fetal cells. In one experiment, fetal lymph node samples did not react against cells from the pregnant woman unless the fetal regulatory T cells were removed. Children retain these cells for several years. It may be possible to stimulate production of T regs later in life to help a recipient's body accept an organ transplant.

Systemic Sclerosis

In systemic sclerosis (also called scleroderma), the skin hardens into an armorlike texture, described by one patient as "the body turning to stone" (**figure 17B**). Symptoms usually begin in middle age, and the condition affects mostly women. It was long thought

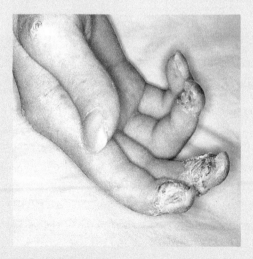

Figure 17B An autoimmune disease—maybe. Some cases of systemic sclerosis appear to be caused by a long-delayed immune response to cells retained from a fetus decades earlier. *Courtesy of Dr. Maureen Mayes*

(Continued)

to be autoimmune, but discovery of Y chromosomes in skin cells from patients who are mothers of sons revealed a very different source of the illness—lingering cells from a fetus. Cells from long-ago female fetuses can presumably have the same effect but cannot be distinguished from cells of the mother on the basis of a sex chromosome check.

The degree of genetic difference between a mother and a son may play a role in development of systemic sclerosis. Mothers who have the condition tend to have cell surfaces that are more similar to those of their sons than mothers who do not have the condition. Perhaps the similarity of cell surfaces enabled the fetal cells to escape destruction by the mother's immune system.

Rh Incompatibility

"Rh," the rhesus factor discovered in rhesus monkeys, defines a blood group. A person is Rh⁺ if red blood cells have a surface molecule called the RhD antigen. Rh type is important when an Rh⁺ man and an Rh⁻ woman conceive a child who is Rh⁺ (**figure 17C**). The woman's immune system manufactures antibodies against the few fetal cells that enter her bloodstream. Not enough antibodies form to harm the fetus that sets off the reaction, but the number of antibodies continues to increase. If she carries another Rh⁺ fetus, the antibodies can attack the fetal blood supply, causing potentially fatal hemolytic disease of the fetus and newborn. It can be treated at birth with a transfusion of Rh⁻ blood.

Fortunately, natural and medical protections make this complication rare today. If a woman's ABO blood type is O and the fetus is A or B, her anti-A or anti-B antibodies attack the fetal cells in her circulation before her immune system produces anti-Rh antibodies. Also, if a pregnant woman alerts her health care provider to a potential incompatibility, she can be given RhoGAM, which is antibody against the Rh antigen. It shields fetal cells so her system does not manufacture the harmful antibodies. When she becomes pregnant again, fetal DNA in her circulation can be tested to see if it is Rh⁻ or Rh⁺. If the second fetus is Rh⁻, she does not need RhoGAM. A first Rh⁺ fetus developing in an Rh⁻ mother can be affected if her blood has been exposed to Rh⁺ cells in amniocentesis, a blood transfusion, an ectopic (tubal) pregnancy, a miscarriage, or pregnancy termination.

Questions for Discussion

1. Explain why a maternal immune system might attack a fetus.
2. How do regulatory T cells help a pregnant woman's body accept the fetus?
3. Explain how systemic sclerosis may result from retained fetal cells in the maternal body.
4. What is Rh incompatibility?

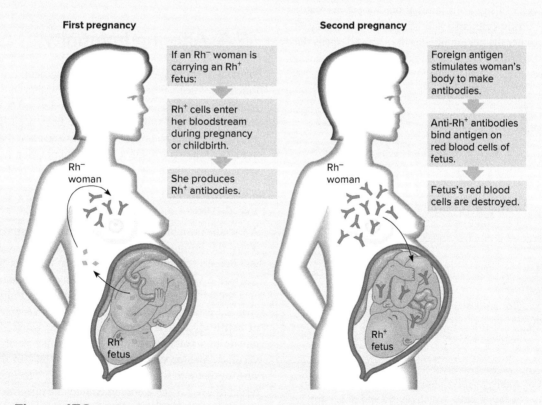

First pregnancy

If an Rh⁻ woman is carrying an Rh⁺ fetus:

Rh⁺ cells enter her bloodstream during pregnancy or childbirth.

She produces Rh⁺ antibodies.

Rh⁻ woman

Rh⁺ fetus

Second pregnancy

Foreign antigen stimulates woman's body to make antibodies.

Anti-Rh⁺ antibodies bind antigen on red blood cells of fetus.

Fetus's red blood cells are destroyed.

Rh⁻ woman

Rh⁺ fetus

Figure 17C **Rh incompatibility.** Fetal cells entering the pregnant woman's bloodstream can stimulate her immune system to make anti-Rh antibodies, if the fetus is Rh⁺ and she is Rh⁻. A drug called RhoGAM prevents attacks on subsequent fetuses.

Both humoral and cellular immune responses take part in an allergic reaction (**figure 17.13**). Antibodies of class IgE bind to mast cells, sending signals that open the mast cells, which release allergy mediators such as histamine and heparin. Allergy mediators cause inflammation, with symptoms that may include runny eyes from hay fever, narrowed airways from asthma, rashes, or the overwhelming body-wide allergic reaction called anaphylactic shock. Allergens also activate a class of helper T cells that produce cytokines.

A mutation in one specific gene may cause several common allergies. Half the normal amount of a skin protein called filaggrin can lead to development of atopic dermatitis (a type of eczema), asthma, peanut allergy, and hay fever (**figure 17.14**). Filaggrin is a gigantic protein that binds to the keratin proteins that form most of the outermost skin layer, the epidermis. Filaggrin normally breaks down, releasing amino acids that rise to the top of the skin and keep it moisturized. The epidermis forms a barrier that keeps out irritants, pathogens, and allergens.

People with the rare inherited disease ichthyosis vulgaris have two mutations in the gene for filaggrin, and experience severe skin flaking. Many more individuals—possibly 1 in 10 of us—has a mutation that causes ichthyosis that is so mild that we just treat it with skin lotion. When researchers realized that people with either form of ichthyosis very often also have the itchy red inflamed skin of atopic dermatitis, they realized there could be a connection—when the epidermis is cracked, allergens enter and reach deeper skin layers, where they activate dendritic cells to signal inflammation. Atopic dermatitis results. The dendritic cells also activate immune memory, so that when years later the person inhales the same allergens that once crossed the broken skin, an immune response ensues in the airways, causing a form of asthma due to a skin barrier

deficiency. Perhaps treating breaks in the skin early in life can prevent atopic dermatitis, asthma, and other allergies.

Inherited allergies can be rather strange. In familial vibratory urticaria, hives arise from vibration, such as from clapping hands, a bumpy car ride, running, and even toweling off after a shower. A single mutation in a gene called *ADGRE2* causes the condition. Mast cells orchestrate vibratory urticaria by releasing histamine, as they do when hives arise in response to other stimuli. Blood levels of histamine spike as the hives arise and diminish within an hour. ADGRE2 protein has two subunits, one embedded in the plasma membrane and the other extending from the cell surface. In people with familial vibratory urticaria, the two subunits are not close enough to each other to enable the mast cell to resist releasing histamine upon even light mechanical stimulation.

Key Concepts Questions 17.3

1. What are some causes of inherited immune deficiencies?

2. Describe the effects of HIV on the human immune system.

3. Explain how the immune system malfunctions in autoimmune disorders and in allergic reactions.

17.4 Altering Immunity

Medical technologies can alter or augment immune system functions. Vaccines trick the immune system into acting early. Antibiotic drugs, which are substances derived from organisms such as fungi and soil bacteria, have been used for decades to assist an immune response. Cytokines and altered antibodies are used as drugs. Transplants require suppression of the immune system so that the body will accept a nonself replacement part.

Vaccines

A **vaccine** is an inactive or partial form of a pathogen that stimulates the immune system to alert B cells to produce antibodies. People thought about developing vaccines a millennium ago (see A Glimpse of History). When a vaccinated person then encounters the natural pathogen, a secondary immune response ensues, even before symptoms arise. Vaccines consisting of entire viruses or bacteria can, rarely, cause illness if they mutate to a pathogenic form. This was a risk of the smallpox vaccine. A safer vaccine uses only the part of the pathogen's surface that elicits an immune response. Vaccines against different illnesses can be combined into one injection, or the genes encoding antigens from several pathogens can be inserted into a harmless virus and delivered as a "super vaccine."

Most vaccines are injections. New delivery methods include nasal sprays (flu vaccine) and genetically modified fruits and vegetables. A banana as a vaccine makes sense in theory, but in

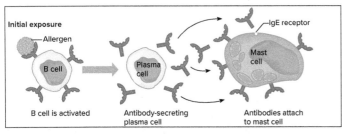

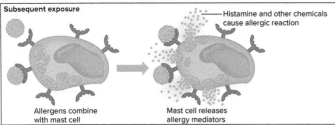

Figure 17.13 Allergy. In an allergic reaction, an allergen such as pollen activates B cells, which divide and give rise to antibody-secreting plasma cells. The antibodies attach to mast cells. When the person encounters the allergen again, the allergen particles combine with the antibodies on the mast cells, which then burst, releasing the chemicals that cause itchy eyes and a runny nose.

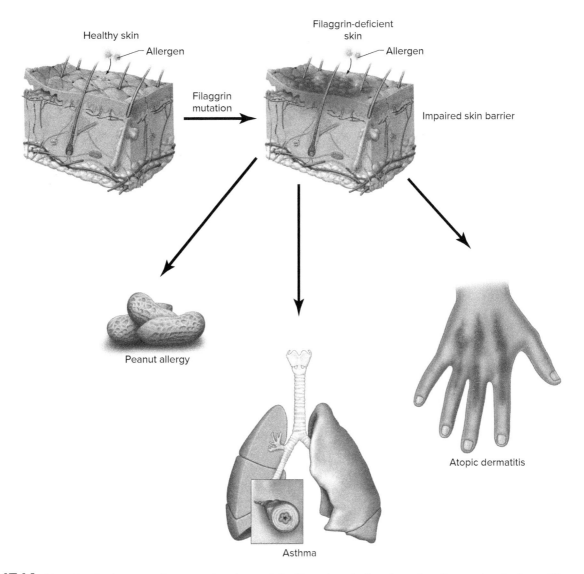

Figure 17.14 **Mutation in the gene that encodes the protein filaggrin sets the stage for allergy.** Atopic dermatitis, peanut allergy, asthma, and hay fever may result from lack of filaggrin protein to protect the skin.

practice it is difficult to obtain a uniform product. Edible plants are grown from cells that are given genes from pathogens that encode the antigens that evoke an immune response. When the plant vaccine is eaten, the foreign antigens stimulate phagocytes beneath the small intestinal lining to "present" the antigens to nearby T cells. From here, the antigens go to the bloodstream, where they stimulate B cells to divide to yield plasma cells that produce antibodies that coat the small intestinal lining, protecting against pathogens in food. Potatoes and tomatoes have also been genetically modified to function as vaccines. Edible vaccines are still experimental.

Whatever the form of vaccine, it is important that a substantial proportion of a population be vaccinated to control an infectious disease. This establishes "herd immunity"—that is, if unvaccinated people are rare, then if the pathogen appears, it does not spread, because so many people are protected. If the population includes unvaccinated individuals who come into contact with each other, the disease can spread.

An infectious disease such as flu that is mild or harmless to most people can kill a person who has a compromised immune system. Diseases that had been nearly eradicated thanks to vaccination have returned in areas where people either refuse to have their children vaccinated, or cannot due to war. Pertussis, measles, and polio have returned to certain parts of the world due to lapses in vaccination.

Immunotherapy

Immunotherapy amplifies or redirects the immune response. It originated in the nineteenth century to treat disease. Today, a few immunotherapies are in use, with more in clinical trials.

Monoclonal Antibodies Boost Humoral Immunity

When a B cell recognizes a single foreign antigen, it manufactures a single, or monoclonal, type of antibody. A large amount

Vaccine technology dates back to the eleventh century in China. Because people saw that those who recovered from smallpox never got it again, they crushed scabs from pox into a powder that they inhaled or rubbed into pricked skin. In 1796, the wife of a British ambassador to Turkey witnessed the Chinese method of vaccination and mentioned it to English country physician Edward Jenner. Intrigued, Jenner was vaccinated the Chinese way, and then thought of a different approach.

It was widely known that people who milked cows contracted a mild illness called cowpox but did not get smallpox. The cows became ill from infected horses. Because the virus seemed to jump from one species to another, Jenner wondered whether exposing a healthy person to cowpox lesions might protect against smallpox. A slightly different virus causes cowpox, but Jenner's approach worked, leading to development of the first vaccine (the word comes from the Latin *vacca,* for "cow").

Jenner tried his first vaccine on a volunteer, 8-year-old James Phipps. Jenner dipped a needle in pus oozing from a cowpox sore on a milkmaid named Sarah Nelmes, then scratched the boy's arm with it. He then exposed the boy to people with smallpox. Young James never became ill. Eventually, improved versions of Jenner's smallpox vaccine eradicated a disease that once killed millions (**figure 17D**). By the 1970s, vaccination became unnecessary. Several nations have resumed smallpox vaccination in case the virus is ever used as a bioweapon. Smallpox has not naturally infected a human since 1977. Because many doctors are unfamiliar with smallpox, and people are no longer vaccinated, an outbreak could be a major health disaster.

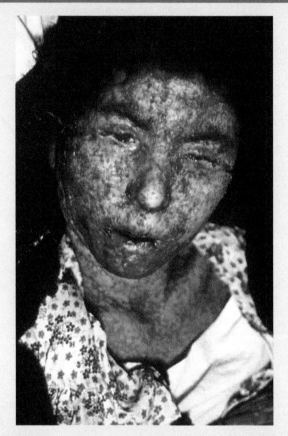

Figure 17D Smallpox. This woman had such a severe case of smallpox that the lesions joined. *CDC/Carl Flint, Armed Forces Institute of Pathology*

of a single antibody type could target a particular pathogen or cancer cell because of the antibody's great specificity.

In 1975, British researchers Cesar Milstein and George Köhler devised monoclonal antibody (MAb) technology, which mass-produces a single B cell, preserving its specificity and amplifying its antibody type. First, they injected a mouse with a sheep's red blood cells (**figure 17.15**). They then isolated a single B cell from the mouse's spleen and fused it with a cancerous white blood cell from a mouse. The fused cell, called a hybridoma, had a valuable pair of talents. Like the B cell, it produced large numbers of a single antibody type. Like the cancer cell, it divided continuously.

MAbs are used in basic research, veterinary and human health care, agriculture, forestry, and forensics. They can diagnose everything from strep throat to turf grass disease. In a home pregnancy test, a woman places drops of her urine onto a paper strip containing a MAb that binds hCG, the "pregnancy" hormone. The color changes if the MAb binds its target. In cancer diagnosis, if a MAb attached to a fluorescent dye and injected into a patient or applied to a sample of tissue or body fluid binds its target—an antigen found mostly or only on cancer cells—fluorescence indicates disease. MAbs linked to radioactive isotopes or to drugs deliver treatment to cancer cells. The MAb drug trastuzumab (Herceptin) blocks receptors on certain breast cancer cell surfaces, preventing them from receiving signals to divide. MAbs used in humans are made to more closely resemble natural human antibodies than the original mouse-derived products, which caused allergic reactions. More than 30 MAb-based treatments are available for human conditions.

Cytokines Boost Cellular Immunity

As coordinators of immunity, cytokines are used to treat a variety of conditions. However, it has been difficult to develop these body chemicals into drugs because they act only for short periods. They must be delivered precisely where they are needed, or overdose or side effects can occur.

Interferons (IFs) were the first cytokines tested on a large scale, and today are used to treat cancer, genital warts, multiple sclerosis, and some other conditions. Interleukin-2 (IL-2)

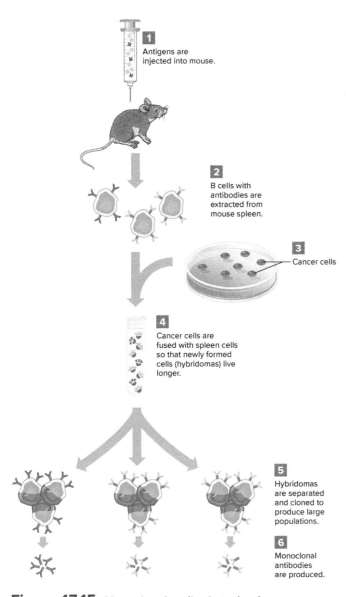

Figure 17.15 **Monoclonal antibody technology.**
Monoclonal antibodies are pure preparations of a single antibody type that recognize a single antigen type. They are useful in diagnosing and treating disease because of their specificity. *Source: CDC/ Carl Flint; Armed Forces Institute of Pathology*

1. Antigens are injected into mouse.

2. B cells with antibodies are extracted from mouse spleen.

3. Cancer cells

4. Cancer cells are fused with spleen cells so that newly formed cells (hybridomas) live longer.

5. Hybridomas are separated and cloned to produce large populations.

6. Monoclonal antibodies are produced.

is a cytokine that is administered intravenously to treat kidney cancer recurrence. Colony-stimulating factors, which cause immature white blood cells to mature and differentiate, are used to boost white blood cell levels in people with suppressed immune systems, such as individuals with AIDS or those receiving cancer chemotherapy. Treatment with these factors enables a patient to withstand higher doses of a conventional drug.

Because excess of another cytokine, tumor necrosis factor (TNF), underlies some diseases, blocking its activity treats some conditions. One drug consists of part of a receptor for TNF. Taking the drug prevents TNF from binding to cells that line joints, relieving rheumatoid arthritis pain. Excess TNF in rheumatoid arthritis prevents cells lining the joints from secreting lubricating fluid.

Transplants

When a car breaks down, replacing the damaged part often fixes the trouble. The same is sometimes true for the human body. The heart, kidneys, liver, lungs, corneas, pancreas, skin, and bone marrow are routinely transplanted, and sometimes several organs at a time. Although transplant medicine had a shaky start, many problems have been solved. Today, thousands of transplants are performed annually and recipients gain years of life. The challenge to successful transplantation lies in genetics, because individual inherited differences in cell surfaces determine whether a particular recipient will accept tissue from a particular donor.

Transplant Types

Transplants are classified by the relationship of donor to recipient (**figure 17.16**):

1. An autograft transfers tissue from one part of a person's body to another. A skin graft taken from the thigh to replace burned skin on the chest, or a leg vein that replaces a coronary artery, are autografts. The immune system does not reject the graft because the tissue is self. (Technically, an autograft is not a transplant because it involves only one person.)
2. An isograft is a transplant of tissue from a monozygotic twin. Because the twins are genetically identical, the recipient's immune system does not reject the transplant. Ovary isografts are used to treat infertility.
3. An allograft comes from an individual who is not genetically identical to the recipient, but is a member of the same species. A kidney transplant from an unrelated donor is an allograft.
4. A xenograft transplants tissue from one species to another. (See **Bioethics.**)

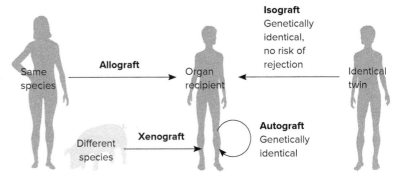

Figure 17.16 **Transplant types.** An autograft is a transfer of tissue within an individual. An isograft is between identical twins. An allograft is between members of the same species, and a xenograft is between members of different species.

Pig Parts

In 1902, a German medical journal reported an astonishing experiment. A physician, Emmerich Ullman, had attached the blood vessels of a patient dying of kidney failure to a pig's kidney set up by her bedside. The patient's immune system rejected the attachment almost immediately.

Nearly a century later, in 1997, a similar experiment took place. Robert Pennington, a 19-year-old suffering from acute liver failure and desperately needing a transplant, survived for 6.5 hours with his blood circulating outside of his body through a living liver removed from a 15-week-old, 118-pound pig named Sweetie Pie. The pig liver served as a bridge until a human liver became available. But Sweetie Pie was no ordinary pig. She had been genetically modified and bred so that her cells displayed a human protein that controlled rejection of tissue transplanted from an animal of another species. Because of this bit of added humanity, plus immunosuppressant drugs, Pennington's body tolerated the pig liver's help for the few crucial hours. Baboons have also been organ donors (**figure 17E**).

Successful xenotransplants would help alleviate the organ shortage. A possible danger of xenotransplants is that people may acquire viruses from the donor organs. Viruses can "jump" species, and the outcome in the new host is unpredictable. For example, a pig virus called PERV—for "porcine endogenous retrovirus"—can infect human cells in culture. However, several dozen patients who received implants of pig tissue did not show evidence of PERV years later. That study, though, looked only at blood. We still do not know what effect pig viruses can have on a human body. Because many viral infections take years to cause symptoms, introducing a new infectious disease in the future could be the trade-off for using xenotransplants to solve the current organ shortage.

Stem cell technology may in the future enable human organs to grow in non-human animals whose organs are about the same size as ours. In a technique called interspecies blastocyst complementation, pluripotent stem cells (PSCs) from one species are introduced into blastocysts of another species for which gene editing has resulted in an organ not developing. (Recall from figure 3.15 that the blastocyst is a hollow ball of cells surrounding the inner cell mass, which develops into the embryo.) As the embryo develops, the donor cells form the replacement organ. Early experiments created mice that grew rat pancreases, for example. Researchers have created such chimeric early embryos in which human PSCs are part of the inner cell mass of pig blastocysts. This is a first step to growing human organs in pigs.

Questions for Discussion

1. Pig parts as transplants may become necessary due to the human organ shortage. Discuss the pros and cons of the following systems for rationing human organs:

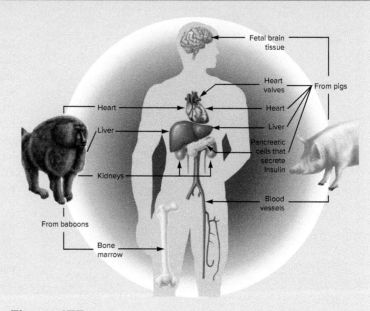

Figure 17E Baboons and pigs can provide tissues and organs for transplant.

   a. first come, first served
   b. closeness of match of cell surface antigens
   c. ability to pay
   d. the importance of the recipient
   e. the youngest
   f. the most severely ill, who will soon die without the transplant
   g. the least severely ill, who are strong enough to survive a transplant
   h. people who are not responsible for their condition, such as a nonsmoker with hereditary emphysema versus a person who has emphysema caused by smoking

2. In the novel and film *Never Let Me Go,* Kazuo Ishiguro describes a society in which certain people are designated as organ donors. They know that at a certain age, their organs will be removed, one by one, until they die, to provide transplants for wealthy recipients. The film *The Island* has a similar plot, except that the donors do not know their fate. In Robin Cook's novel *Chromosome Six,* a geneticist places HLA genes into fertilized ova from bonobos (pygmy chimps), and the animals are raised to provide organs for wealthy humans.

   Choose a book or film with a transplant plot and discuss the source of the transplants from the points of view of the donor, the recipient, the families of both, and the government.

3. Discuss the issues that people might find disturbing about growing human organs in pigs.

Rejection Reactions—or Acceptance

The immune system recognizes most donor tissue as nonself and launches a tissue rejection reaction in which T cells, antibodies, and activated complement destroy the foreign tissue. The greater the difference between recipient and donor cell surfaces, the more rapid and severe the rejection. An extreme example is the hyperacute rejection reaction against tissue transplanted from another species. Donor tissue from another type of animal is usually destroyed in minutes as blood vessels blacken and cut off the blood supply.

Physicians use several approaches to limit rejection to help a transplant recipient survive. These include closely matching the HLA types of donor and recipient and stripping donor tissue of antigens. Gene expression microarrays can be used to better match donors to recipients.

Immunosuppressant drugs inhibit production of the antibodies and T cells that attack transplanted tissue. If recipients get bone marrow stem cells from the donors, they need immunosuppressant drugs for only a short time, because along with the new organ, the bone marrow stem cells help the recipient's body accept the transplanted tissue. Gene expression profiling can identify transplant recipients unlikely to reject their new organs, sparing some people from the need to take immunosuppressants, which have side effects.

Rejection is not the only problem that can arise from an organ transplant. Graft-versus-host disease can develop when bone marrow transplants are used to correct certain blood deficiencies and cancers. The transplanted bone marrow, which is actually part of the donor's immune system, attacks the recipient—its new body—as foreign. Symptoms include rash, abdominal pain, nausea, vomiting, hair loss, and jaundice.

Sometimes a problem arises if a bone marrow transplant to treat cancer is too closely matched to the recipient. If the cancer returns with the same cell surfaces as it had earlier, the patient's new bone marrow is so similar to the old marrow that it is equally unable to fight the cancer. The best tissue for transplant may be a compromise: different enough to control the cancer, but not so different that rejection occurs.

Key Concepts Questions 17.4

1. How does a vaccine protect against an infectious disease?
2. How are monoclonal antibodies and cytokines used clinically?
3. Describe the types of transplants.
4. What happens in organ rejection?

17.5 Using Genomics to Fight Infection

Immunity against infectious disease arises from interactions of two genomes—ours and the pathogen's. Human genome information is revealing how the immune system halts infectious disease. Information from pathogen genomes reveals how they make us sick.

Researchers use genomic information to better understand how an infection affects the body and how infections spread, causing outbreaks and epidemics. Genomic information can inspire new treatment approaches. The DNA sequence for *Streptococcus pneumoniae*, for example, revealed genetic instructions for a huge protein that enables the bacterium to adhere to human cells. Potential drugs could dismantle this adhesion protein. Sequencing the genomes of pathogens can also help to halt the spread of the diseases they cause, as the following two examples illustrate.

Reverse Vaccinology

Older vaccines consisted of parts of pathogens that were detected using standard microbiological approaches. In a strategy called reverse vaccinology, researchers consult genome sequence information to identify genes that encode "hidden" antigens that might serve as the basis for a vaccine. If the proteins that correspond to the genes induce immunity in experimental animals such as mice, the preparation is tested in humans.

The first reverse vaccine targeted meningococcus B, which causes more than half of all cases of meningococcal meningitis, a bacterial infectious disease that inflames the membranes around the brain and/or spinal cord. Meningococcal meningitis can be fatal, or cause deafness or brain damage. Using part of the bacterium as a vaccine is difficult, because some molecules on the surface resemble human proteins so much that a vaccine might direct an immune response against the body. Proteins unique to the bacterium are too varied to easily use as the basis for a vaccine.

To develop a vaccine against meningococcus B, researchers used bioinformatics (computer analysis of DNA sequences) to identify hundreds of bacterial antigens, and tested the antigens in mice to see if they would prevent the infection. A few antigens that were highly specific to the bacterium and protected mice became the basis of the vaccine that is now used for humans. Vaccines against severe acute respiratory syndrome (SARS) and several varieties of influenza were also developed using information from pathogen genomes. Reverse vaccines do not consist of DNA, but rather are based on finding protein antigens that elicit a human immune response using pathogen genome information.

Genome Sequencing to Track Outbreaks

DNA is a changeable molecule, and so over time, even over short periods, mutation alters genome sequences. Epidemiologists can follow the rapidly changing genome sequences in pathogens taken from body fluids or on objects to identify the source of an outbreak and trace the spread of infectious disease. This approach is an example of a strategy called **metagenomics**, which sequences random pieces of DNA, or entire genomes, that are present in a particular environment, such as a drop of lake water—or a blood or stool sample. Following are two examples of how a genome sequencing approach helped protect public health.

Hospital-Acquired Pneumonia

In June 2011, infection by the drug-resistant bacterium *Klebsiella pneumoniae* killed six patients at the National Institutes of Health's (NIH) Clinical Center in Bethesda, Maryland, and caused life-threatening pneumonia in five others (**figure 17.17**). *K. pneumoniae* is normally found on the skin and in the mouth and intestines, but certain strains, when inhaled, can severely damage the lungs. Researchers from the nearby National Human Genome Research Institute teamed pathogen genome sequencing with classic epidemiological sleuthing to quickly identify the source of the outbreak at the clinical center, halting the spread and saving lives.

K. pneumoniae lung infections kill half of the people they infect and are resistant to many antibiotic drugs; also the bacteria exist as variants that are so alike genetically that standard microbiological techniques to identify pathogenic variants miss subtle genetic changes. Genome sequencing, however, checks every nucleotide. When researchers sequenced bacterial genomes from the first ("index") case in the outbreak, a 43-year-old woman who recovered, they discovered distinguishing single nucleotide polymorphisms (SNPs) at 41 sites in the 6-million-base genome of *K. pneumoniae*. The bacteria were evolving—mutating—so rapidly that secretions sampled from different parts of the patient's body yielded *K. pneumoniae* with slightly different SNP patterns.

Using an algorithm to compare bacterial sequences, the researchers found that some of the sick people shared SNP patterns with bacteria from the lungs and groin of the index patient, and some infected individuals had SNP patterns like those from the first patient's throat. One patient got the infection from a contaminated ventilator, which provided a clue to transmission of the infection in the hospital.

To see who might be infected but not yet show symptoms, the researchers next sequenced the genomes of all 1,115 patients in the hospital at the time. This part of the investigation showed that five people carried the bacteria without developing symptoms, and they had spread the infection to six others who hadn't developed symptoms but would. Identifying all of these people, and discovering objects that they had touched, enabled the investigators to control the outbreak.

Using pathogen genome sequencing revealed two new facts about the infection: *K. pneumoniae* survives on inanimate objects and can be transmitted by carriers who never become ill. Comparing pathogen genomes may help to prevent some of

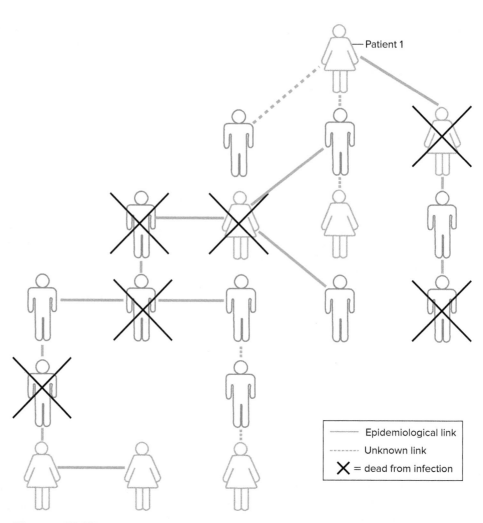

Figure 17.17 **On the trail of a pathogen.** Researchers compared SNPs at 41 sites in the genome of *Klebsiella pneumoniae* to assemble the chain of infection to trace an outbreak at the National Institutes of Health Clinical Center.

the 1.7 million hospital-acquired infections that occur in the United States each year, killing nearly 100,000 people.

Toxic *Escherichia coli*

Pathogen genome sequencing helped epidemiologists control a more widespread outbreak than that of pneumonia at the NIH Clinical Center. It involved *Escherichia coli*, a bacterium that is a normal part of the human intestinal microbiome (see figure 17.1). A strain of the bacterium, called "STEC" for Shiga-toxigenic *Escherichia coli*, produces a toxin that causes severe diarrhea and, in some individuals, a form of kidney failure that can be lethal.

In June 2011, an outbreak in Germany sickened nearly 4,000 people and resulted in 52 fatalities. The specific strain of *E. coli* had never been seen in an epidemic. Identifying the pathogen was a life-and-death race against time, and standard microbiological culturing techniques were too slow. So researchers conducted a metagenomic analysis on stool samples from people who suffered from diarrhea during the outbreak.

Each sample contained DNA from thousands of microorganisms, as well as a great deal of DNA from the patient.

Of the 45 samples that the researchers analyzed, 40 included *E. coli* and 27 of those 40 carried Shiga toxin genes. The investigators then used traditional epidemiological methods to determine that the affected individuals had all eaten raw bean sprouts in a particular town in Germany, which were traced to a farm where runoff from cattle feces had contaminated the sprouts. Five of the patients' stool samples had genomes from other types of bacteria, including *Salmonella*, *Campylobacter*, and *Clostridium* species. Unfortunately the infection does not respond to antibiotics, and the only treatment is to keep the patient hydrated.

Since the 2011 cases, epidemiologists have used genomic sleuthing to contain other outbreaks. Because it is now possible to sequence a microbial genome in just hours, it will soon be possible to much more accurately diagnose an infectious disease in a health care provider's office. As researchers continue to identify variants of human immune system genes that protect against infection as well as gene variants in the pathogens that provoke the immune response, we will be better able to prevent, contain, and treat infectious diseases.

Key Concepts Questions 17.5

1. How does reverse vaccinology use pathogen genome information?

2. How are pathogen genome sequences used to identify the source of an outbreak of an infectious disease and to trace its spread?

Summary

17.1 The Importance of Cell Surfaces

1. The cells and biochemicals of the immune system distinguish self from nonself, protecting the body against infections and cancer.

2. Genes encode immune system proteins and may confer susceptibilities to certain infectious diseases.

3. An **antigen** is a molecule that elicits an immune response. Patterns of cell surface proteins and glycoproteins determine blood types. **Human leukocyte antigens** (HLAs) are cell surface antigens that bind foreign antigens that **antigen-presenting cells** display to the immune system. **Dendritic cells** and **macrophages** are antigen-presenting cells.

17.2 The Human Immune System

4. If a pathogen breaches the body's physical barriers, **toll-like receptors** on macrophages and dendritic cells bind proteins that several types of pathogens share, triggering the **innate immune response**, which produces the redness and swelling of **inflammation**, plus **complement**, **collectins**, and **cytokines** such as **interferons** and **interleukins**. The innate immune response is broad and general.

5. The **adaptive immune response** is slower and specific, and it remembers antigens it has responded to.

6. The **humoral immune response** begins when macrophages display foreign antigens near HLAs. This activates **T cells**, which activate **B cells**, which give rise to **plasma cells** that secrete specific **antibodies**. Some B cells give rise to **memory cells**. The **primary immune response** is the first reaction to encountering a nonself antigen, and the **secondary immune response** is a reaction to subsequent encounters.

7. An antibody subunit is Y-shaped and has four polypeptide chains, two **heavy chains** and two **light chains**. Each antibody molecule has regions of constant amino acid sequence and regions of variable sequence.

8. The tips of the Y of each antibody subunit form **antigen binding sites**, which include the more specific **idiotypes** that bind foreign antigens at their **epitopes**.

9. Antibodies bind antigens to form immune complexes large enough for other immune system components to detect and destroy. Antibody genes are rearranged during early B cell development, providing instructions to produce a great variety of antibodies.

10. T cells carry out the **cellular immune response**. Their precursors are selected in the thymus to recognize self cells. Helper T cells secrete cytokines that activate other T cells and B cells. A helper T cell's CD4 antigen binds macrophages that present foreign antigens. Cytotoxic T cells release biochemicals that kill bacteria and destroy cells covered with viruses.

17.3 Abnormal Immunity

11. Mutations in antibody or cytokine genes, or in genes encoding T cell receptors, cause inherited immune deficiencies. Severe combined immune deficiencies (SCIDs) affect both branches of the immune system.

12. HIV binds to the coreceptors CD4 and CCR5 on macrophages and helper T cells, and, later in infection, triggers apoptosis of cytotoxic T cells. As HIV replicates, it mutates, evading immune system attack. Falling CD4 helper T cell numbers allow opportunistic infections and cancers to flourish. People who cannot produce a complete CCR5 protein resist HIV infection.

13. In **autoimmunity**, the body manufactures **autoantibodies** against its own cells.

14. In people who are susceptible to allergies, allergens stimulate IgE antibodies to bind to mast cells, which causes the cells to release allergy mediators. Certain helper T cells release selected cytokines.

17.4 Altering Immunity

15. A **vaccine** presents a disabled pathogen, or part of one, to elicit a primary immune response.

16. Immunotherapy enhances or redirects immune function. Monoclonal antibodies are useful in diagnosing and treating some diseases because of their abundance and specificity. Cytokines are used to treat various conditions.

17. Transplant types include autografts (within oneself), isografts (between identical twins), allografts (within a species), and xenografts (between species). A tissue rejection reaction occurs if donor tissue is too unlike recipient tissue.

17.5 Using Genomics to Fight Infection

18. Knowing the genome sequence of a pathogen provides more information than can be obtained from traditional microbiological classification.

19. Reverse vaccinology uses pathogen genome sequence information to identify antigens that can form the basis of an effective vaccine.

20. Comparing SNP patterns among samples of bacteria taken from patients in an outbreak or epidemic can reveal the spread of the infection, including inanimate objects and carriers. It can also identify bacterial strains that produce toxins. Tracking outbreaks is an application of **metagenomics**, which is sequencing all the DNA in a particular environment.

Review Questions

1. Match the cell type to the type of biochemical it produces.

   1. mast cell
   2. T cell
   3. plasma cell
   4. macrophage and dendritic cell
   5. all cells with nuclei

   a. antibodies
   b. class II human leukocyte antigens
   c. interleukin
   d. histamine
   e. interferon
   f. heparin
   g. tumor necrosis factor
   h. class I human leukocyte antigens
   i. toll-like receptors

2. What does "nonself" mean? Give an example of a nonself cell in your own body.

3. Distinguish between viruses and bacteria.

4. What is the physical basis of a blood type?

5. Distinguish between using serology or genotyping to type blood.

6. Explain why an HLA-disease association is not a diagnosis.

7. Explain how mucus, tears, cilia, and earwax are part of the immune response.

8. Distinguish between the following:
   a. a T cell and a B cell
   b. innate and adaptive immunity
   c. a primary and a secondary immune response
   d. a cellular and a humoral immune response
   e. an autoimmune condition and an allergy
   f. an inherited and an acquired immune deficiency

9. Which components of the human immune response explain why we experience the same symptoms of an upper respiratory infection (a "cold") when many different types of viruses can cause these conditions?

10. State the function of each of the following immune system biochemicals:
    a. complement proteins
    b. collectins
    c. antibodies
    d. cytokines
    e. filaggrin

11. What does HIV do to the human immune system?

12. Cite three reasons why developing a vaccine against HIV infection has been challenging.

13. What would be the consequences of lacking each of the following?
    a. helper T cells
    b. cytotoxic T cells
    c. B cells
    d. macrophages

14. Explain how the immune system can respond to millions of different nonself antigens using only a few hundred antibody genes.

15. How are SCID and AIDS similar and different?

16. What part do antibodies play in allergic reactions and in autoimmune diseases?

17. What do a plasma cell and a memory cell descended from the same B cell have in common? How do they differ?

18. Why is a deficiency of T cells more dangerous than a deficiency of B cells?

19. Cite two explanations for why autoimmune diseases are more common in females.

20. How do each of the following illnesses disturb immunity?
    a. graft-versus-host disease
    b. SCID
    c. systemic sclerosis
    d. AIDS
    e. atopic dermatitis

21. Why is a polyclonal antibody response valuable in the body, but a monoclonal antibody valuable as a diagnostic tool?

22. State how each of the following alters immune system functions:
    a. a vaccine
    b. an antibiotic drug
    c. a cytokine-based drug
    d. an antihistamine drug
    e. a transplant

23. Explain how a "reverse" vaccine is similar to but also different from a traditional vaccine.

24. How can knowing the genome sequence of a pathogen be useful in fighting an outbreak of infectious disease?

Applied Questions

1. Explain how the famous case of Timothy Brown is leading to development of a vaccine against HIV.

2. "Winter vomiting disease," a form of gastroenteritis sometimes called "stomach flu," is caused by a virus called norovirus. It makes a person miserable for a day or two. Why do some people get the illness every year?

3. Rasmussen's encephalitis causes 100 or more seizures a day. Affected children have antibodies that attack brain cell receptors that normally bind neurotransmitters. Is this condition most likely an inherited immune deficiency, an autoimmune disease, or an allergy? State a reason for your answer.

4. In people with a certain HLA genotype, a protein in the joints resembles an antigen on the bacterium that causes Lyme disease. This infection is transmitted in a tick bite and causes flulike symptoms soon and joint pain (arthritis) a few weeks or months later. When the individuals with the HLA genotype become infected, their immune systems attack the bacteria and their joints. Explain why antibiotics treat the early phase of the disease, but not the arthritis.

5. A person exposed for the first time to Coxsackie virus develops a painful sore throat. How is the immune system alerted to the exposure to the virus? When the person encounters the virus again, why don't symptoms develop?

6. A young woman who has aplastic anemia will soon die as her lymphocyte levels drop sharply. What type of cytokine might help her?

7. Tawanda is a 16-year-old with cystic fibrosis. She receives a lung transplant from a woman who has just died in a car accident. Tawanda and the donor share four of the six HLA markers commonly used to match donor to recipient. What action can the transplant team take to minimize the risk that Tawanda's immune system will reject the transplanted lung?

Case Studies and Research Results

1. Researchers infected human neural progenitor cells with Zika virus and carried out three types of experiments.

   - Genome editing technology removed one human gene at a time from the neural progenitor cells and identified nine genes that the virus uses to infect human cells. The proteins that the genes encode participate in how the endoplasmic reticulum (ER) processes viral proteins.

   - "Transcriptomics" catalogued the mRNAs that the infected cells produce as infection ensues and compared the mRNAs to those made in brain cells from children who died from microcephaly due to genetic syndromes. Many of the gene variants were overexpressed in both the neural progenitor cells infected with Zika virus in the laboratory culture and from the children who had microcephaly.

   - Transcriptomics revealed activation of genes encoding toll-like receptors, interleukins, tumor necrosis factor, CXCR4, an interferon, and a colony-stimulating factor, in the neural progenitor cells.

   a. One of the roles of the human ER in Zika virus infection is to cut two viral proteins in half, which activates them. What would you have to know in order to target this activity in developing a drug? (See chapter 2.)

   b. How might the transcriptomics findings for microcephaly be used to determine the critical period for Zika virus infection? (See chapter 3.)

   c. Does Zika virus infection activate innate immunity, adaptive immunity, or both, and how do you know?

   d. Which of the experimental results explains why adults who develop symptoms of Zika virus infection experience fever, joint pain, a rash, and conjunctivitis?

2. State whether each of the following situations involves an autograft, an isograft, an allograft, or a xenograft.

   a. A man donates part of his liver to his daughter, who has a liver damaged by cystic fibrosis.

   b. A woman with infertility receives an ovary transplant from her identical twin.

   c. A man receives a heart valve from a pig.

   d. A woman who has had a breast removed has a new breast built using her fatty thigh tissue.

3. Researchers have successfully treated mice for type 1 diabetes with transplants of beta cells from the pancreases of pigs. What type of transplant is this?

4. Mark and Louise are planning to have their first child, but they are concerned because they think that they have an Rh incompatibility. He is Rh$^-$ and she is Rh$^+$. Will there be a problem? Why or why not?

5. Twenty-four children and teens at a summer camp for people with cystic fibrosis contract severe lung infections from the multidrug-resistant bacterium *Mycobacterium abscessus*. Describe a technology, based on genetics, that infectious disease experts might use to discover exactly how the infection spread at the camp.

6. In a Dutch family, a father, son, and daughter frequently develop fungal infections (candidiasis) of the mouth, throat, and skin of the feet. Each of the three family members has at least one autoimmune condition. The mother is unaffected. Investigation of the immune responses of the father and children found that they do not make sufficient helper T cells, nor do they produce enough interferon. They also make autoantibodies against two types of interleukins.

   a. What is the likely mode (or modes) of inheritance of the underlying immune system dysfunction that affects the father and two children?

   b. Explain how the immune systems of the father and children are abnormal.

   c. How can autoimmunity indirectly increase the risk of infection?

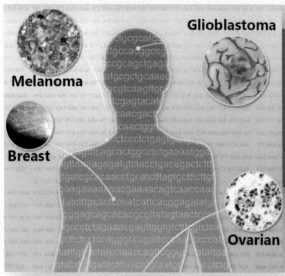

Melanoma

Glioblastoma

Breast

Ovarian

Source: Jonathan Bailey, National Human Genome Research Institute

CHAPTER 18

Cancer Genetics and Genomics

Decisions on how to treat cancer are increasingly considering both location (body part) and mutation (genetic changes that occur as a cancer starts, progresses, and responds to treatment).

Learning Outcomes

18.1 Cancer Is an Abnormal Growth That Invades and Spreads
1. List the characteristics of a cancerous tumor.
2. Explain how loss of cell cycle control causes cancer.
3. Explain how most cancers are not inherited, but are genetic.

18.2 Cancer at the Cellular Level
4. Describe cancer cells.
5. Explain how cancer cells can arise from stem cells.

18.3 Cancer Genes and Genomes
6. Distinguish between driver and passenger mutations in cancer.
7. Discuss how mutations in several genes drive cancer.
8. Describe what can happen to chromosomes in cancer cells.
9. Explain how mutations in oncogenes and tumor suppressor genes cause or increase susceptibility to cancer.
10. Discuss environmental factors that increase risk of or cause cancer.

18.4 Diagnosing and Treating Cancer
11. Explain how cancer diagnosis and treatment have become more specific and increasingly based on genetics and genomics.

 ## The BIG Picture

A complex chain reaction of changes at the gene, chromosome, and genome levels causes and propels the common and diverse collection of diseases that we call cancer.

Treating Cancer by Location and Mutation

In the past, we have classified the more than 100 types of cancer according to the cell type, tissue type, or body part initially affected. Treatment was to remove the tumor and/or use drugs that kill all rapidly dividing cells. Today, cancers are being increasingly classified by, and treatment decisions based on, mutations in cancer cells.

Attacking cancer based on mutations is particularly helpful for forms of the disease that are so rare that there is little research and no specific treatments. This is the case for Erdheim-Chester disease, a cancer of macrophages that causes widespread organ failure and bone pain. Only about 500 cases have been identified since the 1930s. About half of patients with the cancer have a mutation, V600, in a gene called *BRAF*. These patients are benefitting from a targeted drug (Zelboraf, or vemurafenib) developed to treat the skin cancer melanoma, which is also caused by the V600 mutation.

When the head physician at a major cancer center tested the drug on Erdheim-Chester patients with the V600 mutation, results were astonishing. "The first patient we treated had been on the way to hospice. She couldn't walk. Two weeks after joining the clinical trial she came back, walking and feeling fine." She and others have recovered from a cancer that had no treatment—by considering the underlying mutation.

Choosing a drug based on one mutation oversimplifies the genetic complexity of cancer cells, which may have dozens of mutations. The National Cancer Institute's Molecular Analysis for Therapy Choice (MATCH) program is assigning thousands of cancer patients to receive combinations of targeted drugs tailored to the subsets of 143 well-studied mutations in their cancer cells.

18.1 Cancer Is an Abnormal Growth That Invades and Spreads

Cancer is a type of disease in which certain cells become able to divide more often, leading to an abnormal growth (a tumor) or disruption of the proportion of blood cell types (a "liquid" tumor). One in three people develop cancer. A person may learn of cancer after noticing symptoms and reporting them to a health care provider, who then orders diagnostic tests. A change in bowel habits, lower abdominal discomfort, blood-tinged stools, and fatigue are warning signs of colon cancer, for example. Or, a cancer diagnosis may follow a routine screening test, if the test reveals a biomarker in the blood or shows a growth on a colonoscopy scan. Sometimes cancer diagnosis is secondary, such as discovering a shadow on an X ray taken to diagnose pneumonia and sampling cells to find a tumor.

However a cancer is detected, chances are it has been present for years, perhaps decades. A cancer takes time to grow because it is the culmination of a series of genetic and genomic changes—mutations—that enable certain cells to divide more frequently than normal, forming a new growth that enlarges, eventually crowding healthy tissue.

Eradicating cancer is challenging because of its great diversity. Although a few "driver" mutations initiate a cancer by affecting a small set of cellular processes, this can happen in so many ways that the cancers of no two patients are exactly alike, nor are multiple tumors within the same body genetically alike, or even the cells within a single tumor.

The underlying derangement of the cell cycle that causes and sustains a cancer may be set into motion by inherited cancer susceptibility genes, or, more commonly, by environmental exposure to factors such as ultraviolet radiation in sunlight or toxins in cigarette smoke. The observation that smokers who develop lung cancer have 10 times as many mutations in their cancer cells as nonsmokers who develop lung cancer indicates the powerful role of the environment in causing cancer.

Researchers have sequenced the genomes of thousands of tumors to better understand the genetic changes that cause and accompany a cancer. Sequences of mutations that occur over time in somatic cells, the shattering of chromosomes, and changes in gene expression underlie the progression of cancer as it spreads. Mutations may affect the expression of other genes, but so may epigenetic influences, such as DNA methylation and chromatin remodeling (see section 11.2). One researcher calls the accumulating DNA changes that lie behind cancer "genomic scars."

This chapter explores cancer at the cellular, genetic, and genomic levels. **Figure 18.1** summarizes the major concepts. The depiction can be read from either direction, but the central portion—the cellular level—is the way of looking at cancer that may ultimately prove the most useful. All of the genes in which mutations increase susceptibility to, or cause, cancer affect three basic cellular pathways: cell fate, cell survival, and genome maintenance. Cell fate refers to differentiation (specialization). Cell survival refers to oxygen availability and preventing apoptosis. Genome maintenance refers to the abilities to survive in the presence of reactive oxygen species and toxins, to repair DNA, to maintain chromosome integrity and structure, and to correctly splice mRNA molecules.

Cancer is a complication of being a many-celled organism. Our specialized cells must follow a schedule of mitosis—the cell cycle—so that organs and other body parts grow appropriately during childhood, stay a particular size and shape throughout adulthood, and repair damage by replacing tissue. If a cell in solid tissue escapes normal controls on its division rate, it forms a tumor (**figure 18.2**). In the blood, cancer cells divide more frequently than others, taking over the population of blood cells.

A tumor is benign if it grows in place but does not spread into, or "invade," surrounding tissue. A tumor is cancerous, or malignant, if it infiltrates nearby tissue. Pieces of a malignant tumor can enter the bloodstream or lymphatic vessels and travel to other areas of the body, where the cancer cells "seed" the formation of new tumors. The process of spreading is termed **metastasis**, which means "not standing still." Metastasis can make a cancer deadly, because the new growth may be in an inaccessible part of the body, or genetically distinct enough from the original, or primary, tumor that drugs that were effective early in the illness no longer work. Metastases are difficult to detect. If a few sites of metastases appear on a medical scan, there may actually be dozens of growths in the body.

An early hint at the genetic nature of cancer was the observation that most substances known to be carcinogens (causing cancer) are also mutagens (damaging DNA). Researchers first discovered genes that could cause cancer in humans in 1976. These genes were versions of genes from certain viruses that had been associated with tumors in birds in studies dating from the mid-nineteenth century. In the 1980s and 1990s, research to identify cancer-causing genes began with rare families that had many young members who had the same type of cancer and specific unusual chromosomes. Finding genes in the affected chromosome regions whose protein products could alter cell cycle control led to the discovery of more than 100 **oncogenes**. An oncogene is a gene that causes cancer when it is expressed when it wouldn't be in healthy cells, or is overexpressed. In a sense, oncogenes are inappropriately activated.

Family studies also identified more than 30 **tumor suppressor genes**, which cause cancer when they are deleted or inactivated. The normal function of a tumor suppressor gene is to keep the cell cycle running at the appropriate rate for a particular cell type under particular conditions.

Most mutations that cause cancer are in oncogenes or tumor suppressor genes. The effects of mutations in oncogenes

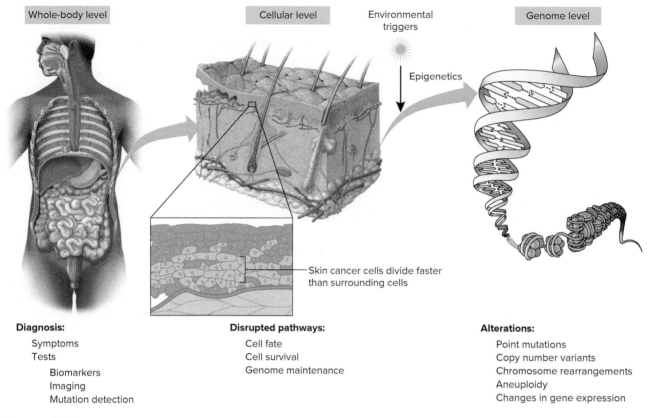

Whole-body level **Cellular level** Environmental triggers **Genome level**

Epigenetics

Skin cancer cells divide faster than surrounding cells

Diagnosis:

Symptoms
Tests
    Biomarkers
    Imaging
    Mutation detection

Disrupted pathways:

Cell fate
Cell survival
Genome maintenance

Alterations:

Point mutations
Copy number variants
Chromosome rearrangements
Aneuploidy
Changes in gene expression

Figure 18.1 **Levels of cancer.** It takes years for a cancer to produce symptoms, as more and more cells lose their specializations and divide more frequently than the cells from which they descend. Mutations—from single-base changes to large-scale chromosomal upheavals—drive the disease, accompanied by changes in gene expression, some of which are responses to environmentally induced epigenetic effects.

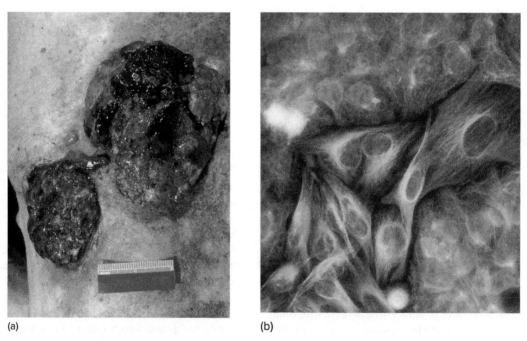

(a) (b)

Figure 18.2 **Cancer cells stand out.** **(a)** A cutaneous melanoma is a cancer of the pigment-producing cells (melanocytes) in the skin. This cancer may have any or all of five characteristics, abbreviated abcde: it is <u>a</u>symmetric, has <u>b</u>orders that are irregular, <u>c</u>olor variations, a <u>d</u>iameter of more than 5 millimeters, and <u>e</u>levation. **(b)** These cutaneous melanoma cells stain orange. The different staining characteristics of cancer cells are due to differences in gene expression patterns between the normal and cancerous states. *(a): © McGraw-Hill Higher Education; (b): © Nancy Kedersha/Science Source*

are typically dominant, and those of tumor suppressor genes recessive. A third category of cancer genes includes mismatch mutations in DNA repair genes (see section 12.7) that allow other mutations to persist. When these other mutations activate oncogenes or inactivate tumor suppressor genes, cancer can result. Most DNA repair disorders are inherited in a single-gene fashion, and are quite rare. They typically cause diverse and widespread tumors, often beginning at a young age. We return to the genes behind cancer in section 18.3.

Loss of Cell Cycle Control

The most fundamental characteristic of cancer is the underlying disruption of the cell cycle. **Figure 18.3** repeats the cell cycle diagram from chapter 2. Cancer begins when a cell divides more frequently, or more times, than the noncancerous cell it descended from (**figure 18.4**). Mitosis in a cancer cell is like a runaway train, racing along without signals and control points.

The timing, rate, and number of mitoses a cell undergoes depend on protein growth factors and signaling molecules from outside the cell, and on transcription factors from within. Because these biochemicals are under genetic control, so is the cell cycle. Cancer cells probably arise often, because mitoses are so frequent that an occasional cell escapes control. However, the immune system destroys most cancer cells after recognizing tumor-specific antigens on their surfaces.

The discovery of the checkpoints that control the cell cycle revealed how cancer can begin. A mutation in a gene that normally halts or slows the cell cycle can lift the constraint, leading to inappropriate mitosis. Failure to pause long enough to repair DNA can allow a mutation in an oncogene or tumor suppressor gene to persist.

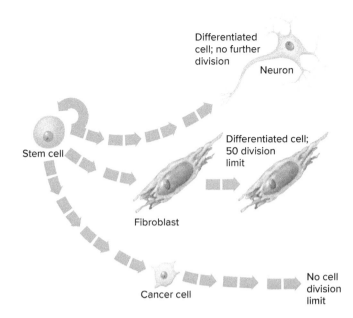

Figure 18.4 **Cancer sends a cell down a pathway of unrestricted cell division.** Cells may be terminally differentiated and no longer divide, such as a neuron, or differentiated yet still capable of limited cell division, such as a fibroblast (connective tissue cell). Cancer cells either lose specializations or never specialize; they divide unceasingly. (Arrows represent some cell divisions; not all daughter cells are shown.)

Loss of control over telomere length may also contribute to cancer by affecting the cell cycle. Recall that telomeres, or chromosome tips, protect chromosomes from breaking (see figure 2.16). Human telomeres consist of the DNA sequence TTAGGG repeated thousands of times. The repeats are normally lost from the telomere ends as a cell matures, from 15 to 40

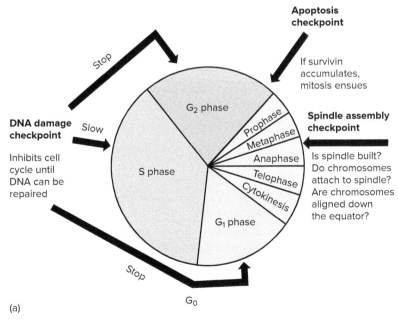

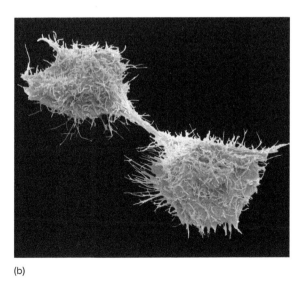

Figure 18.3 **Cell cycle checkpoints. (a)** Checkpoints ensure that mitotic events occur in the correct sequence. Many types of cancer result from faulty checkpoints. **(b)** These fibrosarcoma cancer cells descend from connective tissue cells (fibroblasts) in bone. The photo captures them in telophase of mitosis. *(b): © Steve Gschmeissner/SPL/Getty Images*

nucleotides per cell division. The more specialized a cell, the shorter its telomeres. The chromosomes in skin, nerve, and muscle cells, for example, have short telomeres. Chromosomes in a sperm cell or oocyte, however, have long telomeres. This makes sense—as the precursors of a new organism, gametes must retain the capacity to divide many times.

Gametes keep their telomeres long using an enzyme, telomerase, that consists of RNA and protein. Part of the RNA—the sequence AAUCCC—is a template for the 6-DNA-base repeat TTAGGG that builds telomeres. Telomerase moves down the DNA like a zipper, adding six "teeth" (bases) at a time. Mutation in the gene that encodes telomerase, called *TERT*, causes some cancers.

In normal, specialized cells, telomerase is turned off and telomeres shrink, signaling a halt to cell division when they reach a certain size. In cancer cells, telomerase is turned back on. Telomeres extend, and this releases the normal brake on rapid cell division. As daughter cells of the original abnormal cell continue to divide uncontrollably, a tumor forms, grows, and may spread. Usually the longer the telomeres in cancer cells, the more advanced the disease. However, turning on telomerase production in a cell is not sufficient in itself to cause cancer. Many other things must go wrong for cancer to begin.

Cancer cells can divide continuously if given sufficient nutrients and space. Cervical cancer cells of a woman named Henrietta Lacks, who died in 1951, vividly illustrate the hardiness of these cells. Her cells persist today as standard cultures in many research laboratories. These "HeLa" cells divide so vigorously that when they contaminate cultures of other cells, they soon take over.

Cells vary greatly in their capacity to divide. Cancer cells divide more frequently or more times than the cells from which they arise. Yet even the fastest-dividing cancer cells, which complete mitosis every 18 to 24 hours, do not divide as often as some cells in a normal human embryo do. A tumor grows more slowly at first because fewer cells divide. By the time the tumor is the size of a pea—when it is usually detectable—billions of cells are actively dividing. A cancerous tumor eventually grows faster than surrounding tissue because a greater proportion of its cells are dividing.

Inherited versus Sporadic Cancer

Cancer is *genetic*, because it is caused by changes in DNA, but it is not usually *inherited*. Only about 10 percent of cases result from inheriting a cancer susceptibility allele from a parent. The inherited allele is a germline mutation, meaning that it is present in every cell of the individual, including the gametes. Cancer develops when a second mutation occurs in the other allele in a somatic cell in the affected body part (**figure 18.5**).

The majority of cancers are sporadic, and caused by somatic mutations, which affect only non-sex cells (see figures 12.1 and 12.2). A sporadic cancer may result from a single dominant mutation or from two recessive mutations in copies of the same gene. The cell loses control of its cell cycle, and accelerated division of its daughter cells forms the tumor. Eventually, tumor cells may contain dozens of mutations that are not in neighboring, healthy cells.

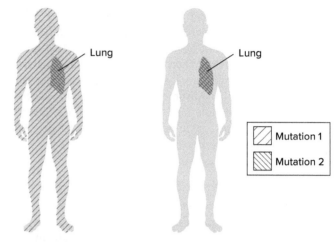

(a) Germline (inherited) cancer   (b) Sporadic cancer

Figure 18.5 **Germline versus sporadic cancer.** **(a)** In germline cancer, every cell has one gene variant that increases cancer susceptibility and a somatic mutation occurs in cells of the affected tissue. This type of predisposition to cancer is inherited as a single-gene trait and is due to an initial germline mutation. **(b)** A sporadic cancer forms when a dominant mutation occurs in a somatic cell or two recessive mutations occur in the same gene on homologous chromosomes in a somatic cell. An environmental factor can cause the somatic mutations of cancer.

Germline cancers are rare, but they have high penetrance and tend to strike earlier in life than do sporadic cancers. Germline mutations may explain why some heavy smokers develop lung cancer, but many do not; the unlucky ones may have inherited a susceptibility allele. Years of exposure to the carcinogens in smoke eventually cause a mutation in a tumor suppressor gene or oncogene of a lung cell, giving it a proliferative advantage. Without the susceptibility gene, two recessive somatic mutations are necessary to trigger the cancer. This, too, can be the result of an environmental insult, but it takes longer for two events to occur than one.

Key Concepts Questions 18.1

1. Distinguish between a benign and a cancerous tumor.
2. What is the relationship between genes and cancer?
3. What are the three cellular processes that cancer disrupts?
4. Distinguish between oncogenes and tumor suppressor genes.
5. Distinguish between inherited and sporadic cancers.

18.2 Cancer at the Cellular Level

Cancer begins at the genetic and cellular levels. If not halted by the immune system or treatment, it may spread through tissues and eventually take over organs and organ systems.

Characteristics of Cancer Cells

A cancer cell looks different from a normal cell. Some cancer cells are rounder than the cells they descend from because they do not adhere to surrounding normal cells as strongly as other cells do. Because the plasma membrane is more fluid, different substances cross it. A cancer cell's surface may sport different antigens than are on other cells or different numbers of the antigens that are also on normal cells. The "prostate specific antigen" (PSA) blood test that indicates increased risk of prostate cancer, for example, detects elevated levels of this protein that may come from cancer cell surfaces.

When a cancer cell divides, both daughter cells are cancerous, because they inherit the altered cell cycle control. Therefore, cancer is said to be heritable because it is passed from parent cell to daughter cell. A cancer is also transplantable, because a cancer cell injected into a healthy animal of the same species will proliferate.

A cancer cell is **dedifferentiated**, which means that it is less specialized than the normal cell types near it that it might have descended from. A skin cancer cell, for example, is rounder and softer than the flattened, scaly, healthy skin cells above it in the epidermis, and is more like a stem cell in both appearance and division rate.

Cancer cell growth is unusual. Normal cells in a container divide to form a single layer; cancer cells pile up on one another. In an organism, this pileup would produce a tumor. Cancer cells that grow all over one another are said to lack contact inhibition—they do not stop dividing when they crowd other cells.

Cancer cells have surface structures that enable them to squeeze into any space, a property called **invasiveness**. They anchor themselves to tissue boundaries, called basement membranes, where they secrete enzymes that cut paths through healthy tissue. Unlike a benign tumor, an invasive malignant tumor grows irregularly, sending tentacles in all directions. A cancer cell can move. Mutations affect the cytoskeleton (see figure 2.11), breaking down actin microfilaments and releasing actin molecules that migrate to the cell surface, moving the cell from where it is anchored in surrounding tissue.

The changes that craft a cancer cell from a healthy cell, and the proliferation of cancer cells and eventual invasion and metastasis, take time. Pancreatic cancer, for example, begins 10 to 15 years before the first abdominal pain, then progresses rapidly if not treated. Smoking-induced lung cancer begins with irritation of the lining tissue in respiratory tubes and may not produce symptoms for two decades. Cancer cells on the move eventually reach the bloodstream or lymphatic vessels, which take them to other body parts. This is metastasis.

Once a tumor has grown to the size of a pinhead, interior cancer cells respond to the oxygen-poor environment by secreting a protein called vascular endothelial growth factor (VEGF). It stimulates nearby capillaries (the tiniest blood vessels) to sprout new branches that extend toward the tumor, bringing in oxygen and nutrients and removing wastes. This growth of new capillary extensions to bring in a blood supply is called **angiogenesis**, and it is critical to a cancer's growth and spread. Capillaries may snake into and out of the tumor. Cancer cells wrap around the blood vessels and creep out upon this

Table 18.1	Characteristics of Cancer Cells
Oilier, less adherent	
Loss of cell cycle control	
Heritable	
Transplantable	
Dedifferentiated	
Lack contact inhibition	
Induce local blood vessel formation (angiogenesis)	
Invasive	
Increased mutation rate	
Can spread (metastasize)	

scaffolding, invading nearby tissue. In addition to attracting their own blood supply, cancer cells may also secrete hormones that encourage their own growth. This is a new ability because the cells they descend from do not produce these hormones.

When cancer cells move to a new body part, the DNA of these secondary tumor cells often mutates, and chromosomes may break or rearrange. Many cancer cells are aneuploid (with missing or extra chromosomes). The metastasized cancer becomes a new genetic entity that may resist treatments that were effective against most cells of the original tumor. **Table 18.1** summarizes the characteristics of cancer cells.

Origins of Cancer Cells

Factors that influence whether or not cancer develops include how specialized the initial cell is and the location of that cell in the tissue. Cancer can begin at a cellular level in at least four ways:

- activation of stem cells that produce cancer cells
- dedifferentiation
- increase in the proportion of a tissue that consists of stem or progenitor cells
- faulty tissue repair

Dedifferentiation is not an all-or-none phenomenon. Most cancer cells are more specialized than stem cells, but considerably less specialized than the differentiated cells near them in a tissue. From which does the cancer cell arise, the stem cell or the specialized cell? A cancer cell may descend from a stem cell that yields slightly differentiated daughter cells that retain the capacity to self-renew, or a cancer cell may arise from a specialized cell that loses some of its features and can divide. Certain stem cells, called **cancer stem cells**, veer from normal development and produce both cancer cells and abnormal specialized cells. Cancer stem cells are found in cancers of the brain, blood, and epithelium (particularly in the breast, colon, and prostate).

Figure 18.6 illustrates how cancer stem cells may cause brain tumors. In figure 18.6a, as cancer stem cells give rise to progenitors and then differentiated cells (neurons, astrocytes, and oligodendrocytes), a cell surface molecule called CD133

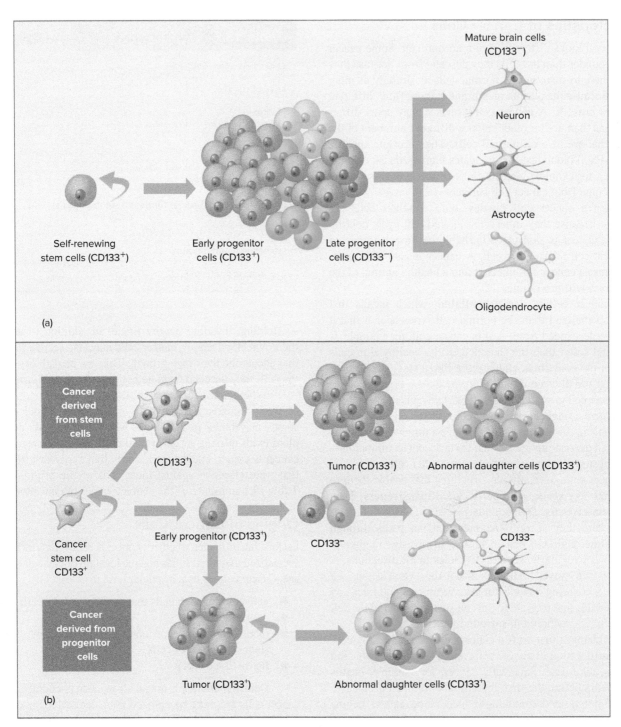

Figure 18.6 **Cancer stem cells.** **(a)** In the developing healthy brain, stem cells self-renew and give rise to early progenitor cells, which divide to yield late progenitor cells. These late progenitor cells lose the CD133 cell surface marker, and they divide to give rise to daughter cells that specialize as neurons or one of two types of supportive cells, astrocytes or oligodendrocytes. **(b)** A cancer stem cell can divide to self-renew and give rise to a cancer cell, which in turn can also spawn abnormal daughter cells (top row). Or, an early progenitor cell can give rise to normal differentiated cells (middle row). Or, cancer-causing mutations occur in the cancer stem cell–derived early progenitor cell. In this case, the early progenitors form the tumor, which may spawn some abnormal daughter cells (bottom row). Note that stem cells, cancer stem cells, early progenitor cells, and abnormal daughter cells all have the CD133+ marker, but the differentiated cells do not.

is normally lost (designated by CD133⁻) at the late progenitor stage. In contrast, in figure 18.6a, cancer cells retain the molecule (designated by CD133⁺). Some progenitor cells that descend from a cancer stem cell can relentlessly divide, and they ultimately accumulate, forming a brain tumor.

Cancer may also begin when cells lose some of their distinguishing characteristics as mutations occur when they divide. Or, cells on the road to cancer may begin to express "stemness" genes that override signals to remain specialized (**figure 18.7**).

Another possible origin of cancer may be a loss of balance at the tissue level in favor of cells that can divide continually or frequently—like a population growing faster if more of its members are of reproductive age. Consider a tissue that is 5 percent stem cells, 10 percent progenitors, and 85 percent differentiated cells. If a mutation, over time, shifts the balance in a way that creates more stem and progenitor cells, the extra cells pile up, and a tumor forms (**figure 18.8**).

Uncontrolled tissue repair may cause cancer (**figure 18.9**). If too many cells divide to fill in the space left by injured tissue, and those cells keep dividing, an abnormal growth may result.

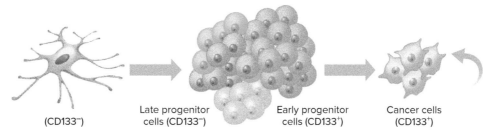

Figure 18.7 **Dedifferentiation reverses specialization.** Mutations in a differentiated cell could reactivate latent "stemness" genes, giving the cell greater capacity to divide while removing some specialization. *Adapted from "Neurobiology: At the root of brain cancer," Michael F. Clarke, Nature 432:281–282. Macmillan Publishers Ltd, November 18, 2004.*

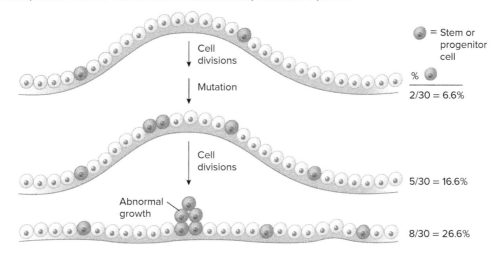

Figure 18.8 **Shifting the balance in a tissue toward cells that divide.** If a mutation renders a differentiated cell able to divide to yield other cells that frequently divide, then over time these cells may take over, forming an abnormal growth.

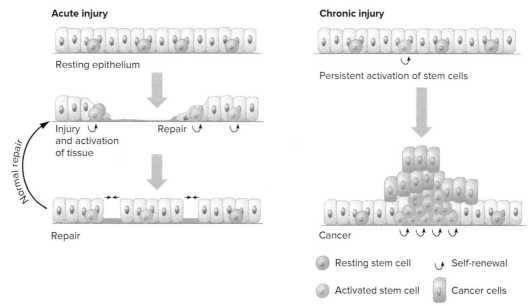

Figure 18.9 **Too much repair may trigger tumor formation.** If epithelium is occasionally damaged, resting stem cells can become activated and divide to fill in the tissue. If injury is chronic, the persistent activation of stem cells to renew the tissue can veer out of control, fueling an abnormal growth.

With so many millions of cells undergoing so many error-prone DNA replications, and so many ways that cancer can arise, it isn't surprising that cancer is so common. Yet most of the time, the immune system destroys a cancer before it progresses very far.

Key Concepts Questions 18.2

1. What are the characteristics of cancer cells?
2. Explain how cancer stem cells produce cancer cells and abnormal specialized cells.
3. Explain how altering the balance between stem and progenitor cells and differentiated cells, or excess tissue repair, can cause cancer.

18.3 Cancer Genes and Genomes

Researchers have sequenced the genomes of cells from thousands of tumors. As a result, a more comprehensive view is emerging of how mutations cause cancer and keep changing the course of the disease.

Driver and Passenger Mutations

A driver of a vehicle takes it to the destination; a passenger goes along for the ride. In cancer genetics, a **driver mutation** provides the selective growth advantage to a cell that defines the cancerous state. A **passenger mutation** occurs in a cancer cell, but does not cause or propel the cancer's growth or spread. Drivers can be oncogenes or tumor suppressor genes, and may be generated from rearranged chromosomes.

Only about 200 genes are known to have driver mutations, but thousands of genes can harbor passenger mutations, which occur in cancerous as well as noncancerous cells. More than 99 percent of the mutations in cancer cells are passengers, just along for the ride. Looked at another way, about 1 percent of all of our genes are involved in the cell cycle or DNA repair and can be implicated in some way in cancer.

Tumors vary greatly in the numbers of each type of mutation. A cancer generally has two to eight driver mutations. The number of passenger mutations increases with age. For a 40-year-old and an 80-year-old with cancer in the same type of tissue, the older person's tumor cells will have many more passenger mutations than the younger person's tumor cells. This makes sense. The passage of time brings DNA replication errors and environmental exposures.

The effect of driver mutations is cumulative. One model describes the accumulation of mutations as a cancer forms and progresses as "three strikes" that correspond to three stages: breakthrough, expansion, and invasion (**figure 18.10**).

The initial mutation ("strike") enables a normal epithelial (lining) cell, for example, to divide slightly faster than cells near it. In this way, a clone (a group of cells that descend from one cell) of faster-dividing cells gradually forms. The second strike occurs when a second driver mutation boosts the division rate in the already mutation-bearing cells, and their proportion within the tissue increases. The cells form a tumor that is benign (not yet cancerous). Even if each driver mutation boosts the cell division rate by only 0.4 percent, and if cells divide only once or twice a week, in several years the tumor will grow to a size that a person might be able to feel, and consist of billions of cells.

In the expansion stage, cancer cells acquire the ability to withstand low levels of nutrients, oxygen, and growth factors, which may happen in interior portions of a solid tumor. In the third strike, invasion, additional driver mutations representing several pathways send tentacles of the tumor into surrounding tissue. **Table 18.2** lists the sequences of mutational events behind four types of cancer. Metastasis is not given its own "strike" stage because mutations that enable a tumor to metastasize may have been present from the beginning of the disease.

Researchers use several techniques that compare tumors to deduce the mutational steps that drive the disease. One approach is to count and compare mutations in tumor cells from people at different stages of the same type of cancer, as classified by body part. The older the tumor, the more genetic changes have accumulated. A mutation present in all stages among several individuals' tumors acts early in the disease process, whereas a mutation seen only in the tumor cells of sicker people acts later. Looked at another way, when a cancer is considered over time in the same individual, even an advanced, widely spread cancer has the same initial driver mutations.

The evolution of a cancer is similar to the evolution of species in that genetic changes accumulate over time. Just as linear changes may represent only part of an evolutionary tree diagram, linear progressions of mutations may be too simple as a way to view the development of cancer. In many tumors, cell lineages branch when they acquire new mutations, which may accelerate or accompany metastasis (**figure 18.11**).

The following subsections look at specific oncogenes and tumor suppressor genes. Chromosome abnormalities can cause these mutations. The chromosomes in cancer cells may be abnormal in number and/or structure. They may bear translocations, inversions, or have extra or missing pieces. A translocation that joins parts of nonhomologous chromosomes can hike expression of a gene enough to turn it into an oncogene. Duplications can increase the number of copies of a particular oncogene from two—one on each of a pair of homologs—to up to 100. A deletion may remove a tumor suppressor gene. A one-time event called chromothripsis shatters several chromosomes and may kill the cell—or trigger cancer. The chapter opener of chapter 13 discusses chromothripsis. Point mutations can also cause cancer.

A Closer Look at Oncogenes

Genes that normally trigger cell division are called **proto-oncogenes**. They are active where and when high rates of cell division are necessary, such as in a wound or in an embryo. When proto-oncogenes are transcribed and translated

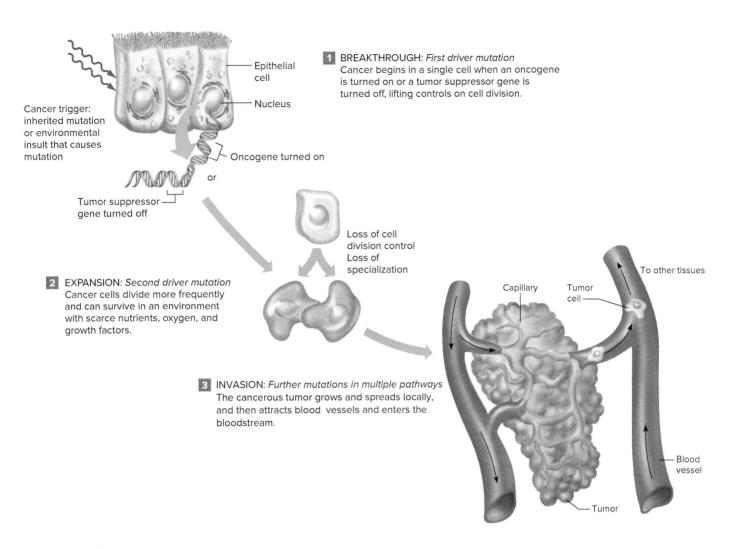

Figure 18.10 The "three strikes" of cancer.

1. BREAKTHROUGH: *First driver mutation*
Cancer begins in a single cell when an oncogene is turned on or a tumor suppressor gene is turned off, lifting controls on cell division.

Cancer trigger: inherited mutation or environmental insult that causes mutation

Oncogene turned on

or

Tumor suppressor gene turned off

Epithelial cell

Nucleus

Loss of cell division control
Loss of specialization

2. EXPANSION: *Second driver mutation*
Cancer cells divide more frequently and can survive in an environment with scarce nutrients, oxygen, and growth factors.

3. INVASION: *Further mutations in multiple pathways*
The cancerous tumor grows and spreads locally, and then attracts blood vessels and enters the bloodstream.

To other tissues

Capillary

Tumor cell

Blood vessel

Tumor

Table 18.2	Three Strikes Drive Cancer			
	Melanoma	**Pancreatic Cancer**	**Cervical Cancer**	**Colorectal Cancer**
Breakthrough	BRAF	KRAS	TP53, RB	APC
Expansion	TERT	CDK2NA	PIK3CA	KRAS
Invasion	CDK2NA	SMAD4	MAPK1	SMAD4
	TP53	TP53	STK11	TP53
	PIK3CA		FBXW7	PIK3CA
				FBXW7

too rapidly or frequently, or perhaps at the wrong time in development or place in the body, they function as oncogenes (*onco* means "cancer"). Usually oncogene activation is associated with a point mutation or a chromosomal translocation or inversion that places the gene next to another that is more highly expressed (transcribed). Oncogene activation causes a gain-of-function. In contrast, a tumor suppressor gene mutation is usually a deletion that causes a loss-of-function (see figure 4.8).

Proto-oncogenes can also become oncogenes by being physically next to highly transcribed genes. Three examples of genes that can activate proto-oncogenes are a viral gene, a gene encoding a hormone, and parts of antibody genes.

Models of tumor evolution

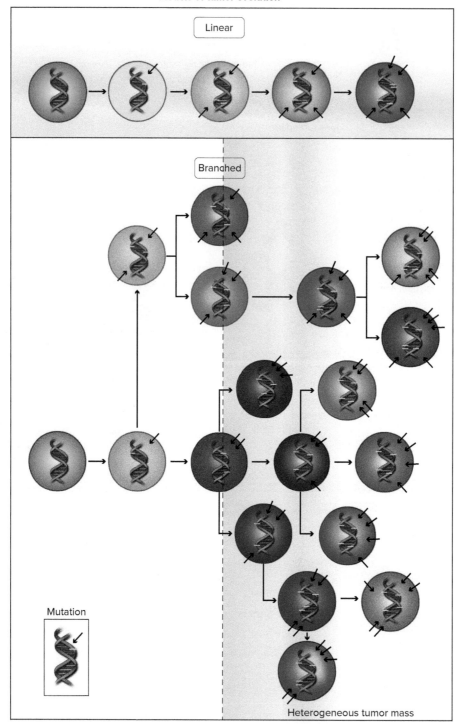

Figure 18.11 The evolution of a cancer. Genetic changes in cancer may occur in a linear and/or branching pattern. A heterogeneous tumor mass evolves as new mutations arise, while old ones remain. Note that the number of mutations increases from left to right.

A virus infecting a cell may insert DNA next to a proto-oncogene. When the viral DNA is rapidly transcribed, the adjacent proto-oncogene (now an oncogene) is also rapidly transcribed. Increased production of the oncogene's encoded protein then switches on genes that promote mitosis, triggering the cascade of changes that leads to cancer. Viral damage to a human genome may be catastrophic, activating and amplifying oncogenes as well as inverting and translocating chromosomes. Viruses cause cervical cancer, Kaposi sarcoma, and acute T cell leukemia.

A proto-oncogene may be activated when it is moved next to a gene that is normally very actively transcribed. This happens when an inversion on chromosome 11 places a proto-oncogene next to a DNA sequence that controls transcription of the parathyroid hormone gene. When the gland synthesizes the hormone, the oncogene is expressed, too. Cells in the gland divide, forming a tumor.

Antibody genes are among the most highly transcribed, so it isn't surprising that a translocation or inversion that places a proto-oncogene next to an antibody gene causes cancer. Cervical cancer and anal cancer following human papillomavirus infection may begin when proto-oncogenes are mistakenly activated with antibody genes. Similarly, in Burkitt lymphoma, a cancer common in Africa, a large tumor develops from lymph glands near the jaw. Epstein-Barr virus stimulates specific chromosome movements in maturing B cells to assemble antibodies against the virus, and a translocation places a proto-oncogene on chromosome 8 next to an antibody gene on chromosome 14. The oncogene is overexpressed, and the cell division rate increases. Tumor cells of Burkitt lymphoma patients have the translocation (**figure 18.12**).

A proto-oncogene may not only move next to another gene, but also be transcribed and translated with it as if they are one gene. The double gene product, called a **fusion protein**, activates or lifts control of cell division. For example, in acute promyelocytic leukemia, a translocation between chromosomes 15 and 17 brings together a gene coding for the retinoic

acid cell surface receptor and an oncogene called *myl*. The fusion protein functions as a transcription factor, which, when overexpressed, causes cancer. The nature of this fusion protein explains why some patients who receive retinoid (vitamin A–based) drugs recover. Their immature, dedifferentiated cancer cells, apparently stuck in an early stage of development where they divide frequently, suddenly differentiate, mature, and die. Perhaps the cancer-causing fusion protein prevents affected white blood cells from getting enough retinoids to specialize, locking them in an embryonic-like, rapidly dividing state. Supplying extra retinoids allows the cells to resume their normal developmental pathway. **A Glimpse of History** describes the fusion protein that led to development of Gleevec, one of the first cancer drugs targeted to cells with specific mutations.

Another way that an oncogene can cause cancer is by excessive response to a growth factor. In about 25 percent of women with breast cancer, affected cells have 1 to 2 million copies of a cell surface protein called HER2 that is the product of an oncogene. The normal number of these proteins is 20,000 to 100,000 per cell.

The HER2 proteins are receptors for epidermal growth factor. In breast cells, the receptors traverse the plasma membrane, extending outside the cell into the extracellular matrix and also dipping into the cytoplasm. They function as a tyrosine kinase, as is the case for the leukemia described in A Glimpse of History. When the growth factor binds to the tyrosine (an amino acid) of the receptor, the tyrosine picks up a phosphate group, which signals the cell to activate transcription of genes whose protein products stimulate cell division. In HER2 breast cancer, too many tyrosine kinase receptors send too many signals to divide.

HER2 breast cancer usually strikes early in adulthood and spreads quickly. However, a monoclonal antibody-based drug called Herceptin (trastuzumab) binds to the receptors, blocking the signal to divide. Herceptin works when the extra receptors arise from multiple copies of the gene, rather than from extra transcription of a single *HER2* gene.

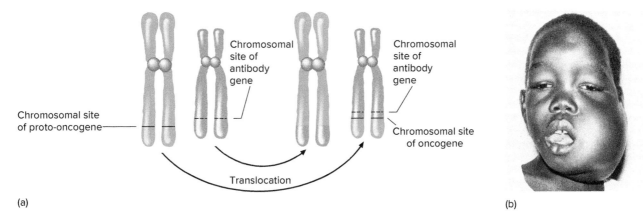

(a)

(b)

Figure 18.12 **A translocation that causes cancer.** **(a)** In Burkitt lymphoma, a proto-oncogene on chromosome 8 moves to chromosome 14, next to a highly expressed antibody gene. Overexpression of the translocated proto-oncogene, now an oncogene, triggers the molecular and cellular changes of cancer. **(b)** Burkitt lymphoma typically affects the jaw. *(a): Adapted from "Tissue repair and stem cell renewal in carcinogenesis," Philip A. Beachy, Sunil S. Karhadkar, and David M. Berman, Nature 432:324–331. Macmillan Publishers Ltd, 2004; (b): Source: Centers for Disease Control and Prevention*

On August 13, 1958, two men entered hospitals in Philadelphia and reported weeks of fatigue. Each had a very high white blood cell count and was diagnosed with chronic myelogenous leukemia (CML). Too many immature white blood cells were crowding the healthy cells. The men's blood samples eventually fell into the hands of pathologist Peter Nowell and cytogeneticist David Hungerford. They had developed ways to stimulate white blood cells to divide in culture, and they probed the chromosomes of both leukemic and normal-appearing white blood cells in the two tired men and five others with CML.

Nowell and Hungerford discovered a small, unusual chromosome that was only in the leukemic cells, the first chromosome abnormality linked to cancer. Later, it would be dubbed "the Philadelphia chromosome" (Ph¹). By 1972, new chromosome stains that distinguished AT-rich from GC-rich regions revealed that Ph¹ is the result of a reciprocal translocation (see figure 13.19). By 1984, researchers had homed in on the two genes juxtaposed in the translocation between chromosomes 9 and 22. One gene from chromosome 9 is called the Abelson oncogene (*abl*), and the other gene, from chromosome 22, is called the breakpoint cluster region (*bcr*). Two different fusion

genes form. The *bcr-abl* fusion gene is part of the Philadelphia chromosome, and it causes CML. (The other fusion gene does not affect health.)

The discovery that the fusion protein, called the BCR-ABL oncoprotein, is a form of the enzyme tyrosine kinase (the normal product of the *abl* gene) led directly to development of the drug Gleevec. A kinase is an enzyme that transfers a phosphate group (PO_4) to another molecule, and is a key part of signal transduction pathways. The cancer-causing form of tyrosine kinase is active for too long, which sends signals into the cell, stimulating it to divide, too many times.

Through the 1980s, drug developers tested more than 400 small molecules in search of one that would block the activity of the errant tyrosine kinase, without derailing other important enzymes. They found what would become Gleevec in 1992. The drug nestles into the pocket on the tyrosine kinase that must bind ATP to stimulate cell division. With ATP binding blocked, cancer cells do not receive the message to divide, and they cease doing so. After passing safety tests, the drug worked so dramatically that it set a new speed record for drug approval—10 weeks. However, after several years of using Gleevec, resistant cancer cells may appear. Other drugs are available to continue targeted treatment for these patients.

A Closer Look at Tumor Suppressor Genes

Some cancers result from loss or silencing of a tumor suppressor gene that normally inhibits expression of genes involved in all of the activities that turn a cell cancerous. Cancer can result when a tumor suppressor gene is deleted or if the promoter region binds too many methyl (CH_3) groups, which blocks transcription.

Wilms' tumor is an example of a cancer that develops from loss of tumor suppression. A gene is deleted that normally halts mitosis in the rapidly developing kidney tubules in the fetus. As a result, an affected child's kidney retains pockets of cells dividing as frequently as if they were still in the fetus, eventually forming a tumor. Following are descriptions of specific tumor suppressor genes.

Retinoblastoma

Retinoblastoma (RB) is a rare childhood eye tumor (**figure 18.13**). **A Glimpse of History** traces recognition of this cancer to many years ago.

About 1 in 20,000 infants develops RB, and half of them have inherited susceptibility to the disease. They have one germline mutant allele for the *RB1* gene in each of their cells, and then cancer develops in a somatic cell where the second copy of the *RB1* gene mutates. Therefore, inherited retinoblastoma requires two point mutations or deletions, one germline and one somatic. In some sporadic (noninherited) cases, two

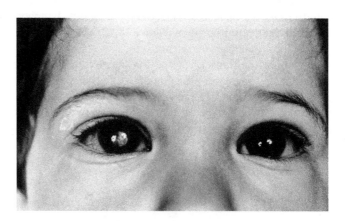

Figure 18.13 **Retinoblastoma type 1 is due to a mutation in a tumor suppressor gene.** The retinal tumor appears opaque in photographs because it reflects light. Fortunately nearly all cases of retinoblastoma are treatable today with surgery and/or chemotherapy. © *Science Source*

somatic mutations occur in the *RB1* gene, one on each copy of chromosome 13. Either way, the cancer usually starts in a cone cell of the retina, which provides color vision. Study of retinoblastoma inspired the "two-hit" hypothesis of cancer causation—that two mutations (germline and somatic or two somatic) are required to cause a cancer related to tumor suppressor deletion or malfunction. (A second form of retinoblastoma is caused by mutation in an oncogene, *MYCN*.)

People have been aware of retinoblastoma for a long time. A 2000 B.C.E. Mayan stone carving shows a child with a bulging eye. A Dutch anatomist provided the earliest clinical description as a growth "the size of two fists." In 1886, researchers noted that the cancer can be inherited, and in those families, secondary tumors sometimes arose, usually in bone. Once flash photography was invented, parents would notice the disease as a white spot in the pupil on a photograph, from light reflecting off a tumor.

The discovery that many children with RB have deletions in the same region of the long arm of chromosome 13 led researchers to the *RB1* gene and its protein product, which linked the cancer to control of the cell cycle. The RB1 protein normally binds transcription factors so that they cannot activate genes that carry out mitosis. It normally halts the cell cycle at G_1. When the *RB1* gene is mutant or missing, the hold on the transcription factor is released, and cell division ensues.

For many years, the only treatment for retinoblastoma was removal of the affected eye. Today, children with an affected parent or sibling, who have a 50 percent chance of having inherited the mutant *RB1* gene, can be monitored from birth so that noninvasive treatment (chemotherapy) can begin early. Full recovery is common.

Mutations in the *RB1* gene cause other cancers. Some children successfully treated for retinoblastoma develop bone cancer as teens or bladder cancer as adults. Mutant *RB1* genes have been found in the cells of patients with breast, lung, or prostate cancers, or acute myeloid leukemia, who never had the eye tumors. Expression of the same mutation in different tissues may cause these cancers.

p53 Normally Prevents Many Cancers

A single gene that causes a variety of cancers when mutant is *p53*. Recall from chapter 12 that the p53 protein transcription factor "decides" whether a cell repairs DNA replication errors or dies by apoptosis. If a cell loses a *p53* gene, or if the gene mutates and malfunctions, a cell with damaged DNA is permitted to divide, and cancer may be the result.

More than half of human cancers arise from a point mutation or deletion in *p53*. This may be because p53 protein is a genetic mediator between environmental insults and development of cancer (**figure 18.14**). A type of skin cancer, for example, is caused by a *p53* mutation in skin cells damaged by an excessive inflammatory response that can result from repeated sunburns. That is, *p53* may be the link between sun exposure and skin cancer.

In most *p53*-related cancers, mutations occur only in somatic cells. However, in the germline condition Li-Fraumeni syndrome, family members who inherit a mutation in *p53* have a very high risk of developing cancer—50 percent do so by age 30, and 90 percent by age 70. A somatic *p53* mutation in the affected tissue results in cancer because a germline mutation in the gene is already present.

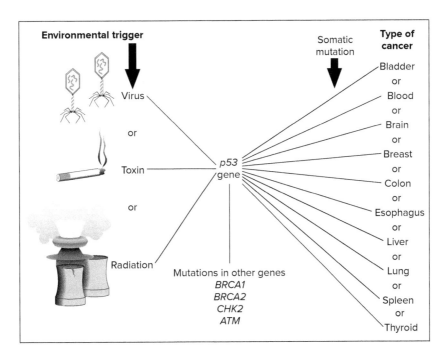

Figure 18.14 *p53* **cancers reflect environmental insults.** The environment triggers mutations or changes in gene expression that lead to cancer. The *p53* gene is a mediator—"the guardian of the genome." The protein products of many genes interact with p53 protein. (*p53* is also called *TP53,* for tumor protein.)

Stomach Cancer Lifts Cellular Adhesion

E-cadherin normally acts as a cellular adhesion protein found in tissue linings, but it is also a tumor suppressor because when its gene is deleted, cancer results. This was the case for the Bradfield family. Golda Bradfield died of stomach cancer in 1960. By the time some of her grown children developed the cancer too, the grandchildren began to realize that their family had a terrible legacy. Genetic testing revealed familial diffuse gastric cancer, caused by an "exon skipping" missense mutation in the E-cadherin gene that deletes an entire exon as the mRNA is transcribed from the gene (see section 12.5).

Golda's grandchildren had genetic tests. Eleven of them had inherited the mutant gene, but scans of their stomachs did not show any tumors. Still, they all had their stomachs removed to prevent the cancer. Most of them already had hundreds of tumors, too tiny to have been seen on medical scans. The cousins without stomachs are doing well. Like people who have their stomachs surgically shrunk to lose weight, the cousins avoid hard-to-digest foods and eat a little at a time, throughout the day. The inconvenience, they say, is a fair trade for eliminating the fear of their grandmother's, parents', and aunts' and uncles' fates—stomach cancer.

BRCA1/BRCA2 Mutations Disrupt Repair

Breast cancer that runs in families may be due to inheriting a germline mutation and then having a somatic mutation occur in a breast cell (a familial form), or two somatic mutations affecting the same breast cell (a sporadic form), as figure 18.5 depicts for lung cancer. However, breast cancer is so common that a family with many affected members may actually have multiple sporadic cases, rather than an inherited form of the disease causing them all.

Only about 5 percent of breast cancers are familial, caused by mutations in any of at least 20 genes. Most of the genes associated with susceptibility to familial breast cancer encode proteins that interact in ways that enable DNA to survive damage. If DNA cannot be repaired, mutations that directly cause cancer can accumulate and persist.

The two major breast cancer susceptibility genes are *BRCA1* and *BRCA2*. Together they account for 15 to 20 percent of the 5 percent of cases that are familial. A *BRCA* mutation is inherited in an autosomal dominant manner, with incomplete penetrance because it increases susceptibility, rather than directly causing cancer.

Inheriting a mutation in *BRCA1*, which stands for "breast cancer predisposition gene 1," greatly increases the lifetime risk of inheriting breast and/or ovarian cancer. This risk, however, varies in different population groups because of the modifying effects of other genes.

The most common *BRCA1* mutation deletes two adjacent DNA bases, altering the reading frame and shortening the protein. Thousands of mutations in the gene are known, most of them rare. Cancer risk depends on the site of the mutation. For *BRCA1*, mutations in the ends of the gene are more likely to increase risk of breast cancer and mutations

in the middle part of the gene are more likely to increase risk of ovarian cancer.

BRCA1 encodes a protein that interacts with many other proteins that counter DNA damage in several ways. One important form of protection is the mending of areas of the genome where both DNA strands are broken at the same site. These double-stranded breaks are particularly dangerous because they cut the chromosomes all the way through, making rearrangements such as deletions and translocations possible. **Figure 18.15** depicts the central role of *BRCA1* in protecting DNA.

BRCA1 mutations have different incidences in different populations. Only 1 in 833 people in the general U.S. population has a *BRCA1* mutation. That figure is more than 1 in 50 among Ashkenazi Jewish people, due to population bottlenecks and nonrandom mating. The *BRCA1* gene was initially discovered in Ashkenazi families in which several members developed the associated breast cancer at very young ages. In this

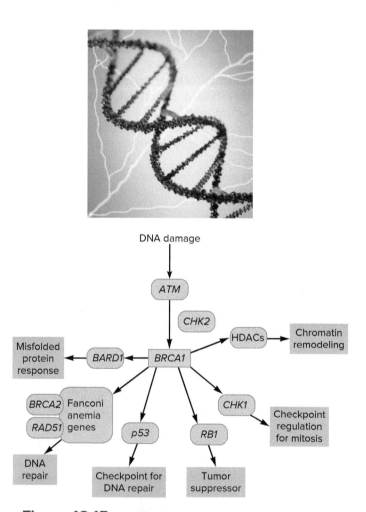

Figure 18.15 **The *BRCA1* gene controls many cellular defense mechanisms.** *BRCA1* functions as a genetic crossroads in handling DNA damage. Inheriting a *BRCA1* mutation increases susceptibility to cancers of the breast, ovary, cervix, uterine tube, uterus, peritoneum, pancreas, and prostate gland. Mutations in any of the genes with which *BRCA1* interacts increases cancer susceptibility too. (HDACs are histone deacetylases.) © *Comstock Images/Jupiter Images*

population, a woman who inherits a *BRCA1* mutation has up to an 87 percent risk of developing breast cancer over her lifetime (five times the general population risk), and a 50 percent risk of developing ovarian cancer. For Ashkenazi families without a strong clinical history of breast cancer, the risks are 65 percent for breast cancer and 39 percent for ovarian cancer.

Some women who learn that they have inherited an allele predisposing to breast or ovarian cancer have their breasts and/or ovaries removed. This action makes more sense for an Ashkenazi woman facing an 87 percent lifetime risk of breast or ovarian cancer than it does for a woman in a population group with a much lower risk. The general population risk of actually developing a *BRCA1* cancer if one inherits a mutation is only about 10 percent, based on empirical (observational) evidence. Women with such mutations born after 1940 have a higher risk than those born earlier, suggesting that the environment also plays a role in whether inheriting a mutation causes cancer.

BRCA2 breast cancer is also more common among the Ashkenazim. Ashkenazi women who inherit a mutation in *BRCA2* face a 45 percent lifetime risk of developing breast cancer and an 11 percent risk of developing ovarian cancer. Men who inherit a *BRCA2* mutation have a 6 percent lifetime risk of developing breast cancer, which is 100 times the risk for men in the general population. Inheriting a *BRCA2* mutation also increases the risk of developing cancers of the colon, kidney, prostate, pancreas, gallbladder, skin, or stomach.

Tests for any cancer susceptibility genes can reveal a "variant of uncertain significance" (see **Bioethics** in section 12.7). This means that the gene sequence is not wild type, but the identified variant has not been associated with increased cancer susceptibility or any other disease phenotype.

Environmental Causes of Cancer

Environmental factors contribute to cancer by mutating or altering the expression of genes that control the cell cycle, apoptosis, and DNA repair. Inheriting a susceptibility gene places a person farther along a particular road to cancer, but cancer can happen in somatic cells in anyone. It is more practical, for now, to identify environmental cancer triggers and develop ways to control them or limit our exposure to them, than to alter genes that cause or predispose a person to develop cancer.

Looking at cancer at a population level reveals the interactions of genes and the environment. For example, researchers examined samples of non-Hodgkin's lymphoma tumors from 172 farmers, 65 of whom had a specific chromosomal translocation. The 65 farmers were much more likely to have been exposed for long times to toxic insecticides, herbicides, fungicides, and fumigants, compared to the farmers with lymphoma who did not have the translocation. These findings associate exposure to the chemicals with developing the cancer.

Determining precisely how an environmental factor such as diet affects cancer risk can be complicated. Consider the cruciferous vegetables, such as broccoli and brussels sprouts, which are associated with decreased risk of colon cancer. These vegetables release chemicals that activate enzymes that detoxify carcinogenic products of cooked meat, called heterocyclic aromatic amines. With a vegetable-poor, meaty diet, these amines accumulate and elevate cancer risk. They cross the lining of the digestive tract and circulate to the liver, where enzymes metabolize them into compounds that cause driver mutations for colon cancer. Therefore, careful combining of foods can lower cancer risk—add brussels sprouts to that steak dinner!

Both environmental and genetic factors contribute to cancer risk. Consider melanoma, a type of skin cancer (see figure 18.2). Sun exposure elevates risk, but certain variants of the melanocortin-1-receptor (*MC1R*) gene, known for imparting red hair, fair skin, and freckles, double the risk of developing melanoma, even if a person avoids intense sunlight.

Key Concepts Questions 18.3

1. Distinguish between driver and passenger mutations.
2. Explain how cancer is a multistep process.
3. Discuss the role of chromosomes in causing cancer.
4. Compare and contrast the mechanisms by which oncogenes and tumor suppressor genes cause cancer.
5. How does the environment contribute to causing cancer?

18.4 Diagnosing and Treating Cancer

The experience of having cancer has changed markedly in recent years, as both diagnosis and treatment have evolved from general measures to highly specific strategies to classify and control the disease. New approaches go beyond identifying oncogenes and tumor suppressor mutations.

The ability to sequence cancer exomes and genomes is providing a new level of clinical specificity. For example, if several members of a family have cancer of the same body part, yet they all test negative for known cancer gene driver mutations, then sequencing their exomes can reveal other genes that are mutant in the cancer cells and have a function that could disrupt the cell cycle or DNA repair. If researchers can then find the same mutations in other families with the same type of cancer, then a new entry can be added to the list of cancer driver genes, and perhaps a new drug target identified—or even a discovery that an existing drug might work.

For a long time, cancer was diagnosed only after a person reported symptoms to a physician, such as abdominal bloating, a lump, bruising, a scaly patch of skin, or fatigue. Follow-up tests then might include scans to distinguish a solid mass from a cyst, then biopsies to check for cancer cells in a tissue sample, using special stains to highlight the cancer cells, as in figure 18.2, or noting cell shape changes and loss of specialized characteristics. In contrast to such a diagnostic test that follows reporting of symptoms, a screening test for cancer is done on groups of people who are at elevated risk, but do not have symptoms. An example of a cancer screening test is to test for *BRCA1* or *BRCA2* mutations in an Ashkenazi woman who has young relatives who have or had breast or ovarian cancer, but who does not have symptoms or signs of the diseases herself. Sometimes large populations are screened if the test is economical, such as screening men over age 50 for prostate specific antigen for prostate cancer, which is a protein biomarker.

The traditional ways of treating cancer are broad. The oldest approach, surgery, is still effective and often the first treatment, if it removes a primary tumor before it has invaded healthy tissue and spread through the bloodstream. The other two standard approaches, chemotherapy and radiation, hit all rapidly dividing cells, causing adverse effects when they harm healthy cells. Newer approaches that are targeted based on genetic information, such as the tyrosine kinase inhibitor Gleevec (see A Glimpse of History in section 18.3) are highly effective if matched to a sensitive tumor cell type. For all of these approaches, however, it takes just a few escaped or resistant cancer cells to sow the seeds of a future tumor.

Following are examples of a new approach to diagnosing cancer (based on gene expression patterns) and to treating cancer (combining immune system components).

Gene Expression Profiling

Gene expression profiling can specifically describe the type of cell that turns cancerous, which informs treatment choices. A striking example is what is now recognized as mixed lineage leukemia (MLL) (**figure 18.16**). The discovery came from the observation that 10 percent of children with acute lymphoblastic leukemia (ALL) do not respond to the chemotherapy that helps the other 90 percent, although all the patients have the same symptoms of fever, fatigue, and bruising. The leukemic cells all look alike under a microscope. However, gene expression analysis using DNA microarrays (chips) revealed that in the 10 percent of patients who did not respond to standard drugs, the cancerous cells make too little of about 1,000 proteins and too much of 200 others, compared to protein production in children who did respond to the standard "cocktail" of chemotherapy drugs. Those 10 percent, who actually have MLL and not ALL, respond to different drugs.

Chimeric Antigen Receptors

The immune system regularly recognizes and destroys cancer cells, but **chimeric antigen receptor technology** improves upon nature. It creates DNA instructions for a hybrid surface protein on T cells that the body does not normally synthesize. The protein, called a chimeric antigen receptor, or CAR, is part T cell receptor and part antibody. (Figure 17.4 shows both T cell receptors and antibodies.) The receptor part is engineered to guide the T cell to a specific target, such as cancer cells. The antibody part then binds the target, alerting the immune system to respond and kill the cancer cells. A CAR is a little like a drone.

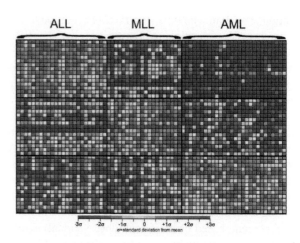

Figure 18.16 **Cancer cells that look alike may be genetically distinct.** These leukemias—ALL, MLL, and AML—differ in gene expression patterns. The columns of squares represent DNA from tumor samples and the rows compare the activities of particular genes. Red tones indicate higher-than-normal expression and blue tones show lower-than-normal expression. The different patterns indicate distinct cancers, although the cells look alike under a microscope.

The first patient to receive CAR-based treatment was a 65-year-old man with leukemia who had exhausted all treatments. He recovered after receiving CARs made from his own cells. In 2012, a 6-year-old with another form of leukemia and who had also tried everything possible received CAR treatment using her own cells. She had been near death when treated that April, but was back in school by September, her hair regrown and energy back. In 2015, another young girl who had undergone chemotherapy and a bone marrow transplant but was still very sick had CAR treatment, but with a key difference from the man and the other child—she had UCAR treatment, the "U" standing for universal. She received CAR T cells from a donor, and recovered within a month. The ability to use donated cells means that more people will be able to be treated. However, fatalities have occurred resulting from use of harsh chemotherapy drugs needed to clear the bone marrow of cancer cells before CAR treatment.

Genome editing tools (see section 19.4) are used to create the chimeric antigen receptors. Typically a patient receives a billion or more altered T cells. In the leukemia that the two girls had, the engineered cells bind CD19 antigens that are abundant on the surfaces of the cancerous B cells. CAR technology is also being tested on multiple myeloma, brain cancer, breast cancer, and soft tissue cancers.

Unfortunately, living with cancer entails not just diagnosis and treatment, but also the possibility of recurrence of the disease. This is why patients are checked frequently, even if they feel well, and even if years have elapsed since the first illness. **Clinical Connection 18.1** discusses how liquid biopsy detects cancer recurrence with a blood test.

A Liquid Biopsy Monitors Cancer Recurrence and Response to Treatment

The fact that cancerous tumors shed DNA into the bloodstream can be used to monitor disease and perhaps even to detect disease in people who do not have symptoms (**figure 18A**). Checking DNA pieces in the blood plasma for oncogene or tumor suppressor mutations is termed a "liquid biopsy." The entire genome sequence of a cancer can be deduced by overlapping DNA pieces from the bloodstream. The simple blood test of a liquid biopsy is much less painful and invasive than a traditional biopsy, which samples cancer cells from a solid tumor, such as in a breast or the liver, or from a lining such as the skin.

The DNA detected in a liquid biopsy is called cell-free tumor DNA, or ctDNA. It is similar to the cell-free fetal DNA that is detected in pregnant women (see figure 13.10). Very rarely, such a prenatal test detects cancer in an unsuspecting pregnant woman.

The first uses of liquid biopsy were in people who already had cancer, and required monitoring. If a liquid biopsy soon after surgery to remove a tumor does not have ctDNA, but 2 years later does, then the cancer likely has recurred, and may have new mutations. Liquid biopsy is also useful for monitoring response to treatment. If a drug is working, the level of ctDNA will decrease. If a cancer has become resistant to a drug, the level of ctDNA will increase.

Using tumor DNA as a biomarker is more specific than using a protein biomarker because a protein biomarker may also be present on healthy cells. Cell-free tumor DNA can also be collected from urine (bladder cancer), sputum (lung cancer), and feces (colorectal cancer). A liquid biopsy is particularly useful when tissue is difficult to obtain or not enough cells are sampled, or when cancer has spread and the initial site is unknown.

More controversial than detection of ctDNA for people who already have cancer is its use to screen high-risk populations, such as people who smoke (for lung cancer) or people with family histories of specific cancers. A finding of ctDNA in a person who does not have cancer symptoms can warn the individual to seek additional types of tests that might diagnose cancer very early.

Liquid biopsy might be used someday on everyone, to detect cancers that are entirely unsuspected. For example, ovarian cancer is sometimes diagnosed at a very late stage because the symptoms of bloating and fatigue are vague and common, and may be attributed to other factors, such as a change

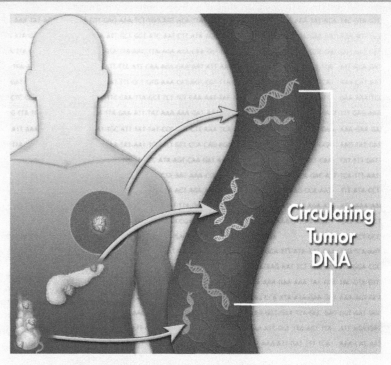

Figure 18A Liquid biopsy. Detecting tumor cell DNA in the blood is a less invasive way to monitor cancer recurrence than a traditional surgical biopsy. *Source: National Human Genome Research Institute.*

in diet or exercise routine. The danger of using liquid biopsy on people of average risk, however, is a false positive finding—an oncogene or tumor suppressor variant that doesn't cause cancer in everyone, due to effects of other genes. Current clinical trials are performing liquid biopsies on thousands of healthy people at low or average risk of cancer to assess the predictive value of the technology.

Questions for Discussion

1. What would you want to know before you have a liquid biopsy?

2. What are the advantages and possible disadvantages of a liquid biopsy?

3. Explain why the results of a liquid biopsy would likely change as a cancer is monitored over many years.

4. When in the course of checking someone for cancer do you think a liquid biopsy should be offered?

Continuing analysis of cancer cell genomes combined with the ongoing annotation of human genomes (discovering what genes do) will tell us more about exactly what happens when a cell becomes malignant. But conquering cancer may be an elusive goal. The DNA of cancer cells mutates in ways that enable cells to pump out any drug sent into them. Cancer cells have redundant pathways, so that if a drug shuts down angiogenesis or invasiveness, the cell completes the task another way. A more realistic goal than eradicating cancer may be to kill enough cancer cells, and sufficiently slow their spread, so that it takes the remainder of a human lifetime for tumors to grow back enough to harm health. In this way, cancer can become a chronic, manageable condition. As long as our cells divide, we are at risk of developing cancer.

Key Concepts Questions 18.4

1. How can exome sequencing help to identify a previously unknown cancer gene in a family?

2. Contrast traditional approaches to diagnosing and treating cancer to newer methods.

3. Explain how testing for mutations and for gene expression patterns can help a physician to select a treatment for a patient.

4. Discuss how chimeric antigen receptor technology treats a cancer.

Summary

18.1 Cancer Is an Abnormal Growth That Invades and Spreads

1. Cancer is a genetically driven loss of cell cycle control, creating a population of highly proliferative cells that invades surrounding tissue.

2. Cancer-causing mutations (activated **oncogenes**, inactivated **tumor suppressor genes**, and mutant DNA repair genes) affect cell fate, cell survival, and genome maintenance. They affect transcription factors, cell cycle checkpoint proteins, growth factors, repair proteins, or telomerase.

3. Spread of a cancer is called **metastasis**.

4. Sporadic cancers result from two recessive somatic mutations in the two copies of a gene. They are more common than inherited cancers that are caused by germline mutations that confer susceptibility in all cells along with somatic mutations in affected tissue.

18.2 Cancer at the Cellular Level

5. A tumor cell divides more frequently or more times than cells surrounding it, has altered surface properties, loses the specializations of the cell type it arose from (**dedifferentiates**), and produces daughter cells like itself.

6. A malignant tumor infiltrates tissues and can metastasize by attaching to basement membranes and secreting enzymes that penetrate tissues and open a route to the bloodstream. **Angiogenesis** establishes a blood supply for a tumor. A cancer cell can travel, establishing secondary tumors.

7. Cell specialization and position within a tissue affect whether cancer begins and persists.

8. **Cancer stem cells** can divide to yield cancer cells and abnormally differentiated cells.

9. A cell that dedifferentiates and/or expresses "stemness" genes can begin a cancer.

10. A mutation that enables a cell to divide continually can alter the percentages of cells in a tissue that can divide, resulting in an abnormal growth.

11. Chronic repair of tissue damage can provoke stem cells into producing an abnormal growth.

18.3 Cancer Genes and Genomes

12. A **driver mutation** provides a selective growth advantage to a cell. A **passenger mutation** occurs in a cancer cell but does not cause or contribute to the disease. Passengers greatly outnumber drivers.

13. Driver mutations hit with three strikes: breakthrough (initial mutation in one cell), expansion (second mutation hikes division rate), and invasion. Mutations that will cause metastasis may be present early.

14. Sequences of driver mutations may be linear or branched.

15. Chromosomes in cancer cells may be highly abnormal (translocations, inversions, deletions, duplications, and chromothripsis).

16. Cancer is often the result when **proto-oncogenes** become oncogenes or when tumor suppressor genes are deleted or silenced. Mutations in DNA repair genes cause cancer by enabling mutations to persist.

17. Proto-oncogenes normally promote controlled cell growth, but are overexpressed because of a point mutation, placement next to a highly expressed gene, or transcription and translation with another gene, producing a **fusion protein**. Oncogenes may also be overexpressed growth factor receptor genes.

18. A tumor suppressor gene normally enables a cell to respond to factors that limit its division.

19. Environmental factors contribute to cancer by mutating genes and altering gene expression.

18.4 Diagnosing and Treating Cancer

20. Traditional cancer treatments are surgery, radiation, and chemotherapy. Newer approaches based on identifying mutations and **gene expression profiling** are used to subtype cancers and better target treatments.

21. **Chimeric antigen receptor technology** creates a cell-surface protein that is part T cell receptor and part antibody. It functions like a drone to highlight cancer to the immune system.

Review Questions

1. Distinguish between a benign and a malignant tumor.

2. Explain how a sporadic cancer differs from a Mendelian disease in terms of predicting risk of occurrence in the relatives of a patient.

3. Name the three basic cellular pathways that cancer disrupts.

4. Why is metastasis dangerous?

5. Explain the connection between cancer and control of the cell cycle.

6. List four characteristics of cancer cells.

7. Describe four ways that cancer can originate at the cell or tissue level.

8. Distinguish between a driver mutation and a passenger mutation.

9. Describe two abnormalities of chromosomes in cancer.

10. Explain how comparing mutations in cells from the same cancer type at different stages can reveal the sequence of genetic changes that drive the cancer.

11. How can classifying cancer genomically differ from classifying cancer histologically or by initial affected body part?

12. Distinguish between a proto-oncogene and an oncogene.

13. Describe two events that can activate an oncogene.

14. Explain why an oncogene is associated with a gain-of-function and a mutation in a tumor suppressor gene is associated with a loss-of-function.

15. How are retinoblastoma type 1, a *p53*-related cancer, inherited stomach cancer, and *BRCA1* breast cancer similar?

16. Explain how a virus can cause cancer, and provide an example.

17. What is the role of poor DNA repair in causing cancer?

18. Describe the three "strikes" of the origin and progression of cancer.

19. Distinguish among mutations, altered gene expression, and epigenetic changes in cancer.

20. Explain why not all cancers affecting the same cell type respond the same way to a particular drug.

21. Describe an alternative to performing surgery to determine whether a metastatic tumor is responding to a new drug.

Applied Questions

1. Herceptin (traztuzumab) is a monoclonal antibody–based drug that binds a receptor present in many copies on the cells of certain aggressive and early-onset breast cancers. (Monoclonal antibodies are discussed in section 17.4.)

    a. How does this treatment approach differ from standard chemotherapy?

    b. Herceptin is now used to treat tumors of the stomach and esophagus. Explain how the same drug can be used for cancers that affect different body parts.

2. Breast cancer can develop from inheriting a germline mutation and then undergoing a second mutation in a breast cell, or from two recessive mutations in a breast cell, one in each copy of a tumor suppressor gene. Cite another type of cancer, discussed in the chapter, that can arise in these two ways.

3. Why is a "cocktail" of several drugs likely to be more effective at slowing the course of a cancer than using a single, powerful drug?

4. von Hippel-Lindau syndrome is an inherited form of cancer. The responsible mutation lifts control over the transcription of certain genes, which, when overexpressed, cause tumors to form in the kidneys, adrenal glands, and blood vessels. Is the von Hippel-Lindau gene an oncogene or a tumor suppressor gene? Cite a reason for your answer.

5. The *BRCA2* gene causes some cases of Wilms' tumor and some cases of breast cancer. Explain how the same tumor suppressor mutation can cause different cancers.

6. Ads for the cervical cancer vaccine present the fact that a virus can cause cancer as startling news, when in fact this has been known for decades. Explain how a virus might cause cancer.

7. A tumor is removed from a mouse and broken up into cells. Each cell is injected into a different mouse. Although all the mice used in the experiment are genetically identical and raised in the same environment, the animals develop cancers with different rates of metastasis. Some mice die quickly, some linger, and others recover. What do these results indicate about the characteristics of the original tumor cells?

8. Explain why receiving a test result of a variant of uncertain significance in an oncogene or tumor suppressor gene might be upsetting.

9. State a genetic factor and an environmental factor that increase risk of developing melanoma.

10. Select a gene from table 18.2 and describe its general function as shown in figure 18.15.

11. If cancer genome sequencing from liquid biopsy becomes a clinical tool, why might it be useful to repeat the sequencing every few years, rather than basing treatment decisions for many years on just an initial sequence?

Case Studies and Research Results

1. Most cases of retinoblastoma (RB) are due to a germline mutation in the tumor suppressor gene *RB1* and typically affect both eyes. Researchers discovered extra copies of an oncogene, called *MYCN*, in children who had a tumor in one eye, but had no affected relatives, and who did not have mutations in *RB1*. *MYCN* sometimes becomes overexpressed in later stages of RB, but in the newly recognized type of disease, called MYCN RB, the oncogene is the initial driver mutation. The mutation is somatic. That is, the *MYCN* mutation originated in an eye cell of the affected child. Which form of retinoblastoma, the one due to the oncogene or the tumor suppressor, is likely to recur in a sibling of the affected child? Explain your answer.

2. In a large family, 15 people over four generations develop the skin cancer melanoma, but none of them have mutations in known melanoma genes. Researchers sequenced the exomes of several affected family members and discovered a mutation in the *TERT* gene. Explain how the cancer likely arises.

3. When I had thyroid cancer in 1993, I had to wait on the operating table while slices of my two tumors were sent to the pathology lab. The growths looked so unusual that the surgeon could not tell whether they were of a type that tended to spread or a less dangerous type. So the pathologist examined the tissue microscopically to determine that one tumor was papillary, and the other follicular—both not dangerous. What newer technology could have distinguished the tumor types at the time of biopsy weeks earlier?

4. DeShawn takes the drug Gleevec to treat his leukemia, and it has worked so well that he thinks he is cured. He stops taking the drug, and 4 months later his leukemia returns. This time, the cancer cells do not have the BCR-ABL mRNA characteristic of the disease. Explain what has happened.

5. The genomes of four patients with acute myeloid leukemia are sequenced and mutations in the following genes noted:

   patient 1   *IDH1* and *NPM1*

   patient 2   just *IDH1*

   patient 3   *IDH1, NPM1,* and *IDH2*

   patient 4   *IDH1, NPM1, IDH2,* and *FLT3*

   Explain how these patients can have the same diagnosis, yet mutations in different genes.

6. Which technology explained in the chapter could be used to investigate different activities in cells at the edges of a tumor compared to cells in the interior?

7. A "polygenic risk score" for breast cancer consists of common variants of 80 genes that collectively increase the risk of developing the cancer by 14 percent. Do you think that this test is useful? Cite a reason for your answer.

8. An epidemiological study found that average cumulative breast cancer risk to age 70 in women with *BRCA1* mutations is 50 percent for women born between 1920 and 1950 but 58 percent if born after 1950. What might account for the difference?

9. Liquid biopsy detected cell-free tumor DNA in some early-stage colon cancer patients, but not in others. How should researchers follow up this study?

CHAPTER 19

© Image Source

DNA Technologies

Pig poop pollutes, so researchers introduced into the pig genome a bacterial gene that encodes an enzyme that dismantles phosphorus. The result of raising genetically modified pigs: less-polluting manure.

Learning Outcomes

19.1 Patenting DNA
1. State the criteria for a patentable invention.
2. Discuss the history of patenting organisms and DNA.

19.2 Modifying DNA
3. Distinguish between recombinant DNA and transgenic organisms.
4. Describe applications of recombinant DNA technology.

19.3 Monitoring Gene Function
5. Explain how a DNA microarray is used to monitor gene expression.

19.4 Gene Silencing and Genome Editing
6. Describe ways to decrease expression of a specific gene.
7. Explain how genes and genomes are edited.

The BIG Picture

Ancient biotechnologies used selective breeding to give us bakeries and breweries, foods and medicines. Modern biotechnologies manipulate DNA to give us new ways to study, monitor, and treat disease, and alter foods and the environment.

Improving Pig Manure

Pig manure presents a serious environmental problem. The animals do not make an enzyme to extract the mineral nutrient phosphorus from a compound called phytate in cereal grains in their feed, so they are given dietary phosphorus supplements. As a result, their manure is full of phosphorus. The element washes into natural waters, contributing to fish kills, oxygen depletion in aquatic ecosystems, algal blooms, and the greenhouse effect. Fortunately, biotechnology solved the "pig poop" problem.

In the past, pig raisers have tried to keep their animals healthy and the environment clean by feeding animal by-products from which the pigs can extract more phosphorus and by giving supplements of the enzyme phytase, which liberates phosphorus from phytate. But consuming animal by-products can introduce prion diseases, and giving phytase before each meal is costly. A "phytase transgenic pig," however, is genetically modified to secrete bacterial phytase in its saliva, which enables the animal to excrete low-phosphorus manure. It is called an Enviropig™.

A transgenic organism has a gene in each of its cells from an organism of another species. The Enviropig has a phytase gene from the bacterium *E. coli,* as well as a promoter DNA sequence from a mouse that controls secretion of phytase from the salivary glands. Enviropig's manure has 75 percent less phosphorus than unaltered pig excrement.

19.1 Patenting DNA

DNA is the language of life, the instruction manual for keeping an organism alive. Yet we also use DNA for other purposes. Manipulating DNA is part of **biotechnology**, which is the use or alteration of cells or biological molecules for specific applications, including products and processes. Biotechnology is an ancient art as well as a modern science, and is familiar as well as futuristic. Using yeast to ferment fruit into wine is a biotechnology, as is extracting and using biochemicals from organisms.

The terminology for biotechnology can be confusing. The popular terms "genetic engineering" and "genetic modification" refer broadly to any biotechnology that manipulates DNA. It includes altering the DNA of an organism to suppress or enhance the activities of its own genes, as well as combining the genetic material of different species. Organisms that harbor DNA from other species are termed **transgenic** and their DNA is called recombinant. The Enviropig described in the chapter opener is transgenic.

Creating transgenic organisms is possible because all life uses the same genetic code—that is, the same DNA triplets encode the same amino acids (**figure 19.1**). Mixing DNA from different species may seem unnatural, but in fact DNA moves and mixes between species in nature—bacteria do it, and it is why we have viral DNA sequences in our chromosomes. But human-directed genetic modification usually gives organisms

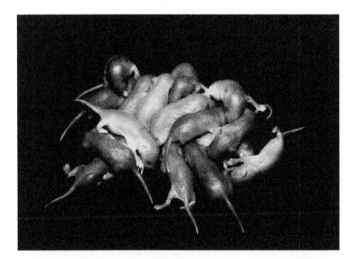

Figure 19.1 **The universality of the genetic code makes biotechnology possible.** The greenish mice contain the gene encoding a jellyfish's green fluorescent protein (GFP). Researchers use GFP to mark genes of interest. The GFP mice glow less greenly as they mature and more hair covers the skin. The nongreen mice are not genetically modified.
© Eye of Science/Science Source

traits they would not have naturally, such as fish that can tolerate very cold water, tomatoes that grow in salt water, and bacteria that synthesize human insulin.

What Is Patentable?

The invention of transgenic organisms raises legal questions, because the design of novel combinations of DNA may be considered intellectual property and, therefore, patentable. To qualify for patent protection, a transgenic organism, as any other invention, must be new, useful, and not obvious to an expert in the field. A corn plant that manufactures a protein naturally found in green beans but not in corn, thereby making the corn more nutritious, is an example of a patentable transgenic organism. A DNA sequence might be patentable if it is part of a medical device used to diagnose an inherited or infectious disease. DNA-based tests, for example, can identify specific mutations that cause cystic fibrosis. Another test can distinguish among bacterial pathogens and identify strains that are resistant to specific antibiotic drugs. DNA is also patentable as a research tool, as are algorithms used to extract information from DNA sequences and databases built of DNA sequences. The Technology Timeline highlights some of the events and controversies surrounding patenting of genetic material.

Patent law has had to keep up with modern biotechnology. In the 1980s, when sequencing a gene was painstakingly slow, only a few genes were patented. In the mid-1990s, with faster sequencing technology and shortcuts to finding the protein-encoding parts of the genome, the U.S. National Institutes of Health and biotech companies began seeking patent protection for thousands of short DNA sequences, even if their functions weren't known. Because of the flood of applications, the U.S. Patent and Trademark Office tightened requirements for usefulness. Today, with entire genomes being sequenced much faster than it once took to decipher a single gene, a DNA sequence alone does not warrant patent protection. It must be useful as a tool for research or as a novel or improved product, such as a diagnostic test or a drug. In the United States, more than one in five human genes is patented in some way, and only a few gene patents have been challenged.

DNA patenting became highly controversial in 2009, when several groups, including the American Civil Liberties Union and the Public Patent Foundation, challenged patents on two breast cancer genes (*BRCA1* and *BRCA2*) held by biotechnology company Myriad Genetics and the University of Utah. The company did not license the full gene sequences to other companies, so some patients were forced to take Myriad's test for increased familial breast cancer risk, costing them more than $3,000 to sequence the entire genes. (Companies could market tests for the most common mutations in the genes though.) The patents discouraged research and prevented patients from getting second opinions. In 2010, a federal judge in the United States ruled seven patents on the genes "improperly granted" because they are based on a "law of nature." In 2011, the court invalidated the patents on the two genes, but a federal appeals court overruled that action, claiming that an isolated gene is not the same as a gene in a cell, which is part of a chromosome.

PATENTING LIFE AND GENES

1790	U.S. patent act enacted. A patented invention must be new, useful, and not obvious.
1873	Louis Pasteur is awarded first patent on a life form, for yeast used in industrial processes.
1930	New plant variants can be patented.
1980	First patent awarded on a genetically modified organism, a bacterium given four DNA rings that enable it to metabolize components of crude oil.
1988	First patent awarded for a transgenic organism, a mouse that manufactures human protein in its milk. Harvard University granted a patent for "OncoMouse," transgenic for human cancer.
1992	Biotechnology company awarded patent for all forms of transgenic cotton. Groups concerned that this will limit the rights of subsistence farmers contest the patent several times.
1996–1999	Companies patent partial gene sequences and certain disease-causing genes for developing specific medical tests.
2000	With gene and genome discoveries pouring into the Patent and Trademark Office, requirements for showing utility of a DNA sequence are tightened.
2003	Attempts to enforce patents on non-protein-encoding parts of the human genome anger researchers who support open access to the information.
2007	Patent requirements must embrace a new, more complex definition of a gene.
2009	Patents on breast cancer genes challenged.
2010	Direct-to-consumer genetic testing companies struggle to license DNA patents for multigene and SNP association tests. Patents on breast cancer genes invalidated.
2011	U.S. government considers changes to gene patent laws.
2013	U.S. Supreme Court declares genes unpatentable.

In 2013 the U.S. Supreme Court ruled that "a naturally occurring DNA segment is a product of nature and not patent eligible merely because it has been isolated." The court did allow patenting of complementary DNA (cDNA), which is synthesized in a laboratory using an enzyme (reverse transcriptase) that makes a DNA molecule that is complementary in sequence to a specific mRNA. The cDNA represents only the exons of a gene because the introns are spliced out during transcription into mRNA. A cDNA is considered not to be a product of nature because its exact sequence is not in the genome of an organism.

1. How do modern applications of biotechnology differ from ancient applications?
2. What are the requirements for a patented invention?
3. Why did the U.S. Supreme Court rule that a gene's DNA sequence is not patentable?

19.2 Modifying DNA

Recombinant DNA technology adds genes from one type of organism to the genome of another. It was the first gene modification biotechnology, and was initially done with bacteria to make them produce peptides and proteins useful as drugs. When bacteria bearing recombinant DNA divide, they yield many copies of the "foreign" DNA, and under proper conditions they produce many copies of the protein that the foreign DNA specifies. Recombinant DNA technology is also known as gene cloning. "Cloning" in this context refers to making many copies of a specific DNA sequence.

A GLIMPSE OF HISTORY

In February 1975, molecular biologists convened at Asilomar, on California's Monterey Peninsula, to discuss the safety and implications of a new type of experiment: combining genes of two species. Would experiments that deliver a cancer-causing virus be safe? The researchers discussed restricting the types of organisms and viruses used in recombinant DNA research and brainstormed ways to prevent escape of a resulting organism from the laboratory. The guidelines drawn up at Asilomar outlined measures of "physical containment," such as using specialized hoods and airflow systems, and "biological containment," such as weakening organisms so that they could not survive outside the laboratory. The containment measures drawn up in the late 1970s are being used today to maximize safety of genome editing technologies (see section 19.4).

Recombinant DNA

Despite initial concerns, recombinant DNA technology turned out to be safer than expected, and it spread to industry faster and in more diverse ways than anyone had imagined. However, recombinant DNA-based products have been slow to reach the marketplace because of the high cost of the research and the long time it takes to develop a new drug. Today, several dozen such drugs are available, and more are in the pipeline (**table 19.1**). Recombinant DNA research initially focused on providing direct gene products such as peptides and proteins. These included the human versions of insulin, growth hormone, and clotting factors. However, the technology can target carbohydrates and lipids by affecting the genes that encode enzymes required to synthesize them.

Table 19.1   Drugs Produced Using Recombinant DNA Technology

Drug	Use
Atrial natriuretic peptide	Dilates blood vessels, promotes urination
Colony-stimulating factors	Help restore bone marrow after marrow transplant; restore blood cells following cancer chemotherapy
Deoxyribonuclease (DNase)	Thins secretions in lungs of people with cystic fibrosis
Epidermal growth factor	Accelerates healing of wounds and burns; treats gastric ulcers
Erythropoietin (EPO)	Stimulates production of red blood cells in cancer patients
Factor VIII	Promotes blood clotting in treatment of hemophilia A
Glucocerebrosidase	Corrects enzyme deficiency in Gaucher disease
Human growth hormone	Promotes growth of muscle and bone in people with very short stature due to hormone deficiency
Insulin	Allows cells to take up glucose in treatment of type 1 diabetes
Interferons	Treat genital warts, hairy cell leukemia, hepatitis B and C, Kaposi sarcoma, multiple sclerosis
Interleukin-2	Treats kidney cancer recurrence
Lung surfactant protein	Helps lung alveoli to inflate in infants with respiratory distress syndrome
Renin inhibitor	Lowers blood pressure
Somatostatin	Decreases growth in muscle and bone in pituitary gigantism
Superoxide dismutase	Prevents further damage to heart muscle after heart attack
Thrombin	Stops postsurgical bleeding
Tissue plasminogen activator	Dissolves blood clots in treatment of heart attack, stroke, and pulmonary embolism

Constructing and Selecting Recombinant DNA Molecules

Manufacturing of recombinant DNA molecules requires three components:

1. **Restriction enzymes**, which are enzymes that cut DNA at specific sequences (They are also called restriction endonucleases because they cut DNA within the molecule ["endo"] rather than from the ends ["exo"].)
2. **Cloning vectors**, which are pieces of DNA used to deliver specific DNA sequences to cells
3. Recipient cells, such as bacteria or cultured single cells

After inserting DNA into vectors, the loaded vectors are delivered to selected cells. The cells then use the added genetic instructions to manufacture the desired protein.

Restriction enzymes are naturally found in bacteria, where they cut DNA of infecting viruses, protecting the bacteria. Methyl (CH_3) groups shield the bacterium's own DNA from its restriction enzymes. Bacteria have hundreds of types of restriction enzymes. Some of them cut DNA at particular sequences of four, five, or six bases that are symmetrical in a specific way—the recognized sequence reads the same, from the 5′ to 3′ direction, on both strands of the DNA. For example, the restriction enzyme EcoR1, whose actions are depicted in **figure 19.2**,

cuts at the sequence GAATTC. The complementary sequence on the other strand is CTTAAG, which, read backwards, is GAATTC. (You can try this with other sequences to see that it rarely works this way.) In the English language, this type of symmetry is called a palindrome, referring to a sequence of letters that reads the same in both directions, such as "Madam, I'm Adam." Unlike the language palindrome, however, palindromic sequences in DNA are on complementary strands.

The cutting action of some restriction enzymes on double-stranded DNA creates single-stranded extensions because the individual cuts are slightly offset, creating overhangs. They are called "sticky ends" because they are complementary to each other, forming hydrogen bonds as their bases pair. Restriction enzymes work as molecular scissors in creating recombinant DNA molecules because they cut at the same sequence in DNA from any source. That is, the same sticky ends result from the same restriction enzyme, whether the DNA is from a mockingbird or a maple.

Another natural "tool" used in recombinant DNA technology is a cloning vector. This structure carries DNA from the cells of one species into the cells of another. A vector can be any piece of DNA into which other DNA can insert. A commonly used vector is a **plasmid**, which is a small circle of double-stranded DNA that exists naturally in some bacteria, yeasts, plant cells, and cells of other types of organisms.

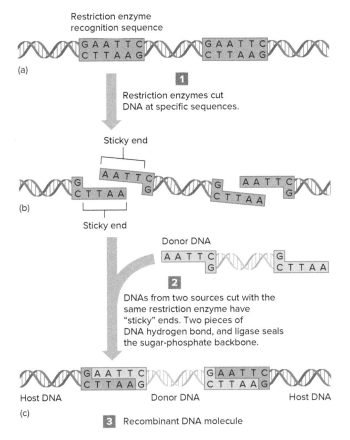

(a)

Restriction enzyme recognition sequence

GAATTC
CTTAAG

GAATTC
CTTAAG

1 Restriction enzymes cut DNA at specific sequences.

Sticky end

(b)

G
CTTAA

AATTC
G

G
CTTAA

AATTC
G

Sticky end

Donor DNA

AATTC
G

G
CTTAA

2

DNAs from two sources cut with the same restriction enzyme have "sticky" ends. Two pieces of DNA hydrogen bond, and ligase seals the sugar-phosphate backbone.

(c)

GAATTC
CTTAAG

GAATTC
CTTAAG

Host DNA   Donor DNA   Host DNA

3 Recombinant DNA molecule

Figure 19.2 **Recombining DNA uses restriction enzymes to insert a foreign DNA sequence.** These enzymes can be used as molecular scissors because they cut DNA from any source at the same base sequence.

Viruses that infect bacteria, called bacteriophages, are another type of vector. They are manipulated to transport DNA, but not cause disease. Disabled retroviruses are used as vectors, too, as are DNA sequences from bacteria and yeast called artificial chromosomes.

Choice of a cloning vector must consider the size of the gene to be delivered, which must be short enough to insert into the vector. Gene size is typically measured in kilobases (kb), which are thousands of bases. Cloning vectors can hold up to about 2 million DNA bases.

To create a recombinant DNA molecule, a restriction enzyme cuts DNA from a donor cell at sequences that bracket the gene of interest (**figure 19.3**). The enzyme leaves single-stranded ends on the cut DNA, each bearing a characteristic base sequence. Next, a plasmid is isolated and cut with the same restriction enzyme used to cut the donor DNA. Because the same restriction enzyme cuts both the donor DNA and the plasmid DNA, the same complementary single-stranded base sequences extend from the cut ends of each. When the cut plasmid and the donor DNA are mixed, the single-stranded sticky ends of some plasmids base pair with the sticky ends of the donor DNA. The result is a recombinant DNA molecule, such as a plasmid carrying the human insulin gene. The plasmid and its human gene can now be transferred into a cell, such as a bacterium or a white blood cell.

Selecting Recombinant DNA Molecules

Much of the effort in recombinant DNA technology is in identifying and separating cells that contain the gene of interest. This entails distinguishing cells bearing the gene from cells that lack plasmids or that have taken up "empty" plasmids that

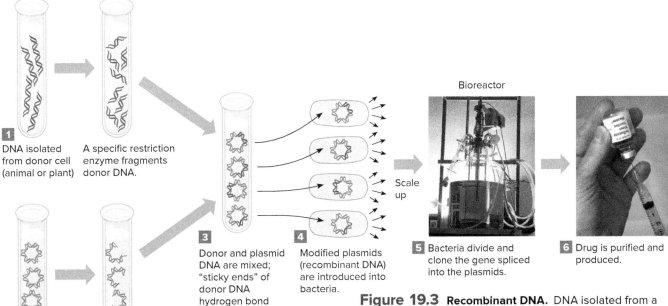

1 DNA isolated from donor cell (animal or plant)

A specific restriction enzyme fragments donor DNA.

2 Plasmid isolated from bacterium

The same restriction enzyme that fragmented donor DNA is also used to open plasmid DNA.

3 Donor and plasmid DNA are mixed; "sticky ends" of donor DNA hydrogen bond with sticky ends of plasmid DNA fragment; recombinant molecule is sealed with ligase.

4 Modified plasmids (recombinant DNA) are introduced into bacteria.

Bioreactor

Scale up

5 Bacteria divide and clone the gene spliced into the plasmids.

6 Drug is purified and produced.

Figure 19.3 **Recombinant DNA.** DNA isolated from a donor cell and a plasmid from a bacterium are cut with the same restriction enzyme and mixed. Sticky ends from the donor DNA hydrogen bond with sticky ends of the plasmid DNA, forming recombinant DNA molecules. When such a modified plasmid is introduced into a bacterium, it is mass produced as the bacterium divides. *(5): © Maximilian Stock Ltd./ SPL/Science Source; (6): Source: CDC/Jim Gathany*

do not contain the gene. Researchers have clever ways of separating the useful cells. One separation strategy uses plasmids that have an antibiotic resistance gene as well as a gene that encodes an enzyme that catalyzes a reaction that produces a blue color. When the antibiotic is applied, only cells that have plasmids survive. If a human gene inserts and interrupts the gene for the enzyme, the bacterial colony that grows is not blue and is, therefore, easily distinguished from the blue bacterial cells that have not taken up the human gene.

When cells containing the recombinant plasmid divide, so does the plasmid. Within hours, the original cell gives rise to many cells harboring the recombinant plasmid. The enzymes, ribosomes, energy molecules, and factors necessary for protein synthesis transcribe and translate the plasmid DNA and its foreign gene, producing the desired protein. Then the protein is separated, collected, purified, and packaged to create a product, such as a new drug.

Products from Recombinant DNA Technology

In basic research, recombinant DNA technology provides a way to isolate genes from complex organisms and observe their functions on the molecular level. Recombinant DNA has many practical uses, too. The first was to mass-produce protein-based drugs.

Drugs manufactured using recombinant DNA technology are pure, and are the human version of the protein. Before recombinant DNA technology was invented, human growth hormone came from cadavers, follicle-stimulating hormone came from the urine of postmenopausal women, and clotting factors were pooled from hundreds or thousands of donors. These sources introduced great risk of infection, especially after HIV and hepatitis C became more widespread.

The first drug manufactured using recombinant DNA technology was insulin, which is produced in bacterial cells (E. coli). Before 1982, people with type 1 diabetes mellitus obtained the insulin that they injected daily from pancreases removed from cattle in slaughterhouses. Cattle insulin is so similar to the human peptide, differing in only two of its 51 amino acids, that most people with diabetes could use it. However, about one in 20 patients is allergic to cow insulin because of the slight chemical difference. Until recombinant DNA technology was developed, the allergic patients had to use expensive combinations of insulin from other animals or human cadavers.

Insulin is a simple peptide and is therefore straightforward to mass-produce in bacteria. Some drugs, however, require that sugars be attached to peptides or proteins, or that the proteins must fold in specific, intricate ways to function. These molecules must be produced in eukaryotic cells that readily carry out these modifications. Yeast cells, Chinese hamster ovary cells, insect cells, and even carrot and tobacco cells have been used to produce human proteins.

Drugs developed using recombinant DNA technology must compete with conventional products. Deciding whether a recombinant drug is preferable to an existing similar drug is often a matter of economics. For example, interferon β-1b treats a type of multiple sclerosis, but this recombinant drug costs close to $70,000 per year. Researchers calculated that more people would be served if funds were spent on improved supportive care for many rather than on this costly treatment for a few.

Tissue plasminogen activator (tPA), a recombinant clot-busting drug, also has cheaper alternatives. If injected within 4 hours of a heart attack, tPA dramatically limits damage to the heart muscle by restoring blood flow. It costs $2,500 a shot. An older drug, streptokinase, is extracted from unaltered bacteria and is nearly as effective, at $300 per injection. Patients who have already received streptokinase and could have an allergic reaction if they were to use it again can benefit from tPA. **Bioethics** considers another drug derived from recombinant DNA technology, erythropoietin (EPO).

Safer vaccines are created using recombinant DNA technology. Recall that a vaccine is a live or killed pathogen or part of a pathogen that stimulates an immune response that protects a person who encounters the actual pathogen. Traditional influenza vaccine, for example, is cultured in eggs, to which some people are allergic, and is a time-consuming process. An alternative flu vaccine consists of the genes that encode the hemagglutinin proteins from two influenza A strains and one influenza B strain (see Clinical Connection 17.1), and is effective against H1N1 and H3N2 infections. The influenza virus genes are introduced aboard a virus called a baculovirus that readily infects the cells of certain insects—the fall armyworm in this case. The new "eggless vaccine" not only avoids the use of eggs, but also does not use antibiotics, preservatives, or live flu viruses, and can be manufactured faster than conventional flu vaccine.

A new vaccine that protects against malaria is based on altering a bacterium (Pantoea agglomerans) that normally inhabits the mosquito gut. Mosquitoes transmit Plasmodium falciparum, the parasite that causes malaria. The bacteria are given genes from E. coli bacteria that enable them to produce proteins that tear apart the insect's intestines. Using recombinant bacteria is easier than genetically modifying mosquitoes to prevent malaria, discussed in section 19.4 (**figure 19.4**).

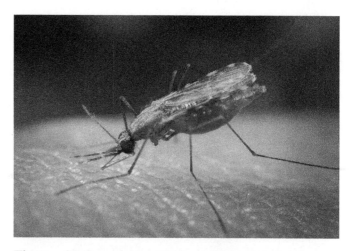

Figure 19.4 **Genetic modification of a gut bacterium in mosquitoes destroys the insect's intestines, halting transmission of malaria.** Section 19.4 considers a more extreme form of genetic modification that could drive malaria-carrying mosquitoes to extinction. *Source: CDC/James Gathany*

EPO: Built-in Blood Booster or Performance-Enhancing Drug?

"Cycle, run, and swim longer and faster than anyone else!" proclaims a website selling a product that supposedly boosts levels of erythropoietin. "EPO" is a glycoprotein hormone that the kidneys produce in response to low levels of oxygen in the blood. The hormone travels to the bone marrow and binds receptors on cells that give rise to red blood cell progenitors. Soon, more red blood cells enter the circulation, carrying more oxygen to the tissues (**figure 19A**). Enhanced stamina results.

The value of EPO as a drug became evident after the 1961 invention of hemodialysis to treat kidney failure. Dialysis removes EPO from the blood, causing severe anemia. But boosting EPO levels proved difficult because levels in human plasma are too low to pool from donors. Instead, in the 1970s, the U.S. government obtained EPO from South American farmers with hookworm infections and Japanese aplastic anemia patients, who secrete abundant EPO into urine. But when the AIDS epidemic came, biochemicals from human body fluids were no longer safe to use. Recombinant DNA technology solved the EPO problem. It is sold under various names

Figure 19A **At least two genes control EPO secretion.** Certain variants of these genes increase the number of red blood cells, increasing endurance but also raising risk of heart attack and stroke. © *Image Source/Getty RF*

to treat anemia in dialysis and AIDS patients and is given with cancer chemotherapy to avoid the need for transfusions.

EPO's ability to increase the oxygen-carrying capacity of blood under low oxygen conditions is why athletes train at high altitudes to increase endurance. Since the early 1990s, athletes have abused EPO to reproduce this effect, at great risk. EPO thickens the blood, raising the risk of a blockage that can cause a heart attack or stroke, especially when intense, grueling exercise removes water from the bloodstream. Excess EPO caused sudden death during sleep for at least 18 cyclists. Olympic athletes now take urine tests that detect recombinant EPO, which has a slightly different configuration of sugars than the form an athlete's kidneys naturally produce.

People with any of four types of familial erythrocytosis get the effects of extra EPO naturally. In one autosomal dominant form, the receptor for EPO is active for too long, leading to large and abundant red blood cells. A member of a family from Finland with this condition won several Olympic medals for skiing thanks to his inborn ability. Autosomal recessive forms of the condition increase the level of EPO in the bloodstream. Erythrocytosis may cause dizziness, headaches, nosebleeds, and shortness of breath, or have no symptoms, but it increases the risk of stroke and heart attack from blocked circulation.

Questions for Discussion

1. Should using a substance made naturally in the body be considered performance enhancement?

2. Should tests be developed to identify athletes whose genes, anatomy, or physiology give them a competitive advantage? Why or why not?

3. When developing drugs that use recombinant DNA technology, should researchers consider how the product could be abused?

Transgenic Organisms

Eukaryotic cells growing in culture are generally better at producing human proteins than are prokaryotic cells such as bacteria. An even more efficient way to express some recombinant genes is in a body fluid of a transgenic animal, such as the saliva of the Enviropig. Getting cells to secrete the human protein from an intact animal more closely mimics the environment in the human body. The genetic change must be introduced into a fertilized ovum so that it is present in every cell of the transgenic organism.

Transgenic sheep, cows, and goats have all expressed human genes in their milk, including genes that encode clotting factors, clot busters, and the connective tissue protein collagen.

Production of human antibodies in rabbit and cow milk illustrates the potential value of transgenic animals. These antibodies are used to treat cancers. Recall from figure 17.8 that antibodies are assembled from the products of several genes. Researchers attach the appropriate human antibody genes to promoters for milk proteins. (Promoters are the short sequences at the starts of genes that control transcription rates.) These promoters normally oversee production of abundant milk proteins. The mammary gland cells of transgenic animals can assemble antibody parts to secrete the final molecules—just as if they were being produced in a plasma cell in the human immune system.

Several techniques are used to insert DNA into animal cells to create transgenic animals. Chemicals and brief jolts of electricity open transient holes in plasma membranes that admit "naked" DNA, or a gunlike device is used to shoot tiny metal particles coated with DNA inside cells. DNA may also cross the plasma membrane in tiny fatty bubbles called liposomes.

Getting foreign DNA into a fertilized ovum is the first step in creating a transgenic organism. It is quite a technical challenge, especially for nonanimal cells, which have cell walls in addition to plasma membranes. The recombinant DNA must enter the nucleus, replicate with the cell's own DNA as part of a chromosome, and be transmitted when the cell divides. Finally, for an animal, an organism must be regenerated from the fertilized ovum, which means gestation in a surrogate mother. If the trait is dominant, the transgenic organism must express it in the appropriate tissues at the right time in development to be useful. If the trait is recessive, crosses between heterozygotes may be necessary to yield homozygotes that express the trait. Then the organisms must pass the characteristic on to the next generation.

Animal Models

Herds of transgenic farm animals supplying drugs in their milk have not become important sources of pharmaceuticals—they are too difficult to maintain. Transgenic animals are more useful as models of human disease (**figure 19.5**). Inserting the mutant human beta globin gene that causes sickle cell disease into mice, for example, results in a mouse model of the disease. Drug candidates can be tested on these animal models and abandoned before being tested in humans if they cause significant side effects.

Transgenic animal models, however, have limitations. Researchers cannot control where a transgene inserts in a genome, and how many copies insert, so that different animals may have different numbers of the gene of interest. The level of gene expression necessary for a phenotype to emerge may also differ in the model and humans. This was the case for a mouse model of familial Alzheimer disease. The transgene has the exact same DNA sequence that disrupts amyloid precursor protein in a Swedish family with the condition, but apparently did nothing to the mice—until researchers increased transcription rate 10-fold. Only then did the telltale plaques and tangles, and neuron cell death, appear in the mouse brains.

Animal models might not mimic the human condition exactly because of differences in rates of development. For example, for some inherited diseases that do not cause symptoms until adulthood in humans, mice simply do not live long enough to have the associated phenotype. For this reason, a transgenic monkey is a more accurate model of Huntington disease than a mouse, because as a primate the monkey is much more similar to a human in life span, metabolism, reproduction, behavior, and cognition.

Genetically Modified Foods

Traditional agriculture is the controlled breeding of plants and animals to select individuals with certain combinations of inherited traits that are useful to us, such as seedless fruits and lean meat. It is a form of genetic modification based on phenotype, such as taste or appearance, and is therefore both subjective and imprecise, affecting many genes. In contrast, DNA-based techniques manipulate one or a few specific genes at a time. Organisms altered to have genes from other species or to over- or underexpress their own genes are termed "genetically modified" organisms, or GMOs. An organism given genes from another species is transgenic.

Golden rice is a well-known GM crop that uses genes from corn and bacteria to produce 23 times as much beta-carotene (a vitamin A precursor) as unaltered rice. It contains no allergens or toxins. Developed by the not-for-profit International Rice Research Institute in the Philippines, the rice was meant to improve human nutrition in the many nations where rice is a dietary staple. In Africa and Asia, 2 million people die each year because lack of vitamin A impairs their immunity to infectious diseases.

Some nations outlaw GM foods, but people in the United States have been eating them for years, and (apparently) safely. A government report found that 75 percent of processed foods in supermarkets contain at least one GM ingredient. From 92 to 94 percent of corn, cotton, and soybean crops in the United States are genetically modified.

People object to GM foods for a variety of reasons (**figure 19.6**). Officials in France and Austria have called such crops "not natural," "corrupt," and "heretical." An enraged consumer declared on a TV news program, "I will not eat food that contains DNA!" GM crops may have inherited traits not normal to them, but DNA is DNA. All foods contain DNA.

Other reasons for objecting to GM foods are practical. Labeling a food that contains a nutrient not normally in it—like a protein from peanuts in corn—could prevent allergic reactions. Many GM crops are designed to resist certain herbicides, requiring use of that herbicide. Bioethical concerns arise when the same company that creates the GM crop also manufactures and markets the herbicide to which it is resistant, forcing farmers to buy the products.

An ecological-based objection to GM plants is that field tests may not adequately predict the effects on ecosystems. Buffer zones of non-GM plants are routinely planted around fields of GM varieties, but wind pollination can take GM plants far. A sampling of 400 canola plants growing along roads in North Dakota found that 86 percent of them had a gene indicating that they descended from GM plants.

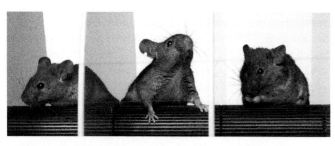

Figure 19.5 Animal models mimic human disease.
Transgenic mice that have a mutation that causes Huntington disease in humans are tested for coordination on a rotating drum with a grooved surface. *Courtesy of MRC Harwell*

Figure 19.6 **Some people fear genetically modified foods.** © Shutterstock/Elnur

Another science-based objection to GM crops is that overreliance on them may lead to genetic uniformity, which is just as dangerous as traditional monoculture. Genetic sameness creates vulnerability in a population should environmental conditions change and no variants are there to survive natural selection. Some GM organisms, such as fish that grow to twice normal size or can survive at temperature extremes, can disrupt natural ecosystems. **Table 19.2** lists some GM organisms.

Bioremediation

Recombinant DNA technology and transgenic organisms provide processes as well as products. In **bioremediation**, bacteria or plants with the ability to detoxify certain pollutants are released or grown in a particular area. Natural selection has

Table 19.2	Some Genetically Modified Organisms
Organism	**Altered Trait**
Grapes	Less sugar
Cassava, papaya, plum	Resist viral infection
Cattle	Resist mad cow disease
Sugar beets, corn, soybeans	Tolerate an herbicide
Bananas, rice	More iron and vitamin A
Canola	Altered fatty acid composition
Salmon	Faster growth

sculpted such organisms, perhaps as adaptations that render them unpalatable to predators. Bioremediation uses genes that enable an organism to metabolize a substance that, to another species, is a toxin. The technology uses unaltered organisms, and also transfers "detox" genes to other species so that the protein products can more easily penetrate a polluted area.

Nature offers many organisms with interesting metabolic characteristics that might benefit other species. A type of tree that grows in a tropical rainforest on an island near Australia, for example, accumulates so much nickel from soil that slashing its bark releases a bright green latex ooze. Genes from this tree can be used to clean up nickel-contaminated soil. Transgenic microorganisms that make proteins that detoxify contaminants are sent into plants whose roots distribute the detox proteins in the soil. In this way, transgenic yellow poplar trees can thrive in mercury-tainted soil thanks to a bacterial gene that encodes an enzyme that converts a highly toxic form of mercury to a less toxic gas. The tree's leaves release the gas.

Cleaning up munitions dumps from wars is another use of bioremediation, such as deploying bacteria that normally break down trinitrotoluene, or TNT, the major ingredient in dynamite and land mines. The enzyme that provides this ability is linked to the *GFP* gene from jellyfish (see figure 19.1). Bacteria spread in a contaminated area glow near land mines, revealing the locations more clearly than a metal detector could. Once the land mines are removed, the bacteria die as their food vanishes.

Key Concepts Questions 19.2

1. Explain the steps of recombinant DNA technology.
2. Discuss applications of recombinant DNA technology.
3. List examples of useful transgenic plants and animals.

19.3 Monitoring Gene Function

We usually can do little about the gene variants that we inherit. Gene expression, in contrast, is where we can make a difference by controlling our environment. Monitoring gene expression requires detecting the mRNAs in particular cells under particular conditions. To do this, devices called gene expression DNA microarrays, or gene chips, detect and display the mRNAs in a cell. The creativity of the technique lies in choosing the types of cells to interrogate. Figure 18.16 depicts use of the technology to distinguish types of leukemia.

Evaluating a spinal cord injury illustrates the basic steps in creating a DNA microarray to assess gene expression. Researchers knew that in the hours after such a devastating injury, immune system cells and inflammatory biochemicals flood the affected area. Gene expression profiling revealed how fast healing begins.

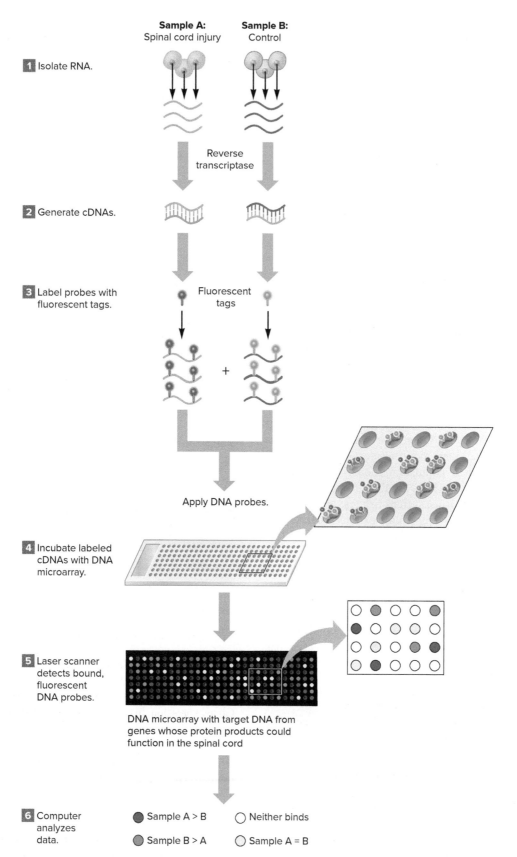

Sample A: Spinal cord injury

Sample B: Control

1 Isolate RNA.

Reverse transcriptase

2 Generate cDNAs.

3 Label probes with fluorescent tags.

Fluorescent tags

+

Apply DNA probes.

4 Incubate labeled cDNAs with DNA microarray.

5 Laser scanner detects bound, fluorescent DNA probes.

DNA microarray with target DNA from genes whose protein products could function in the spinal cord

6 Computer analyzes data.

⬤ Sample A > B   ◯ Neither binds

⬤ Sample B > A   ◯ Sample A = B

Figure 19.7 **A DNA microarray experiment reveals gene expression in response to spinal cord injury.** In this example, the red label represents DNA from a patient with a spinal cord injury and the green label represents control DNA from a healthy person. DNA targets on the microarray that bind red but not green can reveal new points of intervention for drugs.

A **DNA microarray** is a piece of glass or plastic that is about 1.5 centimeters square—smaller than a postage stamp. Many small pieces of DNA (oligonucleotides) of known sequence are attached to one surface, in a grid pattern. The researcher records the position of each DNA piece in the grid. In many applications, a sample from an abnormal situation (such as disease, injury, or environmental exposure) is compared to a normal control. **Figure 19.7** compares cerebrospinal fluid (CSF; the liquid that bathes the spinal cord) from an injured person (sample A) to fluid from a healthy person (sample B). Messenger RNAs are extracted from the samples and complementary DNAs (cDNAs) are made. Researchers make cDNAs from mRNA using an enzyme from a retrovirus, reverse transcriptase. A cDNA includes codons for a mature mRNA, but not sequences for promoters and introns, so it represents the exons of a gene.

The cDNAs from the injury sample are labeled with a red fluorescent dye and the cDNAs from the control sample are labeled with a green fluorescent dye. These labeled DNAs are then applied to the microarray. The microarray might display genes that a researcher hypothesizes are likely to be involved in a spinal cord injury, or the entire human exome or genome, to be certain all possible genes are covered. Considering so many DNA sequences allows for surprises, avoiding the assumption that we know what to look for.

DNA that binds to complementary sequences on the grid fluoresce in place. A laser scanner then detects and converts the results to a colored image. Each position on the microarray can bind DNA pieces from both samples (injured and healthy), either sample, or neither. The scanner also detects fluorescence intensities, which provides information on how strongly the gene is expressed (how much mRNA is in the sample).

A computer algorithm interprets the pattern of gene expression. For the spinal cord example, the visual data mean the following:

- Red indicates a gene expressed in CSF only when the spinal cord is injured (and presumably leaking inflammatory molecules).
- Green indicates a gene expressed in CSF only when the spinal cord is intact.
- Yellow indicates positions where both red- and green-bound dyes fluoresce, representing genes that are expressed whether or not the spinal cord has been injured.
- Black, or a lack of fluorescence, corresponds to DNA sequences that are not expressed in CSF because they do not show up in either sample.

The color and intensity pattern of the microarray provides a glimpse of gene expression following spinal cord injury. The technique is even more powerful when it is repeated at different times after injury. When researchers did exactly that on injured rats, they discovered genes expressed just after the injury whose participation they never suspected. Their microarrays, summarized in **table 19.3**, revealed waves of expression of genes involved in healing. Analysis on the first day indicated activation of the same suite of genes whose protein products heal injury to the deep layer of skin—a total surprise that suggested new points for drugs to intervene.

Table 19.3	Gene Expression Profiling Chronicles Repair After Spinal Cord Injury
Time After Injury (rats)	**Type of Increased Gene Expression**
Day 1	Protective genes to preserve remaining tissue
Day 3	Growth, repair, cell division
Day 10	Repair of connective tissues
	Angiogenesis
Days 30–90	Blood vessels mature
	New type of connective tissue associated with healing

Key Concepts Questions 19.3

1. What do DNA microarrays detect?
2. List the steps of a DNA microarray experiment.
3. How can microarrays that track gene expression be used to reveal the genes that participate in a process?

19.4 Gene Silencing and Genome Editing

Since the 1960s, when the flow of genetic information from DNA to RNA to protein was discovered and described, researchers have been trying to manipulate the process to diminish ("knock down") or silence ("knock out") the expression of specific genes. **Gene silencing** techniques block synthesis of, or degrade, mRNA. **Genome editing** techniques create double-stranded breaks in the DNA double helix, enabling insertion of a desired DNA sequence or removal of a sequence. Different approaches to gene silencing and genome editing have had varying degrees of success. **Figure 19.8** indicates where these techniques intervene in the flow of genetic information. All of these biotechnologies are based on the phenomenon of complementary base pairing and some borrow directly from natural DNA repair pathways.

Gene Silencing

Antisense technology is a form of gene silencing that blocks expression of a gene by introducing RNA that is complementary to the gene's mRNA transcript. The introduced RNA, called antisense RNA, binds to the mRNA, preventing its translation into protein. An early application of antisense technology was a tomato intended to stay fresh longer because the antisense RNA

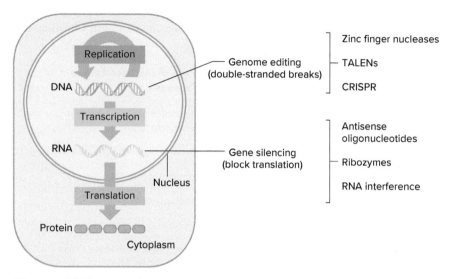

Figure 19.8 Gene silencing and genome editing. Gene silencing techniques prevent mRNA from being translated into protein. Genome editing techniques enable researchers to remove double-stranded segments of DNA, so that they can induce deletions or swap in genes.

squelched activity of a ripening enzyme. Ripening slowed, but the tomatoes still softened and tasted terrible. Nearly 2 million cans of GM tomato paste had made their way to store shelves before consumer complaints led to halted production. Results of genetic modification aren't entirely predictable! The first Case Studies and Research Results question at the end of the chapter returns to the challenges of creating a tasty tomato.

A second variation of antisense technology uses synthetic molecules called **morpholinos** that consist of sequences of 25 DNA bases attached to organic groups that are similar to, but not exactly the same as, the sugar-phosphate backbone of DNA. Morpholinos can block splice-site mutations that would otherwise delete entire exons. For example, a morpholino-based drug targets the dystrophin gene, restoring the function of a skipped exon and enabling muscle cells to produce some dystrophin. The drug has been approved to treat Duchenne muscular dystrophy, but is still being tested to see if it can restore enough dystrophin to provide sustained improvement in mobility.

Another approach to gene silencing uses **ribozymes**. These are RNA molecules that are part of ribosomes (the organelles on which translation occurs) that have catalytic activity, like enzymes. Ribozymes fit the shapes of certain RNA molecules. Because ribozymes cut RNA, they can be used to destroy RNAs from pathogens, such as HIV. Figure 13.14 illustrates yet another way to turn off gene expression—using the XIST long noncoding RNA that normally shuts off one X chromosome in the cells of female mammals to silence the extra chromosome of trisomy 21 Down syndrome.

RNA interference (RNAi) is a gene silencing technique based on the fact that RNA molecules can fold into short, double-stranded regions where the base sequence is complementary. **Figure 19.9** shows how this bonding produces a hairpin shape. Short, double-stranded RNAs introduced into cells can have great effects.

The Nobel Prize in Physiology or Medicine was awarded in 2006 to Andrew Fire and Craig Mello for explaining how RNAi works. They discovered that short, double-stranded RNAs sent into cells separate into single strands, one of which binds its complement in mRNA, preventing it from being translated. The small RNAs that carry out RNA interference are called small interfering RNAs (siRNAs).

Several proteins and protein complexes orchestrate RNAi, and are also part of microRNA function (**figure 19.10** and see section 11.2). First, an enzyme called Dicer cuts long, double-stranded RNAs into pieces 21 to 24 nucleotides long. These pieces contact a group of proteins that form an RNA-induced silencing complex, or RISC. One strand of the double-stranded short RNA, called the guide strand, adheres. Now, as part of RISC, the guide strand finds its complementary RNA and binds. Then another part of RISC, a protein called argonaute, degrades the targeted RNA, preventing its translation into protein. SiRNAs act in the nucleus, too, where they alter methylation and the binding of histones to certain genes.

Shortly after RNA interference was discovered in 1998, biotechnology and pharmaceutical companies began developing the approach to silence genes. Potential applications ranged from creating vaccines by knocking down expression of key genes in viruses that cause diseases (such as AIDS, polio, and hepatitis C), to treating cancer by silencing oncogenes, to creating a better-tasting decaf coffee by silencing an enzyme required for caffeine synthesis in coffee plants. However, problems arose in clinical trials of RNAi-based drugs. In the human body, the synthetic "small interfering RNAs" that carry out RNAi can inflame the liver, rather than reach their intended targets. For now, RNAi may be better suited as a research tool to see what happens when gene expression is turned off in animal models of human disease and in human cells growing in culture.

Figure 19.9 Hairpins. RNA hydrogen bonds with itself, forming hairpin loops. RNAi uses similar molecules to "knock down" expression of specific genes by binding their mRNAs.

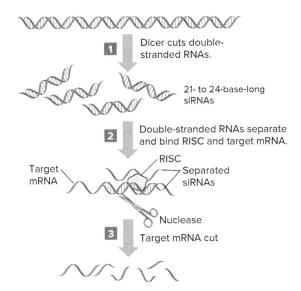

1 Dicer cuts double-stranded RNAs.

21- to 24-base-long siRNAs

2 Double-stranded RNAs separate and bind RISC and target mRNA.

RISC

Target mRNA

Separated siRNAs

Nuclease

3 Target mRNA cut

Figure 19.10 **RNA interference.** Dicer cuts double-stranded parts of RNA molecules, which then associate with RNA-induced silencing complexes (RISCs). The RNA is open, revealing single strands that locate and bind specific mRNAs. Nucleases then break down the targeted mRNAs, preventing their translation into protein.

Genome Editing

Genome editing, like recombinant DNA technology, uses restriction endonucleases to cut and paste DNA molecules in patterns that might not exist in nature. However, the two technologies differ in that in recombinant DNA technology, the enzymes generate offset ends that are "sticky" so that pieces connect by base-pairing. In genome editing, the enzymes generate double-stranded breaks, which natural DNA repair systems mend, removing or even replacing or adding to the targeted area.

Genome editing can be done on somatic cells or on cells of the germline (developing oocytes and sperm). In somatic applications, the change must affect all or enough cells to alter the phenotype, such as treating symptoms of a genetic disease (see section 20.3). Germline genome editing introduces the genetic change to every cell in an organism because it is carried out at the beginning of development.

The three major genome editing techniques are usually referred to by their easier-to-remember abbreviations and acronyms: ZFNs, TALENs, and CRISPR-Cas9 (**table 19.4**). Although ZFNs and TALENs were invented several years before CRISPR-Cas9, they use proteins and can only cut one gene at a time. CRISPR-Cas9 uses RNA and is cheaper and easier to use, and can remove, replace, or add more than one gene at a time. All three genome editing techniques are more precise than the gene therapies described in chapter 20 that deliver genes that do not replace the existing mutant genes.

The restriction enzymes that are used in genome editing function like scissors directed to cut DNA at a specific point. A "DNA binding" domain leads the enzyme to a specific short DNA sequence, while the endonuclease part severs both strands of the double helix at the same point.

Understanding natural DNA repair pathways inspired the development of gene and genome editing in the late 1980s, for which Mario Capecchi, Martin Evans, and Oliver Smithies won the Nobel Prize in Physiology or Medicine in 2007. They adapted one of two DNA repair pathways activated by double-stranded breaks, called homologous recombination, to create "knockout" mice missing key genes. Since then, mice with knocked out genes have provided models for many genetic diseases of humans. The technique of knocking out genes is not used directly on humans because it requires setting up crosses.

The first genome editing tool, zinc finger nucleases, was developed in 2009. CRISPR-Cas9 was first described in 2011 by Emmanuelle Charpentier and Jennifer Doudna and rapidly became widely used in research laboratories.

Zinc finger nuclease technology uses protein motifs (parts of proteins that have characteristic shapes) called zinc fingers that consist of a beta-pleated sheet linked to an alpha helix by a zinc atom (see figure 10.19*b*). Different zinc fingers bind different three-base DNA sequences. If zinc fingers bind, then another nuclease (FokI) cuts the DNA. Zinc finger action is a little like folding a small loop of yarn and snipping it across both strands with one cut. In **transcription-activator-like effector nuclease (TALEN) technology**, a restriction enzyme from a bacterium (*Xanthomonas*) that infects plants cuts DNA on both strands. Chimeric antigen receptor technology to treat cancer, described in section 18.4, at first used ZFNs but switched to the easier-to-use CRISPR-Cas9 technology.

Clustered regularly interspaced short palindromic repeats (CRISPRs) are short sequences of DNA that include several repeats. CRISPRs are natural components of the genomes of certain bacteria, where they provide an action similar to an immune response. Specifically, CRISPRs enable bacteria to deploy a restriction enzyme called Cas9 ("CRISPR-associated protein 9") to recognize and cut out DNA sequences from infecting viruses that have inserted into the bacterial genome. The CRISPR DNA sequences along with the viral sequences are transcribed into CRISPR RNAs. These RNAs

Table 19.4	Genome Editing Techniques
Abbreviation	**Full Name**
CRISPR-Cas9	Clustered regularly interspaced short palindromic repeats-CRISPR-associated protein 9
TALENs	Transcription-activator-like effector nucleases
ZFNs	Zinc finger nucleases

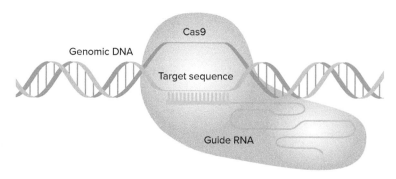

Figure 19.11 CRISPR-Cas9. In the genome editing tool CRISPR-Cas9, a synthetic guide RNA directs the restriction endonuclease Cas9 (purple) to a selected target sequence of DNA. The enzyme makes a double-stranded cut, and the target sequence can be swapped in. *Adapted from Chemical & Engineering News 93(26), 20–21.*

then serve as "guides" that bring Cas9 to other sites where viral DNA has integrated into the bacterial genome (**figure 19.11**). Cas9 then cuts both strands. The particular CRISPR RNAs remain to eject DNA from future viral infections.

The beauty of CRISPR-Cas9 is its great versatility—researchers can design and synthesize guide RNAs to direct which DNA sequences are removed, and even stitch in desired sequences to replace the ones snipped out, or add new genes. It works in any species. While bacteria might remove infecting viral DNA, a human cell might be programmed to replace a mutant gene with the wild type variant. Treating genetic disease and creating animal models are applications of CRISPR-Cas9. **Table 19.5** lists some innovative uses.

Genome editing techniques are powerful, and therefore controversial. One major concern is use of CRISPR-Cas9 to reduce populations of, disarm, or eliminate species that are vectors for human pathogens. Top targets are mosquitoes,

particularly *Aedes aegypti*, which spreads Zika virus disease, dengue fever, and yellow fever, and the *Anopheles* species that transmit malaria.

The application of genome editing to kill, alter, or render infertile a pathogen is called a **gene drive**. It is based on a natural form of DNA repair, called homing, which removes one of a pair of alleles of a selected gene and replaces it with another copy of the remaining allele. Homing genes are found in certain single-celled organisms. The proteins that they encode recognize certain DNA sequences of 15 to 30 bases, cut them out, and replace them with a copy of the DNA sequence on the other strand. In a gene drive, homing proteins are incorporated into CRISPR-Cas9 to delete and then replace a selected gene.

A gene drive counters Mendel's first law, of segregation. **Figure 19.12** compares Mendelian inheritance of a single gene, in which each parent transmits alleles with a frequency of

Table 19.5	Applications of CRISPR-Cas9 Genome Editing
Application	**Example**
GMO containment	Deleting genes that encode hormones that catfish require to reproduce, so that catfish with other genetic modifications can be raised on farms where the hormones are added to the water, but not survive in the wild.
Bringing back extinct species	Replace genes with counterparts from extinct relatives, such as mammoth genes in an elephant.
Limiting spread of infectious diseases	Introduce genes into disease vectors such as mosquitoes and ticks that make them sterile.
Avoiding allergy to vaccines	Remove the gene that encodes albumin, the egg white protein, which makes some people react to vaccines grown in hen eggs.
Adding genes to resist disease	Swapping in a *CCR5* mutation to provide resistance to HIV (see the chapter opener of chapter 17).
Creating organ donors	Introduce human genes into pig genomes to create cell surfaces that the human immune system will not reject.
Adding traits to show animals	Introduce genes for valued traits (such as fur color, body size, and stamina) into pets or show animals in just one or two generations.

Mendelian inheritance

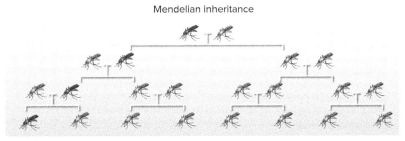

Mutation inherited 50:50

Gene drive

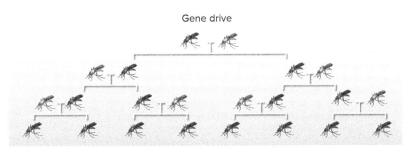

Mutation inherited all the time

Figure 19.12 **A gene drive.** Using a type of natural DNA repair called homing, a gene drive counters Mendel's law of segregation by cutting out an allele and replacing it with a copy of the remaining allele. Incorporating gene variants that confer infertility or interfere with the relationship between an infectious disease vector and a pathogen that it carries can wipe out the vector or destroy its ability to transmit the infectious disease. *Source: Adapted from Dr. Kevin Esvelt, Wyss Institute at Harvard*

50 percent, to a gene drive, in which the parent whose allele is copied essentially transmits it with a frequency near 100 percent. Because gene drives are an incredibly fast way to alter a wild population, research is restricted to highly controlled, laboratory settings. However, a change to the genetic structure of a population resulting from a gene drive may not be permanent because of evolution. New mutations may occur and offer a selective advantage, similar to the way that antibiotic resistance arises.

Gene drives were first demonstrated on laboratory fruit flies, then applied to the mosquitoes that transmit malaria. Developing mosquito oocytes are given the homing gene, guide RNAs corresponding to three mutant genes required for female fertility, and a gene variant giving the altered insects a reproductive advantage. The oocytes are fertilized with sperm wild type for the three genes, and the insects that result are all heterozygotes for the three genes. But when the heterozygotes mate, homing happens and suddenly the next generation of mosquitoes are homozygotes for the three fertility mutations. Population size plummets to below the level necessary to sustain the ability to transmit the protozoan that causes malaria.

Malaria kills more than 1 million people a year, and dengue, yellow fever, and Zika virus sicken and kill many, too. However, drastic alterations to mosquito populations could have unintended consequences on ecosystems by removing a component of food webs, opening up new ecological niches or harming populations of other species.

Gene drives in mosquito populations alter the germline. Many people object to genome editing of the human germline. Although many researchers have agreed not to attempt germline editing of human genomes, it has been done. The first step was germline genome editing of monkeys in 2014. A year later, researchers in China used CRISPR-Cas9 on fertilized human ova that they knew would be unable to develop very far because of extra genetic material. This "proof-of-concept" experiment was widely condemned. Other attempts followed, but objection to germline manipulation in humans remains strong.

Chapter 20 explores how genetic technologies are used to diagnose and treat disease.

Key Concepts Questions 19.4

1. How do antisense technologies, ribozymes, and RNA interference silence gene expression?
2. Explain how the three genome editing techniques work.
3. What are some applications of genome editing?
4. Why is genome editing controversial?

Summary

19.1 Patenting DNA

1. **Biotechnology** alters cells or biochemicals to provide a product or process. It extracts natural products, alters an organism's DNA, or combines DNA from different species.

2. A **transgenic organism** has DNA from a different species. Recombinant DNA comes from more than one type of organism. Both are possible because of the universality of the genetic code.

3. Patented DNA must be useful, novel, and nonobvious. A gene sequence by itself can no longer receive patent protection in the United States.

19.2 Modifying DNA

4. **Recombinant DNA technology** mass-produces proteins in bacteria or other single cells. Begun hesitantly in 1975, the technology has matured into a valuable method to produce proteins that are used as drugs.

5. To construct a recombinant DNA molecule, **restriction enzymes** cut the gene of interest and a **cloning vector** (such as a **plasmid**) at a short palindromic sequence, creating complementary "sticky ends." The DNAs are mixed and vectors that pick up foreign DNA selected.

6. Genes conferring antibiotic resistance and color changes in growth media are used to select cells containing recombinant DNA. Useful proteins are isolated and purified.

7. A multicellular transgenic organism has an introduced gene in every cell. Heterozygotes for a transgene are bred to yield homozygotes. Some transgenic animals are used to model human disease. Transgenic plants are genetically modified to have traits from other species.

19.3 Monitoring Gene Function

8. **DNA microarrays** hold DNA pieces to which fluorescently labeled cDNA probes from samples are applied. They are used in **gene expression profiling**.

19.4 Gene Silencing and Genome Editing

9. **Gene silencing** uses **antisense technology, ribozymes, morpholinos,** and **RNA interference (RNAi)** to block translation of mRNAs.

10. **Genome editing** uses enzymes to create double-stranded breaks in DNA, enabling researchers to add, replace, or delete specific genes. **ZFNs, TALENs,** and **CRISPR-Cas9** are used to edit genomes.

11. Genome editing can correct a genetic defect, create an animal model, or introduce a **gene drive** to change or eradicate a species, among other applications.

12. Germline editing of the human genome is controversial.

Review Questions

1. Cite three examples of a DNA sequence that meets requirements for patentability.

2. Why is a DNA sequence by itself not patentable, but a DNA sequence that is used to diagnose a disease in a medical test is patentable?

3. Describe the functions of each of the following tools used in biotechnology:
   a. restriction enzymes
   b. cloning vectors
   c. DNA microarrays
   d. short nucleic acid molecules with specific sequences
   e. CRISPR sequences

4. Explain how cells containing recombinant DNA are selected.

5. List the components of an experiment to produce recombinant human insulin in *E. coli* cells.

6. Why would recombinant DNA technology and creation of transgenic organisms be restricted or impossible if the genetic code were not universal?

7. What is an advantage of a drug produced using recombinant DNA technology compared to one extracted from natural sources?

8. How are transgenic animals better models of human disease than animal models whose DNA is unaltered? What are limitations of transgenic animal models?

9. How is genetic modification of a crop usually more precise and predictable than using conventional breeding to create a new plant variety?

10. Explain the advantages of using a DNA microarray that covers the exome or genome rather than selected genes whose protein products are known to take part in the disease process being investigated.

11. Describe how a technology to silence a gene or edit a genome can be used to treat a disease.

12. How does the type of enzyme used in genome editing with CRISPR-Cas9 differ from similar enzymes used in recombinant DNA technology?

13. Explain how a gene drive counters Mendel's first law.

14. Discuss why germline genome editing in humans is more difficult and more controversial than somatic cell genome editing in humans.

Applied Questions

1. Phosphorus in pig excrement pollutes aquatic ecosystems, causing fish kills and algal blooms, and contributes to the greenhouse effect. *E. coli* produces an enzyme that breaks down phosphorus. Describe the steps to create a transgenic pig that secretes the bacterial enzyme and therefore excretes less polluting feces.

2. Do you agree with the Supreme Court that a gene is not patentable but a cDNA is? Cite a reason for your answer.

3. Which (if any) objection to GMOs do you agree with, and why?

4. Genetic modification endows organisms with novel abilities. From the following three lists (choose one item from each list), devise an experiment to produce a particular protein, and suggest its use.

Organism	Biological Fluid	Protein Product
Pig	Milk	Human beta globin chains
Cow	Semen	Human collagen
Goat	Milk	Human EPO
Chicken	Egg white	Human tPA
Aspen tree	Sap	Human interferon
Silkworm	Blood plasma	Jellyfish GFP
Rabbit	Honey	Human clotting factor
Mouse	Saliva	Human alpha-1-antitrypsin

5. Collagen is a connective tissue protein that is used in skincare products, shampoos, desserts, and artificial skin. For many years, it was obtained from the hooves and hides of cows collected from slaughterhouses. Human collagen can be manufactured in transgenic mice. Describe the advantages of the mouse source of collagen.

6. People did not object to the production of human insulin in bacterial cells used to treat diabetes, yet some people object to mixing DNA from different animal and plant species in agricultural biotechnology. Why do you think that the same general technique is perceived as beneficial in one situation, yet a threat in another?

7. A human oncogene called *ras* is inserted into mice, creating transgenic animals that develop a variety of tumors. Why are mouse cells able to transcribe and translate human genes?

8. In a DNA microarray experiment, researchers attach certain DNA pieces to the grid. For example, to study an injury, genes known to be involved in the inflammatory response might be attached. How might this approach be limited?

9. Patenting of CRISPR-Cas9 is under dispute. Two researchers from the University of California published the idea of redirecting the underlying bacterial DNA repair system to edit DNA, but a third researcher from the Broad Institute described application of the genome editing technique in mouse and in human cells. Considering the general requirements for a patented invention, what is the basis of the disagreement concerning CRISPR-Cas9?

10. Suggest applications of the following biotechnologies:
    a. recombinant DNA of a single cell
    b. a transgenic organism
    c. gene silencing
    d. gene expression profiling
    e. genome editing

Case Studies and Research Results

1. Commercial varieties of tomatoes tend to be large, hard, and tasteless, traits selected to ease shipping. Researchers asked consumers to identify heirloom tomatoes (which are bred using natural pollination) and wild tomatoes that had pleasing tastes, and then identified the chemicals responsible for the better taste. Genome-wide association studies and whole genome sequencing then revealed gene variants whose encoded proteins affect the identified chemicals in ways that improve tomato flavor. The genes control sweetness, acidity, and synthesis of volatile chemical compounds that we inhale and that contribute to taste. The researchers identified 15 genes that affect production of the volatile compounds and 6 that control overall flavor intensity. One gene, *Lin5*, encodes an enzyme, extracellular invertase, that affects sugar synthesis directly.

A *Lin5* variant rare in commercial tomatoes substitutes an aspartic acid (asp) for an asparagine (asn) at one position in the gene. Tomatoes that overexpress this mutation are much sweeter than other varieties, which are large and firm.

   a. Suggest a specific mutation that would change an asparagine to an aspartic acid (see table 10.3, the genetic code).
   b. Hypothesize how farmers might have hastened the loss of flavor from tomatoes.
   c. Each of the 26 genes that controls volatile compounds has variants that increase or decrease production. How might this information be used to improve tomato taste, either using conventional agriculture or a more recently invented biotechnology?

2. Nancy is a transgenic sheep who produces human alpha-1-antitrypsin (AAT) in her milk. This protein, normally found in blood serum, enables the microscopic air sacs in the lungs to inflate. Without it, inherited emphysema results. Donated blood cannot yield enough AAT to help the thousands of people who need it. Describe the steps taken to enable Nancy to secrete human AAT in her milk.

3. To investigate causes of acne, researchers used DNA microarrays that cover the entire human genome. Samples came from facial skin of people with flawless complexions and from people with severe acne. In the simplified portion of a DNA microarray shown, one sample is labeled green and comes from healthy skin; a second sample is labeled red and represents skin with acne. Sites on the microarray where both probes bind fluoresce yellow. The genes are indicated by letter and number.

    a. Which genes are expressed in skin whether or not a person has acne?

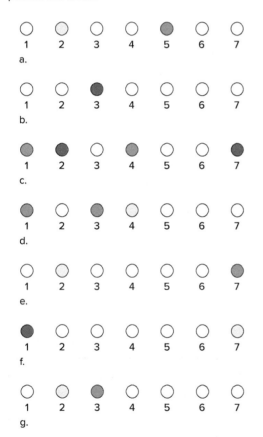

    b. Which genes are expressed only when acne develops?
    c. List three DNA pieces that correspond to genes that are not expressed in skin.
    d. How would you use microarrays to trace changes in gene expression as acne begins and worsens?
    e. Design a microarray experiment to explore gene expression in response to sunburn.

4. Most viral infections of the lower respiratory tract (the lungs and airways) in babies are caused by rhinovirus (the common cold virus), influenza virus, or respiratory syncytial virus (RSV). Little is known about RSV, and treatment is nonspecific—just supportive. A study showed that white blood cells from babies infected with RSV express 2,000 genes at significantly different levels than white blood cells from healthy infants.

    a. What technology discussed in the chapter could be used to assess the differences among the three types of infections?
    b. What further information would reveal more about RSV infection?
    c. How would you develop a diagnostic test to distinguish RSV infection from rhinovirus and flu infections?
    d. Design an experiment to enable physicians to determine which babies with RSV infection will develop symptoms severe enough to require supplemental oxygen and hospitalization.

5. Why was Dolly the sheep, who was cloned from a cell taken from a 6-year-old ewe, not patentable, but a sheep bearing a human gene is?

6. Crohn's disease is inflammation in the intestines due to overactive immune response. At the molecular level, the disease is caused by high levels of a protein called SMAD7, which blocks signals from another protein, transforming growth factor B1 (TGFB1). To restore functioning of TGFB1, researchers developed a drug that complementary base-pairs to SMAD7 mRNA, allowing the TGFB1 signal to increase enough to improve symptoms. Which biotechnology described in the chapter does this research apply?

7. Researchers used CRISPR-Cas9 to knock out one gene at a time in the genomes of human cells growing in culture, then infected the cells with Zika virus, to discover which human genes the virus requires to enter cells. The experiment revealed that the virus uses an antibody-like protein called AXL that transmits signals from the extracellular matrix (the fibers and other material between cells) into the cytoplasm. How can this discovery be used?

8. Describe a technology that could eliminate the deer ticks that transmit the bacteria that cause Lyme disease. Discuss the risks and benefits of doing this.

9. Researchers replaced UAG "stop" codons in E. coli with UAA codons altered to bind and insert a "nonstandard amino acid" (NSAA) into a growing protein. An NSAA is not among the 20 that the natural genetic code specifies. The result, a genomically recoded organism, or GRO, cannot survive without the NSAA. The technology offers "biocontainment," a way to keep organisms from spreading to environments where they are not wanted.

    a. What might a use be of a GRO?
    b. What might a danger be of a GRO?
    c. How does a GRO differ from a transgenic organism?

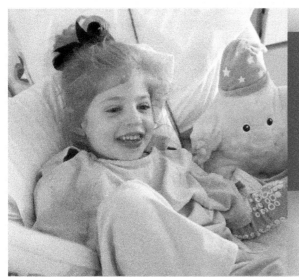

Courtesy of Glenn O'Neill

Genetic Testing and Treatment

Eliza O'Neill had gene therapy to treat Sanfilippo syndrome type A, a lysosomal storage disease.

Learning Outcomes

20.1 Genetic Counseling

1. Describe the services that a genetic counselor provides.

20.2 Genetic Testing

2. Describe types of genetic tests that are done at different stages of human prenatal development and life.
3. Discuss the benefits and limitations of direct-to-consumer genetic testing.
4. Explain how pharmacogenetics and pharmacogenomics personalize drug treatments.

20.3 Treating Genetic Disease

5. Describe three approaches to correcting inborn errors of metabolism.
6. Discuss how gene therapy adds a functional gene to correct or counter symptoms of a single-gene disease.

20.4 CRISPR-Cas9 in Diagnosis and Treatment

7. Explain how CRISPR-Cas9 genome editing may be used to help diagnose and develop treatments for single-gene diseases.

 ## The BIG Picture

DNA-based tests to predict and help to diagnose disease are becoming more common as researchers identify the functions and variants of more genes, and develop faster ways to sequence DNA. Proteins are manipulated to treat certain inborn errors of metabolism. Gene therapy and genome editing are types of interventions that add, delete, or replace genes to correct the faulty instructions behind specific diseases.

From Gene Therapy to Genome Editing

Eliza O'Neill's first symptoms were not very alarming or unusual—slight developmental delay, hyperactivity, recurrent ear infections, and not interacting much with the other children at preschool. After an autism evaluation and diagnosis didn't quite describe the full picture, her pediatrician recommended an MRI. The scan revealed fluid at the back of Eliza's brain and flattened vertebrae in her neck. These findings led to additional tests. A urine test showed the telltale buildup of the sugar heparan sulfate, caused by deficiency of a lysosomal enzyme (see figure 2.6). Then blood tests to detect the enzyme deficiency and the mutant genes led to the diagnosis: mucopolysaccharidosis (MPS) type IIIA, more commonly known as Sanfilippo syndrome type A.

When Eliza was diagnosed at age 3½, her devastated parents dove into fundraising and creating awareness through a nonprofit organization, the Cure Sanfilippo Foundation. They soon learned that a clinical trial to test a gene therapy for the disease was already being planned at a major children's hospital. Eliza was worsening, losing speech and becoming more hyperactive. Shortly after her sixth birthday, she entered the gene therapy trial. A trillion viruses, each bearing a wild type copy of the gene that encodes the missing enzyme, entered her body through a vein in her hand. The viruses should have been able to cross the blood-brain barrier and, if all went well, to allow newly formed enzyme to melt away the buildup of heparan sulfate that caused Eliza's symptoms.

Within a few months, biochemical tests on the first treated children showed that the biochemical buildup had lessened. Six months after gene therapy, Eliza said "mommy" and "daddy"—she hadn't spoken in a long time. Then she began to sing nursery rhymes! It is too soon to tell if the gene therapy has stopped the disease. Without it, symptoms would have worsened and progressed to intellectual disability, loss of mobility, increasing irritability and inability to sleep, seizures, hearing and visual loss, aggression, dementia, and early death.

Gene therapy has experienced dramatic ups and downs. Participants in some clinical trials have died—from their disease, from immune reactions, from drugs used to prepare their bodies, and the unexpected, such as introduced genes inserting into oncogenes. But against the backdrop of these setbacks have been spectacular successes, especially for childhood diseases with few or no other treatment options. Gene therapy adds functional genes. Right behind it is genome editing, which can introduce healing genes and delete the mutant ones. The age of genetic medicine is here.

20.1 Genetic Counseling

Over the past few decades, the field of human genetics has evolved from an academic life science, to a narrow medical specialty, to a growing part of many clinical fields, to a source of personal information that ordinary people can access. This chapter presents the types of genetics services that a health care consumer might encounter: genetic counseling, genetic testing, and protein and gene-based therapies.

Genetic tests provide views of our genomes at several levels. They may detect single-base or copy number variants in individual genes, display or identify abnormal chromosomes or chromosome numbers, assess variability across the genome, or determine the entire protein-encoding part of the genome (the exome) or the whole genome. The U.S. National Institutes of Health maintains information about more than 10,000 DNA-based tests in the Genetic Testing Registry (www.ncbi.nlm.nih.gov/gtr/). Health care consumers, physicians, and researchers use the registry. Such tests identify genotypes that cause, contribute to, or raise the risk of developing a specific disease.

A **genetic counselor** is a health care professional with a master's degree who can help patients and their families understand the inheritance pattern of a specific medical condition, evaluate risk, and possibly navigate the path of genetic testing (**figure 20.1**). Genetic counseling addresses medical, psychological, sociological, and ethical issues, and a genetic counselor has medical, scientific, and communication skills. A counselor

Reasons to seek genetic counseling:

Family history of abnormal chromosomes

Elevated risk of single-gene disease

Family history of multifactorial disease

Family history of cancer

Genetic counseling sessions:

Family history

Pedigree construction

Information provided on specific diseases, modes of inheritance, tests to identify at-risk family members

Testing arranged, discussion of results and treatments

Links to support groups, appropriate services, clinical trials

Follow-up contact

Figure 20.1 **The genetic counseling process.** © mediacolor's/Alamy

can interpret a DNA test; explain uncertainties; and suggest ways to cope with anxiety, fear, or guilt associated with taking genetic tests—or not taking them.

Genetic counseling began in pediatrics and prenatal care (see **A Glimpse of History**). Today it embraces diseases of adults, too, branching into such specialties as cancer, cardiovascular disease, neurology, hematology, and ophthalmology. A genetic counselor can explain the difference between susceptibility from a gene variant that contributes a small degree to risk and a single-gene diagnostic test with a higher risk based on Mendelian inheritance. Education and public policy may be part of the job. Genetic counselors in New York, for example, hold "DNA days" to educate state legislators. Genetic counselors are integral parts of research teams conducting exome and genome sequencing projects, searching the scientific literature and disease databases to assign functions to genes.

A GLIMPSE OF HISTORY

In 1947, geneticist Sheldon Reed coined the term "genetic counseling" for the advice he gave to physician colleagues on how to explain heredity to patients with single-gene diseases. In 1971, the first class of specially trained genetic counselors graduated from Sarah Lawrence College, in Bronxville, New York. Today, 33 programs in the United States offer a master's degree in genetic counseling, and many other countries have programs.

Many genetic counselors work directly with patients as parts of health care teams, typically at medical centers. A consultation may entail a single visit to explore a test result, such as finding that a pregnant woman is a carrier for spinal muscular atrophy, or a several-month-long relationship as the counselor guides a decision to take (or not take) a test for an adult-onset condition, such as Huntington disease or susceptibility to *BRCA* breast cancer.

The knowledge that a genetic counselor imparts is similar to what you have read in this book, but personalized and applied to a disease or other phenotype or concern. A counselor might explain Mendel's laws but substitute a family's condition for pea color. Or, a counselor might explain how an inherited susceptibility can combine with a controllable environmental factor, such as smoking or poor diet, to affect health.

A genetic counseling session begins with a discussion of the family's health history. Using an online tool or pencil and paper, the counselor constructs a pedigree, then deduces and explains the risks of disease for particular family members. The counselor may initially present possibilities and defer discussion of specific risks and options until test results are available. The counselor also explains which second-degree relatives— aunts, uncles, nieces, nephews, and cousins—might benefit from being informed about a test result. The genetic counselor provides detailed information on the condition and refers the family to appropriate medical specialists, including therapists, and to support groups. If a couple wants to have a biological child who does not have the illness, a discussion of assisted reproductive technologies (see chapter 21) might be helpful.

A large part of the genetic counselor's job is to determine when specific biochemical, gene, or chromosome tests are warranted and to arrange for people to take the tests. The counselor then interprets test results and helps the patient or family choose among medical options, in consultation with other medical specialists. Until genetic tests for more conditions became available fairly recently, people most often sought genetic counseling for either prenatal diagnosis or a disease in the family.

Prenatal genetic counseling typically presents population (empiric) and family-based risks, explains tests, and discusses whether the benefits of testing outweigh the risks. The couple, or woman, decides whether noninvasive prenatal testing of cell-free fetal DNA, amniocentesis, chorionic villus sampling, maternal serum screening, ultrasound scans, or no testing is the best option. Part of a prenatal genetic counseling session is to explain that tests that rule out some conditions do not guarantee a healthy baby. If a test reveals that a fetus has a serious medical condition, the counselor discusses possible outcomes, treatment plans, and the option of ending the pregnancy.

Genetic counseling when an inherited disease is in a family is more tailored to the individual situation than general prenatal counseling. For recessive diseases, the affected individual is usually a child. Illness in the first affected child is often a surprise, and recognition of a problem may take months and a diagnosis, years. However, family exome and genome analysis (see Clinical Connection 1.1 and figure 4.18), combined with rigorous mutation database searching, can now diagnose in minutes what once took years, once the DNA information is available.

Communicating the risk concerning other children in the family, future or living, may be difficult. People may think that if one child in a family has an autosomal recessive condition, then the next three will be healthy. Actually, each child has a one in four chance of inheriting the illness. Counseling for subsequent pregnancies requires great sensitivity. Some people will not terminate a pregnancy when the fetus has a condition that already affects their living child, yet some people will see that as the best option. The severity of an illness and available treatments may be important in how people feel about options. Genetic counselors must respect these feelings and tailor the discussion accordingly, while still presenting all the facts and choices.

Genetic counseling for adult-onset diseases does not have the problem of potential parents making important decisions for existing or future children, but presents the conflicting feelings of people choosing whether or not to find out if a disease is likely in their future. Often, they have seen loved ones suffer with the illness. This is the case for Huntington disease (see the opener to chapter 4). Predictive tests are introducing a new type of patient, the "genetically unwell" or those in a "premanifest" state—people with mutant genes but no symptoms (yet). Such a disease-associated genotype indicates elevated risk, but is not a medical diagnosis, which is based on symptoms and results of other types of tests.

When genetic counseling began in the 1970s, it was "nondirective," meaning that the counselor presented options but did not offer an opinion or suggest a course of action. That approach is changing as the field moves from analyzing hard-to-treat, rare single-gene diseases to considering inherited susceptibilities to more common illnesses that are more treatable, and for which lifestyle changes might realistically alter the outcome. A more recent description of a genetic counseling session is "shared deliberation and decision making between the counselor and the client."

Genetic counselors regularly communicate with physicians and other health care professionals. They are important parts of teams at molecular diagnostic testing laboratories, where they guide physicians in ordering and interpreting tests. Before a doctor orders a test, the counselor helps to interpret risks from the patient's pedigree, discusses the pros and cons of the appropriate test, and raises ethical issues that might arise when other family members are considered. While the test is under way, the genetic counselor ensures that time constraints are respected, such as an advancing pregnancy, and updates the physician. Once test results are in, the counselor may request a repeat test if the findings are inconsistent with the patient's symptoms, interpret the results, suggest additional tests, and alert the physician if the patient may be a candidate for participating in a clinical trial or able to use a new treatment.

The United States has about 4,000 genetic counselors, and most of them practice in urban areas. Finding a genetic counselor with a specific expertise can be difficult. For example, only 400 genetic counselors in the United States are specially trained in cancer genetic counseling. Due to the shortage of counselors and demand for their services, other professionals, such as physicians, nurses, social workers, and PhD geneticists, may provide counseling. Dietitians, physical therapists, psychologists, and speech-language pathologists also discuss genetics with their patients, although they may not be specifically trained to do so. Genetic testing companies may connect patients with genetic counselors for phone, online, or email consultations. Some states require that genetic counselors be licensed, requiring specific training.

As genetic testing becomes more commonplace, the need for genetic counselors and other genetics-savvy professionals to help individuals and families best use the new information will increase. Medical schools offer courses in which students interpret their own genetic tests and exome or genome sequences. Physicians attend 2-day workshops to have their genomes sequenced and interpreted, and are given iPads loaded with their results. The goal of this continuing medical education is to learn how to incorporate genetic and genomic testing into clinical practice.

Key Concepts Questions 20.1

1. What services and types of information does a genetic counselor provide?
2. To what issues must a genetic counselor be sensitive?
3. Where do genetic counselors work?
4. How is the medical field preparing for increased use of genetic and genomic testing?

20.2 Genetic Testing

Genetic tests are administered at all stages of human existence, and for a variety of reasons (**table 20.1**). Identifying mutations can help in diagnosis and choosing treatments, such as when a person only has some symptoms that are part of a recognized syndrome. Results of a genetic test are unlike the results of a cholesterol check or an X ray because they can have effects beyond the individual, to family members who share genotypes that affect health.

As the pace of exome and genome sequencing accelerates, and such testing is more widely available, it may become more economical to obtain all the information and parse it for results relevant to an individual, than to do single-gene tests. Cost is based on several factors, including the effort of sequencing and interpreting information. Obtaining a "raw sequence" of a genome may cost under $1,000, yet an analysis of every known variant of the *BRCA1* and *BRCA2* genes costs more than $3,400, for example.

The following subsections consider clinical genetic tests according to a developmental time frame, and then take a closer look at three general types of genetic tests.

Genetic Testing Through the Human Life Cycle

Some prenatal genetic tests have been in use for decades; others are new or in development.

Preconception and Prenatal Testing

When a direct-to-consumer (DTC) genetic testing company was awarded a patent in 2013 for "gamete donor selection based on genetic calculations," many people interpreted the idea as a method to create designer babies. The invention is actually a computer program that predicts the results of meiosis in the gametes of an individual. It analyzes possible gamete genotypes of "the recipient" (presumably a woman) and "a plurality of donors" (presumably men), considering penetrance and epistasis. Recall from chapter 5 that penetrance refers to the frequency that a genotype will manifest as a particular phenotype, and epistasis refers to gene-gene interactions.

The algorithm reports the likelihood that any two people of opposite sex can produce a child with a particular combination of genotypes, interpreted as possible traits. For example, a couple might ask to have their DNA analyzed to predict the likelihood of their having a child with red hair, green eyes, a small nose, and low risk of *BRCA* breast cancer, familial Alzheimer disease, and cystic fibrosis. The computer program also detects if potential parents are close blood relatives. The patent refers to a "hypothetical offspring" rather than a "designer baby" because testing an oocyte or sperm for certain inherited traits would destroy it in the process.

Although we still can't order up a particular baby, it has been possible for many years to collect sperm and separate them into fractions that are enriched for X-bearing or Y-bearing

Table 20.1   Genetic Testing

Test	Description	Chapter
PRENATAL		
Sperm selection	Enriches for X- or Y-bearing sperm, then intrauterine insemination or *in vitro* fertilization	6, 21
Preimplantation genetic diagnosis and sequencing	Sequences genes, exomes, and genomes and examines chromosomes of 1 cell of 8-celled embryo conceived *in vitro*; remaining embryo develops for a few days and is then implanted in the uterus	21
Chorionic villus sampling	Tests DNA and chromosomes of chorionic villus cell	13
Rescue karyotyping	Detects small deletions and duplications in archived cells from past unexplained pregnancy losses	13
Noninvasive prenatal diagnosis	Tests cell-free fetal (placental) DNA	13
Maternal serum markers	Measure levels of biomarkers in pregnant woman's blood	13
Amniocentesis	Tests DNA and chromosomes of amniocytes	13
NEWBORNS		
Screening	Screen metabolites and DNA in heelstick blood sample for 50-plus actionable conditions	20
Genome sequencing	Sequence genomes of many newborns to assess clinical value	20
CHILDREN		
Chromosomal microarray analysis	Detects small deletions, duplications, and other copy number variants associated with certain phenotypes	8
Exome sequencing	Diagnoses unrecognized syndromes or atypical cases; family comparisons distinguish *de novo* from inherited mutations in children	1, 4, 8
ADULTS		
Dor Yeshorim program	Preconception carrier tests for genetic diseases more prevalent among people of Jewish ancestry	15
Sickle cell disease	Tests athletes to identify carriers at risk for symptoms	11, 12
Comprehensive carrier testing	Preconception carrier tests for many single-gene diseases	20
Population carrier screen	Tests for heterozygotes for diseases more prevalent in certain population groups	15
Ancestry testing	Y chromosome and mitochondrial DNA sequences identify paternal and maternal lineages; these and autosomal markers identify distant cousins	16
Forensics testing	Copy numbers of short tandem repeats (STRs) in crime scene or disaster evidence	14
Military	Identify remains; risk for depression, PTSD; rapid infection diagnosis	1
Susceptibility	*BRCA* cancers, Alzheimer disease	5, 18
Pharmacogenetics	Drug efficacy, adverse effects, and dose	20
Genome-wide association studies	Identify genes contributing small degrees to a phenotype	7
Paternity	Half of a child's genome comes from the father	14
Centenarians	Identify gene variants that extremely old people share	3
POSTHUMOUS		
Disease diagnosis	Identify mutations in cells of deceased individuals	2, 22
History	Identify remains	5, 14
Human origins	Compare genomes of modern and archaic humans	16

cells, to attempt to conceive a girl or boy, respectively. Using a technique called flow cytometry, sperm are labeled with fluorescent markers. X-bearing sperm cells glow more intensely than Y-bearing sperm because the X chromosome is so much larger. Flow cytometry assigns each type of sperm a positive or negative charge, and uses the distinction to separate and collect them. The approach is not as specific as selecting a single sperm, but can increase the probability of having, for example, a daughter in a family that has an X-linked condition, rather than a son who would face a 50:50 chance of inheriting the illness.

During the first few days following conception, when the embryo consists of only a few cells, sampling one of them can reveal mutations, and then the rest of the embryo can be placed in the uterus to continue development. This technique, called **preimplantation genetic diagnosis**, is discussed further in chapter 21. Genetic testing techniques used later in pregnancy are discussed in chapter 13. They are chorionic villus sampling, amniocentesis, maternal serum markers, and noninvasive prenatal testing (which tests cell-free fetal DNA).

Another type of chromosome test, called rescue karyotyping, is done on tissue samples that were stored when a woman's uterus was scraped following a miscarriage. In the past, such material was not routinely tested if the woman had experienced fewer than three pregnancy losses. However, the technology to detect microdeletions and microduplications (also called copy number variants) improved as it was used in diagnosing children with developmental and other disabilities (discussed later in this section), giving researchers the idea to seek chromosomal clues in evidence from past pregnancies. Such information can indicate increased risk for future pregnancies—or, more often, alleviate concern by revealing an aneuploid (extra or missing chromosome), which is not likely to repeat.

Genetic Testing of Newborns

The goal of newborn screening is to identify infants who are at high risk of developing certain inherited diseases that are "actionable"—that is, parents and health care professionals can provide treatments and services to improve the quality of life for the child. Testing a few drops of a newborn's blood for metabolites (biochemicals that indicate an inborn error of metabolism) has been routine for decades (**figure 20.2**). For most conditions, an analytical chemistry technique called mass spectrometry detects abnormal metabolites, but DNA testing has been added. The number of conditions screened for in newborns has steadily grown, from the first one tested for, phenylketonuria (PKU), to more than 50 conditions today.

A genetic screening test that indicates a newborn is at increased risk for a specific condition is followed up with diagnostic tests that look for evidence of the disease. Early accurate diagnosis of a genetic disease can lead to treatment or enrollment in a clinical trial to test a new therapy, and it provides information on risks to future pregnancies. Genetics adds precision to a diagnosis. The downside of newborn screening is that follow-up tests may *not* show an abnormality,

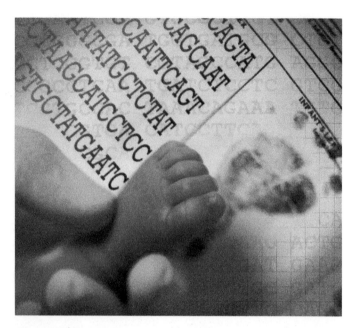

Figure 20.2 **Newborn screening.** In addition to taking a newborn baby's footprint, a few drops of blood are sampled with a heel prick for metabolic and genetic testing. © *National Human Genome Research Institute*

A GLIMPSE OF HISTORY

The field of newborn screening began in 1961, with phenylketonuria (PKU). The Guthrie test sampled blood from a newborn's heel and tested it for the buildup of the amino acid phenylalanine that indicates PKU. In 1963, a specialized diet (legally termed a "medical food") became available, with dramatic positive results. The diet sharply reduces the amount of phenylalanine. It is very difficult to stick to, and must be followed for many years, but it does prevent intellectual disability. After the success of newborn screening for PKU, state testing expanded to include eight genetic conditions and a nongenetic form of hearing loss. Gradually, the offerings have grown. Today, many states screen for 58 conditions.

and the family may then experience a situation called "patients in waiting," when the genotype associated with an illness is present, but the phenotype isn't—yet. The related condition may actually never manifest, perhaps because of effects of other genes. Studies have found that this incomplete knowledge may interfere with parent-child bonding and cause anxiety for years.

The challenges of newborn screening for genetic disease will be amplified with the much more complete information that genome sequencing provides. Five-year pilot projects began in 2013 to assess whether the information from exome and genome sequencing of newborns improves their health care. The programs are also investigating the ethical, social, and legal implications of gathering so much genetic information.

Genetic Testing of Children

Tests on children range from single-gene tests that are done because symptoms match those of a known disease, to chromosome tests, to exome and genome sequencing.

A **chromosomal microarray analysis** detects tiny deletions and duplications that are associated with autism, developmental delay, intellectual disability, behavioral problems, and other phenotypes. If a parent has the deletion or duplication and is healthy, then it is considered harmless. Finding a deletion or duplication that a parent does not also have might be comforting because it may explain the child's problems, although it wouldn't necessarily change treatment.

Since 2011, several projects have been using whole exome or genome sequencing to diagnose children whose symptoms do not match known syndromes, such as the child described in Clinical Connection 1.1. Thousands of children have now received diagnoses. The knowledge helps not only them and their health care providers, but each new diagnosis leads to development of a new genetic test that can help other children.

Exome and genome studies are revealing that some children could not be diagnosed because they actually have *two* genetic diseases. For example, a 9-year-old boy was very weak and had episodes when he would stop breathing. He also had droopy eyelids, difficulty feeding, increased respiratory secretions, and at 8 months developed an enlarged heart. His sister had died at 20 months when she stopped breathing when she had a fever. The boy's exome sequence revealed that he had congenital myasthenic syndrome, in which a fever causes too-rapid breakdown of a neurotransmitter. A drug prevented the boy from dying during a fever, as his sister had. But his exome sequence also revealed mutations in a second gene, *ABCC9*, that enlarged his heart.

Genetic testing of children for diseases that will not cause symptoms for many years is controversial. For Huntington disease, it is generally agreed not to do a presymptomatic test in anyone under age 18, because it can be so upsetting. (Juvenile HD, discussed in the chapter 4 opener, is an exception.) The legality of direct-to-consumer (DTC) DNA testing for health traits is under question, but from 2008 until 2013, even though company websites stated that children should not be tested, parents could and did send in children's DNA under the parents' names. Bioethics in chapter 1 describes one of the first efforts to incorporate genetic testing into a university curriculum.

Genetic Testing of Adults

Like children, adults take single-gene tests as part of diagnostic workups based on symptoms or other test results. They may also take genetic tests to detect increased risk of developing a particular cancer, such as a *BRCA* test. Many adults take direct-to-consumer DNA-based ancestry tests, and DNA testing is commonly done in forensics. In the military, genetic testing is used to identify remains; to detect a disease that might put a soldier at risk, such as sickle cell disease or a susceptibility to develop depression; and to diagnose a communicable disease on the battlefield.

Many adults begin to think about genetic testing when they are considering having children, and wonder what traits and illnesses they might pass on. In the past, carrier tests in adults focused on specific population groups in which a disease is more common, for economic reasons. In the 1970s, carrier testing for sickle cell disease targeted African Americans, while testing for Tay-Sachs disease recruited Ashkenazi Jews (see table 15.2). However, faster DNA sequencing has lowered the cost of carrier tests, at the same time that populations have become much more ethnically mixed. As a result, gynecologists now offer tests that can identify carriers of more than 1,000 recessive disorders to patients considering parenthood, performed on one blood or saliva sample per person. If a woman is a carrier of a specific disease, then her partner can be tested and, if he is a carrier too, genetic counseling offered. The selected diseases affect children, have a carrier frequency of 1 in 100 or greater, cause physical or cognitive impairment requiring treatment, and can be diagnosed prenatally.

Another group of adults who are having their exomes and genomes sequenced are medical students and physicians. The goal is to learn how to provide such information to their patients. **Bioethics** considers a problem that can arise with exome and genome sequencing: detecting and reporting unexpected results, called secondary (or unsolicited) findings, about conditions that were not the reason for the sequencing.

Genetic testing isn't only for the young. We have a lot to learn from the genomes of those who have survived past age 100 with good health. Genetic testing after death is informative, too. The opener to chapter 22 discusses a project to sequence the genomes of the deceased to refine the diagnoses of what had killed them, exploring whether the genetic knowledge might have altered their treatment. The opener to chapter 2 describes a girl with Rett syndrome diagnosed years after her death from the DNA in a saved baby tooth.

A Closer Look at Three Types of Genetic Tests

The genetic tests just discussed are either part of research protocols or are regulated by government agencies. Health care professionals recommend genetic tests to identify people at increased risk of developing an illness or to help diagnose a disease. Other types of DNA-based tests have been offered to consumers by for-profit companies.

Companies have marketed DTC DNA-based tests for traits, susceptibilities, and genetic diseases. In the United States, the Clinical Laboratory Improvement Amendments, or CLIA, control genetic testing of body materials for the prevention, diagnosis, or monitoring response to treatment of a disease or health impairment. The CLIA regulations, instituted in 1988, added "specialty areas" in 1992 to cover very complex tests, such as those involving immunology or toxicology. These did not include genetic testing, which at the time was limited. Since then, the genetics community has asked that genetic tests be included as a specialty area, to no avail. State regulations can override CLIA, but only if they are equally or more stringent. Therefore, the regulation of clinical genetic

Secondary Findings: Does Sequencing Provide Too Much Information?

In medical practice, an "incidentaloma" occurs when a diagnostic work-up for one condition discovers another—such as an X ray to rule out pneumonia revealing lung cancer. A genetic "incidental" (or "secondary") finding arises when sequencing a person's exome or genome to discover a mutation that accounts for one set of symptoms identifies a mutation that indicates a second condition that may or may not already have produced symptoms. The term "secondary finding" is used because "incidental" implies that the information is not useful.

Secondary genetic findings were first identified in children having their exomes sequenced to explain unfamiliar combinations of developmental delay, intellectual disability or other neurological symptoms, and/or birth defects. An early case illustrated how life-saving unexpected information can be. A 2-year-old had severe feeding problems, seizures, failure to thrive, developmental delay, and intellectual disability. Doctors had ruled out infection, Angelman syndrome, Rett syndrome, and mitochondrial disease. Family exome sequencing (see figure 4.18) revealed an autosomal dominant mutation that originated in the boy, in a gene, *SYNGAP1*, which affects synapse formation. That explained the developmental delay, intellectual disability, and seizures. But the exome sequence also revealed a mutation in the connective tissue protein fibrillin, which causes Marfan syndrome, for which the boy had no symptoms (see figure 5.5). An ultrasound of his heart indeed revealed an enlarged aortic root. A burst aorta can be the first, and deadly, symptom of the syndrome. Drug treatment saved the boy's life.

As more people have their exomes and genomes sequenced, secondary findings are inevitable, because we all have mutations. How should a clinician determine which results to report to a patient? Should all of the information in an exome or genome be deciphered?

The American College of Medical Genetics and Genomics recommends reporting 58 conditions that are prevalent, caused by mutation of a single gene, and that are "actionable" (treatment possible). The list includes cancers, connective tissue diseases, and malignant hyperthermia, in which exposure to a certain anesthetic can be lethal. Discussion of what a patient wants to know comes *before* the sequencing. If a patient does not want to know about a particular condition, it isn't included in the exome or genome analysis.

Questions for Discussion

1. What factors should a physician take into account when discussing with a patient which genetic findings to report?

2. Should everyone have their exome or genome sequenced and the information entered into a database, even if they don't want to know the results? The reason for such a database would be to speed gene discovery and development of new drugs and diagnostics.

3. Many people have taken genetic tests for mutations in *APOE4*. Certain variants greatly increase the risk of developing Alzheimer disease. A more recently discovered gene has variants that protect against the disease. How can clinicians prevent harm from the incompleteness of genetic information?

4. How can an overextended health care system that can barely handle people who are ill right now provide for people who have taken genetic tests that indicate they may become sick in the future?

---

tests remains unclear. Availability of exome and genome sequencing will complicate the regulation of testing because of the increased amount and depth of information that sequencing provides.

Tests that offer genetic information, but are not intended to be used to diagnose a disease, do not come under the CLIA regulations. The distinction between information and diagnosis is often based on careful wording. Some company websites use vague phrases such as "genetic risk factors" and a genetic variant "linked to" or "associated with" a particular trait, talent, or health condition. Following are three examples of DNA tests that provide information, but not a diagnosis, and may be offered directly to consumers: tests for inborn athletic ability, nutrigenetics testing, and pharmacogenetic and pharmacogenomic tests.

Inborn Athletic Ability

Several dozen companies worldwide offer DTC genetic tests for athletic ability. Customers mail in saliva or cheekbrush scrapings that yield DNA and receive lists of variants of specific genes a few weeks later. The genes have general physiological functions that are related to fitness and athletic ability, perhaps studied at the population level. **Table 20.2** lists some of the genes that have variants that could impact such characteristics as muscle response to exercise, metabolism, tendency to gain weight on specific diets, endurance, strength, and susceptibility to specific injuries.

Two particular genes have gotten the most attention in predicting athletic ability. The angiotensin-1-converting enzyme gene (*ACE*) has an "insertion/deletion polymorphism,"

Table 20.2 Genes Associated With Athletic Characteristics

Gene Abbreviation	Encoded Protein	Protein Function
ADRBR	Beta-2 adrenergic receptor	Regulates blood glucose level, vasodilation, fat utilization
APOA2	Apolipoprotein A2	Weight gain with high saturated fat diet
COL1A1	Type 1 collagen alpha-1	Increased need for hydration, decreased response to endurance activity
COL5A1	Type 5 collagen alpha-1	Susceptibility to tendon injury
FABP2	Fatty acid binding protein 2	Increases fatty acid uptake in small intestine, increasing blood levels of triglycerides, LDL, and total cholesterol and lowering HDL
GDF5	Growth differentiation factor 5	Increases risk of tendon disease and osteoarthritis
GST	Glutathione S transferase	Detoxification, antioxidant activity, decreased free radicals
IL6	Interleukin-6	Immune response and inflammation in injury repair
PPAR	Peroxisome proliferator-activated receptor gamma	Fat storage and glucose metabolism
TRHR	Thyrotropin-releasing hormone receptor	Controls metabolic rate
VDR	Vitamin D receptor	Degree of muscle growth and bone density increase with strength training
VEGF	Vascular endothelial growth factor	Increased blood vessel extensions with exercise

which means that 250 base pairs are either present or not. Several studies have shown that the "I" (or insertion) allele is more common among endurance athletes, such as people who do triathlons, as well as among elite mountaineers and in populations who live at very high altitudes. The "D" (or deletion) allele, in contrast, is more prevalent among elite swimmers than other groups and is associated with increased power and strength with training.

The second commonly discussed "athletic" gene is *ACTN3* (alpha-actinin 3), which encodes a protein that binds actin, a component of the cytoskeleton in every cell and also one of the two major proteins of muscle. One genotype is more common among elite sprinters and another among endurance athletes. The opener to chapter 7 discusses this gene.

Some parents are testing their young children for gene variants associated with athletic ability and using the results to decide whether the child should pursue a sport that entails sprinting or endurance, for example. While some geneticists warn that DNA-based tests for athletic ability are too simplistic and illustrate genetic determinism, the company websites do have caveats and limitations and stress that gene action combines with environmental factors.

Nutrigenetics Testing

"Nutrigenetics" websites offer DTC genetic tests along with general questionnaires about diet, exercise, and lifestyle habits. The companies return supposedly personalized profiles with dietary suggestions, often with an offer of a costly package of exactly the supplements that an individual purportedly needs to prevent realization of his or her genetic fate.

After the media spread the word of these services a few years ago, the U.S. Government Accountability Office tested the tests. An investigator took two DNA samples—one from a 9-month-old girl and the other from a 48-year-old man—and created 14 lifestyle/dietary profiles for these "fictitious consumers"—12 for the female, 2 for the male. The samples were sent to four nutrigenetics companies, none of which asked for a health history. Here is an example of the information sent to the companies:

- The DNA from the man was submitted as being from a 32-year-old male, 150 pounds, 5′9″, who smokes, rarely exercises, drinks coffee, and takes vitamin supplements.
- The DNA from the baby girl was submitted as being from a 33-year-old woman, 185 pounds, 5′5″, who smokes, drinks a lot of coffee, doesn't exercise, and eats a lot of dairy, grains, and fats.
- The same baby girl DNA was also submitted as that of a 59-year-old man, 140 pounds, 5′7″, who exercises, never smoked, takes vitamins, hates coffee, and eats a lot of protein and fried foods.

The elevated risks found for the three individuals were exactly the same: osteoporosis, hypertension, type 2 diabetes, and heart disease. One company offered the appropriate multivitamin supplements for $1,200, which the investigation found

Figure 20.3 Pharmacogenetics tailors drug treatments to how an individual metabolizes the drug, based on gene action. © Ariel Skelley/Blend Images

to be worth about $35. Recommendations stated the obvious, such as advising a smoker to quit. The advice tracked with the fictional lifestyle/diet information, and *not* genetics. Concluded the study: "Although these recommendations may be beneficial to consumers in that they constitute common sense health and dietary guidance, DNA analysis is not needed to generate this advice." Some of the suggestions could even be dangerous, such as vitamin excesses in people with certain medical conditions.

Matching Patient to Drug

People may react differently to the same dose of the same drug, even accounting for weight, gender, and age, because we differ in the rates at which our bodies respond to and metabolize drugs (**figure 20.3**). Genetic tests can identify these differences. A **pharmacogenetic** test detects a variant of a single gene that affects drug metabolism, and a **pharmacogenomic** test detects variants of multiple genes or gene expression patterns that affect drug metabolism. Pharmaceutical and biotechnology companies use these tests in developing drugs, and physicians are increasingly using these tests in prescribing drugs.

Genetic testing to guide drug selection offers several advantages:

- Identifying patients likely to suffer an adverse reaction to a drug
- Selecting the drug most likely to be effective
- Monitoring response to drug treatment
- Predicting the course of the illness (prognosis)

Many pharmacogenetic tests detect specific variants of genes that encode enzymes called cytochromes that metabolize toxins, including by-products of normal metabolism as well as drugs. The P450 cytochrome enzymes are especially diverse, encoded by 57 genes with many variants. P450 genotyping can be used to predict how quickly a person metabolizes a particular drug. A "rapid metabolizer" may dismantle a drug so fast that a higher dose is required, whereas a "poor metabolizer" may break down a drug so slowly that effects linger and may become toxic. **Table 20.3** lists examples of the effects of certain gene variants on drug metabolism.

One of the first drugs to be described using pharmacogenetics was the blood thinner warfarin (also known as Coumadin). This drug has a small range of concentration in which it keeps blood at a healthy consistency, but people can vary up to 10-fold in the dose required. Too little drug allows dangerous clotting; too much causes dangerous bleeding. In the past, physicians would give an initial standard dose, based on a patient's age, gender, health status, weight, and ethnicity, then monitor the patient for a few weeks to check for too much clotting or bleeding, tweaking the dose until it was about right. This general approach led to hospitalization for abnormal bleeding in 43,000 of the 2 million people prescribed the drug each year.

A "pharmacogenetic algorithm" is now used to prescribe warfarin. It considers two genes: two variants of *CYP2C9* and one variant of *VKORC1,* which are associated with increased

Table 20.3	Pharmacogenetics		
Gene Abbreviation	**Encoded Protein**	**Genotype**	**Phenotype**
CYP2D6	Cytochrome P450	2D6	Poor metabolizer; increased sensitivity to certain beta blocker heart drugs
HTR2A	Serotonin receptor	G allele	Strong response to SSRI drugs to treat depression and anxiety
OPRM1	Opioid receptor	A118G (asp to asn)	Requires higher doses of opiates for pain relief
SLCO1B1	Solute carrier organic anion transporter	T/C (val/ala)	4.5-fold increased risk of muscle pain from statin drugs
		C/C (ala/ala)	17-fold increased risk of muscle pain from statin drugs
VKORC1	Vitamin K epoxide reductase complex subunit 1	1639G>A	Increased sensitivity to warfarin (a blood thinner)

sensitivity to the drug. People with these gene variants require lower doses of warfarin. The genetic test for warfarin response is especially helpful for the 50 percent of patients who fall at the extremes of the range of drug concentration that is effective. However, considering clinical information remains an extremely important part of determining the dose of warfarin in particular, and other drugs in general, because variants of genes other than the ones that are tested for can affect how an individual human body metabolizes a drug.

Key Concepts Questions 20.2

1. On what biological process is "gamete donor selection based on genetic calculations" based?
2. Which genetic tests are used on embryos and fetuses?
3. In what circumstances are genetic tests done on children?
4. Why might adults take genetic tests?
5. What are limitations of direct-to-consumer genetic tests for athletic ability and nutrition?
6. Explain how pharmacogenetics and pharmacogenomics help health care providers select the best drugs for patients.

20.3 Treating Genetic Disease

Tests for genetic diseases greatly outnumber treatments, because treatments are challenging to develop. They must correct the abnormality in enough of the appropriate cells and tissues to prevent or minimize symptoms, while not harming other parts of the body. Treatments for single-gene diseases have evolved through several stages, in parallel to development of new technologies:

- Removing an affected body part
- Replacing an affected body part or biochemical with material from a donor
- Delivering pure, human proteins derived from recombinant DNA technology
- Refolding correctly a misfolded protein
- Blocking gene expression (gene silencing) (see figure 19.11 and figure 19.12)
- Using gene therapy to add wild type alleles without removing mutant alleles
- Using gene and genome editing to replace, delete, or add alleles

Drugs

Preventing a disease phenotype may be as straightforward as adding digestive enzymes to applesauce for a child with cystic fibrosis or giving a clotting factor to a boy with hemophilia. Inborn errors of metabolism are particularly treatable when the biochemical pathways are well understood.

Lysosomal storage diseases are a well-studied subgroup of inborn errors of metabolism. The little girl described in the chapter opener has such a disease, Sanfilippo syndrome type A. In the disease, a deficient or abnormal enzyme leads to buildup of the substrate (the molecule that the enzyme acts on) and deficit of the breakdown product of the substrate. Recall from figure 2.6 that a lysosome is an organelle that dismantles debris. It houses more than 50 types of enzymes, and each breaks down a specific molecule.

Treatment of another lysosomal storage disease, type 1 Gaucher disease, illustrates three general approaches to counteracting an inborn error of metabolism that affects an enzyme, summarized in **table 20.4** and **figure 20.4**. In type 1 Gaucher disease, the enzyme glucocerebrosidase is deficient or absent. As the substrate builds up because there is little or no enzyme to break it down, lysosomes swell, ultimately bursting cells. Symptoms include an enlarged liver and spleen, bone pain, and deficiencies of blood cells. Too few red blood cells cause the fatigue of anemia; too few platelets cause easy bruising and bleeding; and too few white blood cells increase the risk of infection. The disease is variable in age of onset, severity of symptoms, and rate of progression.

Early treatments for Gaucher disease corrected the affected body parts: removing the spleen, replacing joints, transfusing blood, or transplanting bone marrow. In 1991, **enzyme replacement therapy** became available, which supplies the missing or deficient enzyme, in this case recombinant glucocerebrosidase. It is effective but costs about $550,000 a year and the recombinant enzyme takes several hours to infuse, twice a month. Only a few diseases have enzyme replacement therapies.

In 2003 came a different approach: **substrate reduction therapy**. A drug taken by mouth decreases the amount of the substrate, the molecule on which the deficient enzyme acts. In a third approach called **pharmacological chaperone therapy**, an oral drug binds a patient's misfolded enzyme, stabilizing it sufficiently to allow some function.

Less costly than developing new drugs is to "repurpose" drugs already available to treat other conditions. The chapter 3 opener describes a failed cancer drug that helps children who

Table 20.4	Lysosomal Storage Disease Treatments
Treatment	**Mechanism**
Enzyme replacement therapy	Recombinant human enzyme infused to compensate for deficient or absent enzyme
Substrate reduction therapy	Oral drug that reduces level of substrate so enzyme can function more effectively
Pharmacological chaperone therapy	Oral drug that binds to patient's misfolded protein, restoring function

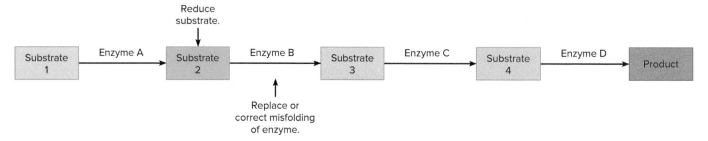

Figure 20.4 **Counteracting a metabolic abnormality.** Treatments of lysosomal storage diseases are based on understanding metabolic pathways, in which a series of enzyme-catalyzed reactions leads to formation of a product. If any enzyme is deficient or its activity blocked, the substrate builds up and the product is deficient. Enzyme replacement therapy delivers an absent or deficient enzyme. Substrate reduction therapy decreases the amount of substrate and pharmacological chaperone therapy corrects misfolded proteins. Points of intervention vary with the specific condition.

have a form of progeria, the accelerated aging condition. A drug that treats erectile dysfunction by increasing blood flow to the penis improves leg muscle function in boys who have Becker muscular dystrophy.

Gene Therapy

Gene therapy delivers working copies of genes to specific cell types or body parts, typically as part of modified viruses that function as carriers, called vectors. More than 2,000 clinical trials of gene therapies have been conducted worldwide since 1990. The first approval of a gene therapy, in Europe, was in 2012 for a rare enzyme deficiency that causes pancreatitis but it was discontinued due to the high cost and low demand.

Adding functional genes to treat an inherited disease may provide a longer-lasting effect than treating symptoms or supplying a protein. A gene therapy may require a single or a few treatments, whereas a traditional drug or even enzyme replacement, substrate reduction, or pharmacological chaperone therapy may have to be administered frequently for the patient's entire life.

Types and Targets of Gene Therapy

Gene therapy approaches vary in the way that healing genes are delivered and where they are sent. **Germline gene therapy** alters the DNA of a gamete or fertilized ovum, so that all cells of the individual have the change and the correction is heritable, passing to offspring. Germline gene therapy is not being done in humans. (The transgenic plants and animals discussed in chapter 19 had germline genetic manipulations.) In contrast, **somatic gene therapy** corrects only the cells that an illness affects. It is nonheritable: A recipient does *not* pass the genetic correction to offspring, unless the cells that give rise to gametes are inadvertently altered. Clearing lungs congested from cystic fibrosis with a nasal spray containing functional *CFTR* genes is an example of somatic gene therapy.

Gene therapy strategies vary in invasiveness (**figure 20.5**). Cells can be altered outside the body and then infused into the bloodstream through a vein. This is called *ex vivo* **gene therapy** ("outside the body"). In the more invasive *in vivo*

gene therapy ("in the living body"), the gene and its vector are introduced directly into the body, such as through a catheter inserted and snaked to the appropriate organ. There, the vector must enter the targeted cells, the human DNA be transcribed into mRNA, the mRNA translated into protein, and then the protein must function enough to halt progression of the disease or improve symptoms.

Researchers obtain therapeutic genes using the polymerase chain reaction, recombinant DNA technologies, and other techniques that cut DNA. In the future, researchers—and, someday, clinicians—may deliver synthetic genes. The origin should not make a difference—DNA is DNA and the genetic code universal.

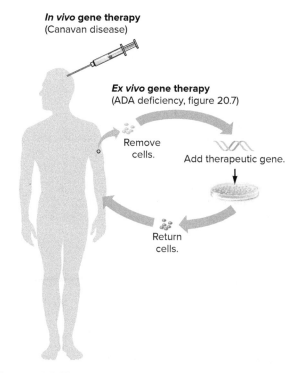

Figure 20.5 **Gene therapy invasiveness.** Therapeutic genes are delivered to cells removed from the body that are then returned (*ex vivo* gene therapy) or delivered directly to an interior body part (*in vivo* gene therapy).

The next step in gene therapy, gene transfer, usually delivers the healing DNA with other DNA that is mobile. To create viral vectors, researchers remove the genes that cause symptoms or alert the immune system to infection and add the corrective gene. Different viral vectors are useful for different treatments. A certain virus may transfer its cargo with great efficiency to a specific cell type, but carry only a short DNA sequence. Another virus might carry a large piece of DNA but enter many cell types, causing "off-target" side effects. Even if a viral vector goes where intended, it must enter enough cells to alleviate symptoms or slow or halt progression of the disease. Finally, a viral vector must not integrate into a gene that harms the patient, such as an oncogene or tumor suppressor gene, which could cause cancer.

Some gene therapies use viruses that normally infect the targeted cells. For example, a herpes simplex virus delivers the gene encoding a pain-relieving peptide to nerve endings in skin. Researchers team parts of viruses to

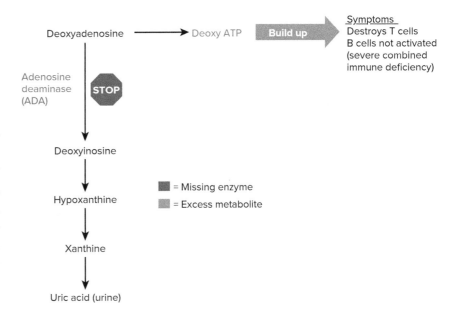

Figure 20.7 **ADA deficiency.** Absence of the enzyme adenosine deaminase (ADA) causes deoxy ATP to build up, which destroys T cells, which then cannot stimulate B cells to secrete antibodies. The result is severe combined immune deficiency (SCID). © *Anna Powers*

target a certain cell type. Adeno-associated virus (AAV), for example, infects many cell types, but adding a promoter from a parvovirus gene restricts it to red blood cell progenitors in bone marrow. Different subtypes of AAV home to different body parts. AAV9 can cross the blood-brain barrier and is given by intravenous infusion, replacing earlier versions that required delivery into the brain through catheters. The girl described in the chapter opener received gene therapy via AAV9 vectors. Fatty structures called liposomes are also used as vectors. **Figure 20.6** shows four targets of somatic gene therapies.

The first gene therapy efforts were aimed at well-studied inherited diseases, even though they are rare. Knowing the biochemical pathway that is disrupted enabled researchers to know where to intervene. **Figure 20.7** illustrates such a biochemical pathway for adenosine deaminase (ADA) deficiency, a form of severe combined immune deficiency that was the first for which gene therapy was tested (see table 17.5 and **Clinical Connection 20.1**). Today, about 8 percent of gene therapy clinical trials are for single-gene diseases. About 65 percent target cancers, and about 8 percent focus on cardiovascular disease.

Endothelium. The tile-like endothelium that forms capillaries can be genetically altered to secrete proteins into the circulation.

Muscle. Immature muscle cells (myoblasts) given healthy dystrophin genes may treat muscular dystrophy.

Liver. To treat certain inborn errors of metabolism, only 5 percent of the liver's 10 trillion cells need to be genetically altered.

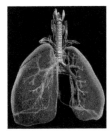

Lungs. Gene therapy can reach damaged lungs through an aerosol spray. Enough cells would have to be reached to treat hereditary emphysema (alpha-1-antitrypsin deficiency) or cystic fibrosis.

Figure 20.6 **Some sites of gene therapy.** © *MedicalRF.com/ Getty Images RF*

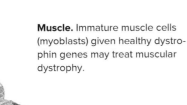

Key Concepts Questions 20.3

1. Explain how the three types of protein-based therapies work.

2. Distinguish between somatic and germline gene therapies.

3. What is the role of viruses in gene therapy?

The Rocky History of Gene Therapy

Gene therapy has had its ups and downs, as does any new medical technology.

Initial Success

In the late 1980s, Ashanthi ("Ashi") DeSilva's parents did not think their little girl would survive her near-continual coughs and colds. She could walk only a few steps before becoming winded. Ashi had severe combined immune deficiency due to ADA deficiency (see table 17.5, figure 20.7, and **figure 20A**). At age 2, she began enzyme replacement therapy, but within 2 years it stopped working. She would likely die of infection. Then her physician heard about a clinical trial of gene therapy that would give her white blood cells functional ADA genes. On September 14, 1990, at the National Institutes of Health in Bethesda, Maryland, Ashi received an intravenous infusion of her own corrected white blood cells. She did well. Today a longer-lasting form of the gene therapy alters stem cells instead of the more specialized T cells. It was approved in Europe in 2016.

Setbacks

In September 1999, 18-year-old Jesse Gelsinger died, 4 days after receiving gene therapy, from an overwhelming immune response to the viruses used to introduce the therapeutic genes. He had ornithine transcarbamylase (OTC) deficiency and could not make a liver enzyme required to break down dietary proteins. In the disease, the nitrogen released from digestion of proteins combines with hydrogen to form ammonia (NH_3), instead of being excreted in urine. The ammonia rapidly accumulates in the bloodstream and travels to the brain, usually causing irreversible coma within 72 hours of birth. Half of affected babies die within a month and another quarter by age 5. The survivors can control their symptoms by following a very-low-protein diet and taking drugs that bind ammonia.

Jesse was diagnosed at age 2. He had a mild case because he was a mosaic—some of his cells could produce the enzyme. At age 10, Jesse's diet lapsed and he became very ill. Years later, after he went into a coma in December 1998 from missing a few days of medications and recovered, he volunteered for a gene therapy trial his doctor had mentioned. On September 13, 1999, a trillion adenoviruses carried functional human *OTC* genes through a tube into Jesse's liver. But the viruses entered not only the hepatocytes as intended, but also the macrophages that alert the immune system. Jesse's organ systems began to fail. Four days later he was brain dead, and the devastated medical team stood by as his father turned off life support.

Also in 1999, researchers in France gave gene therapy for X-linked severe combined immune deficiency (SCID X1) to two babies (see figure 17.10). In SCID X1, T cells do not have certain cytokine receptors, which prevents the immune system from recognizing infection. Most children die in infancy. *Ex vivo* gene therapy removes

Figure 20A **Gene therapy introduces a gene that can compensate for, but not replace, a mutant allele.** *Courtesy of Jane Ades/NHGRI*

a boy's T cell progenitor cells from bone marrow, gives them wild type alleles, and infuses the corrected cells back. The researchers used a retrovirus, which only enters dividing cells. If the healing viruses could infect the T cell progenitor cells, the corrected cells would differentiate into mature T cells that could alert the immune system. However, by 2002, one boy in the trial developed leukemia, followed by several others. The retroviruses had inserted into part of a chromosome that harbors a proto-oncogene, turning it into an active oncogene. Use of a "self-inactivating" vector has made this gene therapy, which is effective against SCID X1, much safer.

More Success

A dramatic gene therapy success story is the treatment of a form of hereditary blindness called Leber congenital amaurosis type 2 due to RPE65 deficiency (LCA2). Several hundred people who lived in dark shadows or total blindness can now see after experimental gene therapy for this condition. One little girl, just days after gene therapy, could navigate a curb using her eyes, not her cane. Another child could see his food in a dimly lit restaurant near the hospital where days earlier he'd had his first eye treated. For years he had needed to feel and smell the items on his plate to identify them.

At the backs of the eyes in a person with LCA2, cells that make up a thin layer called the retinal pigment epithelium cannot make an enzyme necessary to convert vitamin A to a form that the rods and cones, the cells that transmit light energy to the brain, can use. Gene therapy supplies the gene that encodes the enzyme. Figure 5.6 shows the anatomy of the retina and where genes intervene.

Gene therapy also works for adrenoleukodystrophy (ALD). This is the disease of peroxisomes that affected Lorenzo Odone, whose parents devised a dietary oil to help him (see section 2.2). Peroxisomes are tiny sacs inside cells that house enzymes. They have porthole-like openings formed by a protein called ABCD1. A mutation in the *ABCD1* gene prevents the portholes from admitting an enzyme needed to process certain fats used to make myelin, the material that coats neurons, enabling them to send messages. The affected brain cells, called microglia, descend from progenitor cells in the bone marrow.

(Continued)

Gene therapy for ALD alters bone marrow cells outside the body and sends them into the bloodstream, where they go to the brain and give rise to corrected microglia. The viral vector is HIV, stripped of the genes that cause AIDS. The disabled HIV is an excellent vector because it does not insert into proto-oncogenes, is efficient, can carry large genes, and enters many types of cells. For several treated boys, blood levels of the crippling fats fell so greatly and brain neurons gained enough myelin so that the boys could attend school—with only about 15 percent of their microglia corrected! Teaming the gene therapy for ALD with newborn screening may completely prevent symptoms of this otherwise devastating disease.

Questions for Discussion

1. The goal of a clinical trial of a new medicine or other type of treatment, such as gene therapy, is to demonstrate safety and efficacy. Patients are selected so that the trial will yield the most useful information to help as many people as possible—not to help a particular individual. However, different clinical trial protocols select participants in different ways. State the risks and benefits, from different perspectives, for each of the following real clinical trial protocols. What further information do you need to judge whether each procedure is ethical or not?

   a. A clinical trial for metachromatic leukodystrophy enrolls families with more than one affected child. The diagnoses of the older children led to the diagnoses of the younger ones, and the younger children are given the gene therapy. The older children have died. The reasoning is that the treatment can prevent symptoms, whereas it cannot improve or likely reverse symptoms in the older children.

   b. A clinical trial for LCA2 has several families with more than one affected child. The older children are being treated first because their vision is worse and therefore they have more to gain. The continuing deterioration in vision is studied in the younger siblings, who receive the gene therapy a year later.

   c. A clinical trial for Batten disease treated two young daughters of a famous Hollywood producer shortly after they were diagnosed, partly because the case was publicized in the popular media, some people have said. Other children wait years to enter clinical trials or die waiting.

20.4 CRISPR-Cas9 in Diagnosis and Treatment

The genome editing tool CRISPR-Cas9 may one day replace gene therapy because it is more precise (see figure 19.11). CRISPR-Cas9 not only delivers a gene to a specific part of the genome, but can replace a gene, remove a gene, or add a gene, and do so in multiple places or insert, correct, or delete multiple genes. In contrast, genes introduced through older gene therapy methods may integrate into chromosomes at random or remain outside the chromosomes as parts of circles of DNA called episomes. Although genome editing using CRISPR-Cas9 is being developed rapidly, it is largely a research tool, with applications in diagnosis and treatment just beginning.

An Animal Model Identifies a Family's Mutation

CRISPR-Cas9 genome editing can be combined with genome sequencing to engineer an animal model of a family's inherited disease, to refine a diagnosis and reveal how the symptoms arise. This is the case for a family with thoracic aortic aneurysm and dissection (TAAD), in which the wall of the aorta—the largest artery in the body, leading from the heart—stretches, thins, balloons out, and may burst, causing sudden death (**figure 20.8**). If detected early, the ballooning, called an aneurysm, can be repaired. TAAD runs in families and is part of several genetic syndromes that affect connective tissue, including Marfan syndrome (see figure 5.5) and Ehlers-Danlos syndrome (see figure 12.5).

A 35-year-old man came to the attention of researchers at a genome sequencing center. The man had experienced sudden severe chest pain at age 19. When scans revealed that his aorta was on the verge of rupturing, he underwent surgery to replace the blood vessel. The man's mother had also had an aortic repair, and two cousins had aortic aneurysms. Everyone in the family had tests for all mutations known to cause aortic aneurysm, and all were negative. So the researchers sequenced the genomes of the two cousins and identified seven candidate genes. Segregation analysis, based on Mendel's first law, tracked which mutations occurred only in the affected family members and never in the healthy ones. Only one gene fit those criteria, and it encodes an enzyme called lysyl oxidase (LOX).

Next came reasoning: Could mutations in *LOX* cause the symptoms? LOX is necessary for the connective tissue proteins collagen and elastin to link, which they do when they form the walls of arteries. It made sense. To test the hypothesis that *LOX* mutations cause TAAD, the researchers used CRISPR-Cas9 to replace certain DNA nucleotides in the mouse version of the gene to recreate the mutant human sequence. Mice given two copies of the human mutant *LOX* gene died just after birth as their aortas burst. Mice with one copy—similar to the family members—survived but with fragmented and fragile aortic walls. (Many genetic diseases that are autosomal dominant in humans are autosomal recessive in mice.)

A genetic test is now possible for TAAD caused by mutations in *LOX*. Although this is a rare form of the condition, availability of an accurate and precise test can help to diagnose family members who do not yet have symptoms, possibly saving lives.

Somatic Genome Editing Tackles Sickle Cell Disease

The first applications of CRISPR-Cas9 to treat disease are targeting conditions in which the affected cells are easy to reach. Chimeric antigen receptor technology (see section 18.4)

Thoracic Aortic Aneurysm

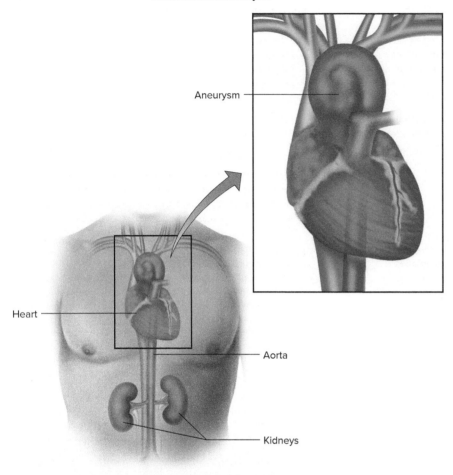

Aneurysm

Heart

Aorta

Kidneys

Figure 20.8 **CRISPR-Cas9 created a mouse model that helped diagnose familial thoracic aortic aneurysm and dissection in one family.** An aortic aneurysm may not produce symptoms until it bursts. The diagnostic test developed from using a mouse model can identify and alert patients.

embryonic and fetal forms of hemoglobin (see figure 11.2).

Reactivating the quiescent globin genes can treat sickle cell disease. One approach uses induced pluripotent stem cells (iPS cells; see table 2.3) that are cultured in laboratory glassware from skin fibroblasts of people who have the disease. Chemicals are added to the iPS cells to direct their development into the cell type that precedes mature red blood cells, called reticulocytes. CRISPR-Cas9 replaces one sickle cell mutation with a wild type copy of the beta globin gene. The reticulocytes normally extrude their nuclei as they mature into red blood cells to make room for as much hemoglobin as possible. The red blood cells that descend from the gene-edited reticulocytes indeed fill with as much normal hemoglobin as a heterozygote—that is, enough to not have the disease. The intervention works, at least on the cellular level.

A second approach to using CRISPR-Cas9 to treat sickle cell disease silences a gene that normally switches fetal hemoglobin production off after birth. Researchers introduced small deletions into the switch gene, called *BCL11A*, into hematopoietic stem cells. The red blood cells that descend from the manipulated cells make fetal hemoglobin, which does not sickle and effectively carries oxygen.

at first used ZFNs and TALENs to treat cancer, and is now being done using CRISPR-Cas9. Other treatments based on genome editing alter cells that descend from the hematopoietic (blood-forming) stem cells in bone marrow because they can be removed, altered, and placed back in the body.

Sickle cell disease is particularly amenable to genome editing. Recall from figure 12.3 that sickle cell disease is caused by a missense mutation in the beta globin gene. It is autosomal recessive. The embryo and fetus use different genes to produce

---

Key Concepts Questions 20.4

1. Distinguish genome editing from gene therapy.

2. Explain how genome editing can help to identify specific mutations that cause a disease in a family.

3. Describe two strategies that use genome editing to treat sickle cell disease.

Summary

20.1 Genetic Counseling

1. **Genetic counselors** provide information on inheritance patterns for specific medical conditions, disease risks and symptoms, and tests and treatments.

2. Prenatal counseling and counseling a family coping with a particular disease pose different challenges.

3. Genetic counselors interpret genetic tests and assist other health care professionals in incorporating genetic information into their practices.

20.2 Genetic Testing

4. Genetic tests are performed before birth and on newborns to detect certain single-gene diseases, on children to assist in diagnosis, on adults for diagnosis and carrier detection, and on human remains. A **chromosomal microarray analysis** detects small deletions and copy number variants and is mostly used for children.

5. The Clinical Laboratory Improvement Amendments regulate some genetic tests.

6. DTC tests may provide inaccurate, inappropriate, or incomplete information on traits such as athletic ability and dietary habits.

7. **Pharmacogenetic** and **pharmacogenomic** tests provide information on how individuals metabolize certain drugs.

20.3 Treating Genetic Disease

8. **Enzyme replacement therapy, substrate reduction therapy,** and **pharmacological chaperone therapy** treat biochemical imbalances at the protein level.

9. Drugs may be repurposed to treat genetic diseases.

10. Gene therapy delivers genes to cells and directs production of a needed substance at appropriate times and in appropriate tissues, in therapeutic (not toxic) amounts.

11. **Germline gene therapy** alters gametes or fertilized ova, affects all cells, and is transmitted to future generations. It is not done in humans. **Somatic gene therapy** affects somatic cells and is not passed to offspring.

12. **Ex vivo** gene therapy is applied to cells outside the body that are then implanted or infused into the patient. **In vivo gene therapy** delivers gene-carrying vectors directly into the body.

13. Several types of vectors deliver therapeutic genes. Viruses are most commonly used. Some gene therapies target stem or progenitor cells, because they can divide and some can move.

14. Gene therapies are attempted on diseases with well understood pathological mechanisms.

20.4 CRISPR-Cas9 in Diagnosis and Treatment

15. Genome editing adds, deletes, or replaces genes at specific loci. Gene therapy introduces genes nonspecifically.

16. Genome editing aids diagnosis by creating animal models that have human mutations.

17. CRISPR-Cas9 can correct the beta globin mutation behind sickle cell disease in induced pluripotent stem cell–derived red blood cell precursors, or it can reactivate fetal hemoglobin genes.

Review Questions

1. How are the services that a genetic counselor provides different from those of a nurse or physician?

2. How are the consequences of a genetic test different from those of a cholesterol check?

3. Compare and contrast the types of information obtained from preconception comprehensive carrier screening, prenatal diagnosis, and newborn screening in terms of the types of actions that can be taken in response to receiving the results.

4. Describe a situation in which a DNA test performed on human remains provides interesting or helpful information.

5. Using information from this or other chapters, or the Internet, cite DNA-based tests given to a fetus, a newborn, a young adult, and a middle-aged person.

6. Select a gene from table 20.2 and discuss how its variants might impact the ability to succeed at a particular sport or exercise.

7. Explain how a pharmacogenetic test can improve quality of life for a person with cancer.

8. Distinguish among enzyme replacement therapy, substrate reduction therapy, and pharmacological chaperone therapy.

9. State a biological and an ethical reason why germline gene therapy and germline genome editing are not being done in humans.

10. What factors would a researcher consider in selecting a viral vector for gene therapy? In selecting a disease amenable to gene therapy?

11. How do gene therapy and genome editing differ?

12. Explain how researchers combined genome sequencing, application of Mendel's first law, and CRISPR-Cas9 to diagnose familial thoracic aortic aneurysm and dissection.

13. Explain why blood conditions will likely be among the first diseases to be treated with CRISPR-Cas9 genome editing.

14. Describe two ways that CRISPR-Cas9 genome editing is used experimentally to correct the mutation underlying sickle cell disease.

Applied Questions

1. Suggest a treatment other than gene therapy that might have helped Eliza O'Neill, the child described in the chapter opener.

2. Explain why high penetrance is an important criterion for including a disease in newborn screening.

3. Discuss the challenges that a genetic counselor faces in explaining to parents-to-be a prenatal diagnosis of trisomy 21 in a fetus with no family history, compared to a prenatal diagnosis of translocation Down syndrome in a family with a reproductive history of pregnancy loss and birth defects. (Chapter 13 discusses Down syndrome.)

4. Should an online dating service include analysis based on the patent for "gamete donor selection based on genetic calculations" to predict the genetic characteristics possible for offspring from two individuals? Cite a reason for your answer.

5. Lysosomal acid lipase deficiency causes hepatocytes (the most abundant type of liver cell) to accumulate triglycerides and low density lipoproteins (LDL), but have low levels of high density lipoproteins (HDL). The resulting "fatty liver" becomes hard, or sclerotic, and its functions become impaired. Certain genotypes are lethal by the age

of 1 year. The condition is a lysosomal storage disease and is autosomal recessive. The statin drugs that millions of people take to combat this unhealthy lipid profile are ineffective, because the drugs target a different pathway (endogenous cholesterol synthesis) than the one that is affected in lysosomal acid lipase deficiency.

a. Compare how the mechanisms of enzyme replacement therapy and gene therapy to treat this disease differ.

b. Who would be good candidates for a genetic test for the deficiency?

c. Lysosomal acid lipase deficiency is frequently misdiagnosed as familial hypercholesterolemia (see figures 5.2 and 12.9). The two conditions have the same phenotype, but different genotypes. How do the diseases differ in the underlying abnormal physiology?

6. Select a disease described in an earlier chapter and explain how any of the technologies discussed in this chapter might be applied to create a treatment. Possible diseases include Rett syndrome (chapter 2 opener), lactase deficiency (Clinical Connection 2.1), progeria (chapter 3 opener), giant axonal neuropathy (Clinical Connection 2.2), Huntington disease (chapter 4 opener), and cystic fibrosis (Clinical Connection 4.1). The technologies are enzyme replacement therapy, substrate reduction therapy, pharmacological chaperone therapy, gene therapy, and genome editing.

Case Studies and Research Results

1. If you were the genetic counselor for the following patients, how would you answer their questions or address their concerns? (See other chapters for specific information.)

a. A couple in their early forties is expecting their first child. Amniocentesis indicates that the fetus is XXX (see table 13.5). When they learn of the abnormality, the couple asks to terminate the pregnancy, fearing severe birth defects caused by the extra chromosome.

b. Two people of normal height have a child with achondroplastic dwarfism, an autosomal dominant trait. Will future children have the condition (see figure 5.1)?

c. A couple has received results from a direct-to-consumer genetic testing company. Tests based on genome-wide association studies (see figure 7.11) indicate that they each have inherited susceptibility to asthma as well as gene variants that in some populations are associated with autism (see table 8.4). Both also have several gene variants that are found in lung cancers (see figure 18.14). Each has a few recessive mutant alleles, but not in the same genes. On the basis of these results, they do not think that they are "genetically healthy" enough to have children.

2. Jill and Scott had thought 6-month-old Dana was developing just fine until Scott's sister, a pediatrician, noticed that the baby's abdomen was swollen and hard. Knowing that the underlying enlarged liver and spleen could indicate an inborn error of metabolism, Scott's sister suggested the child undergo several tests.

  Dana had inherited sphingomyelin lipidosis, also known as Niemann-Pick disease type A. Both parents were carriers, but Jill had tested negative when she took a Jewish genetic disease panel during her pregnancy because her mutation was rare and not part of the test panel. Dana was successfully treated with a transplant of umbilical cord blood cells from a donor. She caught up developmentally and became more alert. Monocytes, a type of white blood cell, from the cord blood traveled to her brain and manufactured the deficient enzyme. Dietary or enzyme replacement therapy do not work for this condition because the enzyme cannot cross from the blood to the brain. Monocytes, however, can enter the brain.

a. Explain the effect of Dana's treatment on her phenotype and genotype.

b. Why did the transplant have to come from donated cord blood, and not from Dana's own, which had been stored?

c. If you were the genetic counselor, what advice would you give this couple if they conceive again?

d. What is the probability that Dana can pass on the disease to a child if her partner is a carrier?

3. A clinical trial of gene therapy for the clotting disorder hemophilia B introduced the gene for factor IX via AAV into the livers of several men. Researchers halted the trial when they detected the altered gene in semen of treated men. Why did they take this precaution?

4. Tanisha and Jamal were concerned that their son Perry did not make eye contact or respond as his older siblings had, but the pediatrician assured them that the boy was well, just easily distracted. In school, though, Perry could not focus, and had angry outbursts. Tanisha, a nurse, learned about chromosomal microarray analysis in a continuing education class. She had Perry tested, which revealed a small duplication. The report stated that the finding was a "variant of uncertain significance" (see chapter 12 Bioethics). The duplication was abnormal, but it hadn't been associated with a syndrome. Tanisha wanted to include the test results in Perry's school records, in case a situation should arise in which a medical explanation or special services might be helpful, but Jamal feared that the genetic test result might stigmatize the boy. Do you agree with either parent? Cite a reason for your answer.

5. In 2004, in the months following approval and initial marketing of a type of arthritis drug called a COX-2 inhibitor, cases of heart damage in some patients who took the drug began to be reported. It is a rare adverse effect that had not shown up in the clinical trials, but did after millions of people began taking the drugs. Several drugs were discontinued or their use restricted because of the risk of heart damage, preventing many people who would not have developed heart problems from obtaining relief for their arthritis. What technology discussed in the chapter might help in determining which arthritis patients can safely take COX-2 inhibitors?

© medicalRF.com

CHAPTER 21

Reproductive Technologies

In vitro *fertilization coupled with genetic testing of preimplantation embryos enables parents-to-be to select embryos for completing development that have the normal number of chromosomes and have not inherited specific disease-causing gene variants. This is a stylized view of sperm meeting oocyte.*

Learning Outcomes

21.1 Savior Siblings and More

1. Explain how a child can be conceived to provide tissue for an older sibling.
2. Define *assisted reproductive technology (ART)*.

21.2 Infertility and Subfertility

3. Distinguish infertility from subfertility.
4. Describe causes of infertility in the male.
5. Describe causes of infertility in the female.
6. List infertility tests.

21.3 Assisted Reproductive Technologies

7. Describe ARTs that donate sperm, uterus, or oocyte.
8. List the steps of *in vitro* fertilization.
9. Explain how preimplantation genetic diagnosis enables people to select embryos conceived *in vitro* that have normal chromosomes and do not have certain mutations.
10. Explain how testing a polar body can reveal information about a genotype of a fertilized ovum.

21.4 Extra Embryos

11. Discuss uses for extra embryos resulting from ARTs.

The BIG Picture

Assisted reproductive technologies provide intriguing and sometimes complex variations on the processes of conceiving, selecting embryos with or without specific gene variants, and carrying offspring to term. Testing early embryos is evolving from visualizing chromosomes to identifying specific single gene variants to exome and genome sequencing.

Choosing an Embryo

The couple met when they were in their early forties. When they began discussing starting a family, they had genetic testing and discovered that they each are carriers of a serious neurological condition, Pompe disease, and that the man has a variant of the factor V Leiden gene that is consistent with his family history of deep vein thrombosis and pulmonary embolism. The couple decides to use *in vitro* fertilization (IVF) with preimplantation genetic diagnosis (PGD) to increase the chances that they have a healthy child. They are concerned because of their family medical histories and their own mutations, and the elevated risk of an abnormal number of chromosomes in an oocyte due to the woman's age. It might take much longer to conceive and carry a pregnancy to term the natural way.

For several weeks the woman injects hormones to prepare her body to ovulate. At the right time, a physician retrieves the largest oocytes bulging from an ovary and mixes them with the man's sperm. Twelve fertilized ova result. They divide in the lab dish over the next 5 days, reaching the stage just before they would implant in the uterus. A few cells are sampled from the nine embryos that remain and look robust, and they are tested for chromosome number as well as for the Pompe disease and factor V Leiden mutations in the families.

Four embryos have one extra or missing chromosome and one embryo has two missing chromosomes. They are donated for research. Of the four embryos with the normal 46 chromosomes, two are XX and two are XY. One XX embryo has inherited the factor V Leiden gene variant but is wild type for Pompe disease. One XY embryo does not have the factor V Leiden mutation but has one copy of the Pompe disease mutation. One XX and one XY embryo inherited neither of the parents' mutations.

The couple chooses the XY embryo that inherited neither mutation. They didn't select the XX with the factor V Leiden mutation because that individual would one day be at higher risk for inappropriate clotting, although the condition would not affect the current pregnancy because the man has the mutation, not the pregnant woman. The XY embryo that is a carrier for Pompe disease will not have the condition, but someday his partner will need to be tested to avoid having an affected child. The couple freezes the XX embryo that does not have the family mutations. Perhaps when the couple is ready to have another child, genome sequencing of embryos will be routine so they can further check that their daughter will be healthy. Genome sequencing would reveal *de novo* mutations—those that originate in the fertilized ovum and are therefore not inherited.

The chapter 15 Bioethics, "Designer Babies: Is Prenatal Genetic Testing Eugenic?" discusses possible population consequences of many people choosing the genetic characteristics of their offspring.

21.1 Savior Siblings and More

A couple in search of an oocyte donor advertises in a college newspaper seeking an attractive, bright young woman from an athletic family. A cancer patient has ovarian tissue removed and frozen before undergoing treatment. Two years later, a strip of the frozen tissue is thawed and oocytes separated and fertilized in a laboratory dish with her partner's sperm. Several cleavage embryos develop and two are implanted in her uterus. She becomes a mother—of twins. A man paralyzed from the waist down has sperm removed and injected into his partner's oocyte. He, too, becomes a parent when he thought he never would.

Lisa and Jack Nash wanted to have a child for an unusual reason. Their daughter Molly, born on July 4, 1994, had Fanconi anemia. This autosomal recessive condition would destroy her bone marrow and her immunity and likely cause leukemia; the median age at death of patients with this disease is 30. An umbilical cord stem cell transplant from a sibling could cure her, but Molly had no siblings. Nor did her parents wish to have another child who would have a one in four chance of

Figure 21.1 Savior siblings. Preimplantation genetic diagnosis removes one cell from an early embryo and tests it for an inherited disease that is in a particular family. The test cannot be done on a fertilized ovum because the cell would be destroyed. © *Science Photo Library/Alamy Stock Photo*

inheriting the disease, as Mendel's first law dictates. Technology offered a way for them to alter the odds.

In late 1999, researchers mixed Jack's sperm with Lisa's oocytes in a laboratory dish. After allowing 15 of the fertilized ova to develop to the 8-cell stage, researchers separated and applied DNA probes to one cell from each embryo (**figure 21.1**). A cell that had wild type Fanconi anemia alleles and that matched Molly's human leukocyte antigen (HLA) type was identified and its 7-celled remainder implanted into Lisa's uterus to develop. This technology, revolutionary at the time, is **preimplantation genetic diagnosis (PGD)**. The result was Adam, born in late summer. His umbilical cord stem cells saved Molly's life. At first the Nashes were sharply criticized for intentionally conceiving a "savior sibling," but as other families followed their example, conceiving and selecting a child to provide cells for a sibling became more accepted.

Increased knowledge of how the genomes of two individuals come together and interact has spawned several novel ways to have children. **Assisted reproductive technologies (ARTs)** replace the source of a male or female gamete, aid fertilization, or provide a uterus. These procedures were developed to treat infertility, but are increasingly including genetic testing, too, so that parents-to-be can select offspring-to-be, like the couple in the chapter opener.

The U.S. government does not regulate ARTs, but the American Society for Reproductive Medicine provides voluntary guidelines. The United Kingdom has pioneered ARTs and its Human Fertilisation and Embryology Authority has served as a model for government regulation. A great advantage of the British regulation of reproductive health services and technologies is that databases include success rates for the different procedures. Another advantage is that access to reproductive technology is not limited to those who can afford it.

Key Concepts Questions 21.1

1. Explain how preimplantation genetic diagnosis saved Molly Nash.
2. What are assisted reproductive technologies?

21.2 Infertility and Subfertility

Infertility is the inability to conceive a child after a year of frequent sexual intercourse without the use of contraceptives. Some specialists use the term *subfertility* to distinguish those individuals and couples who can conceive unaided, but for whom this may take longer than average. On a more personal level, infertility can be a seemingly endless monthly cycle of raised hopes and crushing despair.

As a woman ages, fertility declines. By the start of menopause at about age 51, only a thousand or so oocytes remain in her ovaries. Spontaneous abortion occurs in 15 percent of women under 35, in 20 to 35 percent of women between ages 35 and 45, and in up to 50 percent of women over 45 who try to become pregnant. The incidence of pregnancy-related problems rises with maternal age, including chromosomal anomalies, fetal deaths, premature births, and low-birth-weight babies. Older men are at increased risk of having children who develop autism or schizophrenia. Sperm motility declines with age of the man.

One in six couples has difficulty conceiving or giving birth to children. According to the American Society for Reproductive Medicine, males and females contribute about equally to infertility, with 25 percent of infertile couples having two factors at play. When a physical problem is not obvious, the cause is usually a mutation or chromosomal aberration that impairs fertility in the male. A common combination that accounts for low fertility is a woman with an irregular menstrual cycle and a man with a low sperm count.

Male Infertility

Infertility in the male is easier to detect but sometimes harder to treat than female infertility. Four in 100 men in the general population are infertile, and half of them do not make any sperm, a condition called azoospermia. Some men have difficulty fathering a child because they produce fewer than the average 15 to 200 million sperm cells per milliliter of ejaculate. This condition, called oligospermia, has several causes. If a low sperm count is due to a hormonal imbalance, administering the appropriate hormones may boost sperm production. Sometimes a man's immune system produces IgA antibodies that cover the sperm and prevent them from binding to oocytes. Male infertility can also be due to a varicose vein in the scrotum. This enlarged vein emits heat near developing sperm, which prevents them from maturing. Surgery can remove a scrotal varicose vein. The heat from using laptop computers for long periods of time can raise scrotal temperature, damaging sperm.

Most cases of male infertility are genetic. About a third of infertile men have small deletions of the Y chromosome that remove the only copies of key genes whose products control spermatogenesis. Other genetic causes of male infertility include mutations in genes that encode androgen receptors or protein fertility hormones, or that regulate sperm development or motility (movement). **Clinical Connection 21.1** describes a type of autosomal recessive male infertility that is unusual in that it is not part of a syndrome.

To speed conception, a man with a low sperm count can donate several semen samples over a period of weeks at a fertility clinic. The samples are kept in cold storage, then pooled. Some of the seminal fluid is withdrawn to leave a sperm cell concentrate, which is then placed in the woman's body. It isn't very romantic, but it is highly effective at achieving pregnancy. Test kits that provide a sperm count are available over-the-counter, but they do not assess sperm motility. Men who cannot ejaculate due to a spinal cord injury can use a vibrating device applied to the tip of the penis or inserted into the rectum that stimulates release of sperm. The sperm are collected and transferred to the partner using a syringe.

Sperm quality is more important than quantity. Sperm cells that are unable to move or are shaped abnormally cannot reach an oocyte. Inability to move may be due to a hormone imbalance and abnormal shapes may be due to impaired apoptosis (programmed cell death) that normally removes such sperm. The genetic package of an immobile or abnormally shaped sperm cell can be injected into an oocyte and sometimes this leads to fertilization. However, even sperm that look and move normally may be unable to fertilize an oocyte.

Female Infertility

Abnormalities in any part of the female reproductive system can cause infertility (**figure 21.2**). Many women with subfertility or infertility have irregular menstrual cycles, making it difficult to pinpoint when conception is most likely. In an average menstrual cycle of 28 days, ovulation usually occurs around the 14th day after menstruation begins. At this time a woman is most likely to conceive.

For a woman with regular menstrual cycles who is under 30 years old and not using birth control, pregnancy typically happens within 3 or 4 months. A woman with irregular menstrual periods can tell when she is most fertile by using an ovulation predictor test, which detects a peak in the level of luteinizing hormone that precedes ovulation by a few hours. An older way to detect the onset of ovulation is to record body temperature each morning using a digital thermometer with subdivisions of hundredths of a degree Fahrenheit, which can indicate the 0.4 to 0.6 rise in temperature when ovulation starts. Several apps track a woman's menstrual cycle, enabling her to predict the time of ovulation. Sperm can survive in a woman's body for up to 5 days but the oocyte is only viable for 24 to 48 hours after ovulation.

The hormonal imbalance that usually underlies irregular ovulation has various causes. These include a tumor in the ovary or in the pituitary gland in the brain that controls the reproductive system, an underactive thyroid gland, or use of steroid-based drugs such as cortisone. If a nonpregnant woman produces too much prolactin, the hormone that promotes milk production and suppresses ovulation in new mothers, she will not ovulate.

Fertility drugs can stimulate ovulation, but they can also cause women to "superovulate," producing and releasing more than one oocyte each month. A commonly used drug, clomiphene, raises the chance of having twins from 1 to 2

The Case of the Round-Headed Sperm

In fewer than a tenth of a percent of men who are infertile, sperm cells lack the tip, called the acrosome, which contains the enzymes that break through the layers surrounding an oocyte. This condition is called "globozoospermia" (**figure 21A**). An Ashkenazi Jewish family led researchers to a gene that, when mutant, causes an autosomal recessive form of male infertility due to sperm with round rather than oval heads.

The family went to a center for reproductive medicine in the Netherlands. Of the six sons, three were infertile (**figure 21B**). Four daughters were fertile. The affected sons' sperm were misshapen. The mode of inheritance was recessive, because the parents were fertile.

Researchers suspected consanguinity—a shared ancestor would increase the risk of inheriting a rare autosomal recessive condition if the mutation is in the family. But the family denied knowing a relative who had married a relative. Reasoning that perhaps DNA could reveal consanguinity that the family did not know about, researchers scanned the genomes of all six sons for regions of homozygosity that would indicate they had relatives marrying relatives not too far back in the family tree (see figure 4.14).

A region of homozygosity in this case was defined as 25 consecutive SNPs that were homozygous. The genomes of all six sons were riddled with these regions, suggesting that at some point, a cousin married a cousin or an aunt/uncle wed a nephew/niece. One region of homozygosity was seen in all three infertile brothers, but was heterozygous in two of the three fertile brothers. The remaining brother was homozygous wild type for the region.

Next, the researchers scrutinized the part of the long arm of chromosome 3 where the region of homozygosity lay in the three infertile brothers. It houses 50 genes, only one of which is expressed in the testes. This gene encodes "spermatogenesis-associated protein 16" and is called *SPATA16*. It has 11 exons, and the mutation in the Ashkenazi family is a single base change, from G to A, at the 848th position in the gene, near the end of exon 4. The mutation affects the splicing out of introns as the gene is transcribed.

The wild type protein product of the *SPATA16* gene is transported from the Golgi apparatus into vesicles that take it to the acrosome as it telescopes out of the front end of a sperm cell. By attaching the gene for the jellyfish's green fluorescent protein (see figure 19.1) to the wild type *SPATA16* gene in cells growing in culture, researchers visualized the protein being transported to the forming acrosome in immature healthy sperm.

Questions for Discussion

1. What is the evidence that the infertility in the family is inherited as an autosomal recessive trait?

2. What is the evidence for inheritance of the mutation from shared ancestors?

3. Discuss how researchers discovered the gene that is mutant in the family.

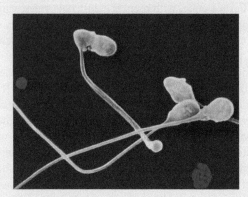

Figure 21A **A sperm cell's streamlined form facilitates movement.** A misshapen sperm cannot fertilize an oocyte. © *Tony Brain/SPL/Science Source*

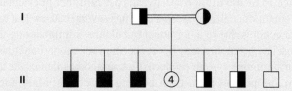

Figure 21B **Inheriting infertility.** In a family with autosomal recessive globozoospermia, three of six sons are infertile.

percent to 4 to 6 percent. If a woman's ovaries are completely inactive or absent (due to a birth defect or surgery), she can become pregnant only if she uses a donor oocyte. Some cases of female infertility are due to "reduced ovarian reserve"—too few oocytes. This is typically discovered when the ovaries do not respond to fertility drugs. Signs of reduced ovarian reserve are an ovary with too few follicles (observed on an ultrasound scan) or elevated levels of follicle-stimulating hormone on the third day of the menstrual cycle.

The uterine tubes are a common site of female infertility because fertilization usually occurs in open tubes. Blockage can prevent sperm from reaching the oocyte, or entrap a

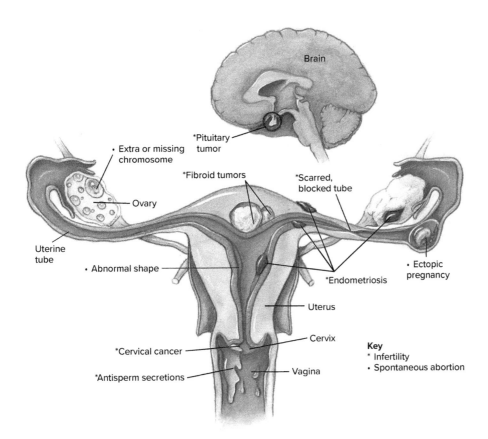

Figure 21.2 Sites of reproductive problems in the female.

Too little mucus can prevent conception, too; this is treated with low daily doses of oral estrogen. Sometimes mucus in a woman's body has antibodies that attack sperm. Infertility may also result if the oocyte does not release sperm-attracting biochemicals.

One reason the incidence of female infertility increases with age is that older women are more likely to produce oocytes that have an abnormal chromosome number, which often causes spontaneous abortion because defects are too severe for development to proceed for long. The cause is usually misaligned spindle fibers when the second meiotic division begins, causing aneuploidy (extra or missing chromosomes). Perhaps the longer exposure of older oocytes to harmful chemicals, viruses, and radiation contributes to the risk of meiotic errors. Losing very early embryos may appear to be infertility because the bleeding accompanying the embryo as it leaves the body resembles a heavy menstrual flow.

fertilized ovum, keeping it from descending into the uterus. If an embryo begins developing in a blocked tube and is not removed and continues to enlarge, the tube can burst and the woman can die. This is called a tubal or ectopic pregnancy.

Uterine tubes can also be blocked due to a birth defect or, more likely, from an infection such as pelvic inflammatory disease. A woman may not know she has blocked uterine tubes until she has difficulty conceiving and medical tests uncover the problem. Surgery can open blocked uterine tubes.

Excess tissue growing in the uterine lining may make it inhospitable to an embryo. This tissue can include benign tumors called fibroids or areas of thickened lining from a condition called endometriosis. The tissue can grow outside of the uterus, too, in the abdominal cavity. In response to the hormonal cues to menstruate, the excess lining bleeds, causing cramps. Endometriosis can hamper conception, but curiously, if a woman with endometriosis conceives, the cramps and bleeding usually disappear after the birth.

Secretions in the vagina and cervix may be hostile to sperm. Cervical mucus that is thick or sticky due to infection can entrap sperm, keeping them from moving far enough to encounter an oocyte. Vaginal secretions may be so acidic or alkaline that they weaken or kill sperm. Douching daily with an acidic solution, such as acetic acid (vinegar) or an alkaline solution, such as bicarbonate, can alter the pH of the vagina so that, in some cases, it is more receptive to sperm cells.

Infertility Tests

A number of medical tests can identify causes of infertility. The man is checked first, because it is easier, less costly, and less painful to obtain sperm than oocytes.

Sperm are checked for number (sperm count), ability to swim (motility), and shape (morphology). An ejaculate containing up to 40 percent of unusual forms of sperm is still considered normal, but many more than this can impair fertility. A urologist performs sperm tests. A genetic counselor can evaluate Y chromosome deletions associated with oligospermia. If a cause of infertility in the male is not identified, a gynecologist checks the woman to see that reproductive organs are present and functioning.

Some cases of subfertility or infertility have no clear explanation. Psychological factors may be at play, or it may be that inability to conceive results from consistently poor timing. Sometimes a subfertile couple adopts a child, only to conceive one of their own shortly thereafter; many times, the causes of infertility remain a mystery.

Key Concepts Questions 21.2

1. What are causes of male infertility?

2. What are causes of female infertility?

3. Describe medical tests used to identify the causes of infertility.

21.3 Assisted Reproductive Technologies

Many people with fertility problems who do not choose to adopt children use alternative ways to conceive. Several of the ARTs were developed in nonhuman animals (see the **Technology Timeline**). In the United States, about 2 percent of the approximately 4 million births a year result from ARTs, and worldwide ARTs account for about 250,000 births a year.

ART procedures can be performed on material from the parents-to-be ("nondonor") or from donors, and may be "fresh" (collected just prior to the procedure) or "frozen" (preserved in liquid nitrogen). Except for intrauterine insemination, ARTs cost thousands of dollars and are not typically covered by health insurance in the United States.

Donated Sperm—Intrauterine Insemination

The oldest ART is **intrauterine insemination (IUI)**, in which a doctor places donated sperm into a woman's cervix or uterus. (It used to be called artificial insemination.) The success rate is 5 to 15 percent per attempt. The sperm are first washed free of seminal fluid, which can inflame female tissues. A woman might seek IUI using donor sperm if her partner is infertile or has a mutation that the couple wishes to avoid passing to their child. Women also undergo IUI to be a single parent without having sex, or a lesbian couple may use it to have a child.

The first documented IUI in humans was done in 1790. For many years, physicians donated sperm, and this became a way for male medical students to earn a few extra dollars. By 1953, sperm could be frozen and stored and IUI became much more commonplace. Today, donated sperm are frozen and stored in sperm banks, which provide the cells to obstetricians who perform the procedure. IUI costs on average $865 per cycle, according to the American Society of Reproductive Medicine, with higher charges from some facilities for sperm from donors who have professional degrees because those men are paid more for their donations. Additional fees are charged for a more complete medical history of the donor, for photos of the man at different ages, and for participation in a "consent program" in which the donor's identity is revealed when his offspring turns 18 years old. If ovulation is induced to increase the chances of success of IUI, additional costs may exceed $3,000.

A couple who chooses IUI can select sperm from a catalog that lists the personal characteristics of donors, such as blood type, hair and eye color, skin color, build, educational level, and interests. Some traits have nothing to do with genetics. If a couple desires a child of one sex—such as a daughter to avoid passing on an X-linked disease—sperm can be separated into fractions enriched for X-bearing or Y-bearing sperm.

Problems can arise in IUI if a donor learns that he has an inherited disease after he has already donated sperm that has been used. For example, a man developed cerebellar ataxia, a movement disorder, years after he donated sperm. Eighteen children conceived using his sperm face a one in two risk of having inherited the dominant mutant gene. Too-popular sperm donors can lead to problems. One man, listed in the Fairfax Cryobank as "Donor 401," was quite attractive, and 45 children were conceived with his sperm. When a few of the families appeared on a talk show, several other families tuning in were shaken to see so many children who resembled their own. When cases came to light of men fathering more than 150 offspring, sperm banks began to limit sales of a particular man's sperm cells. An online "donor sibling registry" has enabled thousands of half-siblings who share sperm-donor fathers to meet.

A male's role in reproductive technologies is simpler than a woman's. A man can be a genetic parent, contributing half of his genetic self in his sperm, but a woman can be both a genetic parent (donating an oocyte) and a gestational parent (loaning her uterus).

A Donated Uterus—Surrogate Motherhood

If a man produces healthy sperm but his partner's uterus cannot maintain a pregnancy, a surrogate mother may help by being inseminated with the man's sperm. When the child is born, the surrogate mother gives the baby to the couple. In this variation of the technology, the surrogate is both the genetic and the gestational mother. Attorneys usually arrange surrogate relationships. The surrogate mother signs a statement signifying her intent to give up the baby. In some U.S. states, and in some nations, she is paid for her 9-month job, but in the United Kingdom compensation is illegal. Outlawing compensation is to prevent wealthy couples from taking advantage of women who become surrogates to earn money.

A problem with surrogate motherhood is that a woman may not be able to predict her responses to pregnancy and childbirth in a lawyer's office months before she must hand over the baby. When a surrogate mother changes her mind, the results are wrenching for all.

Another type of surrogate mother lends only her uterus, receiving a fertilized ovum conceived from a man and a woman who has healthy ovaries but lacks a functional uterus. This variation is an "embryo transfer to a host uterus," and the pregnant woman is a "gestational-only surrogate mother." She turns the child over to the biological parents. About 1,600 babies are born in the United States to gestational surrogates each year.

In Vitro Fertilization

In *in vitro* **fertilization (IVF)**, which means "fertilization in glass," sperm and oocytes join in a laboratory dish. Before the fifth day of development, an embryo or embryos are transferred to a woman's uterus. If all goes well, an embryo implants into the uterine lining and continues development. **Figure 21.3** reviews these early stages of the embryo.

	In Nonhuman Animals	In Humans
1782	Intrauterine insemination (IUI) in dogs	
1790		Pregnancy reported from IUI
1890s	Birth from embryo transplantation in rabbits	IUI by donor
1949	Cryoprotectant enables safe freezing of animal sperm	
1951	First calf born after embryo transplantation	
1952	Live calf born after insemination with frozen sperm	
1953		First reported pregnancy after insemination with frozen sperm
1959	Live rabbit offspring produced from IVF	
1972	Live offspring from frozen mouse embryos	
1976	Intracytoplasmic sperm injection (ICSI) in hamsters	First reported commercial surrogate motherhood arrangement in the United States
1978	Transplantation of ovaries between cows	Baby born after IVF in United Kingdom
1980		Baby born after IVF in Australia
1981	Calf born after IVF	Baby born after IVF in United States
1982	Sexing of embryos in rabbits Cattle embryos split to produce genetically identical twins	
1983		Embryo transfer after uterine lavage
1984		Baby born in Australia from frozen and thawed embryo
1985		Baby born after gamete intrafallopian transfer (GIFT) First reported gestational-only surrogacy arrangement in the United States
1986		Baby born in the United States from frozen and thawed embryo
1989		First preimplantation genetic diagnosis (PGD)
1992		First pregnancies from ICSI
1994		62-year-old woman gives birth from fertilized donated oocyte
1995	Sheep cloned from embryo cell nuclei	Babies born following ICSI
1996	Sheep cloned from adult cell nucleus	
1998	Mice cloned from adult cell nuclei	Baby born 7 years after his twin
1999	Cattle cloned from adult cell nuclei	
2000	Pigs cloned from adult cell nuclei	
2001		First savior sibling born to treat sister for genetic disease Human preimplantation embryo cloned, survives to 6 cells
2003		3,000-plus PGDs performed to date
2004	Woman pays $50,000 to have her cat cloned	First birth from a woman who had ovarian tissue preserved and implanted on an ovary, after cancer treatment
2005	Dog cloned	
2011		First children born free of single-gene disease following sequential polar body analysis
2013		First woman conceives from stored ovarian tissue
2015		First baby born following preimplantation genome sequencing

Louise Joy Brown, the first "test-tube baby," was born in 1978 amid great attention and sharp criticism. A prominent bioethicist said that IVF challenged "the idea of humanness and of our human life and the meaning of our embodiment and our relation to ancestors and descendants." Yet Louise is, despite her unusual beginnings, an ordinary woman. More than 5 million children have been born following IVF.

A woman might undergo IVF if her ovaries and uterus work but her uterine tubes are blocked. Using a laparoscope, which is a lit surgical instrument inserted into the body through a small incision, a physician removes several of the largest oocytes from an ovary and transfers them to a culture dish. If left in the body, only one oocyte would exit the ovary, but in culture, many oocytes can mature sufficiently to be fertilized *in vitro*. Chemicals, sperm, and other cell types similar to those in the female reproductive tract are added to the culture.

An acidic solution may be applied to the zona pellucida, which is the layer around the egg, to thin it to ease the sperm's penetration.

Sperm that cannot readily enter the oocyte may be sucked up into a tiny syringe and microinjected into the female cell. This technique, called **intracytoplasmic sperm injection (ICSI)**, is more effective than IVF alone and has become standard at some facilities (**figure 21.4**). ICSI is helpful for men who have low sperm counts or many abnormal sperm. It makes fatherhood possible for men who cannot ejaculate, such as those who have suffered spinal cord injuries. ICSI has been performed on thousands of men with about a 30 percent success rate.

Two to 5 days after sperm wash over the oocytes in the dish, or are injected into them, a blastocyst is transferred to the uterus. If the hormone human chorionic gonadotropin appears in the woman's blood a few days later, and its level rises, she is pregnant.

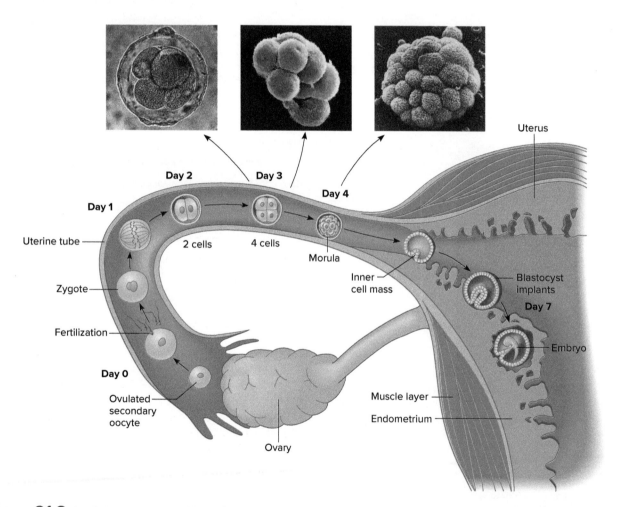

Figure 21.3 **Reviewing the earliest stages of prenatal development.** The zygote forms in the uterine tube when a sperm nucleus fuses with the nucleus of an oocyte at fertilization. The first divisions, called cleavage, proceed while the zygote moves toward the uterus. Preimplantation genetic diagnosis and screening were developed for use on 8-cell cleavage embryos but are now more commonly done on day 5 blastocysts, sampling cells other than those from the inner cell mass (trophoblast cells). By day 7, the blastocyst begins to implant in the uterine lining. *(left):* © *Petit Format/Nestle/Photo Researchers/Science Source; (middle):* © *P.M. Motta & J. Van Bierkom/SPL/Science Source; (right):* © *Petit Format/Nestle/Photo Researchers/Science Source*

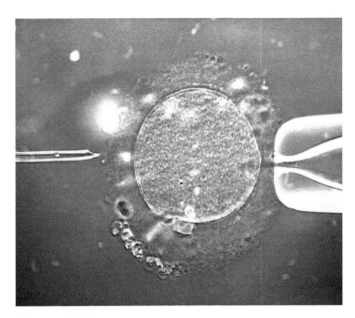

Figure 21.4 Intracytoplasmic sperm injection. ICSI enables some infertile men, such as those with spinal cord injuries or certain illnesses, to become fathers. A single sperm cell is injected into the cytoplasm of an oocyte. © *Science Photo Library/Getty Images RF*

IVF costs on average $12,000 per cycle. Medications can add $3,000 to $5,000. ICSI adds another $1,500. Children born following IVF have a slight increase in the rate of birth defects (about 4 to 7 percent) compared to children conceived naturally (about 3 to 5 percent), but the elevation may be due to the medical problems that caused the parents to seek IVF. Children born after IVF on average have slightly higher birth weights, which is thought to be due to the time the early embryos spend in laboratory culture.

In the past, several embryos were implanted to increase the success rate of IVF, but this led to many multiple births, which are riskier than single births. In some cases, physicians had to remove embryos to make room for others to survive. To avoid the multiples problem, and because IVF has become more successful as techniques have improved, guidelines now suggest transferring only one embryo.

Embryos resulting from IVF that are not soon implanted in the woman can be frozen in liquid nitrogen ("cryopreserved" or "vitrified") for later use. Cryoprotectant chemicals are used to prevent salts from building up or ice crystals from damaging delicate cell parts. Freezing takes a few hours; thawing about a half hour. The "oldest" pregnancy using a frozen embryo occurred 20 years after the freezing.

Overall, the chances of a live birth following IVF are about 20 to 35 percent per cycle, but this rate varies greatly, depending on certain risk factors that lower the likelihood of success. These include:

- maternal age—success is 30 to 40 percent for women (oocyte donors) under age 34, but only 5 to 10 percent for women over 40;

- increased time being infertile;
- number of previous failed IVF attempts;
- number of previous IVF attempts;
- use of a woman's own oocytes rather than a donor's; and
- infertility with a known cause.

Several websites assess risks and predict IVF success. In one example, a couple had been infertile for 11 years. They attempted IVF four times that resulted in two failures and two spontaneous abortions. They had used the woman's eggs and ICSI because too many sperm were abnormal. The chance of success per IVF attempt is about 8 percent, but if they use a donor oocyte, it doubles.

Gamete and Zygote Intrafallopian Transfer

IVF may fail because of the artificial environment for fertilization. A procedure called **gamete intrafallopian transfer (GIFT)** improves the setting. (Uterine tubes are also called fallopian tubes.) Fertilization is assisted in GIFT, but it occurs in the woman's body rather than in glassware.

In GIFT, several of a woman's largest oocytes are removed. The man submits a sperm sample, and the most active cells are separated from it. The collected oocytes and sperm are deposited together in the woman's uterine tube, at a site past any obstruction that might otherwise block fertilization. GIFT is about 22 percent successful but depends upon a woman's age.

A variation of GIFT is **zygote intrafallopian transfer (ZIFT)**. In this procedure, an IVF ovum is introduced into the woman's uterine tube. Allowing the fertilized ovum to make its own way to the uterus increases the chance that it will implant. ZIFT is also 22 percent successful.

GIFT and ZIFT are done less frequently than IVF. These procedures may not work for women who have scarred uterine tubes. The average cost of GIFT or ZIFT is $15,000 to $20,000.

Bioethics considers the unusual situation of collecting gametes from a person shortly after the person has died.

Oocyte Banking and Donation

Oocytes can be stored, as sperm are, but the procedure is more challenging. Because an oocyte is the largest type of cell, it contains a large volume of water. Freezing can form ice crystals that damage cell parts. Reasons to freeze oocytes include:

- Cancer treatment (radiation or chemotherapy) that can harm oocytes and cause early menopause
- Exposure to toxins or teratogens in the workplace
- Disease of the ovaries
- Premature ovarian failure, which may be due to XO syndrome or fragile X syndrome
- Ovary removal to prevent ovarian cancer due to a *BRCA1* or *BRCA2* mutation

Removing and Using Gametes After Death

A gamete is a packet containing one copy of a person's genome. If, after death, gametes are collected and combined with an opposite gamete type, the deceased person can become a parent. This "postmortem gamete retrieval" has happened for years, but nearly always for men.

One of the first cases of postmortem sperm removal affected a couple in their early thirties who had delayed becoming parents, confident that their good health would make pregnancy possible later. Then the man suddenly died of an allergic reaction to a drug. His wife knew how much he had wanted to be a father, so she asked the medical examiner to collect her husband's sperm. The sample was sent to a cryobank, where it lay deeply frozen for more than a year. In the summer of 1978, the sperm were defrosted and used to fertilize one of the woman's oocytes. The couple's daughter was the first case of postmortem sperm retrieval in which the father did not actively participate in the decision. Since 1990, U.S. servicemen who feared infertility from exposure to chemical or biological weapons have taken advantage of sperm bank discounts to the military, preserving their sperm before deploying.

Postmortem sperm retrieval raises legal and ethical issues based on timing. In one case, a woman conceived twins using her husband's preserved sperm 16 months after he died of leukemia; he had stated his wishes for her to do so. The Social Security Administration refused to provide survivor benefits to their daughters, claiming that the husband was not a father, but a sperm donor. The Massachusetts Superior Court reversed this decision. In New Jersey, a mother claimed Social Security benefits for twins conceived after her husband's death. An appeals court upheld the denial of benefits, claiming that the children had to have been dependents at the time of their father's death.

Postmortem oocyte retrieval was considered in a case reported in 2010. A 36-year-old woman stood up on a plane following many hours of sleeping in one position, and her heart stopped. By the time a doctor on board restarted it, the woman's brain had been robbed of oxygen for several precious minutes. The plane made an emergency landing and she was taken to a hospital, where she was placed on a respirator. Scans showed blood clots in her lungs that had caused the collapse. By the fourth day, the woman's brain was dangerously swelling, and

by the ninth day, her brain activity was nearly nil, although she could still open her eyes and move spontaneously. Her husband, parents, and in-laws asked that the tubes keeping her alive be withdrawn. Then, several hours after this was done, they changed their minds, and asked that the breathing tube be reinserted so that the woman's oocytes could be retrieved. They had no idea how difficult this would be.

The physicians, knowing the complexities of the medical situation, wanted to know one other person's opinion—the patient's. So they consulted with the young woman's gynecologist, who had no record or recollection of the patient stating she wished to have children. The young woman was not completely brain dead, so the decisions would not be the same as for donating an organ after death. If her oocytes were to be used to give her husband a child, IVF and a surrogate mother would obviously be necessary. Before that could happen, though, the woman would have to undergo 2 weeks of hormone treatments to ovulate, when she would have to lie flat, which could kill her. For these practical reasons, and the fact that the woman had never stated that she wished to be a parent, the family elected to turn off life support, and she quickly died.

Like other ARTs, postmortem gamete retrieval is not regulated at the federal level in the United States. Bioethicists have identified situations to avoid:

- Someone other than a spouse wishing to use the gamete
- A too-hasty decision based on grief
- Use of the gamete for monetary gain

Questions for Discussion

1. How does the case of the 36-year-old woman whose oocytes were to be retrieved following her brain death differ from that of a pregnant woman in a coma who is kept alive for several weeks so that her baby can be born?

2. The people described in this essay did not have other children. How might the situation differ for a couple who already has children?

3. Do you think that Social Security or another benefit system should cover fetuses, embryos, or gametes?

4. How might postmortem gamete retrieval be abused?

- Inability to obtain viable sperm on the day of IVF
- Extra oocytes remain after IVF
- Donation
- Delaying motherhood for personal reasons

Oocytes are frozen in liquid nitrogen at temperatures below −40°C, when they are at metaphase of the second meiotic

division (**figure 21.5**). At this time, the chromosomes are aligned along the spindle, which is sensitive to temperature extremes. If the spindle comes apart as the cell freezes, the oocyte may lose a chromosome, which would devastate development. Another problem with freezing oocytes is retention of a polar body, leading to a diploid oocyte.

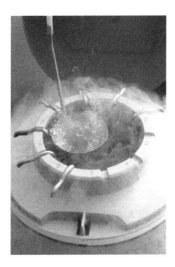

Figure 21.5 **Freezing oocytes is safe.** © *Science Photo Library/Alamy Stock Photo*

To minimize risk of damage to oocytes when freezing, they are first treated with cryoprotectants that remove water and prevent ice crystal formation. When the frozen oocytes are to be used, they are slowly warmed and rehydrated by gradually removing the cryoprotectants. Frozen oocytes are fertilized *in vitro* using ICSI because the outermost layer (the zona pellucida) may be altered. Studies show that frozen and fresh oocytes from the same woman are equally likely to be fertilized, and there is no difference in live birth rate. Thousands of babies have been born following oocyte freezing.

An alternative to freezing oocytes is freezing strips of ovarian tissue that can be stored, thawed, and reimplanted at various sites, such as under the skin of the forearm or abdomen or in the pelvic cavity near the ovaries. The tissue ovulates and the oocytes are collected and fertilized *in vitro*. The first child resulting from fertilization of an oocyte from reimplanted ovarian tissue was born in 2004. The mother, age 25, had been diagnosed with advanced Hodgkin's lymphoma. The harsh chemotherapy and radiation cured her cancer, but destroyed her ovaries. Before her cancer treatment, five strips of tissue from her left ovary were frozen. Later, several pieces of ovarian tissue were thawed and implanted in a pocket that surgeons crafted on one of her shriveled ovaries, near the entrance to a uterine tube. Menstrual cycles resumed, and shortly thereafter, the woman became pregnant with her daughter, who is healthy. Freezing ovarian tissue may become routine for cancer patients of childbearing age.

Women who have no oocytes or wish to avoid passing on a mutation can obtain oocytes from donors, who are typically younger women. Some women become oocyte donors when they undergo IVF and have "extras." The potential father's sperm and donor's oocytes are placed in the recipient's uterus or uterine tube, or fertilization occurs in the laboratory and a blastocyst is transferred to the woman's uterus. A program in the United Kingdom funds IVF for women who cannot afford the procedure if they donate their "extras." The

higher success rate of using oocytes from younger women confirms that it is the oocyte that age affects, and not the uterine lining.

Embryo donation is a variation on oocyte donation. A woman with malfunctioning ovaries but a healthy uterus carries an embryo that results when her partner's sperm is used in intrauterine insemination of a woman who produces healthy oocytes. If the woman conceives, the embryo is gently flushed out of her uterus a week later and inserted through the cervix and into the uterus of the woman with malfunctioning ovaries. The child is genetically that of the man and the woman who carries it for the first week, but is born from the woman who cannot produce healthy oocytes. Embryo donation includes use of IVF "leftovers."

In another technology, cytoplasmic donation, older women have their oocytes injected with cytoplasm from the oocytes of younger women to "rejuvenate" the cells. Although resulting children conceived through IVF appear to be healthy, they are being monitored for a potential problem—heteroplasmy, or two sources of mitochondria in one cell (see chapter 5 Bioethics). Researchers do not yet know the health consequences, if any, of having mitochondria from the donor cytoplasm plus mitochondria from the recipient's oocyte. These conceptions also have an elevated incidence of XO syndrome, which often causes spontaneous abortion.

Because oocytes are harder to obtain than sperm, oocyte donation technology has lagged behind that of sperm banks, but is catching up. One IVF facility that has run a donor oocyte program since 1988 has a brochure that describes 120 oocyte donors of various ethnic backgrounds, like a catalog of sperm donors. The oocyte donors are young and have undergone extensive medical and genetic tests. Recipients may be up to 55 years of age.

Preimplantation Genetic Diagnosis

Preimplantation genetic diagnosis (PGD) is often teamed with IVF to detect genetic and chromosomal abnormalities *before* pregnancy starts, as described in the chapter opener and in section 21.1. The procedure is called "diagnosis" if it detects a specific gene variant known to be in one or both parents-to-be, and "screening" if it is used to construct a karyotype (that is, check chromosome number), without looking for a specific chromosomal anomaly. The couple selects a very early "preimplantation" embryo that tests show has not inherited a specific detectable genetic condition. "Preimplantation" refers to the fact that the embryo is tested at a stage prior to when it would naturally implant in the uterus. PGD has about a 29 percent success rate, and it adds on average $3,500 to the cost of IVF.

PGD is possible because one cell, or blastomere, can be removed for testing from an embryo consisting of at least 8 cells, and the remaining cells can complete development normally in a uterus, even after being frozen. Before the embryo is transferred into the woman, the sampled single cell is karyotyped, or its DNA is amplified and probed for genes that the parents carry, or the exome or genome is sequenced. Embryos that have normal chromosomes and do not have disease-causing genotypes in the family are selected to complete development or are stored. Testing cells from a day-5 embryo, when it is 80 to 120 cells, is more successful than testing a cell from an 8-celled embryo, as was once done, because several cells can be studied. Testing at the 8-cell stage is called a "blasto*mere* biopsy" and the later one a "blasto*cyst* biopsy" (**figure 21.6**). A blastocyst biopsy tests cells of the trophectoderm, which is the surrounding tissue that does not develop into the embryo. Accuracy in detecting a mutation or abnormal chromosome in PGD is about 97 percent. Errors generally happen when a somatic mutation affects the sampled cell but not the rest of the embryo. Amplification of the selected blastomere DNA may cause such a somatic mutation.

PGD is not new. The first children were born following PGD in 1989. In these first cases, probes for Y chromosome-specific DNA sequences were used to select females, who could not have inherited the X-linked conditions their mothers carried. The alternative to PGD would have been to face the 25 percent chance of conceiving an affected male.

In March 1992, the first child was born who underwent PGD to avoid a specific inherited disease. She was checked as an 8-celled preimplantation embryo to see if she had escaped the cystic fibrosis that affected her brother. Since then, PGD

has helped to select thousands of children free of many types of single-gene conditions.

Like most ARTs, use of PGD has expanded as it has become more accurate and more familiar. Today it has taken on a quality-control role in addition to being a tool to detect and prevent inherited diseases. PGD is increasingly being used to screen early embryos derived from IVF for normal chromosome number before implanting them into women, even if there is no family history of single-gene diseases, because the technology can detect *de novo* (new) mutations. In this way, PGD increases the success rate of IVF because it eliminates many embryos that are not likely to complete development. It is also more precise than examining early embryos to see which ones appear most robust—for years researchers who worked with human embryos did this, but then studies revealed that the best-looking embryos are not necessarily the most genetically healthy ones.

PGD can introduce a bioethical "slippery slope" when it is used for reasons other than to ensure that a child is free of a certain disease. Some people may regard gender selection using PGD as a misuse of the technology. The chapter 6 opener explores the population effects of sex selection. The American Society for Reproductive Medicine endorses PGD for sex selection only to avoid passing on an X-linked disease, which was the first application of the technology. Yet even using PGD to avoid disease can be controversial. In the United Kingdom, where the government regulates reproductive technology, inherited cancer susceptibility is an approved indication for PGD. These cancers do not begin until adulthood, the susceptibility is incompletely penetrant (not everyone who inherits the disease-associated genotype will actually develop cancer), and the cancer may be treatable.

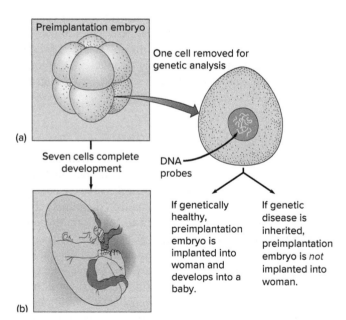

Figure 21.6 **Preimplantation genetic diagnosis (PGD) probes disease-causing genes or chromosome aberrations in an 8-celled preimplantation embryo or from a later stage, the blastocyst.** (a) A single cell is separated and tested to see if it contains a disease-causing genotype or atypical chromosome. (b) If it doesn't, the remaining seven cells divide a few more times and are transferred to a woman's uterus—typically the oocyte donor—to complete development.

Sequential Polar Body Analysis

A technique called **sequential polar body analysis** may substitute for PGD and provides genetic information even earlier in development. The approach is based on the fact that meiosis completes in the female only as a secondary oocyte is fertilized (see figure 3.12).

If a woman is heterozygous for a mutation, then an oocyte will contain the mutation and its associated polar bodies will have the wild type allele, or vice versa. This is in accordance with Mendel's first law, gene segregation. The timetable of female meiosis is important, too. The first polar body forms as the developing oocyte leaves the ovary. That polar body is not accessible, and it would not show the effects of crossing over. A second polar body, however, which forms at fertilization, can be tested (**figure 21.7**). Researchers can sequence linked markers or even entire genomes of second polar bodies to determine whether crossing over has occurred, to be certain that the fertilized ovum has not inherited the mutation.

Because sequential polar body analysis is still experimental, researchers follow it up with PGD to test their predictions

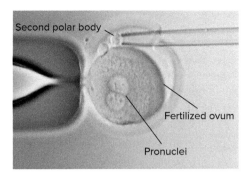

Second polar body

Fertilized ovum

Pronuclei

Figure 21.7 **Deducing a fertilized ovum's genotype by testing a second polar body.** Because of Mendel's law of segregation, if a second polar body associated with a fertilized ovum has a mutation that the woman carries, then the fertilized ovum is inferred to have the wild type allele. *Courtesy of Dr. Anver Kuliev*

and ensure that IVF embryos transferred to the woman's uterus to develop are free of the family's mutation. The idea of probing polar bodies dates from the 1980s, but the technique first led to results in 2011. So far mutations behind more than 150 single-gene diseases have been detected at this initial stage of prenatal development—the very first cell.

Table 21.1 summarizes the ARTs.

Table 21.1	Some Assisted Reproductive Technologies
Technology	**Procedures**
GIFT	Deposits collected oocytes and sperm in uterine tube.
IVF	Mixes sperm and oocytes in a dish. Chemicals simulate intrauterine environment to encourage fertilization.
IUI	Places or injects washed sperm into the cervix or uterus.
ICSI	Injects immature or rare sperm into oocyte, before IVF.
Oocyte freezing	Oocytes retrieved and frozen in liquid nitrogen.
PGD	Analyzes chromosomes and gene variants in early embryos, *in vitro*. Selected embryos are implanted in the uterus and child will be free of tested-for conditions.
Sequential polar body analysis	Genetic testing of polar body attached to just-fertilized ovum enables inference that fertilized ovum is free of a family's mutation.
Surrogate mother	Woman carries a pregnancy for another.
ZIFT	Places IVF ovum in uterine tube.

Key Concepts Questions 21.3

1. What is intrauterine insemination?
2. Distinguish between a genetic and gestational surrogate mother and a gestational-only surrogate mother.
3. List the steps of IVF.
4. Describe GIFT and ZIFT.
5. What is embryo adoption?
6. Explain how PGD optimizes the outcome of IVF.
7. How can genetic testing of a polar body reveal information about a fertilized ovum?

21.4 Extra Embryos

ARTs can result in "extra" oocytes, fertilized ova, or very early embryos. **Table 21.2** lists the possible fates of frozen embryos.

In the United States, nearly half a million embryos derived from IVF sit in freezers; some have been there for years. Most couples who donate their frozen embryos to others do so anonymously, with no intention of learning how their genetic offspring are raised. However, some couples keep track of their embryos, helped by social media to make connections with potential other parents. One couple who had twins from two of 18 IVF embryos chatted online with and selected another couple to receive the remaining embryos. That couple also had twins, and they then shipped the remaining frozen embryos to a third couple, who also had twins. Extra frozen embryos are an excellent resource for same-sex couples to have children, with the help of a surrogate mother if necessary.

Another alternative to disposing of fertilized ova and embryos is to donate them for use in research. The results of experiments sometimes challenge long-held ideas, indicating that we still have much to learn about early human prenatal development. This was the case for a study from Royal Victoria Hospital in Montreal. Researchers examined the chromosomes

Table 21.2	Fates of Frozen Embryos

1. Store indefinitely.
2. Store and destroy after a set time.
3. Donate for embryonic stem cell derivation and research.
4. Thaw later for use by biological parents.
5. Thaw later for use by other parents.
6. Discard.

of sperm from a man with XXY syndrome. Many of the sperm would be expected to have an extra X chromosome, due to nondisjunction (see figure 13.12), which could lead to a preponderance of XXX and XXY offspring. Surprisingly, only 3.9 percent of the man's *sperm* had extra chromosomes, but five of his 10 spare *embryos* had an abnormal X, Y, or chromosome 18. That is, even though most of the man's sperm were normal, his embryos weren't. The source of reproductive problems in XXY syndrome, therefore, might not be in the sperm, but in early embryos—a finding that was previously unknown and not expected, and was only learned because of observing early human embryos.

In another study, Australian researchers followed the fates of single blastomeres that had too many or too few chromosomes. They wanted to see whether the abnormal cells preferentially ended up in the inner cell mass, which develops into the embryo, or the trophectoderm, which becomes extra-embryonic membranes. The study showed that cells with extra or missing chromosomes become part of the inner cell mass much more frequently than expected by chance. The finding indicates that the ability of a blastomere sampled for PGD to predict health may depend on whether it is fated to be part of the inner cell mass. The later blastocyst biopsy samples the trophectoderm, which is the part of the embryo that becomes extra-embryonic membranes, and not the embryo.

Using fertilized ova or embryos designated for discard in research is controversial. Without regulations on privately funded research, ethically questionable experiments can happen. For example, researchers reported at a conference that they had mixed human cells from male embryos with cells from female embryos, to see if the normal male cells could "save" the female cells with a mutation. Sex was chosen as a marker because the Y chromosome is easy to detect, but the idea of human embryos consisting of chromosomally female and male cells caused a public outcry.

ARTs introduce ownership and parentage issues (table 21.3). Another controversy is that human genome information is revealing more traits to track and perhaps control in coming generations. When we can routinely scan the human genome in gametes, fertilized ova, or early embryos, who will decide which traits are worth living with, and which aren't?

Table 21.3	Assisted Reproductive Disasters

1. A physician used his own sperm to perform intrauterine insemination on 15 patients, telling them that he had used sperm from anonymous donors.

2. A plane crash killed the wealthy parents of two early embryos stored at −320°F (−195°C) in a hospital. Adult children of the couple were asked to share their estate with two 8-celled siblings-to-be.

3. Several couples planning to marry discovered that they were half-siblings. Their mothers had been inseminated with sperm from the same donor.

4. Two Rhode Island couples sued a fertility clinic for misplacing several embryos.

5. Several couples sued a fertility clinic for implanting their oocytes or embryos in other women without donor consent. One woman requested partial custody of the resulting children if her oocytes were taken, and full custody if her embryos were used, even though the children were of school age and she had never met them.

6. A man sued his ex-wife for possession of their frozen fertilized ova. He won, and donated them for research. She had wanted to be pregnant.

7. The night before *in vitro* fertilized embryos were to be implanted in a 40-year-old woman's uterus after she and her husband had spent 4 years trying to conceive, the man changed his mind and wanted the embryos destroyed.

ARTs operate on molecules and cells, but affect individuals and families. Ultimately, by introducing artificial selection, these interventions may affect the human gene pool.

Key Concepts Questions 21.4

1. What can be done with extra fertilized ova and early embryos?

2. What can we learn from using early embryos in research?

Summary

21.1 Savior Siblings and More

1. **Assisted reproductive technologies (ARTs)** replace what is missing in reproduction, using laboratory procedures and people other than the infertile couple.

21.2 Infertility and Subfertility

2. **Infertility** is the inability to conceive a child after a year of frequent unprotected intercourse. Subfertile individuals or couples manufacture gametes, but take longer than usual to conceive.

3. Causes of infertility in the male include low sperm count, a malfunctioning immune system, a varicose vein in the scrotum, structural sperm defects, drug exposure, vasectomy reversal, heat to the testes, and abnormal hormone levels. Mutation may impair fertility.

4. Causes of infertility in the female include absent or irregular ovulation, blocked uterine tubes, an inhospitable or abnormally shaped uterus, secretions that are hostile to sperm, or lack of sperm-attracting biochemicals. Early pregnancy loss due to abnormal chromosome number is more common in older women and may appear to be infertility.

21.3 Assisted Reproductive Technologies

5. In **intrauterine insemination (IUI)**, donor sperm are placed into a woman's reproductive tract in a clinical setting.

6. A gestational and genetic surrogate mother provides her oocyte. Then IUI introduces sperm from a man whose partner cannot conceive or carry a fetus. The surrogate provides her uterus for 9 months. A gestational-only surrogate mother receives a fertilized ovum that resulted from **in vitro fertilization (IVF)** of a secondary oocyte by a sperm that came from the couple who asked her to carry the fetus. She is not genetically related to the fetus.

7. In IVF, oocytes and sperm join in a dish, fertilized ova divide a few times, and embryos are placed in the woman's body, circumventing blocked tubes or malfunctioning sperm. **Intracytoplasmic sperm injection (ICSI)** introduces immature or nonmotile sperm into oocytes.

8. Embryos can be frozen and thawed and then complete development when placed in a uterus.

9. **Gamete intrafallopian transfer (GIFT)** introduces oocytes and sperm into a uterine tube past a blockage; fertilization occurs in the woman's body. **Zygote intrafallopian transfer (ZIFT)** places an early embryo in a uterine tube.

10. Oocytes can be frozen and stored. In embryo donation, a woman undergoes IUI. A week later, the embryo is washed out of her uterus and put into the reproductive tract of the woman whose partner donated the sperm.

11. **Preimplantation genetic diagnosis (PGD)** checks chromosomes or specific gene variants that are in a family, and may sequence exomes and genomes in cells from embryos conceived *in vitro*. PGD was developed on 8-celled embryos but is more often done today on trophectoderm cells of day-5 blastocysts.

12. **Sequential polar body analysis** infers absence of a mutation in a fertilized ovum by checking its associated second polar body.

21.4 Extra Embryos

13. Extra fertilized ova and early embryos generated in IVF are used, donated to couples, stored, donated for research, or discarded.

14. Donated embryos enable researchers to study aspects of early human development that they could not investigate in other ways.

Review Questions

1. Which assisted reproductive technologies might help the following couples? (More than one answer may fit some situations.)
   a. A woman is born without a uterus but manufactures healthy oocytes.
   b. A man has cancer treatments that damage his sperm.
   c. A genetic test reveals that a woman will develop Huntington disease. She wants a child but does not want to pass on the disease.
   d. Two women wish to have and raise a child together.
   e. A man and woman are carriers of sickle cell disease. They do not want to have an affected child or terminate a pregnancy to avoid the birth of an affected child.
   f. A woman's uterine tubes are scarred and blocked.
   g. A young woman must have radiation and chemotherapy to treat ovarian cancer but wishes to have a child.

2. Why are men typically tested for infertility before women?

3. What are causes of infertility among older women?

4. Cite a situation in which both man and woman contribute to subfertility.

5. How does ZIFT differ from GIFT and IVF?

6. Explain how PGD is similar to and different from the prenatal diagnosis techniques described in chapter 13.

7. How do each of the following ARTs deviate from the normal biological process?
   a. IVF
   b. GIFT
   c. embryo donation
   d. gestational-only surrogacy
   e. intrauterine insemination
   f. cytoplasmic donation

8. Explain how PGD works, and list two events in early prenatal development that might explain cases in which the PGD result is inaccurate.

9. Describe a scenario in which each of the following technologies is abused:
   a. surrogate mother
   b. gamete donation
   c. IVF
   d. PGD

10. Explain how sequential polar body analysis is based on Mendel's first law.

Applied Questions

1. A company that offers PGD terms an embryo "normal and healthy" if it has 46 chromosomes. Why is this statement misleading?

2. Two famous male actors had twins using a surrogate mother who carried two embryos that had been fertilized *in vitro,* one with one man's sperm and the other with the

other man's sperm. In terms of genetics, how closely are the babies, a boy and a girl, related to each other if they have different fathers?

3. Some ARTs were invented to help people who could not have children for medical reasons or to avoid conceiving a child with a genetic disease in the family. As the technologies became more familiar, in the United States people with economic means began to use ARTs for other reasons, such as a celebrity who wants to avoid pregnancy. Remembering that the U.S. government does not regulate ARTs, do you think that any measures should be instituted to select candidates for ARTs? How can these technologies be made more affordable?

4. A man reads his medical chart and discovers that the results of his sperm analysis indicate that 22 percent of his sperm are shaped abnormally. He wonders why the physician said he had normal fertility if so many sperm are abnormally shaped. Has the doctor made an error?

5. The American Society for Reproductive Medicine recommends oocyte freezing for women who have cancer and want to avoid teratogenic effects of toxic therapies but not for women who want to delay pregnancy but avoid the risks of older oocytes. Do you agree with these limits?

6. An embryo bank offers IVF leftover embryos, which would otherwise remain in the deep freeze or be discarded, to people wanting to have children, for $2,500 each. The bank circumvents bioethical concerns by claiming that it sells a service, not embryos. People in favor of the bank claim that purchasing an embryo is not different from paying for sperm or eggs, or an adopted child. Those who object to the bank claim that it makes an embryo a commodity.

   a. Do you think that an embryo bank is a good idea, or is it unethical?

   b. Whose rights are involved in the operation of the embryo bank?

   c. Who should be liable if a child that develops from the embryo has an inherited disease?

   d. Is the bank elitist because the cost is so high?

7. Many people spend thousands of dollars pursuing pregnancy. What is an alternative solution to their quest for parenthood?

8. Big Tom is a bull with valuable genetic traits. His sperm are used to conceive 1,000 calves. Mist, a dairy cow with exceptional milk output, has many oocytes removed, fertilized *in vitro,* and implanted into surrogate mothers. With their help, Mist becomes the genetic mother of 100 calves—many more than she could give birth to naturally.

Which two ARTs are based on these examples from agriculture?

9. State who the genetic parents are and who the gestational mother is in each of the following cases:

   a. A man was exposed to unknown burning chemicals and received several vaccines while serving in the U.S. military in Iraq, and abused drugs for several years before and after that. Now he wants to become a father, but he is concerned that past exposures to toxins have damaged his sperm. His wife undergoes intrauterine insemination with sperm from the husband's brother, who has led a calmer and healthier life.

   b. A 26-year-old woman has her uterus removed because of cancer. However, her ovaries are intact and her oocytes are healthy. She has oocytes removed and fertilized *in vitro* with her husband's sperm. Two resulting embryos are implanted into the uterus of the woman's best friend.

   c. Max and Tina had a child by IVF and froze three extra embryos. Two are thawed years later and implanted into the uterus of Tina's sister, Karen. Karen's uterus is healthy, but she has ovarian cysts that often prevent her from ovulating.

   d. Forty-year-old Christopher wanted children, but did not want a partner. He donated sperm, which were used for intrauterine insemination of a woman who carried the resulting fetus to term for a fee, and gave birth to a daughter. Christopher adopted the child.

10. An IVF attempt yields 12 more embryos than the couple who conceived them can use. What could they do with the extras?

11. What do you think children born as a result of an ART should be told about their origins and when should this happen?

12. A program in India offers IVF with PGD to help couples who already have a daughter to conceive a son. The reasoning is that because having a male heir is of such great importance in this society, offering PGD can enable couples to avoid aborting second and subsequent female pregnancies. Do you agree or disagree that PGD should be used for sex selection in this sociological context?

13. Men who have small deletions in their Y chromosomes that stop sperm from maturing can use ICSI to conceive a child, but this passes on their infertility. Suggest an ART that they could use to prevent male infertility.

14. The novel and film *My Sister's Keeper* describes the steps of PGD and calls the technique "genetic engineering." Is this correct? Why or why not?

Case Studies and Research Results

1. Ridge Forrester, a major heartthrob on the daytime drama "The Bold and the Beautiful," had a vasectomy because he had so many children he couldn't remember all of them. Many months later, Ridge's urologist performed a test that showed Ridge was azoospermic. Then Ridge suddenly fell in love with the decades-younger Caroline, who wanted children immediately. Ridge complied, telling her that despite having had a vasectomy, the doctor said he had azoospermia and therefore he was still making some sperm cells. Was Ridge correct?

2. Doola is 28 years old and is trying to decide if she and her husband are ready for parenthood when she learns that her 48-year-old mother has Alzheimer disease. The mother's physician tells Doola that because of the early onset, the Alzheimer disease could be inherited through a susceptibility gene. Doola is tested and indeed has the same dominant allele. She wants to have a child right away, so that she can enjoy many years as a mother. Her husband David feels that it wouldn't be fair to have a child knowing that Alzheimer disease likely lies in Doola's future. He suggests adoption.

a. Who do you agree with, and why?

b. David is also concerned that Doola could pass on her Alzheimer gene variant to a child. Which technology might help them avoid this?

c. Is Doola correct in assuming that she is destined to develop Alzheimer disease?

3. Two large, well-known companies offered female employees a $20,000 benefit to freeze their eggs. Why might the companies have done this, and why might many people have objected to the offer?

4. An Oregon man anonymously donated sperm that were used to conceive a child. The man later claimed, and won, rights to visit his child. Is this situation for the man more analogous to a genetic and gestational surrogate mother or to an oocyte donor who wishes to see the child she helped to bring into existence?

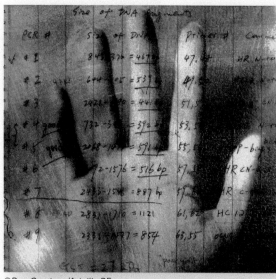
©Don Carstens/Artville RF

CHAPTER 22

Genomics

What can we learn from our genome sequences?

Learning Outcomes

22.1 From Genetics to Genomics

1. Explain how linkage studies, positional cloning experiments, and gene mapping led to the idea to sequence human genomes.
2. Distinguish between the two approaches used to sequence the first human genomes.

22.2 Analysis of Human Genome Content

3. Define *reference genome*.
4. Explain why genome sequencing requires multiple copies of a genome.
5. Identify limitations of exome and genome sequencing.

22.3 Personal Genome Sequencing

6. Discuss the types of information that the first sequenced human genomes provided.
7. Discuss practical aspects of implementing exome and genome sequencing in health care.

 ## The BIG Picture

Nearly two decades ago, the first human genomes were sequenced. Today the cost has plummeted to the point that personal genome sequencing is not only possible, but may soon become widely used to help diagnose disease, inform treatment choices, and reveal our ancestry. Thousands of people have had their genomes sequenced. Will you have your genome sequenced? Getting the most useful information as possible from our genome sequences will depend upon many people permitting their information to be collected and analyzed, even if it remains anonymous. Genomics may impact the future for many of us.

Sequencing the Genomes of the Deceased

Genomes transcend time. DNA sequences provide clues to health, from a cell plucked from the earliest of embryos to the centenarian that embryo might become. The chapter 1 opener envisioned the use of genome information from a newborn over her lifetime. This final chapter opener considers the value of sequencing the genomes of people who have died. Would knowing genome sequences during their lifetimes have changed their health care?

The Genomic Postmortem Research Project of the Marshfield Clinic Research Foundation in Wisconsin is sequencing the genomes of 300 deceased individuals and comparing the information to data in the electronic medical records. The researchers are searching through more than 27 million gene variants, compiled from databases and recommendations from genetics organizations, that could have been used to prevent, detect, and treat specific diseases. Gene variants are classified as pathogenic, likely pathogenic, "variant of uncertain significance," likely benign, and benign (see the Bioethics in chapter 12). To start, the researchers are focusing on certain chronic diseases, cancers, susceptibility to infections, and pharmacogenetics of three common drug classes (see section 20.2).

Practical examples of the value of genome sequencing quickly emerged:

- A woman who died from breast and ovarian cancer at age 59 had no family history of cancer and had never had a mammogram. She had a rare pathogenic variant of *BRCA1*. Had the mutation been detected early, the cancers could likely have been prevented or treated.

- A man with deep vein thrombosis and tangles of abnormal blood vessels in his kidneys, brain, and lungs had hereditary hemorrhagic telangiectasia type 1. If the mutation had been detected, the man could have taken either an old drug (thalidomide) or a newer one (Avastin) to prevent dangerous blood clots.

- Of the 300 deceased individuals, 281 had gene variants that affected their response to drugs, including antidepressants, cancer drugs, painkillers, antipsychotics, blood thinners, and cholesterol-lowering drugs. The information could have guided physicians to initially prescribe the drugs most likely to work.

22.1 From Genetics to Genomics

Genetics is a young science, genomics younger still. As one field has evolved and led to the other, milestones have come at oddly regular intervals. A century after Gregor Mendel announced and published his findings, the genetic code was deciphered; a century after his laws were rediscovered, the first human genomes were sequenced.

Geneticist H. Winkler coined the term *genome* in 1920. A hybrid of "gene" and "chromosome," genome then denoted a complete set of chromosomes and genes. The modern definition refers to all the DNA in a haploid set of chromosomes. The term *genomics,* credited to T. H. Roderick in 1986, indicates the study of genomes. Thoughts of sequencing genomes echoed through much of the twentieth century, as researchers described the units of inheritance from several different perspectives.

As the twenty-first century dawned, people began having their genomes sequenced. First came two superstars in genetics, J. Craig Venter and James Watson. For a time papers were published adding different ethnic groups to the roster of the sequenced. Then a few wealthy actors, musicians, politicians, scientists, and entrepreneurs funded their personal genome sequencing, interested in diseases in their families or themselves. Curious journalists did and still do have their genomes sequenced and then report on their collection of gene variants in books, blogs, podcasts, TED talks, and articles. People long dead have had their genomes sequenced, including Egyptian mummies, Ötzi the Ice Man, and Neanderthals and Denisovans.

At the same time that a few prominent and representative individuals were having their genomes sequenced, the first large-scale genome sequencing research projects began, with an ultimate goal of incorporating genome information into electronic health records and routine health care. The cost of genome sequencing has dropped from an initial $1 billion to under $1,000, not including the all-important interpretation. Researchers project that by 2025, 100 million to 2 billion people will have had their genomes sequenced (**table 22.1**).

The idea of sequencing a human genome was once nearly unfathomable in its complexity; soon it will likely be routine, and span gametes to embryos to the oldest old. Genome sequence information can already be stored on smartphones. **Table 22.2** lists some of the uses of exome and genome sequencing covered in other chapters. Our genomes are nearly identical, yet with

Table 22.1	Cost of Sequencing a Human Genome
Year	**Cost**
1990	$1 billion
2001	$100 million
2003	$30 million
2005	$22 million
2008	$1 million
2009	$100,000
2010	$50,000
2011	$10,000
2012	$8,000
2014	$1,000–$5,000
2018	<$1,000

Table 22.2	Genomics Coverage in Other Chapters
Topic	**Location**
Eve's genome	Chapter 1 opener
Precision medicine initiative	Figure 1.8
Genome sequencing ends a child's diagnostic odyssey	Clinical Connection 1.1
Genomes of very old people	Section 3.6
Exome and genome sequencing clarify pedigrees	Section 4.4
Parent and child trios for exome and genome sequencing	Figure 4.18
The human genome sequence adds perspective	Section 5.1
Exome sequencing led to discovery of autism gene variants	Figure 8.10
The louse genome	Chapter 10 opener
Vincent's diagnostic odyssey	Chapter 12 opener
Dog genomes	Section 15.6
Khoisan genomes	Chapter 16 opener
Neanderthal, Denisovan, and Ötzi the Ice Man genomes	Section 16.1
Native American genomes	Section 16.3
Evolution of human genomes	Section 16.4
Using genomics to fight infection	Section 17.5
Cancer genes and genomes	Section 18.3

enough variability—we differ at millions of single-base sites—to make life and society interesting.

Beginnings in Linkage and Positional Cloning Studies

Sequencing what was thought of as "the" human genome unofficially began in the 1980s with deciphering signposts along the chromosomes. Many of the initial steps and tools grew from existing technology. Linkage maps and studies of rare families that had chromosome abnormalities and specific syndromes enabled researchers to assign a few genes to their chromosomes. Automated DNA sequencing took genetic analysis to a new level—information in the form of a living language.

The evolution of increasingly detailed genetic maps is similar to zooming in on a geographical satellite map (**figure 22.1**). A cytogenetic map (of a chromosome) is like a map of California within a map of the United States, highlighting only the largest cities. A linkage map is like a map that depicts the smaller cities and large towns, and a physical map is similar to a geographical map indicating all towns in an area. Finally, a sequence map is

the equivalent of a Google map showing all of the buildings in a specific town.

Before the first human genomes were sequenced, researchers matched single genes to specific diseases using an approach called positional cloning. The technique began with examining a particular phenotype corresponding to a Mendelian disease in large families. The phenotype was easily matched to a chromosome segment if all the affected individuals shared a chromosome abnormality that relatives without the phenotype did not have. But abnormal chromosomes are rare. Another way that positional cloning located medically important genes on chromosomes used linkage maps that showed parts of a chromosome shared by only the individuals in a family who had the same syndrome. Then researchers isolated pieces of the implicated chromosome, identified short DNA sequences corresponding to the region of interest, and overlapped the pieces, to gradually find the gene behind a particular phenotype.

Positional cloning was an indirect method that narrowed a gene's chromosomal locus without necessarily knowing the protein product. Throughout the 1980s and 1990s, the method

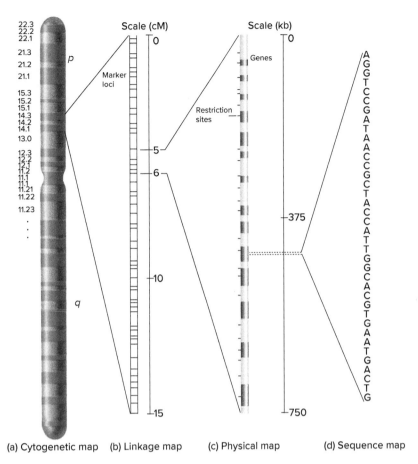

(a) Cytogenetic map   **(b) Linkage map**   **(c) Physical map**   **(d) Sequence map**

Figure 22.1 **Different levels of genetic maps are like zooming up the magnification on a geographical map.** **(a)** A cytogenetic map, based on associations between chromosome aberrations and syndromes, can distinguish DNA sequences at least 5,000 kilobases (kb) apart. **(b)** A linkage map derived from recombination data distinguishes genes hundreds of kilobases apart (cM = centimorgans). **(c)** A physical map constructed from overlapped DNA pieces distinguishes genes tens of kilobases apart. **(d)** A sequence map depicts the order of nucleotide bases.

enabled researchers to discover the genes behind Duchenne muscular dystrophy, cystic fibrosis, Huntington disease, and many other single-gene conditions. So slow was the process that it took a decade to go from discovery of a genetic marker for Huntington disease to discovery of the gene.

Sequencing the First Human Genomes

The idea to sequence the first human genome occurred to many researchers at about the same time, but they had different goals (see the **Technology Timeline**). It was first brought up at a meeting held by the Department of Energy (DOE) in 1984 to discuss the long-term population genetic effects of exposure to low-level radiation. In 1985, researchers convening at the University of California, Santa Cruz, called for an institute to sequence the human genome, because sequencing of viral genomes had shown that it could be done. The next year, virologist Renato Dulbecco proposed that the key to understanding the origin of cancer lay in knowing the human genome

sequence. Later that year, scientists packed a room at the Cold Spring Harbor Laboratory on New York's Long Island to discuss the feasibility of a project to sequence the human genome. At first those against the project outnumbered those for it five to one. The major fear was the shifting of goals of life science research from inquiry-based experimentation to amassing huge amounts of data—ironic considering the importance of bioinformatics today.

A furious debate ensued. Detractors claimed that the project would be more gruntwork than a creative intellectual endeavor, comparing it to conquering Mt. Everest just because it is there. Practical benefits would be far in the future. Some researchers feared that such a "big science" project would divert government funds from basic research and developing treatments for the still-new AIDS epidemic. Finally, the National Academy of Sciences convened a committee representing both sides to debate the feasibility, risks, and benefits of the project. The naysayers were swayed to the other side. In 1988, Congress authorized the National Institutes of Health (NIH) and the DOE

Technology Timeline

EVOLUTION OF GENOME PROJECTS AND RELATED TECHNOLOGIES

1977	Two methods of DNA sequencing invented. The Sanger method persists for detailed sequencing of individual genes.
1985–1988	Idea to sequence human genome suggested at several scientific meetings.
1988	Congress authorizes the Department of Energy and the National Institutes of Health to fund the Human Genome Project.
1989	Researchers at Stanford and Duke universities invent DNA microarrays.
1990	Human Genome Project officially begins. Part of International Consortium with the United Kingdom, Canada, Japan, France, Germany, and China.
1991	Expressed sequence tag (EST) technology identifies protein-encoding DNA sequences.
1992	First DNA microarrays available.
1993	Need to automate DNA sequencing recognized.
1994	U.S. and French researchers publish preliminary map of 6,000 genetic markers, one every 1 million bases along the chromosomes.
1995	Emphasis shifts from gene mapping to sequencing. First genome sequenced: *Hemophilus influenzae*.
1996	Resolution to make all data public and updated daily at GenBank website. First eukaryote genome sequenced—yeast.
1998	International consortium releases preliminary map of pieces covering 98 percent of human genome. Millions of sequences in GenBank. Instructions for creating DNA microarrays posted on Internet. First multicellular organism's genome sequenced—roundworm.
1999	Rate of filing of new sequences in GenBank triples. International consortium and two private companies race to complete sequencing. First human chromosome sequenced—22.
2000	Completion of first draft human genome sequence announced. Microarray technology flourishes. First plant and fungal genomes sequenced.
2001	Two versions of draft human genome sequence published.
2003	Finished human genome sequence announced to coincide with 50th anniversary of discovery of DNA structure.
	Human exome sequence available on DNA microarrays. HapMap project identifies linkage patterns.
2004	Final version of human genome sequence published.
2005	Annotation of human genome sequence continues. Number of species with sequenced genomes soars.
	High-throughput, next-generation DNA sequencing introduced.
2007	Detailed analysis of 1 percent of the human genome (ENCODE project) reveals that most of it is transcribed. First individual human genome sequenced. Human Microbiome Project begins.
2008	First genome synthesized—*Mycoplasma genitalium*.
2010	Several human genomes sequenced. First synthetic genome supports a bacterial cell. By now, 4,000 bacterial and viral and 250 eukaryotic species' genomes have been sequenced.
2012	1000 Genomes Project completes cataloging of human genetic diversity.
2014	Tens of thousands of human genomes and hundreds of thousands of human exomes sequenced.
2015	Precision Medicine Initiative announced, to include human genome sequence information.
2017	Health care providers learn how to incorporate DNA sequence–based screens and tests into their medical practices.
2020	5 million or more people will have had their genomes sequenced.
2025	100 million to 2 billion people will have had their genomes sequenced.

to fund the $3 billion, 15-year Human Genome Project, which began in 1990 with James Watson at the helm. The project set aside 3 percent of its budget for the Ethical, Legal and Social Implications (ELSI) Research Program. It has helped ensure that genetic information is not used to discriminate. Eventually, an international consortium as well as a private company, Celera Genomics, sequenced the first human genomes. The groups worked separately, and one effort finished a few months before the other, but the accomplishment is referred to historically as "the human genome project."

A series of technological improvements sped the genome project. In 1991, a shortcut called expressed sequence tag (EST) technology enabled researchers to quickly pick out genes most likely to be implicated in disease. This was a foreshadowing of future efforts to focus on the exome, the part that encodes protein and is responsible for 85 percent of single-gene diseases. ESTs are cDNAs made from the mRNAs in a cell type that is abnormal in a particular illness, such as an airway lining cell in cystic fibrosis. ESTs therefore represent gene expression. Also in 1991, researchers began using DNA microarrays to display short DNA molecules. Microarray technology became important in DNA sequencing (tiling arrays) as well as in assessing gene expression (expression arrays).

Computer algorithms assembled many short pieces of DNA with overlapping end sequences into longer sequences (**figure 22.2**). When the project began, researchers cut several genomes' worth of DNA into overlapping pieces of about 40,000 bases (40 kilobases), then randomly cut the pieces into small fragments. The greater the number of overlaps, the more complete the final assembled sequence. The sites of overlap had to be unique sequences, found in only one place in the genome. Overlaps of repeated sequences found in several places in the genome could lead to more than one derived overall sequence—a little like searching a document for the word "that" versus searching for an unusual word, such as "dandelion." Searching for "dandelion" is more likely to lead to a specific part of a document, whereas "that" may occur in several places—just like repeats in a genome.

The use of unique sequences is why the human genome project did not uncover copy number variants. For example, the sequence CTACTACTA would appear only as CTA. Researchers did not at first appreciate the fact that repeats are a different form of information and source of variation than DNA base sequences. A balance was necessary between using DNA pieces large enough to be unique, but not so large that the sequencing would take a very long time.

Two general approaches were used to build the long DNA sequences to initially derive the sequence of the human genome (**figure 22.3**). The "clone-by-clone" technique the U.S. government-funded group used aligned DNA pieces one chromosome at a time. The "whole-genome shotgun" approach Celera Genomics used shattered the entire genome, then used an algorithm to identify and align overlaps in a continuous sequence. Whole-genome shotgun sequencing can be compared to cutting the binding off a large book, throwing it into the air and freeing every page, and reassembling the dispersed pages in order. A "clone-by-clone" dismantling of the book would divide it into bound chapters. Whole-genome shotgunning is faster, but it misses some sections (particularly repeats) that the clone-by-clone method detects.

Technical advances continued. In 1995, DNA sequencing was automated, and software was developed that could rapidly locate the unique sequence overlaps among many small pieces of DNA and assemble them, eliminating the need to gather large guidepost pieces. In 1999, the race to sequence the human genome became intensely competitive. The battling factions finally called a truce. On June 26, 2000, Craig Venter from Celera Genomics and Francis Collins, representing the International Consortium, flanked President Clinton in the White House Rose Garden to unveil the "first draft" of the human genome sequence. The milestone capped a decade-long project involving thousands of researchers, culminating a century of discovery. The historic June 26 date came about because it was the only opening on the White House calendar! In other words, the work was monumental; its announcement, somewhat staged. **Figure 22.4** is a conceptual overview of genome sequencing.

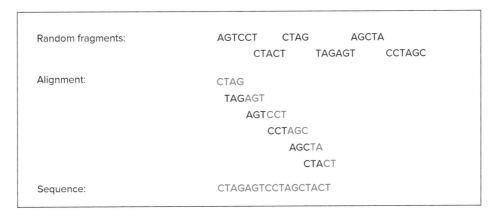

Figure 22.2 **Deriving a DNA sequence.** Automated DNA sequencers first determine the sequences of short pieces of DNA, or sometimes of just the ends of short pieces. Then algorithms search for overlaps. By overlapping the pieces, the software derives the overall DNA sequence.

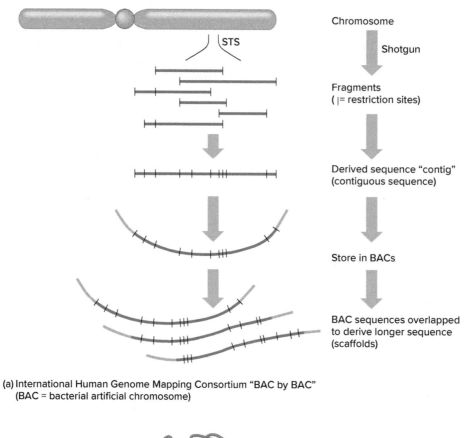

Chromosome

↓ Shotgun

Fragments
(|= restriction sites)

↓

Derived sequence "contig"
(contiguous sequence)

↓

Store in BACs

↓

BAC sequences overlapped
to derive longer sequence
(scaffolds)

(a) International Human Genome Mapping Consortium "BAC by BAC"
    (BAC = bacterial artificial chromosome)

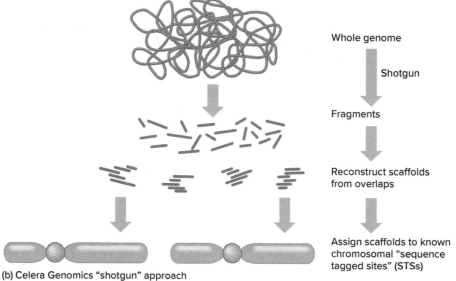

Whole genome

↓ Shotgun

Fragments

↓

Reconstruct scaffolds
from overlaps

↓

Assign scaffolds to known
chromosomal "sequence
tagged sites" (STSs)

(b) Celera Genomics "shotgun" approach

Figure 22.3 **Two routes to sequencing the human genome.** **(a)** The International Consortium began with known chromosomal sites and overlapped large pieces, called contigs, that were reconstructed from many small, overlapping pieces. STS stands for "sequence tagged site," which refers to specific known parts of chromosomes. A BAC (bacterial artificial chromosome) is a cloning vector that uses bacterial DNA. **(b)** Celera Genomics shotgunned several copies of a genome into small pieces, overlapped them to form scaffolds, and then assigned scaffolds to known chromosomal sites. They used some Consortium data.

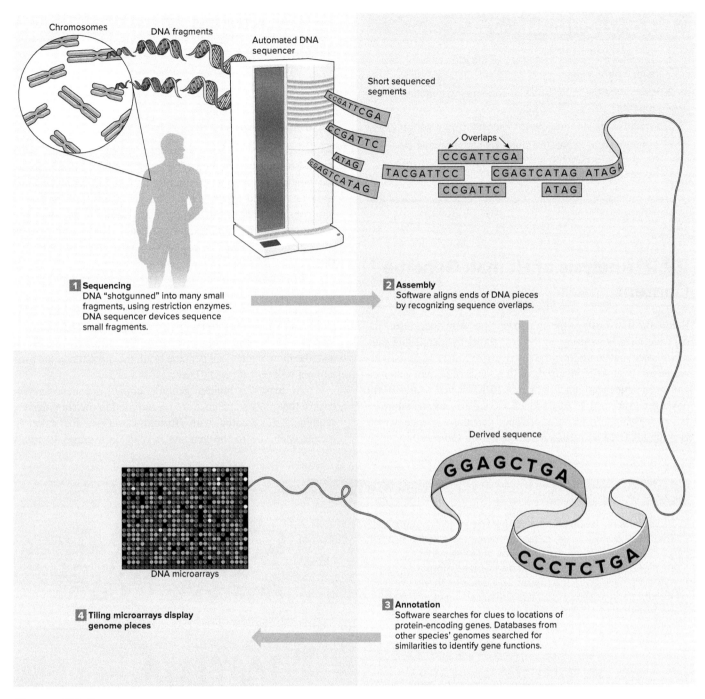

Figure 22.4 Sequencing genomes. The first-generation, whole-genome "shotgun" approach to genome sequencing overlapped DNA pieces cut from several copies of a genome, then assembled the overall sequence. The "next-generation" genome sequencing techniques used today are based on microfluidics and nanomaterials and are much faster.

The labels in the figure read:

Chromosomes

DNA fragments

Automated DNA sequencer

Short sequenced segments

CCGATTCGA

CCGATTC

ATAG

CGAGTCATAG

Overlaps

CCGATTCGA

TACGATTCC   CGAGTCATAG   ATAGA

CCGATTC   ATAG

1 Sequencing
DNA "shotgunned" into many small fragments, using restriction enzymes. DNA sequencer devices sequence small fragments.

2 Assembly
Software aligns ends of DNA pieces by recognizing sequence overlaps.

Derived sequence

GGAGCTGA

CCCTCTGA

3 Annotation
Software searches for clues to locations of protein-encoding genes. Databases from other species' genomes searched for similarities to identify gene functions.

DNA microarrays

4 Tiling microarrays display genome pieces

Key Concepts Questions 22.1

1. Describe the research that led to the idea to sequence human genomes.

2. What were the initial goals of sequencing a human genome?

3. What are the general steps in sequencing a genome?

4. Distinguish between the clone-by-clone and whole-genome shotgunning approaches to sequence the human genome.

22.2 Analysis of Human Genome Content

Before the first drafts of the human genome were done, researchers were already working on the next stages by identifying sites of variation and continuing the discovery of functions of individual genes. Even after hundreds of thousands of human genomes had been sequenced, each new one still revealed nearly 9,000 novel gene variants! To ease sequence comparisons and interpretations, researchers derive a reference genome sequence, which is a digital DNA sequence assembled from the most common base at each point in many sequenced genomes. It is haploid (one copy). Comparing reference genome sequences for different types of organisms and identifying DNA sequences that they share provides views of evolution, discussed in **A Glimpse of (Pre) History**.

Improving Speed and Coverage

Sequencing the first human genomes took 6 to 8 years; today it can be done within a day. The cost has fallen dramatically, too, beating an estimate commonly used in the computer hardware field called Moore's law. The law states that the rate of improvement in a technology (such as number of transistors on integrated circuits) doubles every 2 years. The cost of genome sequencing has fallen at a much sharper rate, as table 22.1 shows.

Improvements in sequencing technology enabled researchers to work with many more copies of an individual's genome, which is termed "coverage." Recall from section 9.4 that DNA pieces cut from several genomes must be sequenced and overlapped to derive the overall sequence. Because some pieces are lost, the more copies of a genome used, the more likely the overlapping will pick up every base.

At least 28 human genome copies are necessary to ensure that most sequences are represented in the final derived sequence. A genome with 40-fold coverage, for example, means each site in the genome is read on average 40 times.

A GLIMPSE OF (PRE)HISTORY: COMPARATIVE GENOMICS

Thousands of species have had their genomes sequenced. The first were viruses and bacteria because they have small genomes. Next came the genomes of familiar animals, such as mice, rats, chimps, cats, and dogs. Most informative, however, have been the genomes of species that represent evolutionary crossroads. These are organisms that introduced a new trait or were the last to have an old one.

Comparing genomes of modern species enables researchers to infer evolutionary relationships from DNA sequences that are conserved (shared) and presumably selected through time. **Figure 22A** shows one way of displaying short sequence similarities, called a pictogram. DNA sequences from different species are aligned, and the bases at different points indicated. A large letter A, C, T, or G indicates, for example, that all species examined have the same base at that site. A polymorphic site, in contrast, has different bases for different species.

Comparative genomics uses conserved sequences to identify biologically important genome regions, assuming that persistence means evolutionary success. But there are exceptions—the human genome has some conserved sequences with no apparent function. Either we haven't discovered the functions or genomes include "raw material" for future functions. About 6 percent of the human genome

Human
Mouse
Rat
Chicken

(a) Not highly conserved

(b) Highly conserved

Figure 22A A pictogram indicates conservation of the DNA sequence. These pictograms are for short sequences in corresponding regions of the human, mouse, rat, and chicken genomes. A large letter means that all four species have the same base at that site. If four letters appear in one column, then the species differ. Pictogram **(a)** shows a sequence that is not highly conserved; **(b)** shows one that is.

(Continued)

sequence is highly conserved. The following examples illustrate the types of information inferred from conserved DNA sequences.

The Minimum Gene Set Required for Life

The smallest microorganism known to be able to reproduce is *Mycoplasma genitalium*. It infects cabbage, citrus fruit, corn, broccoli, honeybees, and spiders, and causes respiratory illness in chickens, pigs, cows, and humans. Researchers call its tiny genome the "near-minimal set of genes for independent life." Of 480 protein-encoding genes, 265 to 350 are essential. Considering how *Mycoplasma* uses its genes reveals the fundamental challenges of being alive. *M. genitalium* was the first organism to have its genome synthesized. Researchers created a "synthetic genome" consisting of the 582,970 bases, in four pieces, of the tiny bacterium's genome and delivered it to *E. coli* or yeast cells that had their genomes removed. The microorganism serves as a model for attempting to create even simpler genomes, discussed in Bioethics at the end of this chapter.

Fundamental Distinctions Among the Three Domains of Life

Methanococcus jannaschii is a microorganism that lives at the bottom of 2,600-meter-tall "white smoker" chimneys in the Pacific Ocean, at high temperature and pressure and without oxygen. Like bacteria, the archaea lack nuclei but they replicate DNA and synthesize proteins in ways similar to multicellular organisms, and so they are considered a third form of life. The genome sequence confirmed this designation.

The Simplest Organism with a Nucleus

The yeast *Saccharomyces cerevisiae* is single-celled with only about 6,000 genes, but a third of them have counterparts among mammals, including at least 70 genes implicated in human diseases. Yeast cells are eukaryotic. Understanding what a gene does in yeast can provide clues to how it affects human health. For example, counterparts of mutations in cell cycle control genes in yeast cause cancer in humans.

The Basic Blueprints of an Animal

The genome of the tiny, transparent, 959-celled nematode worm *Caenorhabditis elegans* is packed with information on what it takes to be an animal. The worm's signal transduction pathways, cytoskeleton, immune system, apoptotic pathways, and brain proteins are similar to our own. Curiously, the fruit fly (*Drosophila melanogaster*) genome has 13,601 genes, fewer than the 18,425 in the much simpler worm. Of 289 disease-causing genes in humans, 177 have counterparts in *Drosophila*. The fly is a model for testing new treatments.

From Birds to Mammals

The sequencing of the chicken genome marked a number of milestones. It was the first agricultural animal, the first bird, and, as such, the first direct descendant of dinosaurs. The genome of the red jungle fowl *Gallus gallus,* the ancestor of the domestic chicken, is remarkably like our own, minus many repeats, but its genome organization is intriguing. Like other birds, fishes, and reptiles, but not mammals, the chicken genome is distributed among very large macrochromosomes and tiny microchromosomes. Repeats may have been responsible for the larger sizes of mammalian chromosomes.

From Chimps to Humans

Most comparisons of the human genome to those of other species seek similarities. Comparisons of our genome to that of the chimpanzee, however, seek genetic *differences,* which may reflect what makes humans unique (see section 16.4). The human and chimp genomes differ by 1.2 percent, equaling about 40 million DNA base substitutions. Within those differences, as well as copy number distinctions, may lie the answers to compelling medical questions. Why do only humans get malaria and Alzheimer disease? Why is HIV infection deadlier than the chimp version, SIV?

High coverage is needed to detect a rare DNA sequence, such as the genome sequencing that helped to diagnose the child described in Clinical Application 1.1. Even today, genomes are not completely sequenced, because the technology is not perfect. The first human genome sequences published, in 2001, had about 150,000 small gaps; current versions have only a few hundred tiny missing pieces.

Sequencing genomes provides much more information than sequencing exomes, which are the exon (coding) parts of protein-encoding genes. Knowing the sequences surrounding the exons can detect "structural variants" such as inversions and reciprocal translocations (see figures 13.16 and 13.19), which flip or move DNA but do not alter the base sequence. Exome sequencing arrays must be designed to distinguish between pairs of SNPs on the same homologous chromosomes (in *cis* configuration) or on different homologs (in *trans* configuration). The *cis* configuration indicates linkage, which is important in predicting transmission of a genotype to offspring.

The Ongoing Goal: Annotation

Just as a book written in a foreign language is meaningless unless translated, knowing the sequence of a human genome is not useful unless we know what the information means. "Annotation" in linguistics means "a note of explanation or comment added to a text or diagram." In genomics, annotations are descriptions of genes, and what the significance of a particular gene variant is likely to be. Researchers are accomplishing this enormous task by meticulously consulting the published scientific literature, DNA sequence databases for every identified gene, and SNPs in noncoding regions that might be associated with specific disease risks. Genetic counselors and people with doctorates in genetics are doing the annotation, and are called curators, variant scientists, or curation specialists.

Annotation of a gene variant might include:

- the normal function of the gene;
- mode of inheritance;
- genotype (heterozygote, homozygote, compound heterozygote); and
- frequency of a variant in a particular population.

Knowing the frequency of a gene variant is important for logical reasons. A variant that is common—which means that many people live with it—is less likely to cause a serious illness than one that is rare. If a third of a population has a gene variant that is associated with hypertension, for example, that gene probably contributes only slightly to the overall risk for this multifactorial condition. Otherwise a third of the population would have severe hypertension. However, the reverse is not true—a rare variant can be harmless, and just unusual. Many annotations take these types of data into consideration to estimate how likely the gene variant or genotype is to cause disease—such as "likely pathogenic," "elevates risk by 4 percent," or a "variant of uncertain significance." Bioethics in chapter 12 discusses the clinical situation of a "VUS," which will become less likely as more gene functions are discovered. **Table 22.3** lists some findings among a few of the first people to have their genomes sequenced.

Table 22.3	A Gallery of Genomes
Gender	**Findings**
XY	Hypertrophic cardiomyopathy (enlarged heart), dominant mutation of "uncertain pathogenicity"
	Peroxisome disease similar to adrenoleukodystrophy, asymptomatic carrier
	Age-related macular degeneration (ARMD; visual loss), mutation "likely pathogenic"
	Hypertension, homozygous recessive for common mutation that increases risk by 8%
	Increased plasma triglycerides due to *ApoA5* mutation, not associated with increased risk of coronary artery disease
XX	Biotinidase deficiency, asymptomatic carrier
	Adenosine deaminase deficiency, asymptomatic carrier
	Stiff arteries, lifetime risk of heart attack increased 0.5–3% above population risk of 15%
	Nephropathy (kidney disease) mutation, risk increased <0.1%
	Hypertension, increased risk associated with variant in noncoding DNA
XY	Baldness (confirmed with use of mirror)
	Ichthyosis vulgaris. Heterozygous for filaggrin mutation, has 30–50% risk of atopic dermatitis (eczema), which individual already has
	Polydactyly mutation, but known in only one family
	Age-related macular degeneration mutation, "likely pathogenic"
XX	Esophageal cancer, dominant mutation increases risk fourfold
	Crohn's disease, mutation in *NOD2* gene "likely pathogenic"
	Down syndrome risk to offspring 0.4% due to mutation in *MTRR* gene (age is more important risk factor)
	Hypertension, 4% increase in risk due to SNPs in noncoding region of angiotensin II gene
XY	Galactosemia deficiency, asymptomatic carrier
	Intellectual disability, dominant variant in *CDH* gene, "not statistically significant" increase in risk
	Lumbar disc disease risk increased from 4% in general population to 11%; individual already has it

To be most valuable, an annotated genome sequence is compared to as much health and family history information as a person can provide. The old-fashioned pedigree is still an invaluable tool and the starting point for many investigations, for this is the information that most people already know—who has what in the family. Genome annotations are also including microbiome data (see section 2.5), because the genes of the organisms that live in and on us can affect expression of our own genes.

Limitations of Genome Sequencing

Genome sequencing does not provide a complete picture of health. Limitations are technical and conceptual. In a technical sense, genome sequencing cannot detect repeats without additional types of tests because having multiples of a sequence does not alter the sequence. Such repeats include copy number variants, triplet repeat mutations, and even the complete extra chromosomes of trisomies, as well as deletions. Genome sequencing is also blind to uniparental disomy, the rare situation in which a child inherits two alleles of a gene from one parent (see figure 13.24) rather than one from each. The two copies are identical in DNA sequence, and so "count" only once in the overall sequence. In addition, exons (protein-encoding gene parts) buried within highly repeated introns may not be detected. Genome sequencing also does not include mitochondrial genes.

In a conceptual sense, genome information must be interpreted to be useful. This means not just identifying the protein that a gene encodes, but deciphering all interactions of genes and the gene networks that they form. Imagine a novel that is "read" one letter at a time, so that it is a string of thousands of copies of 26 letters, rather than a nuanced, coherent story with clues and connections conveyed in words. So, too, is a sequenced but unannotated human genome not informative.

Gene interactions are intricate and complex. Figure 11.4 illustrates just a small sampling of diseases that share genes that have altered expression. One gene's activity affecting the expression of another can explain why siblings with the same single-gene disease suffer to a different extent. For example, a child with severe spinal muscular atrophy (SMA), in which an abnormal protein shortens axons of motor neurons, may have a brother who also inherits the disease but has a milder case thanks to inheriting a variant of a second gene that extends axons (see section 5.1). Computational tools are used to sort out networks of interacting genes, sometimes called "connectomes." What were once regarded as simple epistatic interactions—gene affecting gene—may in reality be the tip of an iceberg of complex networks of genes that influence each other.

Finally, epigenetic changes induced by environmental factors provide a layer of information that must be applied to DNA information. These are the influences that are perhaps the most important, because we can act on many of them.

Key Concepts Questions 22.2

1. What is a reference genome?
2. Why are many copies of a genome sequenced?
3. What types of information might be part of a genome annotation?
4. What are limitations of genome sequencing?

22.3 Personal Genome Sequencing

Nearly three decades ago, when "the human genome project" was just an idea, probably the most important word, in hindsight, was "the." Today, with increasing focus on how we differ genetically from each other, the age of personal genome sequencing is here.

Genome sequencing can provide a canvas on which other types of information can be painted, to give a fuller picture of how our bodies function and malfunction. Although genome sequencing is important for investigating our ancestry and our diversity today, it will perhaps be most practically valuable in health care, which will have to adapt to a new type of medical information—and a deluge of it. One analysis considers human genome information a "big data" science, similar to astronomy, YouTube, and Twitter in terms of complexity and the magnitude of data acquisition, storage, distribution, annotation, and analysis.

Practical Medical Matters

The field of human genetics for many years was strictly an academic discipline, a biological science. Until the acceleration of gene discoveries in the 1990s and the sequencing of the first human genomes in the early 2000s, genetics as a medical specialty was a small field, with a few knowledgeable physicians helping families with rare diseases. With the introduction of direct-to-consumer (DTC) genetic testing in 2008, the possibility of testing genes not only for disease-causing mutations, but for variants that indicate only risk, was suddenly available to anyone—without requiring medical expertise. The Food and Drug Administration intervened in 2013 to forbid marketing of tests for genotypes that directly cause disease and for which consumers might attempt to diagnose themselves, but reversed the decision in 2017 to allow carrier testing.

Genetic and genomic testing as part of health care must meet certain practical criteria. The most important requirement for a regulated DNA test or treatment is clinical utility. Does benefit outweigh risk? Is it as accurate or effective as an existing, approved test or treatment? Will it help people who cannot use the existing test or treatment? Efficacy must be demonstrated. For example, molecular evidence may indicate that people with a particular genotype might respond better to a certain drug than people with different genotypes. A clinical trial must evaluate the drug in both groups of people to demonstrate that prescribing-by-genotype—pharmacogenetics—is helpful on the whole-person level. The challenge of validating genetic testing arose with newborn screening (see section 20.2), when clinicians realized the importance of following up initial identification of infants at high risk of developing a genetic disease with definitive diagnostic tests.

A DNA test result alone is not sufficient to diagnose a disease, but may support a clinical diagnosis based on symptoms and the results of other types of tests. For example, a person might be a heterozygote for familial hypercholesterolemia, but is not diagnosed with the condition unless the serum cholesterol level becomes elevated or a cardiovascular condition develops (figure 5.2). However, knowing that a mutation is present can motivate a person to seek further testing. This is the case for the family described in chapter 18 with stomach cancer. Relatives who learned that they had inherited the mutation had scans that revealed tiny tumors—they had cancer already but didn't know it.

The uncertainty in genetic and genomic testing that makes further diagnostic testing necessary arises from the complications of Mendel's laws discussed in chapter 5. A DNA test alone is not sufficient for diagnosis because of

- incomplete penetrance (genotype does not always foretell phenotype),
- variable expressivity (different severities in different individuals),
- epistasis (gene-gene interactions),
- genetic heterogeneity (mutation in more than one gene causing a phenotype), and
- environmental influences (epigenetics).

How will electronic medical records handle the nuances of genomic data? Will records include entire sequences of the DNA bases A, T, C, and G, or just the diagnostic report that includes gene variants classified as pathogenic? How will the records embrace future discoveries that impact the stored data or diagnoses? How will diagnostic codes work? While these matters are under discussion, the medical profession has had to catch up to both the profusion of new genetic tests and genomic technologies, and the fact that many patients are knowledgeable about DNA—in general and sometimes their own. Genomics is becoming incorporated into more medical specialties, and it already is part of oncology and pediatrics. Medical students analyze their own genomes, and physicians are attending continuing medical education programs to learn genomics. Clinical geneticists, genetic counselors, and molecular pathologists are the specialists who are leading the way in the new genomic medicine.

Types of Information in Human Genomes

The human genome sequenced by the Public Consortium was actually composites of different individuals. The first two genomes from specific individuals to be sequenced and the findings published, of genome research pioneers J. Craig Venter and James Watson, yielded few surprises. Instead, they showed that we had greatly underestimated genetic variation by focusing only on the DNA base sequence. The numbers of copies of short sequences—copy number variants, or CNVs—contribute significantly to genetic variation, too.

Venter learned that he has gene variants associated with increased risk of Alzheimer disease and cardiovascular disease. His genome sequence confirmed that he has genotypes that cause or are associated with dry earwax, blue eyes, lactose intolerance, a preference for activities in the evening, and tendencies toward antisocial behavior, novelty seeking, and substance abuse. He metabolizes caffeine fast, which he also knew.

James Watson, according to his genome sequence, carries a dozen rare recessive diseases that would affect glycogen storage, vision, and DNA repair if homozygous, and he is at elevated risk for 20 other diseases. He elected not to learn his status for the *ApoE4* gene variant that increases risk of Alzheimer disease, which a grandparent had, but people inferred the result from the surrounding published parts of his genome (a deduction called imputation that compares the surrounding sequence to that of other people's genomes). He is a slow metabolizer of beta blockers and antipsychotic medications, indicating that he could overdose on normal weight-based dosages of these medications.

The third person to have his genome sequenced was called, simply, "YH." He is Han Chinese, an East Asian population that accounts for 30 percent of modern humanity. He has no inherited diseases in his family, but his genome includes 116 gene variants that cause recessive diseases, as well as many risk alleles. He shares with Craig Venter a tendency to tobacco addiction and high-risk alleles for Alzheimer disease.

An overall comparison of the first three genome sequences of individuals provides a peek at our variation. Each man has about 1.2 million SNPs, but a unique collection. Each has only 0.20 to 0.23 percent of SNPs that are nonsynonymous, meaning that they alter an encoded amino acid, and the men share only 37 percent of these more meaningful SNPs. The math indicates, therefore, that about 0.07 percent of our SNPs may affect our phenotypes.

After the first three human genome sequences were published, others followed from specific ethnic groups. Then came the genomes of people who could afford the initially high cost: journalists who were paid to write about the experience and celebrities. Reasons varied, as they still do. The late Steve Jobs (founder of Apple) and late journalist Christopher Hitchens had their cancer genomes sequenced to guide drug choices. Scholar Henry Louis Gates Jr. had his genome sequenced to trace his African roots. An actress and a rock star reportedly did it to better understand mental illness in their families.

One geneticist had his genome sequenced to serve as a control for a project to sequence the genomes of all the citizens of Qatar. He discovered that he has gene variants for baldness (which he knew from looking in the mirror), a recessive disease that affects children, Viking ancestors, and most important, a blood clotting disease that explained why he bleeds profusely when injured. Despite learning interesting information, the researcher voiced fears: his family learning things they didn't want to know and even someone using his DNA sequence to frame him for a crime.

Do You Want Your Genome Sequenced?

Many of us will face the choice of knowing our own genome sequences (**figure 22.5**). Here are some final thoughts to consider in deciding whether or not, or when, to have your own genome sequenced, and some more general questions:

- Who should decide which gene variants are reported to a patient—the patient, or a health care provider?
- What criteria should a health care provider consider in deciding which genome findings to report to a patient?

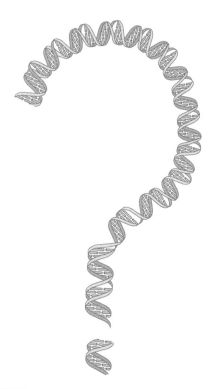

Figure 22.5 Will you have your genome sequenced?

- Should health care providers inform a patient of a gene variant that adds only slightly to the risk of developing a particular multifactorial disease, such as hypertension?
- Should only actionable genome findings be reported to patients?
- How can a health care consumer stay up to date on research results that might be relevant?
- Does a patient have an ethical obligation to share genome results with relatives who may also be at elevated risk of developing an illness?
- Should a parent have a child's genome sequenced?
- Should all newborns have their genomes sequenced?
- Under what circumstances is genetic ancestry information useful?

Other concerns about genome sequencing are societal. Who will pay for genome sequencing? How can the 4,000 genetic counselors and 1,200 clinical geneticists in the United States handle routine genome sequencing and interpretation? Will societal pressure drive people to have their genomes sequenced or will it fail to compete with more traditional health care costs? Could governments compel genome sequencing and could the practice introduce new ways to discriminate against people?

What will the coming flood of genetic and genomic information ultimately mean? Will it tell us where we came from more than family lore and documents? Will physicians consult strings of A, C, T, and G to determine how best to treat their patients, or will signs, symptoms, family history, and a patient's observations turn out to be more valuable types of information? Only time will tell.

While many researchers are busy annotating human genomes, which are highly complex, others are using chemical and microbiological methods to synthesize simple genomes. The effort is part of a field called synthetic biology. **Bioethics** discusses synthetic genomes.

I hope that this book has offered you glimpses of the future and stimulated you to think about the choices that genetic and genomic technologies will present.

Key Concepts Questions 22.3

1. What can personal information from genome sequences add to health care, and what complications might it introduce?
2. What types of information have people learned from having their genomes sequenced?
3. What societal issues does genome sequencing raise?

Should We Create Genomes?

A genome is actually a chemical, and so chemists can synthesize one, then transfer it into a receptive cell whose own genome has been removed. Researchers introduced the first synthetic genome in 2010, and have been updating it in new, smaller versions ever since. They began with the DNA sequence of the tiny genome of *Mycoplasma genitalium* [see A Glimpse of (Pre)History] and whittled it down. The resulting synthetic "hypothetical minimal genome" is about 531,000 DNA base pairs organized into 473 genes. It is called JCVI-syn3.0, or syn3.0 for short. The researchers even stitched their names into the genome using DNA triplets that they assigned to correspond to letters of the alphabet, functioning like a "watermark" on documents to distinguish synthetic life from the natural kind. Syn3.0 "lives" as the genetic headquarters of two species of *Mycoplasma*.

As a short cut, the researchers tested eighths of the *Mycoplasma* genome at a time, using DNA-cutting enzymes to destroy one gene at a time. If a cell can't survive with a specific gene harpooned, then that gene is deemed essential for life. Genes are classified as essential, not essential, or quasi-essential (required for the organism to grow well enough to study, but not to live for very long). **Figure 22B** illustrates the general approach to creating this simplest of synthetic genomes.

The synthetic genome in its host cell is somewhat coddled, given all the small molecules it could possibly require in the laboratory, compared to the natural niche of *Mycoplasma* in goats. The researchers chose *Mycoplasma* as the model because their natural hosts supply nearly all nutrients, enabling them to survive with minimal genomes.

Nearly half of the genes in syn3.0 are involved in making proteins, with many contributing to cell membrane structure and function. Most intriguing and unexpected was that the functions of nearly a third of the genes aren't known. Many genes, however, are highly conserved (found in other organisms), suggesting that they are essential.

Creating a synthetic genome may provide clues to how life began, as complex collections of self-replicating and changeable chemicals coalesced and linked, perhaps using clays or minerals as templates, then knitting themselves fatty protective coverings. In a practical sense, the ability to design a simple microorganism may ultimately offer new energy sources and ways to develop new drugs and vaccines.

Questions for Discussion

1. Is a synthetic genome really creating life, or is it more like a hermit crab taking up residence in an abandoned shell?

2. Do we know enough to use synthetic life technology to create cells that can improve the world?

3. How might the ability to create genomes be misused?

4. Should the synthetic life research community police itself, like the pioneers of recombinant DNA technology did in establishing the containment procedures that persist today?

5. Compare and contrast possible dangers of synthesizing genomes and using genome editing to fashion gene drives that make species extinct (see figure 19.12).

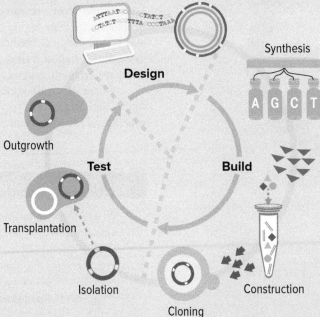

Figure 22B   A synthetic minimal bacterial genome. "Syn3.0" consists of about 531,000 DNA base pairs. It was synthesized in the laboratory from containers holding each of the four types of DNA bases, then mass-produced (cloned) inside yeast cells, and finally individual pieces were isolated and specific genes destroyed, one at a time. Then the structure was transplanted into cells to test "gene essentiality." *From C. Hutchison III et al, Science 351:6280 [2016]*

Summary

22.1 From Genetics to Genomics

1. Genetic maps increase in detail and resolution, from cytogenetic and linkage maps to physical and sequence maps.

2. Positional cloning was used to discover individual genes by beginning with a phenotype and gradually identifying a causative gene, localizing it to part of a chromosome.

3. The Human Genome Project began in 1990 under the direction of the DOE and NIH. Technological advances sped the sequencing.

4. Several copies of a genome are cut and the pieces sequenced, overlapped, and aligned to derive the continuous sequence. For the human genome, the International Consortium used a chromosome-by-chromosome approach and Celera Genomics used whole-genome shotgunning.

22.2 Analysis of Human Genome Content

5. After sequencing genomes became possible, attention turned to refining the process, cataloging human variation, and discovering gene functions.

6. The cost of sequencing decreased while the speed increased, as researchers annotated genes with information on gene function, mode of inheritance, genotype, and frequency of variants.

7. Genome sequencing does not detect copy number variants, mitochondrial DNA, uniparental disomy, or gene-gene and gene-environment interactions.

22.3 Personal Genome Sequencing

8. Genome information used in medical tests and treatments must be validated for clinical utility (be novel or at least as useful as an existing test or treatment).

9. The nuances of Mendel's laws discussed in section 5.1 affect the interpretation of genome sequence data.

10. An individual human genome sequence can confirm what is known from family history, detect disease-associated recessive alleles, detect gene variants that contribute to risk of having or developing a trait or illness, and predict drug responses.

11. Widespread availability of genome sequencing will raise personal as well as societal questions.

Review Questions

1. Explain why the phrase "the human genome project" is inaccurate.

2. How did family linkage patterns and chromosomal aberrations lay the groundwork for sequencing the first human genomes?

3. Describe how the four levels of genetic maps differ, and what the newer types of maps depict.

4. Explain how positional cloning was used to identify disease-causing genes.

5. Why did researchers initially disagree over whether to sequence human genomes?

6. Why is it important to use many copies of a genome when deriving the sequence?

7. Explain why a gene variant that is present at a high frequency in a population is less likely to be harmful than a rare variant in that population.

8. Why is it helpful to include microbiome information in a genome annotation document?

9. Give an example of evolutionary information deduced from comparative genomics.

10. List three types of DNA sequence variation that genome sequencing cannot detect.

11. List three criteria for genome information to have clinical utility.

12. Give examples of types of information that people can learn from knowing their genome sequences.

13. Why is it possible to synthesize a genome?

Applied Questions

1. Suggest a way that the Genomic Postmortem Research Project could help living people.

2. What are uses and limitations of making a clinical decision based on a reference human genome sequence?

3. How should genetic or genomic information be incorporated into clinical diagnosis? Discuss the significance of the differences between Mendelian and multifactorial traits and between the effects of penetrance and epistasis.

4. Compare and contrast DNA testing to an X ray or other standard medical diagnostic test.

5. Some researchers are suggesting that to cut costs and make genome technology available to more people, testing should only look for genes most likely to be mutant in a particular person. What is an advantage and a disadvantage of this strategy?

6. Explain why protein-encoding DNA sequences that are identical in nearly everyone tested are interpreted to be essential to survival.

7. Select a person from table 22.3 and describe health risks to the individual and to her or his offspring.

8. Discuss a medical situation, perhaps in your own family or from this book, that might benefit from genome sequencing.

9. What difficulties would be encountered in trying to synthesize a human genome?

Case Studies and Research Results

1. The Exome Aggregation Consortium (ExAC) sequenced the exomes of 60,706 people with diverse geographic ancestries, coming from Latino, European, African, South Asian, and East Asian populations. The consortium identified about 7.4 million genetic variants. If a human genome is 3.2 billion bases, with the exome 1.5 percent of that, explain how so many variants are possible.

2. After the tsunami that devastated Japan in 2011, many types of fish never before seen washed up on shore, thrown from the deep sea; some were bizarre in appearance. Researchers collected specimens and sequenced DNA to try to classify the animals. Consider the following 8-base sequence that is similar among the species:

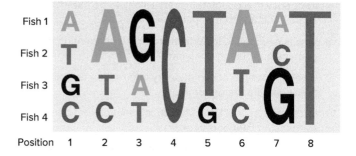

a. Write the DNA sequences for the two most closely related fishes.

b. Which position(s) in the sequence are highly conserved?

c. Which position(s) in the sequence are the least conserved?

d. Which site is probably not essential, and how do you know this?

e. A coelacanth has C T A C T G G T for this section of the genome. Which of the mystery fishes is the coelacanth's closest relative? (A coelacanth is an ancient fish thought to have become extinct, but every few decades one is discovered.)

3. Some researchers have claimed that healthy lifestyle habits (not smoking, following a healthy diet, exercising regularly, having regular wellness exams, and diagnosing and treating common diseases early) would save more lives than sequencing everyone's genome and predicting diseases from the information. Suggest an experiment to evaluate these two approaches to staying healthy.

4. From an analysis of 10,000 human genomes, it was found that, on average, Europeans had 7,214 novel gene variants, admixed individuals had 10,978 novel gene variants, and people of African ancestry had 13,530 novel gene variants. Draw a conclusion about human evolution from this information.

5. Sequencing the genomes of 2,636 people in Iceland to an average depth of 20-fold revealed 20 million SNPs, 1.5 million "indels" (insertions and/or deletions), and more homozygosity than seen in many other populations.

a. What does "depth" refer to and why is it important to the accuracy of findings?

b. How do SNPs differ from indels?

c. What is the significance of the increased homozygosity seen among the Icelanders compared to other human populations?

Glossary

acrocentric (ăk′rō-sĕn′trĭk) A chromosome in which the centromere is near one end.

adaptive immune response (ĭ-myōō′n) A slow, specific type of immune response following exposure to a foreign antigen.

adenine (ăd′en-ēn′) One of two purine nitrogenous bases in DNA and RNA.

admixture New combination of gene variants that arises when individuals from two previously distinct populations have children together.

affected sibling pair study A gene identification approach that looks for SNPs that siblings with a particular trait share, but that siblings who do not share the trait do not share.

allele (ə-lēl′) An alternate (variant) form of a gene.

allelic diseases (ə-lēl ik) Different diseases caused by mutations in the same gene.

allergen (al′er-jen) A substance that can provoke an allergic response.

allergy (al′er-jē) A response of the immune system to a substance that is not dangerous.

alternate splicing Building different proteins by combining exons of a gene in different ways.

amino acid (ə-mē′nō) A small organic molecule that is a protein building block.

amniocentesis (ăm′nē-ō-sĕn-tē′sĭs) A prenatal diagnostic technique that examines fetal chromosomes and biochemicals in amniotic fluid.

anaphase (ăn-ə-fāz′) Stage of mitosis when the centromeres of replicated chromosomes part.

aneuploid (ăn′yū-ploid′) A cell with one or more extra or missing chromosomes.

angiogenesis (an″je-o-jen′ə-sis) Extension of blood vessels.

antibody (ăn′tē-bŏd′ē) A multisubunit protein, produced by B cells, that binds a specific foreign antigen, alerting the immune system or destroying the antigen.

anticodon (ăntē-kō′dŏn) A three-base sequence on one loop of a transfer RNA molecule that is complementary to an mRNA codon, and joins the appropriate amino acid and its mRNA.

antigen (ăn′tē-jən) A molecule that elicits an immune response.

antigen binding sites (ăn′tē-jən) Specialized ends of antibody chains.

antigen-presenting cell (ăn′tē-jən) A cell displaying a foreign antigen.

antiparallelism The head-to-toe orientation of the two nucleotide chains of the DNA double helix.

antisense technology A form of gene silencing that blocks expression of a specific gene using an RNA or similar molecule that is complementary to the gene's mRNA transcript.

apoptosis (āpō-tō′sis) A form of cell death that is a normal part of growth and development.

assisted reproductive technologies (ARTs) Procedures that replace a gamete or provide a uterus to help people with fertility problems have children.

autoantibodies (ô′tō-ăn′tē-bŏdēz) Antibodies that attack the body's own cells.

autoimmunity (ô′tō-ĭ-myōō′nĭ-tē) An immune system attack against the body's own cells.

autophagy (au-toph′ -a-gy) A process in which a cell dismantles its own debris.

autosomal (ôtə-sōməl) **dominant** The mode of inheritance in which one autosomal allele causes a phenotype. Such a trait can affect males and females and does not skip generations.

autosomal (ôtə-sōməl) **recessive** The mode of inheritance in which two autosomal alleles are required to cause a phenotype. Such a trait can affect males and females and can skip generations.

autosome (ôtə-sōm) A chromosome that does not have any genes that determine sex.

B cell A type of lymphocyte that secretes antibody proteins in response to nonself antigens displayed on other immune system cells.

balanced polymorphism (pŏl′ē-môr′fizəm) Maintenance of a harmful recessive allele in a population because the heterozygote has a reproductive advantage.

base excision repair Replacement of one to five contiguous DNA nucleotides to correct oxidative damage.

bioethics A field that addresses personal issues that arise in applying medical technology and genetic information.

bioremediation Use of plants or microorganisms to detoxify environmental pollutants.

biotechnology The alteration of cells or biochemicals for a specific application.

blastocyst (blăs′tō-sĭst′) A fluid-filled ball of cells descended from a fertilized ovum.

blastomere (blăstō′-mēr′) A cell of a blastocyst.

cancer (kăn′sər) A group of disorders resulting from loss of cell cycle control.

cancer stem cells Stem cells that divide and yield cancer cells and abnormal specialized cells.

carbohydrate (kar″bo-hī′-drăt) An organic compound that consists of carbon, hydrogen, and oxygen in a 1:2:1 ratio. Can be a sugar or a starch.

carcinogen (kar-sĭnə-jən) A substance that causes cancer.

cell (sel) The fundamental unit of life.

cell (sel) **cycle** A cycle of events describing a cell's preparation for division and division itself.

cell-free fetal DNA Small pieces of fetal DNA in a woman's bloodstream used for testing for genetic disease.

cellular adhesion A precise series of interactions among the proteins that connect cells.

cellular immune response (ĭ-myōōn′) Process in which T cells release cytokines that stimulate and coordinate an immune response.

centriole (sĕn′-trē-ohl) A structure in cells that organizes microtubules into the mitotic spindle.

centromere (sĕn′-trō-mîr) The largest constriction in a chromosome, located at a specific site in each chromosome type.

centrosome (sĕn′-trō-sōm) A structure built of centrioles and proteins that organizes microtubules into a spindle during cell division.

chaperone protein A protein that binds a polypeptide and guides folding.

chimeric antigen receptor technology (ki-meer-ik ăn′tē-jən re″sep′tor) A cancer treatment that combines DNA instructions for a T cell receptor with those for an antibody, creating a cell surface molecule that guides T cells to cancer cells, alerting the immune system.

chorionic villus sampling (CVS) (kôrē-ŏn′ĭk vĭl-us) A prenatal diagnostic technique that analyzes chromosomes in chorionic villus cells, which, like the fetus, descend from the fertilized ovum.

chromatid (krō′ mə-tĭd) A single, very long DNA molecule and its associated proteins, forming a longitudinal half of a replicated chromosome.

chromatin (krō′ mə-tĭn) DNA and its associated proteins.

chromatin remodeling (krō′ mə-tĭn) Adding or removing chemical groups to or from histones, which can alter gene expression.

chromosomal microarray analysis A technique that detects small deletions and copy number variants.

chromosome (krō′ mə-sōm′) A highly wound continuous molecule of DNA and the proteins associated with it.

chromothripsis A rare event that shatters chromosomes.

cleavage (klēvĭj) A series of rapid mitotic cell divisions after fertilization.

cline (klīn) An allele frequency that changes from one geographical area to a neighboring one.

cloning vector A piece of DNA used to transfer DNA from a cell of one organism into the cell of another.

clustered regularly interspaced short palindromic repeats (CRISPRs) A genome editing tool that uses bacterial DNA and RNA to direct double-stranded cuts in target DNA.

coding strand The strand of the DNA double helix for a particular gene from which RNA is not transcribed.

codominant A form of gene expression in which both alleles are fully expressed in a heterozygote.

codon (kō′dŏn) A continuous triplet of mRNA that specifies a particular amino acid.

coefficient of relatedness The proportion of genes that two people related in a certain way share.

collectins (ko-lek′tinz) Immune system molecules that detect viruses, bacteria, and yeasts.

comparative genomic hybridization (CGH) A technique using fluorescent labels to detect small copy number variants.

comparative genomics (jə-nō′mĭks) Identifying conserved DNA sequences among genomes of different species.

complement Plasma proteins that have a variety of immune functions.

complementary base pairs The pairs of DNA bases that hydrogen bond; adenine bonds to thymine and guanine to cytosine.

complementary DNA (cDNA) A DNA molecule that is the complement of an mRNA, copied using reverse transcriptase.

compound heterozygote An individual with two different recessive alleles in the same gene.

concordance (kən-kôr′dens) A measure indicating the degree to which a trait is inherited; percentage of twin pairs in which both members express a trait.

conditional mutation (myōō-tā′shən) A genotype that is expressed only under certain environmental conditions.

conformation The three-dimensional shape of a molecule.

consanguinity (kŏnsăn-gwĭn′ĭ-tē) Blood relatives having children together.

copy number variant (CNV) A DNA sequence present in different numbers of copies (repeats) in different individuals.

CRISPR A genome editing technology that uses a bacterial enzyme and RNA to make double-stranded breaks at selected sites in a genome.

critical period The time during prenatal development when a structure is sensitive to damage from a mutation or an environmental intervention.

crossing over An event during prophase I when homologs exchange parts.

cytogenetics (sītō-jə-nĕt′ĭks) Matching phenotypes to detectable chromosomal abnormalities.

cytokine (sītō-kīn′) A biochemical that a T cell secretes that controls immune function.

cytokinesis (sī-tō-kin-ē′-sis) Division of the cytoplasm and its contents.

cytoplasm (sī′-tō-plăzm) Cellular contents other than organelles.

cytosine (sī′-tō-sēn) One of the two pyrimidine nitrogenous bases in DNA and RNA.

cytoskeleton (sī-tō-skĕl′ĭ-tn) A framework of protein tubules and rods that supports the cell and gives it a distinctive form.

dedifferentiated A cell less specialized than the cell it descended from. A characteristic of a cancer cell.

deletion A mutation that removes part of a DNA sequence or part of a chromosome.

dendritic cells Immune system cells that reside in places where the body contacts the environment, serving as antigen-presenting cells to stimulate an immune response.

deoxyribonucleic (dē-ŏksē-rībō-nōō-klā′ĭk) **acid (DNA)** The genetic material; the biochemical that forms genes.

deoxyribose (dē-ŏksē-rī′bōs) A 5-carbon sugar that is part of a DNA nucleotide.

differentiation Cell specialization, reflecting differential gene expression.

dihybrid cross Breeding individuals heterozygous for two traits.

diploid (dĭp′ loid) A cell containing two sets of chromosomes.

dizygotic (dīzī-gŏt′ĭk) **(DZ) twins** Twins that originate as two fertilized ova; fraternal twins.

DNA *See* **deoxyribonucleic acid.**

DNA microarray A set of target genes embedded in or attached to a glass or plastic chip, to which labeled cDNA pieces from a clinical sample bind and fluoresce. Microarrays show patterns of gene expression and gene variants present.

DNA polymerase (DNAP) (pə-lim′ər-ās) An enzyme that adds new bases to replicating DNA and corrects mismatched base pairs.

DNA probe A labeled short sequence of DNA that binds its complement in a biological sample.

DNA profiling A biotechnology that detects differences in the number of copies of certain DNA repeats among individuals. Used to rule out or establish identity.

DNA replication Construction of a new DNA double helix using the information in parental strands as a template.

dominant A gene variant expressed when present in one copy.

driver mutation A mutation that provides the selective growth advantage of a cancer cell.

duplication An extra copy of a DNA sequence, usually caused by misaligned pairing in meiosis.

ectoderm (ĕktō-dûrm) The outermost primary germ layer of the primordial embryo.

embryo (ĕm′brē - ō′) In humans, prenatal development until the end of the eighth week, when all basic structures are present.

embryonic (ĕmbrē-ŏn′ĭk) **stem (ES) cell** A cell derived in laboratory culture from inner cell mass cells of very early embryos that can self-renew, and some of its daughters differentiate as any cell type.

empiric risk Probability that a trait will recur based upon its incidence in a population.

endoderm (ĕn′-dō-dûrm) The innermost primary germ layer of the primordial embryo.

endoplasmic reticulum (ĕn′-dō-plaz-mik rĕ-tik′-u-lum) An organelle consisting of a labyrinth of membranous tubules on which proteins, lipids, and sugars are synthesized.

endosome A vesicle that buds inward from the plasma membrane.

enzyme (ĕnzīm) A type of protein that speeds the rate of a specific biochemical reaction.

enzyme replacement therapy Infusion of recombinant human enzyme to compensate for deficient or absent enzyme.

epigenetic (ĕpē-jə-nĕt′ĭk) Any effect that modifies gene expression without changing the DNA sequence, such as methylation.

epistasis (ĕpē-stā-sis) A gene masking or affecting the expression of another gene.

epitope (ep′ĭ-tōp) Part of an antigen that an idiotype of an antibody binds.

equational division The second meiotic division, producing four cells from two.

euchromatin (yōō-krō′mə-tĭn′) Parts of chromosomes that do not stain and that contain active genes.

eugenics (yōō-jĕn′ĭks) The control of individual reproductive choices to achieve a societal goal.

eukaryotic cell (yōō-kar′ē-ŏt′ĭk sel) A complex cell containing organelles, including a nucleus.

euploid (yōō′-ploid) A somatic cell with the normal number of chromosomes for the species.

ex vivo **gene therapy** Replacing mutant genes in cells growing in the laboratory and then introducing the cells into a patient.

excision repair Enzyme-catalyzed removal of pyrimidine dimers in DNA.

exome (x-ōm) The approximately 1.5 percent of the genome that encodes proteins.

exon (x-on) A part of a gene and the resulting mRNA that encodes an amino acid.

exon skipping A protein that is missing contiguous amino acids because a missense mutation creates an intron splicing site.

exosome (x-ō-sōm) A vesicle that carries molecules from cell to cell.

expanding repeat A short DNA sequence that is present in a certain range of copy numbers in wild type individuals but, when expanded, causes a disease phenotype.

expressivity Degree of severity of a phenotype.

fetus (fē′təs) The prenatal human after the eighth week of development, when structures grow and specialize.

founder effect An accelerator of genetic drift in which a few individuals found a new settlement, perpetuating a subset of alleles from the original population.

frameshift mutation (myōō-tā′shən) A mutation that alters a gene's reading frame.

fusion protein A protein that forms from translation of transcripts from two genes.

G$_0$ phase An offshoot of the cell cycle in which the cell remains specialized but does not replicate its DNA or divide.

G$_1$ phase The stage of the cell cycle following mitosis in which the cell resumes synthesis of proteins, lipids, and carbohydrates.

G$_2$ phase The stage of the cell cycle following S phase but before mitosis, when certain proteins are synthesized.

gamete (găm′ĕt) A sex cell.

gamete intrafallopian transfer (GIFT) (găm′ĕt ĭntra-fə-lōpē-ən) An infertility treatment in which sperm and oocytes are placed in a woman's uterine tube.

gastrula (găstrə-lə) A three-layered embryo.

gene (jēn) A section of a DNA molecule whose sequence of building blocks instructs a cell to produce a particular protein.

gene drive An application of genome editing that can eliminate transmission of a selected gene variant, being used to eliminate certain pathogens.

gene expression Transcription of a gene's DNA into RNA.

gene expression profiling Use of DNA microarrays to detect the types and amounts of cDNAs reverse transcribed from the mRNAs in a particular cell source.

gene flow Movement of alleles between populations when individuals migrate and mate.

gene pool All the alleles in a population.

gene silencing Blocking transcription of or degrading mRNA, preventing translation into protein.

gene therapy Introducing a functioning gene to compensate for the effects of a mutation.

genetic code (jə-nĕt′ĭk) The correspondence between specific mRNA triplets and the amino acids they specify.

genetic counselor A medical specialist who calculates risk of recurrence of inherited disorders in families, applying the laws of inheritance to pedigrees and interpreting genetic test results.

genetic determinism Attributing a trait to a gene or genes.

genetic drift Changes in allele frequencies from generation to generation in a population, due to chance.

genetic heterogeneity A phenotype that can be caused by variants of any of several genes.

genetic load The collection of deleterious recessive alleles in a population.

genetic marker A DNA sequence near a gene of interest that is co-inherited unless separated by a crossover.

genetics The study of inherited traits and their variation.

genome (jē′nōm) The complete set of genetic instructions in the cells of a particular type of organism.

genome editing Creating double-stranded breaks in the DNA double helix, enabling insertion or removal of a specific DNA sequence.

genome-wide association study (GWAS) A study in which millions of variants (single nucleotide polymorphisms or copy number variants) that form haplotypes are compared between people with a condition and unaffected individuals to identify parts of the genome that might contribute to a phenotype.

genomic imprinting (jē nō′m ĭk) The covering of a gene or several linked genes by methyl groups, which prevents expression and affects the phenotype, depending upon which parent transmits a particular allele.

genomics (jē nōm ĭks) The field that analyzes and compares genomes.

genotype (jē n′ə- tīp) The allele combinations in an individual that cause traits or diseases.

genotypic (jēn′ə- tīp′ĭk) **ratio** The ratio of genotype classes expected in the progeny of a cross.

germ cells Sperm and oocytes.

germline gene therapy Genetic alteration of a gamete or fertilized ovum, which perpetuates the change throughout the organism and transmits it to future generations.

germline mutation A mutation that is in every cell in an individual because it was present in the fertilized ovum.

Golgi (gōl′jē) **apparatus** An organelle, consisting of flattened, membranous sacs, that packages secretion components.

gonadal mosaicism Having two or more genetically distinct cell populations in an ovary or testis.

gonads (gō′-nadz) Paired structures in the reproductive system where sperm or oocytes are manufactured.

growth factor A protein that stimulates mitosis.

guanine (gwa′ nēn) One of the two purine nitrogenous bases in DNA and RNA.

haplogroup (hăp′ lō-groōp) An extensive set of markers on a chromosome that are inherited together and therefore can be used to trace ancestry.

haploid (hăp′ loid) A cell with one set of chromosomes.

haplotype (hăp′ lō tīp) A series of known DNA sequences or single nucleotide polymorphisms linked on a chromosome, less extensive than a haplogroup.

Hardy-Weinberg equilibrium An idealized state in which allele frequencies in a population do not change from generation to generation.

heavy chains The two longer polypeptide chains of an antibody subunit.

hemizygous (hĕm′ ē-zī′ gəs) The sex that has half as many X-linked genes as the other; a human male is hemizygous.

heredity Transmission of inherited traits from generation to generation.

heritability An estimate of the proportion of phenotypic variation in a population that is due to genetic differences.

heterochromatin (hĕtə-rō-krō′mətĭn) Dark-staining chromosome parts that have few protein-encoding genes.

heterogametic (hĕt′ə-rō-gə-mē′tĭk) The sex with two different sex chromosomes; a human male.

heteroplasmy (hĕt′ə-rō-plăz-mē) Mitochondria in the same cell having different alleles of a particular gene.

heterozygous (hĕtə-rō-zī′gəs) Having two different alleles of a gene.

histone (hĭs′tōn) A type of protein around which DNA coils in a regular pattern.

hominins (hŏm′ə-nĭnz) Animals ancestral to humans only.

homogametic (hō′mō-gə-mē′tĭk) The sex with two of the same sex chromosomes; a human female.

homologous (hō-mŏl′ə-gəs) **pair** Two chromosomes with the same gene sequence.

homozygosity mapping (hōmō-zī-gəs′-ĭ-tē) An approach to gene discovery that correlates stretches of homozygous DNA base sequences in the genomes of related individuals to certain traits or disorders.

homozygous (hōmō-zī′ gəs) Having two identical alleles of a gene.

hormone (hor′ mōn) A biochemical produced in a gland and carried in the blood to a target organ, where it exerts an effect.

human leukocyte antigens (HLAs) (loōkə-sīt′ ăn′tĭ-jən) Cell surface proteins that are important in immune system function and are encoded by closely linked genes on the short arm of chromosome 6.

humoral immune response (yoō′ mər-əl) Process in which B cells secrete antibodies into the bloodstream.

identical by descent Linked DNA sequences that are transmitted together over generations.

idiotype (id′ēo-tīp) Part of an antibody molecule that binds an antigen.

in vitro **fertilization (IVF)** (ĭn vē′trō) Placing oocytes and sperm in a laboratory dish with appropriate biochemicals so that fertilization occurs, then, after a few cell divisions, transferring the embryo(s) to a woman's uterus.

in vivo **gene therapy** Introduction of vectors carrying therapeutic human genes directly into the body part where they will act.

incidence The number of new cases of a disease during a certain time in a population.

incomplete dominance A form of gene expression that results in a heterozygote intermediate in phenotype between either homozygote.

indels Insertions and deletions of genetic material in a genome.

independent assortment Inheritance of a gene on one chromosome does not influence inheritance of a gene on a different chromosome (Mendel's second law) because of the random arrangement of homologous chromosome pairs, in terms of maternal and paternal origin, down the center of a cell in metaphase 1.

induced pluripotent stem (iPS) cells Somatic cells that are reprogrammed toward an alternative developmental fate by altering their gene expression.

infertility The inability to conceive a child after a year of frequent unprotected intercourse.

inflammation Part of the innate immune response that causes an infected or injured area to swell with fluid, turn red, and attract phagocytes.

innate immune response (ĭ-myoō′n) Components of the immune response that are present at birth and do not require exposure to an environmental stimulus.

inner cell mass A clump of cells on the inside of the blastocyst that will continue developing into an embryo.

insertion mutation (myoō-tā′-shən) A mutation that adds DNA bases.

insertional translocation A rare type of translocation in which a part of one chromosome is part of a nonhomologous chromosome.

interferon (in″tər-fēr′on) A type of cytokine.

interleukin (in″tər-loo′kin) A type of cytokine.

intermediate filament A type of cytoskeletal component made of different proteins in different cell types.

interphase (ĭn′tər-fāz′) Stage of the cell cycle when a cell is not dividing.

intracytoplasmic sperm injection (ICSI) (ĭn′trə-sītō-plăzmĭk) An infertility treatment that injects a sperm cell nucleus into an oocyte, to overcome poor or absent sperm motility.

intrauterine insemination (IUI) (ĭn′trə-yōō′tər-ĭn) An infertility treatment that places donor sperm in the cervix or uterus.

introgression Entry of a specific gene variant into a genome from an individual of a different species or subspecies.

intron (ĭn trŏn) Part of a gene that is transcribed but is excised from the mRNA before translation into protein.

invasiveness The tendency of cancer cells to squeeze into surrounding spaces.

isochromosome A chromosome that has two copies of one arm but none of the other, as a result of cell division along the wrong plane.

isoform An alternate version of a protein that arises from a certain combination of translated exons.

karyotype (kă′rē-ō-tīp) A size-order chart of chromosome pairs.

lethal allele (ə-lēl′) An allele that causes death before reproductive maturity or halts prenatal development.

ligase (lī′gās) An enzyme that catalyzes the formation of covalent bonds in the sugar-phosphate backbone of a nucleic acid.

light chains The two shorter polypeptide chains of an antibody subunit.

linkage Genes on the same chromosome.

linkage disequilibrium (LD) A consequence of extremely tight linkage between DNA sequences in which two genes or DNA sequences are very often inherited together.

linkage maps Diagrams that show gene order on chromosomes, determined from crossover frequencies between pairs of genes.

lipid (lĭp′ĭd) A type of organic molecule that has more carbon and hydrogen atoms than oxygen atoms. Can be a fat or an oil.

long noncoding RNA An RNA molecule that associates with chromatin in the cell nucleus and likely helps control gene expression, especially in the brain.

lysosome (lī′sō-sōm) A saclike organelle containing enzymes that degrade debris.

macroevolution (măk′rō-ĕv′ə-lōōshən) Genetic change sufficient to form a new species.

macrophage A large, wandering cell that engulfs and digests debris, including cells, and presents nonself antigens on its surface, alerting other immune system cells.

manifesting heterozygote (hĕt′ə-rō-z īgōt) A carrier of an X-linked recessive trait who expresses the phenotype because the normal allele is inactivated in some tissues.

mast cells Cells that release histamine, which dilates blood vessels near infected or injured areas. Misdirected mast cell activity causes allergic reactions.

meiosis (mī-ō′sĭs) A form of cell division that halves the number of chromosomes, forming haploid gametes.

memory cells B or T cell descendants that carry out a secondary immune response.

mesoderm (mĕz-ō-dûrm) The middle primary germ layer.

messenger RNA (mRNA) A molecule of RNA complementary in sequence to the template strand of a gene that specifies a protein product.

metacentric (mĕta-sĕn′trĭk) A chromosome with a centromere that divides it into two arms of approximately equal length.

metagenomics Sequencing all of the genomes present in a sample of a particular environment.

metaphase (mĕta-fāz) The stage of mitosis when chromosomes align along the center of the cell.

metastasis (mĕta-stā′-sĭs) Spread of cancer from its site of origin to other parts of the body.

microbiome (mahy-kroh-bahy′-ohm) All of the organisms that live in and on another organism.

microdeletion A very small chromosomal deletion (missing part); a type of copy number variant.

microduplication A very small second copy of a DNA sequence; a type of copy number variant.

microevolution Change of allele frequency in a population.

microfilament A solid rod of actin protein that forms part of the cytoskeleton.

microRNA A 21- or 22-base-long RNA that binds to certain mRNAs, blocking their translation into protein.

microtubule (mīkrō-tōōbyōōl) A hollow structure built of tubulin protein that forms part of the cytoskeleton.

mismatch repair Proofreading of DNA for misalignment of short, repeated segments.

missense mutation (mĭs′sĕns) A single base change (point mutation) that alters a codon so that it specifies a different amino acid.

mitochondrion (mītō-kŏn′drē-ən) An organelle consisting of a double membrane that houses enzymes that catalyze reactions that extract energy from nutrients.

mitosis (mī-tōsĭs) Division of somatic (non-sex) cells.

mode of inheritance The pattern in which a gene variant passes from generation to generation. It may be dominant or recessive, autosomal, or X- or Y-linked.

molecular evolution Analysis of changes in genomes, protein and DNA sequences, and chromosome banding patterns over time to estimate how recently species diverged from a shared ancestor.

monohybrid (mŏn′ō-hībrĭd) **cross** A cross of two individuals who are heterozygous for a single trait.

monosomy (mŏn′ō-sō′mē) An abnormality in which a person's cells have forty-five (one missing) chromosomes.

monozygotic (mŏnō-zī-gŏt′ĭk) **(MZ) twins** Twins that originate as a single fertilized ovum; identical twins.

morpholinos Synthetic molecules that resemble 25 linked DNA nucleotides that block splice-site mutations, restoring deleted exons and function in some genes.

morula (môr′ yə-lə) The very early prenatal stage that resembles a mulberry.

multifactorial trait A characteristic determined by the actions of one or more genes and the environment.

mutagen (myōō′tə-jən) Something that causes mutation by changing, adding, or deleting a DNA base. Certain chemicals and radiation are mutagens.

mutant (myōōt′nt) An allele that differs from the normal or most common allele in a population and alters the phenotype.

mutation (myōō-tā′shən) A change in a DNA sequence that affects the phenotype and is rare in a population.

natural selection Differential survival and reproduction of individuals with particular phenotypes in particular environments, which may alter allele frequencies in subsequent generations.

neural (nōōr′əl) **tube** A structure in the embryo that develops into the brain and spinal cord.

neurexin A protein in the presynaptic membrane of a neuron that uses the neurotransmitter glutamate; when misfolded, it is involved in autism.

neuroglia Several types of cells in the nervous system that support and nurture neurons.

neuroligin A protein in the postsynaptic membrane of a neuron that uses the neurotransmitter glutamate; when misfolded, it is involved in autism.

neuron (nōrin′) A nerve cell.

neurotransmitter A molecule that transmits messages in the nervous system across synapses.

nitrogenous (nī-trŏj′ə-nəs) **bases** The nitrogen-containing bases that are the information-carrying part of DNA or RNA.

nondisjunction (nŏndĭs-jŭngk′shən) The unequal partitioning of chromosomes into gametes during meiosis.

noninvasive prenatal testing A general term that includes ultrasound imaging and testing cell-free fetal DNA in maternal plasma.

nonsense mutation (myōōtā′shən) A point mutation that changes an amino-acid-coding codon into a stop codon, prematurely terminating synthesis of the encoded protein.

nonsense suppression Action of a drug that enables protein synthesis to ignore a nonsense mutation (stop codon not at the end of a gene).

nonsense-mediated decay A response that destroys mRNAs in which nonsense mutations encode shortened proteins that could have toxic effects on the cell.

nonsynonymous codon (kō′don) A codon that encodes a different amino acid from another codon.

nucleic (nōō-klē′ĭk) **acid** DNA or RNA.

nucleolus (nōō-klē′ə-ləs) A structure in the nucleus where ribosomes are assembled from ribosomal RNAs and proteins.

nucleosome (nōō′-klē-ō-sōm) A unit of chromatin structure consisting of DNA coiled around a histone.

nucleotide (nōō-klēō-tīde) The building block of a nucleic acid, consisting of a phosphate group, a nitrogenous base, and a 5-carbon sugar.

nucleotide excision repair (nōō-klēō-tīd) Replacement of up to thirty nucleotides to correct DNA damage of several types.

nucleus (nōō-klēəs) A large, membrane-bounded region of a eukaryotic cell that houses DNA.

oncogene (ŏn′kə-jēn) A gene that normally controls the cell cycle, but causes cancer when overexpressed.

oocyte (ō′ə-sīt) The female gamete (sex cell). An egg.

oogenesis (ōə-jĕn′ĭ-sĭs) Oocyte (egg) formation.

oogonium (o-o-go′-ni-um) A cell in the ovary that gives rise to an oocyte, in meiosis.

open reading frame A sequence of DNA that does not include a stop codon.

organelle (ŏr′gə-nĕl′) A specialized structure in a eukaryotic cell that carries out a specific function.

ovaries (o′və-rēz) The female gonads.

paracentric inversion (para sĕn′-trīk) A chromosomal inversion in which the inverted section does not include the centromere.

passenger mutation A mutation in a cancer cell that does not provide selective growth advantage. It doesn't cause cancer.

pedigree A chart of symbols connected by lines that depicts the genetic relationships of and transmission of inherited traits in related individuals.

penetrance Percentage of individuals with a genotype who have an associated phenotype.

pericentric inversion (pər-ē sĕn-trik) A chromosomal inversion in which the inverted part includes the centromere.

peroxisome (pə-rŏk′sĭ-sōm) An organelle with a single outer membrane that is studded with proteins and that houses enzymes with various functions.

pharmacogenetics (farm a kō jə-nĕt-ĭks) Testing for a single gene variant that affects metabolism of a specific drug.

pharmacogenomics (farm a kō jə-nōm-ĭks) Testing for variants of many genes or gene expression patterns that affect metabolism of a specific drug.

pharmacological chaperone therapy Use of an oral drug that binds to a misfolded enzyme, stabilizing it sufficiently to permit some level of function.

phenocopy (fē′ nō-kŏp′ē) An environmentally caused trait that occurs in a familial pattern, mimicking inheritance.

phenotype (fē′ nō-tīp) The expression of a gene in traits or symptoms.

plasma cell A cell descended from a B cell that produces abundant antibodies of a single type.

plasma membrane (plăz′mə mĕm′brăn) The selective barrier around a cell, consisting of proteins, glycolipids, glycoproteins, and lipids on or in a phospholipid bilayer.

plasmid (plăz′ mĭd) A small circle of double-stranded DNA found in some bacteria, yeasts, and plant cells. Used as a vector in recombinant DNA technology.

pleiotropic (plēə-trōpĭk) A single-gene disease with several symptoms or a gene that controls several functions or has more than one effect, causing different symptom subsets in different individuals.

point mutation (myōō-tā′ shən) A single base change in DNA.

polar body A product of female meiosis that contains little cytoplasm and does not continue to develop into an oocyte.

polygenic (pŏlē-jēn′ ĭk) A trait determined by more than one gene.

polymerase chain reaction (PCR) (pŏlə′-mə-r′ās) A nucleic acid amplification technique in which a DNA sequence is replicated in a test tube to rapidly produce many copies.

polymorphism (pŏlē-môr′ fīz əm) A DNA base or sequence at a certain chromosomal locus that varies in a small percentage of individuals in a population.

polypeptide A long chain of amino acids. A protein consists of one or more polypeptides.

polyploid (pŏl′ē-ploid) A cell with one or more extra sets of chromosomes.

population A group of interbreeding individuals.

population bottleneck A large decrease in population size resulting from an event that kills many members of a population, followed by restoration of population numbers; it accelerates genetic drift.

population genetics (jə-nĕt′ĭks) The study of allele frequencies in biological populations.

population study Comparison of disease incidence in different groups of people.

preimplantation genetic diagnosis (PGD) (jə-nĕt′ĭk) Removing a cell from an 8-celled embryo and testing it for a mutation to deduce the genotype of the embryo.

prevalence The number of cases of a disease in a population at a particular time.

primary (1°) structure The amino acid sequence of a protein.

primary germ layers The three layers of an embryo.

primary immune response Immune system's response to initial encounter with a nonself antigen.

prion An infectious protein.

progenitor cell A cell whose descendants can follow any of several developmental pathways, but not all.

prokaryotic cell (prō-kārē - ŏt′ik sĕl) A cell that does not have a nucleus or other organelles. Members of the domains Bacteria or Archaea.

promoter A control sequence that signals the start of a gene.

pronuclei (prō-nōō′klēī) DNA packets in the fertilized ovum, one from each parental gamete.

prophase (prō′fāz) The first stage of mitosis or meiosis, when chromatin condenses.

prospective study A study that follows two or more groups.

proteasome (prō-tēə-sōm) A multiprotein cellular structure with a tunnel-like shape through which misfolded or unneeded proteins pass and are dismantled.

protein A type of macromolecule that is the direct product of genetic information; a chain of amino acids.

proteomics (prō′tē-ōm-ix) Cataloging all of the proteins a specific cell type makes under specific conditions.

proto-oncogene (prōtō-ŏn′kə-jēn) A gene that normally triggers cell division, but when overexpressed causes cancer.

pseudogene (sōō′ dō jēn) A DNA sequence that does not encode a protein, but whose sequence very closely resembles that of a coding gene.

Punnett square A diagram used to follow parental gene contributions to offspring.

purine (pyōō r′ēn) A nucleic acid base with a two-ring structure; adenine and guanine are purines.

pyrimidine (pǐ-rǐm′ǐ-dēn) A nucleic acid base with a single-ring structure; cytosine, thymine, and uracil are pyrimidines.

quantitative trait loci (QTLs) DNA sequences that contribute to polygenic traits.

quaternary (4°) structure A protein that has more than one polypeptide subunit.

reading frame The point in a DNA sequence from which contiguous triplets encode amino acids of a protein. A DNA sequence has three reading frames.

recessive An allele that must be present on both chromosomes of a pair to be expressed.

reciprocal translocation A chromosome aberration in which two nonhomologous chromosomes exchange parts, conserving genetic balance but rearranging genes.

recombinant (rē-kŏm′bə-nənt) A series of alleles on a chromosome that differs from the series of either parent.

recombinant DNA technology (rē-kŏm′bə-nənt DNA) A DNA sequence from one species that includes a DNA base sequence from a gene in another species.

reduction division The first meiotic division, which halves the chromosome number.

reference genome A digital nucleic acid sequence derived from many sequenced genomes that establishes a representative genome sequence for a species.

replication fork Locally opened portion of a replicating DNA double helix.

restriction enzyme An enzyme, typically from bacteria, that cuts DNA at a specific sequence. Used to create recombinant DNA molecules and in genome editing.

retrovirus A virus that has RNA as its genetic material.

reverse transcriptase An enzyme originally from viruses that copies an RNA sequence into a DNA sequence. Used in several biotechnologies.

reverse vaccinology Creating a vaccine using pathogen genome sequence information to identify antigens that stimulate a human immune response.

ribonucleic (rī́bō-nōō-klē′ĭk) **acid (RNA)** A nucleic acid whose bases are A, C, U, and G.

ribose (rī′bōs) A 5-carbon sugar that is part of RNA.

ribosomal RNA (rRNA) (rī′bōs-ō′məl) RNA that, with proteins, comprises ribosomes.

ribosome (rī′bō sōm) An organelle consisting of RNA and protein that is a scaffold and catalyst for protein synthesis.

ribozyme An RNA that is the catalytic part of a ribosome.

risk factor A characteristic or experience associated with increased likelihood of developing a particular medical condition.

RNA interference (RNAi) Introduction of a small interfering RNA molecule that binds to and prevents translation of a specific mRNA.

RNA polymerase (RNAP) (pŏl′ə-mə-rās) An enzyme that adds nucleotides to a growing RNA chain.

Robertsonian translocation (Răb-ərt-sō′-nē-ən) A chromosome aberration in which two short arms of nonhomologous chromosomes break and the long arms fuse, forming one unusual, large chromosome.

run of homozygosity Regions of the genome in which contiguous SNPs (single nucleotide polymorphisms) are homozygous, indicating a shared ancestor with another person with the same pattern.

S phase The stage of interphase when DNA replicates.

secondary (2°) structure Folds in a polypeptide caused by attractions between amino acids close together in the primary structure.

secondary immune response Immune system activation in response to a second or subsequent encounter with a foreign antigen.

segregation The distribution of alleles of a gene into separate gametes during meiosis. (Mendel's first law)

self-renewal Defining property of a stem cell; the ability to yield a daughter cell like itself.

semiconservative replication DNA synthesis along each separated strand of the double helix.

sequential polar body analysis Testing the DNA of a second polar body to infer the genotype of its associated fertilized ovum.

sex chromosome (krō′mə-sōōm) A chromosome containing genes that specify sex.

sex ratio Number of males divided by number of females multiplied by 1,000, for people of a certain age in a population.

sex-influenced trait Phenotype caused when an allele is recessive in one sex but dominant in the other.

sex-limited trait A trait that affects a structure or function present in only one sex.

short tandem repeats (STRs) Repeats of two to ten DNA bases that are compared in DNA profiling.

signal transduction A series of biochemical reactions and interactions that pass information from outside a cell to inside, triggering a response.

single nucleotide polymorphism (SNP) (nōōklēō-tīd pŏlē-môr′ fīz′əm) A site in the DNA that has a different base in at least 1 percent of a population.

sirtuin An enzyme that regulates energy use and may help to maintain health in the elderly.

somatic cell (sō-măt′ĭk sĕl) A non-sex cell, with 23 pairs of chromosomes in humans.

somatic gene therapy (sō-măt′ĭk) Genetic alteration of a specific cell type, not transmitted to future generations.

somatic mosaicism A condition in which a somatic mutation causes a phenotype to affect parts of the body descended from the original cell in which the mutation occurred.

somatic mutation (sō-măt′ĭk myōō-tā′shon) A genetic change in a non-sex cell.

sperm The male sex cell (gamete).

spermatogenesis (spər-măt′ə-jĕn′ĭ-sis) Sperm cell formation and differentiation.

spermatogonium (sper″mah-to-gō′ ne-um) An undifferentiated cell in a seminiferous tubule that can give rise to a sperm cell following meiosis and maturation.

spermatozoon (spər-măt′ə-zō′ŏn) (sperm) A mature male reproductive cell (meiotic product).

spindle A structure composed of microtubules that pulls sets of chromosomes apart in a dividing cell.

splice-site mutation A point mutation at a site in a gene that controls intron removal, resulting in extra or absent amino acids in the protein product.

spontaneous mutation (myōō-tā′sheən) A genetic change that results from mispairing when the replication machinery encounters a base in its rare tautomeric form.

SRY **gene** The sex-determining region of the Y. If the *SRY* gene is activated, the gonad develops into a testis; if not, an ovary forms under direction of other genes.

stem cells Cell that can divide to yield another stem cell (self-renew) and a cell that differentiates.

submetacentric (sŭb mĕt-ə-sĕn′trĭk) A chromosome in which the centromere establishes a long arm and a short arm.

substrate reduction therapy Use of an oral drug that reduces the level of substrate so that an enzyme can function more effectively.

sugar-phosphate backbone The "rails" of a DNA double helix, consisting of alternating deoxyribose and phosphate groups, oriented opposite one another.

synapse The space between two neurons that a neurotransmitter must cross to transmit a message.

synonymous codons (kō d ŏnz) DNA triplets that specify the same amino acid.

synteny (sĭn′tə-nē) Correspondence of genes on the same chromosome in two or more species.

T cell A type of lymphocyte that produces cytokines and coordinates the immune response.

tandem duplication A duplicated DNA sequence next to the original sequence.

telomerase (tə-lŏm′ə-rās) An enzyme, including a sequence of RNA, that adds DNA to chromosome tips.

telomere (tĕl′ə-mîr) A chromosome tip.

telophase (tĕlə-fāz) The stage of mitosis or meiosis when daughter cells separate.

template strand The DNA strand carrying the information to be transcribed.

teratogen (tə-răt′ə-jən) A substance that causes a birth defect.

tertiary (3°) structure Folds in a polypeptide caused by interactions between amino acids and water. This draws together amino acids that are far apart in the primary structure.

test cross Breeding an individual of unknown genotype with a homozygous recessive individual to deduce the unknown genotype from observing the traits of the offspring.

testes (tes′tēz) The male gonads.

thymine (thī′mēn) One of the two pyrimidine bases in DNA.

tissue Aggregate of cells with a shared function.

toll-like receptors Proteins embedded in the plasma membranes of immune system "sentinel" cells (macrophages and dendritic cells) that bind proteins common to many pathogens, stimulating both innate and adaptive immune responses.

transcription Manufacturing complementary RNA from a gene on a strand of DNA.

transcription-activator-like effector nuclease (TALEN) technology A genome editing method that uses a bacterial restriction enzyme to create double-stranded cuts in DNA.

transcription factor A protein that activates the transcription of certain genes.

transcriptomics Cataloging all of the messenger RNA molecules (transcripts) made in a specific cell type under specific conditions.

transfer RNA (tRNA) A type of RNA that connects mRNA to an amino acid during protein synthesis.

transgenic organism (trăns-jĕn′ĭk) An individual with a genetic modification, typically introduction of a gene from another species, in every cell.

transition A point mutation that replaces a purine with a purine or a pyrimidine with a pyrimidine.

translation Assembly of an amino acid chain according to the sequence of base triplets in a molecule of mRNA.

translocation carrier An individual with a large, translocated chromosome who may not have any symptoms. The person often has the usual amount of genetic material, but it is rearranged.

translocation Exchange of genetic material between nonhomologous chromosomes.

transposon (trăns-pōzŏn) A transposable element, or sequence of DNA that can move within the genome.

transversion A point mutation that replaces a purine with a pyrimidine, or vice versa.

trisomy A human cell with one extra chromosome, for a total of 47.

tumor necrosis factor (ne-kro-sis) A type of cytokine.

tumor suppressor gene (tōōmər səprĕs′ər jēn) A recessive gene whose normal function is to limit the number of divisions a cell undergoes.

uniparental disomy (UPD) (yû-ni-pə′rent-əl dīsō mē) Inheriting two chromosomes or chromosome segments from one parent.

uracil (yōōr′ə-sĭl) One of the four types of bases in RNA; a pyrimidine.

vaccine (vak-sē′n) An inactive or partial form of a pathogen that stimulates antibody production.

vesicles (ves-ə-kulz) Bubble-like membrane-bounded organelles that participate in secretion.

virus (vīrəs) An infectious particle built of nucleic acid in a protein coat.

wild type The most common phenotype in a population for a particular gene.

X inactivation A mechanism that turns off most of the genes on one X chromosome in each cell of a female mammal, occurring early in embryonic development.

X-linked Genes on an X chromosome.

X-Y homologs (hŏm′ə-lôgz) Y-linked genes that are similar to genes on the X chromosome.

Y-linked Genes on a Y chromosome.

zinc finger nuclease (ZFN) technology A genome editing technique that uses protein motifs called zinc fingers to bind specific DNA triplets, enabling a nuclease to cut the DNA across both strands.

zygote (zī′gōt) A prenatal human from the fertilized ovum stage until formation of the primordial embryo, at about 2 weeks.

zygote intrafallopian transfer (ZIFT) (zī′ gōt ĭn′trə-fə-lō′ pē-ən) An assisted reproductive technology in which an ovum fertilized *in vitro* is placed in a woman's uterine tube.

Nonself antigens, 332
Nonsense-mediated decay, 215
Nonsense mutations, 208f, 215, 215t, 216f, 314
Nonstandard amino acid (NSAA), 382
Nonsynonymous codons, 184
Noonan syndrome, 121
Noradrenaline, 149
Norovirus, 343
Notochord, 56
NSAA (nonstandard amino acid), 382
NTDs (neural tube defects), 56–57, 129, 236
Nuchal ligament, in *H. erectus*, 313
Nuchal translucency, 236
Nuclear envelope, 17, 19f–21f
Nuclear lamina, 17, 19f
Nuclear pores, 20f, 21f
Nucleic acids, 17, 18cc
Nucleolus, 17, 19f
Nucleosomes, 165, 166f
Nucleotide(s), 163, 163f
Nucleotide excision repair, 222, 222f, 223f, 224
Nucleus (nuclei), 17, 19f, 20f, 23t
    configuration of DNA in, 165, 166f, 167, 167f
    missing in prokaryotes, 17
Nutrients, birth defects caused by, 59
Nutrigenics testing, 391–392

O

"Oasel" (Neanderthal), 298–299
Obesity, 138, 140
OCA2 gene, 71–72, 72f
Odone, Lorenzo, 396cc–397cc
Odorant receptor (OR) proteins, 314
Office of Malaria Control in War Areas, 285
Okazaki fragments, 168f, 169
"Old Man" of La Chapelle-aux-Saints, 299
Old Order Amish, 271. *See also* Plain Populations
    "Amish cerebral palsy," 278cc
    bipolar disorder in, 149
    family records of, 79
Olfactory sense, 314
Oligospermia, 403
Oncogenes, 345, 347, 352–353, 355, 355f
Oncology, genetics in, 2
One-child policy, in China, 106–107
100,000 Genomes Project, 8
O'Neill, Eliza, 383
Oocyte(s), 41, 49f
    banking and donation of, 410–411, 411f, 413t
    postmortem retrieval of, 410
    primary, 47, 48f
    secondary, 47f, 47–48, 48f
Oocyte banking and donation, 410–411, 411f, 413t
Oogenesis, 47–48, 47f–49f
Oogonium(a), 41, 47
Open reading frame, 184
Operon, 179
Opium, 146
OPRM1 gene, pharmacogenetics of, 392t
Opsin genes, 113cc
Orexin, 144
Organelles, 17, 19f, 19–23, 20f, 23t
    energy production in, 23, 23f
    intracellular digestion in, 21–23, 22f
    secretion in, 20–21, 21f
Organogenesis, 56, 57
Organ systems, 6
Ornithine transcarbamylase (OTC)
    deficiency, 396cc
OR (odorant receptor) proteins, 314
Orrorin tugensis, 295, 296f
Osteoarthritis, 209t
Osteogenesis imperfecta ("brittle bone disease"),
    61, 90, 93t, 211

Osteogenesis imperfecta type 1, 209t
Osteoporosis, susceptibility to, 5
Ötzi the Ice Man, 302f, 302–303, 419
Ovarian cancer, detecting, 362cc
Ovarian tissue, reimplantation of, 411
Ovaries, 42, 42f
Ovulation, 48
Ovulation predictor tests, 403
Ovum (ova), injecting foreign DNA into, 371
Oxygen binding, 220t

P

Paabo, Svante, 317
PAH gene, 152t
Pain syndromes, 24
Painter, Theophilus, 234
Palindromic DNA sequences, 108, 368
Pancreas
    pancreatic cancer, 349, 353t
    pancreatic islets, 195, 196f
    structure of, 195, 196f
Panthers, endangered, 7f
Pantoea agglomerans, 370
Paracentric inversion, 249f, 249–250
Parental allele configuration, 100f
Parent-of-origin effects, 118–121, 122
    genomic imprinting, 119f, 119–120, 120f
    imprinting diseases, 120f, 120–121
    sperm and egg production timetables, 121
Parkinson disease, 39, 189, 211t
    early-onset, 192
    misfolded proteins in, 189t
    twin studies of, 140
Passenger mutations, 352, 353t, 353f, 354f
Pasteur, Louis, 367
Patau syndrome (trisomy 13), 239t, 242–243
Patent(s)
    breast cancer genes, 366
    DNA technologies, 366f, 366–367, 380
    transgenic organisms, 366, 366f, 367
Patent and Trademark Office, U.S., 366, 367
Paternal age effects, 48, 49, 49t, 121
Paternity testing, 387t
Patient selection, in GWAS studies, 135
Pattern baldness, 41
Pauling, Linus, 161, 209
PCR (polymerase chain reaction), 169t,
    169–170, 170f
PCSK9 gene, 140
Peanut allergies, 372
Pea plants
    discovery of linkage in, 98, 98f, 99f
    Mendel's experiments with, 67–69, 68f, 69f
    parent-of-origin effects in, 118
Pediculus humanus (louse), 175–176, 318
Pedigree(s), 6, 428
    analysis of, 78f, 78–81, 82
        conditional probability problem, 80, 81f
        exome and genome sequencing and,
            80–81, 81f
        history of, 79, 79f
        Mendel's laws and, 79f, 80, 80f
    for autosomal recessive disorders, 80, 80f
    components of, 78, 78f
    for Hemophilia B, 111t, 112, 114f
    for heritability of traits, 131, 131f
Peeling skin syndrome, 84
Penetrance
    in gene expression, 90, 91, 93t, 120, 430
    of HNPCC, 224
    incomplete, 120, 430
    of single-gene diseases, 67
Penicillamine, 18cc
Pennington, Robert, 338

Peptide(s), 176
Peptide bonds, 185, 186f
Peptide fingerprints, 209
Perchloroethylene exposure, 140
Perforin, 327, 328f
Pericentric inversion, 250, 250f
Period 2 gene, 144
Periodic fever, familial, 279t
Peripheral neuropathy, 211t
Peroxisomes, 22f, 22–23, 23t
Personal genome sequencing, 429–432, 433
    criteria for choosing, 430–432, 431t
    practical matters in, 429–430
    types of information in, 430
PERV (porcine endogenous retrovirus), 338
Pfeiffer syndrome, 49t
PGD. *See* Preimplantation genetic diagnosis
p53 gene, 140, 357, 357f
P29 gene mutation, 317–318
Ph¹ ("Philadelphia chromosome"), 356
Phagocytes, 31f, 324–325
Phagocytosis, 324–325
Pharmacogenetics, 8, 149, 429
Pharmacogenetics testing, 387t, 392t, 392f,
    392–393
Pharmacogenomics testing, 392
Pharmacological chaperone therapy, 393, 393t
Phenocopies, 91, 93t
Phenotype(s), 6, 69
    effect of heteroplasmy on, 95
    effect of X inactivation, 115f, 117–118, 118f
    Neanderthal, 299, 299f
    sex and, 108–110, 109f
Phenotypic frequencies, 257, 257t
Phenotypic sex, 110t
Phenylketonuria (PKU), 152t, 284t
    genetic testing for, 388, 388f
    misfolded proteins in, 189t
    mutations of, 280–281
    phenotypic frequencies for, 257, 257t
"Philadelphia chromosome" (Ph¹), 356
Phipps, James, 336
Phocomelia, 58
Phospholipid(s), 24, 24f
Phospholipid bilayer, 24, 24f
Phosphorus, in pig manure, 365–366, 381
Photolyases, 222
Photoreactivation, 222
Physical barriers, in immunity, 323, 324f
Physical maps, 420, 421f
Phytase, 365–366
Pictogram, 426, 426f, 434f
Pierson syndrome, 279t
Pigs
    "enviropig," 365, 371
    improving pig manure, 365–366, 381
    as organ donors, 338, 338f
Pima Indians, obesity among, 138
Pingelapese blindness, 277
piRNA (piwi-interacting RNA), 202t
Pitchfork, Colin, 262
PITX1 gene, 55cc
Piwi-interacting RNA (piRNA), 202t
PKU. *See* Phenylketonuria
Placenta, formation of, 53
Placental DNA, testing, 233
Plain Populations
    founder effect in, 278f, 278cc–279cc, 279t
    genetic diseases among, 277
Plasma cells, 325, 325f
Plasma membrane, 19, 19f, 21f, 23, 24
Plasmids, 283, 368, 369f, 369–370
Plasmodium falciparum, 284, 284f, 370
Plastin 3 gene, 90
Pleiotropy, 90, 91f, 92, 93t